Silicon Nitride Ceramics

Scientific and Technological Advances

MATERIALS RESEARCH SOCIETY SYMPOSIUM PROCEEDINGS VOLUME 287

Silicon Nitride Ceramics
Scientific and Technological Advances

Symposium held November 30-December 3, 1992, Boston, Massachusetts, U.S.A.

EDITORS:

I-Wei Chen
University of Michigan
Ann Arbor, Michigan, U.S.A.

Paul F. Becher
Oak Ridge National Laboratory
Oak Ridge, Tennessee, U.S.A.

Mamoru Mitomo
National Institute for Research in Inorganic Materials
Tsukuba, Japan

G. Petzow
Max-Planck Institut für Metallforschung
Stuttgart, Germany

Tung-Sheng Yen
Shanghai Institute of Ceramics
Shanghai, China

MATERIALS RESEARCH SOCIETY
Pittsburgh, Pennsylvania

Single article reprints from this publication are available through University Microfilms Inc., 300 North Zeeb Road, Ann Arbor, Michigan 48106

CODEN: MRSPDH

Copyright 1993 by Materials Research Society.
All rights reserved.

This book has been registered with Copyright Clearance Center, Inc. For further information, please contact the Copyright Clearance Center, Salem, Massachusetts.

Published by:

Materials Research Society
9800 McKnight Road
Pittsburgh, Pennsylvania 15237
Telephone (412) 367-3003
Fax (412) 367-4373

Library of Congress Cataloging in Publication Data

Silicon nitride ceramics : Scientific and technological advances / editors, I-Wei Chen, Paul F. Becher, Mamoru Mitomo, G. Petzow, Tung-Sheng Yen

    p. cm.—(Materials Research Society symposium proceedings, ISSN 0272-9172 ; v. 287)
    Proceedings of a symposium held November 30-December 3, 1992 in Boston, Massachusetts.
    Includes bibliographical references and index.
    ISBN 1-55899-182-4
    1. Silicon nitride—Congresses. 2. Ceramics—Congresses. I. Chen, I-Wei II. Becher, Paul F. III. Mitomo, Mamoru IV. Petzow, G. V. Yen, Tung-Sheng VI. Series: Materials Research Society symposium proceedings ; v. 287.

TP245.N8S45   1993   93-6876
666—dc20   CIP

Manufactured in the United States of America

Contents

PREFACE xi

MATERIALS RESEARCH SOCIETY SYMPOSIUM PROCEEDINGS xii

PART I: RETROSPECT AND PROSPECT

*MICROSTRUCTURAL DESIGN OF Si_3N_4 BASED CERAMICS 3
  M.J. Hoffmann and G. Petzow

*SIALON CERAMICS: RETROSPECT AND PROSPECT 15
  Kenneth H. Jack

*PROGRESS IN SILICON NITRIDE CERAMICS IN JAPAN 29
  K. Komeya

*PHASE RELATIONSHIP STUDIES OF SILICON NITRIDE SYSTEM—
A KEY TO MATERIALS DESIGN 39
  T.S. Yen and W.Y. Sun

*SILICON NITRIDE CERAMICS—ALLOY DESIGN 51
  Tseng-Ying Tien

*HIGH-RESOLUTION ELECTRON MICROSCOPY OBSERVATIONS OF
GRAIN-BOUNDARY FILMS IN SILICON NITRIDE CERAMICS 65
  H.-J. Kleebe, M.K. Cinibulk, I. Tanaka, J. Bruley, R.M. Cannon,
  D.R. Clarke, M.J. Hoffmann, and M. Rühle

*NEW GRAIN-BOUNDARY PHASES FOR NITROGEN CERAMICS 79
  Derek P. Thompson

*OXYNITRIDE GLASSES AND GLASS CERAMICS 93
  Stuart Hampshire

*SILICON NITRIDE POWDER FROM CARBOTHERMAL REACTION 105
  Heinrich Hofmann, U. Vogt, A. Kerber, and F. van Dijen

*SIALON CERAMICS SINTERED WITH YTTRIA AND RARE EARTH
OXIDES 121
  Thommy Ekström

*THE INFLUENCE OF POWDERS AND PROCESSING METHODS ON
MICROSTRUCTURE AND PROPERTIES OF DENSE SILICON NITRIDE ... 133
  G. Woetting, H. Feuer, and E. Gugel

*THE INFLUENCE OF MICROSTRUCTURE ON THE MECHANICAL
BEHAVIOR OF SILICON NITRIDE CERAMICS 147
  P.F. Becher, H.T. Lin, S.L. Hwang, M.J. Hoffmann, and I-Wei Chen

*SIALONS AND SILICON NITRIDES; MICROSTRUCTURAL DESIGN
AND PERFORMANCE 159
  M.H. Lewis

*STATUS OF RESEARCH AND DEVELOPMENT ON MATERIALS FOR
CERAMIC GAS TURBINE COMPONENTS 173
  Minoru Matsui

*Invited Paper

*DEVELOPMENT OF HIGH PERFORMANCE SILICON NITRIDE
CERAMICS AND THEIR APPLICATIONS 189
   Yo Tajima

*APPLICATIONS OF SILICON NITRIDE BASED CERAMICS IN THE U.S. 197
   R. Nathan Katz

*SUPERPLASTIC SiAlON—A BIRD'S EYE VIEW OF SILICON NITRIDE
CERAMICS 209
   I-Wei Chen and Shyh-Lung Hwang

PART II: RECENT PROGRESS

PART IIA. POWDER SYNTHESIS AND PROCESSING ROUTES

SYNTHESIS OF Si_3N_4 POWDER BY THERMAL DECOMPOSITION OF
$Si(NH)_2$ 227
   Silvia Ampuero, Paul Bowen, and Terry A. Ring

STRUCTURAL CHARACTERIZATION OF SILICON CARBONITRIDE
CERAMICS FROM POLYMERIC PRECURSORS USING NUCLEAR
MAGNETIC RESONANCE 233
   Corine M. Gerardin, F. Taulelle, and J. Livage

SPECTROSCOPIC CHARACTERIZATION OF THE NITRIDATION
PROCESS OF POLYMERIC PRECURSORS TO Si-M-N-O SYSTEMS
(M=Ti,Zr,Al) 239
   F. Babonneau and G.D. Soraru

FABRICATION AND CHARACTERIZATION OF β-SiAlON
COMPONENTS FROM POLYMERIC PRECURSORS 245
   G.D. Soraru, V.M. Sglavo, F. Vulcan, and F. Babonneau

Mg-Si-Al-O-N GLASSES PREPARED BY NITRIDING CORDIERITE
POWDERS DERIVED FROM POLYMER PRECURSORS 251
   Richard M. Laine, Clint R. Bickmore, Kurt F. Waldner,
   Brian L. Mueller, and Hal W. Estry

EFFECT OF POWDER PROCESSING ON SLIP CASTING PARAMETERS
AND SINTERING OF Si_3N_4-SiO_2-Y_2O_3 257
   Gary Gilde and George Gazza

TAPE CASTING OF SILICON NITRIDE 265
   Young-Wook Kim and June-Gunn Lee

RHEOLOGICAL BEHAVIOR OF SILICON NITRIDE WHISKERS 271
   Patricio Toro and Oscar Wittke

MICROWAVE NITRIDATION OF SILICON COMPACTS UTILIZING
A TEMPERATURE GRADIENT 277
   Jeffrey J. Thomas, Hamlin M. Jennings, and D. Lynn Johnson

SINTERED REACTION-BONDED SILICON NITRIDE BY MICROWAVE
HEATING 283
   Terry N. Tiegs, James O. Kiggans, Jr., and Kristin L. Ploetz

MICROSTRUCTURAL CHARACTERISATION OF MICROWAVE
SINTERED SILICON NITRIDE CERAMICS 289
   Kevin P. Plucknett and David S. Wilkinson

*Invited Paper

STRENGTH OPTIMIZATION THROUGH POWDER MODIFICATION 295
  Arvid E. Pasto, F. Avella, S. Natansohn, and W.J. Rourke

INFILTRATION/PYROLYSIS PROCESSING OF FIBER-REINFORCED
SILICON NITRIDE 303
  Stuart T. Schwab, Renee C. Graef, Cheryl R. Blanchard, Yi-Ming Pan,
  and David L. Davidson

A NOVEL PROCESSING ROUTE FOR THE FABRICATION OF
MONOLITHIC AND COMPOSITE SILICON NITRIDE 309
  R.V. Raman and S.V. Rele

NUCLEATION AND GROWTH OF CVD POLYCRYSTALLINE Si_3N_4
FILMS AT LOW TEMPERATURES 315
  Frederick S. Lauten, Janet Rankin, and Brian W. Sheldon

PART IIB. GRAIN BOUNDARY PHASES AND GLASSES

INTERGRANULAR MICROSTRUCTURE AND OXIDATION BEHAVIOUR OF
Si_3N_4 CERAMICS FORMED WITH Y_2O_3, Al_2O_3 AND ZrO_2 323
  L.K.L. Falk, E.U. Engström, and K. Rundgren

ELEMENTAL ANALYSIS OF MATRIX GRAIN BOUNDARIES IN SiC
WHISKER REINFORCED Si_3N_4 BASED COMPOSITES 329
  J. Liu, K. Das Chowdhury, R.W. Carpenter, and W. Braue

INTERFACE STRUCTURE OF Si_3N_4 MATRIX COMPOSITE WITH
NANO-METER SCALE SiC PARTICLES 335
  Gen Sasaki, Katsuaki Suganuma, Teruaki Fujita, Kenji Hiraga,
  and Koichi Niihara

CHARACTERIZATION OF HIP'ED, HIGH PURITY Si_3N_4 GRAIN
BOUNDARIES 341
  Ping Lu, S.C. Danforth, and W.T. Symons

PARTIAL DEVITRIFICATION OF SINTERED SILICON NITRIDE
DURING STATIC FATIGUE TESTING 347
  W. Braue and G.D. Quinn

FABRICATION AND PROPERTIES OF Si_3N_4 WITH RARE EARTH
APATITE GRAIN BOUNDARY PHASES 353
  Terry N. Tiegs, Stephen D. Nunn, Kristin L. Ploetz,
  Paul A. Menchoffer, and Claudia A. Walls

SILICON NITRIDE CONTAINING RARE EARTH SILICATE INTER-
GRANULAR PHASES 359
  Stephen D. Nunn, Terry N. Tiegs, Kristin L. Ploetz,
  Claudia A. Walls, and Nelson Bell

CHEMICALLY INDUCED DEFECTS IN OXYNITRIDE GLASSES 365
  Donald R. Messier and Parimal J. Patel

PART IIC: MICROSTRUCTURE—DESIGN, DEVELOPMENT AND CHARACTERIZATION

REACTION SINTERING OF β-Si_3N_4/α'-SIALON CERAMICS 373
  S. Boskovic, K.J. Lee, and T.Y. Tien

TRANSIENT VISCOUS PHASE REACTION SINTERED (TVPRS)
SILICON OXYNITRIDE CERAMICS   381
   Kevin P. Plucknett and David S. Wilkinson

FORMATION AND DENSIFICATION OF R-α'-SiAlONS
(R = Nd, Sm, Gd, Dy, Er AND Yb)   387
   P.L. Wang, W.Y. Sun, and T.S. Yen

PARAMETERS AFFECTING PRESSURELESS SINTERING OF α'-
SIALONS WITH LANTHANIDE MODIFYING CATIONS   393
   K.P.J. O'Reilly, M. Redington, S. Hampshire, and M. Leigh

NON-OXIDE ADDITIVES AS SINTERING AIDS FOR Si_3N_4-BASED
CERAMICS   399
   C. Ge, Y. Xia, and L. Chen

MICROSTRUCTURAL DESIGN BY SELECTIVE GRAIN GROWTH OF
β-Si_3N_4 405
   Naoto Hirosaki, Yoshio Akimune, and Mamoru Mitomo

THE EFFECT OF GLASS CHEMISTRY ON THE MICROSTRUCTURE
AND PROPERTIES OF SELF REINFORCED SILICON NITRIDE   411
   Aleksander J. Pyzik, Daniel F. Carroll, and C. James Hwang

ROLE OF SINTERING PARAMETERS ON MICROSTRUCTURE
DEVELOPMENT AND MECHANICAL PROPERTIES OF SINTER/HIP
SILICON NITRIDE   417
   Arnd Kühne, Rainer Oberacker, and Georg Grathwohl

GRAIN MORPHOLOGY AND INTERGRANULAR STRUCTURE OF
Si_3N_4 BASED CERAMICS FORMED BY HIP   423
   H. Björklund, L.K.L. Falk, J. Wasén, J.E. Adlerborn, and
   H.T. Larker

CHARACTERIZATION OF HOT-PRESSED SILICON NITRIDE
CERAMICS WITH ALKOXIDE-DERIVED OXIDE MIXTURES AS THE
SINTERING AID   429
   Y. Sato, C. Sakurai, M. Ueki, and K. Sugita

DIRECT OBSERVATION OF MICROSTRUCTURE CHANGE DURING
DENSIFICATION OF SILICON NITRIDE CERAMICS WITH A
NOVEL CHARACTERIZATION METHOD   435
   Y. Iwamoto, H. Nomura, I. Sugiura, J. Tsubaki, H. Takahashi,
   K. Ishikawa, N. Shinohara, M. Okumiya, Y. Yamada, H. Kamiya,
   and K. Uematsu

ACOUSTIC EMISSION STUDY OF Si_3N_4 441
   E.C. Subbarao, V. Srikanth, J.C. Walck, and C.A. Tarry

PART IID: MECHANICAL PERFORMANCE

DEFORMATION AND TOUGHNESS OF α-SILICON NITRIDE SINGLE
CRYSTALS   449
   H. Suematsu, J.J. Petrovic, and T.E. Mitchell

EVALUATION OF TENSILE STATIC, DYNAMIC, AND CYCLIC
FATIGUE BEHAVIOR FOR A HIPed SILICON NITRIDE AT
ELEVATED TEMPERATURES   455
   Chih-Kuang Jack Lin, Michael G. Jenkins, and Mattison K. Ferber

HIGH TEMPERATURE FATIGUE PROPERTIES OF SILICON NITRIDE
IN NITROGEN ATMOSPHERE 461
    Yasuhiro Shigegaki, Takashi Inamura, Akihiko Suzuki, and
    Tadashi Sasa

CAVITY EVOLUTION DURING TENSILE CREEP OF Si_3N_4 467
    William Luecke, S.M. Wiederhorn, B.J. Hockey, and G.G. Long

DAMAGE RESISTANCE OF *IN SITU* REINFORCED SILICON NITRIDE 473
    Chien-Wei Li, Charles J. Gasdaska, Jeffrey Goldacker, and
    Siu-Ching Lui

CRACK GROWTH RESISTANCE OF CERAMIC COMPOSITE 481
    Seijiro Hayashi, H. Baba, and A. Suzuki

MECHANICAL PROPERTIES OF SiC WHISKER REINFORCED β-SiAlON
COMPOSITES 487
    C. Yamagishi, J. Hakoshima, S. Nakajoh, N. Miyata, and
    K. Tsukamoto

MECHANICAL PROPERTIES OF SILICON CARBIDE WHISKER-
REINFORCED SILICON NITRIDE MATRIX COMPOSITES 493
    Yong Huang, Huirong Le, Jianbao Li, Longlie Zheng, and
    Jianguang Wu

THE MECHANICAL PROPERTIES OF A NOVEL Si_3N_4-AMORPHOUS
Si_3N_4 COMPOSITE 499
    Ivar E. Reimanis, J.J. Petrovic, H. Suematsu, T.E. Mitchell, and
    O.S. Leung

PART IIE: APPLICATIONS, OXIDATION, AND CORROSION

STUDIES ON Si_3N_4 CERAMIC CUTTING TOOL MATERIALS AND
THEIR APPLICATIONS 507
    He-Zhuo Miao, Long-Hao Qi, De-Jin Ma, and Zuo-Zhao Jiang

DEVELOPMENT OF ADVANCED SILICON NITRIDE VALVES FOR
COMBUSTION ENGINES AND SOME PRACTICAL EXPERIENCE ON
THE ROAD 513
    Rainer Hamminger and Juergen Heinrich

OXIDATION KINETICS OF Si_2N_2O CERAMICS 521
    Jeanette Persson, Per-Olov Käll, and Mats Nygren

OXIDATION BEHAVIOUR OF ZIRCONIA-SIALON COMPOSITES 527
    Yibing Cheng and Derek P. Thompson

CORROSION OF SILICON NITRIDE CERAMICS BY NITRIC ACID 533
    Kunihiko Kanbara, N. Uchida, K. Uematsu, T. Kurita,
    K. Yoshimoto, and Y. Suzuki

STRUCTURE AND CORROSION PROPERTIES OF PVD COATINGS IN
THE SYSTEM Si-Al-O-N 539
    Otto Knotek, Frank Löffler, and Wolfram Beele

AUTHOR INDEX 547

SUBJECT INDEX 549

Preface

In the past decade, silicon nitride ceramics have been developed for a number of applications, including those with quite stringent reliability and economic requirements (e.g., automotive turbocharger rotors). These applications serve as graphic examples of the real advances that have been achieved in this field. The improvement in these materials has been based on advances in the fundamental understanding of processing-phase equilibria-microstructure-mechanical properties relationships, coupled with the development of manufacturing and engineering practices capable of producing highly reliable components. These advances are reflected in this proceedings volume, whose papers were authored by an international group of researchers and engineers. These authors have also indicated the direction research is heading in the further development of silicon nitride ceramics. Areas such as innovative alloy design based on phase relationships, basic understanding of microstructure and phase evolution and elucidation of the connections between microstructure (including interfaces) and various mechanical properties will continue to make impressive advances. Progress will also be seen in improving reliability and, hopefully, dramatically lowering cost.

During the symposium, held at the 1992 MRS Fall Meeting in Boston, Massachusetts, Professor T.Y. Tien of the University of Michigan was recognized for his contributions to the field of silicon nitride ceramics. The spirit of the symposium was one of scientific exchange, discussion of ideas, and being with old friends and making new ones. That is a fitting tribute to an international teacher, an innovative materials technologist, and above all, a gentleman. Many of his ideas and contributions have been reflected in the papers of this proceedings which will be presented to Professor Tien as a "festschrift" on his 70th birthday.

No symposium can succeed without the contributions of a number of people—first among these being the efforts of the authors. However, others have played a significant role also. The session chairs helped keep us on track and promoted some excellent discussions. The entire process of organizing the meeting and preparing the proceedings would not have been possible without the assistance of Ms. Bonni Viets of the University of Michigan. Finally, we are most grateful for the financial support of the University of Michigan and the U.S. Department of Energy, Division of Materials Science, Office of Basic Energy Sciences.

I-Wei Chen
Paul F. Becher
Mamoru Mitomo
G. Petzow
Tung-Sheng Yen

January 1993

MATERIALS RESEARCH SOCIETY SYMPOSIUM PROCEEDINGS

Volume 258—Amorphous Silicon Technology—1992, M.J. Thompson, Y. Hamakawa, P.G. LeComber, A. Madan, E. Schiff, 1992, ISBN: 1-55899-153-0

Volume 259—Chemical Surface Preparation, Passivation and Cleaning for Semiconductor Growth and Processing, R.J. Nemanich, C.R. Helms, M. Hirose, G.W. Rubloff, 1992, ISBN: 1-55899-154-9

Volume 260—Advanced Metallization and Processing for Semiconductor Devices and Circuits II, A. Katz, Y.I. Nissim, S.P. Murarka, J.M.E. Harper, 1992, ISBN: 1-55899-155-7

Volume 261—Photo-Induced Space Charge Effects in Semiconductors: Electro-optics, Photoconductivity, and the Photorefractive Effect, D.D. Nolte, N.M. Haegel, K.W. Goossen, 1992, ISBN: 1-55899-156-5

Volume 262—Defect Engineering in Semiconductor Growth, Processing and Device Technology, S. Ashok, J. Chevallier, K. Sumino, E. Weber, 1992, ISBN: 1-55899-157-3

Volume 263—Mechanisms of Heteroepitaxial Growth, M.F. Chisholm, B.J. Garrison, R. Hull, L.J. Schowalter, 1992, ISBN: 1-55899-158-1

Volume 264—Electronic Packaging Materials Science VI, P.S. Ho, K.A. Jackson, C-Y. Li, G.F. Lipscomb, 1992, ISBN: 1-55899-159-X

Volume 265—Materials Reliability in Microelectronics II, C.V. Thompson, J.R. Lloyd, 1992, ISBN: 1-55899-160-3

Volume 266—Materials Interactions Relevant to Recycling of Wood-Based Materials, R.M. Rowell, T.L. Laufenberg, J.K. Rowell, 1992, ISBN: 1-55899-161-1

Volume 267—Materials Issues in Art and Archaeology III, J.R. Druzik, P.B. Vandiver, G.S. Wheeler, I. Freestone, 1992, ISBN: 1-55899-162-X

Volume 268—Materials Modification by Energetic Atoms and Ions, K.S. Grabowski, S.A. Barnett, S.M. Rossnagel, K. Wasa, 1992, ISBN: 1-55899-163-8

Volume 269—Microwave Processing of Materials III, R.L. Beatty, W.H. Sutton, M.F. Iskander, 1992, ISBN: 1-55899-164-6

Volume 270—Novel Forms of Carbon, C.L. Renschler, J. Pouch, D. Cox, 1992, ISBN: 1-55899-165-4

Volume 271—Better Ceramics Through Chemistry V, M.J. Hampden-Smith, W.G. Klemperer, C.J. Brinker, 1992, ISBN: 1-55899-166-2

Volume 272—Chemical Processes in Inorganic Materials: Metal and Semiconductor Clusters and Colloids, P.D. Persans, J.S. Bradley, R.R. Chianelli, G. Schmid, 1992, ISBN: 1-55899-167-0

Volume 273—Intermetallic Matrix Composites II, D. Miracle, D. Graves, D. Anton, 1992, ISBN: 1-55899-168-9

Volume 274—Submicron Multiphase Materials, R. Baney, L. Gilliom, S.-I. Hirano, H. Schmidt, 1992, ISBN: 1-55899-169-7

Volume 275—Layered Superconductors: Fabrication, Properties and Applications, D.T. Shaw, C.C. Tsuei, T.R. Schneider, Y. Shiohara, 1992, ISBN: 1-55899-170-0

Volume 276—Materials for Smart Devices and Micro-Electro-Mechanical Systems, A.P. Jardine, G.C. Johnson, A. Crowson, M. Allen, 1992, ISBN: 1-55899-171-9

Volume 277—Macromolecular Host-Guest Complexes: Optical, Optoelectronic, and Photorefractive Properties and Applications, S.A. Jenekhe, 1992, ISBN: 1-55899-172-7

Volume 278—Computational Methods in Materials Science, J.E. Mark, M.E. Glicksman, S.P. Marsh, 1992, ISBN: 1-55899-173-5

MATERIALS RESEARCH SOCIETY SYMPOSIUM PROCEEDINGS

Volume 279—Beam-Solid Interactions—Fundamentals and Applications, M.A. Nastasi, N. Herbots, L.R. Harriott, R.S. Averback, 1993, ISBN: 1-55899-174-3
Volume 280—Evolution of Surface and Thin Film Microstructure, H.A. Atwater, E. Chason, M. Grabow, M. Lagally, 1993, ISBN: 1-55899-175-1
Volume 281—Semiconductor Heterostructures for Photonic and Electronic Applications, D.C. Houghton, C.W. Tu, R.T. Tung, 1993, ISBN: 1-55899-176-X
Volume 282—Chemical Perspectives of Microelectronic Materials III, C.R. Abernathy, C.W. Bates, D.A. Bohling, W.S. Hobson, 1993, ISBN: 1-55899-177-8
Volume 283—Microcrystalline Semiconductors—Materials Science & Devices, Y. Aoyagi, L.T. Canham, P.M. Fauchet, I. Shimizu, C.C. Tsai, 1993, ISBN: 1-55899-178-6
Volume 284—Amorphous Insulating Thin Films, J. Kanicki, R.A.B. Devine, W.L. Warren, M. Matsumura, 1993, ISBN: 1-55899-179-4
Volume 285—Laser Ablation in Materials Processing—Fundamentals and Applications, B. Braren, J. Dubowski, D. Norton, 1993, ISBN: 1-55899-180-8
Volume 286—Nanophase and Nanocomposite Materials, S. Komarneni, J.C. Parker, G.J. Thomas, 1993, ISBN: 1-55899-181-6
Volume 287—Silicon Nitride Ceramics—Scientific and Technological Advances, I-W. Chen, P.F. Becher, M. Mitomo, G. Petzow, T-S. Yen, 1993, ISBN: 1-55899-182-4
Volume 288—High-Temperature Ordered Intermetallic Alloys V, I. Baker, J.D. Whittenberger, R. Darolia, M.H. Yoo, 1993, ISBN: 1-55899-183-2
Volume 289—Flow and Microstructure of Dense Suspensions, L.J. Struble, C.F. Zukoski, G. Maitland, 1993, ISBN: 1-55899-184-0
Volume 290—Dynamics in Small Confining Systems, J.M. Drake, D.D. Awschalom, J. Klafter, R. Kopelman, 1993, ISBN: 1-55899-185-9
Volume 291—Materials Theory and Modelling, P.D. Bristowe, J. Broughton, J.M. Newsam, 1993, ISBN: 1-55899-186-7
Volume 292—Biomolecular Materials, S.T. Case, J.H. Waite, C. Viney, 1993, ISBN: 1-55899-187-5
Volume 293—Solid State Ionics III, G-A. Nazri, J-M. Tarascon, M. Armand, 1993, ISBN: 1-55899-188-3
Volume 294—Scientific Basis for Nuclear Waste Management XVI, C.G. Interrante, R.T. Pabalan, 1993, ISBN: 1-55899-189-1
Volume 295—Atomic-Scale Imaging of Surfaces and Interfaces, D.K. Biegelson, D.S.Y. Tong, D.J. Smith, 1993, ISBN: 1-55899-190-5
Volume 296—Structure and Properties of Energetic Materials, R.W. Armstrong, J.J. Gilman, 1993, ISBN: 1-55899-191-3

Prior Materials Research Society Symposium Proceedings available by contacting Materials Research Society

PART I

Retrospect and Prospect

MICROSTRUCTURAL DESIGN OF Si_3N_4 BASED CERAMICS

M.J. HOFFMANN AND G.PETZOW
Max-Planck-Institut für Metallforschung, Institut für Werkstoffwissenschaft, Pulvermetallurgisches Laboratorium, Heisenbergstr. 5, D-7000 Stuttgart 80, Germany.

ABSTRACT

Parameters controlling the size and aspect ratio of elongated Si_3N_4 grains are discussed, based on the assumption that only pre-existing β-Si_3N_4 particles of the starting powder grow. Powder mixtures of α-rich and β-rich Si_3N_4 were prepared in order to study the microstructural development. The resulting microstructures were analyzed by quantitative microstructural analysis determining the distribution of the length and aspect ratio of the Si_3N_4 grains. Subsequently, the influence of the sintering conditions on grain growth was analyzed in relation to mechanical properties. A high Weibull modulus and the non-catastrophic failure during thermal shock of coarse-grained materials is attributed to an R-curve behaviour. Finally, the influence of sintering additives on the mechanical properties was studied. The importance of phase relationships between the matrix and the grain boundary phase is discussed for Si_3N_4 with Yb_2O_3 additives. It is demonstrated that the oxygen content of Si_3N_4 powder must been taken into account in order to devitrify defined secondary phases and to achieve a high degree of crystallization. A reduction in the amount of additives does not necessarily improve the properties as high temperature strength and creep data indicate.

INTRODUCTION

Silicon nitride ceramics exhibit excellent mechanical properties, good oxidation resistance and thermal shock behaviour at both room and high temperatures. The high wear resistance and mechanical properties of these ceramics are interesting for several applications such as cutting tools, valves for automotive engines or gas turbine components. However, the properties of silicon nitride ceramics depend on their microstructure. Because of the highly covalent bonding, Si_3N_4 has to be densified with sintering additives. Today, it is common practice to densify Si_3N_4 by pressureless sintering, gas pressure sintering, hot-pressing or hot isostatic pressing. The sintering aids are usually metal oxides such as MgO, Al_2O_3 and most of the rare earth oxides [1-3].

The densification is described as a liquid phase sintering process. At higher temperatures, SiO_2, which is always present at the surface of the Si_3N_4 particles [4], reacts with the oxide additives to form an oxide melt and, with increasing temperature, an oxynitride melt by dissolving Si_3N_4. While the α-Si_3N_4 particles dissolve and supersaturate the liquid phase, β-Si_3N_4 is reprecipitated [5]. Depending on the composition of the sintering aids, the liquid phase can form an amorphous or a crystalline grain boundary phase during cooling, both of which degrade the mechanical properties of the Si_3N_4 at temperatures > 1000°C because of the softening of the grain boundary regions [6,7].

The first part of this paper is focussed on the possibilties of controlling the microstructural development to achieve high strength materials for low temperature applications (≤ 1000°C), and the second part discusses the influence of grain boundary phase on high temperature behaviour.

MICROSTRUCTURAL DEVELOPMENT

Nucleation and Grain Growth

In 1979, F.F. Lange [8] investigated hot-pressed Si_3N_4 with 5 wt.% MgO and found that the aspect ratio of the grains and, therefore, the fracture toughness, was determined by the ratio of α- and β-Si_3N_4 in the starting powder. The strong influence of the α/β ratio on pressureless and gas pressure sintered samples has also been reported by several other research groups [9,10].

Figure 1 shows the relationship between the number of β-particles in the initial powder (β-nuclei density) and the particle density after complete densification and transformation into the β-phase for 5 commercial Si_3N_4 powders densified by pressureless sintering with the same amount of Y_2O_3/Al_2O_3 additives [11,12]. The β-nuclei density was calculated from the β-content determined by X-ray diffraction and the measured particle size distribution of the starting powder under the assumption of the same particle size distribution for α- and β-particles. The particle density was determined by quantitative microstructural analysis without taking into account the 3-dimensional random distribution of prismatic Si_3N_4 grains. Therefore, the measured particle density is somewhat lower than the actual particle density [12].

The hatched area on the left side of the diagram would represent powders with a particle density higher than the initial β-nuclei density. This case would imply an additional nucleation either homogeneously or heterogeneously on α-particles. Powders located on the 1:1-line have the same β-nuclei and particle density. In other words, all β-particles present in the starting powder grow by dissolution of α-Si_3N_4-particles and reprecipitation as β.

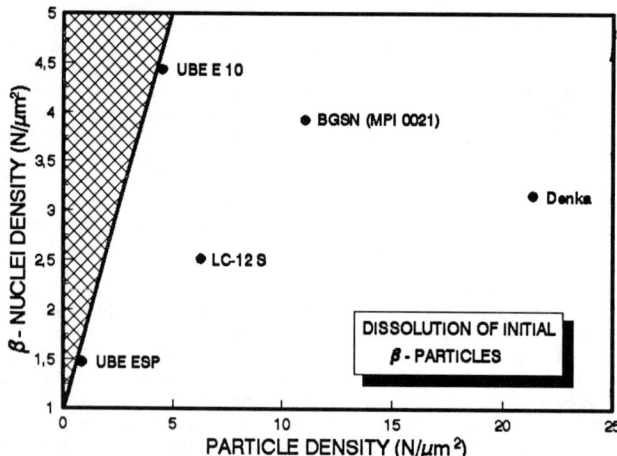

Fig. 1. Particle density after pressureless sintering as a function of the calculated β-nuclei density for 5 Si_3N_4 powders.

The diagram shows that all investigated powders had an equal or lower particle density in comparison to the initial β-nuclei density. While the

UBE-powders had approximately the same particle and β-nuclei density, the particle density of the BGSN (Bayer AG, Germany) and LC12-S (H.C. Starck, Germany) is smaller than the β-nuclei density indicating that smaller β-nuclei dissolve during densification. The Denka powder is a β-rich Si_3N_4-powder (97.5%) and complete densification of this powder occurs only by dissolution of the smaller β-particles instead of dissolution of α-grains and reprecipitation on pre-existing β-grains.

A low amount of initial β-nuclei results in a larger interparticle distance of β-grains in the green compact and the grains were able to grow in the first stage of densification without a steric hindrance. Therefore, the powder with the lowest β-nuclei density (UBE ESP) developed the coarsest microstructure (lowest particle density) with large elongated grains. With increasing initial β-nuclei density, the steric hindrance of grain growth increases, resulting in an increasingly equiaxed microstructure (Denka) as indicated in Fig. 2.

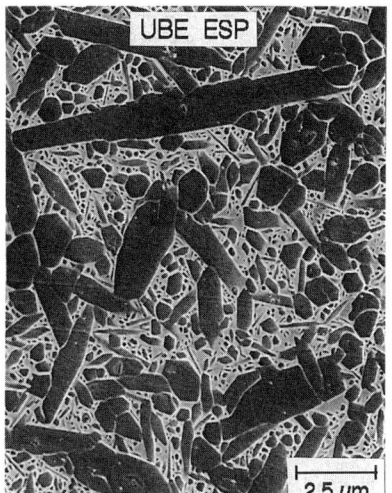 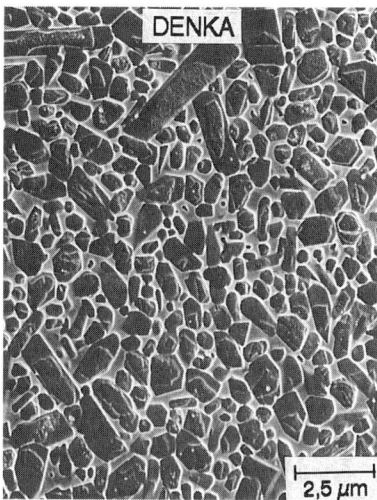

Fig. 2. SEM-micrograph of a plasma-etched sample of UBE ESP and Denka after complete transformation and densification by pressureless sintering.

To obtain the influence of the β-particle size distribution in the starting powder, mixtures of α-rich Si_3N_4 powder and β-rich powder were prepared. The α-rich reference powder had a mean β-Si_3N_4 particle size of 0.12 μm and an β-content of 4.1 v/o. Subsequently, 4 and 20 v/o β-Si_3N_4 (Denka) with a mean particle size of 0.28 μm were added to the reference sample. All specimens were pressureless sintered at 1780°C until phase transformation was complete. The microstructures were characterized by quantitative microstructural analysis taking into account the 3-dimensional random distribution of the elongated, hexagonal Si_3N_4 prisms. The 3-dimensional distribution of the grain size was calculated by a stereological transformation of the data taken from a 2-dimensional cross section. Finally, the results of microstructural characterization of the reference sample UBE E-10 and the powder mixtures were compared with microstructure data from a pressureless sintered sample with pure Denka-powder after 30 min sintering at 1780°C.

Figure 3 shows the 3-dimensional plots of the grain size and the aspect ratio distribution for the 4 investigated powders. The plot of the reference powder (UBE E-10) reveals a bimodal grain size distribution with a maximum grain length of 4.5 μm and a corresponding aspect ratio between 7 and 8. A small addition of 4 v/o β-particles (E-10/Denka 96:4) leads to an increase in grain length with a simultaneous decrease in aspect ratio for the larger grains, indicating the growth of the added larger diameter β-particles. The observation of a smaller steric hindrance of small amounts of large diameter grains has been already described in a previous paper [11]. However, with an increasing content of additional β-Si_3N_4 particles (E-10/Denka 80:20), grain growth of large diameter grains is hindered by an increasing impingment and the maximum grain length is reduced. Figure 3 shows a more uniform grain length and aspect ratio distribution with maximum aspect ratios of 5. In the case of the Denka powder with an initial β-content of 97.5 v/o, the mean grain length is even more reduced and the grains become more equiaxed, as already shown in Fig. 2.

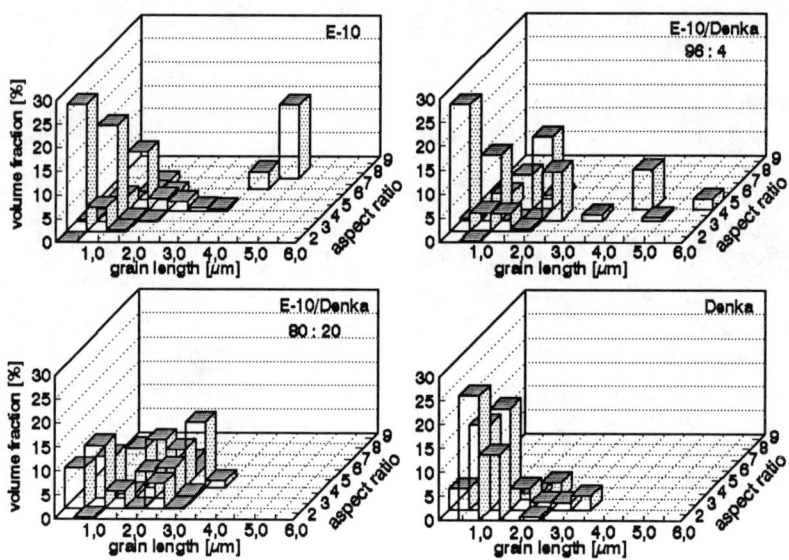

Fig. 3. 3-dimensional grain length and aspect ratio distribution for various Si_3N_4 powders.

Another important parameter for the microstructural development is the sintering temperature and time. After complete phase transformation grain growth starts by dissolution of smaller β-grains, as concluded from Fig.4. Mitomo et al. [13] observed an increasing aspect ratio with increasing sintering temperatures up to 1950°C and a decrease with further temperature increases. However, comparing α- and β-rich Si_3N_4 powders he found only small differences in the mean aspect ratio after sintering at 2000°C for 1 h.

Some grains with a large initial diameter can grow in length direction with a minor steric hindrance more minor than the smaller ones, as pointed out earlier. Their maximum length could reach 100-200 μm. The amount of such large grains, which could reduce the bending strength, depends on the initial β-particle size distribution of the starting powder [12,14].

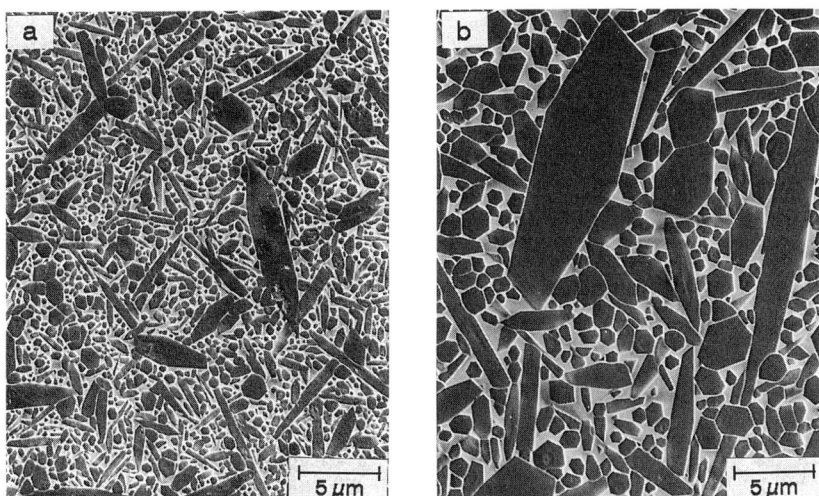

Fig. 4. SEM-micrograph of gas pressure sintered Si3N4 after 35 mins at 1835°C (a) and 360 mins at 1900°C (b), respectively.

Mechanical Properties

The measurement of the fracture toughness of specimens with different microstructures, prepared with the same amount of sintering additives and Si3N4 powder, has revealed a toughness increase with increasing mean grain size. The sample sintered at 1850°C for 35 mins has a toughness of 7.2 MPa$\sqrt{m}$ and the sample sintered for 360 mins at 1900°C has a toughness of 8.4 MPa$\sqrt{m}$. The corresponding strength distribution of both materials is shown in Fig. 5.

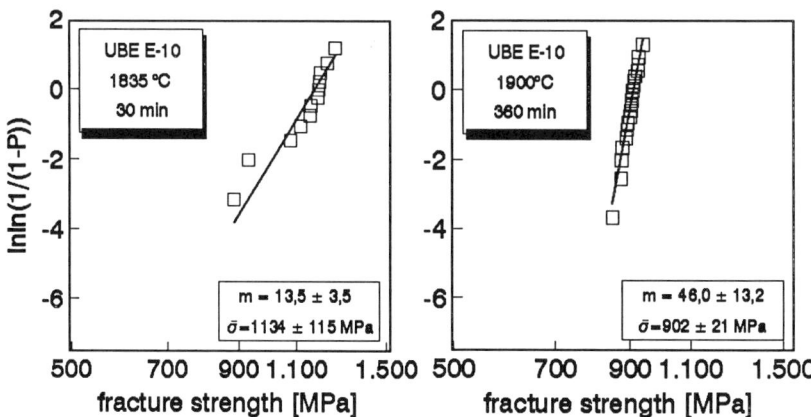

Fig. 5. Strength distribution for a fine- and coarse-grained Si3N4 material.

The fine-grained material with the lower toughness exhibits the higher RT-strength of 1134 MPa with an Weibull modulus of 13.5. With increasing grain size a decrease in strength is detectable, but the Weibull modulus increased from m = 13.5 to m = 46. This significant improvement in reliability is unlikely due to a very narrow flaw size distribution. According to theoretical considerations of Schneider et al. [15], the high Weibull modulus can be easily explained by an R-curve behaviour. The model of Schneider is schematically shown in Fig. 6. Under the assumption of a Weibull distribution of flaw sizes and a constant fracture toughness of the material, the failure probability depends linearly on the fracture strength and a typical Weibull modulus for a material prepared in a laboratory of m = 10 is achieved. A ceramic with the same flaw size distribution and a higher fracture toughness, K_{Ic}^2 in Fig. 6, would have the same Weibull modulus, but the strength data are shifted to higher values. However, if a material with an R-curve behaviour is considered, the fracture toughness increases with crack extension, as shown schematically in the insert of Fig. 6. Under the assumption of the same flaw size distribution for both the material with a constant fracture toughness and the one with an R-curve behaviour, the Weibull modulus is not the same. Figure 6 illustrates the general shape of the strength distribution curve for a material with an initial fracture toughness of K_{Ic}^1 and a plateau value of K_{Ic}^2. Of course, the change in the slope is strongly dependent on the shape of the R-curve. For Si3N4 ceramics with larger diameter grains, as is the case for the material sintered for 360 mins at 1900°C, the expected R-curves should reach the plateau value after approximately 100 μm crack extension [15].

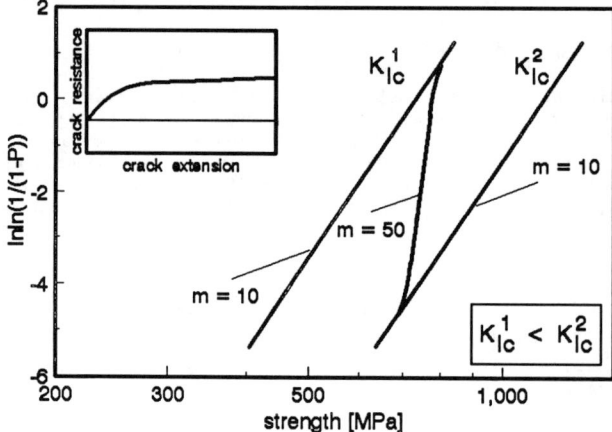

Fig. 6. Schematic diagram of the Weibull modulus change for materials with an R-curve behaviour as compared to materials with a constant fracture toughness.

Thermal Shock Behaviour

Another important advantage of materials exhibiting an R-curve behaviour is the improvement of the thermal shock behaviour. Normally, high strength ceramics with strength determining flaw sizes of 50 μm show a catastrophic failure at a certain quench temperature, Figure 7 left side. The non-catastrophic behaviour is only observed for low strength materials, like porous refractories. However, the right diagram of Figure 7 shows two Si3N4 materials sintered with the same sintering additives, but exhibiting different microstructures.

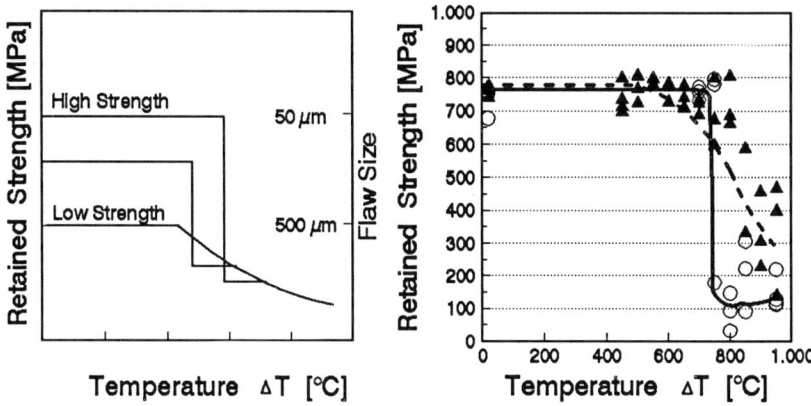

Fig. 7. Retained strength after thermal shock by a water quench test. Left diagram illustrates schematically the typical strength degradation for high strength and low strength materials. Right diagram shows the thermal shock behaviour of Si_3N_4 with different microstructures [16].

The fine-grained material (Fig. 7, open circles) reveals the expected catastrophic failure at a quench temperature of 750°C, while the coarse-grained Si_3N_4 (triangles) shows a strength degradation over a temperature range of 500°C. At the maximum quench temperature of 1000°C, the mean retained strength of the coarse-grained material is still approximately twice that of the fine-grained one. Again, an R-curve behaviour is expected for the coarse-grained material, leading to the observed behaviour during a thermal shock test [16].

INFLUENCE OF THE GRAIN BOUNDARY PHASE

Tajima [17] investigated Si_3N_4 ceramics with similar microstructures and found a dependency of the toughness on the sintering additives. This observation is attributed to a change from a transgranular fracture mode for the low toughness material to an intergranular one for the high toughness Si_3N_4 caused by different interfacial properties [14]. Notwithstanding, the importance of the sintering additives becomes larger in the case of high-temperature properties.

Phase Relationships

The development of specific microstructures with well-defined crystalline grain boundary phases is difficult without the knowledge of phase relationships of the phases present after sintering. Unfortunately, most Si_3N_4 ceramics are quaternary or even quinary systems due to the necessity of oxide additions as sintering additives. However, if the elements do not change their valency, the concentrations can be expressed in equivalent percent (eq.%) and the systems are reduced to pseudoquaternary and pseudoternary ones, respectively [18].

Figure 8 shows, as examples, the pseudoternary phase diagrams for Y- and Yb-containing Si_3N_4 ceramics. The figure on the left side reveals a subsolidus diagram of the system Yb_2O_3-YbN-SiO_2-Si_3N_4 with the two pseudobinary compounds, $Yb_2Si_2O_7$ and Yb_2SiO_5, both coexisting with Si_3N_4. In the Y-based system only $Y_2Si_2O_7$ is in equilibria with Si_3N_4. The Yb-system has one pseudoternary compound ($Yb_4Si_2O_7N_2$) in comparsion to the 4 phases of the yttrium related diagram on the right side of Fig. 8. The reason for the instability of the Apatit- ($Y_{10}(SiO_4)_6N_2$), Wollastonit- ($YSiO_2N$), and Melilitephase ($Y_2Si_3O_3N_4$) is the small cation radius of the ytterbium (85.8 pm) in comparsion to yttrium (89.3 pm). An analysis of the different crystal structures of the four pseudoquaternary phases in the Y-related system shows that the Wöhlerite ($Yb_4Si_2O_7N_2$) is the only structure, tolerating a wide variation in the radii of interstitial cations. The Melilite structure is based on SiO_4-tetrahedron layers connected by cations (e.g. Y^{3+}) and the Wollastonite consists of SiO_4-chains connected by cations. In the case of very small cations causing the narrow distance between the SiO_4-layers and -chains, respectively, repulsive forces between the SiO_4-tetrahedron and both structures became unstable. A pure nitrogen containing Yb-apatite is also not stable due to the small cation radius. Nevertheless, the Yb-apatite phase can be stabilized by substituting less than 20 at.% of the Yb^{3+}-Ion by a larger cation like Ca^{2+} [19]. All pseudoternary compounds contain nitrogen and can cause a severe oxidation of the ceramic at relatively low temperatures (800-1000°C) due to the difference in molar volume between the oxidizing phases and the oxidation products [20].

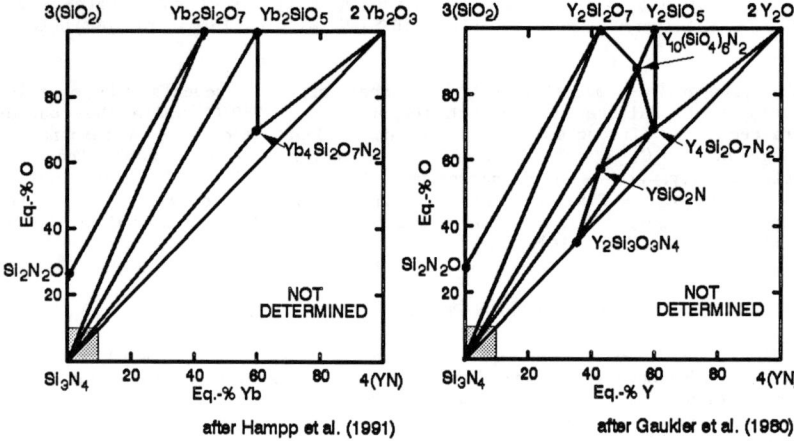

Fig. 8. Comparison of the subsolidus phase relationships in the systems Si_3N_4-Yb_2O_3- SiO_2-YbN [21] and Si_3N_4-Y_2O_3-SiO_2-YN [22].

Most of the Si_3N_4 ceramics have an additive content smaller than 10 eq.% and the overall compositions are located in the marked squares in the Si_3N_4-rich corner. For the final phase relationship between the Si_3N_4 matrix and the grain boundary phase, the SiO_2-content of the Si_3N_4 powder and the oxygen pick-up during powder preparation has to be taken into account because small changes in oxygen content can cause changes in phase relationships.

High Temperature Strength

In contrast to the room temperature properties, the high-temperature strength is dominated by the additives determining the softening temperature of the grain boundary phase. The crystallization of the amorphous grain boundary phase is one opportunity for improving these properties. However, the devitrified materials with crystalline grain boundaries often exhibit improvements in the temperature range > 1200°C, but a simultaneous strength decrease at room temperature.

Figure 9 shows the temperature dependency of the bending strength of Yb-doped Si_3N_4 containing 5 (SN5Yb(Al)) and 10 vol.% Yb_2O_3 (SN10Yb(Al)), respectively, and additionally 0.5 vol.% Al_2O_3. The as sintered SN5Yb(Al) sample with an amorphous grain boundary phase has a bending strength of 1050 MPa at room temperature and a continous decrease above 800°C to 380 MPa at 1350°C. During post heat treatment $Yb_2Si_2O_7$ crystallizes in the triple junctions, but the total degree of crystallization of the grain boundary region, determined by TEM, is only approximately 80%. The reason for the incomplete grain boundary crystallization might be attributed to the Yb_2O_3-SiO_2 ratio. The overall composition is located in the compatibility triangle Si_3N_4-Si_2N_2O-$Yb_2Si_2O_7$, but X-ray investigations showed only $Yb_2Si_2O_7$ and Si_3N_4. At all temperatures, the crystallization of the secondary phase in the grain boundary causes a strength degradation in comparison to the as-sintered material.

The SN10Yb(Al) samples with the higher additive content already had a partially crystalline grain boundary after sintering, causing a strength degradation at room temperature in comparison to the SN5Yb(Al) samples. However, at temperatures > 1000°C both materials exhibit a similar high temperature strength behaviour. The secondary phase ($5Yb_2O_3 \cdot Si_3N_4 \cdot Al_2O_3$) present in the as-sintered SN10Yb(Al) samples is not in equillibrium with Si_3N_4 and reacts during post heat treatment at 1200°C with the residual amorphous phase (approximately 30% of the grain boundary) to form the Yb-silicate phases (Yb_2SiO_5 and $Yb_2Si_2O_7$), as expected from the phase diagram. The nearly complete crystallization of the grain boundary leads to a strength degradation in the intermediate temperature range, but the strength between 1200 and 1350°C remains constant.

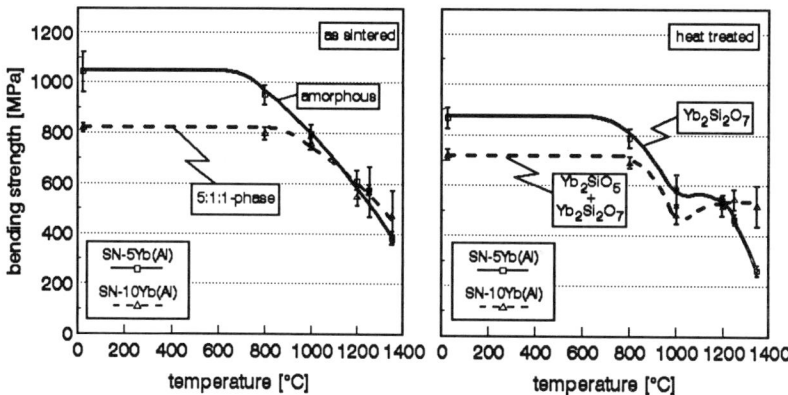

Fig. 9: Bending strength as a function of temperature for Yb-containing Si_3N_4 ceramics.

The often observed strength degradation at room temperature for devitrified Si_3N_4 ceramics can be attributed to the generation of internal stresses caused by the volume misfit between the crystalline and the amorphous grain bondary phase and the misfit in thermal expansion coefficient between matrix and secondary phase. However, a crystallization of the grain boundary region does not necessarily improve the strength at higher temperature, as shown in the case of the SN5Yb(Al)-composition.

Additionally, HRTEM investigations of Kleebe et al. [23] have shown that the complete crystallization of the grain boundary phase is not possible. All investigated Si3N4 ceramics had revealed an amorphous layer in the 2-grain junctions with a thickness between 1 and 2 nm, depending on the additive system. The influence of this grain boundary film on the mechanical properties of Si3N4-ceramics is not well understood.

Creep Behaviour

The creep resistance of the two compositions SN5Yb(Al) and SN10Yb(Al), described in the last chapter was measured in 4-point bending at 1200°C in air. The creep strain as a function of time is plotted in Fig. 10.

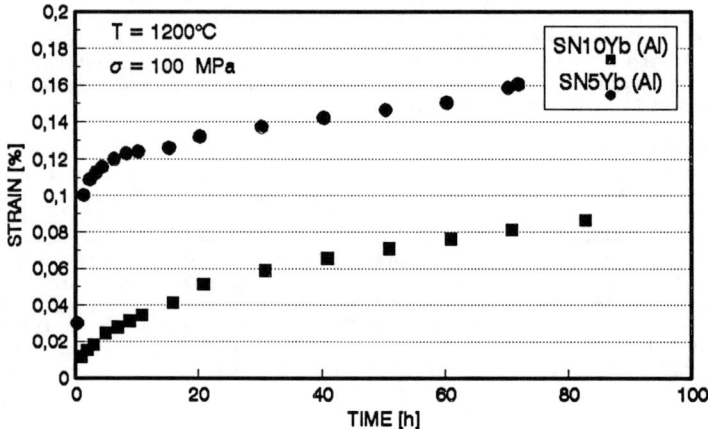

Fig.10: Creep resistance of gas pressure sintered and annealed Yb-containing Si3N4 ceramics.

The SN10Yb(Al) sample with the higher additive content reveals a much better creep resistance in the primary creep stage in comparison to SN5Yb(Al), although the strength of both materials is the same at 1200°C (Fig.9). The stationary creep rate in the secondary creep regime is approximately equal. However, the smaller creep deformation reflects the higher strength of SN10Yb(Al) above 1200°C in comparison to SN5Yb(Al), Fig. 9. It should be pointed out that both materials reveal the same amorphous grain boundary film thickness of 1 nm, but the sample SN10Yb(Al) has a larger grain size and a nearly complete crystallization of the triple junctions.

CONCLUSIONS

Microstructural development of Si3N4 based ceramics has been discussed with respect to amount and size distribution of β-Si3N4 particles in the starting powder under the assumption that only initial β-particles grow. The calculated β-nuclei density and the measured final particle density showed a dissolution of a certain amount of smaller β-nuclei in the case of powders with a high β-content or a wide particle size distribution. The observations were confirmed by model experiments with defined powder mixtures and quantitative microstructural analysis. After complete transformation and densification grain growth occurs and large diameter β-nuclei grow mainly in length direction due to their minor steric hindrance. When the maximum grain length exceeds 100-200 μm, the grain size becomes strength determining.

High strength Si_3N_4 ceramics with a mean bending strength > 1100 MPa exhibit a fine-grained microstructure and residual pores or agglomerates determine the strength. However, the fracture toughness and the reliability of Si_3N_4 could be significantly improved by a precisely controlled grain growth. For the coarse-grained microstructures an increase in fracture resistance with increasing crack extension (R-curve) is expected. The existance of steep R-curves, reaching the plateau value after 100 µm explain the observed high Weibull modulus as well as an improvement in thermal shock behaviour.

The high temperature properties of Si_3N_4 ceramics are mainly controlled by the chemistry of the grain boundaries. The crystallization of the amorphous triple junctions does not always improve the high-temperature strength and creep resistance. Generally, the room temperature strength of Si_3N_4 ceramics with a crystalline grain boundary phase is lower than that for the materials with an amorphous one due to the formation of internal stresses during crystallization. However, for a well-defined devitrification of the grain boundary by a post heat treatment, the phase relationships between the Si_3N_4-matrix and the sintering additives have to be known and the oxygen content of the Si_3N_4 powder as well as the oxygen pick up during processing has to be taken into account in order to crystallize the desired grain boundary phase. A reduction in the total amount of sintering additives does not necessarily result in an improvement in high-temperature properties.

Finally, it has to be mentioned that Si_3N_4 ceramics represent a whole class of materials. Depending on the application, different Si_3N_4 ceramics have to be developed for qualities such as high strength (>1200 MPa up to 1000°C), good thermal shock resistance, or improved high-temperature properties.

Acknowledgement

The authors would like to thank W. Dressler, E. Hampp, and G.A. Schneider for their helpful discussions. The work was supported by the German Ministry of Science and Technology (BMFT) and the companies Bayer AG, Daimler-Benz AG, Hoechst AG, and MTU under contract number 03 M 2012.

REFERENCES

[1] G.R. Terwilliger and F.F. Lange, J.Mater.Sci., 10, (1975), 1169.
[2] S. Boskovic, L.J. Gaukler, G. Petzow and T.Y. Tien, Pow.Met., 9, (1977), 185.
[3] M. Mitomo, J. Mat. Sci., 11, (1976), 1103.
[4] M. Peukert and P. Greil, J. Mat. Sci., 22, (1987), 213.
[5] J. Weiss and W.A. Kaysser, in F.L. Riley (ed.), Progress in Nitrogen Ceramics, (1983), 169.
[6] J. Bressani, Ph.D. Thesis, University Stuttgart, (1984).
[7] E. Butler, R.J. Lumby, A. Szweda and M.Lewis, in: S. Somiya et al. (eds.), Proc. Int. Symp. on Ceramic Component for Eng., Tokyo, (1983).
[8] F.F. Lange, J. Am. Ceram. Soc., 62, (1979), 428.
[9] M. Mitomo, M. Tsutsumi, H. Tanaka, S. Uenosono, F. Salto, "Grain Growth During Gas-Pressure Sintering of β-Silicon Nitride", J. Am. Ceram. Soc., 73, [8], (1990), 2441-2445.
[10] W. Dressler, M.J. Hoffmann, and G. Petzow, "Analysis of Microstructural Development in Si_3N_4-Ceramics", Presentation at the 44th Pacific Coast Regional Meeting of ACS, San Diego (CA/USA), November 1991.
[11] G. Petzow und M.J. Hoffmann, "Grain Growth Studies in Si_3N_4-Ceramics", Materials Science Forum Vols. 113-115, Trans. Tech. Publications, Switzerland, (1993), 91-102.

[12] W. Dressler, "Microstructural Development and Mechanical Properties of Si_3N_4 Ceramics", Ph.D. Thesis, University Stuttgart, (1993).
[13] M. Mitomo, S. Uenosono, "Microstructural Development During Gas-Pressure Sintering of α-Silicon Nitride", J. Am. Ceram. Soc., **75**, [1], (1992), 103-108.
[14] P.F. Becher, H.T. Lin, S.L. Hwang, M.J. Hoffmann, and I.W. Chen, "The Influence of Microstructure on the Mechanical Behaviour of Silicon Nitride Ceramics", MRS-Proceedings, this volume, (1993).
[15] G.A. Schneider, K.D. Debschütz, and G. Petzow "Strength Variability of Ceramics with Rising Crack Growth Resistance", submitted to Acta Met. (1992).
[16] M.J. Hoffmann, G.A. Schneider, and G. Petzow, "The Potential of Si_3N_4 for Thermal Shock Applications", to be published in Thermal Schock and Thermal Fatigue Behaviour of Advanced Ceramics, G.A. Schneider and G. Petzow (eds.), NATO ASI Series, (1993).
[17] Y. Tajima, K. Urashima, M. Watanabe, and Y. Matsuo, "Fracture Toughness and Microstructure Evaluation of Silicon Nitride Ceramics", in Ceramics Transactions, Vol.1: Ceramic Powder Science-IIB, G.L. Messing, E.R. Fuller, Jr., and H. Hausner (editors), Am. Ceram. Soc., Westerville, OH, (1988), 1034-41.
[18] L.J. Gaukler and G. Petzow, "Representation of Multicomponent Silicon Nitride Based Ceramics", in Nitrogen Ceramics, F.R. Riley (ed.), NATO Advanced Study Institute, Canterbury, UK, (1976).
[19] M.J. Hoffmann, J. Gröbner, E. Hampp, and G. Petzow, to be published.
[20] J.K. Patel and D.P. Thompson, "The Low-temperature Oxidation Problem in Yttria-densified Silicon Nitride Ceramics", Brit. Ceram. Trans. J., **87**, (1988), 70.
[21] E. Hampp, "Phase Relationships and Sintering Behaviour of Yb-Doped Si_3N_4", Ph.D. Thesis, University Stuttgart, 1993.
[22] L.J. Gaukler, H. Hohnke and T.Y. Tien, "The System $Si_3N_4-SiO_2-Y_2O_3$", J. Am. Ceram. Soc., **63**, (1973), 35.
[23] H.J. Kleebe, M.J. Hoffmann and M. Rühle, "Influence of Secondary Phase Chemistry on Grain-Boundary Film Thickness in Silicon Nitride", Z. Metallkde, 83, [8], (1992), 610-17.

SIALON CERAMICS: RETROSPECT AND PROSPECT

KENNETH H. JACK
University of Newcastle upon Tyne and the Cookson Group plc,
Wallsend, Tyne & Wear, United Kingdom

ABSTRACT

The future for sialon ceramics is bright. Engine components are slowly increasing in numbers and variety, and applications for wear parts and molten metal handling are set to flourish.

Two major developments have been: (i) the production of useful sialon composites; and (ii) the ability to join pieces to make large and complex components.

In (i), properties of $\alpha':\beta'$ sialon composites can be varied by processing modifications in ways that are impossible for single-phase materials. Also, densifying additives such as yttria and rare-earth oxides can subsequently be incorporated into the α'-structure to minimize the amount of intergranular glass. The potentialities of other composites including $O':\beta'$, $O':ZrO_2$, $\beta':TiN$ and $\beta':BN$ are assessed.

In (ii), the joining of sialons follows from an understanding of phase relationships, glass formation and glass-ceramics in the Y-Si-Al-O-N system. Its exploitation has greatly widened the field of sialon applications.

INTRODUCTION

The major incentive for national programs on engineering ceramics has always been the ceramic gas turbine. The British program, starting in the early 1950s, produced two engineering ceramics: first, reaction-bonded silicon nitride (RBSN); and then hot-pressed silicon nitride (HPSN) using MgO as a densifying additive. RBSN is up to 25% porous and so is not strong enough for most engineering applications. On the other hand, HPSN is limited to simple shapes which are so hard that components must be produced from them by prohibitively expensive diamond machining. Just when the British activity was running down, silicon nitride was resuscitated in 1971 for the first American gas turbine program. Running at a target temperature of 1370°C the ceramic engine promised more efficiency, fuel saving, and less environmental pollution.

American research and development produced pressureless sintered silicon nitride (SSN) and also replaced MgO by Y_2O_3 as a densifying additive.

All additives degrade properties; smaller amounts of additive give less degradation but require higher nitrogen gas pressure to prevent nitride decomposition; this gas-pressure sintering (GPS) is now widely used. Where one component in a mixed additive is alumina, e.g. 5w/o Y_2O_3+5w/o Al_2O_3, the distinction between SSN and a sialon is doubtful.

Finally, densification by hot-isostatic pressing (HIP) uses the least amount of additive and should therefore show the least degradation of properties. Balanced against this, it is arguable that HIP is too expensive, particularly for the production of small numbers of components.

After nearly forty years and several thousand million dollars spent on successive national programs, there is still no commercial ceramic engine. The most recent Japanese project, due for completion by 1996, is for three types of 300kW industrial gas turbines and a 100kW automotive engine, all with a gas inlet temperature of 1350°C. I said five years ago [1] "...the ceramic turbine ... is probably the most difficult application that could

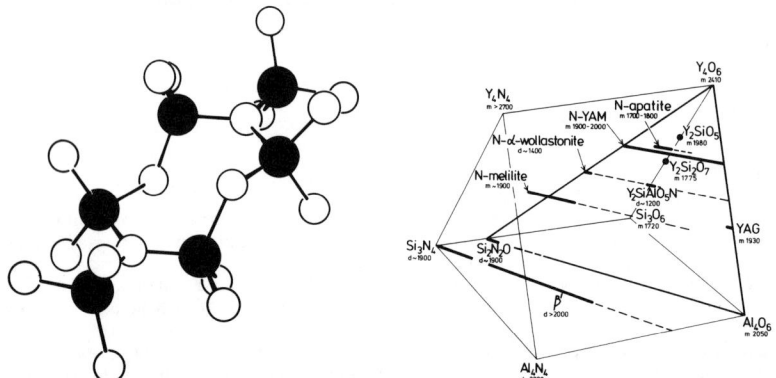

Fig. 1 Crystal structure of
        β-Si$_3$N$_4$ and β'-(Si,Al)$_3$(O,N)$_4$

Fig. 3 The Y-Si-Al-O-N system

have been proposed. To pay for continuing efforts towards this goal, less exacting 'bread and butter' applications must be successfully pursued. At present (i.e. in 1987), the market penetration of these wear parts, bearings, turbochargers and small engine components is almost negligible. Judged by previous experience of materials substitution, a substantial take-over of current automotive engines by ceramic gas turbines cannot be expected before the year 2050".

The present paper attempts to assess any changes in the ceramic engine prospects and in the "bread and butter" applications.

THE SIALONS

One of the offshoots of the British effort was the idea of "ceramic alloying" [2]. Both α and β silicon nitrides are built up of SiN$_4$ tetrahedra joined in a three-dimensional network by sharing nitrogen corners in the same way that SiO$_4$ units are joined in a silicate. In fact, the atomic arrangement in β-Si$_3$N$_4$ (see Figure 1) is the same as in Be$_2$SiO$_4$ and Zn$_2$SiO$_4$ and so the same principles of silicate crystal chemistry are applicable. Thus, up to two-thirds of the Si in β-Si$_3$N$_4$ can be replaced by Al without change of structure provided that an equivalent concentration of

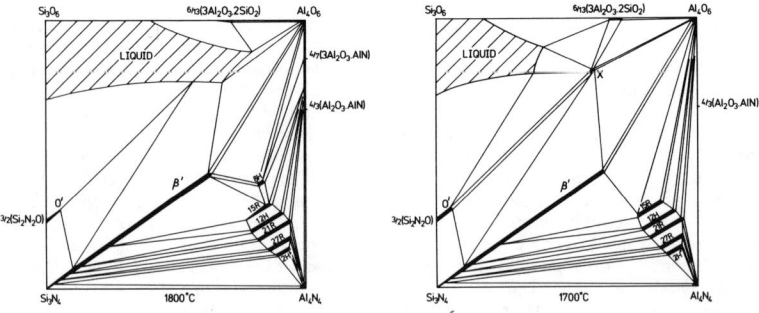

Fig 2. The Si-Al-O-N behaviour diagram at (a) 1800°C; and (b) 1700°C

N is replaced by O [3,4]:

$$Si^{4+} \quad N^{3-} \rightleftharpoons Al^{3+} \quad O^{2-} \tag{1}$$

In the Si-Al-O-N behaviour diagram of Figure 2 the β'-sialon phase extends over a range of hexagonal unit-cell contents:

$$Si_{6-z}^{(24-4z)} Al_z^{(3z)} O_z^{\overline{(2z)}} N_{8-z}^{\overline{(24-3z)}}$$

with 0<z<4. Because of its atomic arrangement, β'-sialon has mechanical and physical properties similar to those of β-Si_3N_4, but it is a solid solution and so, compared with silicon nitride, it forms more liquid at lower temperatures with oxide additives such as Y_2O_3. This allows the sialon to be densified by pressureless sintering like a traditional ceramic.

Yttrium β'-Sialons

The Y-Si-Al-O-N system (Figure 3) is represented by a Jänecke triangular prism the base of which is the Si-Al-O-N square of Figure 2. At 1800°C, Y-sialon liquid extends from the rear oxide face into the middle region of the prism.

Starting with a powder mix of Si_3N_4, Al_2O_3, AlN and Y_2O_3, these dissolve and react in the first-formed liquid to give β'-sialon. The liquid cools to give an intergranular glass but, by subsequent heat-treatment at ∼ 1400°C, this glass reacts further with the sialon;

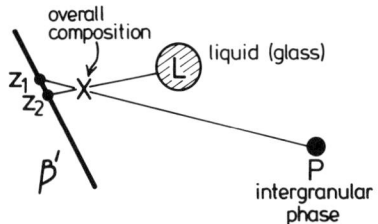

Fig. 4 Reaction of β' with liquid (glass) at 1440°C

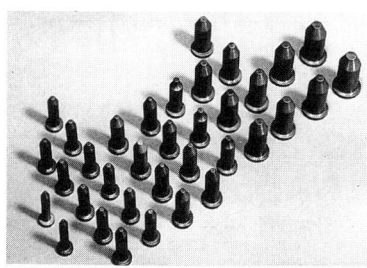

Fig. 6 Sialon weld location pins

Fig. 5 Swarf from cutting Ni-base superalloy for 10sec with (left) β'-sialon and (right) WC:Co

Idealised Si-N layers

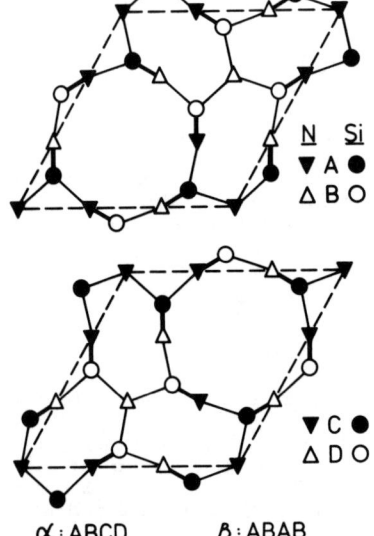

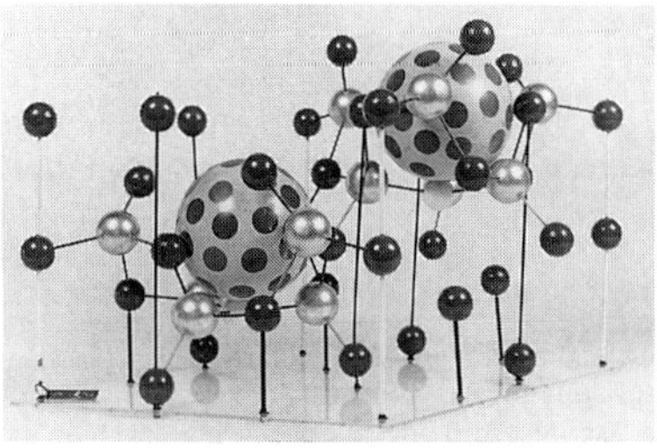

Fig. 8 Interstices in the α-$Si_{12}N_{16}$ unit cell

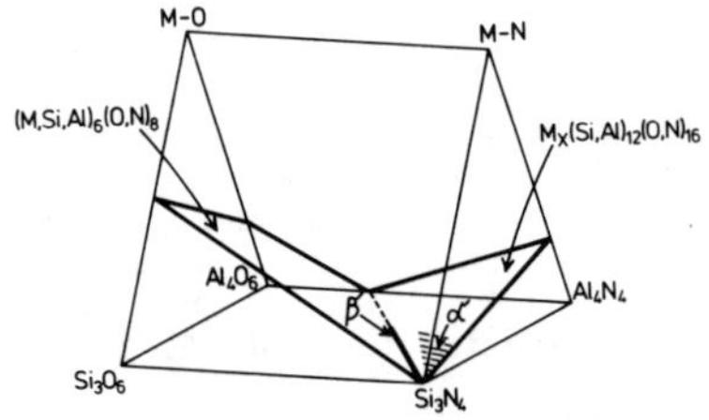

Fig. 7 Si-N layers in α and β silicon nitrides

Fig. 9 Relationship between α' and β' phases

see figure 4. The Si and N from the glass go back into the β'-sialon giving it a small decrease in z-value, while the remaining glass components Y, Al and O crystallise as a grain-boundary yttrium-aluminum garnet, $Y_3Al_5O_{12}$ (YAG):

$$Si_5AlON_7 + Y\text{-}Si\text{-}Al\text{-}O\text{-}N \longrightarrow Si_{5+x}Al_{1-x}O_{1-x}N_{7+x} + Y_3Al_5O_{12}$$

β'-sialon   glass β'-sialon   YAG   (2)

Based on these principles, two groups of sialons - β'+glass and β'+YAG - were commercially developed in the late 1970s. However, the reaction represented by equation (2) never goes to completion. Differences in composition between the liquid and crystalline phases make it impossible to produce a two-phase β'+YAG ceramic without some residual intergranular Y-sialon glass that degrades the properties. Despite this, the materials were and are successful; they have properties of HPSN with the advantages of near-net shaping and cheap, pressureless densification.

The first application of β'+glass was as a cutting tool. Its high hot-hardness allows it to machine metals at very high speeds where the tool-tip temperature exceeds 1000°C. Figure 5 compares the amounts of swarf from machining a nickel-base superalloy for 10 seconds with WC:Co and with sialon. Figure 6 shows sialon locating pins - another early application - used in the resistance welding of captive nuts onto automobile bodies. A conventional pin of hardened steel in a protective alumina sleeve lasts about 7,000 operations - an 8-hour working shift; a sialon pin lasts at least one year!

α'-Sialons

The two Si_3N_4 structures represent a stacking of Si-N layers in either an ABAB.... β-sequence or an ABCD.... α-sequence; see Figure 7. In the hexagonal β unit cell containing Si_6N_8 this gives continuous c-axis channels. In α, CD layers are identical with AB layers but are inverted and translated with respect to them by a c-glide plane. The continuous channels of β are thereby replaced in the $\alpha\text{-}Si_{12}N_{16}$ unit cell by two closed interstices that are large enough to accommodate other atoms; see Figure 8.

As in the "stuffed quartz" silicate structures, Si can be replaced by Al if valency compensation is made by "stuffing" additional cations into these two interstitial sites [5], for example:

$$\alpha\text{-}Si_{12}N_{16} \longrightarrow \alpha'\text{-}Ca_2[Si_8Al_4N_{16}] \quad \text{using } Ca_3N_2 \quad (3)$$

$$\text{or} \quad \alpha\text{-}Si_{12}N_{16} \longrightarrow \alpha'\text{-}Ca[Si_9Al_3ON_{15}] \quad \text{using } CaO \quad (4)$$

α'-Sialons can accommodate Li, Mg, Ca, Y and all the lanthanide elements except La, Ce, Pr and Eu the ionic radii of which (>1.0A) are too large. They have the same small coefficient of thermal expansion as β'-sialons - and hence good thermal shock resistance - and are no less strong, but they are very much harder. The two phases, α' and β', are completely compatible and an α':β' composite can be prepared by a single-stage process by heating together the appropriate mix of nitrides and oxides; see Figure 9. A 50α':50β' composite is more than twice as hard at 1000°C as pure β' and so commercial sialon cutting tool tips for metal machining have for several years been α':β' composites; these still contain some intergranular glass.

α':β' SIALON CERAMIC COMPOSITES

A further advantage of an α':β' composite is that its microstructure can be varied in a way that is impossible for a single-phase sialon. The

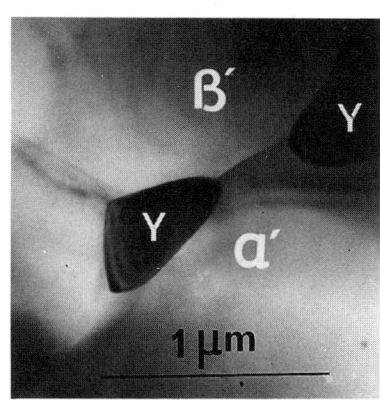

Fig. 10 TEM microstructure of α':β' ceramic after post-sintering heat-treatment; after Jasper and Lewis [6]

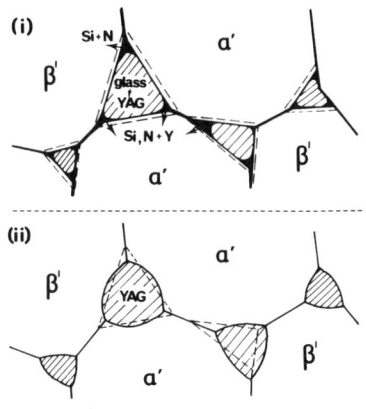

Fig. 11 The transport of non-stoichiometric elements during YAG crystallisation; (i) before and (ii) after heat-treatment. After Jasper and Lewis [6]

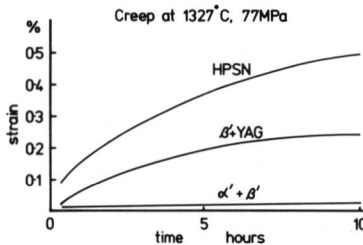

Fig. 12 Creep resistance of α':β' compared with β'+YAG and HPSN; after Cookson Syalons

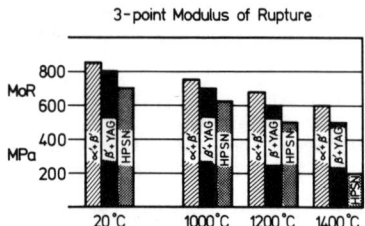

Fig. 13 3-point MoR for α':β' compared with β'+YAG AND HPSN; after Cookson Syalons

flexibility of the β' composition allows it to accommodate excess Si+N during the formation of β'+YAG ceramic from β'+glass. However, the α'-phase accommodates Al, N and O and also Y. Thus, within limits, an α':β' composite can function as a sink for all the elements of the Y-sialon liquid phase that is necessary for densification. This has been explored successfully by the Cookson Group in collaboration with Warwick University [6,7].

Powder mixes of α-Si_3N_4, 21R polytypoid and Y_2O_3 with overall composition on the tie-line between α':β' and the Y-sialon liquid are isostatically compacted and pressureless sintered at up to 1800°C to give an α':β' product with only 3-4v/o glass. This small volume of glass is then reacted and devitrified by a two-stage nucleation and growth heat-treatment to give crystalline YAG only at triple points; see Figure 10. The transport of non-stoichiometric elements during crystallisation is illustrated schematically by Figure 11. TEM lattice images show no intergranular glass [6].

Figures 12 and 13 show the superior creep properties and strength of this new α':β' composite. The absence of intergranular glass also causes slower grain-boundary diffusion of metallic ions and so improves oxidation resistance; after 50h at 1300°C in air only a 10μm oxidised layer is formed compared with 40μm for β'+YAG. The improvement in wear resistance is strikingly demonstrated by the performance of wire-drawing and extrusion dies. Figure 14 shows that for brass rod reduced in cross-section by 12% until the diameter reaches its upper specification limit due to die wear, the yield is 50% higher than with a WC die.

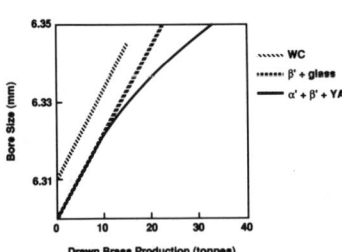

Fig. 14 Wear performance during brass wire-drawing: after Jasper and Lewis [6]

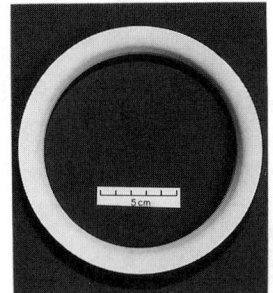

Fig. 15 β':15w/oBN break ring; after Cookson Vesuvius

Fig. 16 β':50w/oTiN composites; after Cookson Syalons

Fig. 17 Commutator of conducting Si_3N_4:TiN segments and insulating Si_3N_4:Al_2O_3 wedges [8]

For their AGT engine, Toyota [8] are developing a similar α':β' composite except that its grain size is much smaller. Acicular β' grains <1μm long combined with equi-axed 0.1μm α' give remarkable bend strengths of 1300 MPa at 20°C and 1000 MPa at 1400°C; $K_{1C} \sim$ 7MPa m$^{\frac{1}{2}}$.

Li_2O and CaO are avoided as densifying additives for nitrogen ceramics because they produce large volumes of low-viscosity glass. However, Li and Ca are accommodated more readily than Y and RE cations in the α' structure; the maximum x-value in the α' composition $M_x(Si,Al)_{12}(O,N)_{16}$ is higher for Ca (x $\sim$ 1.8) than for any other element. Thus, Li_2O and CaO are effectively cheap and "transient" additives that should be explored for densifying α'-sialons and α'-containing composites.

OTHER SIALON COMPOSITES

Composites other than α'-β' are in production or are under development. Thus, incorporation of boron nitride in β'-sialon increases thermal shock resistance by introducing fine porosity that accommodates strain by microcracking. Strength and elastic modulus are somewhat reduced but not critically for molten metal handling applications. Figure 15 shows a β':15w/oBN break ring used in the horizontal continuous casting of steel.

In the β':50w/oTiN composites shown by Figure 16, the TiN forms a continuous net-work. The material is therefore electrically conducting and so can be machined by spark erosion to produce precise and intricate shapes. Hitachi make similar RBSN:TiN composites by nitriding mixed Si+TiN powder shapes; the same is done with Si+Al_2O_3 to give an electrically insulating RBSN:Al_2O_3 composite. Finally, by alternating regions of TiN and Al_2O_3 in the initial Si mix, a complex composite is produced in one nitriding operation; Figure 17 shows an example - a commutator with conducting segments separated by insulating wedges [9].

O':β' Composites

Silicon oxynitride, Si_2N_2O, has a lower coefficient of thermal expansion and a higher oxidation resistance than the α and β nitrides. Its formation by the reaction of Si_3N_4 and SiO_2 (see Figure 2) is slow and incomplete in the absence of a liquid in which the reactants can dissolve. However, it forms O'-sialons with up to 10w/oAl_2O_3 in solid solution.

Additions of Y_2O_3+Al_2O to mixes of Si_3N_4+SiO_2 provide at $\sim$ 1600°C a Y-Si-Al-O-N liquid that allows rapid reaction and then densification to give O'-sialon and glass. Post-preparative heat-treatment replaces the

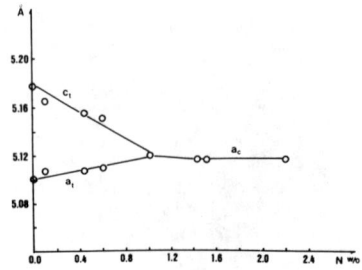

Fig. 18 Variation of TZ-3Y ZrO_2 unit-cell dimensions with nitrogen content [15]

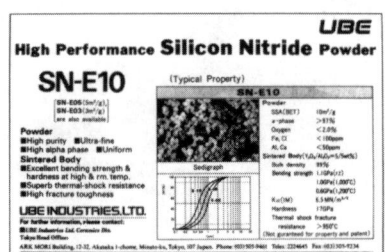

Fig. 19 Advertisement of di-imide Si_3N_4 powder; after Ube Industries

glass by intergranular $Y_2Si_2O_7$ and the product shows good oxidation resistance up to 1300°C [10]. With appropriate changes in the mix, O':ß' composites are obtained [11] but further work is required before the potentialities of these O'-sialons and composites can be assessed.

O':ZrO_2 Composites

Over the past few years the possibility of producing a transformation toughened nitrogen ceramic has been explored by the Cookson Group [12] and in Sweden [13]. Although ∼20w/oZrO_2 dispersed in other ceramics increases their toughness, a Si_3N_4:ZrO_2 composite is unstable at high temperature. ZrO_2 oxidises the nitride and is itself reduced to ZrN but, as shown by equation (5), the oxidation proceeds only as far as silicon oxynitride:

$$2Si_3N_4 + 1.5ZrO_2 \longrightarrow 3Si_2N_2O + 1.5ZrN + 0.25N_2 \quad (5)$$

ZrO_2 and Si_2N_2O are therefore compatible and do not interact chemically.

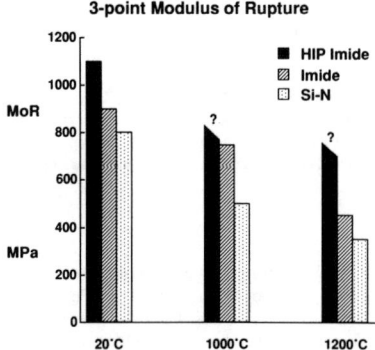

Fig. 20 MoR of ß'-sialon produced from different Si_3N_4 powders; after Nippon Steel Corporation

An intimate mixture of these is produced by the exchange reaction of Si_3N_4 with zircon [14]:

$$Si_3N_4 + ZrSiO_4 \longrightarrow 2Si_2N_2O + ZrO_2(25v/o) \qquad (6)$$

With additions of Al_2O_3, O'-sialon is formed instead of oxynitride and, by pressureless sintering mixtures of Si_3N_4, $ZrSiO_4$, Al_2O_3 and Y_2O_3 at $\sim 1700°C$ in N_2, fully dense $O':ZrO_2$ composites are obtained. Similar composites have been made at the Swedish Ceramic Institute by their NPS nitriding of mixes of Si metal powder, $ZrSiO_4$, Al_2O_3 and Y_2O_3. Both groups claim good strength and excellent high-temperature oxidation resistance but the expected increase in fracture toughness is not observed. Cheng and Thompson [15] explain this by showing that partially stablized tetragonal $ZrO_2:3m/oY_2O_3$ reacts above 1400°C with nitrides, or even molecular N_2, with conversion to the non-transformable t' and cubic forms of zirconia as the N-content increases. Figure 18 shows that conversion to cubic ZrO_2 is complete at $\sim 1.0w/oN$ so that nitrogen is a much more effective zirconia stabilizer than Mg, Ca or Y. It therefore seems unlikely that transformation-toughened composites containing ZrO_2 can be produced with any N-ceramic.

RAW MATERIALS AND PROCESSING

During the past five years there have been remarkable improvements in the properties of nitrogen ceramics, particularly in Japan. This is due to the better processing of purer raw materials and the routine use of GPS and HIP.

The purity of Si_3N_4 produced by nitriding Si or by carbothermally nitriding SiO_2 depends largely on the purity of the initial solids. The best Si_3N_4 is now obtained by the di-imide process from liquid NH_3 and $SiCl_4$, both of which can be purified to an almost unlimited extent.

Figure 19 advertises that imide powder gives, when sintered with $5w/oY_2O_3 + 5w/oAl_2O_3$, a product with a bend strength of 1,100MPa at room temperature and 600MPa at 1,200°C; K_{1C} is 6.5MPa $m^{\frac{1}{2}}$.

Figure 20 compares the strengths of ß'+glass sialon made from (i) nitrided silicon powder, (ii) imide powder, and (iii) imide powder and HIP; the advantages of pure powder and clean processing are obvious.

APPLICATIONS

The only automotive components at present in commercial production, mainly in Japan, are glow plugs, precombustion swirl chambers, rocker arm pads, tappet shims and turbocharger rotors. One-half are of GPS ß'-sialon with glass or YAG as intergranular phases; the remainder are ß'-Si_3N_4 densified with $RE_2O_3+SiO_2$ or with a variety of mixed oxides.

About 30,000 turbocharger rotors are produced per month, and each one is proof-tested at 120,000rpm (20% overspeed) and 1030°C. The negligible rejection rate of <0.1% and no reported failure-in-use have increased confidence in the eventual success of the ceramic gas turbine. Meanwhile, there is a huge market for other automotive components, particularly in valves, seat inserts, guides, push rods and tappets. Publications show that these are technically successful but are not yet commercially viable. Indeed, no automotive applications are profitable, and the ceramic turbocharger is still subsidised. A recent analysis [16] suggests that even if raw material costs are reduced to one-fourth of their present value the cost of a turbocharger rotor or other advanced ceramic component will never be lower than that of its metallic counterpart.

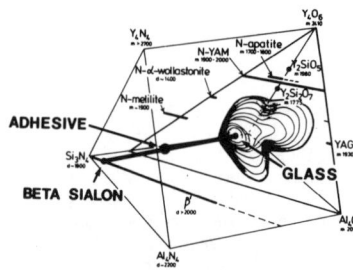

Fig 21. Relative positions of β'-sialon, adhesive, and glass-forming regions in the Y-Si-Al-O-N systems at 1600°C [17]

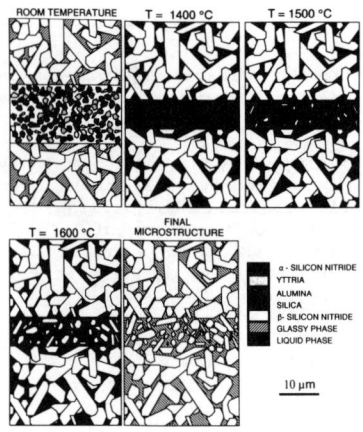

Fig 22 Schematic representation of the joining process [17]

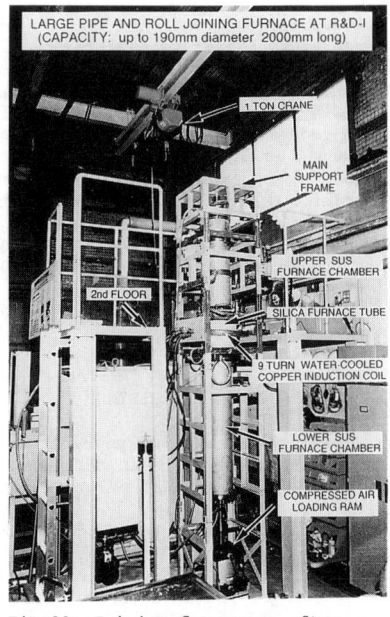

Fig 23 Joining furnace; after Nippon Steel Corporation

The engineering market that promises profitability in the near future is for wear parts like extrusion dies, rolls for producing metal foils, ball and roller bearings, and components for molten metal handling (particularly in aluminum die-casting). These are applications at medium temperature where the high strength and toughness of nitrogen ceramic, its good tribological properties, its non-reactivity with metals, hardness, high elastic modulus and low coefficient of thermal expansion, all give it a combination of properties superior to those of SiC, Al_2O_3 and ZrO_2.

As part of a diversification program, Nippon Steel set up in 1984 an Engineering Ceramics Centre to explore production of the above four "fine ceramics". Now, after eight years, the predominance of sialons (55%) has been made possible by the development of a joining technique [17] that allows the fabrication of large and complex shapes.

Fig 24. Sialon pipe, 95 cm x 17 cm o.d. in pre-joined state (wall thickness, 10mm); after Nippon Steel Corporation

Fig. 25 Sialon rolls, 1.4m x 6cm diam; after Nippon Steel Corporation

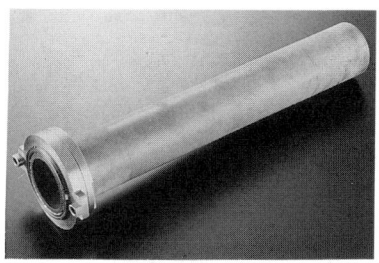

Fig. 26 Stalk, 1m x 10cm diam., for pumping molten Al; after Nippon Steel Corporation

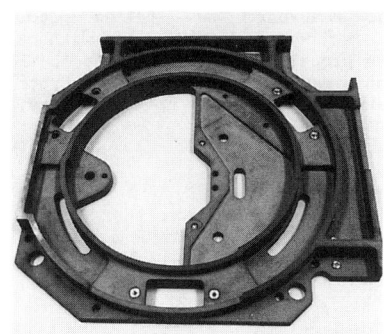

Fig. 27 Sialon X-Y machine table for VLSI semi-conductor manufacture; after Nippon Steel Corporation

Fig. 28 Sialon vacuum jig 20cm diam. for holding Si wafer during machining; after Nippon Steel Corporation

JOINING SIALON CERAMICS

In joining, Figure 21 shows that the adhesive is a powder mix of Si_3N_4, Al_2O_3, Y_2O_3 and SiO_2 which at 1600°C gives 60v/o β'-sialon + 40v/o Y-sialon liquid; the powder slurry in ethanol is sprayed onto the two surfaces, dried, and the pieces then pressed together at 2MPa in a graphite jig for 10min at 1600°C under a nitrogen pressure of 1atm. Sub-micron acicular β' grains are precipitated in the joint (Figure 22) and these intergrow with the sialon on each side. Failure never occurs at the joint since it is stronger than the joined pieces.

Figure 23 shows the jig and furnace for producing joined pipes and rolls up to 2m long and 19cm diameter, while Figure 24 shows pipe pieces of 10mm wall-thickness before joining.

The 1.4m sialon rolls for cold-rolling Cu, Al, Ti, W and stainless steel strip and foil (see Figure 25) are made in two pieces and then joined. Sialon gives twenty times higher wear resistance than a bearing-steel roll, thinner foils and a better finish due to its higher hardness, higher modulus, lower friction and non-adhesion to metals; its small thermal expansion also ensures that the foil keeps a uniform thickness during production. Figure 26 illustrates a stalk, 1m long x 10cm diameter, used in pumping molten aluminum into the mould of a die-casting machine.

The underside of a 30cm diameter sialon X-Y table is shown in Figure 27; it is used in an expanding niche-market within the semi-conductor industry for machining VLSI circuits. Sialon is rigid, does not distort or vary in dimensions with changes of ambient temperature; it is wear resistant and its low inertia allows the table to be positioned rapidly and precisely for each operation. Finally, Figure 28 shows the 20cm diameter sialon vacuum jig for holding the silicon wafer during its precise machining.

With these relatively large components Nippon Steel's Engineering Ceramics Centre is now showing a steadily increasing profitability - unique for any engineering ceramics manufacturer.

CONCLUSIONS

The future for nitrogen ceramics generally, and for sialons in particular, is bright. The ceramic gas turbine is still another decade away and automotive components are not yet commercially viable, but non-automotive applications, especially for wear-parts and molten metal handling, are set to flourish. Successful methods of densification without degradation of properties allow the unique properties of sialons and their composites to be exploited more fully than ever before, and the ability to join pieces to make large and complex components has opened up new markets.

ACKNOWLEDGEMENTS

For permission to reproduce figures, I am greatly indebted to: The Cookson Group (Nos. 6, 10, 11, 12, 13, 14, 15, 16); Hitachi Limited (No. 17); Nippon Steel Corporation (Nos. 20, 21, 22, 23, 24, 25, 26, 27, 28); Dr. D.H. Jack (No. 5); Professor M.H. Lewis (Nos. 10, 11, 14); Dr. D.P. Thompson (No. 18); and Dr. P.A. Walls (Nos. 21, 22).

REFERENCES

1. K.H. Jack, in Ceramics and Civilization, Volume III, High-Technology Ceramics - Past, Present, and Future, edited by W.D. Kingery (The American Ceramic Society, Inc., Westerville OH, 1987) pp. 259-288.

2. S. Wild, P. Grieveson and K.H. Jack, The Crystal Chemistry of Ceramic Phases in the Silicon-Nitrogen-Oxygen and Related Systems, Progress Report No. 1, Ministry of Defence Contract N/CP.61/9411/67/4B/MP.387, 1968.

3. Y. Oyama and O. Kamigaito, Jpn. J. Appl. Phys. **10**, 1637 (1971).

4. K.H. Jack and W.I. Wilson, Nature Phys. Sci. **238**, 28 (1972).

5. S. Hampshire, H.K. Park, D.P. Thompson and K.H. Jack, Nature (London) **274**, 880 (1978).

6. C.A. Jasper and M.H. Lewis in <u>Proc. 4th Int. Symp. on Ceramic Materials and Components for Engines</u>, edited by R. Carlsson, T. Johansson and L. Kahlman (Elsevier Science Publishers, London, 1992) pp. 424-431.

7. C.A. Jasper and Vesuvius Zyalons Midlands Ltd., European Patent Application No. 89303592.3 (12 April 1989).

8. Y. Ukyo, N. Sugiyama and S. Wada, Centennial International Symposium, Ceramic Society of Japan, Yokohama 1991; Exposition RD-33.

9. O.Y. Yasutomi, Centennial International Symposium, Ceramic Society of Japan, Yokohama 1991; Exposition RD-52.

10. M.B. Trigg and K.H. Jack, J. Mater. Sci. **23**, 481 (1988).

11. W-Y. Sun, D.P. Thompson and K.H. Jack, Mater. Sci. Res. **20**, 93 (1986).

12. D.B. Hoggard, H.K. Park, R. Morrison and S. Slasor, Ceram. Bull. **69**, 1163 (1990).

13. R. Pompe and R. Carlsson, "Reaction sintered $Si_2N_2O-ZrO_2$ composite", paper presented at the 90th Am. Ceram. Soc. Annual Meeting, Cincinnati, 1988 (unpublished).

14. S. Slasor and M.R. Anseau, Cookson Group 1988 (unpublished).

15. Y. Cheng and D.P. Thompson, J. Am. Ceram. Soc. **74**, 1135 (1991).

16. S. Das and T.R. Curlee, Ceram. Bull. **71**, 1003 (1992).

17. P.A. Walls and M. Ueki, J. Am. Ceram. Soc. **75**, 2491 (1992).

PROGRESS IN SILICON NITRIDE CERAMICS IN JAPAN

K. KOMEYA
Department of Materials Chemistry, Yokohama National University, 156 Tokiwadai, Hodogayaku, Yokohama, 240, Japan

ABSTRACT

Progress in silicon nitride ceramics in Japan is reviewed. It is historically divided into three stages. Through these stages, basic experimental research and innovations have progressed along with industrial applications, and the government project on fine ceramics in 1981-1992 has contributed much to the acceleration in the development of silicon nitride ceramics. Focus in this paper is mainly on materials development including raw powder synthesis and exploration for applications. The future prospect of utilizing silicon nitride as an engineering material, however, is seen to depend on cost reduction and reliability improvement.

INTRODUCTION

The beginning of silicon nitride development was probably in 1955 by Collins[1] in the UK. In the last 30 years silicon nitride development has consistently progressed year by year through worldwide effort. In particular, much research has been generated since 1971 when the U.S. government project on ceramic gas turbines started. In Japan, research on this material began in the 1960's with fundamental studies. This continued until 1978 when a major government project was initiated. Increasing activities, including material development, component manufacturing, product evaluation, and expansion of applications have flourished since then. In this paper, attention is directed to the history of materials development and steps toward applications in Japan.

STAGES OF DEVELOPMENT

Table I illustrates developmental stages of silicon nitride ceramics in Japan. The division into Stages I, II and III was based upon the author's personal view.
Research in Japan on silicon nitride started around 1963. In 1978, a major milestone, symbolized by the then initiated government sponsored "Moon Light Project," was reached. This project was aimed at the development of high efficiency gas turbine for power generation. Although the activities on ceramics were quite limited in this project, some companies and research organizations started their R & D on ceramic materials around this time. In 1981, the "Fine Ceramics Project" began and then, with the introduction of the ceramic glow plug, practical applications of silicon nitride progressed. At the same time, many issues on further material development and applications were identified.
According to the original plan, 1990 was to be the year when the "Fine Ceramics Project" would have ended. Actually, the project was extended for two more years to allow integration of the developed technologies. By then, however, the boom of ceramic activities had subsided and the remaining R & D effort continued at a more subdued but steady fashion.
As shown in Fig. 1, R & D of structural materials in Japan is now quite diversified. It includes nano-composites, whisker and fiber reinforced materials, functionally gradient materials, and other concepts. Although silicon nitride still figures prominently in the list, the author believes more concentration to this material, with all its excellent characteristics, is warranted. On the other hand, while the importance of cost reduction has been pointed out, there has not been much progress in this area for silicon nitride. Also important is the improvement of mechanical properties. In this respect, we are encouraged by the recent utilization of the self-reinforcing mechanism to toughen silicon nitride. It is interesting to note that the original patent on yttria-added silicon nitride (of 1969) was entitled "Heat Resistant Strengthened Composites."[2] Reliability, hopefully, can be enhanced at the same time by self-reinforcement.

Table I. Development Stages in Silicon Nitride Ceramics in Japan.

Stages	Periods	Activities	Government Projects
I	1963 - 1978	·Seed oriented research ·Fundamental study	
II	1978 - 1990	·Materials development ·Component fabrication processes ·Practical applications development ·Evaluation	Moon Light Proj (1978-85) Fine Ceramics Project (1981-92)
III	1990 -	·More reliable materials ·Cost reduction processes ·Expansion of applications	CGT Proj(300kw) (1988-96) CGT Proj(100kw) (1990-96) (Post Fine Ceramics Proj)

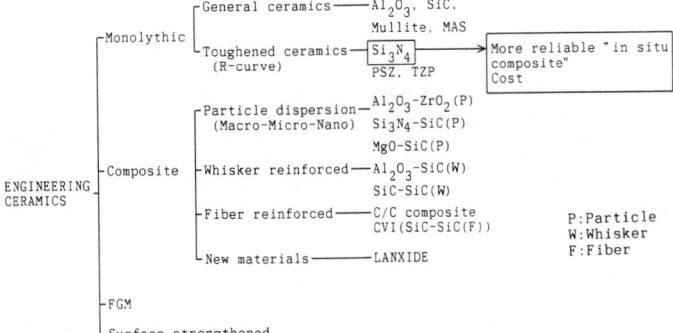

Fig. 1. Engineering Ceramics

MATERIALS DEVELOPMENT IN EACH STAGE

Stage I

During the period of Stage I, seed research on materials was conducted. Some of the highlights, which have contributed to today's progress on silicon nitride ceramics, are reviewed below.

In 1969, it was found that rare earth oxides, especially Y_2O_3, were the optimum additives for densification. Simultaneously, strengthening through the formation of elongated silicon nitride grains (Fig. 2) was reported with these additives.[3] Figure 3 shows high strength data obtained from the hot pressed Si_3N_4-Y_2O_3-Al_2O_3 by Tsuge et al.[4] In this experiment, AME CP-85 silicon nitride powder was used. During this period, the effect of free silica in silicon nitride powder on the strength at RT and 1200°C was also recognized. The results are shown in Fig. 4.[5] It is interesting to note that in this study, the newly available α silicon nitride powder from the silica reduction/nitridation method was used. While this work already suggested the importance of pre-existing oxygen in silicon nitride powder, it was through TEM analysis that the role of crystallization of the grain boundary phase was first demonstrated. In such work, it was shown that an amorphous layer remained in the grain boundary which caused strength degradation at higher temperatures.

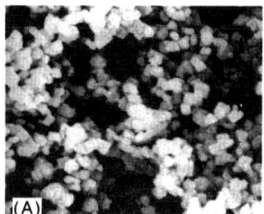 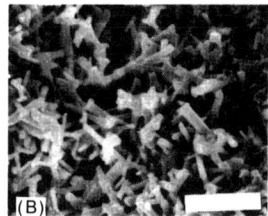

Fig. 2. Changes in grain morphology on powder compacts of α-Si_3N_4 with Y_2O_3 and Al_2O_3. (A) Raw powder and (B) Sintered compact (bar = 5 μm)

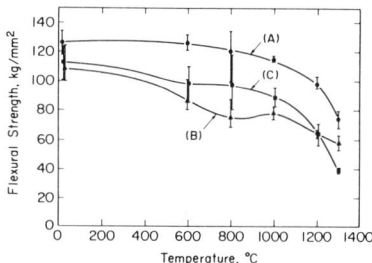

Fig. 3. Flexural strength of hot pressed α-Si_3N_4 with Y_2O_3 and Al_2O_3 as a function of temperature. (A) Compact presintered at 1700°C in AlN powder bed; (B) Compact presintered at 1700°C in Si_3N_4 powder bed; (C) No presintering treatment. All specimens were hot pressed at 1750°C under 490 kg/cm² for 90 min.

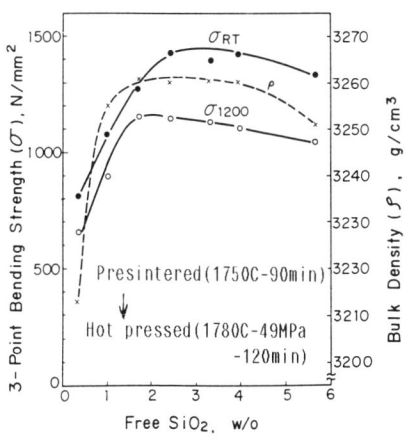

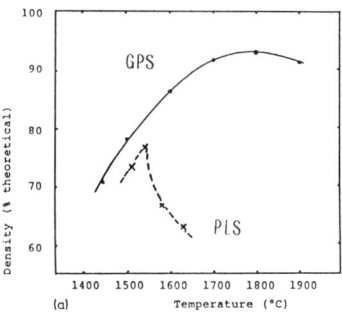

Fig. 5. Densification by N_2 gas pressure sintering of Si_3N_4-5 wt% MgO.

Fig. 4. Effect of free silica in hot pressed Si_3N_4-5 wt% Y_2O_3-2 wt% Al_2O_3 on the strength.

Table II. Schedule for Implementation of R & D in the "Fine Ceramics Project."

	1981	'82	'83	'84	'85	'86	'87	'88	'89	'90	'91	1992
1. Process Technology												
(1) Process Fundamentals												
(2) Material Powder Synthesis	(Laboratory scale)				(Bench-plant scale)							
(3) Forming and Sintering												
·1· Mechanism of Forming & Sintering												
·2· Materials for Three Fields (Hi-Str., Corr'n-Res., Wear-Res.)	(Test pieces)				(First-stage models)				(Gas-turbine element models)			
·3· Materials of Increased Toughns.												
(4) Machining and Joining												
·1· Mechanism of Machining												
·2· Grinding Machine Development												
·3· Joining												
·4· Surface Strengthening												
2. Evaluation Technology												
(1) Property Evaluation												
·1· Material Powder												
·2· Sintered Bodies												
(2) Non-Destructive Testing												
(3) Proof Testing Technique				(Theoretical experiments)					(Appliance study)			
·1· Mechanism of Fracture												
·2· Life Prediction												
3. Application Technology												
(1) Designing Technology												
·1· Designing Criteria												
·2· Fracture Data Accumulation (Combined Str., Impact Stress, Mech. & Thermal Fatigue, Corr'n)												
(2) Testing of Developed Models												
·1· First-Stage Models												
·2· Gas-Turbine Element Models												
4. Integration & Evaluation												

- - - - Survey

Oyama et al.[6] found new silicon nitride related compounds, known as Sialons, in 1971, about the same time as Jack.[7] They called the new compounds silicon nitride solid solutions. This has proven to be an important discovery in the field of ceramics and especially in silicon nitride, as later development amply attests.

Original work on gas pressure sintering was carried out in Si_3N_4-MgO by Mitomo in 1976 (Fig. 5).[8] Very important improvements in properties were obtained through this innovation. This technology was later applied by NGK Spark Plug Co. to manufacture ceramic components with high reliability.

Stage II

During this stage, government projects began and so did the ceramic fever. Many technologies made significant progress including materials development, component manufacturing, product evaluation, and practical applications.

1. Government Programs:—"Moonlight Project" in 1978 and "Fine Ceramics Project" in 1981: The "Fine Ceramics Project" will end in March, 1993, and the report on the project will become available later. However, the main development in this project is briefly described here. Table II[9] shows the schedule for implementation in the project, which is further divided into four periods (Period 1: 1981-83, 2: 1984-87, 3: 1988-90, 4: 1991-92). Four technologies, process technology, evaluation technology, application technology and integration and evaluation of silicon nitride and silicon carbide have been identified as key items. Beginning with Period 3, toughened ceramics such as particle, whisker and fiber reinforced composites have also been added to the project. The success of the activities of the project will be evaluated by their attainment of technology targets (specified properties, etc.). Integration and evaluation is being carried out by the fabrication and testing of the model rotor and stator with the design shown in Fig. 6.[9]

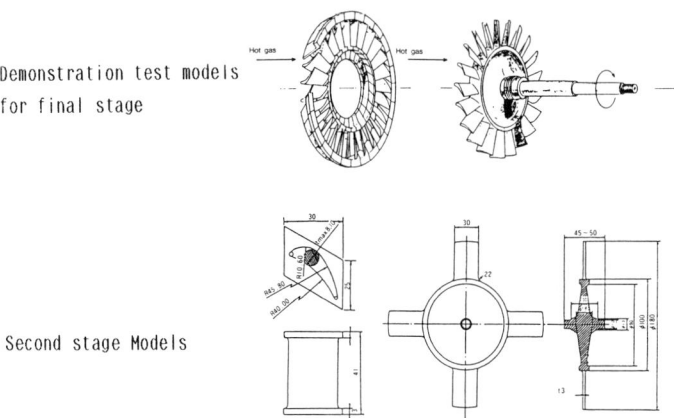

Fig. 6. Model components for evaluation in Fine Ceramics Project.

Standardization activities are also part of the project. Domestically, several Japanese Industrial Standards, JIS, have been established. International collaboration with organizations such as VAMAS and IEA has been progressing. In April, 1992, a conference on international standardization activity was held in Nagoya. Another significant step toward international standardization, the ISO activity on fine ceramics, has also been proposed and started.

2. Powder Synthesis: Processes for powder production of major Japanese powder suppliers are shown in Table III. The different grades of powders which are used for different purposes are also shown. The oxygen content is recognized as an important characteristic. Reproducibility is another important issue. Although progress has been made in this area, it must be admitted that, strictly speaking, reproducibility is still a problem. Further cost reduction is also needed to expand applications.

Table III. Available Silicon Nitride Powders in Japan.

Method	Main supplier
Direct nitridation	Denka(9, 9S, 9F, 9FW, P21,β), Shinetsu Chemicals(K-10SU, 10SP, 10TX), Onoda Cement(α, β)
Silica Reduction	Toshiba Ceramics
Imide decomposition	Ube(E10, E5, ESP)
Vapor phase reaction	Denka

3. Applications: Tables IV and V show fields of applications which are further divided into engine components and industrial parts. Since the realization for practical use of the glow plug in the 1980's, several kinds of engine components such as hot chamber, rocker arm pad, turbocharger rotor, injector link and rocker roller have been put into practical use. The turbocharger rotor had an especially large impact on this field in that it demonstrated in a very meaningful way the feasibility of silicon nitride use in severe, dynamic environments. In the field of industrial applications, anti-friction bearings and cutting tools are the most remarkable. Other kinds of engineering parts are also under development today.

Table IV. Applications for Automotive Engines.

Components	Materials		Periods
	Ceramics	Metals	
Glow plug	Si_3N_4	Ni alloy	1981, 83, 85
Precombustion chamber	Si_3N_4	Al alloy, Steel	1983, 84, 86
Rocker arm pad	Si_3N_4	Al alloy	1984, 87
Turbocharger rotor	Si_3N_4	Ni alloy, Steel	1985, 88
Injector link	Si_3N_4	Steel	1989
Rocker roller	Si_3N_4	Steel	1992

Table V. Applications for Industrial Parts.

Areas	Components
Wear parts	Antifriction bearing, Cutting tool, Balls for powder mixing/grinding, Jigs for semiconductor processing,
Metal casting jigs	Jigs for aluminum diecasting
Heat treatment jigs	Springs for high temperature use, Heat jigs for high temperatures
Others Sport/hobby	Guages, Measuring instruments,

4. Progress in Silicon Nitride Ceramics: During Stage II, many compositions for sintering aids, shown in Table VI, have been studied. These composition systems were investigated to systematically survey the resultant properties with an aim to achieve the technological targets such as strengthening, toughening and oxidation/corrosion resistance. With this in mind, microstructural design and optimal processing to obtain the most desirable combination of characteristics have been emphasized. Some of the major accomplishments are described as follows.

Improvement of fracture toughness was carried out by grain growth. Figure 7 shows examples of microstructures and mechanical properties obtained in the Si_3N_4-Y_2O_3-Al_2O_3 by Kawashima et al.[10] Toughness of 11.3 MPam$^{1/2}$ and 3-point bend strength of 774 MPa were obtained. This was due to the formation of large elongated grains. On the other hand, higher strength (1147 MPa) could be obtained with microstructures composed of homogeneous, small grains. Similar reports were also made by Nakajima et al.[11] who obtained 10.1 MPam$^{1/2}$ toughness by grain size control (Table VII). To understand these microstructures, Mitomo[12] proposed that the grain size distribution during the intermediate stage sintering is important for determining final microstructures (Fig. 8). He found that a homogeneous grain structure was obtained when using β-silicon nitride powder of good uniformity. With the addition of 5 wt%

β-Si₃N₄ nuclei to this powder, large elongated grains appeared in the microstructure. Thus, both the α to β transformation and β seeding may influence the final microstructure and that higher toughness can be obtained with the less uniform microstructure. Contrary to the above conclusion, however, is Urashima et al.'s[13] observation that microstructures with both high and low toughness could be quite similar. This would suggest that the mechanical properties of grain boundary phases and interfacial strength between grain and grain boundary phases are also important factors for toughening.

To improve high temperature strength, grain boundary phase crystallization has been developed as described previously. A mixed phase α + β Sialon material with good high temperature strength was developed by Ukyo and Wada (Fig. 9).[14] Crystallization of the grain boundary phase was also attained by adding HfO_2 to the Si_3N_4-Y_2O_3-AlN system. As a result, a 3-point bend strength of over 900 MPa at 1300°C was obtained (Fig. 10[15]). Another approach is by the addition of other rare earth oxides with low ionic radius of metallic elements.[16] Figure 11 shows weight gain and strength degradation by air oxidation for the densified specimens obtained by two-step sintering of these rare earth oxide-added silicon nitrides (Shimamori et al.[17]). Especially good oxidation resistivity was obtained from the Si_3N_4-Sc_2O_3 composition.

As described previously, properties such as toughness, strength and oxidation resistance are very much affected by microstructures. Microstructures, in turn, are influenced by compositions, starting powders and processing conditions. For such complicated problems, results obtained by many studies in the project, some of which were presented here, have contributed to the improvement of properties either directly or indirectly. In this way, Japanese researchers who have tried to understand phenomena and to determine the optimum processing conditions through much experimental data have made significant contributions.

Table VI. Typical Sintering Aids for Silicon Nitride.

Y_2O_3	MgO
Y_2O_3-Al_2O_3	Al_2O_3
Y_2O_3-Al_2O_3-AlN	MgO-Al_2O_3
Y_2O_3-Al_2O_3-AlN-TiO_2	Al_2O_3-AlN
Y_2O_3-Al_2O_3-MgO	Yb_2O_3
Y_2O_3-Al_2O_3-MgO-ZrO_2	(Si-)Sc_2O_3
Y_2O_3-MgO-ZrO_2	(Si-)Sc_2O_3-Y_2O_3
Y_2O_3-SiO_2	CeO_2-MgO-SrO
Y_2O_3-Cr_2O_3	CeO_2-MgO-SrO-ZrO_2
Y_2O_3-AlN	CeO_2-MgO-Y_2O_3
Y_2O_3-AlN-HfO_2	CeO_2-Al_2O_3
Y_2O_3-AlN-ZrO_2	$BeAl_2O_4$

Table VII. Fracture Toughness for Firing Conditions in the System Si_3N_4-Y_2O_3-Al_2O_3.

Firing temp. (°C)	Density (g/cc)	3P strength(RT) (MPa)	K_{1c} (MPam$^{1/2}$)
1750	3.19	880	7.1
1800	3.21	910	9.4
1850	3.22	800	10.1
2000	3.22	720	10.1

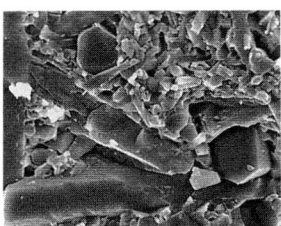

(A) σ_f =774MPa, K_{1c}= 11.3MPam$^{1/2}$

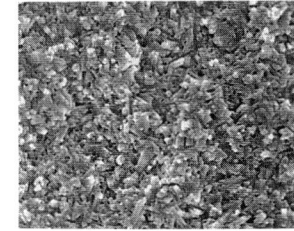

(B) σ_f =1147MPa, K_{1c}= 5.7MPam$^{1/2}$

20 μm

Fig. 7. Grain structures and mechanical properties in sintered Si_3N_4-5 wt% Y_2O_3-2 wt% Al_2O_3.

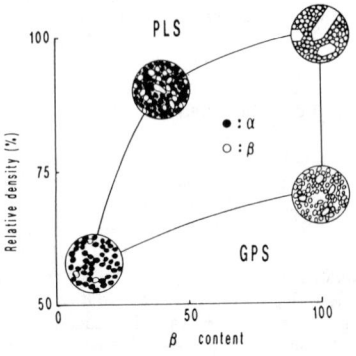

Fig. 8. Relation between densification, phase change and microstructure development.

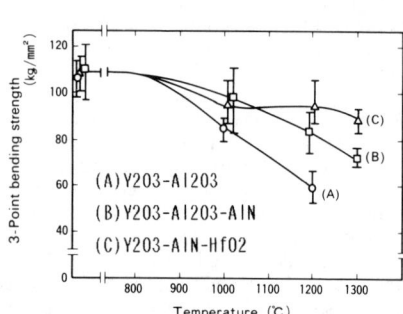

Fig. 10. Bend strength of various sintered Si_3N_4-Y_2O_3 base specimens.

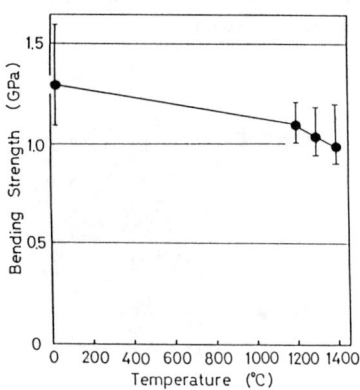

Fig. 9. Bend strength of $\alpha + \beta$ mixed sialon ceramics derived from the Si_3N_4-Y_2O_3-AlN system.

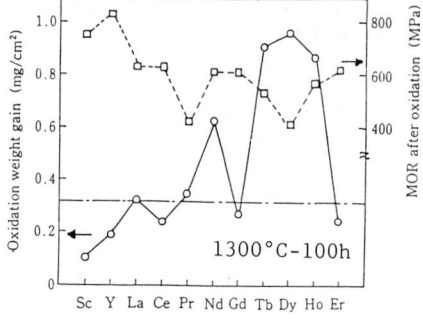

Fig. 11. Weight gain and room temperature strength after air oxidation for several two-step sintered specimens.

Stage III

The planned and initial activities in Step III are shown in Table VIII. More reliable materials and components, cost reduction, and expansion of applications are the main targets. Looking into the future, there is apparently a great deal of expectation for "in situ" toughened silicon nitride composites. However, the biggest obstacle for practical application is probably still the high cost. In this regard, basic research which can lead to cost reduction needs to be continued.

Table VIII. Progress and Expectation in Silicon Nitride Ceramics in Stage III.

Items	Activities & Expectation
More reliable materials/components	·In situ silicon nitride composites ·CMC(Particle dispersion, whisker /Fiber reinforced) ·FGM, SHS, IPC ·Hyper reliable materials
Cost reduction process	(?)
Expansion of applications	Automotives/post-automotives

Some of the step II items have been continued

CONCLUSION

Progress in silicon nitride ceramics in Japan is historically divided into three Stages. In Stage I (1963-78) the activities were seed-research oriented. In Stage II (1978-90) the activities were advancement-oriented, and in Stage III, 1990 and beyond the activities are improvement-oriented. The emphasis is thus now on reliability and cost reduction. The government project, Fine Ceramics (1981-1992), has contributed to the acceleration in the development of silicon nitride ceramics in Japan.

Much advancement has been made through the last 30 years of R & D in silicon nitride. In particular, materials development has been greatly advanced and applications have been constantly expanded. In retrospect, the main activities are numerous and some of them are listed below.
- High quality α silicon nitride powders have been developed.
- High strength, high Weibull modulus, and high toughness have been simultaneously attained in the same material by controlling grain morphology.
- Component fabrication processes and evaluation technologies have been established.
- Important automotive components and industrial parts have been put into practical use.
- New generation, well-controlled "in situ" reinforced silicon nitride composites have been advanced and further improvement is expected.

Nevertheless, despite the above progress, the cost problem remains, which is the most important and critical factor for expanding the application field. Continued attention to this aspect is highly recommended and, perhaps, new breakthrough innovation in materials development or other technologies is required. Beyond that, international collaboration on processing and evaluation technologies is also desirable.

ACKNOWLEDGEMENT

The author would like to thank Dr. M. Mitomo, NIRIM, and Dr. Y. Tajima, NGK Spark Plug Co., for their kind support.

REFERENCES

1. J.F. Collins and R.W. Gerby, J. Metals, **7**, 612 (1955).

2. K. Komeya and H. Inoue, Japanese Pat. No. 703695 (1969).

3. K. Komeya, Am. Ceram. Soc. Bull., **63**, (9) 1193 (1984).

4. A. Tsuge, K. Nishida and M. Komatsu, J. Am. Ceram. Soc., **58**, 323 (1975).

5. A. Tsuge, H. Inoue and K. Komeya, presented at the 81st Am. Ceram. Soc. Annual Meeting, April 1979.

6. Y. Oyama and O. Kamigaito, Japan J. Appl. Phys., **10**, 1637 (1971).

7. K.H. Jack and W.I. Wilson, Nature Phy. Sci. (London), **238**, 28 (1972).

8. M. Mitomo, J. Mater. Sci., **11**, 1103 (1976).

9. Pamphlet, <u>New Energy and Industrial Technology Development Organization</u> (NEDO), March, 1992.

10. K. Kawashima, H. Okamoto, H. Yamamoto, A. Kitamura, J. Ceram. Soc. Japan, **99**, (4) 320 (1991).

11. M. Nakajima, (Proc. of 9th Fine Ceramics Symp. on Engineering Research Association for High Performance Ceramics, 1991) p. 121.

12. M. Mitomo, edited by K. Kimura and K. Niihara (Proc. of 1st Int. Symp. on Science of Engineering Ceramics, 1991) p. 101.

13. K. Urashima, Y. Tajima and M. Watanabe, to be published, (Proc. of 5th Int. Conf. on Fracture Mechanics of Ceramics).

14. Y. Ukyo and S. Wada, J. Ceram. Soc. Japan, **97**, (8) 872 (1989).

15. K. Komeya, M. Komatsu, T. Kameda, Y. Goto and A. Tsuge, J. Mater. Sci., **26**, 5513 (1991).

16. C.A. Anderson and Bratton, Final Technical Report for U.S. Energy Res. Dev. Adm. Contract # EY-76-C-05-5210, August, 1977.

17. T. Shimamori et al., in <u>Progress in Fine Ceramics in Next Generation Research</u>, (Proc. Engineering Research Association for High Performance Ceramics, 1988) p. 115.

PHASE RELATIONSHIP STUDIES OF SILICON NITRIDE SYSTEM—A KEY TO MATERIALS DESIGN

T.S. YEN and W.Y. SUN
Shanghai Institute of Ceramics, Chinese Academy of Sciences, Shanghai, P.R. China.

ABSTRACT

Additions and revisions to several of the most important phase diagrams and phase behavior diagrams in the silicon nitride field are reviewed in this work, with emphasis on the Y-Si-Al-O-N system. This information is further used to make observations on the promising silicon nitride systems containing either highly refractory grain boundary phases or compatible matrix phases of desirable properties. Examples are provided to illustrate the advantage of such a basic approach to materials design. Hardness, toughness, strength at room temperature and elevated temperature and even sinterability can all be improved by adopting such an approach.

INTRODUCTION

Silicon nitride ceramics is a family of the most promising materials for high performance engineering applications. From a structural chemistry point of view, silicon nitride ceramics constitute materials with $[SiN_4]$ and $[(Si,Al)(N,O)_4]$ as the fundamental building units, which, to some extent, is similar to the silicate materials family with $[SiO_4]$ as the basic building unit. One main feature of silicon nitride ceramics is that the Si-N bond is highly covalent in nature with very low diffusivity of either Si or N even at very high temperatures. The excellent properties of silicon nitride ceramics are apparently attributed to this feature, although it also results in difficulty in sintering. Since good mechanical properties of structural ceramics can only be obtained in fully dense materials, sintering additives must be added to silicon nitride. Therefore, the compositional design of nitride ceramics is fundamental to the development of silicon nitride ceramics with desired and improved properties. The significance of phase relationship studies is to provide information for: (a) the selection of effective sintering additives to achieve densification through liquid phase sintering; (b) the control and tailoring of grain boundary phase (or phases); and (c) the design of nitride ceramics with composite Sialon phases.

Y_2O_3 and Al_2O_3 have proven to be effective additives for Si_3N_4 ceramics in the last fifteen years. Correspondingly, much work on the phase relationships in the Si-Al-O-N and Y-Si-Al-O-N systems has been carried out. Most notably are those from the University of Newcastle upon Tyne, the University of Michigan, Max-Planck Institut für Metallforschung and Shanghai Institute of Ceramics.[1-11] Close relationships between these four units have been established and some of the studies were completed through collaboration. In recent years, researchers at the Shanghai Institute have also studied extensively the phase relationships in the M-Si-Al-O-N (M = Na, Li, Mg, Ca and Ln) systems.[12-21] In the Y-Si-Al-O-N or other quinary systems (involving Li, Ca and Ln), several Si_3N_4-based solid solutions, β'-Sialon, α'-Sialon and o'-Sialon, are well-known. These solid solutions have distinct crystalline morphology and physical characteristics. The Shanghai Institute of Ceramics and the University of Michigan have also recently determined the sub-solidus phase relationships in the nitrogen-rich region bound by YN, Si_3N_4, AlN, Al_2O_3 and Y_2O_3 and have also summarized the phase relationships in the whole system.[10-11] Based on the information of these phase relationships, some materials design work has been carried out in Shanghai and the preliminary results show the possibility to develop some rational design philosophy. This report will concentrate on the Y-Si-Al-O-N system and its bearing on materials design.

PHASE RELATIONSHIPS IN THE Y-Si-Al-O-N SYSTEM

The Sub-Systems: Si-Al-O-N, Al-Y-O-N, Si-Y-O-N, Si-Al-Y-O, and Si-Al-Y-N

The phase diagram of the Si-Al-O-N system was first reported by L.J. Gauckler et al.[1] and K.H. Jack.[2] The early behavior diagrams were established on relatively few experimental data and there is still room for revision. Shown in Fig. 1 is the revised Si-Al-O-N behavior diagram by Newcastle.[3] The main argument between the new version and the old one is concentrated on AlN-polytypoids. The early version showed that all polytypoids had tie lines joining with 2H[6] and only 15R was compatible with β'-Sialon. In the Al-Y-O-N system,[15] the formation of aluminum oxynitride spinel occurs at high temperatures (about 1850°C) and the single phase Y_3O_3N is difficult to obtain. Therefore, the tie lines involving these two compounds are sometimes omitted or represented by dash lines in the diagrams. The sub-solidus phase relationships in the Si-Y-O-N system, as represented in Fig. 2(a), are deduced from the observations by the Newcastle group at 1700°C.[8] There does not exist a tie line joining Y_2O_3 and melilite as reported by the Max-Planck Institute.[4] The sub-solidus phase relationships in the Si-Al-Y-N system were first studied by Newcastle,[8] where three compounds on the Si_3N_4-YN join and an α'-Sialon solid solution with a compositional range between m = 1.8 and m = 3.4 ($Y_{1/3m}Si_{12-m}Al_mN_{16}$) were identified. The Shanghai Institute and the University of Michigan[10] have recently reinvestigated this system and observed only one compound which occurred at $Y_2Si_3N_6$ and had the same x-ray diffraction pattern as published by Newcastle for $Y_6Si_3N_{10}$.[8] The pure nitride α'-Sialon solubility region was determined to be from m = 1.3 to m = 2.4.[10] The difference in α'-Sialon compositions between these reports can probably be attributed to the purity of the YN powder. YN is very sensitive to the moisture in air and extreme precaution should be taken during its preparation and follow-up experiments. The revised sub-solidus phase relationships in the systems Si-Y-O-N and Si-Al-Y-N are given in Fig. 2(b). Figure 3 shows the phase relationships of Si-Al-Y-O system reproduced from Michigan's work.[4]

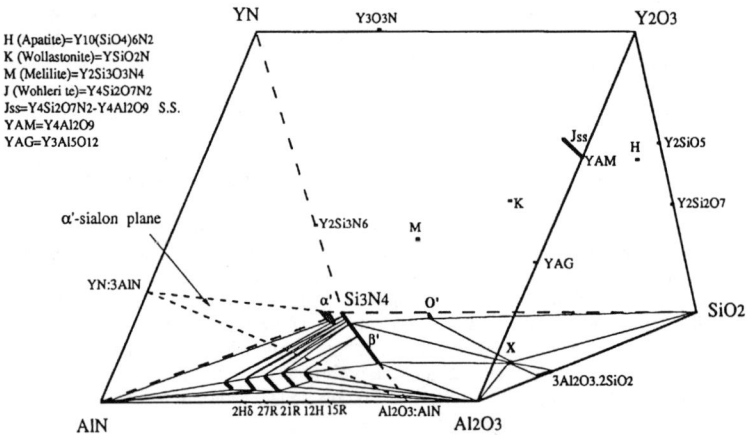

Fig.1. Representation of Si-Al-Y-O-N system showing phases occurring in the system and the revised Si-Al-O-N behaviour diagram (after Newcastle[3]).

The Oxygen-rich Part of the System

The oxygen-rich part of the system has been intensively studied since it involves β'-Sialon and o'-Sialon. β'-Sialon ($Si_{6-x}Al_xO_xN_{8-x}$) has been developed into a good engineering material with high strength. o'-Sialon ($Si_{2-x}Al_xN_{2-x}O_{1+x}$), which has good oxidation resistance, occurs along the line Si_2N_2O and Al_2O_3. YAG has proven to be a good grain boundary phase because of its good oxidation resistance and refractoriness. Figure 4 shows the compatibility relationships between β'-Sialon and YAG, thus indicating the possibility to design β'-YAG ceramics which has already been achieved and found to have good high temperature properties. The sub-solidus phase relationships involving o'-and β'-Sialon were

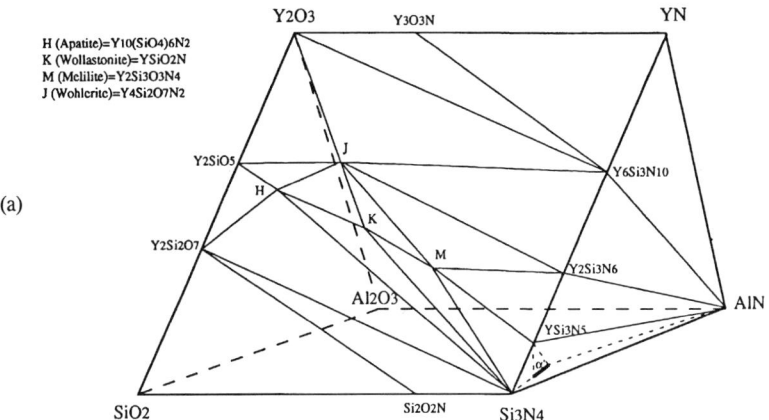

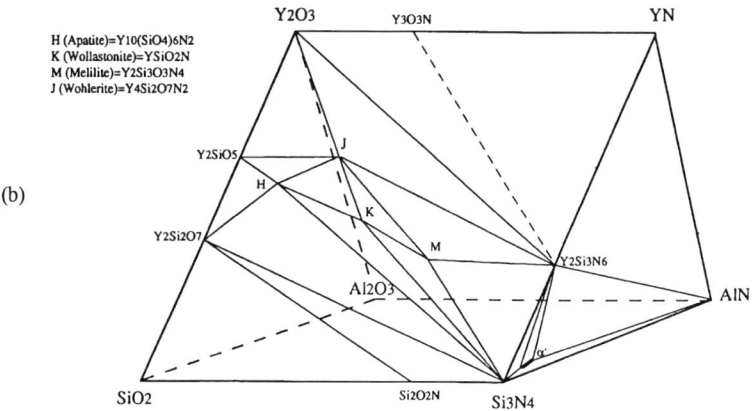

Fig.2. Sub-solidus phase relationships in the systems Si-Al-Y-N and Si-Y-O-N: (a) after Newcastle[8]; (b) after authors' recent work[10].

revised by Shanghai,[11] as shown in Fig. 5. At sub-solidus temperatures the tie lines between o'-Sialon and β'-Sialon do not run parallel and three compatibility tetrahedra exist instead of just one compatibility region, o'-β'-$Y_2Si_2O_7$. The coexistence relationships between β' and o'-Sialons and compounds with high melting points provide the knowledge for designing composite Sialon ceramics with refractory grain boundary phases. $Y_2Si_2O_7$ and/or YAG have been used as the crystalline grain boundary phases in β'-Sialon ceramics which have been developed into commercial materials. Such crystalline compounds have also been used as the grain boundary phases in o' and o'-β'-Sialon ceramics.[22-23] However, the reaction

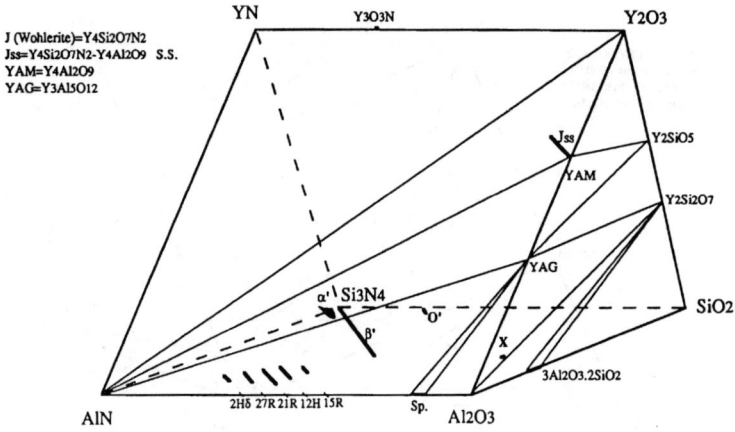

Fig.3. Sub-solidus phase relationships in the Al-Y-O-N system and Si-Al-Y-O system (after Naik[4]).

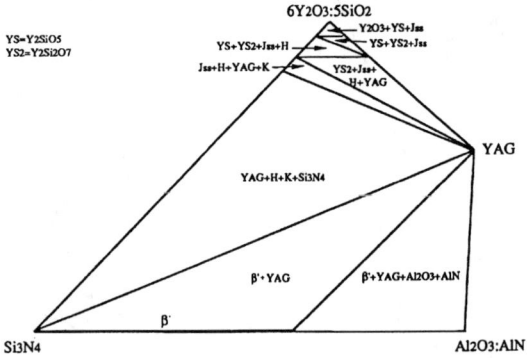

Fig.4. Sub-solidus phase relationships on the β'-YAG plane devitrified at 1350°C (after Newcastle[8]).

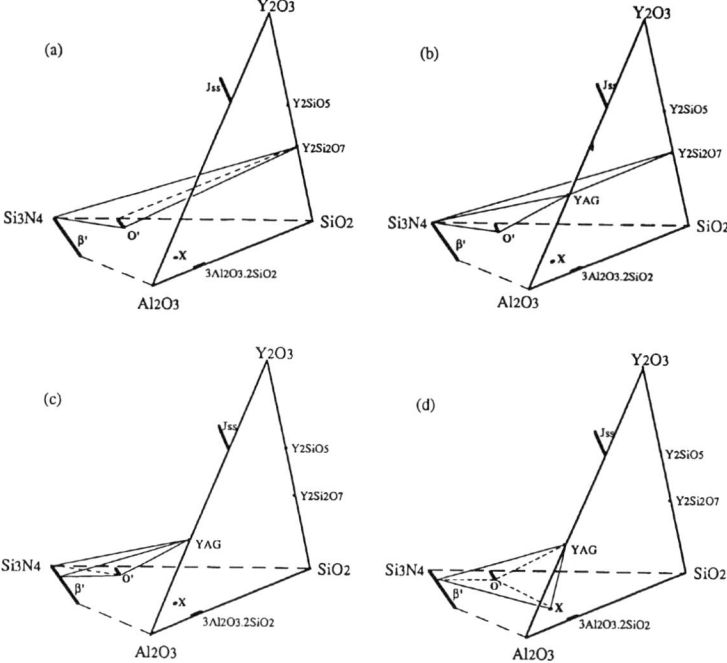

Fig.5. Four compatibility tetrahedra existing in the o'-β' and o'-β'-X-phase regions: (a) Si_3N_4-Si_2N_2O-o'-$Y_2Si_2O_7$; (b) Si_3N_4-o'-$Y_2Si_2O_7$-YAG; (c) Si_3N_4-o'-YAG-β_{10}; (d) β_{10}-o'-YAG-X.

kinetics are too complicated during high temperature processing to obtain the desired o'/β'-Sialon ratio. For this purpose, region (b) of Fig. 5 with $Y_2Si_2O_7$ and YAG as the double grain boundary phases is recommended.[24]

The Nitrogen-rich Part of the System

The nitrogen-rich part of the system has been clarified. Fifty-two compatibility tetrahedra exist in the space bound by the points Si_3N_4, β_{60}, Al_2O_3, AlN, YN and Y_2O_3.[10] Among these, the phase relationships involving α'-Sialon are of particular interest for the development of promising materials. The α'-Sialon region starts at the pure nitride side (n = 0) with m = 1.3 to m = 2.4 and expands towards the alumina side with a gradual decreasing limit of m until m = 1.0 ($Y_{1/3m}Si_{12-(m+n)}Al_{m+n}N_{16-n}$). The maximum oxygen content is approximately represented by n = 1.7, as shown in Fig. 6. The boundary of the α'-Sialon region (from m = 1.3, n = 0 to m = 1, n = 1.7) facing β'-Sialon is compatible with β'-Si_3N_4. The Al_2O_3-rich composition of α'-Sialon (m = 1, n = 1.7) is compatible with β'-Sialon (from β_0 to β_{10}). These two compatibility triangles have tie lines joining melilite and YAG, respectively, as shown in Fig. 7. The compatibility tetrahedra α' (m = 1, n = 1.7)-β_0-β_{10}-YAG provides the possibility to fabricate an α'-β' composite Sialon ceramic with YAG as the grain boundary phase. α'-Sialon coexists with all AlN-

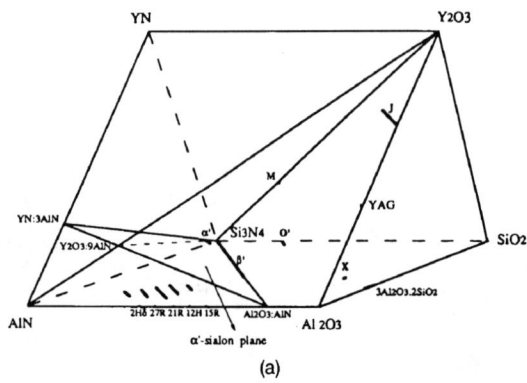
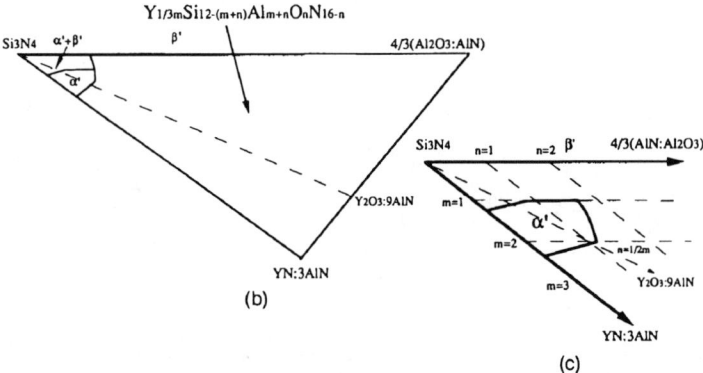

Fig.6. (a) Representation of α'-Sialon plane in Y-Sialon system; (b) α'-Sialon plane; (c) enlarged corner of α'-Sialon plane.

polytypoids (except 15R) at the oxygen-rich terminal composition, m = 1, n = 1.7, as shown in Fig. 8. An important tetrahedron also exists, namely α' (m = 1, n = 1.7)-β_{10}-12H-YAG, as shown in Fig. 9.

NITRIDE MATERIALS DESIGN—PHILOSOPHY, CRITERIA AND TYPICAL RESULTS

Multi-phase Sialon Ceramics

The thorough study of the phase relationships of the Y,Si,Al/O,N system made clear the phase relationships and phase boundaries of all the sialons—o'-Sialon, β'-Sialon, α'-Sialon, and AlN-polytypoids. This information made it possible to design multi-phase ceramics with the proper choice of other accompanying phases. There are good reasons for investigating the multi-phase Sialon ceramics. Two-phase ceramics offer more flexible choices in starting compositions and tailoring of properties. Many of the properties of a two-phase material can be additive or complementary, thus enhancing the overall performance. Since grains of different phases often grow at different rates or develop into different morphologies, they offer new possibilities for microstructural tailoring to gain additional advantages. Lastly, a composite Sialon may sometimes improve the sinterability. (For example a dense single phased α'-Sialon material is rather difficult to fabricate because of its high nitrogen content. However, a dense α'-β' composite Sialon can be achieved with less sintering additives.) From what we have shown in the Y-Si-Al-O-N system, we believe the following five possible combinations are worth considering: o'-β', α'-β', β'-AlN-polytypoids (15R, 12H, 21R, 27R, and 2H$^\delta$), α'-AlN-polytypoids (12H, 21R, 27R and 2H$^\delta$) and α'-β'-12H. The characteristics of these Sialon ceramics can be briefly summarized as follows: o'-Sialon possesses better oxidation resistance; β'-Sialon, high fracture strength and toughness and α'-Sialon, high hardness and good thermal shock resistance. The properties of AlN and polytypoids are not quite clear, but their fiber-like morphology will most probably offer an in-situ strengthening effect. The composites of these phases are all possible and may have superior microstructures, properties, and sinterability.

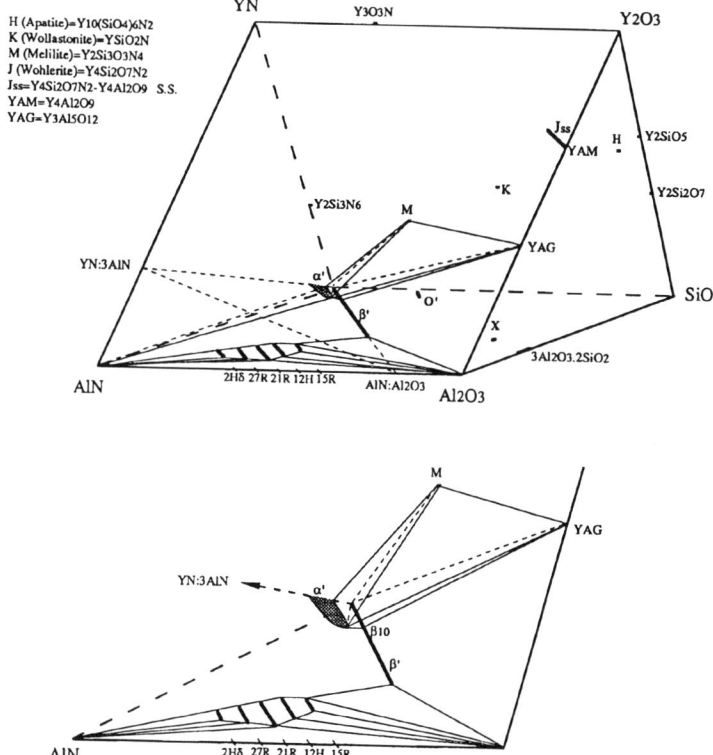

Fig.7. α'-β' two phase region being compatible with melilite(M) and YAG, forming three compatibility tetrahedra: β-Si$_3$N$_4$-α'(boundary facing β')-M; β-Si$_3$N$_4$-α'(m=1,n=1.7)-M-YAG and β-Si$_3$N$_4$-α'(m=1,n=1.7)-β$_{10}$-YAG.

It should be clear from the above discussion that Sialon system ceramics are actually a family of compositions or alloys. Better materials can be designed and fabricated as long as the basic phase-relationship information is used with the additional aid of processing studies, which unfortunately are usually rather complicated. Our work on α'-β'[25] and α'-12H composite Sialon ceramics[26] gave some evidence of the strengthening effect by incorporating β'-Sialon or 12H into α'-Sialon ceramics. Table I shows our preliminary data on the α'-12H composite material. The starting powders used were Si_3N_4 (Starck LC12), AlN (laboratory made, containing 1.2 wt% O), Al_2O_3 (99.9% pure) and Y_2O_3 (99.9% pure). All the compositions listed in Table I contained extra 2.5 wt% YAG and were hot pressed at 1750°C for 1 h under 250 kg/cm^2. The differences in bend strength and fracture toughness (determined by SENB) between the monolithic ceramics and composite ceramics are apparent. The α'-12H composite Sialon ceramics are superior to both monolithic α'-Sialon ceramics and the 12H material. Further studies on these materials are on-going in our laboratory, and the mechanical properties are expected to be improved with compositional adjustment and processing improvement.

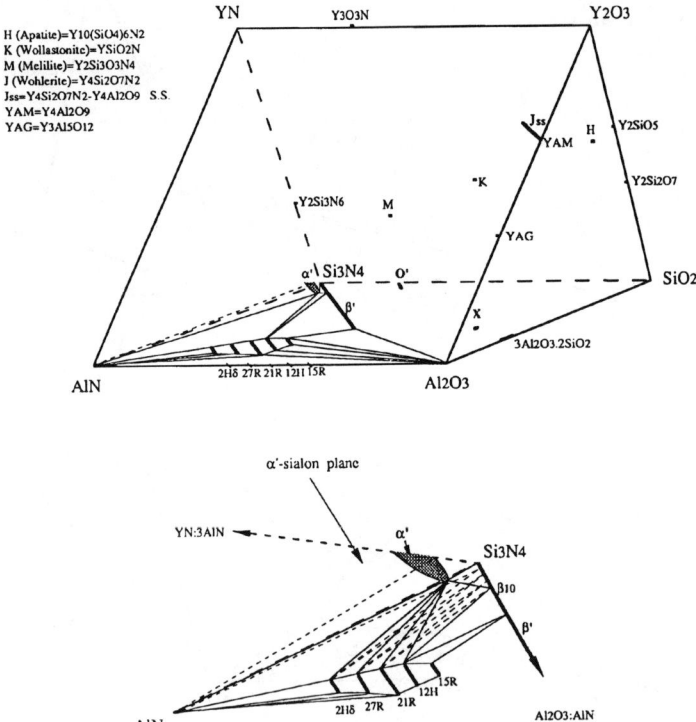

Fig.8. α'-Sialon being compatible with all AlN-polytypoids(except 15R)

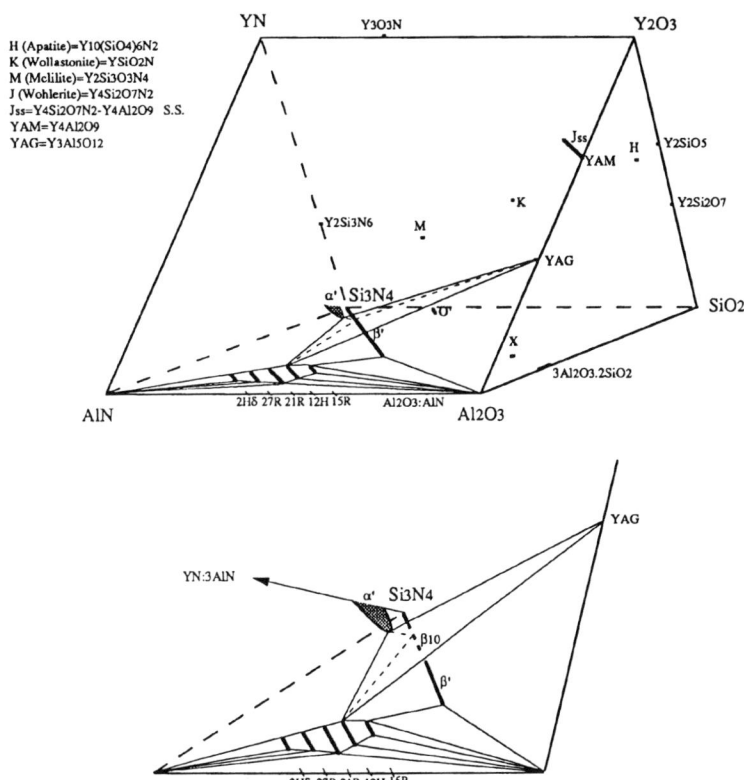

Fig.9. Compatibility tetrahedron YAG-α'($m=1, n=1.7$)-12H-β_{10}.

Table I. Mechanical properties and phase compositions of α'-12H composite ceramics

Nominal Composition	σ_f (MPa)	K_{1c} (MPa.m$^{1/2}$)	12H wt%	α' wt%	β' wt%
α'-Sialon	471	6.2	nil	89	8
90wt% α'+10wt% 12H	641	7.4	19	71	7
75wt% α'+25wt% 12H	525	6.1	41	34	20
12H	364	3.8	95	nil	2

Grain Boundary Engineering and Microstructural Control

One obvious approach to improving high temperature properties is to induce a controlled structural change to the intergranular phase. If the composition of the amorphous phase at the grain boundaries after sintering can be manipulated to conform to that of a stable compound (or compounds) compatible with the main phases such as α-Sialon, β-Si$_3$N$_4$ or β'-Sialon, then follow-up annealing treatments will result in grain boundary phase crystallization. In the case of liquid phase sintered β'-Sialon, the final glass composition will also be depleted of Al by the amount of Al/O substitution incorporated during the α-Si$_3$N$_4$ to β'-Sialon transformation. This reaction can, in fact, be advantageous in eliminating or reducing the glassy phase and allow an additional degree of freedom in compositional

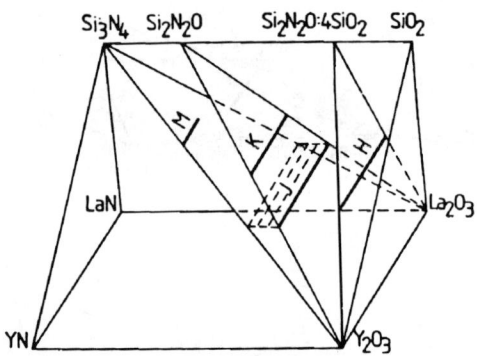

Fig. 10. Sub-solidus phase relationships in the system Si-Y-La-O-N

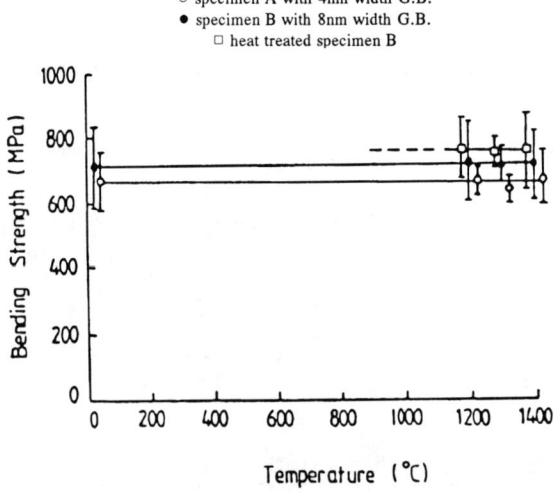

Fig. 11. Bending strength vs temperature relationship of specimens with various G.B. width.

tailoring. In a similar way, for the sintering and processing of α'-Sialon ceramics, more advantages can be gained by eliminating the glassy phase which contains Y and Al during the formation of the α'-Sialon phase. The crystallization products of post sintering annealing depend on the equilibrium phase relationships at the reaction temperature and on the kinetics. The optimum choice of oxide compounds will be based on their mechanical and thermal properties and the chemical compatibility of the phases. Garnet ($3Y_2O_3 \cdot 5Al_2O_3$) is an example of a compound that has been successfully crystallized at the grain boundaries of β-Si_3N_4 and β'-Sialon ceramics with an improvement in high temperature strength of the materials.

Another alternative approach to improving high temperature properties is to design a highly refractory glassy grain boundary phase. Mixed rare earth oxides (including Y_2O_3) have proven to be good sintering additives for Si_3N_4-based ceramics to achieve such an aim.[27] Our previous work[28] indicated that the simultaneous use of Y_2O_3 and La_2O_3 as complex additives is especially advantageous. In order to understand these complex additives, the sub-solidus phase relationships in the system Si_3N_4-Y_2O_3-La_2O_3 have been determined in our recent work,[20] as shown in Fig. 10. β-Si_3N_4 materials processed with these complex additives possess a highly refractory grain boundary phase in either the glassy state or the partially crystallized state. With the grain boundary thickness controlled by adjusting the additive amount, these materials can retain mechanical strength up to 1300°C or even up to 1400°C, as shown in Fig. 11. The $LaYO_3$ crystalline phase, along with other N-containing phases, can be readily recrystallized at the grain boundaries under ordinary atmosphere by heat treatment, and this has further enhanced creep properties.[29]

REFERENCES

1. L.J. Gauckler, H.L. Lukas and G. Petzow, J. Am. Ceram. Soc., **58**, 346 (1975).

2. K.H. Jack, J. Mat. Sci., **11**, 1135 (1976).

3. S. Slasor, PhD thesis, University of Newcastle upon Tyne, 1986.

4. I.K. Naik and T.Y. Tien, J. Am. Ceram. Soc., **62**, 642 (1979); L.J. Gauckler, H. Hohnke and T.Y. Tien, ibid., **63**, 35 (1980).

5. W.Y. Sun, Z.K. Huang and J.X. Chen, Brit. Ceram. Soc., **82**, 173 (1983).

6. Z.K. Huang, P. Greil and G. Petzow, J. Am. Ceram. Soc., **66**, C96 (1983).

7. S. Slasor and D.P. Thompson, in Non-Oxide Technical and Engineering Ceramics, edited by S. Hampshire (Proc. Inter. Conf., Limerick, Ireland, 1985) pp. 223-30.

8. D.P. Thompson, in Tailoring Multiphase and Composite Ceramics, edited by R.E. Tressler et al. (Proc. 21st Univ. Conf. on Ceram. Sci., Pennsylvania State Univ., PA, 1985) pp. 79-91.

9. G.Z. Cao, Z.K. Huang, X.R. Fu and D.S. Yan, Int. J. High Tech. Ceram., **1**, 119 (1985); W.Y. Sun, Z.K. Huang, G.Z. Cao and D.S. Yan, ibid., **3**, 277 (1987).

10. W.Y. Sun, T.Y. Tien and T.S. Yen, J. Am. Ceram. Soc., **74**, 2547 (1991); ibid., **74**, 2753 (1991).

11. W.Y. Sun, T.S. Yen and T.Y. Tien, Sci. in China (Series A), **35**, 877 (1992).

12. W.Y. Sun and T.S. Yen, Mater. Letters, **8**, (5) 145 (1989); ibid., **8**, (5) 150 (1989).

13. W.Y. Sun, L.T. Ma and D.S. Yan, Chine. Sci. Bull., **35**, 1189 (1989).

14. Z.K. Huang, D.S. Yan and T.Y. Tien, J.Solid State Chem., **85**, 51 (1990).

15. W.Y. Sun, Z.K. Huang and T.Y. Tien, Mater. Letters, **11** (3-4) 67 (1991).

16. W.Y. Sun, T.S. Yen and T.Y. Tien, J.Solid State Chem., **95**, 424 (1991).

17. Z.K. Huang, W.Y. Sun and D.S. Yan, J. Mat. Sci. Letters, **4**, 255 (1985).

18. Z.K. Huang, T.Y. Tien and D.S. Yan, J. Am. Ceram. Soc., **69**, C241 (1986).

19. W.Y. Sun, D.S. Yan and T.Y. Tien, Ceram. Inter., **14**, 199 (1988).

20. G.Z. Cao, Z.K. Huang and D.S. Yan, Sci. in China (Series A), **32**, 429 (1989).

21. S.F. Kuang, Z.K. Huang, W.Y. Sun and T.S. Yen, J. Mat. Sci. Letters, **9**, 69 (1990); ibid., **9**, 72 (1990).

22. M.B. Trigg and K.H. Jack, in <u>Ceramic Components for Engines</u>, edited by S. Somiya, E. Kanai and K. Ando (Proc. of the 1st Int. Symp., Hakone, Japan, 1983) pp. 343-49.

23. W.Y. Sun, D.P. Thompson and K.H. Jack, in <u>Tailoring Multiphase and Composite Ceramics</u>, edited by R.E. Tressler et al. (Proc. 21st Univ. Conf. on Ceram. Sci., Pennsylvania State Univ., PA, 1985) pp. 93-101.

24. W.Y. Sun, W.L. Li and D.S. Yan, J. Eur. Ceram. Soc., **5**, 99 (1989).

25. W.Y. Sun, L.J. Tong and T.S. Yen, in <u>Ceramic Materials and Components for Engines</u>, edited by R. Carlsson, T. Johansson and L. Kahlman (Proc. of the 4th Int. Symp., Gothenburg, Sweden, 1991) pp. 292-99.

26. H.X. Li, on-going PhD thesis work, Shanghai Institute of Ceramics.

27. K. Ueno and T. Toibana, Yogyo-Kyokai-Shi, **91**, 409 (1983).

28. Y.R. Xu, L.P. Huang, X.R. Fu and T.S. Yen, Scientia Sinica (Series A), **28**, 556 (1985).

29. T.S. Yen, Z.K. Huang, Z.H. Lin, Y.R. Xu and J.L. Shi, in <u>Ceramics Today and Tomorrow's Ceramics</u>, Part B, edited by P. Vincenzini (7th CIMTEC—World Ceramics Congress, Montecatini Terme, Italy, 1990) pp.701-19.

SILICON NITRIDE CERAMICS—ALLOY DESIGN

TSENG-YING TIEN
Materials Science and Engineering, The University of Michigan, Ann Arbor, MI 48109

ABSTRACT

Properties of silicon nitride ceramics depend on the phases present and their microstructure. In order to obtain ceramics with optimum properties, knowledge of phase relationships in silicon nitride-metal oxides systems and the mechanism and kinetics of microstructure development are necessary.

There are four major stable silicon nitride phases in the Si_3N_4-metal oxides systems: β-Si_3N_4, α'-SiAlON, AlN-polytypoids and silicon oxynitride. The morphologies of these phases are different. The β-Si_3N_4 grains are elongated hexagonal rods; the α'-SiAlON grains are equiaxed; the AlN-polytypoids are platelets; and the silicon oxynitride are equiaxed. Ceramics with combinations of these phases would have different microstructures and properties. The system Si_3N_4-SiO_2-AlN-Al_2O_3-YN-Y_2O_3 will be used to demonstrate the concept of the alloy design. Kinetics of microstructural development will also be discussed.

INTRODUCTION

Silicon nitride is one of the most promising materials for high temperature structural applications. The superior thermal and mechanical properties of this material are due to the highly covalent nature of their chemical bonds, which makes densification of silicon nitride by solid state diffusion impossible. However, when a liquid phase is introduced into the system, high density ceramics can be obtained. This liquid phase is produced by the addition of sintering additives, usually metal oxides, which form a low melting eutectic liquid with the surface layer of the silicon nitride powder. During sintering, the starting silicon nitride powder transforms from α-Si_3N_4 to β-Si_3N_4 by solution-precipitation. The newly formed β-Si_3N_4 grains have an elongated hexagonal rod morphology. These rod-shaped grains form an interlocking microstructure which are responsible for the superior mechanical properties of these materials.

Although densification is facilitated by the use of oxide additives, upon cooling, the high silica-containing liquid becomes glass and envelops the acicular β-Si_3N_4 grains. This intergranular glassy phase will become soft above the glass transition temperature which results in an inferior creep resistance. The glass layer can be crystallized or reacted with the major phase. The crystallization of the grain boundary glass will improve the high temperature mechanical properties of the silicon nitride ceramics.

There are a number of sintering additives used for densifying silicon nitride ceramics. Different additives result in different compositions of the liquid which affect the growth kinetics of the silicon nitride grains, the microstructural development, and the phases present at the grain boundaries after crystallization.

For structural applications, silicon nitride ceramics should have high flexural strength and high fracture toughness at room temperature and good creep resistance at elevated temperatures. The properties of these ceramics are determined by the phases present and their microstructures. For developing silicon nitride ceramics with optimum mechanical properties, knowledge of the phase relationships in silicon nitride-metal oxide systems and the kinetics of microstructural development are essential.

PHASES AND MICROSTRUCTURES

There are four major stable silicon nitride phases in the Si_3N_4-metal oxide systems. These four silicon nitride phases are: β-Si_3N_4, α'-SiAlON, AlN-polytypoids and silicon oxynitride. The morphology of these phases are different: β-Si_3N_4 grains are elongated hexagonal rods; the α'-SiAlON grains are equiaxed; the AlN-polytypoids are platelets; and the silicon oxynitride grains are equiaxed. Ceramics consisting of a combination of these phases would have different microstructures, and hence, different properties. Using knowledge of phase equilibrium, the following ceramics can be designed:

1. Acicular β-Si₃N₄ grains with grain boundary phases
2. Acicular β-Si₃N₄ grains with equiaxed α'-SiAlON
3. Acicular β-Si₃N₄ grains with polytypoid platelets
4. Equiaxed α'-SiAlON with polytypoid platelets
5. Acicular β-Si₃N₄ grains with equiaxed silicon oxynitride

The Si₃N₄-SiO₂-AlN-Al₂O₃-YN-Y₂O₃ system can be used as a model to demonstrate the design of ceramics with these different assemblies of phases. The sub-solidus phase relationships of the system Si₃N₄-SiO₂-AlN-Al₂O₃-YN-Y₂O₃ have been reported by Sun et al.[1] as shown in Fig. 1. Ceramics with different phases and microstructures are discussed in the following sections.

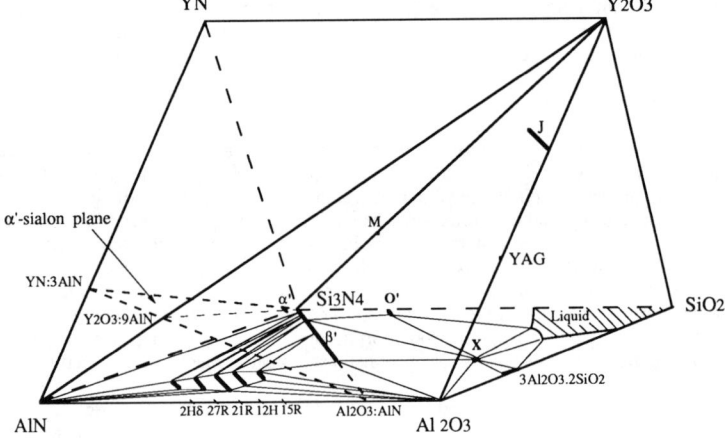

Fig. 1 Representation of Y-sialon system.

Acicular β-Si₃N₄ Grains with Grain Boundary Phases

Glassy Grain Boundary Phase—The most common sintering additives for silicon nitride ceramics are MgO, Al₂O₃, and Y₂O₃ or combinations of these oxides. During processing, mixtures of the starting materials, α-Si₃N₄ with metal oxide sintering aids, were mixed, compacted and sintered in a nitrogen atmosphere. During the first stage of sintering, the oxide additives and the surface layer of the silicon nitride particles formed a eutectic liquid during sintering. This reactive liquid helps the compact to densify by particle rearrangement. A sintering shrinkage curve[2] is given in Fig. 2. After the first stage of sintering, the α-Si₃N₄ particles dissolved in the reactive liquid and re-precipitated as β-Si₃N₄. The growth kinetics of these grains are different in different crystallographic directions. With prolonged heating of the ceramics containing β-Si₃N₄ and liquid phase, β-Si₃N₄ grains grow anisotropically, following the empirical grain growth law:

$$D^n - D_0^n = K_D t$$

The rate constant "K" and growth exponent "n" are different for the length and width directions. Hence, the β-Si$_3$N$_4$ grains will have an acicular morphology and the aspect ratio increases with temperature and time. Lai et al.[3] shows that the exponent "n" equals 3 in the length direction and 5 in the width direction.

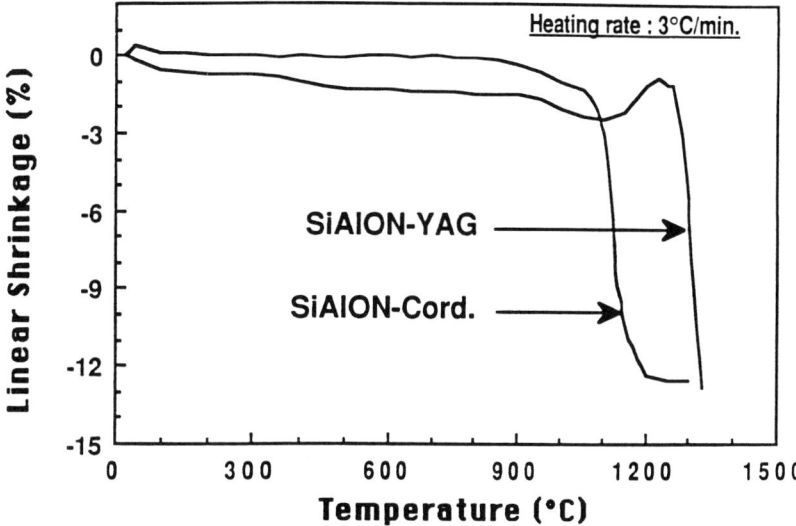

Fig. 2. Continuous sintering curves of SiAlON-YAG and SiAlON-cordierite ceramics conducted at a heating rate of 3°C/min.

Figure 3 shows our results in grain growth studies.[3] Figure 4 shows the aspect ratio as a function of sintering time. The aspect ratio vs. time curves sintered at different temperatures are parallel straight lines with a slope equal to 2/15. This slope is equivalent to the difference in reciprocal growth exponents in both directions, 1/3-1/5 = 2/15. Since mechanical properties of the silicon nitride ceramics depend on their microstructure, knowledge of the kinetics of grain growth are very important in microstructural design of silicon nitride ceramics.

Lai has found that the aspect ratio of β-Si$_3$N$_4$ grains depends on the starting silicon nitride powder. Powders with different β-Si$_3$N$_4$ content (5, 10, 20% of β-Si$_3$N$_4$ in the starting powder) sintered at the same temperature for the same length of time resulted in the same aspect ratio. However, starting powders with the same β-Si$_3$N$_4$ content but of different average particle size sintered to different aspect ratios under the same conditions. The results are given in Figs. 5 and 6.

Lai et al.[4] have shown that the mechanical properties of silicon nitride ceramics depend on the aspect ratio of the β-Si$_3$N$_4$ grains (Fig. 7). These acicular β-Si$_3$N$_4$ grains are the cause for the high toughness and good creep resistance of these ceramics.

The solid-liquid reactions in the system β-SiAlON-YSi$_3$Al$_5$O$_{12}$ (yttrium-aluminum-garnet-YAG) have been determined by Wisnudel[5] and the diagram is shown in Fig. 8. This diagram can be used to determine the amount of intergranular glass for specimens heat treated at different temperatures. This diagram also shows the composition of the intergranular glass. When glass composition is known, the property of the glassy phase will also be known. Hence, the property of the ceramics can then be predicted.

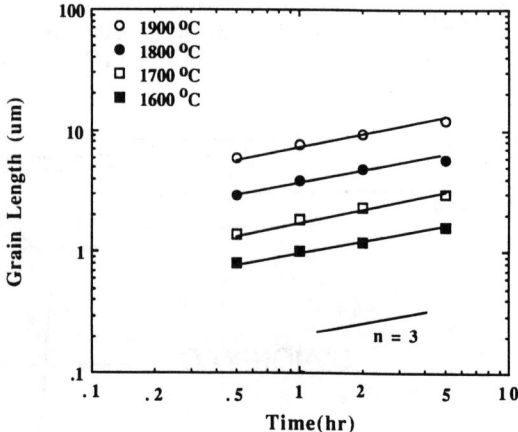

Fig. 3(a). Grain growth behavior of B10Y-Si$_3$N$_4$ in length direction. Samples were sintered under 10 atm N$_2$.

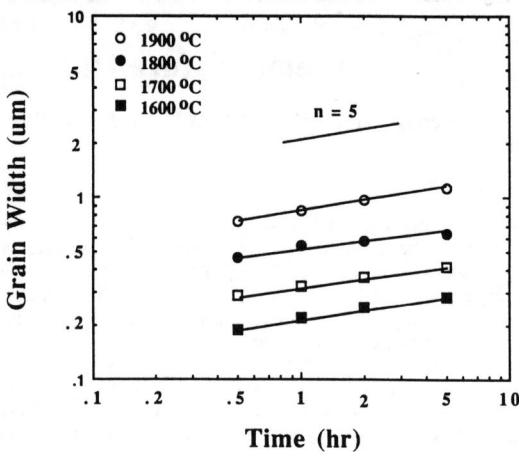

Fig. 3(b). Grain growth behavior of B10Y-Si$_3$N$_4$ in width direction. Samples were sintered under 10 atm N$_2$.

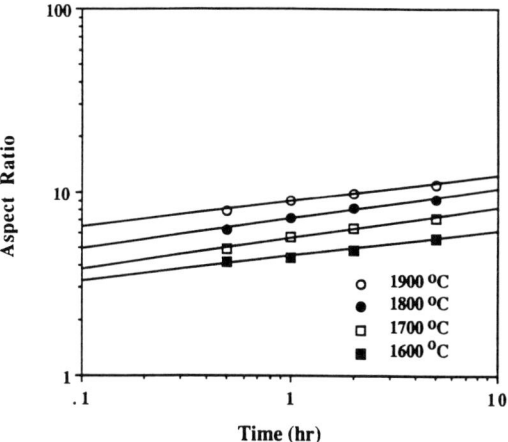

Fig. 4. Aspect ratios of B10Y-Si$_3$N$_4$ as a function of sintering time at various isothermals.

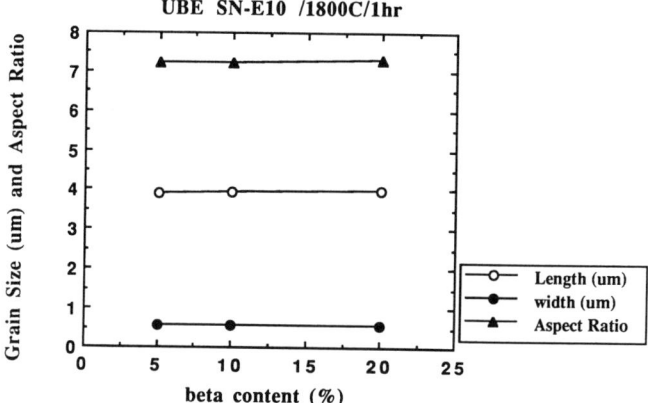

Fig. 5. Aspect ratio, grain length and width among the Si$_3$N$_4$ ceramics sintered from powders consisting of various β contents.

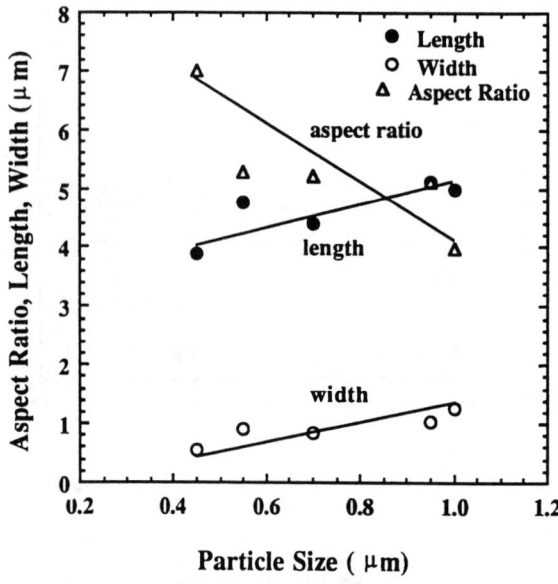

Fig. 6. Grain size and aspect ratio of β-Si$_3$N$_4$ as a function of the particle size of the starting powders.

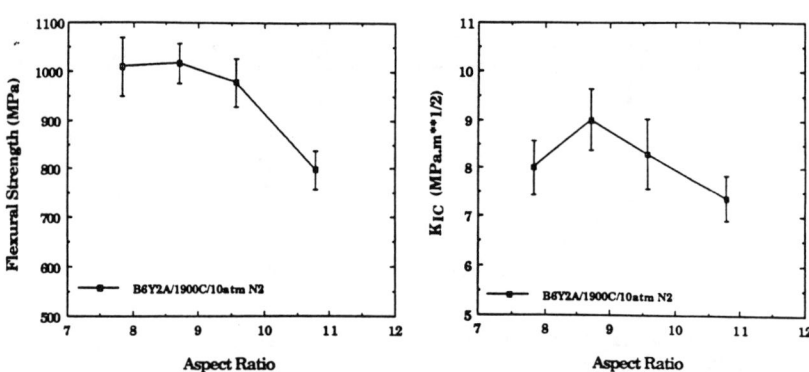

Fig. 7. Microstructure dependence of flexural strength and fracture toughness of silicon nitride ceramics. The compositions of these specimens were 90 wt% Si$_3$N$_4$-10 wt% Y$_3$Al$_5$O$_{12}$.

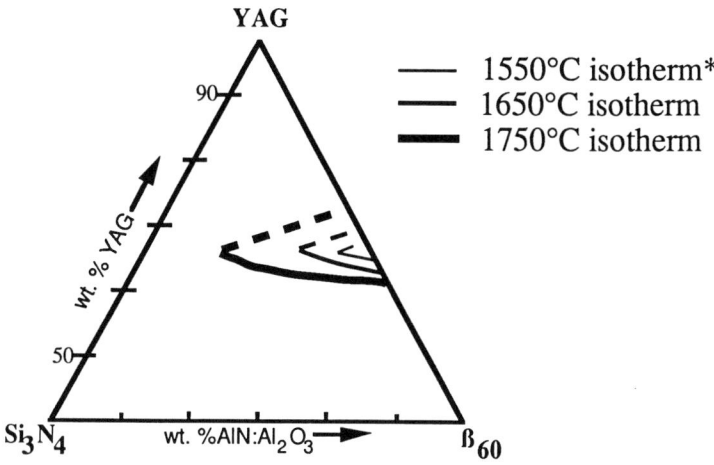

Fig. 8. Liquidus isotherm of the system SiAlON-YAG.

Crystalline Grain Boundary Phases—Chen[6] has shown that creep resistance increased after the grain boundary glassy phase was crystallized (Fig. 9). There are three curves in this figure, each representing a different heat treatment. The first specimen was hot pressed at 1750°C for 60 minutes under a pressure of 20 MPa. This specimen failed in 5 hours under a load of 133 MPa at 1170°C. The second specimen was also hot pressed under the same conditions as the first, but was annealed at 1250°C for 50 hours allowing the intergranular glass to crystallize at the grain boundaries. This sample failed after 150 hours under load. A third sample was hot pressed using the same conditions but was held for 150 minutes and then annealed. This specimen did not fail after 150 hours under load. This specimen has larger β-Si_3N_4 grains. These results suggest that the creep resistance of silicon nitride ceramics depends on their microstructure and the nature of their grain boundary phase; the larger the β-Si_3N_4 grains, the slower the creep rate. Improved creep resistance at high temperatures was also proven by the crystalline materials at the grain boundary.

Figure 10 relates the room-temperature fracture toughness to composition and state of the grain boundary phase (crystalline or amorphous) as shown by Bonnell[7] in the systems β-SiAlON-Garnet and β-SiAlON-cordierite. A decrease in toughness with increasing volume of the grain boundary glass was seen in the hot-pressed samples in both systems. More dramatic differences were observed between samples with amorphous and crystalline grain boundary phase. After cordierite crystallized at the grain boundaries, the toughness of the specimen increased, whereas toughness decreased when the garnet crystallized. This phenomenon illustrates that the thermal expansion mismatch affects the mechanical properties of silicon nitride ceramics. Crystalline garnet has a thermal expansion coefficient of 8×10^{-6} which is higher than that of silicon nitride (3×10^{-6}). In the β-SiAlON-cordierite system, however, the grain boundary crystalline phase has a thermal expansion coefficient of 2×10^{-6} which is lower than that of silicon nitride. The grain boundary phase will be under tension in the β-SiAlON-garnet system but will be under compression in the β-SiAlON-cordierite system.

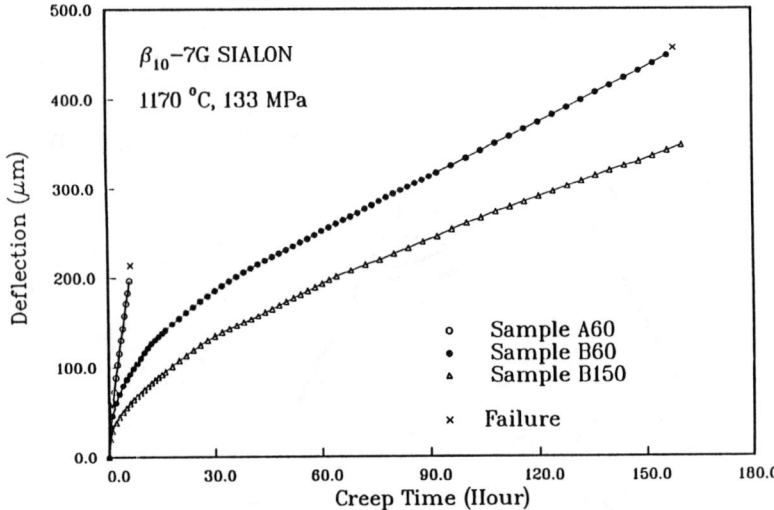

Fig. 9. Creep curves obtained at 1170°C, 133 MPa for (i) sample A60, (ii) sample B60, and (iii) sample B150.

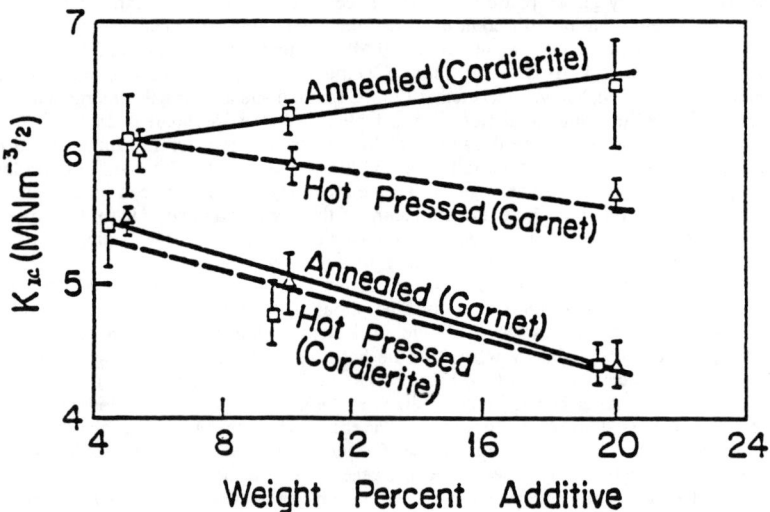

Fig. 10. Comparison of room temperature fracture toughness illustrates the increase in toughness when cordierite crystallizes at the grain boundaries and the decrease with garnet crystallization.

Composites Consisting of Two Major Si3N4 Phases

Sun[8] has shown that β-Si$_3$N$_4$ solid solutions, α'-SiAlON and AlN polytypoids form a series of compatibility triangles (see Fig. 11). Using different combinations of these phases, ceramics with different microstructures can be designed. Melting points of these phases are very high and ceramics with these compositions are expected to have superior high temperature mechanical properties.

Acicular β-Si$_3$N$_4$ Grains with Equiaxed α'-SiAlON—As shown in Fig. 12, the line connecting β-Si$_3$N$_4$ and the composition Y$_2$O$_3$:9AlN intersect the α'-SiAlON phase field. This was also demonstrated by Huang et al.[9] Compositions on this line between the β-Si$_3$N$_4$ and α'-SiAlON contain two phases below sub-solidus temperatures. Since the melting points of both of these

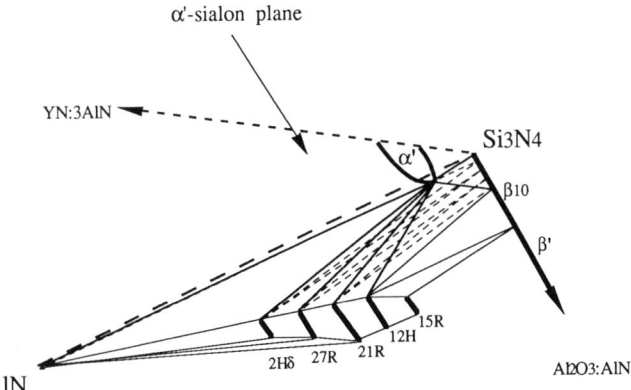

Fig. 11. α'-SiAlON is compatible with polytypoids (from 2Hδ to 12H), AlN and β' forming compatibility tetrahedra.

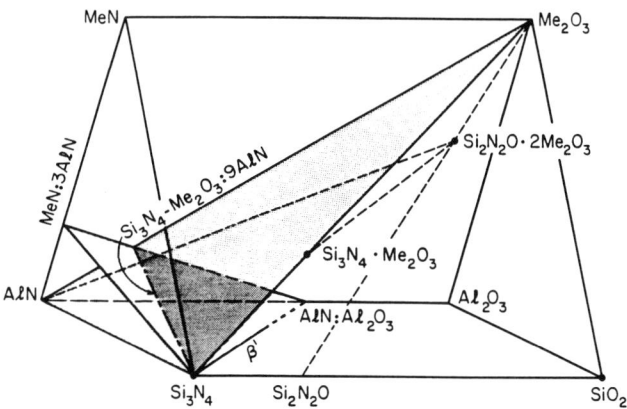

Fig. 12. The system Si$_3$N$_4$-SiO$_2$-AlN-Al$_2$O$_3$-MeN-Me$_2$O$_3$ showing one line Si$_3$N$_4$-Me$_2$O$_3$:9AlN cut through one α'-SiAlON phase field.

two phases are very high, ceramics containing β-Si₃N₄ and α'-SiAlON will have better creep resistance at moderate temperatures. Specimens containing these two phases were hot pressed by Sheu.[10] The composition containing 70% β-Si₃N₄ and 30% α'-SiAlON showed a flexural strength of 1200 MPa at room temperature and 800 MPa at 1400°C. The results are shown in Fig. 13.

α'-SiAlON with Polytypoid Platelets—As shown in Fig. 11, α'-SiAlON is compatible with all of the AlN polytypoids. Ceramics with equiaxed α'-SiAlON and polytypoid platelets should have high toughness.[11] Since α'-SiAlON has a high hardness value, it is anticipated that such ceramics would have superior mechanical properties. This system is currently being studied in the author's laboratory. Preliminary results are shown in Fig. 14.

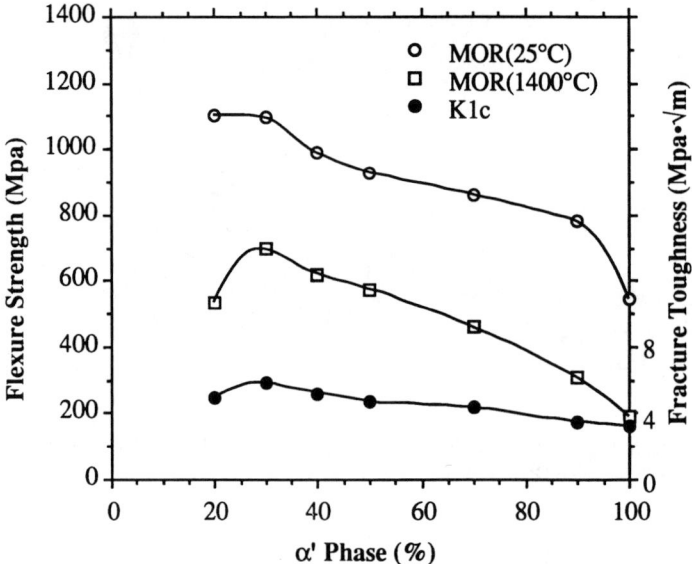

Fig. 13. Mechanical properties of α'-SiAlON plus β-Si₃N₄ two-phase material being tested at 25°C and 1400°C in the air.

Acicular β-Si₃N₄ Grains with Polytypoid Platelets—High density ceramics containing β-Si₃N₄ grains with polytypoid platelets are very difficult to synthesize because the liquid forming temperature is very high. Scientists at Shanghai Institute of Ceramics[12] have synthesized two-phase ceramics consisting of β-Si₃N₄ grains and polytypoid platelets with La₂O₃ as a sintering aid. Their results showed a modest increase in toughness. Because a sintering aid was used, a glassy grain boundary phase was observed.

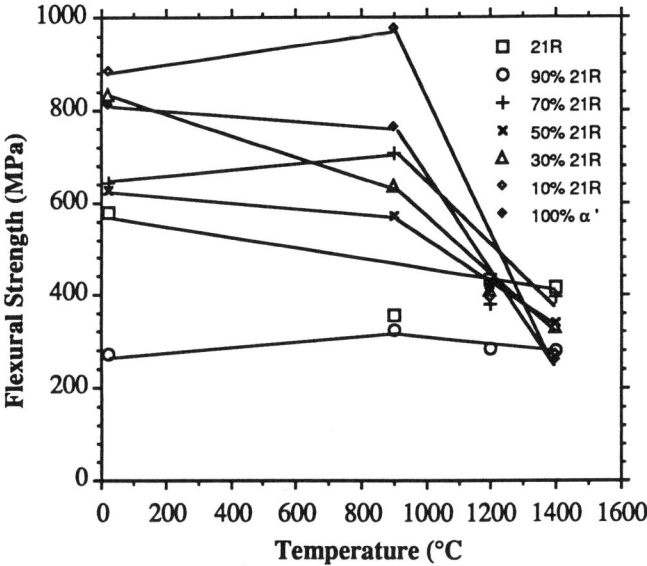

Fig. 14. Flexural strength of a two-phase composite, α + 21R, at temperatures 25°C-1400°C.

Acicular β-Si$_3$N$_4$ Grains with Silicon Oxynitride—Trigg and Jack[13] prepared ceramics containing oxynitride and β-Si$_3$N$_4$. As described before, the oxynitride grains are equiaxed and β-Si$_3$N$_4$ grains are elongated hexagonal rods. It is anticipated that these microstructures should produce ceramics with good mechanical properties. Figure 15 shows a compatibility relationship between oxynitride and β-Si$_3$N$_4$. Note that the tetrahedron β-Si$_3$N$_4$-β-SiAlON-Y$_2$Si$_2$O$_7$-Y$_3$Al$_5$O$_{12}$ separate the silicon oxynitride from α'-SiAlON and AlN polytypoids. This indicates that ceramics with silicon oxynitride and α'-SiAlON and AlN polytypoids cannot be formed.

Three-phase Composites

As shown in the phase diagram (Fig. 11), the phases β-Si$_3$N$_4$, α'-SiAlON and different AlN polytypoids form compatibility triangles. Compositions containing these three phases have been prepared in the author's laboratory by hot pressing and the preliminary results are encouraging.

Multi-phase Ceramics with Grain Boundary Liquid

Compositions containing β-Si$_3$N$_4$, α'-SiAlON and different AlN polytypoids can only be densified by transient liquid phase sintering. During transient liquid phase sintering, the amount of liquid changes with time and diminishes when the reactions are completed. Therefore, it is not possible to develop the acicular β-Si$_3$N$_4$ grains. In order for the silicon nitride grain to develop, equilibrium liquid has to be present during sintering. As shown in Fig. 16, the compounds YAG and melilite form compatibility tetrahedrons with β-Si$_3$N$_4$ and α'-SiAlON. A low melting liquid field should be present in the four phase mixture. Therefore, compositions inside the tetrahedron can be used to form ceramics with controlled microstructures.

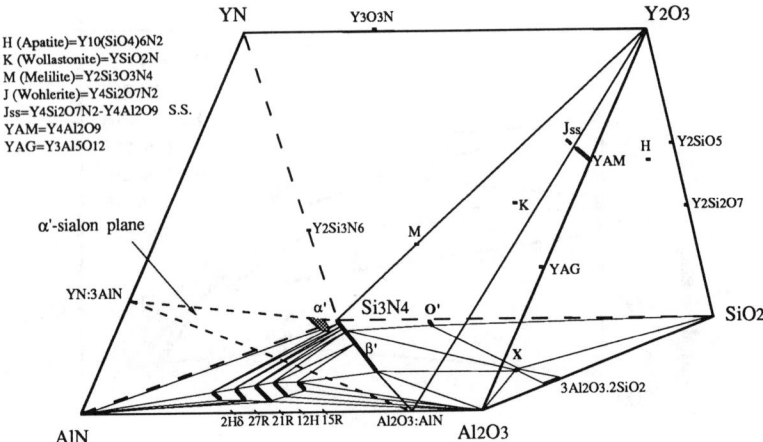

Fig. 15. Subsolidus phase relationships in the Si_3N_4-Y_2O_3-Al_2O_3:AlN system.

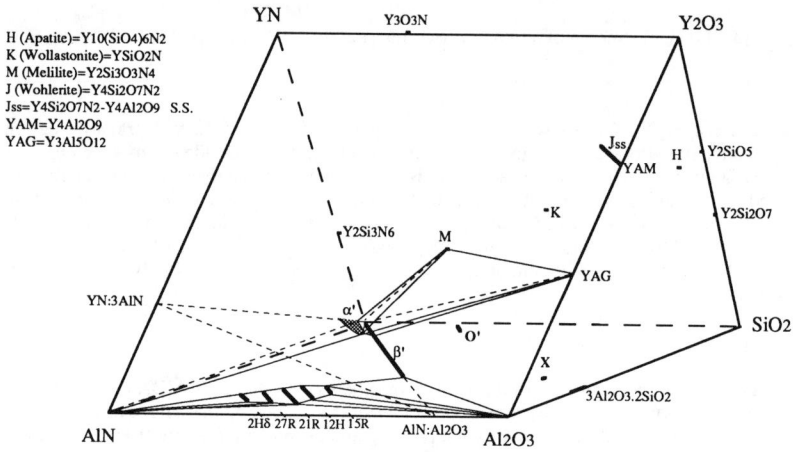

Fig. 16. α'-b' two-phase region is compatible with melilite (M) and YAG forming three compatibility tetrahedra: Si_3N_4-α' (boundary facing to β')--M, Si_3N_4--α'(m = 1, n = 1.7)-M--YAG and Si_3N_4--α' (m = 1, n = 1.7)--β_{10}-YAG.

REFERENCES

1. W-Y. Sun, T-Y. Tien and T-S. Yen, J. Am. Ceram. Soc. **74** (11) 2753-58 (1991).

2. C.M. Hwang, PhD thesis, The University of Michigan, 1988.

3. K R. Lai and T.Y. Tien, J. Am. Ceram. Soc., in press.

4. K.R. Lai and T.Y. Tien, unpublished data.

5. M. Wisnudel, MS thesis, University of Michigan, 1991.

6. C.F. Chen and T.Y. Tien, Ceram. Eng. and Sci Proc. **8** (7-8) 778-95 (1987).

7. D.A. Bonnell, T-Y. Tien and M. Rühle, J. Am. Ceram. Soc., **70** (7) 460-65 (1987).

8. W-Y. Sun, T-Y. Tien and T-S. Yen, J. Am. Ceram. Soc., **74** (10) 2547-50 (1991).

9. Z-K. Huang, T-Y. Tien and T-S. Yen, J. Am. Ceram. Soc., **69** (10) C241-2 (1986).

10. T.S. Sheu and T.Y. Tien, Bimonthly Report, Ceramic Technology for Advanced Heat Engine Project, Oak Ridge National Laboratory, October 31, 1991, p. 69.

11. T.S. Sheu and T.Y. Tien, Bimonthly Report, Ceramic Technology for Advanced Heat Engine Project, Oak Ridge National Laboratory, August 28, 1992, p. 70.

12. H.R. Zhung, W.L. Li, J.W. Feng, Z.K. Huang and D.S. Yan, Europ. Ceram. Soc., **7**, 329-333 (1991).

13. M.B. Trigg and K.H. Jack, <u>Proc. First International Symposium on Ceramic Components for Engines</u>, 1983, Hakone, Japan, edited by S. Somiya, E. Kanai and K. Ando (KTK Scientific Publishers, Tokyo, Japan, 1984) pp. 343-349.

HIGH-RESOLUTION ELECTRON MICROSCOPY OBSERVATIONS OF GRAIN-BOUNDARY FILMS IN SILICON NITRIDE CERAMICS

H.-J. KLEEBE, M. K. CINIBULK, I. TANAKA*, J. BRULEY, R. M. CANNON[†],
D. R. CLARKE[§], M. J. HOFFMANN, and M. RÜHLE
Max-Planck-Institut für Metallforschung, Institut für Werkstoffwissenschaft,
D-7000 Stuttgart 1, Germany

ABSTRACT

Characterization of silicon nitride ceramics by transmission electron microscopy (TEM) provides structural and compositional information on intergranular phases necessary to elucidate the factors that can influence the presence and thickness of grain-boundary films. Different TEM techniques can be used for the detection and determination of intergranular-film thickness, however, the most accurate results are obtained by high-resolution electron microscopy (HREM). HREM studies were applied, in conjunction with analytical electron microscopy, to investigate the correlation between intergranular-phase composition and film thickness. Statistical analyses of a number of grain-boundary films provided experimental verification of a theoretical equilibrium film thickness. Model experiments on a high-purity Si_3N_4 material, doped with low amounts of Ca, suggest the presence of two repulsive forces, a steric force and a force produced by an electrical double layer, that may act to balance the attractive van der Waals force necessary to establish an equilibrium film thickness.

INTRODUCTION

The high-temperature performance of silicon nitride ceramics strongly depends on secondary phases that are present at multiple-grain junctions and grain boundaries. Following liquid-phase sintering, which is required to achieve complete densification, the residual liquid forms an intergranular amorphous phase. Although studies[1-5] have shown that the bulk of this intergranular phase can be crystallized by an appropriate post-sintering heat treatment, there remains a thin amorphous film at the grain boundaries. Bulk material properties at room temperature such as fracture toughness also have been shown to be related to the presence or absence of a thin amorphous grain-boundary film in Si_3N_4 and SiC ceramics.[6]

Although the presence of amorphous phases is known to control mechanical properties, few studies have focused on the existence of amorphous intergranular films. Lange[7] first applied a squeeze-film analysis to predict that a finite film thickness will remain for kinetic reasons between compressed particles when the liquid perfectly wets the solid. Clarke[8]

---

*On leave from: ISIR, Osaka University, Ibaraki, Osaka 567 Japan
[†]On leave from: Lawrence Berkeley Laboratory, University of California, Berkeley, CA 94720
[§]On leave from: Department of Materials, University of California, Santa Barbara, CA 93106

subsequently provided a theoretical basis for the existence of a stable equilibrium film thickness, in terms of a balance between the van der Waals attraction of two grains and a repulsive structural force due to distortions of SiO_4^{4-} tetrahedra in the silicate liquid film. For siliceous films, an equilibrium thickness on the order of 10-20 Å was proposed, depending ultimately on the precise composition of the liquid and the bounding grains. Marion et al.[9] suggested that the two-grain boundary films were more highly ordered than the glass at multiple-grain junctions, based on an electron diffraction study of sintered alumina-anorthite; this supported the theory of "stress-supporting entities" that could provide a repulsive force. Clarke et al.[10] recently proposed another model to explain the stability of these films, which includes the presence of an electrical double layer at the interface to provide the repulsive force. In all of the proposed models the thickness of the grain-boundary film is strongly dependent on the chemical composition of the system, i.e., the composition of both the grains and the intergranular film itself.

Transmission electron microscopy (TEM) allows for a comprehensive characterization of microstructural features such as grain boundaries and intergranular phases. Structural information and elemental composition of these features can be obtained using the appropriate techniques and instrumentation. This paper discusses the characterization of silicon nitride ceramics by TEM to obtain structural and compositional information on intergranular phases in order to clarify the various factors that can influence the presence and thickness of grain-boundary films. A review of the techniques that can be used to detect and quantitatively measure the thickness of the grain-boundary film is presented along with experimental verification of the existence of an equilibrium thickness. The influence of secondary-phase chemistry on the equilibrium film thickness will be described, along with a model experiment that demonstrates the possible dependencies of the equilibrium film thickness, as proposed by theoretical models.

TEM ANALYSIS OF AMORPHOUS GRAIN-BOUNDARY FILMS

The detection of thin (~10 Å) intergranular films over the past 15 years by transmission electron microscopy has been accomplished with a number of specialized techniques, of which the most widely used are diffuse dark-field imaging, defocus Fresnel-fringe imaging, and lattice-fringe imaging by high-resolution electron microscopy (HREM), as illustrated schematically in Fig. 1. Clarke[11] first presented an overview of these techniques and they subsequently have been used for locating and identifying thin amorphous films in ceramics. However, the question as to which one of these techniques is the most accurate for quantitatively determining the thickness of amorphous intergranular films is often raised. The first quantitative comparison of the various imaging techniques showed that only high-resolution lattice imaging gave accurate and consistent results of intergranular film thickness in a sintered silicon nitride.[12]

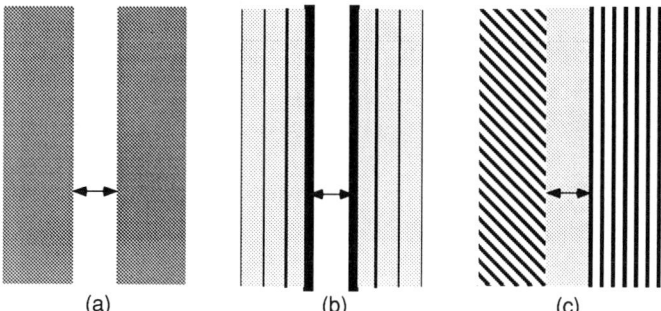

Fig. 1. Schematic illustration of images of a grain boundary containing an intergranular film, produced by (a) diffuse dark-field, (b) defocus Fresnel-fringe, and (c) high-resolution lattice-fringe imaging. Arrows indicate apparent intergranular film thickness.

The dark-field method in which the diffusely scattered electrons from the amorphous intergranular phase are directly imaged is perhaps the most widely used method for detecting amorphous phases because of the experimental ease associated with this technique. A portion of the halo formed by the diffusely scattered electrons is centered and selected by the smallest objective aperture, while eliminating contributions of Bragg diffracted electrons. The intergranular film is then imaged directly in bright contrast relative to the bounding grains, due to its contribution of diffusely scattered electrons as shown in Fig. 1(a). This method can be used to detect intergranular films irrespective of grain orientation and boundary inclination, although the low intensities prevent imaging very thin films unless the interface is exactly edge-on. This technique was found to be the most inaccurate for determining intergranular film thickness, overestimating the film thickness by up to a factor of 2, compared with HREM.[12] Factors that contribute to the artificial increase in film thickness include the nature of diffusely scattered electrons, leading to poorly defined crystal/glass interfaces, and specimen drift stemming from the long exposure times necessary to obtain adequate contrast images at high magnifications.

The defocus Fresnel-fringe technique has been widely used as well to detect amorphous intergranular films.[13] When the mean inner potential of the intergranular film is lower than that of the grains, imaging using an under-focused objective lens produces a bright fringe at the intergranular film with alternating dark and bright Fresnel fringes on either side as shown in Fig. 1(b). The over-focused image yields a reversal in contrast. The spacing of the fringes decreases with decreasing defocus until zero defocus is reached, at which point the fringe spacing equals the film thickness. Therefore, by extrapolating the data obtained from a series of out-of-focus images to zero defocus the thickness of the amorphous film can be obtained. This method can be helpful in aligning a boundary edge-on since the Fresnel fringes produced on defocusing are only symmetric on both sides of the boundary when the interface is exactly parallel to the electron beam. The use of this method for intergranular film thickness determination resulted in values 20-35% greater than those determined by HREM.[12] A complete analysis of the Fresnel fringes produced at an interface is complicated by a number of factors including the difference in projected potential between the grains and the amorphous film, geometry of the potential discontinuity (interface), and specimen thickness.

The measurement of intergranular films by high-resolution lattice (or structure) imaging consists of orienting the grain boundary edge-on while obtaining good diffracting conditions for both adjacent grains such that interference fringes are produced on either side of the film. The thickness of the intergranular film corresponds to the area of discontinuity in these fringes at the interface, shown schematically in Fig. 1(c). The spacing of the interference fringes (equal to the corresponding interplanar spacing) can be used as an internal scale to determine intergranular film thickness most accurately. To obtain interference fringes the foil must be of a thickness at which the fringes have high contrast in both grains. In general this limits the area of the specimen where suitable boundaries can be obtained to only the very thinnest. Interfaces must be edge-on and images should be taken at close to zero defocus, in addition to Scherzer defocus, to ensure that the lattice fringes do not extend into the intergranular film. The use of high-resolution lattice fringe imaging has been found to be the best method of determining the grain-boundary film thickness in Si_3N_4 materials, to within ± 1 Å.[12,14] Hence, this method is used by the authors for accurately determining intergranular film thicknesses as demonstrated in the following sections.

EXPERIMENTAL VERIFICATION OF AN EQUILIBRIUM FILM THICKNESS

Relating bulk material properties, such as creep behavior, to the grain-boundary structure and its corresponding interface chemistry requires a detailed characterization of the intergranular films. Even small changes in film chemistry will affect both the properties of the intergranular film, as well as the film thickness. These changes in film thickness are small, typically less than 5 Å. Hence, a reliable and reproducible technique for measuring such small deviations is required. Although the HREM-imaging technique is the most accurate method to determine sub-nanometer-scale features of intergranular films, evaluating grain-boundary film thicknesses from single HREM images may lead to misinterpretations as these micrographs in general may not be quantitative and representative; e.g., usually only very small regions of the interfaces are imaged. By statistically evaluating grain-boundary film thicknesses along the entire length of a grain boundary and for a large number of grain boundaries, it was possible to verify both the validity of determining intergranular film thicknesses from single HREM micrographs and the assumption of an equilibrium film thickness in these materials.[14]

Four Si_3N_4 materials were investigated and the results, obtained by measuring the intergranular film widths over the entire length of the interface, were compared with the film thicknesses determined from single HREM images, as summarized in Table I. The results of two of these materials (MgO-doped Si_3N_4 and Yb_2O_3-doped Si_3N_4) will be discussed. Statistical measurements were performed by printing HREM images over the entire length (up

TABLE I. SUMMARY OF INTERGRANULAR FILM THICKNESSES

Material	Additive	Intergranular Film Thickness		
		Single HREM Image [Å]*	Statistical Evaluation [Å]	Standard Deviation [Å]
A: NC 132	~2 wt% MgO	8 ± 1	8.3 7.9	0.58 0.65
B	5.0 vol% Yb_2O_3	10 ± 1	10.0 10.7 10.5[†] 10.2[§]	0.69 0.70 1.13[†] 0.49[§]
C	5.0 vol% Yb_2O_3 +0.25 vol% CaO	13 ± 1	14.2	0.94
D	5.0 vol% Yb_2O_3 +0.25 vol% CaO annealed	14 ± 1	15.1	0.95

*Average of three data points from a single HREM image. †Faceted interface.
§Fifteen grain boundaries; each measured at four locations.

to 1 μm) of a given grain boundary between two triple pockets. Every 2 cm of the enlargedprints, corresponding to about 20 nm of the interface length, a thickness data point was evaluated giving 40 to 50 data points for each boundary. The distributions of the film-thickness values were plotted as histograms (with a bar width of 0.5 Å), and the standard deviation was determined. For one of the Si_3N_4 materials, fifteen grain boundaries with different misorientations of the adjacent Si_3N_4 grains at the interface were also studied; the film thickness from single HREM micrographs was measured by averaging three or four measurements from each print.

MgO-doped Si_3N_4

The MgO-doped Si_3N_4 (material A: NC-132, Norton Company, Northboro, MA) contains an amorphous intergranular phase present at both triple-grain junctions and grain boundaries. An amorphous intergranular film can clearly be seen in Fig. 2(a) connecting both triple-point regions. The HREM micrograph in Fig. 2(b) depicts one small region of the grain boundary shown in Fig. 2(a). The grain-boundary film thickness estimated from the single HREM image in Fig. 2(b) is 8 ± 1 Å, as indicated in the micrograph. For the statistical measurements HREM images of the entire length of the interface were printed as four micrographs (indicated by the boxed areas in Fig. 2(a)) and the film-thickness data obtained from 50 locations, as previously described. Histograms of two different grain boundaries and the corresponding standard

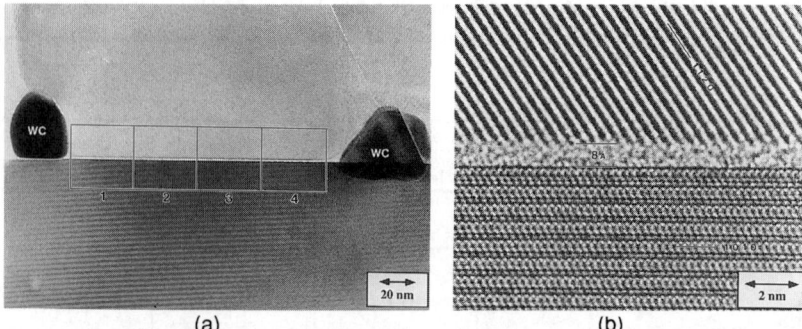

Fig. 2. HREM images of a grain-boundary in an MgO-doped Si_3N_4 (material A: NC-132). (a) Entire length of the grain boundary shown terminating at two triple-grain junctions (containing WC inclusions). Image was taken at a defocus of -200 nm. (b) Enlargement of the central region of boxed area 3 in (a) indicating the film thickness to be 8 ± 1 Å. Four such enlargements were used for statistically measuring the film thickness over the entire length of the grain boundary at 50 locations.

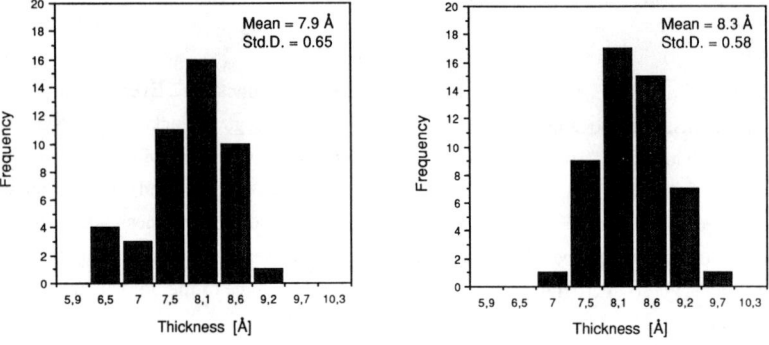

Fig. 3. Histograms of the intergranular film-thickness distributions of two interfaces in material A (NC-132). The mean thickness and standard deviation for the measurements of each interface are given.

deviations are given in Fig. 3. Statistical evaluations of the two interfaces resulted in mean film-thickness values of 7.9 Å and 8.3 Å, respectively. Comparing both mean values with the value of 8 Å, evaluated from the single HREM image in Fig. 2(b), a good agreement was achieved. The standard deviations for both statistical measurements were 0.58 Å and 0.65 Å, which is less than the estimated error of ± 1 Å from a single HREM image. Moreover, the difference between the average thicknesses determined by statistical evaluation and the single HREM image is 0.1 Å and 0.3 Å for the two interfaces, respectively, which lies well within the bounds of the standard deviations.

Yb_2O_3-doped Si_3N_4

The intergranular film thickness in material B, sintered with 5 vol% Yb_2O_3, was determined from single HREM images of fifteen different grain boundaries. Fig. 4(a) depicts a characteristic interface in this material. The film thickness estimated from the single HREM images was 10 ± 1 Å. The histograms of the film thickness of three different interfaces in material B, and the corresponding standard deviations, are given in Fig. 5. Comparing the statistical evaluations with the film-thickness values obtained from single HREM images, a good agreement was achieved (see Table I). The average of the statistical mean values of the film thickness in material B is 10.4 Å. Hence, the difference between thickness values obtained from a single HREM image and the statistical approach is only 0.4 Å. This value also lies well

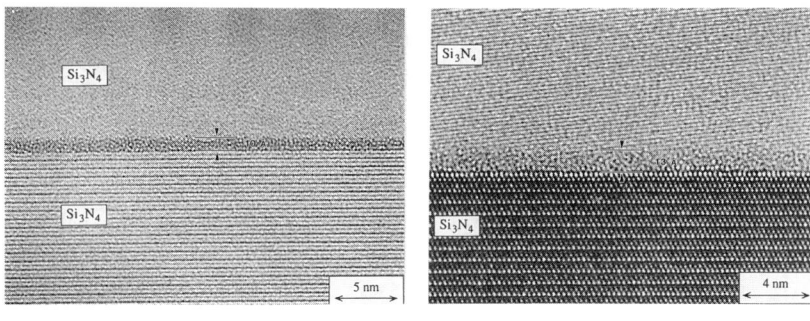

(a)　　　　　　　　　　　　　(b)
Fig. 4. HREM images of grain-boundaries in (a) a Yb_2O_3-doped Si_3N_4 (material B) with an intergranular film thickness of 10 ± 1 Å and (b) a Yb_2O_3-doped Si_3N_4 containing 0.25 vol% CaO (material C) with a film thickness of 13 ± 1 Å.

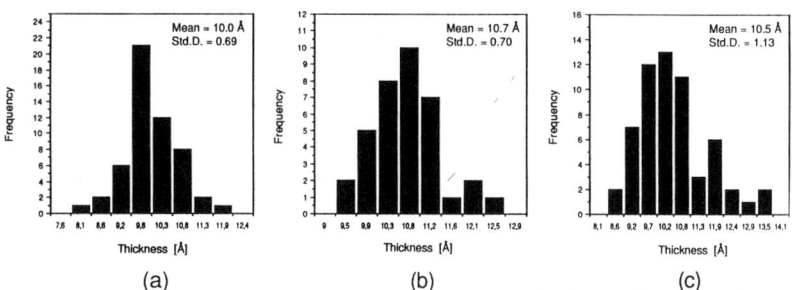

(a)　　　　　　　　(b)　　　　　　　　(c)
Fig. 5. Histograms of the intergranular film-thickness distributions of three interfaces in material B. The mean thickness and standard deviation for the measurements of each interface are given. Note that the standard deviation in (c) is ~60% higher than in (a) and (b) due to faceting of one of the grains at the interface.

within the evaluated standard deviations of 0.69 Å and 0.70 Å for boundaries in Figs. 5(a) and (b), respectively.

One boundary in material B (see Fig. 5(c)) showed a nearly bimodal film-thickness distribution and an increase in standard deviation of ~60%. A detailed characterization of the grain-boundary structure revealed the step-like nature of this interface. Close to the interface step a higher film thickness was measured, while in the central region between two grain-boundary facets a film width close to the average mean value of 10.4 Å was determined. This seems to suggest that the inclination of the grain-boundary plane out of a low-energy configuration can be compensated by grain-boundary faceting, lowering the overall boundary energy. At structurally disturbed regions, such as the surface on one side of the crystal-glass interface, the intergranular film width is locally increased; however, the average film width remains the characteristic film thickness of the material.

The cause for differences in the thicknesses determined by the statistical method and the single HREM measurements is twofold: (i) a real variation of the boundary thickness as a material characteristic and (ii) a variation due to difficulties in determining the surfaces of the two grains adjacent to the amorphous film in the micrograph. The measured scatter of the data of materials A and B is attributed to the latter. However, if grain boundaries show faceting, as observed in one boundary of material B, a variation in film widths is expected and the scatter of the data is related to the characteristic grain-boundary structure and to difficulties in defining the crystal surfaces at the boundary.

To further verify the existence of an equilibrium film thickness in this material, fifteen grain boundaries were selected and the HREM imaging technique applied to each boundary. The corresponding intergranular-film thicknesses were measured at four locations at each interface and plotted as a histogram, shown in Fig. 6. A Gaussian-like distribution was observed, similar to those observed of measurements of the film thickness from a single grain boundary. A mean value of 10.2 Å with a standard deviation of 0.49 Å is in good agreement with values obtained from single HREM micrographs and those values determined statistically from single grain boundaries. This implies that the thickness of the intergranular film is indeed, not only constant along a given grain boundary, but the same at all high-angle grain boundaries in this material. Previous studies of the intergranular film thickness in other Si_3N_4 materials by the authors[5,13,15-19] have led to the same conclusions; however, the present observations are the strongest evidence of the existence of an equilbrium film thickness in Si_3N_4 materials.

The experimental results also showed no dependence of film thickness on grain orientation. All grain boundaries studied in each of the materials revealed a constant film width, although the boundaries observed showed distinct differences in misorientation of the adjacent Si_3N_4 grains. It should be noted that this result (the independence of the equilibrium thickness from grain orientation) is unexpected and not predicted by theory. Moreover, it was shown that single HREM images of a number of interfaces suffice to quantitatively determine the film thickness in Si_3N_4 ceramics.

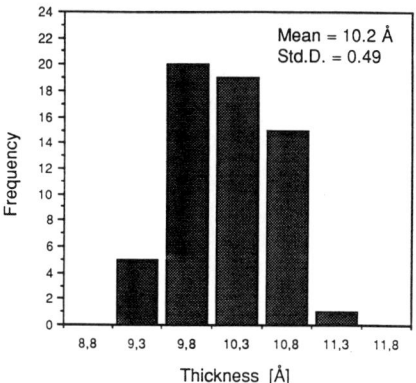

Fig. 6. Histogram of the intergranular film-thickness distribution of fifteen grain boundaries, each measured at four locations, in material B. The mean thickness and standard deviation for the measurements of each interface are given.

DEPENDENCE OF FILM THICKNESS ON INTERGRANULAR-PHASE CHEMISTRY

The verification of an equilibrium-film thickness implies that each material, with its given secondary-phase composition, shows a characteristic film thickness. Previous studies by the authors on Si_3N_4 sintered with different additives, in fact, revealed a strong dependence of intergranular-film thickness on chemistry. For example the two materials studied in the previous section (MgO-doped Si_3N_4 and Yb_2O_3-doped Si_3N_4) were densified with different additives and therefore have a different grain-boundary film chemistry. A consequence of this great difference in amorphous-film chemistry is the difference in film thickness observed by HREM in these materials as can be seen by comparing Figs. 2(b) and 4(a). Minor amounts of impurities are expected to have an effect on the intergranular film thickness as well.

The effect of a small amount of impurity on film thickness was studied by preparing two Yb_2O_3-doped Si_3N_4 materials containing a small amount of CaO (0.25 vol%) as the controlled impurity.[19] In one material the secondary phase was completely amorphous, while in the other a post-sintering heat treatment resulted in crystallized secondary phases at multiple-grain junctions. Two crystalline secondary phases were observed, $Yb_2Si_2O_7$ and $CaYb_9(SiO_4)_6ON$ (Ca- and N-stabilized apatite). Both materials, sintered and annealed, revealed wider grain-boundary films than the undoped material (see Fig. 4). The annealed material revealed a slightly wider film than the as-sintered material (14 Å vs. 13 Å), which is attributed to a higher Ca-concentration at the grain boundaries, as crystallization at triple junctions reduces the overall glass volume in the sintered body and, hence, Ca segregates to the boundary. Obviously, the formation of a Ca-, N-stabilized apatite did not completely compensate for the chemical shift in the specimen during annealing. Therefore, a slightly higher Ca-concentration is expected in the amorphous grain-boundary films in the annealed material than in those in the CaO-doped Si_3N_4 that did not undergo the heat treatment. A similar dependence of film thickness on impurity concentration was also observed in two materials undergoing oxidation.[18] During oxidation the intergranular phase is essentially purified as additive and impurity cations diffuse to the surface oxide layer,[20] resulting in a compositonal change that is reflected by a decrease in grain-boundary film thickness.

A MODEL SYSTEM: LOW Ca-DOPED AMORPHOUS SiO_2 FILMS IN SILICON NITRIDE

As discussed in the previous section, small amounts of added CaO result in a significant increase in the equilibrium thickness of the intergranular film. The thickness variation may reflect the modification of the atomic structure of the film by the change in chemical composition of the glass. A model system consisting of Si_3N_4 with a pure SiO_2 intergranluar phase that was doped with low amounts of Ca was chosen to obtain a better understanding of the influence of impurities on grain-boundary films. Dedicated scanning transmission electron microscopy (STEM) is required for the determination of elemental composition with a high degree of spatial and energy resolution. The 8-Å probe size of the VG HB501 STEM was necessary for elemental analysis of intergranular films of the order of 10 Å in thickness in these materials.

High-purity α-Si_3N_4 powder (SN-E10, Ube Corp., Tokyo, Japan) can be sintered without additives by hot-isostatic-pressing using the glass-encapsulation technique.[21] This powder contains ~1.3 wt% oxygen as surface oxide, but less than 50 ppm total cation impurities, resulting in a densified Si_3N_4 ceramic with an SiO_2 amorphous grain-boundary phase. Four materials were prepared in this manner, but with controlled amounts of Ca (450 ppm, 220 ppm, 80 ppm, and undoped) added to the α-Si_3N_4 powder. In a previous study, Tanaka et al.[22] found that the long-term strength at 1400°C was significantly reduced by doping with only 80 ppm Ca, although the general microstructure was the same as in the undoped material.

HREM images of the materials with the low levels of CaO dopant are shown in Fig. 7. Grain-boundary films in these materials also revealed an equilibrium thickness. The grain-boundary film thickness in Si_3N_4 containing a pure SiO_2 amorphous phase was also examined in another undoped Si_3N_4 (prepared from a different Si_3N_4 powder) and in a Si_3N_4-SiO_2 material with higher SiO_2 content; the film thickness was found to be 10 ± 1 Å in all of the materials.[23] The presence of Ca was detected in the grain-boundary films of all materials doped with Ca.[24] Assuming that all Ca and O are confined in the intergranular phase the composition of the film can be estimated from the intensity ratio of the Ca and O peaks, using electron energy loss spectroscopy. In the 80 ppm Ca-doped material, 1.5 ± 0.3 mol% of CaO was found in the amorphous SiO_2 phase. This value shows good agreement with the simple estimate of the glass containing 1.0 mol% CaO, assuming that all Ca atoms are localized at the intergranular glass which is 3.6 vol% of the material. No impurities were detected in the film in the undoped material, indicating the CaO concentration was less than the detection limit of 0.08 mol% CaO. Film thickness as a function of Ca content is plotted in Fig. 8. It is interesting to note that a decreasing film thickness in the very dilute region (to minimum value of 7 Å at 80 ppm Ca) is followed by an increase in thickness as a function of Ca content.

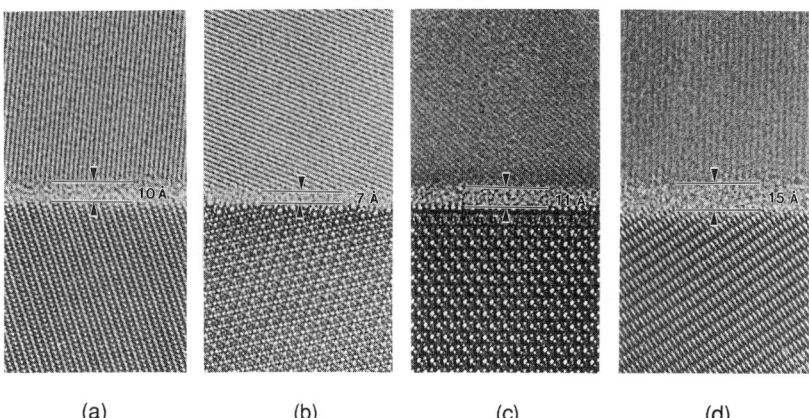

(a) (b) (c) (d)
Fig. 7. HREM images of grain boundaries in a high-purity Si_3N_4 containing an SiO_2 intergranular phase (a) and doped with (b) 80 ppm Ca, (c) 220 ppm Ca, and (d) 450 ppm Ca. Equilibrium film thicknesses were 10 ± 1 Å, 7 ± 1 Å, 11 ± 1 Å, and 15 ± 1 Å, respectively.

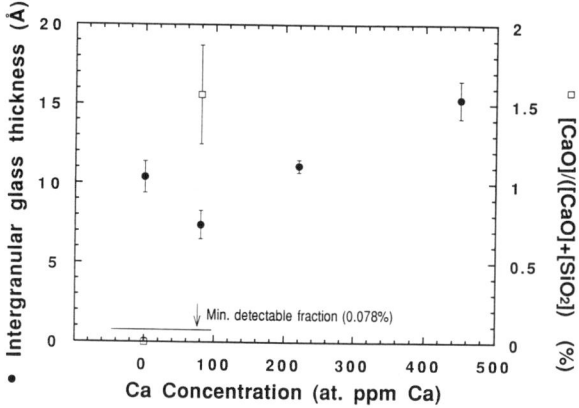

Fig. 8. Plot of equilibrium film thickness in high-purity Si_3N_4 as a function of Ca content. The experimentally determined CaO concentration in the intergranular film was determined by high-resolution electron-energy loss spectroscopy.

The observed variation in thickness of the intergranular film with concentration of added CaO can be understood in terms of the forces discussed by Clarke,[8,10] namely the attractive van der Waals force and the repulsive steric and electrical double layer forces. In the absence of CaO, we can assume that there is no electrical double layer at the grain boundaries, and so the equilibrium thickness is controlled by a balance between the van der Waals and the steric forces; the resulting grain-boundary structure is shown schematically in Fig. 9(b) to consist of a pure silica glass intergranular film. The addition of calcia has a dual effect. It is expected to disrupt the network structure of the silica intergranular phase, decreasing the structural correlation length and hence the magnitude of the steric repulsive force, as shown in Fig. 9(c). The addition of CaO also provides a charged species to the material which may result in the formation of an electrical double layer, as shown in Fig. 9(d). Since CaO is known to be a potent silica network modifier, causing a decrease in viscosity of several orders of magnitude for additions of less than 1 mol%, it might be expected that the strongest effect of minor calcia additions will be to decrease the steric force and cause a reduction in the film thickness. Further additions of CaO can be expected to increase the repulsive electrical double layer force, leading to further increases in film thickness. A minimum in thickness of the intergranular film at a low concentration of a network modifier, as observed at ~80 ppm Ca, is thus consistent with a competition between the effects of disrupting the network structure and the development of a force due to an electrical double layer. While more detailed calculations need to be performed to quantify these competing effects, it is nevertheless interesting to note that if all the CaO in the 450 ppm Ca-doped material were adsorbed to the surface of the Si_3N_4 grains, ~40% of the Si_3N_4 surface would be covered by a 2-Å thick monolayer, assuming 1-μm diameter grains. If the charge of each Ca^{2+} ion were uncompensated, this would correspond to a rather high surface charge density.

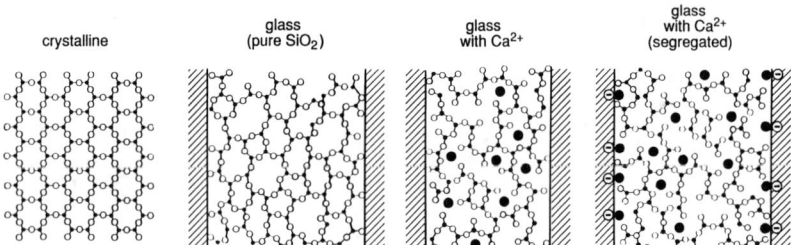

Fig. 9. Schematic illustration of changes occuring in the glass structure of the intergranular phase with increasing impurity cation (Ca^{2+}: ●) concentration, and its influence on the equilibrium film thickness. A combination of steric and electrical double-layer repulsive forces is proposed to balance the attractive van der Waals dispersion force.

CONCLUSIONS

Recent experimental studies on intergranular films in silicon nitride by transmission electron microscopy have been presented. The characterization of silicon nitride ceramics by TEM to obtain structural and compositional information on intergranular phases in order to elucidate the various factors that can influence the presence and thickness of grain-boundary films has been discussed. HREM was used, in conjunction with STEM, to investigate the influence of intergranular-phase composition on film thickness. Statistical analyses of HREM images of grain-boundary films have shown that they attain a constant thickness (to within 1 Å) along a grain boundary, as well as obtaining the same constant thickness along all high-angle grain boundaries for a given material composition, thereby providing experimental verification of an equilibrium film thickness. Moreover, the film thickness was observed to be independent of grain misorientation. Model experiments on a specially prepared, high-purity Si_3N_4 material, doped with low amounts of Ca, have shown that the film thickness is dependent on Ca content. STEM reveals the presence of Ca at the grain-bounary films in these materials. These observations suggest that a combination of steric and electrical double layer repulsive forces act to balance the attractive van der Waals force, resulting in an equilibrium film thickness.

ACKNOWLEDGMENTS

D.R. Clarke and I.Tanaka thank the Alexander von Humboldt Foundation for support during their stay in Stuttgart. D.R. Clarke was also supported by the NSF. This work was supported by the BMFT under contract number NTS 0230/0.

REFERENCES

1. D.A. Bonnell, T.-Y. Tien, and M. Rühle, *J. Am. Ceram. Soc.* **70** [7] 460 (1987).
2. L.K.L. Falk and G.L. Dunlop, *J. Mater. Sci.* **22** 4369 (1987).
3. M.K. Cinibulk, G. Thomas, and S.M. Johnson, *J. Am. Ceram. Soc.* **73** [6] 1606 (1990).
4. M.K. Cinibulk, G. Thomas, and S.M. Johnson, *J. Am. Ceram. Soc.* **75** [8] 2037 (1992).
5. J.S. Vetrano, H.-J. Kleebe, E. Hampp, M.J. Hoffmann, R.M. Cannon, and M. Rühle, *J. Mater. Sci.*, submitted.
6. H.-J. Kleebe, *J. Eur. Ceram. Soc.* **10** 151 (1992).
7. F.F. Lange, *J. Am. Ceram. Soc.* **65** [2] C-23 (1982).
8. D.R. Clarke, *J. Am. Ceram. Soc.* **70** [1] 15 (1987).
9. J.E. Marion, C.H. Hsueh, and A.G. Evans, *J. Am. Ceram. Soc.* **70** [10] 708 (1987).
10. D.R. Clarke, T.M. Shaw, A.P. Philipse, and R.G. Horn, *J. Am. Ceram. Soc.*, in press.

11. D.R. Clarke, *Ultramicroscopy* **4** 33 (1979).
12. M.K. Cinibulk, H.-J. Kleebe, and M. Rühle, *J. Am. Ceram. Soc.*, in press.
13. J.N. Ness, W.M. Stobbs, and T.F. Page, *Philos. Mag. A* **54** [5] 679 (1986).
14. H.-J. Kleebe, M.K. Cinibulk, R.M. Cannon, and M. Rühle, *J. Am. Ceram. Soc.*, submitted.
15. H.-J. Kleebe and M. Rühle, *Mat. Res. Soc. Symp. Proc.* **238** 859 (1992).
16. H.-J. Kleebe, M.J. Hoffmann, and M. Rühle, *Z. Metallkd.* **83** [8] 610 (1992).
17. H.-J. Kleebe and M.K. Cinibulk, *J. Mater. Sci. Lett.* **12** 70 (1993).
18. M.K. Cinibulk and H.-J. Kleebe, *J. Mater. Sci.*, submitted.
19. H.-J. Kleebe, J. Bruley, and M. Rühle, *J. Mater. Res.*, submitted.
20. D.R. Clarke and F.F. Lange, *J. Am. Ceram. Soc.*, **63** [9-10] 586 (1980).
21. I. Tanaka, G. Pezzotti, T. Okamoto, Y. Miyamoto, and M. Koizumi, *J. Am. Ceram. Soc.* **72** [9] 1656 (1989).
22. I. Tanaka, K. Igashira, T, Okamoto, K. Niihara, and R.M. Cannon, *J. Am. Ceram. Soc.*, submitted.
23. I. Tanaka, H.-J. Kleebe, M.K. Cinibulk, J. Bruley, and M. Rühle, *Philos. Mag. A*, submitted.
24. I. Tanaka, H.-J. Kleebe, M.K. Cinibulk, J. Bruley, D.R. Clarke, and M. Rühle, *J. Am. Ceram. Soc.*, submitted.

NEW GRAIN-BOUNDARY PHASES FOR NITROGEN CERAMICS

DEREK P. THOMPSON
Materials Division, Dept. of Mechanical, Materials & Manufacturing Engineering,
The University of Newcastle, Newcastle upon Tyne, NE1 7RU, U.K.

ABSTRACT

The drive for improved refractoriness in nitrogen ceramics, which has motivated the removal of residual glass from grain-boundaries by heat-treatment at sub-solidus temperatures, has resulted in the discovery of a wide range of crystalline oxynitride phases. Generally, these phases are oxygen-rich, and can be classified in the same way as the mineral silicates, in which oxygen atoms are coordinated to a maximum of two silicon atoms, with SiO_4 tetrahedra joined at either 0,1,2,3 or 4 corners to other tetrahedra. However, the field of oxynitride crystal chemistry is wider than this, because nitrogen in SiN_4 tetrahedra commonly occurs coordinated to three silicon atoms, and an additional range of structures exist with nitrogen in this coordination.

Oxynitride analogues of mineral silicates are well known in neso-, soro- and cyclo-structure types; new structures have recently been reported which belong to the pyroxene family (ino-silicates), and information on their preparation and crystal chemistry is reported here. The sialon U-phase (typical composition $Ln_3Si_3Al_3O_{12}N_2$, Ln = La,Ce,Nd,Sm), is an example of an oxynitride with a structure intermediate between layer (phyllo-) and framework (tecto-) types. Sialon W-phase (approximate composition $Ln_4Si_9Al_5O_{30}N$, Ln = La,Ce,Nd) has not been completely characterised, but appears to have a structure related to the amphibole group of double-chain silicates.

Comments are made on the many oxynitride structures which are still uncharacterised. The suitability of all these oxynitrides as grain-boundary phases in silicon nitride and sialon ceramics is discussed.

INTRODUCTION

Since the earliest years in the development of silicon nitride as an engineering ceramic, the presence of intergranular glassy phase in fully dense materials has been accepted as a necessary evil, because without the addition of glass-forming metal oxides to the starting mix, it has proved impossible to obtain fully-dense, high-strength ceramics. The discovery that the glassy phase could be converted into refractory, crystalline oxides and oxynitrides by an appropriate post-preparative heat-treatment, has resulted in more refractory products, but generally at some cost to mechanical performance; also, it must be admitted that the upper temperature limits even for heat-treated materials are still low compared with the decomposition temperature of silicon nitride.

Post-preparative heat-treatment studies have ushered in a wide new field of ceramic oxynitrides, and this has attracted interest in their crystal chemistry. It had previously been recognized [1] that ceramic nitrides and oxynitrides had crystal structures analogous to the mineral silicates and aluminosilicates, in which oxygen was replaced by nitrogen and valency balance was achieved by substituting silicon for aluminium, other cations possibly being incorporated as well. However, it was only when yttrium oxide was used as a densification additive for silicon nitride that a significant number of oxynitrides were produced with varied types of silicate structures. Thus the yttrium nitrogen apatite phase, $Y_{10}Si_6O_{24}N_2$, provided the first example of an oxynitride containing individual $Si(O,N)_4$ tetrahedra, the yttrium nitrogen wollastonite phase, $YSiO_2N$, was the first example of an oxynitride containing three-membered $[Si_3O_6N_3]$ rings, and the yttrium nitrogen melilite phase, $Y_2Si_3O_3N_4$, which by analogy with oxide melilites might have been expected to contain $Si_2(O,N)_7$ groups, in fact has a layer structure, in which all the silicon atoms occur in $Si(O,N)_4$ tetrahedra which form a continuously linked two-dimensional network in the (001) plane. The other yttrium silicon oxynitride, J-phase, of composition $Y_4Si_2O_7N_2$ is structurally analogous to the group of $Ln_4Al_2O_9$ aluminates and also to the mineral cuspidene, $Ca_4Si_2O_7F_2$. Early work on rare earth oxide densified nitrogen ceramics concentrated on the elements lanthanum, cerium and neodymium, and in these Ln-Si-O-N systems, similar oxynitride phases occurred, with certain differences which are discussed in more detail later.

The classical method of categorising silicate structures is in terms of the Si:O ratio as shown in Table I. In this classification, apatites are neso-silicates, J-phase is a soro-silicate,

Table I. Classification of silicate structures

Si:O ratio	Structural units	Type	Oxide example	Oxynitride example
1 : 4.00	SiO_4 groups	Neso-	Mg_2SiO_4	N-apatite, $Y_{10}(SiO_4)_6N_2$
1 : 3.50	Si_2O_7 groups	Soro-	$Sc_2Si_2O_7$	J-phase, $Y_4Si_2O_7N_2$
1 : 3.00	Rings	Cyclo-	$Be_3Al_2Si_6O_{18}$	N-wollastonite, $YSiO_2N$
	Chains	Ino-	$CaMgSi_2O_6$	
1 : 2.75	Double chains		$Mg_7Si_8O_{22}(OH)_2$	
1 : 2.50	Sheets	Phyllo-	$K_2Al_4Si_6Al_2O_{20}(OH)_4$	N-melilite, $Y_2Si_3O_3N_4$
1 : 2.00	Frameworks	Tecto-	$NaAlSi_3O_8$	

wollastonite is a 3-membered ring cyclo-silicate, and melilite is more correctly classified as a phyllo-silicate than as a soro-silicate (which the oxide melilites are). Noticeable absentees from this classification in the oxynitride field are pyroxene-type chain silicates, amphibole type double-chain silicates and framework silicates. However, it should be mentioned that α and β silicon nitride are special examples of framework silicates, in which nitrogens at the corners of SiN_4 tetrahedra are linked to three tetrahedra instead of two as in oxide systems.

The main factor restricting the scope of oxynitride crystal chemistry has been the thermodynamic limitations governing the stability of metal oxides in contact with silicon nitride. This has been discussed in detail by Negita [2]. Most transition metal oxides react with silicon nitride to form nitrides or (more commonly) silicides. Group I elements (with the exception of lithium) are too volatile to survive silicon nitride sintering temperatures, and hence the only stable metal oxides are those from metals in Groups II and III plus the rare

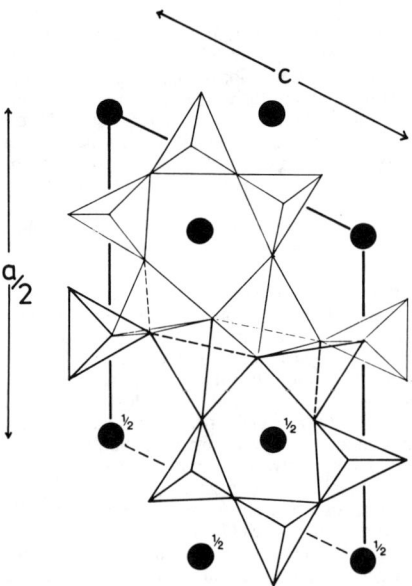

Figure 1. (010) projection of the crystal structure of $Ln_3Si_5O_4N_{11}$.

earths. The latter two groups have been discussed briefly above, and in Group II, Be-Si-O-N oxynitrides have not been pursued because of toxicity problems, MgO does not form stable Mg-Si-O-N oxynitrides, the Ba-Si-O-N and Sr-Si-O-N systems are relatively unexplored and the Ca-Si-O-N system contains several new oxynitrides, which are still not fully characterized. This latter system is a candidate for receiving more attention, but in previous years researchers have been discouraged from carrying out detailed investigations of this system because calcium impurity in silicon nitride starting mixes results in the formation of low-melting Ca-Si-O-N glasses with serious consequences for high-temperature properties.

Aluminium plays a similar role in the crystal chemistry of nitrogen ceramic systems as in mineral systems, but with certain additional features. Not only does it dissolve in the matrix nitride phases to form the well known α-, ß- and O'-sialons, but it also increases the range and stability of liquid phases in these systems, and also provides more scope for allowing a wider range of crystalline oxynitrides to occur.

A particularly important factor when assessing the suitability of a crystalline oxynitride as a grain-boundary phase for a silicon nitride or sialon ceramic, is the O:N ratio. In many cases, the crystallization of the grain-boundary glass in a nitrogen ceramic results in an oxide being produced ($Y_2Si_2O_7$ in yttria-densified silicon nitride; $Y_3Al_5O_{12}$ in yttria-densified sialons). Even though oxides are generally refractory, they exhibit low eutectic temperatures in equilibrium with nitride matrices. Similarly, oxynitrides with high O:N ratios are often close to the eutectic composition in a system, and have limited refractoriness. Oxynitrides with a low O:N ratio are therefore more desirable, because not only are they more refractory, but the higher nitrogen content makes them more similar in composition to the matrix phase and therefore the eutectic temperature is higher. In fact the most desirable situation is when the grain boundary glass has exactly the same composition as a high-melting crystalline oxynitride. There is then no eutectic between the crystalline grain-boundary phase and the matrix phase, and the limit of refractoriness is determined by the melting point of the grain-boundary phase.

Oxynitrides richer in nitrogen than the maximum solubility of nitrogen in the densifying liquid are seldom useful. They behave as additional components of the matrix but with (generally) worse oxidation resistance. If they have formed by crystallizing out of the grain boundary glass, there will be other more oxygen-rich phases also present which will impair overall refractoriness. Examples of this type of phase are the "$Ln_2O_3.2Si_3N_4$" series of compounds which occur in low atomic number (Ln=La,Ce,Nd) Ln-Si-O-N systems. Recent X-ray diffraction studies have shown that the correct formulation is $Ln_3Si_8O_4N_{11}$, and that the crystal structure (Figure 1) has strong similarities with that of $LaSi_3N_5$, and is of interest because the non-metal atoms are coordinated to 1, 2 or 3 silicon atoms in the same structure. ^{15}N NMR spectra show 2 bands corresponding to the latter two types of coordination, confirming that the non-metal atoms coordinated to one Si plus 3 Ln are in all probability oxygen [3].

The present review of recent work on new oxynitride ceramics builds on previous work described in the reviews by Leng-Ward & Lewis [4] and Thompson [5] and describes attempts made during the last five years to prepare new crystalline oxynitride phases by crystallization of M-Si-O-N, M-Si-Al-O-N, M_1-M_2-Si-O-N glasses. These compounds are candidate grain-boundary phases for silicon nitride or sialon ceramics and are therefore assessed in this light.

EXPERIMENTAL

Most oxynitride phases described in this paper have been described in previous publications and readers are referred to these papers for details of their preparation. In general, they have been produced initially in the grain boundaries of a silicon nitride or sialon ceramic either during cooling from the sintering temperature or by subsequent devitrification of the grain-boundary glass. EDAX methods have then been used to estimate cation ratios, and glass compositions corresponding to these cation ratios but with various nitrogen contents have been melted and subsequently crystallized. By proceeding in an iterative manner, pure phases have been produced. In most cases, subsequent crystallographic studies confirm the correctness of the composition deduced by the above preparative and analytical procedures.

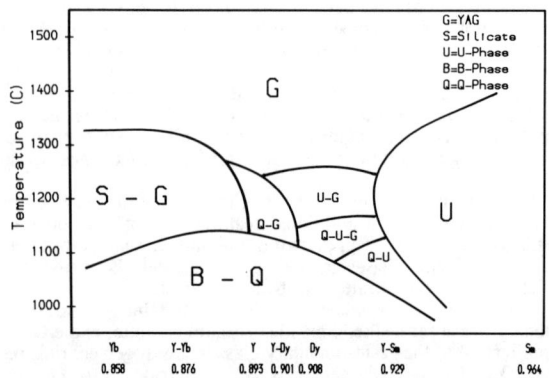

Figure 2. Crystalline phases observed in heat-treated 1700°C HIPped rare-earth densified ß-sialon ceramics.

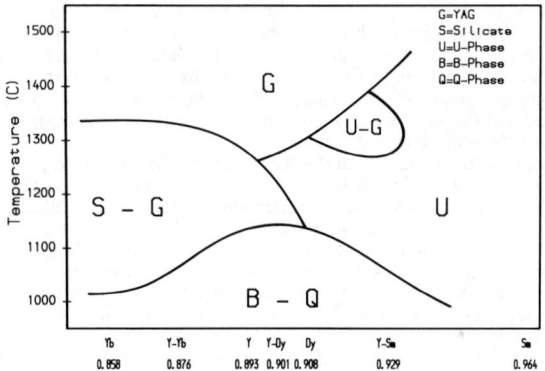

Figure 3. As Figure 2, but samples HIPped at 1775°C.

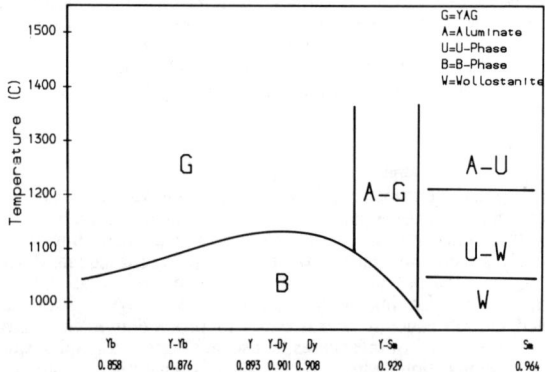

Figure 4. Crystalline phases observed in heat-treated 1700 or 1775°C HIPped rare-earth oxide densified α-ß sialon ceramics.

OXYNITRIDE GRAIN-BOUNDARY PHASES IN RARE-EARTH OXIDE DENSIFIED SIALON CERAMICS.

Mandal et al. [6] described heat-treatment studies carried out on typical ß-sialon and α-ß sialon compositions densified with either 6 w/o Y_2O_3 or equivalent amounts of various rare earth oxides and yttria-rare earth oxide combinations. The aim of this work was to identify how the grain-boundary phase assemblage changed as a function of rare earth cation and heat-treatment temperature. The inital firing was carried out by HIPping to mimimize weight loss and also to ensure fast enough cooling to avoid crystallization of the grain-boundary glass on cooling. Details of the experimental procedure are fully described elsewhere [6]. Figures 2-4 show the results obtained. In most sialon systems, ß-sialon - liquid and α-sialon - liquid tie lines run approximately parallel, with the latter at higher yttria levels (i.e. further away from the Si-Al-O-N base plane). As a result, Ln-Si-Al-O-N liquids in equilibrium with ß-sialon are richer in Si and O than the liquid phases which are in equilibrium with α-sialon containing matrices, which are therefore richer in Ln,Al and N. On devitrification, the higher atomic number rare earth glasses give only a YAG-type phase for α-ß sialon compositions, whereas a $Ln_2Si_2O_7$ silicate phase is also produced in ß-sialon compositions. In contrast, the low atomic number rare earths give mainly U-phase, with additional Ln-rich phases coming in for the α-ß sialon compositions. There are slight differences in the results obtained for ß-sialon samples HIPped at either 1700 or 1775 °C, due possibly to the different compositions of the liquid phase at these two temperatures. The phases B-, Q- and U- observed in these studies are discussed in later sections.

SIALON B-PHASE

The designation *B-phase* was given to the compound of approximate composition Y_2SiAlO_5N, mid-way along the $YSiO_2N$-$YAlO_3$ join, which forms when Y-Si-Al-O-N glasses present in ß-sialon - YAG compositions are heat-treated in the temperature range 950-1100°C [7]. The similarity between the composition of the glass and B-phase is believed to facilitate easy crystallization at these low temperatures. Subsequently, researchers showed that B-phase could be produced with different compositions in the range 60%$YSiO_2N$.40%$YAlO_3$ - 30%$YSiO_2N$.70%$YAlO_3$ by crystallization from glasses of different compositions, and often a range of B compositions occurred within the same specimen. There is still some doubt as to whether this phase is as stoichiometric as might be deduced from the 1 : 1 : 3 Y:(Si+Al)+(O+N) ratio required for an α-wollastonite type of structure [8]. In contrast to $MSiO_2N$-type wollastonites (M = Y,Ln), B-phase exhibits perfect hexagonal symmetry with no stacking faults.

As discussed above, B phases are stable at the high atomic number (i.e. small cation radius) end of the rare earth series (see Figure 5). A major disadvantage of all B-phase compositions is that they are unstable above 1200°C. This is not untypical for compositions close to the eutectic in these systems. Above the decomposition temperature, the major crystalline phase is a YAG phase plus either wollastonite or silicate phases.

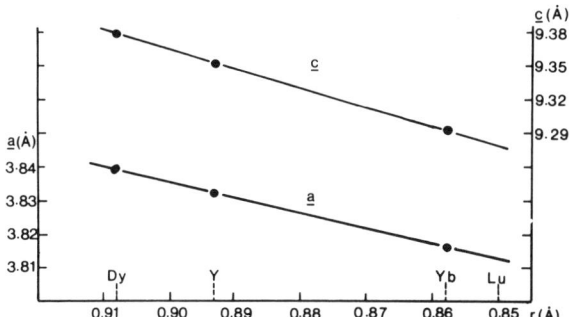

Figure 5. Unit cell dimensions for rare earth sialon B-phases.

Table II. X-ray diffraction data for Q-phase

Orthorhombic: a = 8.22Å, b = 9.42Å, c = 5.53Å

hkl	d_{calc}	d_{obs}	I_{hkl}
100	8.22	8.24	ms
011	4.768	4.740	vw
020	4.712		
101	4.587	4.586	w
111	4.124		
200	4.110	4.116	w
210	3.767	3.761	mw
201	3.298	3.317	ms
220	3.097	3.097	ms
130	2.934	2.942	mw
002	2.764	2.755	ms
031	2.731	2.731	m
310	2.631	2.629	ms
102	2.620		
131	2.592	2.595	ms
301	2.455	2.460	w
.	.	.	.
.	.	.	.

Q-Phase

Table II shows diffraction data for the phase Q which was observed during low-temperature heat-treatment of the sialon compositions described above. Despite preparing a range of samples close to possible compositions for Q, it was impossible to increase the amount of Q in the products. In fact the maximum amount of Q-phase was ≈30%, which occurred in ß or α-ß compositions densified with mixed Y_2O_3 + Dy_2O_3 additions.

Table II shows that the Q-phase diffraction pattern can be indexed on the basis of an orthorhombic cell of dimensions:
a = 8.22Å, b = 9.42Å, c = 5.53Å,
but no known silicate or aluminosilicate was found with similar cell dimensions. Clearly more work is needed to identify more precisely the conditions under which this phase occurs before further characterization can proceed.

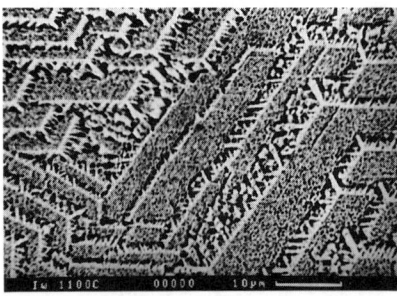

Figure 6. Scanning electron micrograph of I_w phase.

I_w-Phase

A phase designated I_w was reported by Leng-Ward and Lewis [4] as a result of heat-treatment of Y-Si-Al-O-N glasses. Similar preparations at Newcastle gave products with very characteristic microstructures (see Figure 6), but the X-ray diffraction pattern obtained always showed the major constituent to be B-phase (sometimes several B-phases coexisted together), and this is consistent with Leng-Ward and Lewis's EDAX composition which showed a Y : (Si+Al) ratio of 1 : 1. However, the existence of a spare set of X-ray diffraction lines, not correlated with any other known phase in the system, provides strong evidence that another Y-Si-Al-O-N phase is present in these compositions.

U-PHASE

U-phase is a common devitrification product of La-, Ce- and Nd- sialon glasses, and readily occurs in the grain boundaries of α- and ß-sialons densified with these rare earth oxides. Detailed preparative and structural work has now been carried out on rare earth U-phases [9], and it has now been established that this phase is only stable for rare earth cations between La and Dy (see Figure 7), with La U-phase being the most stable (melting point $\approx 1400\,°C$). The unit cell dimensions of the hexagonal cell decrease with increasing atomic number of rare earth cation as expected, and Figure 8 shows the crystal structure projected down the x and z crystallographic axes. The structure is isostructural with the rare earth gallogermanate compounds of the type $Ln_3Ga_5GeO_{14}$, and has a comparable formula of the type $Ln_3Si_{3-x}Al_{3+x}O_{12+x}N_{2-x}$ where $0 < x < 1$. Figure 8(b) clearly shows how the 5 tetrahedra in the unit cell link together to form a plane which is then linked to adjacent planes by the origin-centred AlO_6 octahedron. A complete description of the structure of U-phase has been given elsewhere [10].

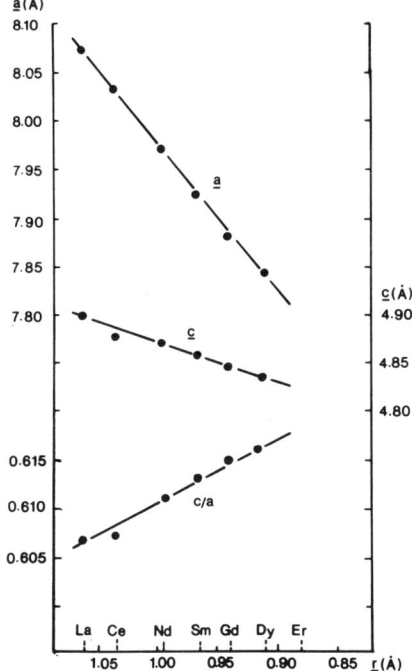

Figure 7. Unit cell dimensions for sialon U-phase.

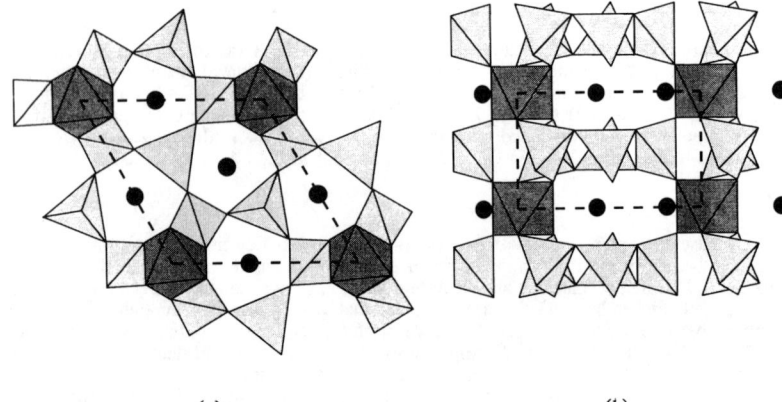

(a) (b)

Figure 8. Crystal structure of sialon U-phase projected down (a) the z-axis and (b) the x-axis.

Table III. X-ray diffraction pattern for Nd W-phase, $Nd_4Si_9Al_5O_{30}N$

Unit cell: a=12.27Å, b=5.14Å, c=10.88Å; ß=107.8°.

hkl	d_{calc}	d_{obs}	I_{obs}
100	11.69	11.62	s
001	10.37	10.42	mw
$10\bar{1}$	9.28	9.35	mw
101	6.80	6.80	mw
$20\bar{1}$	5.920	5.936	mw
200	5.844	5.846	m
$10\bar{2}$	5.382	5.380	mw
010	5.137	5.148	w
011	4.605	4.614	vw
201	4.535	4.534	w
$11\bar{1}$	4.496	4.418	vw
102	4.280	4.296	vw
111	4.100	4.105	ms
$21\bar{1}$	3.881	3.881	ms
210	3.859	3.862	ms
.	.	.	.
.	.	.	.

W-PHASE

Fernie et al. [11] reported uncharacterized crystalline phases in the Nd-Si-Al-O-N system at compositions of the type $Nd_{3.5}Si_4Al_{2.5}(O,N)_x$ and $Nd_5Si_9Al_6(O,N)_y$, where x and y were both small. These compositions are in fact very close to each other in the Jänecke prism and similar preparative work at Newcastle failed to produce two distinct phases, but instead confirmed the presence of one unique crystalline phase at compositions very close to the second Fernie et al. composition. EDAX analyses on a range of X-ray pure W-phase compositions showed that the most likely composition for W-phase was approximately $Ln_4Si_9Al_5O_{30}N$. An indexed X-ray diffraction pattern of W-phase is given in Table III, which shows that the unit cell is monoclinic, with dimensions as shown in Figure 9. The microstructure (Figure 10) shows a characteristic acicular morphology. Preparative work on a range of rare earth systems showed that it was only stable in La,Ce and Nd systems.

Attempts to correlate W-phase with existing silicate systems showed only the mineral *latiumite* to have an approximately similar unit cell. This crystallizes in space group $P2_1$ with unit cell dimensions:

$$a = 12.06 \text{Å}, \; b = 5.08 \text{Å}, \; c = 10.81 \text{Å}; \; \beta = 106.0°,$$

and an overall composition:

$$KCa_3(Si,Al)_5O_{11}([SO_4],[CO_3])_{1.0}.$$

The potassium atom is interstitial and this site might be unoccupied in W-phase; equally the ionic groups might be replaced by an SiO_4 group. The resulting $Ca_3(SiAl)_6O_{15}$ formulation could then be correlated with the observed $Ln_4(Si,Al)_{14}O_{31}$ W-phase composition if some aluminium joined Ln in the calcium sites, which (dividing by 2) would result in a W-phase formulation of $Ln_2Al(Si,Al)_6(O,N)_{15}$. This structure is of interest (see Figure 11) because the $(SiAl)_5O_{11}$ double chains are characteristic of the amphibole group of minerals, and this would be the first time that a nitrogen analogue (albeit with a very low nitrogen content) had

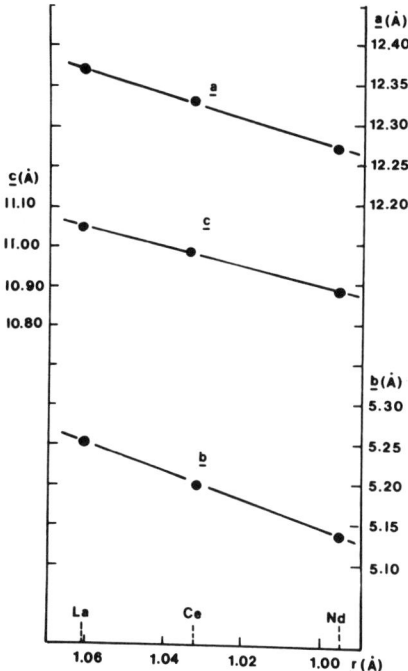

Figure 9. Unit cell dimensions of sialon W-phases.

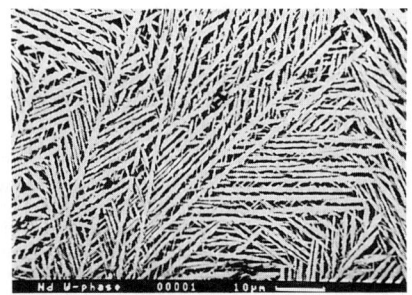

Figure 10. Microstructure of sialon W-phase.

been observed of this type. Further structural work is needed to confirm that W-phase is indeed of this type.

NITROGEN PYROXENES

Work at Newcastle on Si_3N_4 densified with mixed MgO (0-2w/o) and Y_2O_3 (6-12 w/o) additions showed that on heat-treatment at 1200-1400 °C, a new phase occurred in addition to the apatite $(MgY_4(SiO_4)_3O)$ and garnet $(Y_3Mg_{2.5}Si_{2.5}O_{12})$ phases and that this was the major phase in most heat-treated compositions. The new phase indexed on the basis of a monoclinic unit cell of dimensions:

$$a = 9.734\text{Å}, \quad b = 8.750\text{Å}, \quad c = 5.313\text{Å}, \quad ß = 105.6°,$$

which compared very favourably with those of the silicate mineral diopside, $CaMgSi_2O_6$. Careful EDAX analysis showed that Y, Mg and Si were present in the atomic ratio 1:1:2, corresponding to an overall formulation $YMgSi_2O_5N$. Table IV shows the unit cell dimensions and Figure 12 illustrates the structure projected down the **b**-axis. An important feature of the diopside structure is that Ca and Mg are ideal in size to occupy the large and small cation sites in this structure. In the nitrogen analogue, yttrium is slightly smaller than

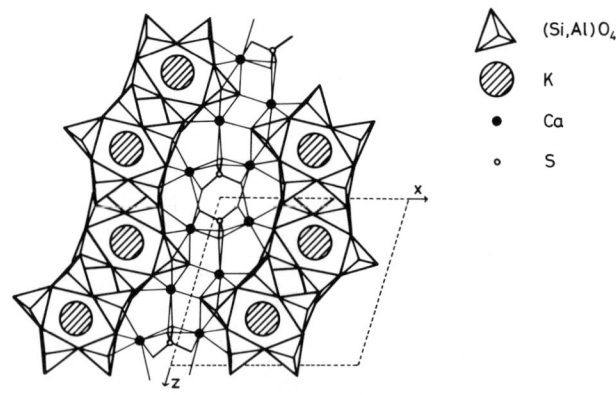

Figure 11. (010) projection of latiumite, $KCa_3(Si,Al)_5O_{11}([SO_4,[CO_3])_{1.0}$

Table IV. Unit cell dimensions of some $Mg(Y,Ln)Si_2O_6$ nitrogen pyroxenes.

	Yb	Y	Dy	Sm	Nd	Diopside
a (Å)	9.724	9.737	9.739	9.749	9.763	9.761
b (Å)	8.634	8.750	8.779	8.908	8.988	8.926
c (Å)	5.308	5.313	5.319	5.343	5.351	5.258
ß (°)	105.65	105.63	105.67	105.79	105.70	105.79

calcium, but easily substitutes into the calcium sites.

Further preparative work has shown that the yttrium in $MgYSi_2O_5N$ can be substituted by all rare earth elements as shown in Table IV. Substitution of Mg by other cations is more difficult because although several divalent cations are of similar size to magnesium, most of them react with silicon nitride at sintering temperatures. However, scandium can substitute for magnesium, but the additional valency requires that an additional oxygen atom must be replaced by nitrogen. The resulting compound, $YScSi_2O_4N_2$, also crystallizes with a pyroxene structure, showing that the extent of nitrogen substitution in these compounds (just as with ß-sialon) is mainly determined by cation size and valency balance considerations.

In common with most of the oxynitrides discussed in this paper, the nitrogen pyroxene phases are all glass ceramics which melt congruently, and can be quenched to form glasses, which on subsequent devitrification crystallize to form fine-grained microstructures. Figure 13 shows this in the case of $YScSi_2O_4N_2$. As stated earlier, the refractoriness of most oxynitrides is determined by their O:N ratio, and DTA analysis showed melting temperatures for most of the $LnMgSi_2O_5N$ pyroxenes to be close to 1400 °C (see Table V). In contrast, $YScSi_2O_4N_2$ remains solid to at least 1550 °C, and is believed to melt just above 1600 °C. This is in noticeable contrast to most mineral pyroxenes, which exhibit melting points below 1300 °C.

CONCLUSIONS

Heat treatment studies carried out on ß-sialon and α-ß sialon ceramics densified with rare earth oxide additions showed that different combinations of oxynitride phases were produced at the low atomic number end of the range compared with the high atomic number end. In particular, the behaviour of ytterbium (Z = 70) was very similar to yttrium, and the current

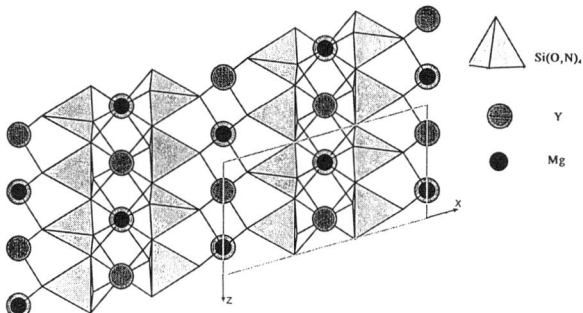

Figure 12. (010) projection of the diopside-type pyroxene structure of $MgYSi_2O_5N$.

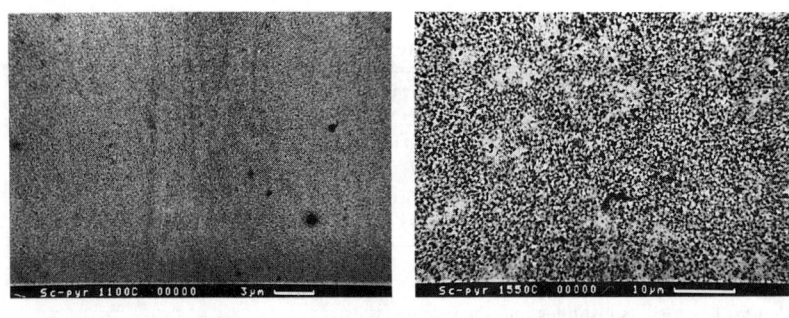

(a)   (b)

Figure 13. Micrographs of (a) $ScYSi_2O_4N_2$ glass and (b) $ScYSi_2O_4N_2$ glass ceramic.

Y-Si-Al-O-N prism (Figure 14) is applicable also to the high Z rare earth elements. In contrast (Figure 15), the low Z rare earths exhibit different aluminate phases and also different oxynitride phases.

Work in all these systems has shown that the use of mixed cations in the starting mix improves the likelihood of finding new crystalline phases. This principle has been established by other researchers, and for example the mixed Mg,Nd apatite phase $(MgNd_4(SiO_4)_3O)$ [12] is more stable than when magnesium is absent. In the present work, the W phase is stabilized by as little as 2 a/o nitrogen, and no compound of this type has ever been reported in the Y_2O_3-Al_2O_3-SiO_2 system. Not only is the presence of mixed cations advantageous for permitting a particular stoichiometry to be achieved, or valencies to be balanced, but as seen with the nitrogen pyroxenes, the size of the cation may advantageously stabilize a new structure.

Most of the structures described in this work are not stable to temperatures in excess of $\approx 1450\,°C$ (the $YScSi_2O_4N_2$ phase being exceptional here) and in contact with nitrogen ceramic matrices, eutectic temperatures are barely in excess of $\approx 1350\,°C$. These compounds are not likely therefore to find application as refractory grain-boundary phases for silicon nitride and sialon ceramics, but may be of greater interest as nitrogen glass ceramics, where the presence of nitrogen offers advantages of improved refractoriness and mechanical performance compared with oxide analogues; the $YScSi_2O_4N_2$ phase clearly shows that fine-grained microstructures can be produced for these materials.

The present studies show that the range of crystalline oxynitride ceramics has not been exhausted, and further work should be carried out using a wider range of cations, mixed in more varied proportions; it should also be extended to include more nitrogen-rich compositions.

Table V. Melting points of some $LnMgSi_2O_5N$ nitrogen pyroxenes

	Yb	Y	Dy	Sm	Nd
Melting Temp (°C)	1455	1442	1435	1403	1422

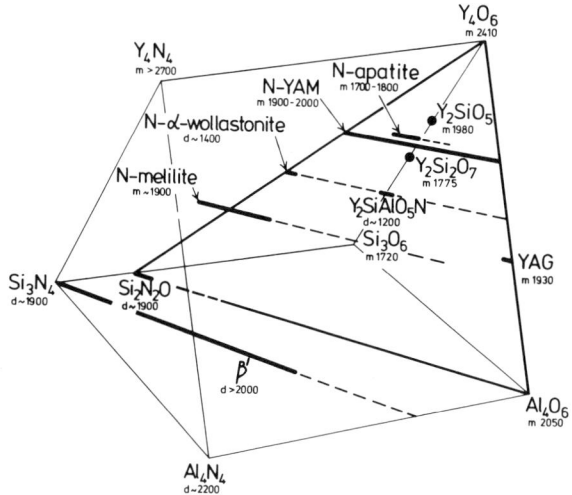

Figure 14. Crystalline phases in the Y-Si-Al-O-N system.

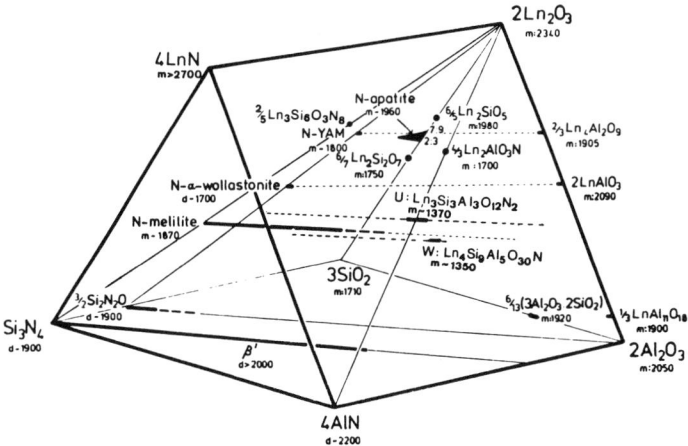

Figure 15. Crystalline phases in Ln-Si-Al-O-N systems for low atomic number Ln.

ACKNOWLEDGEMENTS

It is a pleasure to thank numerous past and present colleagues who have contributed to some aspect of this work. In particular I would like to thank Dr Hasan Mandal, who carried out most of the sialon heat treatment studies and characterized the nitrogen pyroxenes, and also Professors Ekström and Nygren at Stockholm University, who assisted in the preparation and characterization of several of the phases reported here. I would also like to thank the Cookson Technology Centre for financial support.

REFERENCES

1. K.H. Jack, Trans. & J. Brit. Ceram. Soc., 72, 376 (1973).

2. K. Negita, J. Mat. Sci. Lett., 4, 755 (1985).

3. D.P. Thompson in Advanced Structural Materials, edited by Y.Han (C-MRS '90, Peking China, 1990) pp. 435-440.

4. G. Leng-Ward, M.H. Lewis in Glasses and Glass-Ceramics, edited by G. Leng-Ward and M.H. Lewis, Chapman & Hall, 1990), p. 106.

5. D.P. Thompson in Preparation and properties of silicon nitride based ceramics, edited by T.Y. Tien and D.A. Bonnell (Trans Tech, Zurich, 1989), p. 21.

6. H. Mandal, D.P. Thompson, T. Ekström in Materials for Advanced Technology Applications, edited by M. Buggy and S. Hampshire (Trans Tech, Zurich, 1991), p. 187.

7. P. Korgul, D.P. Thompson, Brit. Ceram. Proc. 42, 69 (1989).

8. P. Korgul, private communication (1988).

9. P.-O. Käll, J. Grins, K. Liddell, P. Korgul, D.P. Thompson in New Materials and their Applications, edited by D. Holland (IOP Publishing Ltd., Bristol,1990), p. 427.

10. P.-O. Käll, J. Grins, P.-O. Olsson, K. Liddell, P. Korgul, D.P. Thompson, J. Mater. Chem., 1, 233 (1991).

11. J.A. Fernie, G. Leng-Ward, M.H. Lewis, J. Mat. Sci. Lett., 9, 29 (1989).

12. S. Hampshire, M. Leigh, V. Morrissey, M.J. Pomeroy, B. Saruhan in 3rd Int. Symp. on Ceramic Materials and Components for Engines, edited by V.J. Tennery (American Ceramic Society, Westerville, Ohio, 1989), p. 442.

OXYNITRIDE GLASSES AND GLASS CERAMICS

STUART HAMPSHIRE
Materials Research Centre, University of Limerick, Ireland

Silicon nitride based ceramics contain oxynitride glass phases at the grain boundaries which can impair subsequent high temperature properties. Investigations of bulk glasses in various M-Si-Al-O-N systems have been carried out and it has been shown that up to 15 atomic % N can be incorporated into these oxynitride glasses. Studies have revealed that nitrogen increases the viscosity, hardness and glass transition temperature of the glasses. Heat treatments of the glasses to form crystalline phases have been reported but further improvements are possible if glass-ceramic processes using two-stage heat treatments are introduced. The development of oxynitride glasses and the effects of nitrogen on properties are reviewed and the optimisation of glass-ceramic heat-treatments are reported.

1. INTRODUCTION

The occurrence of oxynitride glasses as grain-boundary phases in silicon nitride based ceramics combined with the ease of shaping glasses and the possibility of producing glass-ceramics containing refractory oxynitride crystalline phases gave the impetus for a number of investigations[1-11] on oxynitride glass formation and properties. Originally, small concentrations of nitrogen in oxide glasses were reported to increase their softening temperature, viscosity and resistance to devitrification[12,13].

Crystallization of selected oxynitride glasses has been investigated[6,8,9,11], principally to complement more extensive studies of phase equilibria in M-Si-Al-O-N systems and the effects of vitreous phases on high-temperature mechanical properties of silicon nitride based ceramics. Though a U.S. patent[14] describes a series of alumino-silicate glass compositions containing nitrogen that may have potential as glass-ceramics, studies of the glass-ceramic process as applied to oxynitride glasses have been a more recent development. This paper reviews the background to and the development of oxynitride glasses and glass-ceramics.

2. PREVIOUS INVESTIGATIONS OF OXYNITRIDE GLASSES

2.1 <u>Solubility of nitrogen in glasses</u>

Mulfinger[15,16] was one of the first investigators to study the solubility of nitrogen in glasses and found that the physical solubility of nitrogen in glasses was very low after bubbling nitrogen gas through the glass melt. However by bubbling ammonia gas through the melt for five hours at a temperature of 1400°C, the chemical solubility of nitrogen in the glass reached a value 10^5 times higher than that of the physical solubility. Using this method, 0.33 w/o nitrogen was introduced into soda-lime-silica glass. Mulfinger suggested that the substitution of nitrogen for oxygen must lead to a higher than average coordination of non-metal atoms and that increased crosslinking should produce a more rigid glass network as follows:

$$Si-O-Si \quad \rightarrow \quad Si-N-Si$$
$$\qquad\qquad\qquad\qquad\qquad |$$
$$\qquad\qquad\qquad\qquad\qquad Si$$

Elmer and Nordberg[12] observed that devitrification of certain glasses could be induced electrolytically. They showed that incorporation of nitrogen into these glasses inhibited the electrolytically induced devitrification and they attributed this to increased viscosity, due to the presence of (=NH) and (=N-) groups in the glass. This was one of the first observations of an improvement in some physical property of a glass resulting from the incorporation of nitrogen into the structure. In this case, ammonia was again used as the nitriding agent and nitrogen contents of the order of 3 ^{w}/o, or ten times that reported by Mulfinger, were obtained.

Davies and Meherali[17] concluded from their investigations that the solubility of nitrogen in glass melts was chemical rather than physical and they found that severe reducing conditions had to be imposed in order to dissolve significant amounts of nitrogen in the glass melts. They discovered that the solubility of nitrogen increased with increasing basicity, indicating that bridging rather than non-bridging oxygen atoms were involved in the dissolution reaction.

Dancy and Janssen[18] investigated the solubility of nitrogen in $CaO-SiO_2-Al_2O_3$ slags. They compared physical and chemical methods of dissolving nitrogen in these melts and found that under one atmosphere of nitrogen, an equilibrium solubility of 0.25 to 2.5 ^{w}/o nitrogen was achieved after 24 hours. By contrast when Si_3N_4 was added to the melt, again under an atmosphere of nitrogen, nitrogen incorporation was very rapid and reached significantly higher levels (4 ^{w}/o). They suggested that most oxide melts would not be significantly reducing to dissolve N_2 or NH_3 to any great extent. More recent studies on nitrogen additions to silicate melts have mainly used Si_3N_4 or AlN.

2.2 Formation of M-Si-Al-O-N glasses and their properties

It is now well established[19-22] that both Si_3N_4 and β'-sialons require an oxide additive for liquid phase densification. The silicate liquid formed dissolves some nitrogen and cools to form a grain boundary glass, sometimes in conjunction with other oxynitride or sialon phases. Jack[1] observed the close similarity between the building units for the structure of silicate glasses (SiO_4 tetrahedra) and those in silicon nitride (SiN_4 tetrahedra) and also the similarity between the lengths of Si-N, Si-O and Al-O bonds, and proposed that nitrogen could be incorporated in the network of silicate and alumino-silicate glasses.

Jack[2] reported preparing oxynitride glasses in the following systems: $Si_3N_4-Al_2O_3-SiO_2$, $Si_3N_4-MgO-SiO_2$, and $AlN-Y_2O_3-SiO_2$, with nitrogen levels up to 10 ^{a}/o. Changes in physical properties due to incorporation of nitrogen were not reported at this point. Subsequently, considerable investigations[6-11,23] have been carried out on glass formation and properties in a wide range of M-Si-O-N and M-Si-Al-O-N systems where M=Y, Mg, Ca, Al or Nd and the effects of increasing nitrogen content on properties of these glasses have also been reported.

Both Shillito et al.[3] and Loehman[4,5] were among the first to report correlations between amounts of nitrogen incorporated into oxynitride glasses and changes in their physical properties. Shillito et al. reported a linear increase in the Knoop hardness of a Y-Si-Al-O-N glass as the nitrogen content increased. Loehman produced more detailed results of changes in physical properties due to incorporation of up to 7 ^{a}/o nitrogen in glasses in the same system. Glass transition temperature (T_g) (see figure 1), microhardness and relative fracture toughness, all increased with increasing nitrogen content, while the thermal expansion coefficient decreased. IR spectroscopic analysis carried out by Loehman indicated that the incorporated nitrogen became chemically bonded to silicon in the glass network, and by substitution for oxygen, produced a more tightly and highly linked structure. Loehman reported that typical glass forming systems for oxynitride glasses were the M-Si-O-N and M-Si-Al-O-N systems where M=Ca, Li, Mg or Y, and that glasses with up to 10 a/o nitrogen had been prepared from these compositions. However, while these results did indicate improvements in properties of glasses related to incorporation of nitrogen, these property changes could not be attributed solely to the incorporation of nitrogen, since it is well known that viscosities of glasses will depend on field strength, polarizability and size and coordination requirements of the added cation. Thus for glasses with a constant nitrogen:oxygen ratio, changes in Al or M concentration may cause changes in viscosity, Tg and hardness and these variances remained unaccounted for.

Drew et al.[6,8] carried out extensive studies on nitrogen-containing glasses in M-Si-O-N and M-Si-Al-O-N systems. Glasses with varying nitrogen : oxygen ratios, with a fixed cation composition were prepared, to allow direct comparison between different M-Si-Al-O-N systems and the effect of replacing oxygen by nitrogen within each system. Drew et al.[6,8,9] found that, for glasses with a constant cation ratio, incorporation of nitrogen resulted in increasing T_g (see figure 2), viscosity, resistance to devitrification, refractive index, dielectric constant and a.c. conductivity, in all the Mg-, Ca-, Y- and Nd-sialon glasses.

2.3 Representation of M-Si-Al-O-N glass forming systems

Convenient methods of representing both Si-Al-O-N and M-Si-Al-O-N systems have been developed[24]. Five component metal sialon systems can be represented by Janecke's triangular prism[25]. The Nd-Si-Al-O-N prism is shown in figure 3. The basal plane of the prism is a square which, in this case, represents the Si_3N_4-Al_4N_4-Al_4O_6-Si_3O_6 system, that is, the oxides and nitrides of silicon and aluminium. The bottom right hand corner is Si_3N_4 ($3Si^{4+}$, $4N^{3-}$) and, maintaining 12 positive and 12 negative valency units throughout, the other corners are then Si_3O_6, Al_4O_6 and Al_4N_4. Addition of a fifth component such as Nd, produces a prism with the back triangular face being a ternary oxide system and the front face the nitrides.

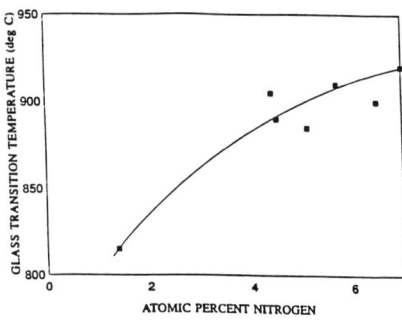
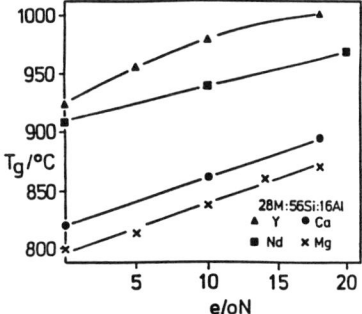

Figure 1. Change in T_g with N content for Y-Si-Al-O-N glasses (after Loehman[4]).

Figure 2. Change in T_g with N content for Mg-, Ca-, Y- and Nd-sialon glasses (after Drew et al.[6,8,9]).

The concentrations of all components are expressed in equivalent units so that any point in the prism again represents a combination of 12 positive and 12 negative valencies. As shown by figure 3, the distance, x, of any point P from the front face represents the concentration (in equivalent units) of oxygen, and the distance y represents the equivalent concentration of nitrogen, that is, the equivalent concentration ratio of nitrogen = y/(x+y) = (3[N])/(2[O]+3[N]) where [O] and [N] are, respectively, the atomic concentrations of oxygen and nitrogen within any composition. The edge of the prism is scaled such that x+y = 12. The vertical plane is scaled such that each division is two valency units. The point P thus has a composition $Nd^{6+}Si^{4+}Al^{2+}O^{6.6-}N^{5.4-}$ in valency units and hence $Nd_{2.0}Si_{1.0}Al_{0.67}O_{3.3}N_{1.8}$ in atomic units. It is convenient to use the concept of equivalent percent (e/o) to express compositions.

This representation was adopted by various investigators[8-10,26] to describe the limits of glass formation in different metal sialon systems. The limits of the metal alumino-silicate glass regions were plotted on the oxide face of the prism and it was possible to observe how the glass region extended into the M-Si-Al-O-N prism on replacing oxygen by nitrogen. The three dimensional representation of the complete glass forming regions in both the Mg- and Y-Si-Al-O-N systems are shown in figures 4 [a] and [b] respectively.

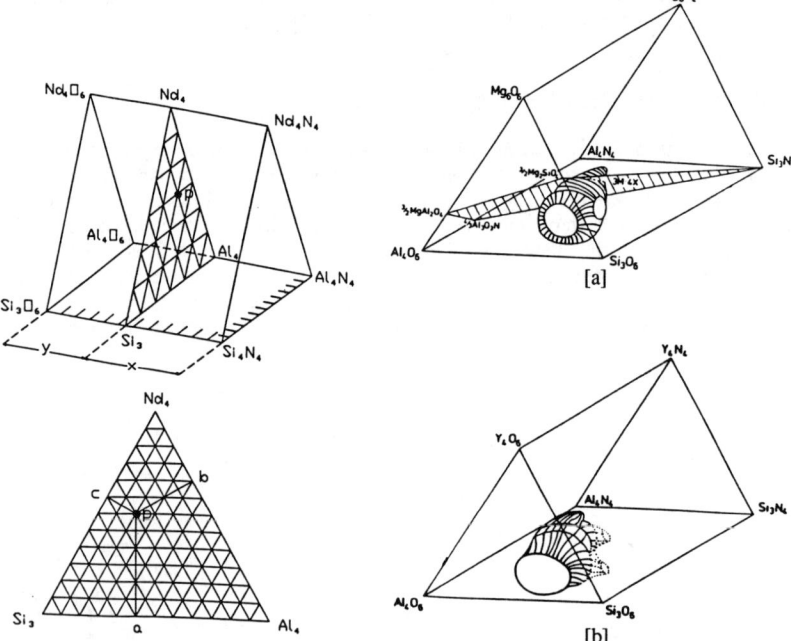

Figure 3. The Janecke[25] prism representation of the Nd-Si-Al-O-N system (see Jack[24]).

Figure 4. Glass forming regions in
(a) Mg-Si-Al-O-N system
(b) Y-Si-Al-O-N system.
(after Drew et al.[6,8,9]).

Prior to this, investigation of these systems had been carried out by Jack[2] and Loehman[4] but adequate exploration of the full extent of these systems was not completed. Hampshire and Jack[22] showed that nitrogen lowers the eutectic temperatures in metal oxide-silica systems and increases the tendency to form glass. Figure 4 [a] shows that the extent of the glass forming region in the Mg-Si-Al-O-N system expands away from the oxide face until 10 e/o nitrogen is incorporated, after which the glass forming region contracts with a simultaneous shift towards slightly more Mg-rich compositions[26]. This suggests that while Mg is a network modifier in oxide systems, in oxynitride glasses it appears to act as a network former or intermediate. In the Y-Si-Al-O-N system (figure 4 [b]) the expansion of the glass forming region is less at 10 e/o N but the maximum nitrogen solubility is much greater than with Mg. Depending on the particular system it was found that a limit of 17-25 e/o of the oxygen could be replaced by nitrogen.

Drew et al.[6,8,9] showed that the corresponding M-Si-O-N systems displayed a much smaller glass-forming region, thus showing the ability of Al_2O_3 to extend the range of glass formation. Shaw et al[23] showed that the miscibility gap in the $MgO-SiO_2$ binary system extends into the Mg-Si-O-N system with resulting phase separation in most of the glasses. Ohashi and Hampshire[27] mapped out the glass forming region in the Ce-Si-O-N system, shown in figure 5, and here also phase separation occurs.

2.4 Nitrogen coordination in glasses

The resulting improvements in glass properties by substitution of nitrogen for oxygen was usually attributed to the replacement of a 2-coordinated bridging oxygen atom, by a nitrogen atom coordinated by 3 silicon ions. Thus, it was assumed that properties were improved due to an increase in the crosslinking of the silicate network due to the tri-coordinated nitrogen. However, originally, there was no direct evidence for this and published IR data[4,28] only suggested the presence of Si-N bonds in the structure. Brow and Pantano[29] carried out more extensive studies on the coordination of nitrogen in oxynitride glasses using Fourier Transform Infrared spectroscopy (FTIR) and X-ray Photoelectron Spectroscopy(XPS).

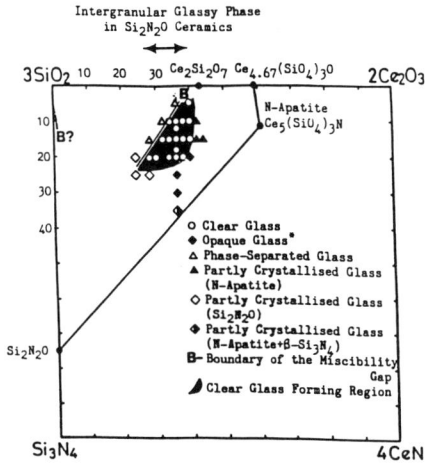

Figure 5. Glass formation in the Ce-Si-O-N system (after Ohashi and Hampshire[27]).

It was concluded that nitrogen was present in the structural network because introduction of nitrogen caused shifting of the position of the Si-O-Si stretching peak towards that of Si-N.

If nitrogen existed as precipitated Si_3N_4 the position of the Si-O-Si peak would not be expected to change. Rand and Roberts[30] also observed a similar shift of the Si-O-Si stretching vibration to lower wavelengths in nitrided silicon thin-films. XPS studies by Brow and Pantano also revealed that nitrogen is usually present in more than one form, and they proposed that non-bridging nitrogen ions may also be present, similar to the following:

$$(1) \equiv Si - N^- - Si \equiv$$
$$(2) \equiv Si - N^{2-}$$

The local charge on the non-bridging nitrogen ions is balanced by the presence of interstitial metal ions in their vicinity. Thus, while it has not been proven beyond doubt that nitrogen is present in oxynitride glasses in a tri-coordinated state, all evidence indicates that this is the symmetry that it adopts.

2.5 Nucleation and crystallization in oxynitride glasses

After formation of various M-sialon glasses, further suitable heat-treatment results in formation of tiny nuclei, upon which crystals then grow. The crystalline phases formed in glasses on heat-treatment and the extent of their formation will determine the properties of the particular material. The phases formed will depend on both the composition of the parent glass and the heat-treatment process. Many glasses require the addition of a nucleating agent to promote the crystallization process, but in general oxynitride glasses appear to be self-nucleating.

Abramovici et al.[31] investigated the effect of nitrogen on nucleation and crystallisation in SiO_2-Al_2O_3-MgO and related glass forming systems. In the SiO_2-Al_2O_3-Li_2O system they found the presence of nitrogen to influence the phase composition of crystallized samples only to a limited extent. In the SiO_2-Al_2O_3-MgO system they reported that in samples with TiO_2 as a nucleating agent the addition of nitrogen leads to a more advanced and finer crystallization. They concluded from this that nitrogen promotes nucleation and in some cases advances crystallization, but they failed to give any explanation for this. Nitrogen is known to be an inhibitor of crystallization because it increases viscosity and a more probable explanation of their observation is that nitrogen does in fact inhibit growth of large crystals, but in some cases this may be compensated for by a more extensive growth of smaller crystals, where less matter transport would be required for their propagation.

Lewis et al.[32] investigated crystallization in the Mg-sialon system and found that forsterite was the main crystallizing phase. They also identified secondary phases and these included a magnesium substituted β'-sialon, designated as β" which was first reported by Drew et al.[6,33]. At higher temperature, this is replaced by a Mg-Si-Al-O-N petalite phase.

Ahn and Thomas[34] carried out preliminary studies on crystallization of Y-sialon glasses. They reported that appreciable crystallization was only effected after glasses were doped with up to 5 w/o ZrO_2 which acted as a nucleating agent. They identified the main crystalline phase as $Y_2Si_2O_7$. Winder et al[35] carried out further work in this system and reported that low nitrogen:oxygen ratios again favour formation of yttrium disilicate ($Y_2Si_2O_7$) while the increased glass viscosity, associated with an increase in the nitrogen:oxygen ratio, favoured suppression of the crystallization of $Y_2Si_2O_7$ and preferential formation at higher temperatures of yttrogarnet ($Y_3Al_5O_{12}$).

More extensive studies of crystallization in Y-sialon glasses were carried out by Lewis et al.[36]. They reported that on heat-treatment at 1250°C the oxide glasses fully crystallized to yttrium disilicate, mullite and Al_2O_3. Again with increasing nitrogen content, they found that the disilicate phase was progressively replaced by yttrium aluminium garnet and nitrogen was mainly incorporated into Si_2N_2O. They also reported that heat-treatment of the nitrogen glasses at 1100°C produced partial crystallization involving intermediate phases related to nitrogen wollastonite.

In a study of crystallization of glasses in the Y_2O_3-SiO_2-AlN system, Dinger et al.[37] found that nucleation occurs heterogeneously on FeSi particles distributed in the glass. In a later study, Besson et al.[38] found a similar phenomenon for Y-sialon glasses. The phases formed after heat treatment at various temperatures and times are shown in Table 1.

Table I. Crystalline phases after heat treatment at various temperatures.
s: strong, m: medium, w: weak.

T(°C)	1050	1100	1200	1250	1300
t(h)	90	36	36	36	10
Phases	FeSi w	$\alpha Y_2Si_2O_7$ s	$\beta Y_2Si_2O_7$ s	$\beta Y_2Si_2O_7$ s	$\beta Y_2Si_2O_7$ s
	$\alpha Y_2Si_2O_7$	$Y_3Al_5O_{12}$ w	$\alpha Y_2Si_2O_7$ m	Si_2N_2O w	Si_2N_2O w
				$AlYO_3$ w	$AlYO_3$ w

At 1050°C, crystallization is very slow resulting in the formation of dendritic α-$Y_2Si_2O_7$. At 1100°C, the β form of yttrium silicate begins to replace the α modification. The dendritic branches have thickened. The nitrogen content increases in the parent glass. At 1250°C and above, α-$Y_2Si_2O_7$ is no longer observed and aluminium is present as $AlYO_3$. As a result of the enrichment in nitrogen of the glass, silicon oxynitride precipitates. Raman studies[40] have revealed that Si_2N_2O is already present after a treatment at 1150°C. The microstructure of the material has lost its dendritic aspect.

Crystallization can also be followed using an ultrasonic technique for investigating Young's modulus, which is very sensitive to structural changes. The change in relative modulus (E/E_o) with temperature for the Y-sialon glass studied by Besson et al.[38] is shown[39] in figure 6. The relative modulus decreases linearly up to 920°C at a rate of $10^{-2}K^{-1}$. Above this temperature, it decreases more quickly up to 1120°C, then rapidly and, finally, shows a plateau from 1160 to 1300°C. From the results of viscosity determination, 920°C corresponds to the viscosity associated with the strain point ($10^{13.5}$Pa.s). The relatively slow decrease of Young's modulus between 920 and 1120°C may be tentatively associated with the decrease in viscosity of the glass above the transition domain, partially hindered by the formation of α-$Y_2Si_2O_7$ crystals. Then, the softening of the remaining glass becomes predominant until the formation of δ-$Y_2Si_2O_7$, which has a high crystallization rate, counteracts it, resulting in the plateau observed above 1160°C.

In figure 6 the change in Young's modulus for silicon nitride sintered with yttria and alumina additions is also shown. The same features are strikingly evident, occurring over the same temperature ranges. The crystallization leads to an irreversible change in the modulus. It has been shown elsewhere[41] that post-sintering heat treatments around 1200°C improve the creep resistance of this material by crystallizing the intergranular vitreous phase.

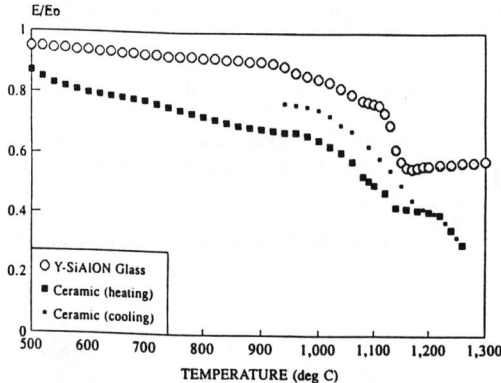

Figure 6 Young's modulus v. temperature for a Y-Si-Al-O-N glass and a SiYAlON ceramic (after Rouxel et al.[39]).

3. OPTIMISATION OF GLASS-CERAMIC PROCESS IN OXYNITRIDE SYSTEMS

The conventional glass-ceramic process involves heating the as-prepared glass to an initial heat-treatment temperature at which internal nucleation occurs followed by heating to a second higher temperature to allow crystal growth of the nuclei. The first temperature is usually about 50°C above the annealing temperature. Morrissey et al.[42] identified process parameters for selected oxynitride glass-ceramics and in a further study[43] the same authors reported the process parameters controlling the preparation of oxynitride glass-ceramics in the Nd-Mg-Si-O-N system. In the absence of any information about annealing temperatures or viscosities, initial heat-treatments were carried out for 2.5 h at temperatures in the range T_g to $T_g + 90$ (°C) based on results from thermal analysis. For the second heat-treatments, temperatures from $T_c - 40$ to T_c (°C) were employed.

Figure 7 shows T_g as a function of N content for two series of Nd-Mg-Si-O-N glasses each containing 64 e/o Si, and respectively, 12 e/o Nd, 24 e/o Mg (12:24:64 composition) and 24 e/o Nd, 12 e/o Mg (24:12:64 composition), both with a range of N contents from 0 to 25 e/o. For the 24:12:64 composition, T_g increases almost linearly with N content, while for the 12:24:64 composition, T_g increases up to 17 e/o N and then levels off. It was concluded that the glass phase present in the 25 e/o N sample had a composition almost identical with that of the 17 e/o N, the excess nitrogen being accommodated in a Nd-N-apatite phase.

The variation of viscosity with temperature for a glass melt is important for various manufacturing operations, particularly in forming, annealing and the controlled crystallization steps necessary for glass-ceramic production. As observed in other systems, nitrogen increases the viscosity of a glass of constant cation ratio. The viscosity changes by three orders of magnitude over 100K in the temperature range 650-750°C. Glasses with high Nd:Mg ratio have higher viscosities than more Mg-rich compositions. Pomeroy et al.[44] noted that, when sintering silicon nitride with mixed magnesia/neodymia additives, compositions containing higher $MgO:Nd_2O_3$ ratios were easier to densify and the microstructural development of the β-silicon nitride grains was consistent with a lower viscosity liquid.

Following the measurement of T_g and crystallization temperatures by thermal analysis, heat treatments were carried out on the glasses. Increases in microhardness after heat-treatment were used as the main criterion for assessment of increases in strength due to crystallization to form glass-ceramics. Morrissey et al.[42] showed that heat-treatment at a single temperature resulted in a small increase in hardness for the 12:24:64 composition but two-stage heat-treatments resulted in much higher increases. Further work[43] has concentrated on optimising the glass-ceramic process for the whole range of glasses within this system.

Figure 8 shows the effect of single heat-treatments on microhardness of the glasses of the 12:24:64 composition after 2.5 hours at 900°C, T_g + 50(°C) and T_g + 90(°C). At a constant temperature of 900°C, there is no clear trend of microhardness with N since the temperature relative to T_g will be different at each N content, whereas, under the other conditions, which themselves vary with N, there is an almost linear increase in microhardness with N content. This confirms the fact that nucleation temperature is directly related to the glass transition temperature. From the figure, it is clear that a heat-treatment of T_g + 90(°C) gives the highest values of hardness.

Initial experiments showed that a second heat-treatment temperature of 1085°C gave glasses with high hardness values. This temperature was investigated for all the glasses but it was noted that this is also 40°C below the crystallization temperature for the 17 e/o N (12:24:64) glass. Glasses were then subsequently treated at T_c - 40(°C) and also T_c - 20(°C) as the upper heat-treatment temperature in a two-stage process.

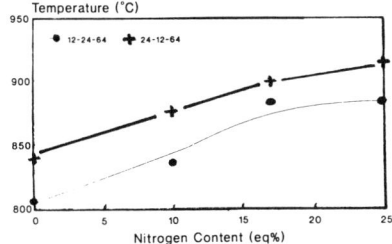
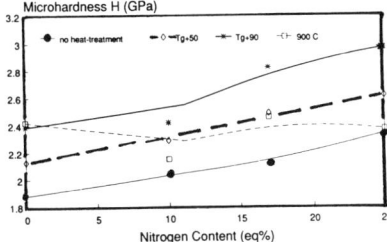

Figure 7 Effect of N on T_g for Nd-Mg-Si-O-N glasses

Figure 8 Variation of microhardness with N content for 12:24:64 (Nd:Mg:Si) composition after single heat treatments.

Figure 9 shows the variation in microhardness with nitrogen for all the double-stage heat-treated samples. For the processes where the nucleation stage is directly related to T_g and the

increases. The optimum treatment giving much higher hardness values is at $T_g + 90(°C)$ followed by $T_c - 40(°C)$.

Figure 10 shows a scanning electron micrograph of 12:24:64 composition following the optimum heat-treatment schedule. As can be seen, this is a fine-grained, closely interlocked microstructure which would be very suitable as a glass-ceramic for use in mechanical applications.

The same composition heat-treated under non-optimum conditions has a larger grain structure than the optimum with more residual glass remaining between the grains.

The crystals are an apatite phase which from X-ray analysis can exist as a solid solution over the range of composition:

$$MgNd_4Si_3O_{13} \rightarrow Nd_{4.67}Si_3O_{13} \rightarrow Nd_5Si_3O_{12}N \rightarrow Mg_{0.5}Nd_{4.5}Si_3O_{12.5}N_{0.5}$$

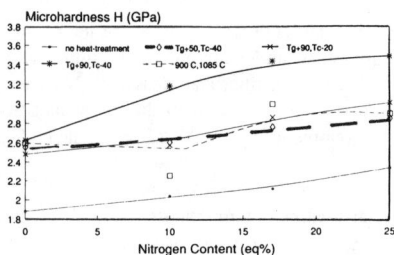
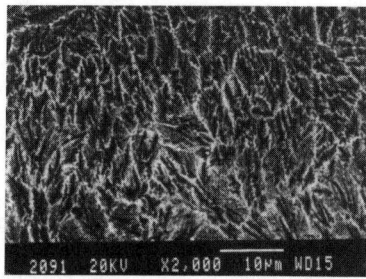

Figure 9 Variation of microhardness with N content for double-stage heat-treated glass-ceramics of 12:24:64 (Nd:Mg:Si) composition

Figure 10 Scanning electron micrograph of 12:24:64 (Nd:Mg:Si) composition with 10 e/o N after optimum heat treatment.

4. SUMMARY

Extensive investigation of oxynitride glass formation and property evaluation has been carried out by several groups of scientists. Glasses have been prepared in M-Si-Al-O-N systems (where M = Mg, Y, Nd, etc.) with up to 25 e/o N by melting appropriate mixtures of oxides and nitrides in nitrogen at 1700°C. When the cation composition is kept constant and only the oxygen/nitrogen ratio is varied, properties such as T_g, microhardness, viscosity and Young's modulus increase with increasing nitrogen content.

The area of oxynitride glasses and glass-ceramics offers encouraging possibilities for developing improved materials. From results of differential thermal analysis, T_g and T_c for Nd-Mg-Si-O-N glasses have been determined. Heat-treatments have been carried out on glasses for 2.5 h at various temperatures involving also two-stage treatments. The optimum schedule as evidenced by increases in microhardness is 2.5 h each at $T_g + 90$ (°C) and $T_c - 40$ (°C). This leads to formation of fine-grained glass-ceramics containing an Nd-N-apatite phase. More detailed property evaluation must be carried out before these materials can be exploited fully. The possibility of developing quality oxynitride glass-ceramics with enhanced properties by suitable heat-treatments makes the future of this field very attractive.

Acknowledgements

Financial support for work on oxynitride glasses at the University of Limerick has been provided by the European Research Office (U.S. Army) and EOLAS (the Irish Science and Technology Agency).

I wish to thank Ms. D. O'Reilly, Mr. D. O'Sullivan and Mr. R. Flynn of the Materials Ireland Ceramics Research Unit at the University of Limerick for their assistance in the preparation of this paper.

REFERENCES

1. K.H. Jack, J. Mater. Sci. **11**, 1135 (1976).
2. K.H. Jack, in Nitrogen Ceramics, edited by F.L. Riley, (Noordhoff, Leyden, 1977) pp.257-267.
3. K.R. Shillito, R.R. Wills and R.E. Bennett, J. Am. Ceram. Soc. **61**, 537 (1978).
4. R.E. Loehman, J. Am. Ceram. Soc. **62**, 491 (1979).
5. R.E. Loehman, J. Non-cryst. Solids **42**, 433 (1980).
6. R.A.L. Drew, S. Hampshire and K.H. Jack, in Special Ceramics 7, edited by D.E. Taylor and P. Popper, (Proc. Brit. Ceram. Soc. **31**, 1981) pp. 119-132.
7. D.R. Messier and A. Broz, J. Am. Ceram. Soc. **65**, C-124 (1982).
8. R.A.L. Drew, S. Hampshire and K.H. Jack, in Progress in Nitrogen Ceramics, edited by F.L. Riley, (Martinus Nijhoff, The Hague, 1984) pp.323-330.
9. R.A.L. Drew, S. Hampshire and K.H. Jack, in Ceramic Components for Engines, edited by S. Somiya, E. Kanai and K. Ando, (Reidel Pub. Co., Dordrecht, 1984) pp. 394-403.
10. S. Hampshire, R.A.L. Drew and K.H. Jack, J. Am. Ceram. Soc. **67**, C46 (1984).
11. T.M. Shaw and G.R. Thomas, in Progress in Nitrogen Ceramics, edited by F.L. Riley, (Martinus Nijhoff, The Hague, 1983) pp. 331-336.
12. T.H. Elmer and M.E. Nordberg, J. Am. Ceram. Soc. **50**, 275 (1978).
13. F.L. Harding and R.J. Ryder, Glass Tech. **11**, 54 (1970).
14. K. Chyung and R.R. Wusirika, U.S. Patent No. 4 070 198, (Jan. 1978).
15. H.O. Mulfinger, J. Am. Ceram. Soc., **49**, 462 (1966).
16. T. Kelen and H.O. Mulfinger, Glasstechn. Ber. **41**, 230 (1968).
17. M.W. Davies and S.G. Meherali, Metall. Trans. **2**, 2729 (1971).
18. E.A. Dancy and D. Janssen, Canad. Met. Quart. **15**, (2) 103 (1976).
19. P. Drew and M.H. Lewis, J. Mater. Sci. **9**, 261 (1974).
20. M.H.Lewis, B.D. Powell, P. Drew, R.J. Lumby, B. North and A.J. Taylor, J. Mater. Sci. **12**, 61 (1977).
21. A.W.J.M Rae, D.P. Thompson and K.H. Jack, in Ceramics for High Performance Applications II, edited by J.J. Burke, E.N. Lenoe, and R.N. Katz, (Proc. 5th Army. Mat. Tech. Conf., Brook Hill Pub. Co., Chestnut Hill, Mass.1977) pp. 1039-1067.
22. S. Hampshire and K.H. Jack, in Special Ceramics 7, edited by D.E. Taylor and P. Popper, (Proc. Brit. Ceram. Soc. **31**, 1981) pp.37-49.
23. T.M. Shaw, G. Thomas and R.E. Loehman, J. Am. Ceram. Soc. **67**, 643 (1984).
24. K.H. Jack, in Non-oxide Technical and Engineering Ceramics, edited by S. Hampshire, (Elsevier - Applied Science, Barking, England, 1986) pp.1-30.

25. E.Z. Janecke, Anorg. Chem. **53**, 319 (1907).
26. S. Hampshire, R.A.L. Drew and K.H. Jack, Phys. & Chem. Glass. **26** (5), 182 (1985).
27. M. Ohashi and S. Hampshire, J. Am. Ceram. Soc. **74**, 2018 (1991).
28. C. Schrimp and G.H. Frischat, J. Non-Cryst. Sol., **56** (1-3), 153 (1983).
29. R.K. Brow and G.G. Pantano, J. Am. Ceram. Soc. **67** (4), C72 (1984).
30. M.J. Rand and J.F. Roberts, J. Electro-chem. Soc. **120** (3), 446 (1973).
31. R. Abramovici and M. Ish-Shalom, Ind. Eng. Chem. Prod. Res. Dev. **24**, 586 (1985).
32. M.H. Lewis, G. Leng-Ward and S. Wild, J. Mater. Sci. **21**, 1647 (1986).
33. R.A.L. Drew, Nitrogen Glasses, Research Reports in Materials Science, edited by P. Evans, Series one, No. 2, (Parthenon Press, England, 1986).
34. C.C. Ahn and G. Thomas, J. Am. Ceram. Soc., **65** (11), C185 (1982).
35. S.M. Winder and M.H. Lewis, J. Mater. Sci. Lett. **4**, 241 (1985).
36. M.H. Lewis and G. Leng-Ward, Mater. Sci. Eng. **71**, 101 (1985).
37. T.R. Dinger, R.S. Raj and G. Thomas, J. Am. Ceram. Soc. **71**, 236 (1988).
38. J.L. Besson, D. Billieres, T. Rouxel, P. Goursat, R. Flynn and S. Hampshire, paper submitted to J. Am. Ceram. Soc. (1992).
39. T. Rouxel, J.L. Besson, C. Gault, P. Goursat, M. Leigh and S. Hampshire, J. Mater. Sci. Lett. **8**, 1158 (1989).
40. T. Rouxel, J.L. Besson, E. Rzepka and P. Goursat, J. Non Cryst. Sol. **122**, 298 (1990).
41. J.L. Besson, T. Chartier, P. Goursat and W. Mustel, in Science of Ceramics 14, edited by D. Taylor (Institute of Ceramics, Stoke-on-Trent, 1988) p. 371.
42. V. Morrissey, J. Lonergan, M.J. Pomeroy and S. Hampshire, in Fabrication Technology, edited by R.W. Davidge and D.P. Thompson (Brit. Ceram. Proc., **45**, 1990) pp. 23-32.
43. J. Lonergan, V. Morrissey and S. Hampshire, in Special Ceramics 9, (Brit. Ceram. Proc. **49**, 1991) pp. 57-66.
44. M.J. Pomeroy, B. Saruhan and S. Hampshire, in Special Ceramics 8, edited by S.P. Howlett and D. Taylor (Brit. Ceram. Proc. **37**, 1986) pp. 21-28.

SILICON NITRIDE POWDER FROM CARBOTHERMAL REACTION

Heinrich Hofmann*, U. Vogt*, A. Kerber** and F. van Dijen**

* Alusuisse-Lonza Services Ltd., CH-8212 Neuhausen am Rheinfall, Switzerland
** Lonza-Werke GmbH, D-7890 Waldshut-Tiengen, Germany

ABSTRACT

The synthesis of silicon nitride powder by the reaction between silica, carbon and nitrogen is the economically most interesting route. Only this route makes possible the use of very cheap raw materials as well as the use of a simple production process. On the other side, the properties of the powders are within the limit of the technical requirements of the ceramic part manufacturer. The aim of this paper is to show that an optimized carbothermal Si_3N_4 powder synthesis leads to a product which fulfills the economical (low price) as well as the technical requirements. After a short introduction regarding the economical requirements, an overview of the carbothermal synthesis will be given. The influence of raw material, conditions of synthesis and removal of excess carbon on the powder properties will be discussed. Also, the properties of the powder as well as the sintering behaviour of Si_3N_4 powders, produced by Si-direct nitridation, Si-diimide process and the carbothermal route, will be compared in detail.

1. INTRODUCTION

Silicon nitride (Si_3N_4) is one of the most extensively studied structural ceramic materials for high temperature applications. Especially applications in automotive engines, such as valves, turbo charger rotors and valve guides are of high interest. This increasing interest in Si_3N_4 ceramics has led to a demand for high quality, low cost powders. Unfortunately, high quality commercially available powders are very expensive whereas low-cost powders are of relatively poor quality [1].

It is generally recognised that high quality structural ceramics can only be produced from submicron raw powders with high purity. Especially the manufacturers of ceramic components for engines have the following requirements for Si_3N_4 powder:

- Narrow grain size distribution and a mean grain size of 0,5 - 0,8 μm. No oversized grain (d > 2 μm)
- Low metallic impurity level (Fe, Ti, etc.)
- Oxygen content < 2 mass-%
- High α-content (> 90 % but lower than 98 %!)
- Price < 35 US$/kg
- Whisker and agglomerate free.

Very fine powder must be used because the sintering temperature is limited by the decomposition of Si_3N_4 which occurs above approx. 1700 °C at 1 bar. The Narrow grain size distribution is important for the development of a homogeneous microstructure. On the other hand, forming by dry pressing, slip casting or by injection molding requires powders with a broad grain size distribution and with different specific surfaces. For application at high temperatures the powders must have a very low level of metallic impurities. Concerning the optimum oxygen level as well as the "best" β-Si_3N_4 content, no definite answers exists.

One of the most important factors is the price of the powder. It is well known, for example, that the cost of a ceramic valve must be equal to or lower than the cost of a valve produced from metallic material. With the assumption that the cost for starting material for valves is 10 %, a maximum price for Si_3N_4 powder of 35 US$ can be estimated. Figure 1 shows the influence of the powder on the market potential for Si_3N_4 powder used for structural parts (valves, turbo charger rotors) [2].

There are several routes available for the production of silicon nitride powders. Commercial powders are produced by direct nitridation, Si-diimide processing and carbothermal reduction. Beside these processes, synthesis by vapor, laser and plasma reactions are under investigation or in a pilot stage.

The overall chemical reactions of these processes are:

Si-nitridation	$3\ Si + 2\ N_2 = Si_3N_4$	(1)
Si-diimide	$SiCl_4 + 6\ NH_3 = Si(NH)_2 + 4\ NH_4Cl$	(2a)
	$3\ Si(NH)_2 \rightarrow Si_3N_4 + N_2 + 3\ H_2$	(2b)
Carbothermal reduction	$3\ SiO_2 + 2\ N_2 + 6\ C \rightarrow Si_3N_4 + 6\ CO$	(3)
Gas phase reaction	$3\ SiCl_4 + 4\ NH_3 \rightarrow Si_3N_4 + 12\ HCl$	(4a)
	$3\ SiH_4 + 4\ NH_3 \rightarrow Si_3N_4 + 12\ H_2$	(4b)

Typical properties of powders produced with the three commercial routes are given in Tab. 1. A very high purity as well as a very small grain size is shown for the Si_3N_4 powder produced by the Si-diimide route. Unfortunately this powder is very expensive. Powders manufactured by the carbothermal route have the highest O_2 and impurity contents.

The commercially important powders are produced by direct nitridation of Si. The values of the typical properties of these powders are between the values of the other two powders.

Beside these chemical and physical properties, from an industrial point of view additional requirements must be considered:

- Cost of investment
- Use of hazardous raw materials
- Waste
- Safety aspects of the process.

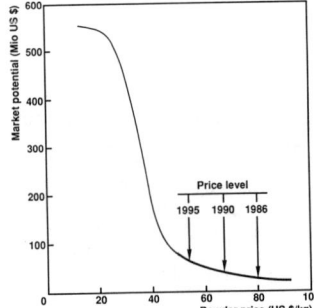

Fig. 1: Influence of the Si_3N_4 powder price on the market potential for structural Si_3N_4 parts (e.g. valves, turbo charger rotors).

Table 1 Typical properties of commercial Si_3N_4 powder for structural ceramics

Property	Production route		
	Si-nitridation	Si-diimide	Carbothermal
O_2 (mass-%)	1.6	1.2	2.1
N_2 (mass-%)	38.7	39.3	39.7
C_{free} (mass-%)	0.04	0.02	0.06
Si_{free} (mass-%)	0.04	0.05	0.04
Metallic impurities (mass-%)	0.04	0.01	0.15
α-Si_3N_4 (%)	96	97	100
spec. surface (m²/g)	15-20	10-12	13
grain size (μm)	0.78	0.38	1.21

Si-direct nitridation uses cheap and harmless raw material but the additional acid wash step after milling with steel balls results in a serious waste problem. The production via the Si-diimide used raw material which required a special infrastructure for $SiCl_4$ and ammonia. The reaction itself is strongly exothermic (safety problem) and there exists some problem with the waste (ammonium chloride). Therefore, the investment costs are also very high, especially if no existing infrastructure can be used for $SiCl_4$ and NH_3. The carbothermal process starts also from very cheap and harmless raw materials (SiO_2, C, N_2). Additionally, the reaction is endothermic and no special infrastructure is necessary. In Europe a potential problem is only the high amount of CO in the exhaust gas.

A detailed investigation of the powder requirements of the Si_3N_4-part producers, as well as of the economical and ecological facts, results in the conclusion that the carbothermal route is the only one which has the potential to fullfill all of these requirements. In this paper we give first an overview of the literature concerning the thermodynamics and experimental work done regarding the carbothermal process. In the second part, the results of our work in this field will be given with special emphasis on the reactor type, the reaction parameters and the typical properties of the powders.

2. LITERATURE REVIEW

2.1 Thermodynamic

The overall reaction between silicon, carbon and nitrogen can be described by eqn. (3). Reaction (3) indicates that 2 moles of gaseous N_2 must be supplied to the SiO_2 sample to produce 1 mole of Si_3N_4 with 6 moles of gaseous CO being given off. The CO content in the powder bed will rise and the N_2 content falls as the reaction proceeds. It is generally thought [3 - 8] that this overall reaction consists of two principal reaction steps:

1. Reduction of silica to silicon monoxide by

$$SiO_2 + C \leftrightarrow SiO + CO \qquad (5)$$
$$K_5 (1700K) = 5,23 \cdot 10^{-4}$$

and/or

$$SiO_2 + CO \leftrightarrow SiO + CO_2 \qquad (6)$$
$$K_6 (1700K) = 7,32 \cdot 10^{-8}$$

$$CO_2 + C \leftrightarrow 2 CO \qquad (7)$$
$$K_7 (1700k) = 7,14 \cdot 10^3 .$$

The reaction (5) is a solid/solid reaction whereas the reaction (6) is a solid/gas reaction.

2. Reaction of $SiO_{(g)}$ to Si_3N_4 by

$$3\ SiO + 3\ CO + 2\ N_2 \longleftrightarrow Si_3N_4 + 3\ CO_2 \qquad (8)$$
K_8 (1700K) $5,46 \cdot 10^{-5}$

and/or

$$76\ SiO + 74\ C + 50\ N_2 \longleftrightarrow Si_{76}N_{100}O_2 + 74\ CO. \qquad (9)$$

Reaction (8) needs three gas species whereas reaction (9) needs two gas and one solid component.

Reaction (8) was proposed by Zhang and Cannon [5]. The morphology of the Si_3N_4-crystal indicated the formation of silicon nitride by this gas phase reaction. Zhang and Cannon observed uniform particle size and shape of the Si_3N_4. They proposed that supersaturation of SiO needed for the nucleation of Si_3N_4 can be achieved only during the initial stage of the reaction, resulting in a burst of nucleation followed by growth at lower supersaturation of reactants. The results of Inoue et al. [9] support this hypothesis. These results indicated the importance of seeding the SiO_2-C mixture for increasing the rate of silicon nitride production and for decreasing the particle size.

Reaction (9) was proposed by Siddiqi and Hendry [6]. These authors assumed $Si_{76}N_{100}O_2$ to be the composition of the α-Si_3N_4 due to the low probability of reacting three gas species in rather complicated stoichiometry. Additionally, they pointed out that the equilibrium partial pressure of CO_2 at reaction temperatures above 1400 °C is very low. Figusch and Licko [10] as well as Shanker et al. [11] show, in disagreement with Siddiqi and Hendry, that once Si_3N_4 has formed by the reaction

$$3\ SiO + 3\ C + 2\ N_2 \longleftrightarrow Si_3N_4 + 3\ CO \qquad (10)$$

the growth reaction (8) might occur in a chemisorbed layer on the surface of α-Si_3N_4 by simple elementary steps [12].

Additionally, in the system SiO_2-C-N the reaction

$$SiO_2 + 3\ C \Rightarrow SiC + 2\ CO \qquad (11)$$

is also possible. It is known from experiments [3, 13, 14] that there exists a lower boundary temperature above which SiC can form.

This temperature is estimated to be between 1390 and 1590 °C depending on the thermodynamic data used.

2.2 Experimental results

The synthesis of Si_3N_4 powder by the carbothermal reaction can be divided into three steps:

- Preparation of the raw material
- Reaction
- Carbon removal.

The first two steps have been very well investigated and many publications exist concerning them. However, regarding the removal of the carbon and the influence of powder characteristics on the sintering and also on the mechanical and chemical properties, very little has been published.

Preparation of the raw material

The following most important parameters were investigated:

- C/SiO_2 ratio in the starting mixture
- SiO_2 modification
- C-modification
- Particle size and specific surface of the starting powder
- Mixing procedure
- Impurities in the starting powder.

The most interesting parameter is the SiO_2/C ratio. The influence of this ratio on the reaction behaviour was the objective of many investigations [4, 5, 9, 10, 13 - 22].

Generally, with an increasing carbon: silicon ratio in the starting mixture the rate of formation of silicon nitride increased significantly. SiO_2/C ratios of 1:2 (moles) to 1:4 was the most investigated range. The most investigations were done with amorphous silica (Aerosil, Cabosil) with a specific surface of 1.2 - 560 m²/g [4, 23, 24, 26, 28]. With increasing specific surface, an increasing reaction rate was observed. Additionally, with increasing specific surface of the carbon source (Carbon black, Lamp black) and increasing homogeneity of the mixture, the reactivity also increases. The possible and most accepted reason for this dependence of the reaction rate on properties of the raw material is that the first reaction step is based on reaction (5). For a high reaction rate, the SiO partial pressure in the powder bed must be also high. Because reaction (5) is a solid/solid reaction, a high specific surface combined with homogeneous mixing is essential for a high reaction rate. Mitomo and Yoshioka support this conclusion with their results [29]. Si_3N_4 powder prepared from an alkoxide derived oxide and carbon mixture heated in flowing N_2 at 1430 - 1500 °C showed 95 % α- and 5 % β Si_3N_4 with a low O_2 content. The influence of impurities has not been investigated in detail. Fe decreases the α/β ratio of the Si_3N_4 powder [14], whereas other authors [6] reported that Fe in the starting mixture increased the amount of SiC in the Si_3N_4 powder formed above 1320 °C.

Reaction

The two most important parameters during the reaction are
Temperature.
Gas flow

These two parameters determine the partial pressure of SiO and CO. From the data reported in the literature for the amount of N_2 required, it is very difficult to estimate a specific N_2-demand. Much more useful are the gas velocity data. The influence of the gas amount on the reaction rate and the Si_3N_4 content of the final powder was determined by several investigaters [24, 30 - 32]. Increasing gas amount or, in other words, increasing gas velocity, increased the reaction rate as well as the content of Si_3N_4 in the final powder. The reason is that at a high N_2 gas velocity the p_{co} will be very low. The most investigated temperature range was that of 1400 - 1500 °C. At lower temperatures, the reaction rate was very low whereas at higher temperatures the formation of SiC was always observed [4, 13, 14, 15, 17, 19, 25, 26, 30, 33].

The influence of seeds has been investigated by different authors [9, 22, 29, 34 - 36]. The addition of 6 - 10 % submicron Si_3N_4 powder enhanced the reaction rate and led to a decreased grain size. These observations also support the assumption that the first reaction step is according to reaction (5), followed by reaction (8).

Carbon removal

This last step in the Si_3N_4 synthesis is very important regarding the quality of the powder. In this step the content of free carbon and oxygen will be fixed. For the carbon removal, temperatures in the range between 600 and 850 °C were used. Normally, air was the oxidizing gas and the reaction time was 1 - 8 h.

2.3 Conclusion from the lierature review

Summarizing the literature review, for the carbothermal systhesis of Si_3N_4 a high reaction rate and a high Si_3N_4 content in the reacted powder can be attained with the following recommended practices:

- SiO_2 with a high specific surface (amorphous SiO_2 like Aerosil, Cabosil)
- Carbon with high spec. surface
- homogeneous mixture of SiO_2 and C
- Partial pressure of CO as low as possible by an increased N_2 partial pressure (high N_2 gas flow)
- Seeding with submicron Si_3N_4 (5 - 15 vol-%)
- low amount of metallic impurities
- Reaction temperature 1450 - 1500°C, Reaction time ca. 5h
- $p_O < 2 \cdot 10^{-20}$.

Powders produced according to these recommendations show the following properties:

α - Si_3N_4 content	: 98 %
O_2 content	: 1.5 - 2 wt-%
C content	: 0.5 wt-%
N content	: 38 wt-%
Grain size	: < 1 µm
Whisker content	: none.

The objective of this study was to investigate the process of carbothermal synthesis of Si_3N_4 in greater detail and to develop a process for the production of low cost Si_3N_4 powder useful for the manufacture of parts for automobile engines.

3. EXPERIMENTAL INVESTIGATION

The aim of this study was to design a process for the carbothermal synthesis of Si_3N_4 based on the following conditions:

- use of cheap raw materials
- production of submicron powder without milling of the Si_3N_4
- the powder quality must satisfy the requirements for production of ceramic parts for automotive engines.

The investigation was split into two parts:

1. Studies of the reaction mechanism. For these experiments, a pressed SiO_2-mixture was reacted in a horizontal furnace

2. Development of a low cost process using a fluidized bed reactor.

3.1 Reaction mechanism

3.1.1 Experiments

Experiments were carried out with gas black (Printex U, Degussa), fumed silica (specific surface of 50 m²/g, Aerosil OX 50, Degussa) with a molar ratio of SiO_2:C of 1:2.5. Additionally, 5 wt-% (relative to SiO_2) Si_3N_4 (LC12 Starck) was added for seeding. The gas black, silica and Si_3N_4 powder were mixed in demineralised water using an attritor lined with polyurethane and using Si_3N_4 milling beads. After mixing, the slurry was filtered, dried and broken into particles with a diameter < 0,5 mm. The reaction was performed in a Carbolite horizontal tube furnace.
The reaction took place at 1470 °C for 0 - 6 h. Details of the experiments as well as of the characterization are described in reference [37].

3.1.2 Results

Table 2 gives the results of the experiments. Upon reaching the reaction temperature, the conversion was 6 % complete. During the isothermal heating the conversion increased linearly with a slope of 23 %/h. Figures 2a - c show TEM pictures of the formation of Si_3N_4 during isothermal heating. Figure 2a shows the composition of the starting powder: spherical carbon black particles with a diameter of 35 nm, spherical amorphous silica particles with a diameter of 160 nm and a silicon nitride seed particle. Figure 2b shows the mixture at 30 % conversion after an hour at reaction temperature. The silica agglomerates are partially sintered, whereas the carbon black particles have not changed in size. Figure 2c show the mixture at 55 % conversion (reaction time 2 h). Crystalline Si_3N_4 particles, silica spheres and carbon black can be observed. Figures 2d and 2e show the mixture after 4 and > 6 h respectively. After full conversion, well crystallized particles of Si_3N_4 and carbon black particles with a diameter of 35 nm are shown.

Table 2 Results of experiments at 1470 °C in N_2, concerning the reaction mechanism

Number	Time (h)	O content (wt%)	N content (wt%)	Conversion (%)	Si_3N_4 Phases
0[a]	-	38.2	1.0	0	α
1	0	35.5	2.6	6	α
2	1	28.6	8.9	30	α
3	2	19.9	15.8	55	α
4	4	2.7	31.7	97	α
5	6	1.7	32.0	98.5	α

[a] Raw material.

3.1.3 Conclusion

The results from TEM investigations indicate that the reaction mechanism is

1. Formation of SiO by a solid/solid reaction [eqn (5)]
2. Formation of Si_3N_4 by a gas/solid reaction [eqn (10)]

The main argument for this is that after synthesis the excess carbon in the products occurs as unreacted carbon black particles. This is in agreement with recent work of different authors [10, 11]. For an optimized synthesis of Si_3N_4, the following conclusions can be drawn:

- Enhancement of the SiO formation (eqn (5)) by excellent mixing of silica and carbon black and the formation of dense pellets.

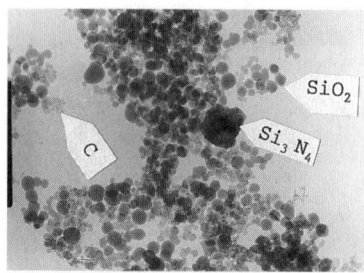

Fig. 2a: Mixture of raw materials. The length of the black bar is 1600 nm. The large particle is Si_3N_4, the 160 nm diameter particles are silica and the 35 nm diameter particles are gas black.

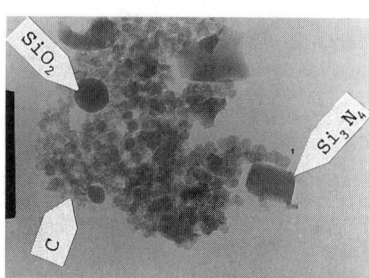

Fig. 2b: The mixture at 30 % conversion (dwell time 1 h). The length of the black bar is 620 nm. Partially crystallised particles and amorphous silica can be seen.

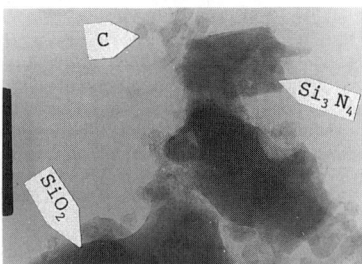

Fig. 2c: The micture at 55 % conversion (dwell time 2 h). The length of the black bar is 450 nm. The figure shows crystalline Si_3N_4 particles, particles of gas black and amorphous silica.

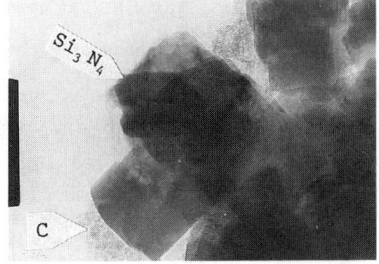

Fig. 2d: The mixture at 97 % conversion (dwell time 4 h). The length of the black bar is 450 nm. Well crystallised Si_3N_4 particles as well as gas black particles are seen.

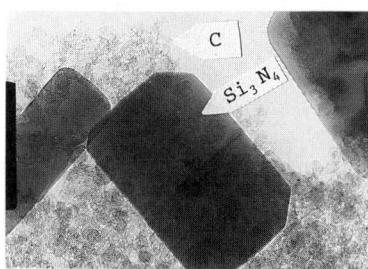

Fig. 2e: The mixture at 100 % conversion (dwell time > 6 h). The length of the black bar is 450 nm. The diameter of the gas black particles is 35 nm. The Si_3N_4 particles are well crystallised.

Fig. 2: Formation of Si_3N_4 during isothermal heating of a SiO_2:C powder mixture with a molar ratio of 1:25 at 1470 °C in flowing N_2.

- Enhancement of the gas exchange (high p_N, low p_{co}) in the pellets by formation of porous granules and high gas velocity.

These two contradictory demands have to be optimized in order to meet the aim of a low cost process route for an adequate Si_3N_4 powder. For large scale production a type of reactor with excellent heat transfer and gas solid interaction is needed. This is due to the endothermic character of the carbothermal synthesis of Si_3N_4 [38] and to the fact that this synthesis is also a gas-solid reaction. A fluidized bed reactor could fulfill the requirements.

3.2 Process development

According to the conclusions of the literature review as well as the experiments regarding the reaction mechanism, a process with a fluidized bed reactor was developed.

3.2.1 Experiments

The schematic of the reactor is shown in Figure 3. A SiC tube with an inner diameter of 76 mm and a length of 1500 mm was used. The length of the constant heat zone was 190 mm. The gas distributor was made of Si_3N_4.
For a first optimization run, the following parameteres were investigated: temperature, reaction time, gas velocity and SiO_2:C - ratio. The investigated ranges of these parameters are listed in Table 3. As raw materials, amorphous silica (Aerosil OX 50) and carbon black (Printex U) were used. The preparation of the raw material was the same as described in 3.1.1. For each reaction, 320 g of SiO_2/C mixture was used.

The granules were characterized by SEM as well as by Hg-porosimetry. During the measurements, the CO content in the outlet gas was determined (URAS 10E, Hartmann & Braun). From the final powder, the contents of O, N, C and free C, as well the phase composition, were measured. To minimize the number of experiments, a statistical planning and evaluation program was used. The excess carbon was removed in a fluidized bed reactor with an air/N_2-gas mixture at temperatures < 700 °C. Finally, a comparative sintering test with different powder was done.

Table 3 Range of investigated parameters for the experiments with a fluidised bed reactor

Parameter	Range
Temperature	1400 - 1500 °C
Reaction time	4 - 8 h
Gas velocity	2,7 - 26,1 cm/s
SiO_2:C ratio	1:2.1 - 1:3

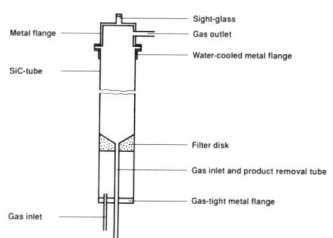

Fig. 3: Schematic of the fluidized bed reactor.

3.2.2 Results

The shapes of the granules before and after the reaction are shown in Figures 4a and 4b. Before the reaction, the granules show the typical shape of broken particles. The diameter of the granules is in the range of 0,2 - 0,8 mm. The granules, after the process in the fluidized bed reactor, have a rounded shape and a size range of 0,2 - 0,6 mm. Also, the size distribution is narrower. The surfaces of the granules before and after the reaction, including the removal of the excess carbon, are shown in Figure 5a - c. The SiO_2/C granule surface is very dense and rough (Figure 5a) whereas the surface after the reaction is smooth and more pores are visible (Figure 5b). After carbon removal, the Si_3N_4 grains are visible. The grain size range is 0,1 - 0,5 µm. The distribution of the pore radius is shown in Figure 6. The as-received granules have a mean pore radius of 20 nm. During the synthesis, the pore radius increases to 100 nm and the pore radius distribution is broad. After the carbon removal, the mean pore radius of the Si_3N_4 granules increases further to 200 nm. In addition, a second peak at the large pore size of 4 µm can be observed.

The CO concentration in the outlet gas is shown in Figures 7 and 8. With increasing reaction temperature, the CO content in the outlet gas increases during the first three hours of the reaction (Figure 7). At 1500 °C, the reaction is finished after 4 - 5 h whereas the reaction at 1400 °C is very slow and not finished during the observed period of 8 h. The influence of the gas velocity is very pronounced. Figure 8 shows the CO content in the outlet gas at a reaction temperature of 1500 °C as a function of the gas flow velocity in the range of 2,6 - 23,2 cm/s. At low gas velocity, the CO content is very high and the reaction rate extremely low. At gas velocities higher than 10 cm/s the CO content is constant after heating up (6 - 7 vol-%) and drops down at the end of the reaction. With increasing gas velocity the reaction is finished after a shorter time. The CO-peak during heating is caused by organic additives used for the preparation of the raw material.

Figure 9 shows the influences of temperature, gas velocity and the reaction time (including heating up) on the O_2 level of the powder. The curves indicate the 2 wt-% O_2 level. At higher temperatures and for higher gas velocities, the O_2 content of the powder is lower.

Further optimization results in the following process parameters:

Temperature	:	1450 °C
Gas velocity	:	20 cm/s
Reaction time	:	8 - 9 h
SiO:C ratio	:	1:2.1.

The properties of the powders produced with these parameters are:

O_2	:	1.81 wt-% before carbon removal
		2.5 wt-% after carbon removal
C	:	0.8 wt-%
C_{free}	:	0.02 wt-%
SiC	:	ca. 2 wt-%
spec. surface	:	12 m²/g.

Figure 5c shows a SEM-picture of the powder after carbon removal. The grain size is << 1 µm and the grain size distribution is narrow.

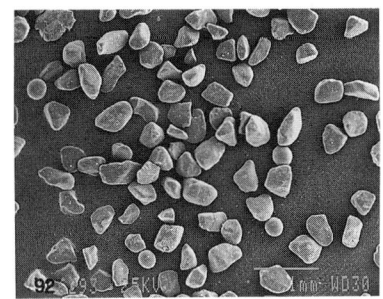

Fig. 4a: Before the reaction

Fig. 4b: After the reaction

Fig. 4: Shape of the SiO_2/C granules before and after the reaction.

Fig. 5a: Surface before the reaction

Fig. 5b: Surface after the reaction

Fig. 5c: Surface after carbon removal

Fig. 5: Surface of the granules

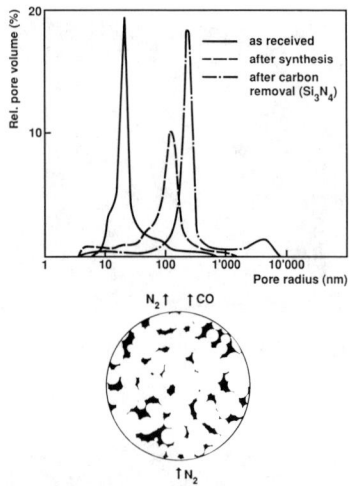

Fig. 6: Distribution of the pore size in the granules.

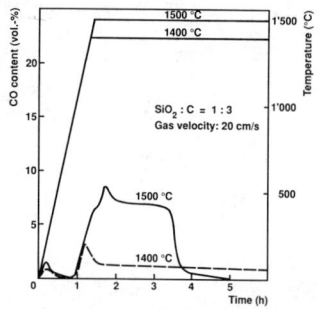

Fig. 7: Influence of the reaction temperature on the CO content in the outlet gas. (SiO_2:C = 1:3, N_2-gas velocity > 20 cm/S).

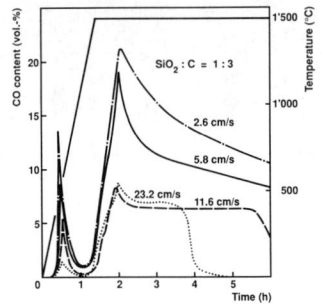

Fig. 8: Influence of the gas velocity on the CO content in the outlet gas. (SiO:C = 1:3, reaction temperature 1500 °C).

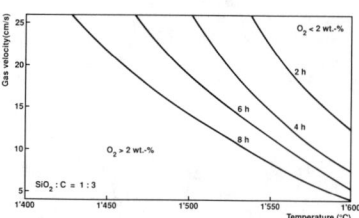

Fig. 9: Influence of temperature, gas velocity and reaction time on the O content in the powder. The lines represent the condition O content = 2 wt-%.

The sintering behaviour of this carbothermal Si_3N_4 powder is shown in Figure 10. For comparison, the sintering behaviour of Si_3N_4 powder produced by direct nitridation and the diimide route, respectively, are also shown. For these experiments, a pressureless sintering process with 5 wt-% Y_2O_3 and 3 wt-% Al_2O_3 sintering additives was used. The sintering behaviour of the carbothermal Si_3N_4 is very similar to the other powders. The begin of the shrinkage is 50 - 100 °C higher, but the sintering rate is also higher. Figures 11a - c show the microstructure of these three samples. All three powders sintered to nearly dense material and the microstructures are very similar.

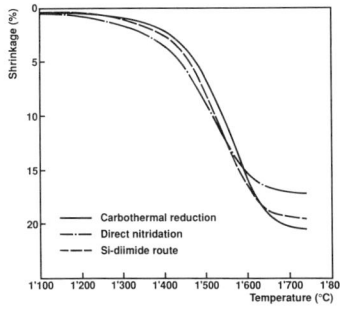

Fig. 10: Sintering behaviour of Si_3N_4 powder by routes:
a) carbothermal
b) direct nitriding
c) Si-diimide process

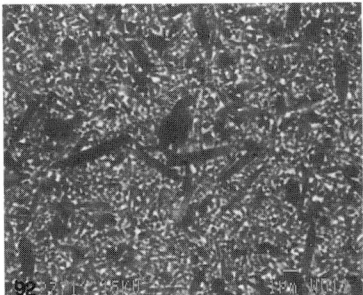

Fig. 11a: carbothermal

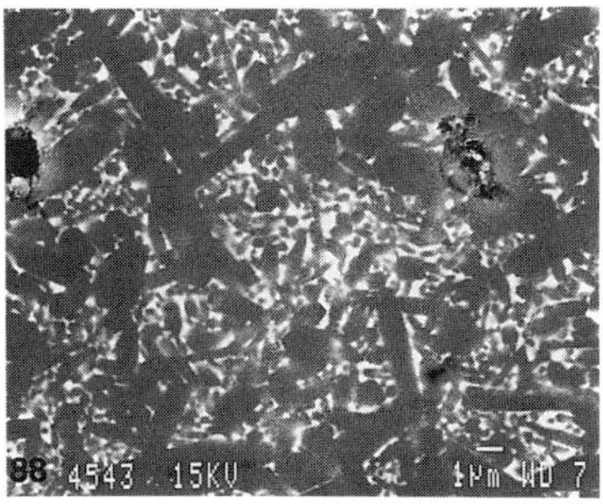

Fig. 11b: direct nitridation

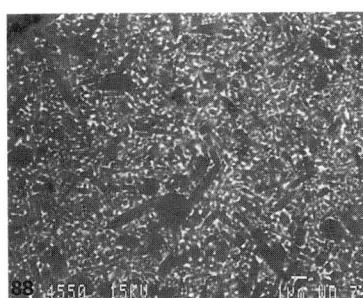

Fig. 11c: Si-diimide process

Fig. 11: Microstructure of pressureless sintered Si_3N_4. Powders were produced by different routes:

3.2.3 Discussion

From the engineering point of view, the most interesting point is the behaviour of the granules in the fluidized bed. The pore radius of the as-received granules is in the range of the grain size of the raw material (20 nm, C; 40 nm SiO_2). This is in agreement with the results of Reyes and Iglesia [39] who simulated the mean pore radius of compacted and sintered fumed silica. During the reaction, the increase of pore radius was due to three reasons:

1. Burn out of the fine carbon increases the pore size because pores between the coarser SiO_2 determined the mean pore size.

2. During heating up, sintering of SiO_2-particles occur and the pore size increases by a factor of two [39].

3. The grain size of the synthesized Si_3N_4 is 0,5 μm, much higher than the grain size of the raw material. This leads to a pore radius of 0,1 - 0,3 μm.

The gas diffusivity in such granules was simulated by Rexes and Iglesia [39]. With an estimated granule porosity of 40 %, the effective diffusivity of N_2 is in the range of 0,02 - 0,04 cm²/s. The time for the diffusion of N_2 to the middle of a granule is therefore in the range of 1 s. Therefore the gas exchange between the inside of the granules and the gas in the furnace is not a serious problem and, compared with the long reaction time is very short.

The observed dependence of the CO content in the outlet gas is in agreement with the literature [40]. The CO content decreases with increasing N_2 gas velocity because a dilution takes place. Additionally, the partial pressure of SiO increases (see eqn. (5)) and the reaction rate also increases. Therefore, the reaction time decreases. These observations support again the assumption that the carbothermal process starts according to reaction (5).

Because the high number of parameters prevent a detailed investigation of the influence of these parameters on the powder properties, a statistical design of the experiments with the important parameters was used. The results show us that only a narrow range of each paramenter can be utilized:

Temperature: Below 1450 °C, the reaction rate is very low and not of industrial interest. Above 1500 °C the amount of SiC in the powder increases very rapidly.

Gas velocity: At low gas velocities, the CO content is too high and therefore the resulting SiO partial pressure and the reaction rate are too low. At higher gas velocities, the yield decreases because the particles will be blown out of the furnace.

SiO:C ratio: This ratio must be close to the stoichiometric value. The optimized ratio of 1:2.1 fulfills this requirement.

Other important parameters, e.g. specific surface of the raw material, homogeneity of the mixture, pore size and pore volume of the starting granules, and kind of raw materials have not been investigated in detail.

CONCLUSION

The carbothermal process is very useful for the production of a high quality Si_3N_4 powder. The application of a fluidized bed reactor using cheap raw material makes it possible to produce a low cost powder with acceptable properties. The carbothermal process is still not understood in detail. The assumption that first SiO develops through a solid/solid reaction and in a second step the SiO, N_2 and CO react at the surface of Si_3N_4 seeds, or Si_3N_4 is formed by a reaction of SiO, N_2 and C, is supported by this work.

Literature

1. M.D. Pugh and R.A.L. Drew, in **Proceedings of the 3rd International Symposium on Ceramic Materials and Components for Engines,** edited by V.J. Tennery (American Ceramic Society, Westerville, OH, 1989) p. 139.
2. Daimler Benz and others, private communication.
3. A. Hendry and K.H. Jack, in **Special Ceramics 6,** edited by P. Popper (British Ceramic Research Association, Stoke-on-Trent, 1975) p. 199.
4. K. Komeya and H. Inoue, J. Mat. Sci. **10,** 1243 (1975).
5. S.-C. Zhang and W.R. Cannon, J. Am. Ceram. Soc. **67,** 691 (1984).
6. S.A. Siddiqi and A. Hendry, J. Mat. Sci. **20,** 3230 (1985).
7. Y.W. Cho and J.A. Charles, Mat. Sci. Tech. **7,** 289 (1991).
8. M. Ekelund and B. Forslund, J. Europ. Ceram. Soc. **9,** 107 (1992).
9. H. Inoue, K. Komeya and A. Tsuge, J. Am. Ceram. Soc. **5,** C-205 (1982).
10. V. Figusch and T. Licko, in **High Technical Ceramics,** edited by P. Vincenzini (Elsevier Science Publisher B.V., Amsterdam, 1987) p. 517.
11. K. Shanker, S. Grenier and R.A.L. Drew, in **Ceramic Powder Science III,** edited by G.L. Messing, S. Hirano and H. Hausner (Am. Ceram. Soc. 1990). p. 321.
12. W.R. Cannon and S-C. Zhang, in **Ceramic Materials & Components for Engines,** edited by V.J. Tennery (American Ceramic Society, Westerville, OH, 1989) p. 86.
13. A. Szweda, A. Hendry and K.H. Jack, in **Special Ceramics 7,** edited by P. Popper (British Ceramic Research Association, Stocke-on-Trent, 1981) p. 107.
14. J.G. Lee and I.B. Cutler, in **Nitrogen Ceramics,** edited by F.L. Riley (Noordhoff, Leyden, 1977) p. 175.
15. M. Mori and N. Takai, Germ. Offenlegungsschrift Nr. 2703354 (28. July 1977).
16. M. Mori, H. Inoue and T. Ochiai, NATO ASI Ser., SerE, 65 (Prog. Nitrogen Ceramic, 1983) p. 149.
17. K. Komeya, H. Inoue, S. Matake and H. Endo, Germ. Offenlegungsschrift Nr. 2814235 (31. August 1978).
18. T. Ishii, A. Sano and I. Imai, in **Silicon Nitride-1,** edited S. Somiya, M. Nitomo and M. Yoshimura (Elsevier Appl. Sci., London, 1990) p. 59.
19. T. Ishii, Japan Patent No. 308193 (26. December 1986).
20. T. Ishii, Japan Patent No. 308190 (26. December 1986).
21. J.G. Lee and I.B. Cutler, in **Nitrogen Ceramics,** edited by F.L. Riley (Noordhoff, Leyden, 1977) p. 275.
22. D.S. Perera, J. Mat. Sci. **22,** 2411 (1987).
23. B.G. Durham, M.J. Murtha, G. Burnet, Advanced Ceramic Materials, **3,** 45 (1988).
24. H. Yoshimatsu, H. Kawasaki, Y. Miura, J. Mat. Sci. **24,** 3280 (1989).
25. K. Komeya, H. Inoue, T. Ohta, Germ. Offenlegungsschrift No. 2622554 (7. April 1977).
26. M. Ekelund, B. Forslund, in **Ceramic Powder Science III,** edited by G.L. Messing, S. Hirano and H. Hausner (American Ceramic Society, Westerville, OH, 1990) p. 337.
27. B.G. Durham, MS-Thesis, Iowa State University (1986).
28. E. Friedrich, L. Sitting, Z. Anorg. Allg. Chem. **143,** 293 (1925).
29. M. Mitomo and Y. Yoshioka, Advanced Ceramic Materials, **2,** 253 (1987).
30. S. Natansohn and G. Czupryna, U.S. Patent No. 4619905 (28. October 1986).
31. M. Ekelund and B. Forslund, in **Ceramic Materials & Componentes for Engines,** edited by V.J. Tennery (American Ceramic Society, Westerville, OH, 1989) p. 101.
32. Y. Tuoshino, R. Laitinen and K. Torkkell, in **Ceramic Powder Science III,** edited by G.L. Messing, S. Hirano and H. Hausner (American Ceramic Society, Westerville, OH, 1990) p. 329.

33. T. Ishii, Japan Patent No. 308192 (26. Dezember 1986).
34. M. Komatsu and T. Miguno, Germ. Offenlegungsschrift No. 3511709 (29. March 1985).
35. T. Ishii, Japan Patent No. 308195 (26. December 1986).
36. H. Inoue, K. Komeya and A. Tsuge, U.S. Patent No. 4368180 (11. January 1983).
37. F.K. van Dijen and U. Vogt, J. Europ. Germ. Soc. **10**, 273 (1992).
38. F.K. van Dijen, PhD Thesis, Eindhoven University of Technology (1986).
39. S.C. Reyes and E. Iglesia, J. Catalysis **129**, 457 (1991).
40. T. Licko, V. Figusch and J. Puchyovà, J. Europ. Cerm. Soc. **9**, 219 (1991).

Sialon Ceramics Sintered with Yttria and Rare Earth Oxides

THOMMY EKSTRÖM

Department of Inorganic Chemistry, Arrhenius Laboratory, University of Stockholm,
S-106 91 Stockholm, Sweden.

ABSTRACT

Dense single-phase α-, β- and O'-sialon cermics or mixed sialon ceramics without a glassy grain-boundary phase can be prepared at high temperatures and pressures, and these materials are well suited for high-temperature use, but they are usually brittle. Additional quantities of oxides of group IIIB metals in the periodic table are often added as sintering aids to achieve pressureless sintering and thereby to allow more complicated shapes to be manufactured directly and at lower costs. The most common additive is yttria, but the rare earth oxides are also of interest. All these oxides will promote the growth of elongated β crystals in the microstructure, and the fracture toughness will be improved considerably. Low-cost oxides like Nd_2O_3, La_2O_3 or CeO_2 may replace Y_2O_3 without significantly impairing the mechanical properties at room temperature. The expensive rare-earth oxides like Sm_2O_3, Dy_2O_3 or Yb_2O_3 have been found to be as good additives as yttria, or even better, but improvements in mechanical properties are generally small and do not justify the use of these additives in large-scale production. The residual intergranular glassy phase usually found in the microstructure of metal-oxide-doped sialons will deteriorate the properties at very high temperatures, and this type of material is best suited for use at operation temperatures below 900-1000°C.

1. INTRODUCTION

It is well known that silicon nitride, Si_3N_4, is difficult to sinter to full density without the use of additives that form an eutectic liquid well below the used sintering temperature. The sintering aids react with the ever-present silicon oxide layer on the surface of the Si_3N_4 particles and also with other cation or anion impurities to form a liquid via which mass transport and densification occur by solution-reprecipitation of the Si_3N_4 particles. In most Si_3N_4 ceramics it is this liquid that after sintering forms the residual intergranular phase, which is less refractory than the Si_3N_4 grains and hence determines the high-temperature properties.

The sialon ceramics (acronym of Si-Al-O-N) offer a special advantage because the sintering temperature can be reduced and all of the added sintering aid might be taken up in solid solution(s) of Si_3N_4 phase(s), i.e. a transient liquid phase sintering route is allowed. The advantages of using sialon ceramics were early pointed out by Jack[1,2], and a vast number of publications have since then appeared on different aspects of the sialons; more recently some of this work has been reviewed[3]. The three sialon phases that have attracted most attention are the α and β sialons, isostructural with α- and β-Si_3N_4, and also the O'-sialon, which has a narrow solid solution range based on the Si_2N_2O structure. The positions of all these sialon phases in the Y-Si-Al-O-N system are schematically illustrated in Figure 1a.

By addition of appropriate amounts of Al_2O_3/AlN/SiO_2, the liquid phase region that expands rapidly in the Si-Al-O-N system at temperatures above 1650°C can be used to reduce sintering temperatures and times to produce β or O'-sialon ceramics, cf. Figure 1b. Especially the β sialon phase, $Si_{6-z}Al_zO_zN_{8-z}$, which has a very extensive solid solution range (up to z= 4.2) may be used to obtain single-phase β sialon ceramics through transient-liquid-phase sintering[4-7]. The O'-sialon, $Si_{2-x}Al_xO_{1+x}N_{2-x}$, where the x-value only varies from zero to 0.2, is less efficient in consuming the sintering aids[8,9]. The α sialon phase finally forms an extended two-dimensional solid solution range in many metal-sialon systems (10,11). It is found in the plane Si_3N_4 - $4/3(Al_2O_3 \cdot AlN)$ - $YN \cdot 3AlN$ shown for the yttrium sialon system in Figure 1a, and it may form a suitable solid solution for transient-liquid-phase sintering when other metal oxides are used as sintering aids, as described below.

The addition of glass-forming oxides of group IIA (CaO, MgO) or of group IIIB elements (Sc, Y, rare earth metals) to sialon ceramics is a frequently used means to drastically lower both the eutectic temperature and the liquid viscosity. The reaction paths through which the sialon phases prepitate are governed by the composition of the liquid and the overall composition of the sample, but the reaction rate is mainly controlled by the volume and viscosity of the liquid. Hence the metal oxides used as sintering aids and the increase of sintering temperature above the eutectic is of importance for both the densification and the phase transformations. Furthermore, most sialon materials with these metal oxide additives retain some liquid phase, which upon cooling forms a residual intergranular glassy or partly crystallized intergranular phase. The amount and characteristics of this glassy phase will markedly influence the materials properties of the sialon ceramics, especially at very high temperatures. Properties of the glass, such as hardness and glass transition temperature (T_g), are important and depend on the cation composition and the nitrogen content of the residual metal oxynitride glass.

The group IIIB elements are known to form refractory glasses upon cooling from the sintering temperature. Therefore special attention has been paid in the past to the effects of using elements like yttria or more recently the rare-earth oxides in preparing sialon ceramics. This paper will summarize some results obtained at Stockholm University in replacing yttrium with other rare earth elements. The selected rare earth elements are marked by bold face text in the following series of elements with increasing atomic number Z; **La, Ce,** Pr, **Nd,** Pm, **Sm,** Eu, Gd, Tb, **Dy,** Ho, Er, Tm, **Yb** and Lu.

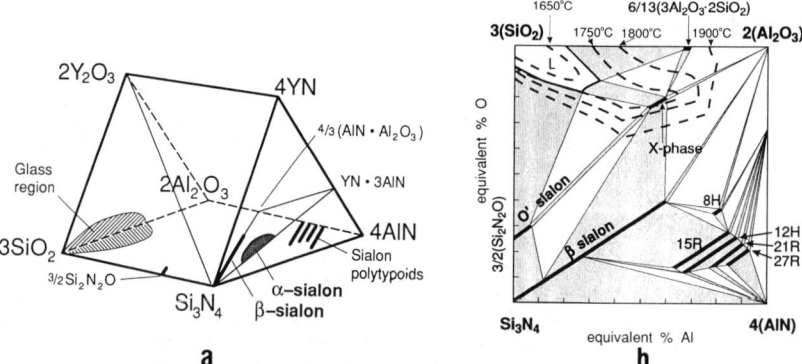

Figure 1. Part of the Y-Si-Al-O-N system, drawn schematically in (a) as a so-called Jänecke prism to illustrate the positions of the α-, β-, and O'-sialon phases. Note that the β- and O'-sialon phases have solid solution ranges only in the "basal plane" representing the Si_3N_4-SiO_2-Al_2O_3-AlN system, which is shown in more detail in (b). In the latter system the liquid phase denoted by L is close to the SiO_2 corner at 1700°C. Its extension is taken from Ref. 12. Calculated liquid-phase regions at different temperatures above 1650°C are marked with dashed lines.

2. MATERIALS AND EXPERIMENTAL PROCEDURE

The selected compositions for this study have used Si_3N_4 as the parent with additions of Al_2O_3/AlN and group IIIB metal oxides. The metal oxides used were Y_2O_3, Nd_2O_3, La_2O_3, CeO_2, Sm_2O_3, Dy_2O_3 and Yb_2O_3 or mixtures of Y_2O_3 with these oxides. The source materials used were silicon nitride (UBE, grade SN10E, or H.C. Starck-Berlin, grade LC1), aluminum oxide (Alcoa, grade A16SG), aluminum nitride (H.C. Starck-Berlin, grade A), yttrium oxide (H.C. Starck-Berlin, grade Finest), neodymium oxide (Molycorp. Inc., 99.5%), cerium oxide (H.C. Starck-Berlin, grade Finest), lanthanum oxide, samarium oxide, dysprosium oxide, ytterbium oxide (all from Johnson Matthey Chem. Ltd, 99.9%).

The relative molar amount of oxide dopant was the same in each series, corresponding to 4.0 or 6.0 wt% Y_2O_3, as to allow direct comparision of results. Two glass-phase-containing sialon materials have been selected in this article to illustrate the effects of replacing yttria with other oxides. The overall compositions of the sialon materials are close to the Si_3N_4 corner of the Me-Si-Al-O-N phase diagram. The first is a β sialon material (equiv. Al= 0.095, equiv. O= 0.070) and the other a mixed α–β sialon material (equiv. Al= 0.095, equiv. O= 0.048) with an α-to-(α+β) ratio around 0.3. The starting powders were carefully weighed, mixed in water-free propanol and milled in a vibratory mill with silicon nitride milling balls. After drying, the powder mixes were dry-pressed (125 MPa) into compacts. The samples were either pressureless-sintered in nitrogen atmosphere in a powder bed of submicron BN, or glass-encapsulated and hot isostatically pressed with 200 MPa of argon. Details of overall compositions and preparations may be found in Refs. 13-16.

The density of the sintered samples was measured using Archimede's principle. Hardness (HV10) and indentation fracture toughness (K_{1C}) at room temperature were obtained with a Vickers diamond indenter using a 98 N (10 kg) load. The method of Anstis et al.[17] was used for calculating the fracture toughness, assuming a value of 300 GPa for Young's modulus. The phase analysis was based on X-ray powder patterns recorded by the Guinier-Hägg film technique and with silicon as internal standard. Scanning electron microscopy was performed on carbon-coated materials, using instruments equipped with EDS analyzers.

3. RESULTS AND DISCUSSION

3.1 Yttria as Sintering Aid for Sialon Ceramics

Sialon materials for very high temperatures (around 1400°C) are only to be found among the single-phase sialon materials or where sialon phases are combined in such a way that no glassy intergranular phase will be present in the materials microstructure. The single-phase α, β or O'-sialon are of obvious interest, but more material combinations are found among multiphase α–β and β-O' sialon ceramics. It should be stressed, however, that only α sialon can accomodate other glass-forming elements in its fairly flexible solid solution range. Hence, to achieve complete transient-liquid-phase sintering in yttria- or some other metal-oxide-doped sialon ceramics, the use of the α sialon phase as a "sink" for the added metal is one possible means.

As mentioned above, the use of a further metal oxide additive to the sialon prior to sintering will greatly promote the densification by forming a eutectic liquid at low temperatures. Addition of Y_2O_3 to sialon will initially give rise to a liquid in the Y_2O_3-SiO_2-Al_2O_3 system at temperatures below 1400°C, see Figure 2a. The amount of liquid phase during sintering and also the amount of unwanted residual glassy phase in sintered β sialon ceramics increases with increased yttria addition. These amounts are also affected by the "alumina content" represented by the z value for the overall β sialon composition. The sialon liquid range expands rapidly with increasing temperature and aluminium-content, cf. Figure 1b. For example, addition of 1 and 3 wt% Y_2O_3 to a β sialon of overall composition z= 0.5 gave rise to 2 and 4.5 vol% residual glassy phase, respectively, and with a β sialon of composition z= 4 about twice as much glass is produced. Analysis of the glass composition in multi-grain pockets of the high-z material indicates that the proportions of the cations are Si : Al : Y = 1 : 1-1.2 : 0.5-0.7, or as weight proportions around SiO_2 : $0.6Al_2O_3$: $0.9Y_2O_3$, which is in fair agreement with the low-temperature eutectic shown in Figure 2a.

The pronounced positive effects on the solidification by simultaneous addition of Y_2O_3 and Al_2O_3 to silicon nitride are exemplified in Figure 2b. At 6 wt% Y_2O_3-addition only around 2 wt% Al_2O_3 is required to obtain nearly full density. This may be compared with a total addition of Al_2O_3 or (Al_2O_3 + AlN) above 8 wt% for β or α–β sialon materials commercially used as metal cutting tools[18]. Thus a good margin is at hand for pressureless sintering, and practical experiments have shown that it is possible either to reduce the sintering temperature to about 1700°C for these ceramics or to reduce the amount of yttria to 4 wt% at 1775-1825°C and still obtain dense materials.

Besides the effect upon sintering, a small addition of yttria will have a positive influence on the fracture toughness of β sialon ceramics[6]. For a pure β sialon material the toughness increases from 2.8 to 3.6 MPam$^{1/2}$ with only 1 wt% Y_2O_3 added. With further increase of the amount of Y_2O_3 up to 6 wt%, the toughness will increase to about 6 MPam$^{1/2}$. Several different toughening mechanisms may play a role for the toughening, but at additions higher than 2-3 wt% yttria the growth of elongated β-grains in the microstructure is probably dominating. However, at low additions the fracture toughness is greatly dependent on the grain-boundary properties. In the absence of a glassy grain-boundary phase the bonding is strong; cracks propagate transgranularly in the material and the toughness is low. The glassy grain-boundary phase that forms upon yttria addition is a preferred crack path, and crack branching will occur, thereby increasing the fracture energy. The interfacial bond strength between the grains and the glassy phase will play a role, but it must also be noted that the glassy phase, after cooling from the sintering temperature, will be in tensile stress. The β-sialon grains has a thermal expansion coefficient of 3.2 x 10^{-6}/°C, but the coefficient for a glass is higher. The thermal expansion coefficient of Y-sialon glasses containing 6 at% of nitrogen is around 4 x 10^{-6}/°C, and this coefficient increases with decreasing nitrogen content so that a pure oxide glass corresponds to 7.5 x 10^{-6}/°C [19].

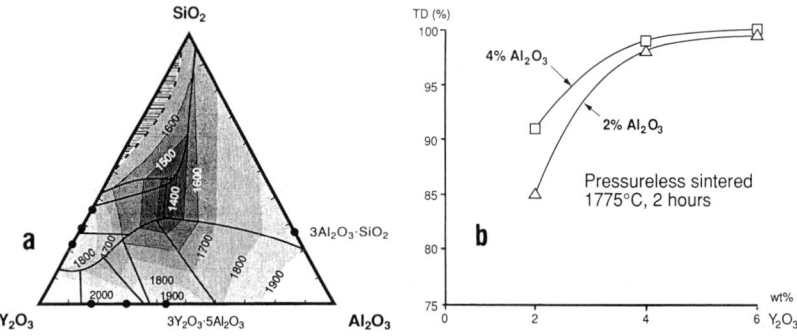

Figure 2. The liquid phase formation in the Y_2O_3-SiO_2-Al_2O_3 system at different temperatures in (a). The diagram is given in weight proportions. Note that the lowest eutectic liquid forms below 1400°C. The effects of (Al_2O_3 + Y_2O_3) sintering aids on the densification of Si_3N_4 by pressureless sintering at 1775°C are illustrated in (b).

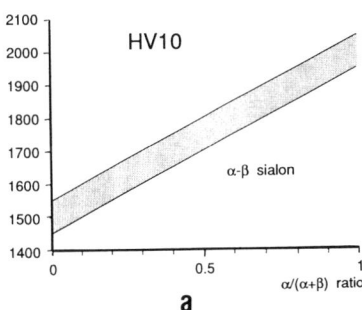
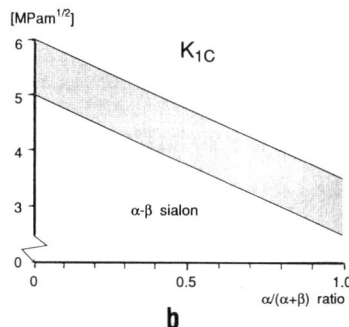

Figure 3. A summary of the variations found in the room-temperature mechanical properties of glass-containing α–β sialon materials. The Vickers hardness, HV10, and the indentation fracture toughness (K_{1C}) are shown as functions of the α sialon content, given as the α/(α+β) phase ratio.

A substantial excess of yttria (6 wt% or higher) is added to most β sialon ceramics, and consequently they contain a considerable amount of residual glassy phase, typically 8-15 vol%. These materials contain well developed, elongated β sialon grains in the microstructures with an aspect ratio (length-to-diameter ratio) typically in the range 4 to 8. They have the highest observed fracture toughness values at room temperature among the different sialon materials, 5-6 MPam$^{1/2}$, but the presence of the glassy phase makes them somewhat soft HV10: 1450-1550.

By changing the overall composition in the Y-sialon phase diagram, α–β sialon ceramics can be prepared with a varying α to β sialon phase ratio and with varying mechanical properties, see Figure 3[18,20]. The possibility of varying the α-to-(α+β) sialon phase ratio by slightly changing the overall composition opens many possibilities to prepare Y-sialon ceramics with desired properties. This is of importance in connection with tailoring properties for specific applications, such as metal cutting tool materials[20].

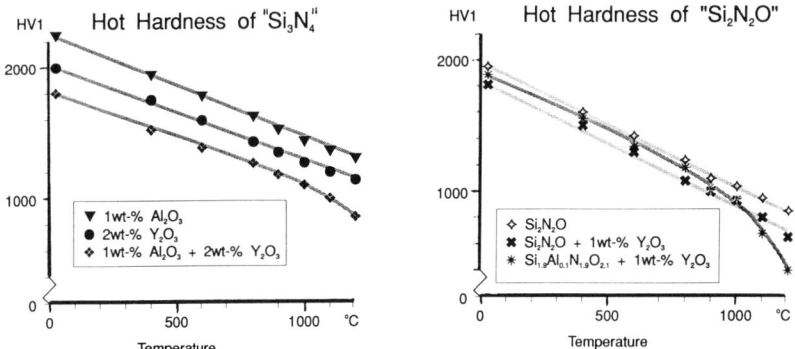

Figure 4. The effects on the Vickers hot hardness (HV1) of adding small amounts of oxides to Si_3N_4-, Si_2N_2O- and O' sialon (x= 0.1)-ceramics. Separate addition of Y_2O_3 or Al_2O_3 yields no significant glass-softening effects up to 1200°C, whereas the high temperature hardness deteriorates at about 900-1000°C when these oxides are combined.

It is possible to achieve much lower contents of residual glassy phase when some or nearly all of the added yttria is used in the formation of the α sialon phase, depending on the selected overall composition in the Y-Si-Al-O-N phase diagram. Pressureless sintering to full density is possible for mixed α–β sialon ceramics with more than 90% of α-phase and with only around 3 vol% of residual glassy phase. The hardness of mixed α–β sialon materials increases markedly with increasing α sialon phase content, whereas the fracture toughness decreases. The α sialon appears as equiaxed grains in the material's microstructure, and this decrease in toughness depends partly on the shape of the α sialon grains but probably also on other factors.

It is not good for the high-temperature properties to add any glass-forming metal oxide to β- or O'-sialon compositions, but this has to be done to make pressureless sintering possible. Addition of only yttria to Si_3N_4 results in fairly refractory glasses, and addition of only alumina to Si_3N_4 will mainly produce β sialon. These two types of materials will not show any glass-phase softening at temperatures up to 1200°C, see Figure 4a. However, if even small amounts of alumina and yttria are combined, then hot-hardness measurements will clearly indicate that the glass becomes soft at around 1000°C. Similar effects of small yttria additions on the hot hardness of different β sialon materials with z= 0.5-3.8 have been reported[21]. Even when only 1 wt% Y_2O_3 is added, both the mechanical and chemical properties at high temperatures are affected, and the effect on the oxidation resistance is demonstrated in Figure 5[22].

Of course, the deteriorating properties of the glassy phase over the glass transition temperature (T_g) cause the observed loss in materials properties, and this temperature is therefore of interesest, but it is dependent on the glass composition. A number of investigations have been done on synthetically prepared Y-sialon glasses, and it is generally accepted that glass transition temperature, hardness, elastic modulus, density, thermal diffusivity and fracture toughness are all higher for oxynitride glasses than for the corresponding oxide glasses and that all these properties tend to improve with increasing N content[23-27]. The substitution of nitrogen for oxygen leads to a higher average coor-dination of the non-metal atoms, resulting in a more rigid glass network, and Y-sialon glass with a very high nitrogen content, corresponding to one in four oxygen atoms replaced by nitrogen, has been reported[23]. Sintered sialon materials contain residual glasses of quite different compositions, however[28,18]. By preparing Y-sialon glasses of a standard cation composition close to that expected to form in sintered sialon ceramics it has been shown that the glass transition temperature rises from 920 to about 1000°C when the nitrogen content increases[23,25]. This temperature interval fits well with the observed loss in hot hardness for Y-sialon materials[21]. Thus, yttria-doped sialon ceramics are not suited for long-time use at operation temperatures above around 950°C.

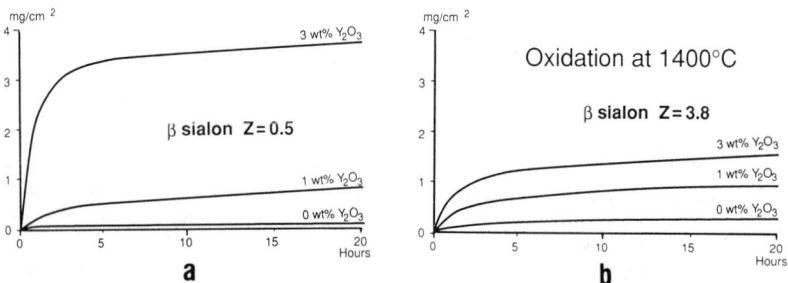

Figure 5. The oxidation resistance as a function of yttria addition (0, 1 and 3 wt%) of two β-sialon materials $Si_{6-z}Al_zO_zN_{8-z}$, where z= 0.5 and 3.8, at 1400°C[22]. The extent of oxidation is given as the weight gain per unit area (mg/cm^2).

3.2 Replacement of Yttria With Low-Cost Rare Earth Oxides

The advantage of replacing Y_2O_3 by other low-cost metal oxides in preparing sialon materials is obvious from an industrial production point of view. The oxides La_2O_3, CeO_2 or Nd_2O_3 have been chosen first for investigation, because these metals belong to the group IIIB elements, which are known to form refractory glasses. However, a number of questions need to be answered: the effects upon densification, the changes in phase compositions or microstructures of sintered materials and, of course, the resulting properties compared to those of Y_2O_3-doped sialon materials.

The solidification process will be dependent on the amount of liquid phase that forms and its viscosity when yttria is partly or fully replaced by the other oxides. A constant molar concentration of added oxide was used here in all preparations, to simplify comparison of the observed properties, and for 6 wt% Y_2O_3 this corresponds to 8.4, 8.9 and 9.2 wt% of the oxides La_2O_3, Nd_2O_3 and CeO_2, respectively. Two specific materials will be used in the examples, a β sialon and an α–β sialon material with about 30% of α sialon phase (when doped with 6 wt% yttria). The sintering temperature and time are factors which influence the densification results, but at sintering temperatures in the range 1750-1825°C, and using 4-6 wt% Y_2O_3 only as an additive, fully dense sialon materials are obtained after 2 hours' sintering time both for the β and α–β sialon ceramics.

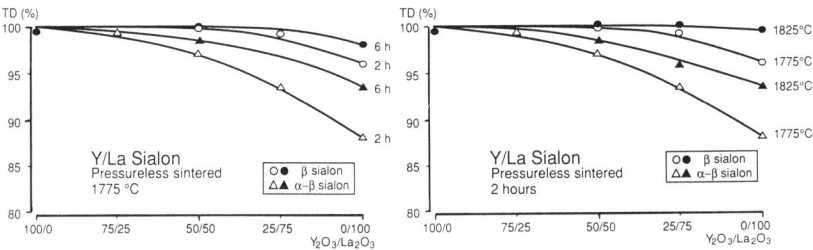

Figure 6. Pressureless sintering in nitrogen atmosphere of sialon materials with La_2O_3 replacing Y_2O_3 as sintering aid. The effects of the Y_2O_3/La_2O_3 ratio, sintering time (2 and 6 hours) and temperature (1775° and 1825°C) are given.

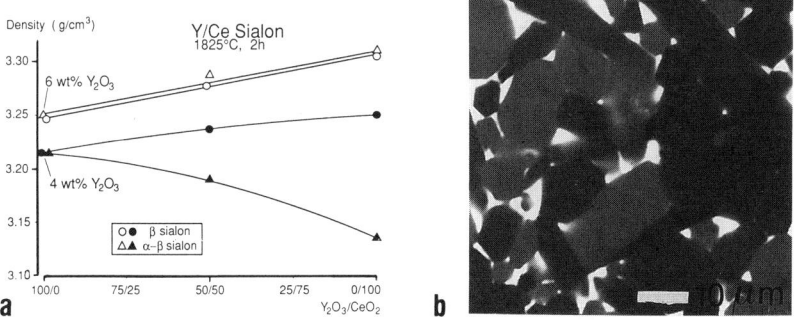

Figure 7. The effects of replacing Y_2O_3 by CeO_2 are illustrated in (a), cf. text. A microstructure of the α–β sialon material with 6 wt% Y_2O_3 added and sintered at 1825°C is shown in (b). The heavy-metal-rich intergranular phase appears bright, the rounded α sialon grains are medium grey, whereas the β sialon grains are dark.

The replacement of Y_2O_3 by Nd_2O_3 or CeO_2 at the "6 wt% Y_2O_3 level" yields dense materials for all Y/Me compositions and for both types of sialons. However, La_2O_3, needed at least 2 hours sintering at 1825°C to produce dense β sialon materials, and dense "α–β sialon" with only La_2O_3 added are difficult to obtain, see Figure 6. This poor ability of La_2O_3 as sintering aid is believed to be a result of the high solubility of nitrogen in the La-sialon liquid at the sintering temperature, with a corresponding increase in viscosity[16]. The β sialon materials, which represent a more oxygen-rich overall composition, were always easier to sinter to full density than the nitrogen-rich α–β sialon materials, and this tendency is even more pronounced when the total concentration of sintering aids is lowered to the "4 wt% Y_2O_3 level". By smaller additions of sintering aids only the β sialon material with Nd_2O_3 or CeO_2 replacing Y_2O_3 will have acceptable densities, as shown for CeO_2 in Figure 7a. The β sialon with La_2O_3 as well as all α–β sialons with all other sintering aids than pure yttria did not reach full density.

The use of different oxides as sintering aids will also affect other properties of the sintered materials, such as the α-to-(α+β) phase ratio. The α sialon phase is not stabilized by cerium and lanthanum, and replacement of yttrium oxide by the oxides of these two elements in the α–β sialon material will therefore diminish the amount of α sialon phase in the materials. Replacement of yttrium oxide by neodymium oxide also gives a slightly lower α/(α+β) phase ratio for the α–β sialon material, being around 0.1. The reduction in the amount of α phase in the α–β sialon materials will affect the materials overall hardness, as exemplified below.

The microstructure will also be affected to some extent by the choice of sintering aids, but it has to be said that the overall microstructures of the β or α–β sialon materials are similar in appearance at a first glance when other rare earth metal oxides replace yttria. However, some changes do appear, and these will subsequently have an influence on the mechanical properties, especially the hardness. The first example is the α sialon phase formation mentioned above, and the α sialon grains have an equi-axed or rounded shape and appear with a medium grey contrast in SEM micrographs, cf. Figure 7b. Another example is the amount and aspect ratio of the elongated β sialon grains, which contain no high-Z metal and appear dark in the SEM micrographs with hexagonal cross-sections. The presence of high-aspect-ratio β sialon grains is of importance for obtaining ceramics of high fracture toughness, and the used metal dopant and the resulting liquid phase viscosity at the sintering temperature are factors that may influence the growth of these elongated crystals. Yttria is a well-known good additive to Si_3N_4 or sialon promoting the growth of elongated β-phase grains. When yttria is replaced by the other rare earth metal oxides in the sialon materials it is observed that the size of the β-grains becomes smaller when the replacement for yttria increases[13,14,16]. The grains also become slightly more needle-like with typical aspect ratios reaching 5-11. Besides the grain framework, the structure is also largely formed by the intergranular phase, which has a direct influence on the properties, especially at high temperatures. The amount of residual glassy phase was found to increase with incresing replacement of yttria with the other oxides, from around 8 vol% to 15-20 vol% in the β sialon material and from around 5 to 10 vol% in the α–β sialon material.

The effects on the Vickers hardness and indentation fracture toughness of replacing yttria with other low-cost rare earth oxides are illustrated in Figure 8a,b and Figure 8c,d for Nd_2O_3 and La_2O_3, respectively[13,16]. The effect of using CeO_2 is very similar and is therefore not shown here, but full details are found in Ref. 14. The decrease in hardness of the β sialon material with increasing Nd_2O_3 or La_2O_3 content is caused by the increasing volume of residual glassy phase, whereas the decrease in hardness of the α–β sialon is dependent also on the diminishing amount of α-sialon.

The oxidation resistance of sialons prepared with Y_2O_3 as sintering aid is better than that for sialons prepared with Nd_2O_3 or Y_2O_3 mixed with Nd_2O_3[29]. The oxidation rates of α–β sialons were generally lower than for the β sialons; this is believed to be a result of lower concentrations of residual glassy phase in the former materials, cf. Figure 9. The lower oxidation resistance of neodymia-doped sialon materials may be compared with the fact that

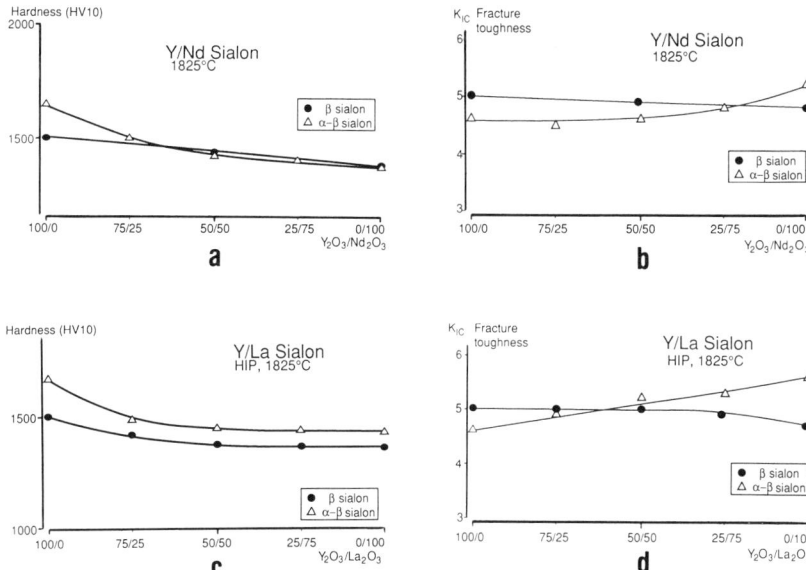

Figure 8. Vickers hardness (HV10) and indentation fracture toughness (K_{1C}) of sialon ceramics as functions of $Y_2O_3:Me_2O_3$ ratio for Me= Nd and La.

Nd-sialon glasses get softer than Y-sialon glasses at lower temperatures[23]. The glass transition temperature of a Nd-sialon glass increases with increasing N content from 910°C to at most 960°C at the highest N content, whereas an N-rich Y-sialon glass has a T_g-value around 1000°C.

Reducing the amount of sintering aids from a level corresponding to 6 wt% down to 4 wt% Y_2O_3 affects the amount of liquid phase available for sintering, with negative effects on the densification as seen above, but the β sialon grains might also be sterically hindered from growing with high aspect ratios. Less residual glassy phase will be present in the microstructures, and these factors might explain the differences in the observed mechanical properties, because the observed Vickers hardness (HV10) increased and the fracture tough-

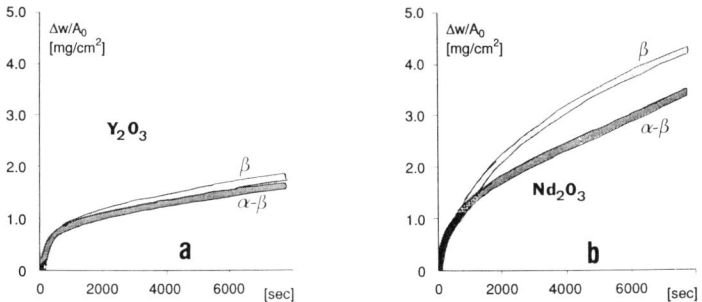

Figure 9. The oxidation rate at 1350°C of sialon ceramics with 6 wt% Y_2O_3 or 8.9 wt% Nd_2O_3 as sintering aids in (a) and (b), respectively[29].

ness decreased for both the β sialon and α–β sialon materials. Typically, the hardness increased by 100-150 and the toughness decreased by 1-1.5 MPam$^{1/2}$ 3.

In summary it can be said that replacement of yttrium oxide at an addition level corresponding to 6 wt% Y_2O_3 by the low-cost La-, Nd- or Ce-oxides preserves the overall good room-temperature properties of the sialon ceramics[13-16]. The fracture toughness is of the same magnitude as, or even better than, found for yttrium-doped sialons, but the hardness of the α–β sialon ceramic predictably decreases in all cases, as the α sialon phase content decreases in the materials. There is no reason why these good mechanical properties should not be retained to temperatures around 900°C, whereas at higher temperatures oxidation results indicate that the yttria-doped sialon has a better oxidation resistance.

3.3 Replacement of Yttria With High-Cost Rare Earth Oxides

The replacment of yttria with more expensive sintering aids in preparing sialon ceramics can only be defended if substantial advantages in processing or in the resulting properties accrue. All the high-cost rare earth oxides (Pm to Lu) have not been investigated, since similarities exist between neighbouring members in the rare earth series. The elements Sm, Dy and Yb were selected in this study to represent rare earth elements of increasing atomic numbers (Z= 62, 66 and 70). The oxides of these latter metals have previously been used to sinter Si_3N_4 ceramics (e.g. 30-32), but fairly few publications have appeared relating to sintering or heat-treatment of sialon ceramics[33-35].

For the oxygen-rich β sialons, fully dense materials were obtained with all three rare earth metal oxides replacing yttria at 1775°C. Sm_2O_3 and Dy_2O_3 give fully dense α–β sialon materials at both sintering temperatures used, 1775 and 1825°C, whereas ytterbium oxide does not give dense α–β sialon materials by pressureless sintering at either temperature, see Figure 10a. Ytterbium oxide is thus found to be a poor sintering aid when the overall nitrogen content increases. This might partly be caused by the higher viscosity of high-Z rare-earth sialon liquids (Yb) compared with the low-Z liquids (Sm,Dy) known to exist for a given level of nitrogen. Another reason for the poor sintering might be that the rate of formation of α sialon is higher in the case of Yb compared with Y, Sm or Dy, thereby diminishing the amount of available liquid phase.

Of the three high-cost rare earth oxides studied , Sm_2O_3 and Dy_2O_3 were found to be as good sintering aids as yttria for preparing β sialon ceramics at low temperatures. Especially Dy_2O_3 was found to be very effective even at fairly low temperatures, see Figure 10b. It is notable that this oxide gives 98% of theoretical density at 1500°C and fully dense materials at 1650°C.

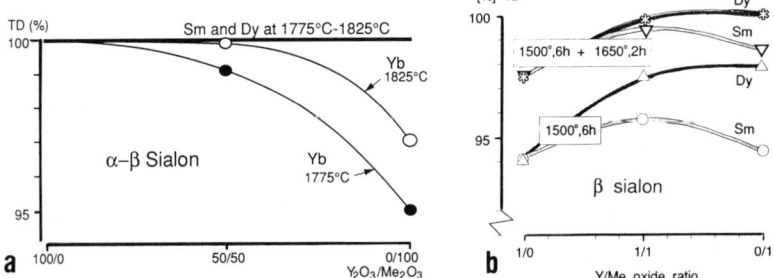

Figure 10. The sintering behaviour of Sm_2O_3, Dy_2O_3 and Yb_2O_3 when replacing Y_2O_3 in pressureless sintering of sialons at different temperatures, cf. text. The densification is given as % of theoretical density (%TD).

In the β sialon ceramics, the replacement of yttria by samarium oxide gives a finer overall grain size in the microstructure and β sialon grains of higher aspect ratio. No significant effects were noted when replacing yttria by dysprosium or ytterbium oxide. However, in the α–β sialon microstructure ytterbium oxide causes formation of significantly more of the α sialon phase compared with the other two oxides. This is in accordance with earlier studies of rare-earth metal-oxide-doped α–β sialon ceramics, where elements of different ionic radii give varying values of the $\alpha/(\alpha+\beta)$ phase ratio[36].

The α–β sialon ceramics were analyzed by point analysis (EDS) to establish any differences in the distribution of the metal dopants in the microstructure. None of the dopants Y, Sm, Dy or Yb enters the β sialon grains, and the analyses of the α sialon grains in the different specimens gave very similar results of the concentration of metal; namely 2.3 at%, which corresponds to x= 0.29 in $M_x(Si,Al)_{12}(O,N)_{16}$. A clear difference emerged, however, in the rare earth concentration of the glassy phase, which was 12.5 at%, 7.9 at% and 6.5 at% in the Sm, Dy and Yb doped α–β sialon, respectively.

The room-temperature mechanical properties when yttrium oxide is replaced by the rare earth oxides are summarized in Figure 11. In the case of samarium, a higher fracture toughness is seen both for the β and α–β sialon ceramics, and in the case of ytterbium a significantly higher hardness is noted for the α–β sialon ceramic. For samarium this result fits well with the evidence of the microstructure containing β sialon grains of a high aspect ratio; for ytterbium the α–β sialon microstructure contained considerably more of hard α sialon grains.

The high-temperature properties of the Sm-, Dy- or Yb-doped sialons have not been investigated yet. However, in the case of Sm_2O_3 the glass-forming region in Sm_2O_3-Al_2O_3-SiO_2 [37] is similar to the same region for Y_2O_3 [38], but Sm-sialon glasses have a significantly lower glass transition temperature than corresponding Y-sialon glasses and even lower than Nd-sialon glasses[39]. This may imply that the high-temperature properties of the Sm-doped sialons are inferior to those of Y-sialon ceramics.

The most notable effect of replacing yttria with high-cost rare earth oxides was that samaria had a clearly positive influence on the room temperature indentation fracture toughness. The highest fracture toughness value in this study, 6.3 MPam$^{1/2}$, was found with samaria. The increase in hardness for the Yb-doped α–β sialon can also be noted, but an increase in hardness is very easy to achieve in the Y-doped α–β sialon as well by only changing the overall composition slightly to obtain a higer content of α sialon (see above).

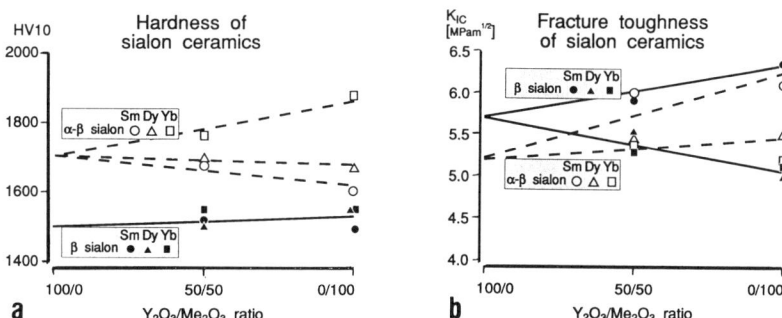

Figure 11. Vickers hardness (HV10) and indentation fracture toughness (K_{1C}) of sialon ceramics as function of the Y_2O_3:Me_2O_3 ratio for Me= Sm, Dy and Yb.

4. CONCLUSIONS

Replacement of yttria with low-cost rare earth oxides in sialon ceramics produces materials of the same good overall properties when used at temperatures below around 900°C.

Replacement of yttria with high-cost rare earth oxides gives equally good materials, but the cost of using these additives is not justified in large-scale production.

REFERENCES

1. K.H. Jack and W.I. Wilson, Nature, Phys. Science, **238**, 28 (1972).
2. K.H. Jack, J. Mater. Sci. **11**, 1135 (1976).
3. T. Ekström and M. Nygren, J. Am. Ceram. Soc. **75**, 259 (1992).
4. S. Umebayashi, K. Kishi, E. Tani and K. Kobayashi, Yogyo Kyokai Shi **92**, 35 (1984).
5. E. Tani, H. Ichinose, K. Kishi, S. Umebayashi and K. Kobayashi, Yogyo Kyokai Shi **92**, 675 (1984).
6. T. Ekström, P.-O. Käll, M. Nygren and P.-O. Olsson, J. Mater. Sci. **24**, 1853 (1989).
7. T. Ekström and P.-O. Olsson, J. Am. Ceram. Soc. **72**, 1722 (1989).
8. M.B. Trigg and K.H. Jack, J. Mater. Sci. Letters **6**, 407 (1987).
9. T. Ekström, P.-O. Olsson and M. Holmström, J. Europ. Ceram. Soc., in print.
10. S. Hampshire, H.K. Park, D.P. Thompson and K.H. Jack, Nature (London) **274**, 880 (1978).
11. D.P. Thompson, Mater. Sci. Forum **47**, 21 (1989).
12. B. Bergman, T. Ekström and A. Micski, J. Europ. Ceram. Soc. **8**, 141 (1991).
13. P.-O. Käll and T. Ekström, J. Europ. Ceram. Soc. **6**, 119 (1990).
14. E. Söderlund and T. Ekström, J. Mater. Sci. **25**, 4815 (1990).
15. T. Ekström and P.-O. Olsson, J. Mater. Sci. Letters **8**, 1067 (1989).
16. P.-O. Olsson and T. Ekström, J. Mater. Sci. **25**, 1824 (1990).
17. G.R. Anstis, P. Chantikul, B.R. Lawn and D.P. Marshall, J. Am. Ceram. Soc. **64**, 533 (1981).
18. C. Chatfield, T. Ekström and M. Mikus, J. Mater. Sci. **21**, 2297 (1986).
19. R.E. Loehman, J. Non-Cryst. Solids **42**, 433 (1980).
20. T. Ekström, Mater. Sci. Engin. **A109**, 341 (1989).
21. T. Ekström and J. Persson, J. Am. Ceram. Soc. **73**, 2834 (1990).
22. J. Persson and M. Nygren, in 11th Risö Symposium on Metallurgy and Materials Science, edited by J.J. Bentzen, J.B. Bilde-Sörensen, N. Christiansen, A. Horsewell and B. Ralph (Risö National Laboratory, Roskilde, Denmark, 1990) p. 451.
23. S. Hampshire, R.A.L. Drew and K.H. Jack, Phys. Chem. Glasses **26**, 182 (1985).
24. D.R. Messier and A. Broz, J. Am. Ceram. Soc. **65**, C-123 (1982).
25. S. Hampshire, R.A.L. Drew and K.H. Jack, J. Am. Ceram. Soc. **67**, C-46 (1984).
26. R.E. Loehman, J. Non-Cryst. Solids **56**, 123 (1983).
27. D.R. Messier, Int. J. High Tech. Ceram. **3**, 33 (1987).
28. S.M. Winder and M.H. Lewis, J. Mater. Sci. Letters **4**, 241 (1985).
29. J. Persson, T. Ekström, P.-O. Käll and M. Nygren, J. Europ. Ceram. Soc., in print.
30. T. Takahashi, Ceram. Trans. **7**, 674 (1990).
31. M.K. Cinibulk, G. Thomas and S.M. Johnson, J. Am. Ceram. Soc. **75**, 2037 (1992); **75**, 2044 (1992); **75**, 2050 (1992).
32. N. Hirosaki, A. Okada and K. Matoba, J. Am. Ceram. Soc. **71**, C-144 (1988).
33. D.P. Thompson, Br. Ceram. Proc. **45**, 1 (1989).
34. H. Mandal, D.P. Thompson and T. Ekström, Key Engin. Mater. **72-74**, 187 (1992).
35. H. Mandal, D.P. Thompson and T. Ekström, Br. Ceram. Proc. **49**, 97 (1992).
36. T. Ekström, K. Jansson, P.-O. Olsson and J. Persson, J. Europ. Ceram. Soc. **8**, 3 (1991).
37. E.M. Erbe and D.E. Day, J. Am. Ceram. Soc. **73**, 2708 (1990).
38. R.A.L. Drew, S. Hampshire and K.H. Jack, Proc. Br. Ceram. Soc. **31**, 119 (1981).
39. S. Hampshire, R. Flynn, J. Lonergan and A. O'Riordan, in Ceramic Materials and Components for Engines, edited by R. Carlsson, T. Johansson and L. Kahlman (Elsevier Applied Science Publishers, London, 1991) p. 157.

THE INFLUENCE OF POWDERS AND PROCESSING METHODS ON MICROSTRUCTURE AND PROPERTIES OF DENSE SILICON NITRIDE

G. WOETTING, H. FEUER AND E. GUGEL
CFI - Cremer Forschungsinstitut, 8633 Roedental, FRG

ABSTRACT

Experiments with commercially available as well as with specially prepared Si_3N_4-powders were performed in order to estimate the extent of influence of the powder properties and the homogeneity of the additive distribution on the sintering behaviour, the microstructure and on mechanical properties of the dense sintered materials.

The quality of the processing steps was examined by dilatometry which also proved to be a useful tool for deducing appropriate sintering conditions for different Si_3N_4-powder/additive systems.

The effects of compositional variations and processing on microstructure and mechanical properties are discussed with respect to strength, reliability (Weibull analysis) and fracture toughness. Some important factors to be considered for the production of high-performance, reliable silicon nitride materials and components are pointed out.

The careful analysis of possible influences on the final ceramic resulted in the production of a high-quality Si_3N_4-material which is used for engine valves, bearing balls, etc.

1. Introduction

For technical applications, Si_3N_4-materials have gained much importance during the last years and are standing before a broad industrial use. One reason for this is the high strength level and the reliability of Si_3N_4-ceramics which have been achieved by using highly advanced raw materials.

A tremendous lot of work has been done to characterize Si_3N_4-powders and to relate their properties to the sintering behaviour, the microstructural development and to resulting properties of the densified material [1 - 4]. The evaluation of these results, however, does not yet result in a consistent picture. Reasons therefore are thought to be various different processing routes and sintering-additives used. Besides this, often the powder characteristics before the processing of the sintering composition are discussed, neglecting changes due to this step.

A very similar situation exists with the microstructural development during sintering and possibilities to influence it [4 - 9]. This question gained increasing importance by finding the beneficial effect of bimodal microstructures on mechanical properties, especially on the fracture toughness and the R-curve behaviour, respectively [4, 10 - 14]. Impressive success is achieved with this so-called "self-" or "in-situ-reinforcement", leading to toughness-values comparable to whisker-reinforced Si_3N_4-composites [15 - 17]. These results are desired very much as they avoid a lot of problems associated with the use of whiskers.

Concerning properties at high temperatures, the lack of knowledge on influences of powder properties and microstructure is still great, though there are reasonable concepts [18 - 20]. This also seems to be due to the high experimental effort to characterize Si_3N_4-materials in this respect. Simplified methods to estimate influences and properties are therefore highly recommended.

This contribution discusses influences of powder characteristics on sintering, microstructure and properties, based on the state-of-the-art in literature as well as on own work. Central points are an extensive analysis of such relations by means of statistical methods [20, 21] and possibilities and mechanisms to tailor the microstructure of dense Si_3N_4. Besides this, technological aspects are outlined, like the influence of processing or composition of the Si_3N_4-additive-mixture on the topics mentioned above. It will be demonstrated that the application of the concepts described allow the fabrication of highly reliable Si_3N_4-components for different applications in an economic way.

2. Si_3N_4-powders

2.1 Si_3N_4-powder properties and their influence on sintering

Most Si_3N_4-materials start with Si_3N_4-powders (the exception is SRBSN) and the choice of the powder grade already determines the processing route applicable and the property level achievable to an essential amount. Thus, the choice of the Si_3N_4-powder should be oriented by the application of the Si_3N_4-material as well as by the processing to be applied.

Criteria to be considered for the selection of a Si_3N_4-powder:
- <u>Granulometric characteristics</u>: fineness, coarse residues, particle size distribution, particle morphology/uniformity
- <u>Chemical characteristics</u> (homogeneous): oxygen, metallic impurities, further anions
- <u>Chemical characteristics</u> (inhomogeneous/particles): Fe_xO_y, $FeSi_x$, superalloy particles, WC, organics like hairs, grinding debris
- <u>Mineral phases</u>: α-, ß-, amorphous Si_3N_4, free Si, SiAlON-phases, polytypes, others
- <u>Processing characteristics</u>: green densification for pressing, dispersibility for slip-casting/injection-moulding, maximum solid content of slips or polymer-ceramic blends, rate of green body formation in slip-casting
- <u>Sintering characteristics</u>: densification kinetics, microstructure
- <u>Economy</u>: consistency, availability, price

Concerning their properties, today's available Si_3N_4-powders cover a rather broad spectra. To evaluate the influence of the main characteristics on sintering and materials' properties a comparative test was performed with various Si_3N_4-powders commercially available as well as laboratory grades, applying constant processing conditions and additives. The variation range of properties of the Si_3N_4-powders of this test are listed in Tab. I. Results were analyzed by means of statistical methods [21, 22], revealing characteristics responsible for the sintering behaviour, the strength at room-temperature and at high temperatures.

Tab. I: Variation range of Si_3N_4-powder properties evaluated

Property	Variation range
Specific surface area	5 - 20 m^2/g
Oxygen content	0.8 - 2.8 wt.%
Carbon content	0.04 - 0.8 wt.%
Impurity content (Fe, Al, Ca ...)	0.005 - 0.3 wt.%
α-Si_3N_4	2 - 96 %

Concerning the first point, the sintering behaviour, results shown in Fig. 1 demonstrate different influences within directly nitrided (DN) and gas-phase (GP) processed Si_3N_4-powders. For the first ones, at lower sintering temperatures the impurity concentration and the specific surface area are of greatest importance for the sintering density. With increasing sintering temperature, however, the oxygen-content gets determining. Over the whole temperature range considered, the degree of description of the result "sintering density" by the powder characteristics and the polynom evaluated, respectively, is about 80 %, indicating a satisfactory degree.

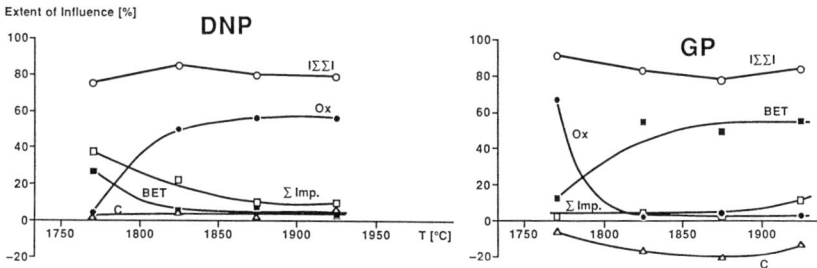

Fig. 1: Influence of powder properties on sintering of DN- and GP-Si_3N_4

The corresponding results of the GP-powders show the oxygen content of greatest importance at lower temperatures, decreasing with raising temperature in favour of the specific surface area.

These differences can be explained by the fact that DN-powders contain most of the oxygen content within the particles while GP-powders often show an oxygen enrichment on the particle surface [1, 2]. The latter can participate in the liquid phase formation, supporting the densification already at rather low temperatures while the oxygen within the particles is available not before they get dissolved in the liquid phase. Thus, the surface oxygen content reveals as a very important characteristic for the sintering behaviour.

2.2 Effects of processing on Si_3N_4-powder properties

The tasks of the processing of the starting composition are:
- Destruction of coarse particles and unfavourable particle morphologies
- Destruction of agglomerates
- Homogeneous distribution of the sintering aids

- Homogeneous distribution of the processing aids
- Yield a confectionated form for further processing respectively moulding:
  * granulate for pressing, injection-moulding
  * high solid loaded suspension for slip-casting or spray-drying

Normally, however, it is not possible to achieve:
- Improvement of the purity (more often purity is reduced due to grinding debris)
- Removement of particulate impurities
- Reduction of oxygen content

It is obvious that the intensity of grinding or comminution determines the extent of change of powder properties. Fig. 2 shows changes in the specific surface area respectively the oxygen content of different powders applying constant grinding conditions (planetary mill, 1,000 min^{-1}, 10 h, propanol; solid : liquid : Si_3N_4-balls (5 mm) = 1 : 2.5 : 2.5). Thus, less fine powders and such low in oxygen are altered much more than fine powders with medium to higher oxygen content.

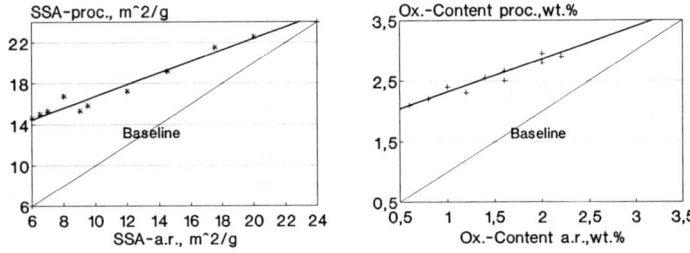

Fig. 2: Change in specific surface area (left) and oxygen content (right) of various Si_3N_4-powders by processing

Rising specific surface areas improve the sintering behaviour, however, they make moulding more difficult. The increase in oxygen has to be considered with respect to the chemical composition of the grain-boundary phase formed. Generally, rising oxygen contents also improve the sintering behaviour, but only up to a certain amount. Above a critical oxygen content, the formation of (transient) Si_2N_2O may occur, which reduces the densification as well as properties [23].

Besides the change of the powder properties, the second important point is the homogeneity of the additive distribution achieved. The process normally used is grinding, however, chemical precipitation methods are discussed increasingly in literature [24, 25].

Concerning mechanical mixing, the rule can be given "the longer the better". However, one has to keep in mind the simultaneous change in powder properties as shown before. As can be seen in Fig. 3, with suitable conditions, homogeneous distributions can be reached.

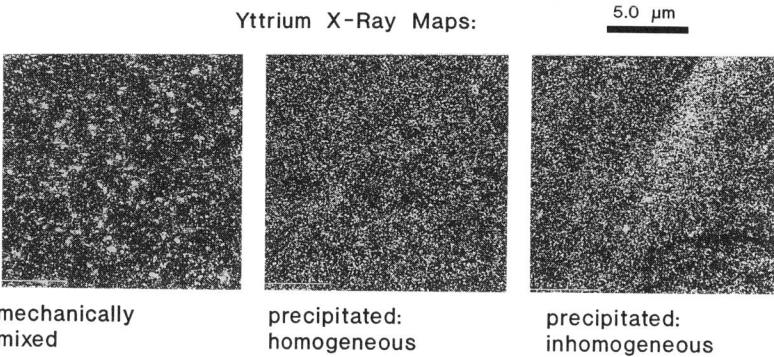

Fig. 3: Additive distribution in differently processed Si_3N_4-compositions:
left: mechanically mixed; middle/right: precipitated homogeneous/inhomogeneous

An even better distribution may be achieved with chemical methods. Besides a nearly ideal homogeneity, however, coarse residues and large inhomogeneities may remain (Fig. 3). Additionally, the precipitation of salts or hydroxides affords a calcination step which again may create new hard agglomerates. This calcined material, of course, does not contain any processing aids, thus a redispersion for deagglomeration and mixing with the processing aids becomes necessary, meaning an additional large effort.

Probably, the application of mechanical mixing will gradually decrease in the near future due to increasing offers of such "ready-to-use" compositions by skilled powder producers. For mechanical mixing, however, one should be aware to use fine and well-dispersable additive grades to avoid the necessity of very intensive processing which may alter the Si_3N_4-powder properties too much, as discussed above.

Attempts to start with crude Si_3N_4-powders and to improve them by intensive grinding [26] is often associated with secondary effects (e. g. oxygen increase) and with unfavourable material properties, e. g. low strength, due to particular impurities and/or grinding debris.

3. Sintering behaviour and sintering mechanisms

It is generally known that without sintering aids only insufficient densification of Si_3N_4 occurs associated with unfavourable properties. One exception of this is capsule-HIP of additive-free Si_3N_4-compacts [27]. This process, however, seems to be still far from a technical use.

The addition of sintering aids, e. g. Y_2O_3 + Al_2O_3, results in a markedly improved densification by initiating a liquid-phase sintering process. Recording the densification by dilatometer results in curves like the ones in Fig. 4.

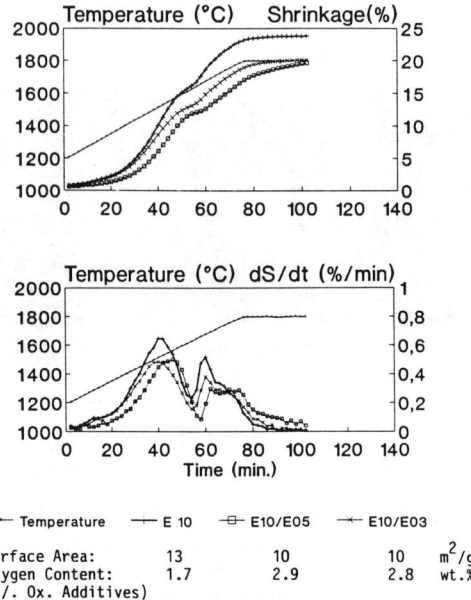

Fig. 4: Dilatometer curves of Si_3N_4-imid powders with different grain-size distribution (sintering additives: 5 wt.% Y_2O_3 + 1 wt.% Al_2O_3)

These curves often show slightly unsteady slopes, which by differentiation divide into two or more densification maxima. According to Kingery's liquid-phase sintering theory, comprising the three mechanisms rearrangement, solution-diffusion-reprecipitation and coalescence, it could be shown that the first maximum results of pure rearrangement of particles due to capillary forces of the liquid phase formed without any considerable contribution of solution [28]. Further experiments showed that temperature and extent of this rearrangement step markedly depend on the amount and properties of the liquid phase as well as on the Si_3N_4-powder characteristics. It is self-evident that coarse grains and/or grains with unfavourable morphologies, like needles, hinder this densification. Moreover, such constituents may lead to bigger defects, as fine-grained areas densify while the coarse grains resist, opening inner cracks/holes which are difficult to remove during sintering and may act as strength-determining defects.

The second (and occasionally further) densification maximum is related to solution-diffusion-reprecipitation processes. The intensity and extent again depends on the amount and properties of the liquid phase and the granular characteristics of the Si_3N_4-powder respectively of the green body. The broader and the more distinct it is, the more inhomogeneous is the powder mixture and the broader is the particle size distribution. This is demonstrated by the examples in Fig. 4, in which the densification behaviour of compositions of pure UBE E10 + 5 Y_2O_3 + 1 Al_2O_3 (wt.%) with mixtures of the coarser grades E5 resp. E3, each of 50 %, are compared. Though, after processing the characteristics are nearly similar, remaining coarser particles or agglomerates retard densification and prevent the achievement of full density. This is reflected in the lower intensity and the delay in the kinetic curve, sometimes also in further small peaks of the densification curves. Such occurrences may also be a result of insufficient deagglomeration or homogenization, giving important hints for process control.

Thus, dilatometry is a valuable tool to characterize the sintering behaviour of Si_3N_4-compositions, to check for the homogeneity of the mixture prepared and to fix the maximum sintering temperature necessary for complete densification. For completeness it should be mentioned that the third densification mechanism, the coalescence, does not contribute to densification and therefore cannot be seen in the dilatometric curve.

4. Microstructural development during sintering

4.1 Relations between α-ß-conversion and densification

For a rather long time it was accepted that the α-ß-conversion of Si_3N_4 occurs simultaneously with densification and supports it [29 - 31]. However, in recent studies it has been demonstrated that there is no relation between densification and the α-ß-transformation [32, 33]. In order to shed light upon this controversy, we performed sintering experiments with improved Si_3N_4-powders of high surface oxygen concentration. Two special powders were prepared, one with a high α-content ($\geq$ 95 % α/(α+ß)) and one with a high ß-content ($\geq$ 98 % ß/(α+ß)). All other properties of the two powders were almost identical and both were sintered with 5 wt.% Y_2O_3 and 1 wt.% Al_2O_3. The densification and the transformation behaviour of the α-material is shown in Fig. 5.

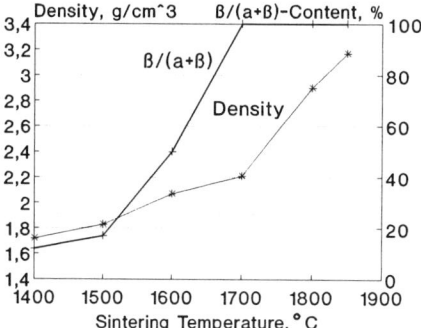

Fig. 5: Sintering density and α-ß-transformation as a function of temperature for GP-Si_3N_4 + 5 wt.% Y_2O_3 + 1 wt.% Al_2O_3

The α-phase has already transformed completely to ß at temperatures below 1,700°C. However, the density at this temperature is still rather low. A clearly defined increase of the density does not occur before 1,700°C. Thus, in this case, the densification is not related to the α-ß-transformation.

Comparing this densification with the one of the ß-material reveals an almost identical behaviour. This proves that the effect of the α-phase on the densification is small and that the densification mechanisms of α- and ß-powders are basically the same. Primarily, the particle size distribution, the oxygen content and the type and amount of additives determine the densification. Therefore, most former reports describing a lower sintering activity of ß-based Si_3N_4-powders seem to be the result of different characteristics of the powders used.

The α-ß-conversion rate depends on the particle size and the particle size distribution: coarse-grained powders and those with a broad particle size distribution transform sluggishly, whereas fine-grained powders transform with a higher rate. The temperature of the transformation is primarily determined by the type and amount of additives.

The microstructures of the two sintered powders, however, are different. The α-based material consists of highly elongated rod-like grains, whereas the ß-based material developed a microstructure with rather globular grains (Fig. 6.). These observed microstructural differences are in agreement with published results [5 -7, 29, 30].

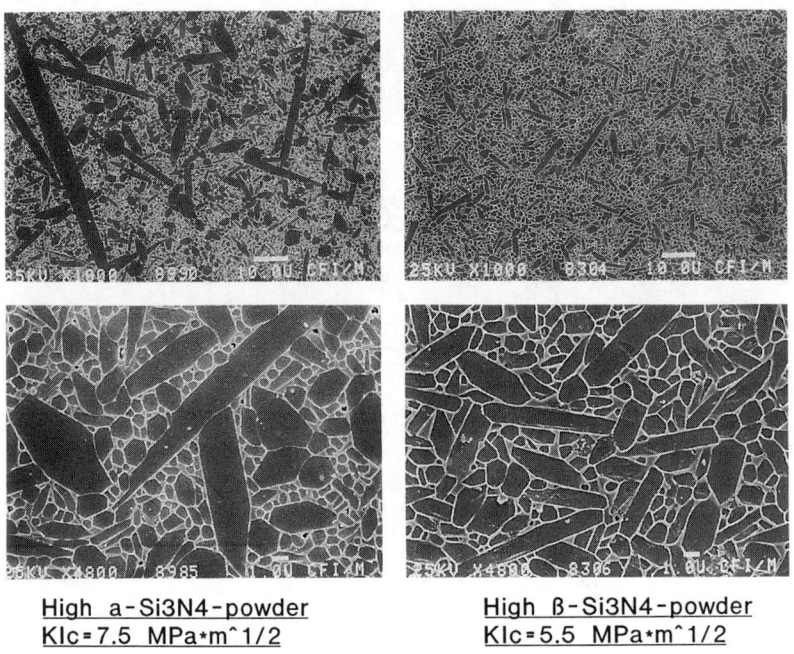

High a-Si3N4-powder   High ß-Si3N4-powder
KIc=7.5 MPa*m^1/2   KIc=5.5 MPa*m^1/2

Fig. 6: Microstructures of Si_3N_4, prepared of high α- (left) resp. high ß-powder (right)

These observations pose the question: Why does ß-Si_3N_4 in the starting powder behave different to ß-Si_3N_4 formed by the α-ß-conversion in α-Si_3N_4 based compacts (before densification occurs)?

To analyze reasons for this, the α- and the ß-based material were both heat-treated at temperatures slightly below and above the α-ß-conversion (Fig. 7). In the α-based material tiny needles form out of clusters of nearly equiaxed former α-grains. The morphology of these metastable needles seems to be determined by homogeneous nucleation and spontaneous growth out of locally supersaturated solutions with respect to ß-Si_3N_4. In the ß-based material, however, the formation of the rather equiaxed globular grains continues.

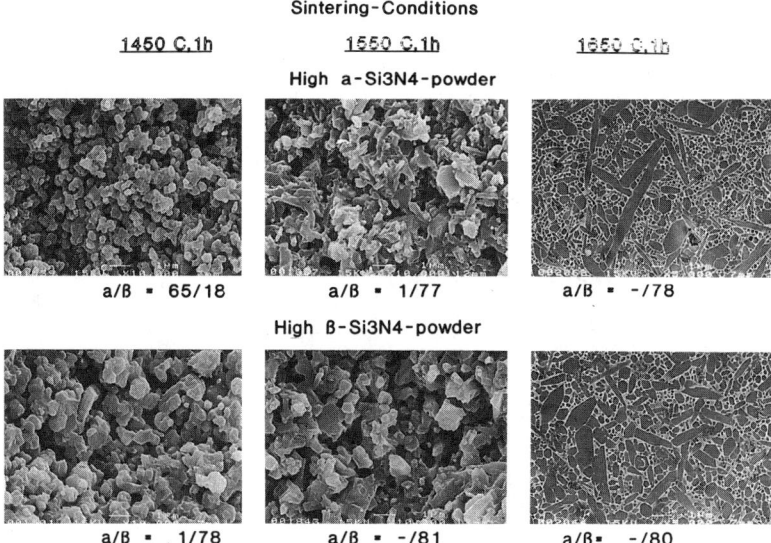

Fig. 7: Microstructural development in Si_3N_4, prepared of high α- resp. high ß-powder

Thus, in the two materials completely different grain morphologies are present at the beginning of sintering, and it is anticipated that these morphologies determine the further grain growth during sintering as demonstrated in the SEM-pictures (Fig. 7). This finally results in the differences shown before (Fig. 6).

4.2 Influences on the microstructural development

One possibility to influence the microstructure, meaning the α/ß-phase ratio, is already given above. Accepting this explanation, there still remains open the role of ß-particles present in highly α-based-Si_3N_4-powders and ß-Si_3N_4-seeds, respectively. Concerning the latter, a series of experiments was performed by different research groups, either to understand the role of ß-particles in α-Si_3N_4-powders or to develop tools to tailor special microstructures, e. g. the so-called bimodal structures or a "self-reinforcement" [8 - 11, 34, 35].

The microstructural features of Si_3N_4-material can be enhanced by plasma-etching of polished sections with CF_4+O_2 gas [36, 37]. The crystalline Si_3N_4 is etched whereas the glassy phases remain nearly unchanged (e. g. Fig. 6). It appears that the etching of the crystalline phases is selective with respect to chemical composition and structural defects, enabling also to reveal features within the grains. By comparing samples prepared by different Si_3N_4-powders, some show nuclei-effects, meaning the contours of small grains inside larger ones and some not. This leads to the conclusion, that ß-particles present in the starting powder only act as nuclei if they have a size greater than the mean size of the ß-grains resulting from the α-ß-conversion of the starting powder at the beginning of the solution-reprecipitation stage. Otherwise, they get dissolved in the liquid phase due to the energy potential between small and big grains.

To verify this hypothesis, powders with high α- or ß-phase content were mixed. The high ß-grades comprised a fine, a coarse and a very crude pure ß-Si_3N_4-powder, the latter prepared by an SHS-process. SEM-micrographs of the sintered samples are shown in Fig. 8, the base material, prepared of pure α-Si_3N_4 is already enclosed in Fig. 6.

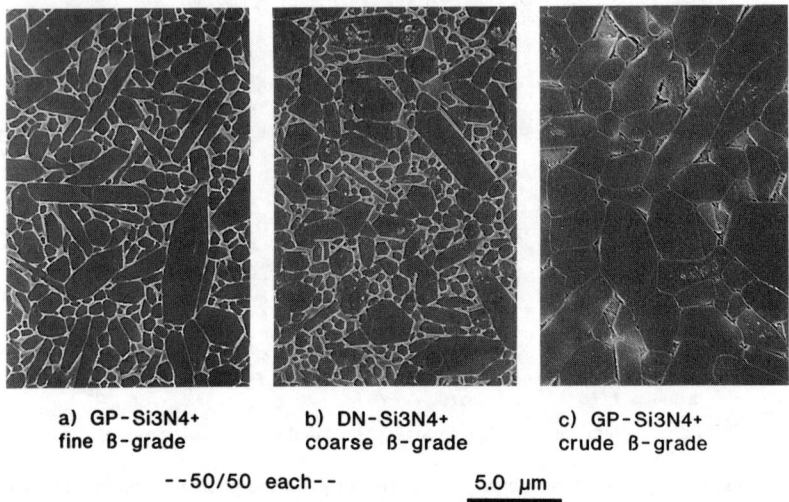

a) GP-Si3N4+ fine ß-grade

b) DN-Si3N4+ coarse ß-grade

c) GP-Si3N4+ crude ß-grade

--50/50 each--   5.0 μm

Fig. 8: Microstructures of Si_3N_4, prepared of mixtures of high α-phase powder and different ß-phase grades (50/50 each)

From Fig. 6, the base-material, it can be seen that very few nuclei or contours are present. In Fig. 8a only a few more nuclei can be seen, in no case about 50 %, corresponding to the mixing ratio. In Fig. 8b, however, in most large grains nuclei are present, proving the above hypothesis concerning the size effect. The same is true with Fig. 8c, where primarily the coarse ß-particles grew to big sizes.

Additionally, it can be seen from these pictures that only a small number of grains with a nuclei contour developed to rod-like grains with high aspect-ratios. The bigger number formed coarse, more equiaxed grains similar to those found in the sample prepared of pure ß-Si_3N_4 (Fig. 6). Thus, besides the size, the morphology of the ß-nuclei-particle seems to be of importance, and only the few ones with a needle-like morphology and suitable crystallographic orientation are obviously able to favour the formation of the rod-like grains in the developing microstructure.

Therefore, in order to tailor a microstructure, the addition of needle-like ß-particles of sizes around 1 μm is necessary which, according to our knowledge, are not available commercially at the moment. It is also questionable if such particles would survive the processing step of homogenization and adding the sintering aids. The more easier and also common route is to use a fine-grained, high α-Si_3N_4-powder with a high α-ß-conversion rate, where the ß-nuclei form itself during sintering in a sufficient amount and a favourable morphology, as demonstrated.

This explanation, however, still leaves open the question how bimodal microstructures with the so-called "self-reinforcement" can be achieved. Among possible routes, we propose the following way:

In a first approximation it can be stated that the two most important Si_3N_4-powder routes, the directly-nitrided Si_3N_4 (DN) and the low- or high-temperature gas-phase processed Si_3N_4 (GP), result in different particle size distributions. The distribution of GP-powders is relatively narrow, compared with the broad distribution of DN-powders.

The broad particle size distribution results in a slightly retarded α-ß-conversion, probably due to the presence of bigger grains. This also seems to be the reason for a broad grain size distribution in the sintered material (Fig. 9). This distribution remains even after extended sintering times, as the energetical difference between bigger and smaller grains gets diminished, but is still active. This is in agreement with data in literature [5, 6] where a constant normalized grain size distribution was found, independent of the sintering temperature. This indicates that normal grain growth takes place. The K_{IC}-value of Si_3N_4-ceramics with such previously favoured microstructures is about $7\ MPa \cdot m^{1/2}$. Consequently, in GP-materials a more narrow grain-size distribution remains, resulting in K_{IC}-values of about $6\ MPa \cdot m^{1/2}$.

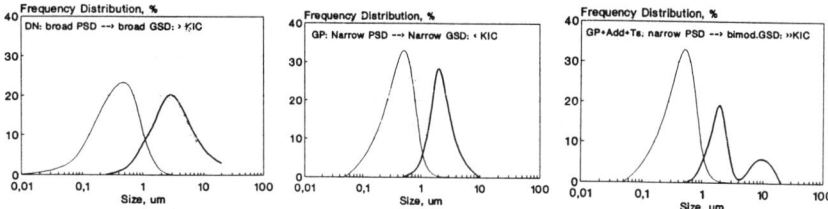

Fig. 9: Schematic representation of relations between the powder particle size distribution (PSD) and the grain-size distribution (GSD) in sintered Si_3N_4

With further heat-treatment, the DN-based microstructure coarsens uniformly, thereby hindering abnormal rod-like grain-growth. This may result even in a reduction of K_{IC}. In the GP-sample, however, by local activation individual grains grow to much bigger size, comparable to exaggerated grain-growth in SiC, favoured by special additive combinations (see e. g. Fig. 6a). These grains give rise for a marked increase in K_{IC}, values up to $12\ MPa \cdot m^{1/2}$ are reported in literature for such microstructures [38]. We could verify with this concept values up to about $10\ MPa \cdot m^{1/2}$ [17].

This concept therefore affords a high-grade, suitable Si_3N_4-powder, special sintering additives and a very careful control of the sintering conditions, however, there is no need of any seeds.

5. Relations between microstructure and mechanical properties

Applying the statistical evaluation method already discussed, a rather bad description of experimental results by the polynom describing relations between Si_3N_4-powder properties and strength at room-temperature. A marked improvement, however, was achieved by the the addition of the sintering density to the polynomial variables.

Thus, features favouring the densification like the oxygen-content and the specific surface area, the sintering density itself and the α-Si_3N_4-phase content determine the resulting strength. However, the degree of description remains below 50 % indicating the effects of further parameters on strength, which are supposed to be microstructural characteristics. All these features, however, are not directly influencing the strength, but primarilly the fracture toughness. The basic relation behind this is the well-known Griffith equation:

$$\sigma = Y \cdot K_{IC}/\sqrt{c}$$

which for a constant defect size c proposes an increasing strength with increasing K_{IC}. Ways to reach this have just been discussed.

The alternative is to reduce c, which for $K_{IC} = 7$ MPa $\cdot$ m$^{1/2}$ and $\sigma = 1,000$ MPa is in the order of 40 μm. Such big defects still determine the strength of "high-performance" Si_3N_4! Various evaluations have shown that defects of such size result from:
- machining of samples
- moulding-defects
- agglomerates/hard granulates in the processed material
- additive-rich areas resulting from insufficient mixing
- metallic inclusions and reaction-zones, resulting from impure starting powder, grinding debris, etc.

With the exception of the last point, all other causes may be eliminated or reduced by an optimized proper processing. As an example, results of a set of bending bars which showed a relatively low mean strength value are shown in Fig. 10 (curve A). Nearly 20 % of the strength determining defects were machining defects. By excluding these samples from the calculations the Weibull-modulus rises from 11 to 22 (Fig. 10, B).

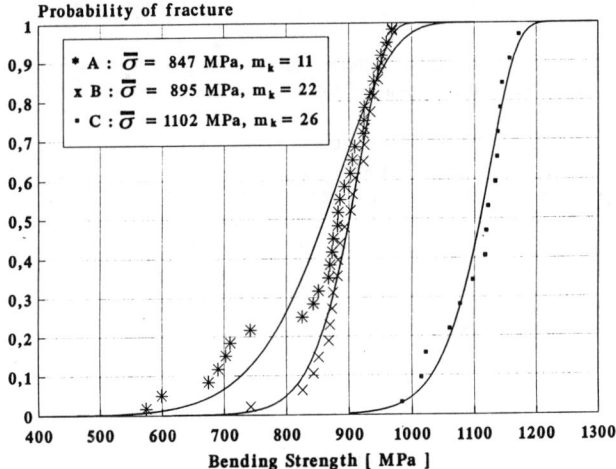

Fig. 10: Weibull-analysis of Si_3N_4-materials: A) basic material; B) corrected for non-specific defects; C) for an improved grade

Further serious strength-determining defects were found to be metallic inclusions and their reaction zones, respectively. Thus, by changing the Si_3N_4-powder source and repeating the analysis, using the same processing, but a more careful machining of the samples, a markedly improved mean strength and a Weibull-modulus of 26 resulted, as shown in Fig. 10 (curve C).

This points out the importance of pure Si_3N_4-powders with respect to particular metallic impurities to produce good and reliable materials even without the application of cost-effective equipment like clean-rooms etc.

Refering again to the statistical analysis cited above, the short-term strength at temperatures above $1,100°C$ primarily depends on the oxygen- and impurity-content, especially Fe and Ca, as well as the absolute ß-phase content in the sintered material. However, the degree of description of the target value "high-temperature strength" by Si_3N_4-powder characteristics is lower than 60 %, indicating a great influence of further factors. These are supposed to be grain-size and -morphology effects as well as the amount and state of secondary phases.

Concerning the overall oxygen content, this can only be reduced by using very active starting powders low in oxygen, water-free processing and non-oxide sintering-additives. Using this concept, materials are obtained which show up to 80 % of the room-temperature strength of about 800 MPa at $1,350°C$. Additionally, with this concept oxidation resistance as well as creep resistance are improved markedly. It seems that there is also a positive effect of rod-like grains on the creep resistance, indicating a beneficial effect of bimodal microstructures with this respect.

6. Summary and conclusions

From recent studies and from our own experiments we draw the following conclusions concerning factors influencing the sintering behaviour, the resulting microstructure and properties of Si_3N_4:
- Si_3N_4-powder: Characteristics not changed by processing are important (particulates). Surface oxygen enrichment promotes densification.
- Processing: Introduction of too much oxygen should be avoided (Si_2N_2O). Improved additive homogeneity can be achieved by chemical methods.
- Sintering: No direct relation exists between α-ß-transformation and densification of highly active Si_3N_4-powders.
- Microstructure: ß-Si_3N_4 seed nuclei must have a prticular size and morphology (needle-like).
Bimodal microstructures can be tailored without the use of added seeds.
- Properties: R.T.-strength is improved by rising K_{IC} and/or reduced defect size. H.T.-strength is determined by the purity of the Si_3N_4-powder, state and amount of secondary phases as well as grain-size resp. morphology.

Thus, it may be summarized that the use of high purity, uniform, α-rich Si_3N_4-powders give better prerequisites to prepare high-performance, dense Si_3N_4-materials than the attempt to improve "cheap" Si_3N_4-powders by intensive processing and/or to influence microstructures by additions of Si_3N_4-seeds or other nuclei.

Based on these concepts, we developed Si_3N_4-materials that are used as engine valves and as bearing components, e. g. balls for Hertzian pressure up to 2,900 MPa. Meanwhile, the high performance of the valve material is demonstrated by 80,000 trouble-free kilometers of our company car (Daimler-Benz 300, 24 valves).

References

1. G. Franz, G. Schwier, in Raw Materials for New Technologies, edited by M. Kürsten (E. Schweizerbart'sche Verlagsbuchhandlung, Stuttgart, 1990), p. 139.
2. M. Peuckert, P. Greil, J. Mater. Sci. **22**, 3717-3720 (1987).

3. M. Mitomo, N. Yang, Y. Kishi, Y. Bando, J. Mater. Sci. **23**, 3413-3419 (1988).
4. G. Ziegler, J. Heinrich, G. Wötting, J. Mater. Sci. **22**, 3041-3086 (1987).
5. M. Mitomo, S. Uenosono, J. Am. Ceram. Soc. **75**, 103-108 (1992).
6. M. Mitomo, M. Tsutsumi, H. Tanaka, J. Am. Ceram. Soc. **73**, 2441-2445 (1990).
7. V. K. Sarin, Mater. Sci. Eng. **A105/106**, 151-159 (1988).
8. M. Herrmann, S. Keßler, Ch. Schubert, presented at the 2nd ECerS Conf., Augsburg, 1991; to be published in Proceedings.
9. M. J. Hoffmann, W. Dreßler, G. Petzow, presented at the 4th Symposium on Ceramic Materials & Components for Engines, Göteborg, Sweden, 1991; to be published in Proceedings.
10. M. Mitomo, in MRS Int'l. Mtg. on Adv. Mats. **5** (Mater. Res. Soc., 1989), p. 69.
11. Y. Tajima, K. Urashima, M. Watanabe, Y. Matsuo, Ceram. Transactions 1/Proc. 1st Int. Conf. "Cer. Powder Proc. Sci.", Orlando, 1987.
12. P. Becher, J. Am. Ceram. Soc. **74**, p. 255 (1991).
13. Ch.-W. Li, J. Yamanis, Ceram. Eng. Sci. Proc. **10**, 632-645 (1989).
14. A. Okada, N. Hirosaki, J. Mater. Sci. **25**, p. 1656 (1990).
15. T. Kawashima et al., J. Ceram. Soc. Jap. (Inter. Ed.) **90**, p. 310 (1991).
16. S. R. Choi, J. A. Salem, J. Mater. Sci. **27**, p. 1491 (1992).
17. G. Wötting, E. Gugel, G. Schwier, H. Lange, cfi/Ber. DKG **69**, 88-90 (1992).
18. P. Greil, Sci. of Ceramics **14**, 645-651 (1987).
19. M. H. Lewis, G. Leng-Ward, C. Jasper, Ceram. Transactions 1/Proc. 1st Int. Conf. "Cer. Powder Proc. Sci.", Orlando, 1987.
20. W. Braue, G. Wötting, G. Ziegler, Proc. Int. Mater. Symp., Berkeley (1986).
21. G. Franz, B. Laubach, U. Wickel, G. Wötting, E. Gugel, in Ceramic Materials and Components for Engines, edited by V. J. Tennery, Am. Ceram. Soc., p. 1 (1988).
22. G. Nietfeld, U. Wickel, G. Wötting, in Ceramic Powder Processing Science II, Berchtesgaden (1988).
23. R. de Jong, R. A. McCauley, in Proc. 1st ECerS, Maastricht, 1150-1154 (1989).
24. J.-S. Kim, H. Schubert, G. Petzow, J. Europ. Ceram. Soc. **5**, 311-319 (1989).
25. F. Riley, presented at the 2nd ECerS Conf., Augsburg, 1991; to be published in Proceedings.
26. T. Quadir, R. W. Rice, J. C. Chakraverty, J. A. Breindel, C. CM. Wu, Ceram. Eng. Sci. Proc. **12**, 1952-1957 (1991).
27. I. Tanaka, G. Pezotti, T. Okamoto, Y. Miyamoto, M. Koizumi, J. Am. Ceram. Soc. **72**, p. 1656 (1989).
28. G. Wötting, G. Peitzsch, H. Hausner, Science of Sintering **17**, 87-95 (1985).
29. J. I. Iskoe, F. F. Lange, Ceramic Microstructures **76**, 669-678 (1977).
30. H. Knoch, G. E. Gazza, Ceramurg. Int. **6**, 51-56 (1980).
31. S. Hampshire, K. H. Jack, Special Ceramics **7**, 37-49 (1981).
32. O. Abe, J. Mater. Sci. **25**, 4018-4026 (1990).
33. D. Sutter, G. S. Fischman, J. Am. Ceram. Soc. **75**, p. 1063 (1992).
34. G. Petzow, M. J. Hoffmann, to be publ. in Proc. of Recrystallization '92, San Sebastian, Spain, 1992, Trans Tech. Pub., CH, 1992.
35. D. E. Wittmer, et al., Ceram. Eng. Sci. Proc. **13**, 907-917 (1992).
36. U. Täffner, M. J. Hoffmann, M. Krämer, Pract. Met. **27**, p. 385 (1990).
37. M. Mitomo, Y.-I. Sato, N. Ayuzawa, J. Am. Ceram. Soc. **74**, 856-858 (1991).
38. Annon., Note in "Advanced Ceramics Report", August 1992.

The Influence of Microstructure on the Mechanical Behavior of Silicon Nitride Ceramics

P. F. Becher, H. T. Lin, and S. L. Hwang
Metals and Ceramics Division
Oak Ridge National Laboratory
Oak Ridge, TN

M. J. Hoffmann
Max-Planck-Institut für Metallforschung
Powder Metallurgy Laboratory
Stuttgart, Germany

I-Wei Chen
University of Michigan
Department of Materials Science
Ann Arbor, MI

BACKGROUND

Mechanisms such as crack bridging (with and without interfacial friction) and pullout that operate in the crack tip wake, Figure 1, can significantly toughen ceramics and greatly improve their overall mechanical performance. The relative contributions of each of these crack wake mechanisms are dependent upon the interface properties, thermo-mechanical properties of matrix and reinforcing phases, and the amount, size, and geometry of the reinforcing phase. The degree of toughening achieved depends upon the energy dissipated by each mechanism as follows:

$$K = [E_c (\Delta J^m + \Delta J^{cb})]^{1/2} \qquad (1)$$

where K is the fracture toughness of the composite system, E_c is its Young's modulus, ΔJ^m is the energy dissipated by crack extension in the matrix and ΔJ^{cb} is the energy dissipated by one of the crack wake bridging processes.[1]

Debonding of the interface between the matrix and the reinforcing phase is, of course, a necessary condition to achieve the toughening effects. One way to describe this condition is that the energy required for interfacial debonding γ_i must be less than the fracture energy of the reinforcing phase Γ_r. However, this does not require that the debond energy or the stress to debond the interface equal or approach zero. In the SiC whisker reinforced alumina system, the radial compressive stresses imposed on the whisker-matrix interfaces due to the thermal expansion mismatch is quite high. This will most likely preclude a very low value of either the interfacial debonding or pullout shear resistance. However, this system exhibits toughening by pullout of the reinforcement. Therefore, one does need to be able to tailor the interface to promote pullout (and debonding).

Note that here we have used the debonding relationship of Budiansky et al.[2]:

$$\gamma_r/\gamma_i = Al_{db}/d \qquad (2)$$

where l_{db} is the length of the debonded interface, γ_r/γ_i is $\propto \sigma_f^r/\tau_{db}$. Additionally, σ_f^r, d, and γ_r, are the tensile strength, diameter, and fracture energy of the reinforcing phase, respectively, γ_i and τ_{db} are the debonding energy of and the shear stress to debond the interface, respectively, and A is a constant with a value of two to three. For a constant γ_r/γ_i ratio (e.g., consistent with that for a composite where only the reinforcing phase dimensions are changed), the amount of debonding increases as the diameter of the reinforcing phase increases. For each mechanism, the toughening contribution increases as the diameter of the reinforcing phase increases (this assumes that the matrix, interface, and reinforcing phase characteristics do not change when the size of the reinforcing phase is changed).

There are a number of examples that reveal that such crack wake processes occur in monolithic ceramics including alumina[3] and silicon nitride[4], as well as in ceramics reinforced with whisker[1] or platelet[5] second phases. Therefore, it is consistent with the theme of this symposium to consider how the microstructure of silicon nitride ceramics might influence the fracture resistance, as well as the other mechanical properties.

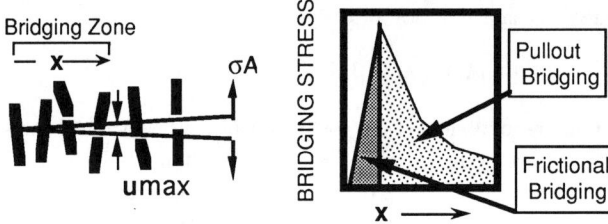

Figure 1. Crack wake processes can increase the fracture resistance by imposing bridging stresses that act to impede crack opening and propagation.

First, let us review the micromechanics descriptions appropriate to ceramics toughened by the actions of such discontinuous reinforcements (e.g., elongated or large grains) in the crack wake. The contribution of the various crack wake bridging mechanisms to the fracture resistance are as follows:[1]

1. Crack Bridging (Frictional Interface) by Elongated Grains

$$\Delta J^{fb} = [A_r(\sigma_f{}^r)^2 \, l_{db} / 9E_r] = [A_r(\sigma_f{}^r)^2 \, (\gamma r/\gamma i)/ 9E_r] \, d \qquad (3)$$

where A_r and E_r are the areal fraction and Young's modulus of the reinforcing phase, respectively.

2. Grain Rotation[6]

$$\Delta J^{gr} = [\pi \, A_{gr} \, \mu \, \sigma_f{}^r /2] \, d \qquad (4)$$

3. Grain Pullout

$$\Delta Jpo = [A_{po} \, (\sigma_f{}^r)^3 /12 \, E \, \tau_i] \, d \qquad (5)$$

where τ_i [$\propto \mu\sigma_r$ where τ_i is the sliding resistance of the interface, μ is the coefficient of friction, and σ_r is the radial (compression) stress imposed on the interface] and A_{gr} and A_{po} are the areal fraction of grains on the crack surface that rotate and pullout, respectively (i.e., $\propto$ volume fraction of elongated grains).

These simplified relationships reveal the basic contributing parameters. With each mechanism, apparently the size and shape of the grains can significantly affect the fracture toughness of silicon nitride ceramics provided that the grain boundary phase promotes debonding. The following concerns, however, need to be clarified. How significant is the size of the elongated grains in the toughening effect, what effect does this have on strength and strength distribution, and how might the presence of silicate-based phases at the grain boundaries coupled with grain size influence fatigue resistance? To this end, observations of the influence of microstructure on the mechanical behavior of silicon nitrides containing elongated grains are discussed in hope of stimulating research on approaches to tailor the microstructure and composition of silicon nitrides to optimize various combinations of mechanical properties.

OBSERVATIONS

Fracture Toughness
Various experimental studies show that the toughening response in silicon nitrides with elongated grains can be described in terms of crack wake mechanisms. Bridging of the crack by intact elongated grains is often observed behind the crack tip. Evidence of pullout of these grains further behind the crack tip is also observed. Most importantly, the observed toughening contribution associated with the introduction of quite large diameter elongated grains is, indeed, found to scale with (diameter of elongated grain)$^{1/2}$, Figure 2. Thus, the toughening behavior observed in silicon nitrides containing elongated grains is consistent with the predictions

based on crack wake mechanisms. Other types of toughening processes (e.g., crack deflection) would not result in this dependence of the toughness upon the diameter of the reinforcing phase.

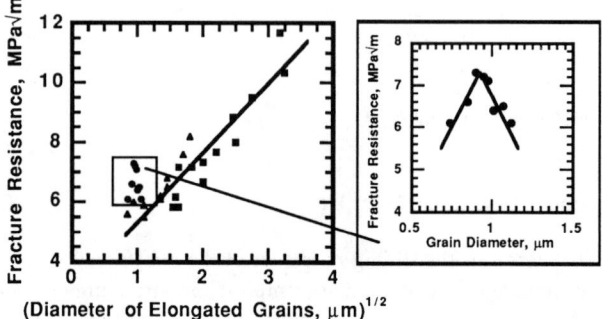

Figure 2. The steady-state fracture toughness of Si_3N_4 ceramics having elongated grain structures increases with increase in grain diameter, d, consistent with various crack bridging models (refs. 7 and 8). Insert illustrates grain size response observed in some materials containing smaller diameter elongated grains (ref. 9).

However, several characteristics that should affect crack wake processes remain to be tested. Note that the descriptions of the contribution of each mechanism indicate that the toughness should increase with an increase in the volume fraction of the larger elongated grains. Yet studies to date have documented the effect of changing the volume content. Another observation suggests that a maximum in the toughness occurs for submicron diameter grains for silicon nitrides fabricated with somewhat different levels of yttrium and aluminum, Figure 2 (insert). These results raise the question why a maximum in toughness might occur with certain additives and/or microstructures.

Finally, Tajima observed that the presence of larger elongated grains was not sufficient to ensure a toughening effect.[10] He found that in the case of two silicon nitrides exhibiting similar elongated grain microstructure but employing different additives that in one the crack propagated through the elongated grains while leaving bridging grains in the crack tip wake in the other. This points out the importance of interface debonding (e.g., intergranular fracture versus transgranular fracture) in crack wake processes. One can speculate that the properties of the intergranular phase(s) (e.g., thermal expansion, fracture resistance, or chemical bonding with silicon nitride) would alter the stress state at the interfaces and the conditions for interfacial debonding. However, very little is known about the properties of the oxynitride glasses that are present. It is known that the cations present in the sintering additives can modify the equilibrium thickness of two grain junctions in silicon nitride,11 but it is not clear how this might effect debonding. Thus, several questions remain as to how to optimize the toughening effects resulting from the presence of elongated grains in silicon nitrides.

Fracture Strength
Another very significant attribute of the toughened large grained material described in the previous figure is the very modest decrease in fracture strength with increase in grain diameter.[7] Both equiaxed grained ceramics and the Si_3N_4 having much smaller diameter elongated grains (and different additives)[12] exhibit a very substantial loss in strength with increase in grain size. To achieve strengths in the range of ≥ 1000 MPa in dense alumina ceramics grain sizes must be much < 1 μm and the strengths decrease to 500 MPa for grain sizes of ~ 2 μm.[13] Similar behavior is shown for the hot pressed Si_3N_4 ceramic in Figure 3. On the other hand, strengths of 800 MPa or greater can be achieved in the sintered Si_3N_4 having elongated grains with diameters of up to ~ 9 μm! These materials also can exhibit quite high Weibull moduli, e.g., m = 26 as shown in Figure 3.

Note that often the larger elongated grains are observed at the fracture origin; this suggests that they might act as flaws. This is not inconsistent with the toughening effect as relatively easy debonding at the silicon nitride interfaces is required for toughening. A large partially debonded elongated grain interface could then contribute to the formation of a larger flaw and a reduction in strength versus that achieved in the less tough finer grained material. This is born out by the strength versus grain diameter behavior as shown in Figure 3. Nonetheless, the toughening effect achieved minimizes the strength reduction.

One might, then, expect that clusters of the larger elongated grains may generate larger defects to further lower the fracture strength. If such clusters are randomly distributed throughout the microstructure, a much broader strength distribution could result. Evidence for such can be seen in the strength distributions for two silicon nitrides prepared with the same additives and sintering schedule but from two different powders, Figure 4. Both materials exhibit the same level of fracture toughness, but ceramic A exhibits a slightly higher strength (~ 900 MPa) and a much higher Weibull modulus (m > 40) versus ~ 810 MPa and 18, respectively, for material B. The main difference in these two materials is the presence of elongated grain clusters in material B often found at the fracture origins. Thus, simply increasing the size of the elongated grains without controlling the volume fraction and the distribution of the largest elongated grains could prove to be quite detrimental to the mechanical performance, especially when a narrow strength distribution along with toughening effects is crucial for the intended application.

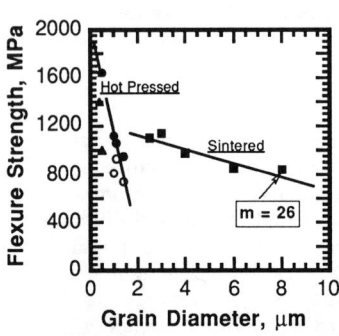 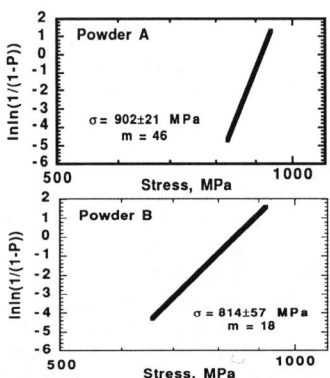

Figure 3. The grain size dependence of the fracture strength of Si_3N_4 can be substantially reduced in some of the systems toughened by the formation of elongated grain structures. Such materials also exhibit quite high Weibull moduli.

Figure 4. Narrow strength distributions attained with well-dispersed large grains are degraded when the large elongated grains tend to form clusters.

Evaluation of the two powders used reveals nearly identical size distributions, phase contents, and contents of oxygen and other impurities. In the case of powder A, the large powder particles consist of friable agglomerates of submicron particles. However, in the case of powder A, the number of larger β-particles in the starting powder is smaller in comparison to powder B. Detailed grain growth experiments show that these large β-particles act as nuclei for the growth of large elongated grains (lengths > 100 μm) during isothermal sintering after completion of the α to β transformation.[14] Thus, attention needs to be given not only to the effects of additives and sintering conditions on the formation of elongated grain microstructures but also, not surprisingly, to the powder characteristics.

R-Curve Response
High fracture toughness coupled with high strengths and narrow strength distributions that can be achieved in certain silicon nitrides suggests that the rise in the fracture resistance with initial extension of a crack (e.g., R-curve response) may be an important factor. However, little information is available on the R-curve behavior of these materials particularly regarding the influence of microstructure. Figure 5 illustrates the R-curves for several Si_3N_4 ceramics having elongated grains but containing different additives. Materials with larger diameter elongated grains tend to exhibit a steeper initial portion of the R-curve in the limited materials studied. The reader is cautioned that these R-curves cannot be compared directly to the results in

the proceeding figures due to differences in additives and processing and thus microstructures. However, if we assume that their microstructures are quite similar, as suggested by preliminary examinations, one could suggest that differences in second phase and grain boundary phase chemistry and content may play a role. Both the microstructural and chemistry and phase content aspects in the toughness, strength, and R-curve behaviors need to be systematically explored.

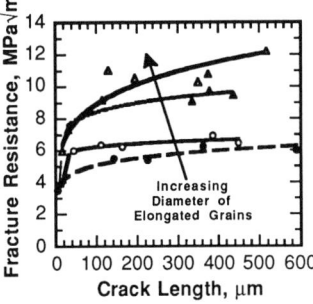 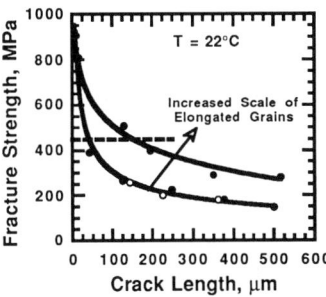

Figure 5. The R-curve behavior of four different Si_3N_4 ceramics illustrate that the initial rate of rise of the fracture resistance with crack length can increase with increase in the diameter of the elongated grains.

Figure 6. High strengths are retained for larger crack sizes in material with the most rapid rise in the R-curve response as compared to the material with both the lower toughness and slower rise in the R-curve (see Figure 5).

To obtain some insight into the influence that the R-curve might have, it is informative to consider the strength-flaw size response of two materials representing the extremes in the R-curve response shown in Figure 5. The importance of the details of the R-curve behavior is indicated by the strength-crack size response in these materials, Figure 6. While the strength of each decreases with increase in crack size, the rate of decrease is reduced as the rate of initial rise in the R-curve and the magnitude of the toughness plateau. In other words, materials with R-curves that rapidly rise to high levels are more damage resistant. This can be seen in Figure 6 where, for example, the strength level can be maintained $\geq$ 500 MPa for crack sizes up to 125 μm in the material that has the steepest and highest R-curve. While in the material with the poorest R-curve response, the same levels of strength can only be maintained for crack lengths of < 50 μm. It is obvious from the findings to date that understanding what factors control the crack bridging mechanisms and related R-curve behavior can also be of significant engineering importance in meeting the demands of various applications.

Fatigue Resistance

With a microstructure consisting of amorphous and crystalline silicate grain boundary phases, one should consider the resistance of these toughened silicon nitride ceramics to fatigue in the form of slow crack growth under static and cyclic loading. It is well documented that silicates and, in particular, glasses are quite susceptible to crack growth at applied stress intensities that are substantially lower than those required for catastrophic fracture.[15] On the other hand, nonoxides with no oxide additives, as typified by sintered silicon carbide containing boron and carbon, can be extremely resistant to slow crack growth.[16] In addition, toughening achieved by crack bridging and pullout may exhibit somewhat poorer crack growth resistance under cyclic versus static loading due to damage of the bridging elements during crack closure on unloading.[17] Also, the presence of amorphous silicate grain boundary (interfacial) phases may substantially alter the elevated temperature response due to softening and viscous flow of the amorphous phases.

At room temperature, slow crack growth is observed in silicon nitrides containing elongated grains and silicate grain boundary phases under static loading, Figure 7a. However, two features should be noted. First, the dependence of the crack velocity on the applied stress intensity is quite strong and the exponent of the stress intensity derived from a power law crack growth description is quite large (e.g., $n > 50$). Second, the high resistance to slow crack growth is maintained when the grain size is increased to further toughen the ceramic. When cyclic loading is applied, crack growth in the toughened silicon nitride is promoted in chemically active environments, Figure 7b, as happens in glasses and oxides.[15] Thus, one can state that the use of glass phases to promote the growth of elongated grains in silicon nitride ceramics may lower their resistance to slow crack growth and make them susceptible to environmental effects at room temperature. While these static fatigue effects may not be very strong, attention should be paid to regulating the amount and the composition of the grain phases to avoid promoting cyclic fatigue.

Besides room temperature fatigue, the presence of silicate grain boundary phases suggests that the fatigue response at elevated temperatures may be a concern as softening of these phases may alter the crack bridging behavior. Static and low cycle fatigue of the silicon nitride with the large elongated grains as shown in Figure 8 at 1400°C in air reveals that slow crack growth is inhibited under cyclic versus static loading. Observations of the crack path indicate that bridging of the crack can be enhanced under cyclic loading. Lin et al. noted similar effects at elevated temperatures in a commercial silicon nitride containing elongated grains where the time to failure increased and the strain rates decreased with cyclic versus static tensile loading.[18]

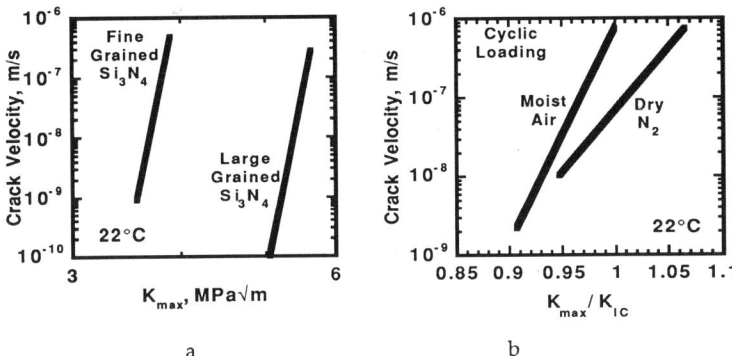

Figure 7. Slow crack growth is observed at room temperature in elongated grain structured silicon nitrides due to the presence of silicate phases (a. static loading; b. cyclic loading in various environments).

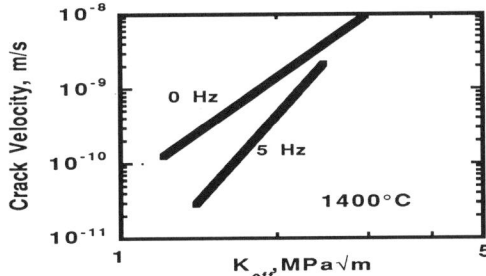

Figure 8. Slow crack growth resistance is enhanced at 1400°C under cyclic loading.

While it is not known why certain mechanisms are activated or suppressed in cyclic loading, it is observed in the experiments by Lin et al. that cavitation is less prevalent.[17] Combined with the observed increase in the contribution of crack bridging mechanisms under cyclic loading, this suggests that stress relaxation during unloading could be a contributing factor. How this depends on the viscous behavior of the grain boundary phase that must influence both bridging and cavitation remains to be resolved.

SUMMARY

The introduction of elongated silicon nitride grains during densification in the presence of a liquid phase can impart considerable improvement to the fracture toughness. This toughening is not universally attained but depends on the activation of intergranular rather than transgranular fracture. This is reminiscent of the requirement of interfacial debonding in whisker-reinforced ceramics. In fact, additional observations such as bridging in the crack wake by elongated grains and pullout of some of these grains further suggest that the crack wake mechanisms that contribute to the toughening of whisker-reinforced ceramics can also operate in silicon nitrides containing elongated grains. Various investigators have found that, consistent with crack wake mechanisms, the fracture toughness of silicon nitrides increases with increase in the diameter of the larger elongated grains. However, little is known about the effects of the grain boundary phase(s) and their properties on the interfacial debonding/intergranular fracture in such silicon nitrides. This is critical as observations show that crack propagation in some systems exhibiting larger elongated grains occurs transgranularly and no toughening occurs.

The fracture strengths of these toughened silicon nitrides containing larger elongated grains are quite impressive, reaching levels of 900 MPa or so. Large grains are frequently found at the fracture origins; this is not surprising if debonding of their interfaces is required for toughening (i.e., the debonding of a large grain surface should act to increase the flaw size). However, some observations reveal that the accompanying improvements in toughness act to minimize the reduction in fracture strength as the size of the larger elongated grains increases. Even with diameters of the larger elongated grains approaching 9 µm, strengths > 800 MPa have been achieved along with high Weibull moduli. However, care must be taken to avoid clustering of these larger elongated grains (i.e., either maintain lower volume fractions or uniform spatial distributions) as such clusters can reduce the strength and the Weibull modulus. The strength-damage response in such materials must also reflect their R-curve behavior where the fracture resistance increases as the crack extends through the microstructure and develops a bridging zone behind the crack tip. The details of how the grain boundary phase(s) and scale of the microstructure influence the R-curve response need to be addressed. Preliminary studies only indicate that large differences in the plateau toughness values can alter the initial rising portion of the R-curve.

The fatigue behavior of these toughened silicon nitrides is influenced by the silicate grain boundary phases. Slow crack growth behavior is observed at room and elevated temperatures. At room temperature, the slow crack growth-environment response is similar to those seen in glasses and oxides indicating moisture-assisted crack growth in the silicate grain boundary phase. At elevated temperatures, cyclic loading appears to enhance the fatigue resistance in several toughened silicon nitride ceramics with evidence for both enhanced crack bridging and reduced cavitation as compared to that under static loading. Obviously, design of such toughened silicon nitrides must consider the effects of the silicate phase(s) composition and content on the fatigue and creep response as well. Further attention to the various microstructural characteristics should provide an approach to designing high strength silicon nitride ceramics with further improvements in toughness and mechanical reliability.

Acknowledgments

W. H. Warwick is recognized for his many technical contributions during these studies as is V. M. Gibson in the preparation of the manuscript. The research of PFB, HTL, and SLH is supported by the Division of Materials Science, U. S. Department of Energy under contract DE-AC05-84OR21400 with Martin Marietta Energy Systems, Inc., that of IWC is supported by the Air Force Office of Scientific Research under grant number AFOSR-91-0094, and that of MJH by the German Ministry of Science and Technology under contract 03 M 2012.

References

1. a. P. F. Becher, C. H. Hsueh, P. Angelini, and T. N. Tiegs, "Toughening Behavior in Whisker Reinforced Ceramic Matrix Composites," J. Am. Ceram. Soc. 71(12) 1050-61 (1988).
   b. P. F. Becher, "Microstructural Design of Toughened Ceramics," J. Am. Ceram. Soc. 74(2) 255-69 (1991).
   c. P. F. Becher, E. R. Fuller, Jr., and P. Angelini, "Matrix-Grain-Bridging Contributions to the Toughening of Whisker-Reinforced Ceramics," J. Am. Ceram. Soc. 74(9) 2131-35 (1991).
2. B. Budiansky, J. W. Hutchinson, and A. G. Evans, "Matrix Fracture in Fiber-Reinforced Ceramics," J. Mech. Phys. Solids 34(2) 167-89 (1986).
3. a. R. F. Cook, "Segregation Effects in the Fracture of Brittle Materials: Ca-Al_2O_3," Acta Metall. 38(6) 1083-1100 (1990).
   b. S. T. Bennison and B. R. Lawn, "Role of Interfacial Grain-Bridging Friction in the Crack-Resistance and Strength Properties of Nontransforming Ceramics," Acta Metall. 37(10) 2659-71 (1989).
4. a. F. F. Lange, "Fracture Toughness of Si_3N_4 as a Function of the Initial a-Phase Content," J. Am. Ceram. Soc. 62 (7-8) 428-30 (1979).
   b. C. W. Li and J. Yamanis, "Super-Tough Silicon Nitride with R-Curve Behavior," Ceram. Sci. and Engr Proc., 10 (7-8) 632-45 (1989).
5. a. K. B. Alexander, P. F. Becher, and S. B. Waters, "Characterization of Silicon Carbide Platelet-Reinforced Alumina," pp. 106-7 in Proc. 12th Int'l Congress for Electron Microscopy, San Francisco Press, San Francisco, CA, 1990.
   b. S. M. Ketchion, G. Leng-Ward, and M. H. Lewis, "SiC Dispersoid-Reinforced Si_3N_4 Composites," pp. 757-63 in Ceramic Trans., Vol. 19, M. D. Sacks (ed.), Am. Ceram. Soc., Westerville, OH. 1990.
6. G. Vekinis, M. F. Ashby, and P. W. R. Beaumont, "R-Curve Behavior of Al_2O_3 Ceramics," Acta Metall. Mater. 38(6) 1151-62 (1990).
7. T. Kawashima, H. Okamoto, H. Yamamoto, and A. Kitamura, "Grain Size Dependence of the Fracture Toughness of Silicon Nitride Ceramics," J. Ceram. Soc. Japan, 99: 1-4, (1991).
8. M. Mitomo, "Toughening of Silicon Nitride Ceramics By Microstructural Control," pp. 101-07 in Proc. Sci. Eng'g Ceram., S. Kimura & K. Niihara, eds., Ceram. Soc. Jpn, Tokyo, 1991.
9. T. Y. Tien, "Silicon Nitride Ceramics-Alloy Design," lectures at Max-Planck-Institut, Institut für Metallforschung, Stuttgart, Germany.

10. Y. Tajima, K. Urashima, M. Watanabe, and Y. Matsuo, "Fracture Toughness and Microstructure Evaluation of Silicon Nitride Ceramics," pp. 1034-41 in Ceramic Transactions, Vol. 1: Ceramic Powder Science-IIB, G. L. Messing, E. R. Fuller, Jr., and H. Hausner (editors), Am. Ceram. Soc., Westerville, OH, 1988.

11. M. Rühle, H. J. Kleebe, R. M. Cannon, and D. R. Clarke, "Atomistic Structure and Composition of Grain Boundaries in Silicon Nitride Ceramics," this volume.

12. A. J. Pyzik and D. R. Beaman, "Self-Reinforced Silicon Nitride," to be published.

13. R. W. Rice, "Microstructure Dependence of Mechanical Behavior of Ceramics," Treatise Mat. Sci. Tech., 11:199-381 (1977).

14. G. Petzow and M. J. Hoffmann, "Grain Growth Studies in Si_3N_4-Ceramics," pp. 91-102 in Materials Science Forum, Vols. 113-115, Trans. Tech. Publns, Switzerland, 1993.

15. S. W. Freiman, "Effect of Environment on the Fracture of Ceramics," Ceramurgia, 2: 111-18 (1976).

16. G. D. Quinn, "Review of Static Fatigue in Silicon Nitride and Silicon Carbide," Ceram. Eng. Sci. Proc. 3(1-2) 77-98 (1982).

17. R. H. Dauskardy, D. Yao, B. J. Dalgleish, P. F. Becher, and R. O. Ritchie, "Cyclic Fatigue-Crack Growth in a Silicon Carbide Whisker-Reinforced Alumina Composite: Role of Load Ratio," to be published.

18. C. K. J. Lin, M. G. Jenkins, and M. K. Ferber, "Evaluation of Tensile Static, Dynamic, and Cyclic Fatigue Behavior for a HIPed Silicon Nitride at Elevated Temperatures," this volume.

SIALONS AND SILICON NITRIDES ; MICROSTRUCTURAL DESIGN AND PERFORMANCE

M.H. LEWIS, Centre for Advanced Materials Technology, University of Warwick, Coventry, CV4 7AL, U.K.

ABSTRACT

A survey is presented of developments in silicon nitride and sialon ceramic microstructures designed for application in differing temperature regimes.

For low temperature (<1000°C) application, pressureless-sinterable sialons with moderate (10-15%) intergranular glass, anisotropic β' grains and high values of MOR and K_c (1 GPa and 6-10MPa$\sqrt{m}$ respectively) are preferred.

Improved hardness and high temperature capability may be achieved by tailoring intergranular phases for crystallisation and further enhanced by the introduction of mixed α'/β' Sialon microstructures. Examples are given of microstructural evolution in β'-$Y_3Al_5O_{12}$(garnet), β'-$Nd_3Si_3Al_3O_{12}N_2$ and α'/β'/garnet ceramics and a comparison of their mechanical behaviour.

Novel Sialon ceramics containing dispersed transition metal compounds (TiN, TiB_2) may be formed by in-situ redox reaction, utilising the β' Sialon phase as an oxygen receptor. The dispersed phase may enhance hardness and toughness and confer electro-discharge machinability.

Oxidation instability of Sialon compositions dictates the use of diphasic $Si_3N_4/M_2Si_2O_7$ microstructures for application above 1300°C. The thermal cycle during pressurised sintering of these non-sialon compositions is critical in avoiding crystallisation of mixed polymorphs of the intergranular disilicate, with consequent microcracking.

INTRODUCTION

Structural ceramics based on the compound Si_3N_4 and its 'Sialon' derivatives have been developed from the 1960s with a view to structural engineering application. The two major factors which have resulted in a disappointingly small commercial uptake have been the high cost of synthetic ceramic fabrication and the limitation to key mechanical properties. For example, valve train components for reciprocating engines have demonstrated clear advantages of low inertia, high wear resistance with adequate strengths and toughness but are too expensive for the automotive market. For critical engine components, such as turbine blades or nozzle guide vanes, the limited fracture toughness, strength variability and susceptibility to surface impact damage are problems which await solution. However, there have been successful efforts to 'design' microstructures with specific property variations and a good understanding has evolved of the micromechanisms which control mechanical and environmental behaviour over a range of temperature.

The early Si_3N_4 ceramics required pressurised sintering in the presence of liquid sintering aids and suffered from poor high temperature (>1000°C) properties due to the liquid residues. A critical improvement was achieved in sinterability and properties with Sialon compositions which enabled pressureless densification, more flexible control of residual liquids and improvements in toughness and high temperature stress rupture behaviour with the suppression of creep-cavitation.

In this paper a survey of the earlier development of diphasic β' Sialons is followed by a review of recent work at Warwick on mixed α'/β' ceramics and on composite microstructures in which transition metal borides and nitrides are dispersed in Sialon matrices by 'in-situ' reaction. Some of the problems in translating from small scale laboratory fabrication to commercial practice are exemplified in a final section concerning microstructural control in Si_3N_4 ceramics designed for applications at temperatures above that for Sialons.

β' SIALON CERAMICS

β' structure

A majority of Sialon microstructures contain the β' solid solution ($Si_{3-x} Al_x O_x N_{4-x}$) as a major phase. Although the substitution level x is generally small (< 0.5) and intrinsic β' properties deviate little from Si_3N_4, there remains an interesting scientific question concerning co-ordination of Si and Al atoms in the crystal structure. The similarity in Si-N and Al-O bond lengths (0.174 and 0.175 nm, respectively) has been used in explanation[1] of the large level of possible substitution of Al for Si and O for N, represented by the β' line on the base of the Janecke prism construction (Fig.1). This explanation infers a preferred coupling of Si-N and Al-O with a deviation from the random solid solution. Diffraction methods have not detected this site selectivity but recent MAS-NMR spectroscopy[2,3] has identified a SiN_4 tetrahedral preference via the constancy of ^{29}Si peak position with varying substitution level x (Fig.2). Peak shifts would be observed for oxygen-substituted SiN_4 tetrahedra, as indicated by the comparison spectra for Si_2N_2O which contains SiN_3O tetrahedra (Fig.2).

NMR data from ^{29}Si and ^{27}Al has been used in support of a 'microdomain' model of preferred SiN_4 and AlO_4 regions[3]. This has been substantiated in recent high field/high spinning rate MAS-NMR which has enabled refinement of spectra from the quadrupolar ^{27}Al nucleus and confirmed the very low AlN_4 population[4].

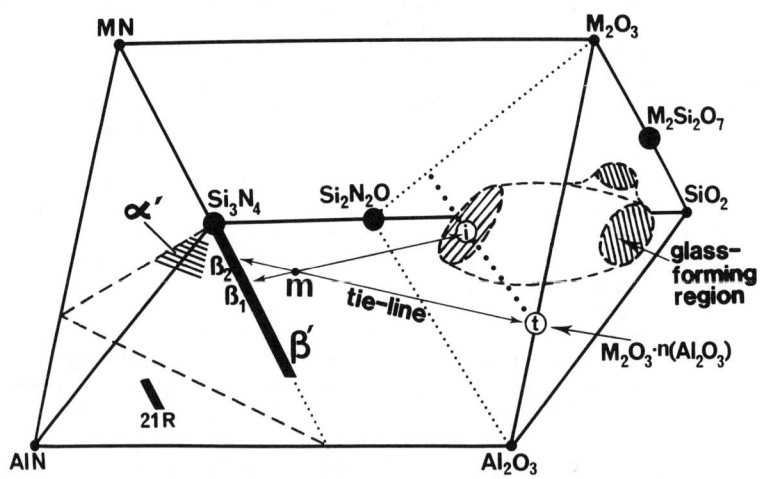

Fig.1 A Janecke prism construction showing the relation between principal Sialon phases (β', α'), sintering liquid and residual glassy and crystalline phases (i or t).

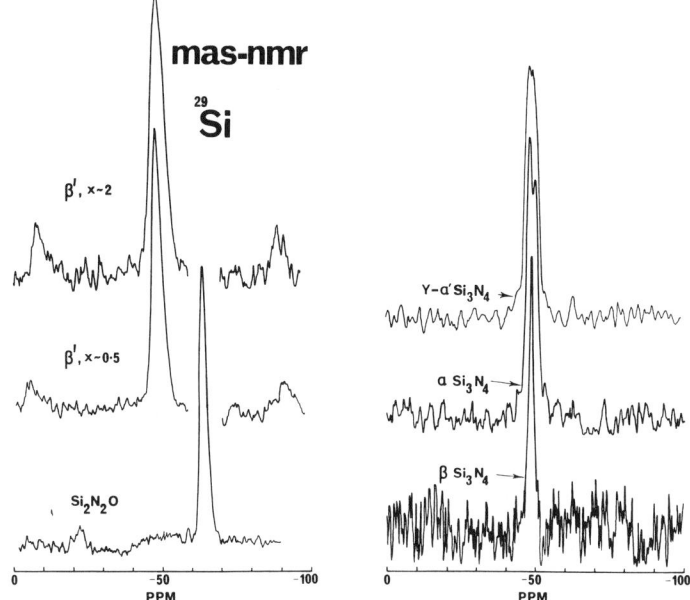

Fig.2 ^{29}Si MAS-NMR spectra from β, and α Si$_3$N$_4$ compared with spectra from β' Sialon with varying substitution level and from the compound Si$_2$N$_2$O.

Diphasic β' microstructures

A recognition of the importance of oxygen/nitrogen ratio in achieving a monophase β' microstructure via transient liquid sintering[5,6] motivated the early development of these Sialons. Their microstructures contained negligible intergranular residues and equiaxed β' grains which resulted in poor fracture toughness and strength levels. This, together with the need for pressurised sintering, inhibited their development as structural ceramics although they have superior high temperature oxidation and creep properties. The monophase ceramics also played an important role as models for high temperature deformation with a first demonstration of diffusional creep and suppression of creep cavitation due to control of grain-boundary microstructure[7,8,9].

Diphasic β' ceramics resulted from an increase in sintering liquid volume which enabled pressureless sintering to theoretical density and the freedom for β' crystals to grow with their natural anisotropic hexagonal prism morphology (Fig.3). Pressureless sintering enables complex component shaping at reduced cost and β' morphology is important in enhancing the K_c level from 3-4 MPa$\sqrt{}$ m (Fig.4). These have been key factors in engineering application as cutting tips, extrusion dies, welding shields, tappet inserts and turbocharger rotors. Typical microstructures (Fig.3) contain 5-15 volume% of the sintering liquid residue, normally in the glassy state, based on M$_2$O$_3$-Al$_2$O$_3$-SiO$_2$ eutectics with dissolved nitrogen (Fig.1). These ceramics (e.g. Syalon 101 - Cookson, UK) are normally prepared from αSi$_3$N$_4$-Al$_2$O$_3$-SiO$_2$-21R mixtures with an Y$_2$O$_3$ additive, the 21R polytypoid being used in preference to AlN as a nitrogen source (Fig.1). The microstructure in Fig.3 is based on a Nd$_2$O$_3$ additive[12] which results in comparable properties to Syalon 101; mean MOR values of 1GPa and $K_c \sim$ 6-8 MPa$\sqrt{}$ m (Fig.4 and 5).

The use of Sialon compositions and, especially, M_2O_3 additives (M = Y, Nd, La) provides for greater flexibility in crystallisation of the sintering liquid residue. The principles of crystallisation to achieve a negligible glass residue have been explained previously [10,12] in relation to the Janecke - prism representation of participating phases (Fig.1). For a diphasic crystalline product the mean composition (m) should lie on the tie-line between β' and minor phase. In the earlier 'Syalon 201' (Cookson) ceramics the minor phase is YAG ($3Y_2O_3.5Al_2O_3$) containing limited Si and N substitution such that the β'_1 composition must change from β'_1 to β'_2 with rotation of the tie-line (Fig.1). In recent Nd-containing compositions[12] the glass crystallises at constant composition (i) to a newly-identified hexagonal phase[13] $Nd_3Si_3Al_3O_{12}N_2$, (Fig.3) which has inferior thermal stability ($\sim 1350°C$) to YAG.

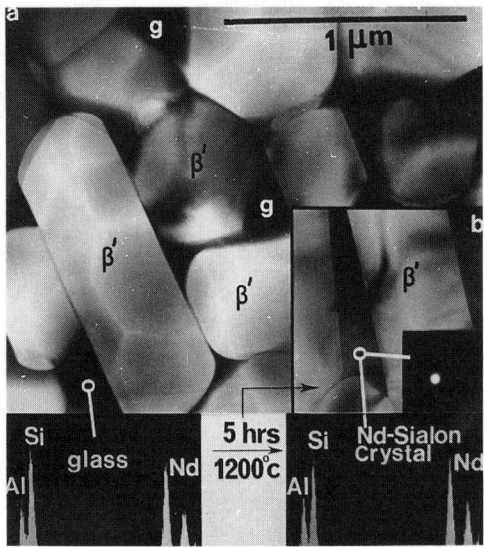

Fig.3 The microstructure of a pressureless-sintered diphasic β' Sialon illustrating the elongated hexagonal prism morphology of β' crystals in a semi-continuous matrix. The M-Si-Al-O-N matrix may be transformed from glass (a) to crystalline state (b) by post-sintering heat-treatment (the example here is for a β' - $Nd_3Si_3Al_3O_{12}N_2$ microstructure).

The variable β' composition is important in accommodating non-stoichiometric species during crystallisation, hence minimising residual intergranular glass and enhancing high temperature properties. MOR shows a useful increment over glass-matrix materials above $\sim 1000°C$ whereas there is a small penalty in low temperature properties (MOR - Fig.5 and K_c - Fig.4) probably due to transformation - induced residual stresses. Of critical importance is the improvement in time-dependent high temperature properties; creep rates are reduced and creep cavitation-induced stress rupture is eliminated[14,15].

Diphasic β' Sialons have an application limit of $\sim 1350°C$ due either to reversion of the crystalline matrix to the liquid state (e.g. in β' - $Nd_3Si_3Al_3O_{12}N_2$) or to a reaction of the matrix oxide with a SiO_2 - rich surface oxidation layer to form a eutectic liquid (e.g. in β'-YAG)[16].

SIALONS CONTAINING α'

$\alpha\ Si_3N_4$ is the main component of initial ceramic powder mixtures used in forming β' Sialons but may be stabilised in sintered products by the simultaneous substitution of Al and O (as in β') together with interstitial metallic atoms such as Y or Ca[17]. The

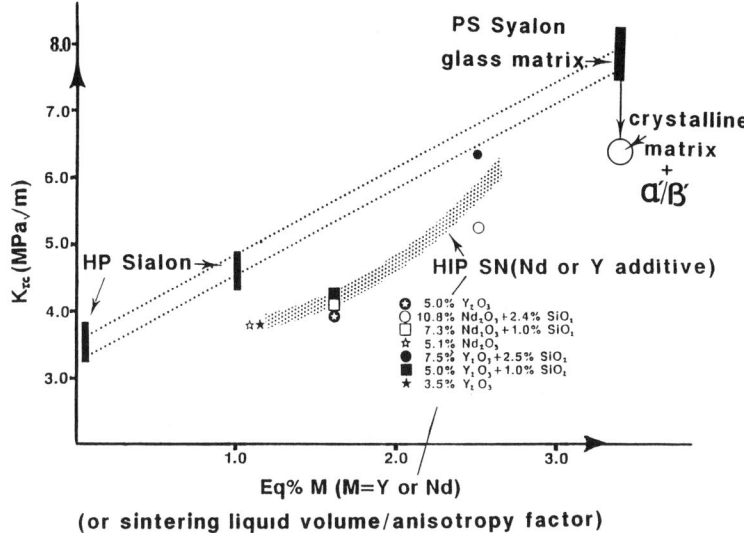

Fig.4 Typical fracture toughness (Kc) increments associated with enhanced morphological anisotropy for β' or βSi_3N_4 crystals, induced by increasing the sintering liquid volume.

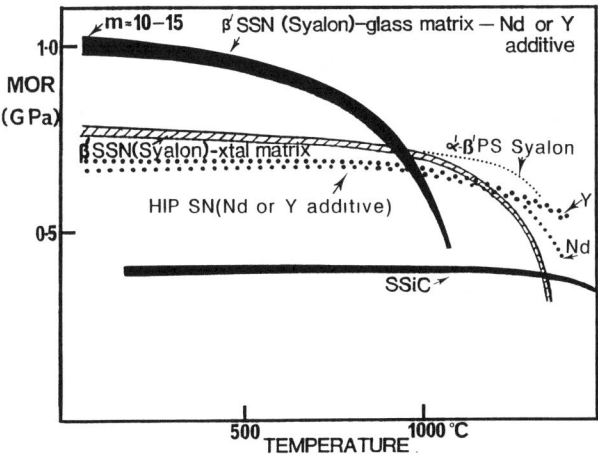

Fig.5 Typical MOR/Temperature data for β' Sialons with glassy or crystalline matrices and for α'/β'/YAG microstructures (Data for a typical sintered SiC and for HIP Si_3N_4 ceramics shown for comparison).

resulting α' solid solution has a limited stability range on a plane (Fig.1) defined by $Y_{m/3} Si_{12-(m+n)} Al_{(m+n)} O_n N_{16-n}$, for the yttrium-stabilised form, with m in the range 1-2 and n< 2.

α' has a similar SiN_4-based tetrahedral network structure to β' but with a double c parameter in the unit cell. This is believed to be a key factor in the enhancement of indentation flow stress (i.e. increased hardness) since the short c axis dislocation Burgers vectors are dominant during indentation plasticity of β' and hence have a lower Peierls stress [18]. In addition to increased hardness the α' phase is capable of more flexible accommodation of residual elements during post-sintering crystallisation, hence reducing the final traces of intergranular glass and improving high temperature properties. This property is exemplified in a recent α'/β'/YAG microstructure[19] and illustrated here.

The development of α' during sintering requires a high N/O ratio, moving the mean composition from the β'/liquid tie-line (m in Fig.1) towards the α' plane by further additions of 21R polytypoid. The specific microstructure is sensitive to thermal cycle; the best properties are associated with 50-70% α' content, a dispersion of prismatic β' crystals (30-40%) and isolated YAG particles (5-10%) in distinctive morphology (Fig.6). A sintering-reaction sequence[19] (Fig.7) illustrates the evolution of α' and β' with temperature and indicates the importance of 2-step sintering cycles in developing the appropriate α'/β'ratio which is influenced by kinetic factors in addition to initial composition. The isolation of YAG, with curved (non faceted) YAG/α' interfaces results from equilibration of triple phase junctions following post-sintering crystallisation of residual glass. It is evidence for the complete removal of non-stoichiometric glass-stabilising elements (Si, Al, O, Y) by solid solution within α'. Similar YAG morphology has been observed at the surface of β' ceramics only after prolonged heat-treatment in oxidising atmospheres[20]. β' is not a receptor for additive ions such as Y, which may be removed only by diffusion to the surface SiO_2 oxidation layer.

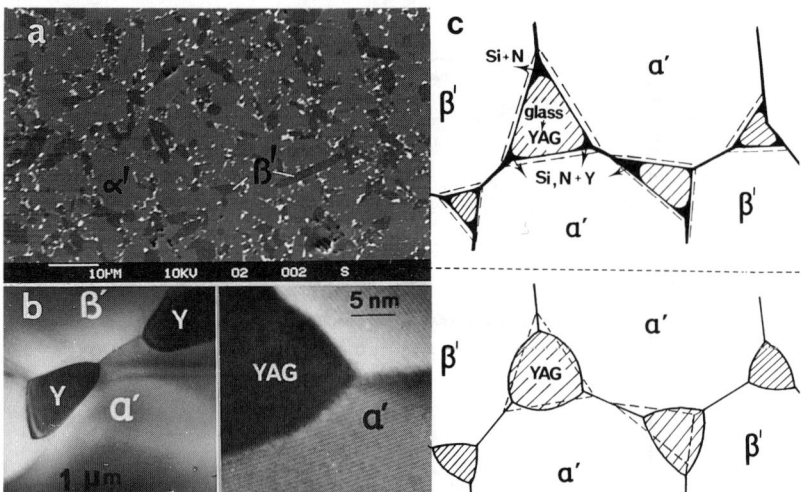

Fig.6 (a) SEM and (b) TEM images of a β'/α'/YAG Sialon ceramic, illustrating prismatic β' (dark grey in SEM) and the isolated YAG distribution with a morphology characteristic of equilibrated solid/solid triple junctions. (c) illustrates the partitioning of elements during YAG crystallisation.

The hardness increment of α'-containing microstructures (Fig.6) is compared for a range of Sialon ceramics in Table 1. This property, together with comparable K_c and improved MOR make this ceramic more suitable for wire-drawing applications than the β'-YAG microstructure (Syalon 201)[15,19]. The isolated YAG morphology and total removal of glassy interfacial residue also improves the creep resistance over the β'-YAG ceramic (Table I); this is due to longer diffusion pathways with reduced diffusivity associated primarily with pure α'/α' and α'/β' interfaces. For similar reasons oxidation resistance is improved; after 50 hours at 1300°C SiO_2-rich oxidation layer thicknesses are typically 30-40μm and 10μm for Syalon 201 and the α'/β'/YAG ceramic, respectively. Although the viscosity-reducing influence (on the passive SiO_2 layer) of Y out-diffusion is suppressed at the crystal boundaries, this is partly compensated by the alternative source of Y within α' crystals. At the substitution limits for defined by the α' phase this does not appear to be a problem.

The oxidative reversion of YAG to the liquid state, which presents a ~1350°C limit to Syalon 201 application, is also observed in α'/β'/YAG ceramics although the kinetics are impaired due to the YAG dispersion. Hence for the highest temperatures, pure α' ceramics, produced by transient liquid sintering may be preferred. However they will suffer the same problem as the earlier monophase β'Sialons, of reduced sinterability and Kc.

Table I - Sample properties of Sialon-based Ceramics

Ceramic type	Hv(kg/mm²)	K_c	% creep strain/100hrs 77MPa.	
			1277°C	1327°C
α'/β'/YAG	2010	6.5	<0.01	<0.1
β'/YAG	1596	6.5	0.07	0.5(20hr)
β'/Nd or Y glass	1350	7-8	-	-
β'/$Nd_3Si_3Al_3O_{12}N_2$	1596	6.5	0.17	~0.3
β'-15% TiB_2	1894	8.9	-	-
β'-26% TiB_2	2204	12.4		

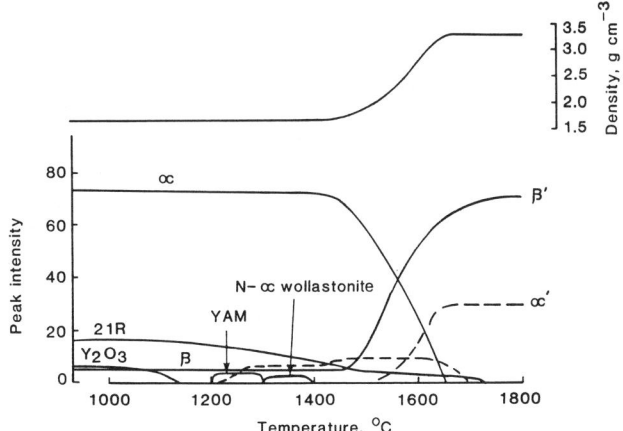

Fig.7 Reaction sequences during the sintering of α'/β' sialons, identified by X-ray diffraction, showing the evolution of α' and β' phases at different temperatures.

SIALON-MATRIX COMPOSITES

A realisation that the microstructural 'design' of monolithic silicon nitride and Sialon ceramics has reached a state of refinement has motivated the study of more radical changes in microstructure not achievable by natural sintering reactions. During the late 1980s there have been numerous studies of the artificial addition of inert dispersions (particles, whiskers and platelets) to nitride matrices and limited studies of fibre additions. The major objective in these 'composite' microstructures is to enhance fracture toughness beyond the level of ~ 10 MPa√m via mechanisms of crack deflection, crack bridging and phase 'pull-out' together with thermally-initiated microcracking.

SiC has been the principal dispersoid but Kc increments (usually <50%) have been limited by dispersion morphology and interface characteristics. These increments have been achieved at the expense of sinterability and, occasionally, MOR[21,22].

Recent research at Warwick[23,24] has been aimed at reduced cost of raw materials and more flexible component fabrication, utilising 'in-situ' sintering reactions to develop the composite dispersions produced by 'redox' reactions from titanium oxides. The benefit of a Sialon 'matrix' is that of flexible accommodation of oxygen within β', to balance the redox reaction. There are a large number of possible sintering reactions, associated with different initial powder mixtures, within the thermodynamic constrains of the redox reactions; examples for TiN and TiB$_2$ dispersions are;

(a) $Ti_2O_3 + 3AlN + 5Si_3N_4 = 3Si_5AlON_7 + 2TiN$

(b) $TiO_2 + 2BN + 2AlN + Si = SiAl_2O_2N_2 + TiB_2$

Reactions (a) and (b) generate β' Sialon matrices with different substitution level and, this is a factor (in addition to Ti-oxide stoichiometry) which determines the volume fraction of dispersion (TiN or TiB$_2$). Higher volume fractions induce higher hardness, Kc and electrical conductivity. The latter property is important for selected applications and an ability to shape components by electro-discharge machining (EDM). Properties, exemplified in Table I, show significant increments in hardness and Kc for the TiB$_2$-containing Sialons and EDM capability above 20 vol.% of TiN, due to an interconnected particle morphology (Fig.8) between β' grains (random dispersion of TiN, as in Syalon 501, requires >30% for EDM).

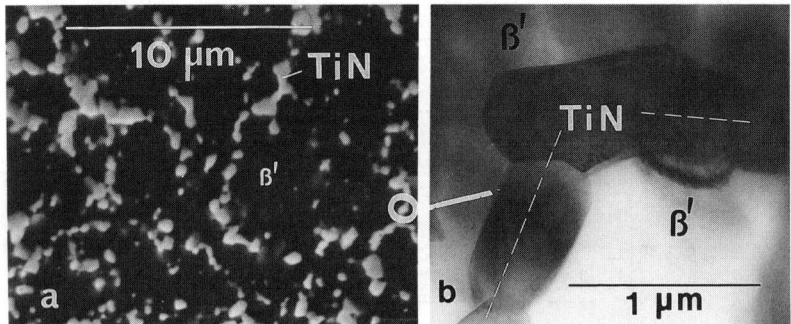

Fig.8 The microstructure of TiN/β' Sialon ceramic produced by 'in-situ' redox reaction during sintering illustrating the interconnected form of TiN particles in (a) SEM and (b) TEM micrographs.

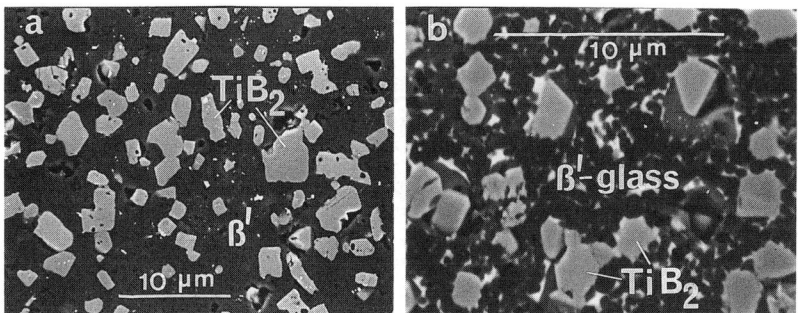

Fig.9 The microstructure of a TiB$_2$/β' Sialon ceramic produced by reaction-sintering; (a) hot-pressed, (b) pressureless-sintered, containing a β'-glass matrix.

The application-potential for those 'in-situ' composites focuses on components in which wear resistance, hardness etc. are important, with less emphasis on high temperature environment due to the impairment to oxidation resistance above ~1000°C.

On-going research is concerned with improvement in densification under pressureless-sintering conditions and greater control over dispersion scale and morphology. A critical factor is initial particle size of reactants; an example for the TiB$_2$ dispersion is illustrated in Fig.9.

HIGH-TEMPERATURE Si$_3$N$_4$ COMPOSITIONS

The temperature limit for diphasic and triphasic Sialon ceramics, imposed by instability of the minor crystalline phases, has motivated the development of compositions without Al$_2$O$_3$ or AlN additions for applications above ~1300°C. These β Si$_3$N$_4$ ceramics

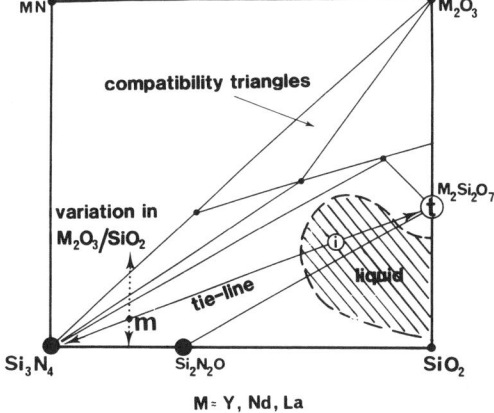

Fig.10 The preferred compatability triangle Si$_3$N$_4$ - M$_2$Si$_2$O$_7$ - Si$_2$N$_2$O from which βSi$_3$N$_4$ ceramic compositions are selected. SiO$_2$ additions are necessary to avoid unstable oxynitride phases when M$_2$O$_3$ additions exceed ~5%.

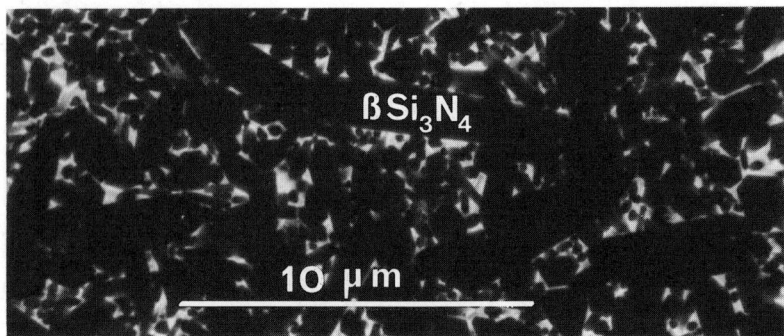

Fig.11 The SEM microstructure of a HIPped Si_3N_4 ceramic, with prismatic βSi_3N_4 crystals and crystalline $Y_2Si_2O_7$ intergranular phase.

are normally based on Si_3N_4 - SiO_2 - M_2O_3 mixtures which fall within the Si_3N_4 - Si_2N_2O - $M_2Si_2O_7$ compatibility triangle (Fig.10) with M = Yttrium. The philosophy underlying the Warwick research has been an improvement in sinterability and fracture toughness using relatively large Y_2O_3 or Nd_2O_3 additions with a SiO_2 balance to maintain a diphasic crystalline microstructure (Fig.11) on the βSi_3N_4 - $M_2Si_2O_7$ tie line[25,26]. Low temperature properties are comparable with crystalline diphasic Sialons (Fig.4,5) with superior strength retention and stress-rupture behaviour. The careful tailoring of composition and sintering/crystallisation cycle is the key to reduced susceptibility to time-dependent failure by creep cavitation to at least 1400°C (Fig.12).

The disadvantage of this class of ceramic, compared to Sialon compositions, is the need for pressurised sintering. This is normally accomplished by glass encapsulated hot-isostatic pressing (HIP) at 100-200 MPa, but reduced pressures are possible with the enhanced sintering liquid volumes used for these compositions compared to parallel developments in the U.S.A.[27].

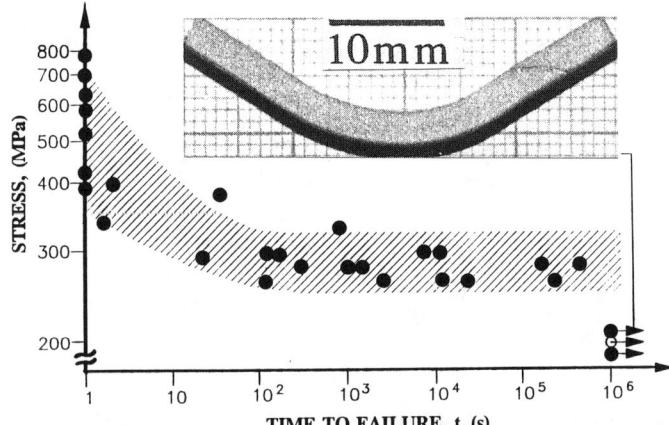

Fig.12 Stress-rupture properties for a βSi_3N_4 - $Y_2Si_2O_7$ ceramic at 1400°C, showing a stress threshold below which diffusional creep occurs to high strain level (inset).

Recent research has focussed on improved processing and scaling-up to component size and sintering furnace content which would be typical of commercial operation. This has revealed significant problems in microstructural control which are not prominant at the small-scale laboratory level and are outlined here.

The heart of the problem resides in the limited thermal cycle rates of large furnaces and the thermal capacity of large ceramic components. For example, it is no longer possible to achieve post-sintering cooling at a rate which would retain a totally glassy residual phase and permit a controlled crystallisation treatment. This is especially prominant for βSi_3N_4 - $M_2Si_2O_7$ ceramics in which the disilicate sintering residue may crystallise to a variety of polymorphic or metastable phases[28]. This is evident in the comparison of microstructures of sintered billets taken from different positions in a large HIP furnace[29]. The differences have been ascribed to cooling rates, with an increasing probability in the following sequence;

(Low rate) β $Y_2Si_2O_7$ - α $Y_2Si_2O_7$ - 'y'$Y_2Si_2O_7$ - glass (High rate)

The problem of slow, continuous, cooling is illustrated by the appearance of large (mm size) islands of β $Y_2Si_2O_7$, with a more general microstructure containing $\alpha Y_2Si_2O_7$, formed during the lower temperature cooling interval. These polymorphs have different atomic volumes (densities p = 3.69 and 3.98 g cm^{-2}, respectively) such that tensile stresses surrounding the β regions induce a network of microcracks[29] (Fig.13). Such inhomogeneous microstructures are not generally recoverable; subsequent heat treatment has little influence on the sinter/cooled state except when the residue is glassy (Table 2). A programmed intermediate cooling rate is the best compromise, to generate a consistant βSi_3N_4 - $\alpha Y_2Si_2O_7$ microstructure which, with other processing improvements, may give room temperature MOR approaching 1GPa with a Weibull modulus of 15-20.

Table II   Intergranular phase combinations in HIP-Si_3N_4 with Y_2O_3/SiO additives for differing cooling rates from the sintering temperature.

as HIPped	post-HIP anneal (8 hrs 1200°C + 16hrs 1400°C)
glass (rapid cool)	y + β
y + α	y + α
glass + α	α
α	α
α + β (slow cool)	α + β

The constraints of cooling rate associated with larger-scale sintering are intrinsically more severe with encapsulated HIP. This may further motivate the move to non-encapsulated, lower pressure, sintering and this is reinforced by the problems of ceramic/encapsulant interaction. Using glass encapsulant there is an observable out-diffusion of yttrium and a SiO_2 enrichment of surface layers to give a βSi_3N_4 - Si_2N_2O microstructure with reduced $Y_2Si_2O_7$ content[29]. This normally extends to a depth of ~0.01mm, but a more significant problem may be reduced Kc within a 1mm surface zone believed to result from a grain-size reduction possibly associated with inhomogeneous densification/reaction kinetics due to the inward motion of a pressure wave.

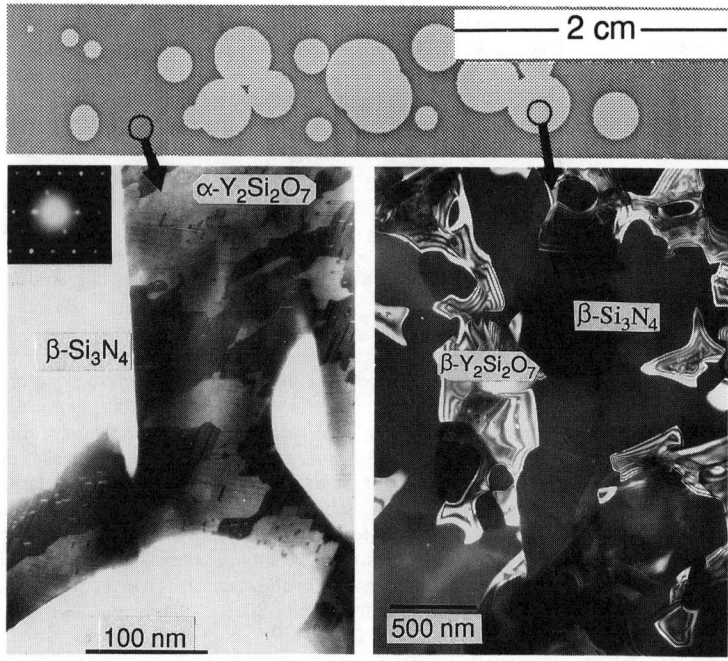

Fig.13 The inhomogeneous microstructure of a slow-cooled Si_3N_4 - $Y_2Si_2O_7$ ceramic with 'islands' of β $Y_2Si_2O_7$ enclosed by the twinned α polymorph in the intergranular phase.

FUTURE DEVELOPMENTS

Developments in Sialon and Silicon Nitride ceramics over the past 20 years have represented a significant achievement in microstructural design and the understanding of micromechanisms limiting mechanical and environmental performance. Microstructural control requires a complex interplay between initial constitution, sintering reactions and post-sintering transformations. The toughness limitation to significant engineering application is unlikely to be removed by further microstructural refinement whereas an expanded market will stem from cost reductions associated with processing.

The potential for innovative microstructural design in the relatively mature Sialon field is now limited. Oxynitride glass-ceramics[30], offered a prospect of glass-forming methods coupled with higher hardness etc. associated with higher covalency phases than in conventional silicates. The properties are limited by the relatively high oxygen content or low volume fraction of the nitrogen-containing phases and glass-preparation requires atmospheric control.

The development of long-fibre silicon-nitride-matrix composites[31] to circumvent the toughness problem will continue for 'critical' components in applications which may tolerate the enhanced cost. There is a continued interest in the aerospace industries where performance and efficiency goals are impossible with metallic systems.

ACKNOWLEDGEMENTS

The research surveyed in this paper has been supported via long collaboration with Lucas-Cookson Syalon (later Vesuvius Zyalons). Recent work on composite microstructures has been supported by SERC, MOD and Rolls Royce plc.

References

1. K.H. Jack, J.Mat.Sci. II, 1135 (1976)

2. R. Dupree, M.H. Lewis, G. Leng-Ward and D.S. Williams, J.Mat.Sci.Lett. 4, 393 1985).

3. R. Dupree, M.H. Lewis and M.E. Smith, J.Appl.Cryst. 21, 109 (1988).

4. M.E. Smith, J.Phys.Chem., 96, 1444 (1992).

5. R.J. Lumby, B. North and A.J. Taylor, Special Ceramics 6, 321, ed. P.Popper, (Brit.Ceram.Res.Assoc. 1974).

6. M.H. Lewis, B.D. Powell, P. Drew, R.J. Lumby, B.North and A.J. Taylor, J.Mat.Sci. 12, 61 (1977).

7. B.S.B. Karunaratne and M.H. Lewis, J.Mat.Sci. 15, 449 (1980).

8. B.S.B. Karanaratne and M.H. Lewis, J.Mat.Sci. 15, 1781 (1980).

9. M.H. Lewis and B.S.B. Karunaratne, 'Fracture Mechanics for Ceramics, Rocks and Concrete', 13, ed. S. Frieman and E. Fuller (ASTM, STP745 Philadelphia 1981).

10. M.H. Lewis, A.R. Bhatti, R.J. Lumby and B.North, J.Mat.Sci 15, 103 (1980).

11. M.H. Lewis, G. Leng-Ward, S.M. Winder and R.J. Lumby, Deformation of Ceramics II, 605, eds. R.E. Tressler & R.C. Bradt (Plenum Press 1984).

12. M.H. Lewis, G.Leng-Ward and C. Jasper, Ceramic Transactions 1; Ceramic Powder Science, 1019, ed. G.L. Messing, E.R. Fuller and H. Hausner, (Am.Cer.Soc., Ohio 1988).

13. J. Fernie, G. Leng-Ward and M.H. Lewis, Materials Letters 9, 29 (1989).

14. M.H. Lewis, S. Mason and A. Szweda, Non-Oxide Technical and Engineering Ceramics, 175, ed. S. Hampshire (Elsevier Applied Science 1986).

15. C. Jasper, Ph.D. Thesis, University of Warwick 1990.

16. M.H. Lewis, B.S.B. Karunaratne, J. Meredith and C. Pickering, 'Creep and Fracture of Engineering Materials and Structures', 365, ed. B. Wilshire and D.R. Owen, (Pineridge Press, U.K., 1981).

17. S. Hampshire, H.K. Park, D.P. Thompson and K.H. Jack, Nature 274, 280 (1977).

18. M.H. Lewis, R. Fung and D.M.R. Taplin, J.Mat.Sci. 16, 3437 (1981).

19. C.A. Jasper and M.H. Lewis, 'Ceramic Materials and Components for Engines', 424, ed. R. Carlsson, T. Johansson and L. Kahlman, (Elsevier Appl. Science, 1992).

20. M.H. Lewis and R.J. Lumby, Powder Metall. 26 73, (1983).

21. S. Hampshire, Y-J Song, D. O'Sullivan and V. Gunay, 'Ceramic Materials and Components for Engines', 707, ed. R. Carlsson, T. Johansson and L. Kahlman (Elsevier Appl. Science, 1992).

22. S.M. Ketchion, G. Leng-Ward and M.H. Lewis, ibid., 609.

23. F. Hong, R.J. Lumby and M.H. Lewis, J. European Ceram.Soc., in press.

24. F. Hong and M.H. Lewis, Ceram.Eng. and Sci. Proceedings 14, (in press., Am. Ceram.Soc., 1993).

25. I.P. Tuersley, G. Leng-Ward and M.H. Lewis, Ceram. Mat. and Components for Engines, 856, ed. V.J. Tennery (Am.Ceram.Soc.,1988).

26. I.P Tuersley, G. Leng-Ward and M.H. Lewis, Advanced Engineering with Ceramics, 231, ed R. Morrell (Inst. of Ceramics 1990).

27. R.L. Yeckley and K.N. Sieben, Ceram.Mat. and Components for Engines, 751, ed. V.J. Tennery (Am.Ceram.Soc. 1988).

28. K. Liddell and D.P. Thompson, Br.Ceram.Trans.J., 85 17 (1986)

29. G. Leng-Ward and M.H. Lewis, in preparation for publication (from EURAM contract MA1E/OO99/C)

30. G. Leng-Ward and M.H. Lewis, 'Glasses and Glass-Ceramics', 106, ed. M.H. Lewis (Chapman and Hall 1988).

31. A.G. Razzell and M.H. Lewis, Ceram. Eng. and Science Proc., 12, 1304, (Am.Ceram.Soc., Ohio 1991).

STATUS OF RESEARCH AND DEVELOPMENT ON MATERIALS
FOR CERAMIC GAS TURBINE COMPONENTS

MINORU MATSUI
NGK Insulators, Ltd., 2-56 Suda-cho, Mizuho-ku, Nagoya, 467 Japan.

INTRODUCTION

Aggressive research and development has raised structural ceramics to a level where they are now of practical use as an alternative to metal. Major applications include use in automobile engine components, such as turbocharger rotors [1].
Recently, because of energy conservation and environmental protection concerns, ceramic gas turbines have become a major research and development theme in many countries [2],[3]. In Japan, too, ceramic gas turbines have been researched for a variety of applications including co-generation [4], automobiles [5], and electric power generation [6]. To design a practical ceramic gas turbine, it is necessary to develop a ceramic that can resist heats of over 1350°C and a special component manufacturing process. It is also necessary to develop material evaluation and component design techniques.
Unlike metals, ceramics are brittle. They are very structure-sensitive, and their strength depends upon the existing flaws. Thus, the strength of ceramic components tends to vary, and reliability in terms of strength and lifetime has been a crucial issue in research. To ensure that ceramics are reliable, it is necessary to establish a method to evaluate strength in various fracture modes, analyze fracture mechanisms, accumulate data about strength and fatigue for use in reliability evaluation, and establish proof test techniques. Once a reliability evaluation technique is established, we will be able to design highly reliable ceramic components by selecting ceramics with properties suitable for the component operating environment or by finding out a shape suitable for the properties of the ceramics.
In this paper, the status of structural ceramics development for gas turbine component, data on the strength and fatigue of ceramics (in particular, silicon nitride (Si_3N_4)), and a component design technique based on Fracture Map are reported on.

STATUS OF STRUCTURAL CERAMICS DEVELOPMENT FOR GAS TURBINE COMPONENT

Two candidate structural ceramics for gas turbine components are under examination. They are Si_3N_4 and silicon carbide (SiC). SiC has excellent heat resistance, and maintains its strength up to 1500°C. However, it has a low fracture toughness and is very brittle. Si_3N_4 is slightly inferior to SiC in terms of heat resistance, but it has a high fracture toughness and is less brittle than SiC. In addition, Si_3N_4 has excellent thermal shock resistance, because its Young's modulus and its thermal expansion coefficient are lower than those of SiC. Through recent research efforts, the heat resistance of Si_3N_4 has been improved, and a Si_3N_4 has been developed that maintains its strength up to 1400°C. Figure 1 shows improvements in the flexural strength of sintered Si_3N_4 at room temperature and 1400°C, confirmed by four-point flexural tests. The strength has been remarkably improved since 1980, at both room temperature and 1400°C. The strength at room temperature was improved by increasing density of the sintered body, and that at 1400°C was improved by promoting crystallization in the grain boundary phase. Figure 2 shows the improvements in the fracture toughness of sintered Si_3N_4, and Figure 3 shows the improvements in the Weibull modulus of sintered Si_3N_4. Fracture toughness

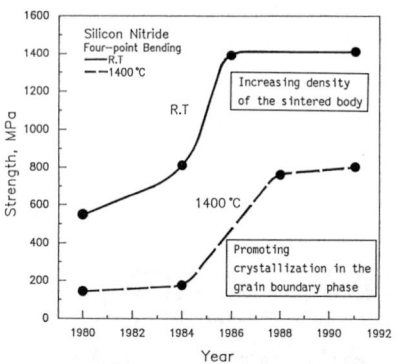

Fig.1. Improvement of Si_3N_4 in strength at room temperature and 1400°C.

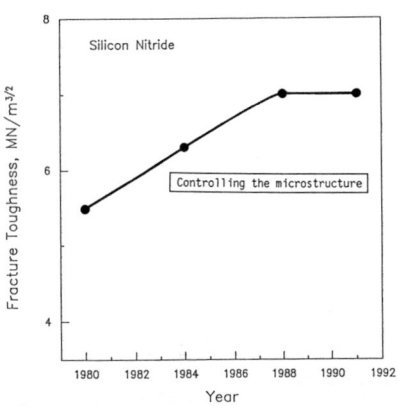

Fig.2. Improvement of Si_3N_4 in fracture toughness.

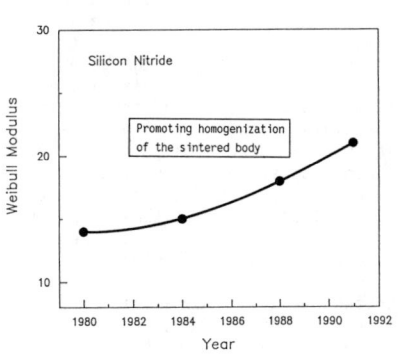

Fig.3. Improvement of Si_3N_4 in Weibull modulus.

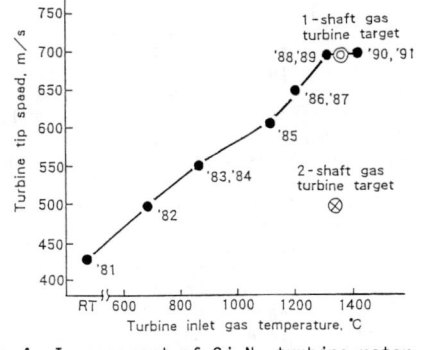

Fig.4. Improvement of Si_3N_4 turbine rotor in the spin test condition.

has been improved every year by controlling the microstructure and the Weibull modulus was improved by promoting homogenization in the sintered body.

Figure 4 shows the improvements in the endurance limit of Si_3N_4 gas turbine rotors, confirmed by hot gas spin tests. Improvements in the strength of Si_3N_4 have raised both the durable tip speed and the turbine inlet temperature of the turbine rotor. In 1991, a turbine tip speed of 700 m/s and a turbine inlet temperature of 1400°C were achieved [7].

EVALUATION OF CERAMICS FOR GAS TURBINE COMPONENTS

To develop highly reliable gas turbine components, materials have been evaluated in multiple aspects including fast fracture strength, static fatigue strength, cyclic fatigue strength, thermal stress fracture strength, thermal shock strength, thermal fatigue strength, oxidation fatigue strength, and foreign object damage, and so on. In designing ceramic components, the most important properties are the fast fracture strength and the static and cyclic fatigue strengths. This section gives an overview of the fast fracture strength and static and cyclic fatigue strengths of ceramic materials, and especially those of Si_3N_4.

Fast Fracture Strength

Volume Effect on Strength

The strength of ceramics depends upon the dimensions of their existing flaws. Therefore, the distribution of strength is dominated by the distribution and dimensions of flaws, and Weibull statistics is generally used for the analysis of the distribution. The failure probability of a component under uniaxial stress is expressed by the Weibull distribution function in Equation (1).

$$F = 1 - \exp\left[-\int_V \left(\frac{\sigma - \sigma_u}{\sigma_0}\right)^m dV\right] \tag{1}$$

σ is the stress at an arbitrary point in the material, σ_u is the location parameter, σ_0 is the scale parameter, and m is the shape parameter. Usually, m is called the Weibull modulus. In general, σ_u is substituted by 0 and a two-parameter Weibull distribution is applied. With the maximum stress on the component, σ_{max}, applied, Equation (1) is transformed into Equation (2).

$$F = 1 - \exp\left\{-\left(\frac{\sigma_{max}}{\sigma_0}\right)^m V_e\right\} \tag{2}$$

$$V_e = \int_V \left(\frac{\sigma(x,y,z)}{\sigma_{max}}\right)^m dV$$

where V_e is an effective volume. The strengths of components with different effective volumes can be represented by the ratio between the effective volumes, as shown in Equation (3):

$$\frac{\sigma_1}{\sigma_2} = \left(\frac{V_{e2}}{V_{e1}}\right)^{1/m} \tag{3}$$

Figure 5 shows the relationship between the effective volumes and mean strengths confirmed in four materials: three sintered Si_3N_4 (SSN-A1, SSN-A2, and SSN-B), and SiC [8]. The difference between SSN-A and SSN-B depends on the applied sintering additives. Tensile strength was tested with rod specimens 6 to 20 mm in gage diameter. Bending strength was tested through three- and four-point bending tests according to JIS (Japanese industrial standard)-R1601. Under uniaxial stress, a linear relationship was observed

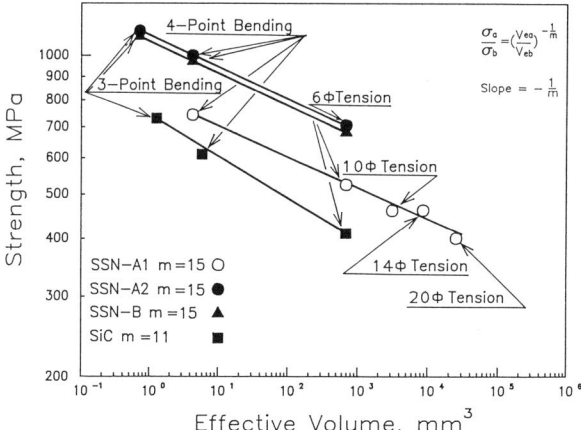

Fig.5. The relationship between the effective volumes and mean strengths for SSN-A1, SSN-A2, SSN-B and SiC under uniaxial stress.

between the effective volume and mean strength in both Si_3N_4 and SiC. The Weibull modulus can be obtained from the gradient of the line. The Weibull modulus obtained from the gradient of the line was 15 for SSN-A1, SSN-A2, and SSN-B, and 11 for SiC. These values agree with the values obtained from the Weibull plot of the four-point bending strengths of the ceramics. Therefore, the strength of ceramics based on two-parameter Weibull distribution can be estimated from the volume effect in Equation (3).

Multiaxial Stress Effect on Strength

Gas turbine components are subject not only to simple uniaxial stresses but also to complex multiaxial stresses, such as tension-compression and tension-tension biaxial stresses. There is a known distribution function for failure probability under multiaxial stresses, as shown in Equation (4) [9]. By using the effective volume shown in Equation (5), the strength under multiaxial stresses can be predicted in the same way as that under uniaxial stress.

$$F = 1 - \exp\left\{ -Ve_m \left(\frac{\sigma_{1max}}{\sigma_0}\right)^m \right\} \qquad (4)$$

$$Ve_m = \frac{2}{\pi \Omega} \int_V \left\{ \int_0^{\pi/2} \int_0^{\pi/2} \left(\frac{Z}{\sigma_{1max}}\right)^m \sin\phi \, d\phi \, d\theta \right\} dV \qquad (5)$$

ϕ and θ are the angles between the principal stress axes and the line normal to the crack plane. Ω is the normalization constant to ensure agreement between the failure probability under a uniaxial tensile stress and that obtained by the Weibull uniaxial distribution function. Z is an equivalent normal stress defined by the unstable brittleness propagation condition of the crack in a mixed mode, and the equation for Z varies with the fracture criterion.

Figure 6 shows the results of testing SSN-A1 rod specimens, which were 6 mm in diameter, under tension-compression biaxial stresses [10]. The tension-compression biaxial stresses were applied in the following sequence: tensile-torsional, compressive-torsional, and torsional tests. The ratios between shear stress τ caused by torsional load and tensile or compressive stress σ (τ/σ) were set as 0.5, 1, 2, 0, -2, and -1. Figure 6 shows the failure curves representing the failure strengths predicted by the uniaxial distribution function of Equation (2) (maximum principal stress theory) and multiaxial distribution functions based on various fracture criteria: Weibull multiaxial theory, G-criterion (coplanar crack extension and non-coplanar

Table I. The fracture criteria and equivalent normal stress Z

Fracture Criterion	:	Equivalent Normal Stress Z
- Shear Insensitive Criterion -		
Weibull Multiaxial Theory	:	$Z = \sigma_n$
- Shear Sensitive Criterion -		
G-criterion (coplanar)	:	$Z = \sqrt{\sigma_n^2 + \frac{4}{(2-\nu)^2} \tau_n^2}$
(non-coplanar)	:	$Z = (\sigma_n^4 + 6\sigma_n^2 \tau_n^2 + \tau_n^4)^{1/4}$
σ_θ-criterion	:	$Z = \sigma_n \cos\left(\frac{\theta_m}{2}\right) \left[\cos^2\left(\frac{\theta_m}{2}\right) + \frac{3}{2} \alpha \sin\theta_m \right]$

σ_n : Normal stress vertical to crack surface
τ_n : Shear stress parallel to crack surface
α : Stress ratio $\alpha = \tau_n / \sigma_n$, $\quad \theta_m$: $\theta_m = \cos^{-1}\left(\frac{3\alpha^2 + \sqrt{8\alpha^2 + 1}}{9\alpha^2 + 1}\right)$

crack extension), and σ_θ-criterion. Table I shows the fracture criteria and equations for equivalent normal stress Z used in strength prediction. σ_1 and σ_2 are the maximum and minimum principal stress causing the test specimen to failure. Both experimental and predicted values for σ_1 and σ_2 are normalized so that the tensile strength is 1. The strengths predicted according to the maximum principal stress theory and the Weibull multiaxial theory are in good agreement with the failure strength. The strengths predicted according to the G-criterion and the σ_θ-criterion are lower than the failure strength. Figure 7 shows the results of similar tests on SiC [8]. Unlike the results from Si_3N_4, the strength predicted according to the G-criterion is in good agreement with the failure strength. Thus, it is considered that, under tension-compression biaxial stresses, the suitable fracture criterion for strength prediction varies with the ceramics. The modified G-criterion, shown as Equation (6), as the fracture criterion under multiaxial stresses are proposed [10].

$$Z = ((\sigma_n^2 + (\beta \tau_n)^2)^{1/2} \qquad (6)$$

With the modified G-criterion, the effect of shear stress on the failure can be specified by changing constant β in the equation. For each of SSN-A1, SSN-B, and SiC, the value for constant β to ensure the best possible agreement between the strength predicted according to the modified G-criterion and the failure strength were obtained. Figure 8 shows the strengths predicted according to the modified G-criterion with the values for β thus obtained, together with the failure strengths of the ceramic [8]. The values

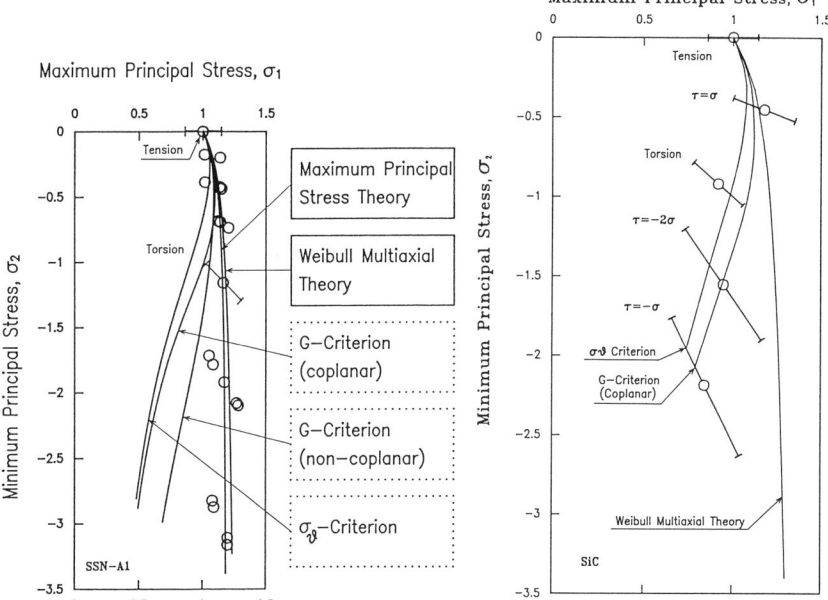

Fig.6. Comparison of the experimental results and predictions according to various fracture Criterion for SSN-A1 under tension-compression biaxial stresses. Plots represent experimental results, lines represent predictions.

Fig.7. Comparison of the experimental results and predictions according to various fracture Criterion for SiC under tension-compression biaxial stresses. Plots represent experimental results, lines represent predictions.

for constant β that resulted in the closest predictions were 0.65 for SSN-A1, 0.8 for SSN-B, and 1.1 for SiC. β probably varies because of differences in the microstructure.

Figure 9 shows the results of testing an SSN-A2 disk specimens under tension-tension biaxial stresses [8]. The tension-tension biaxial stresses were applied by ball-on-ring and ring-on-ring tests. Figure 9 shows the failure curves that represent the strengths predicted according to each fracture criterion. σ_1 and σ_2 are the principal stress causing the specimen to failure, and both experimental and predicted values for σ_1 and σ_2 are normalized so that the tensile strength is 1. The strengths predicted according to the σ_θ -criterion are in good agreement with the failure strength. The strengths predicted according to other multiaxial stress-based criteria are lower than the failure strength, resulting in a strength prediction that is on the safe side from the view point of component design. The strength predicted according to the maximum principal stress theory is higher than the failure strength, resulting in strength prediction on the dangerous side. Similar results were obtained in the test of SiC [8]. Thus, the strength did not vary with the ceramics under tension-tension biaxial stresses, although it did under tension-compression biaxial stresses.

Therefore, when predicting the strength of various ceramics under multiaxial stresses consisting of combined tension-compression and tension-tension biaxial stresses, a fracture criterion which are determined according to experiment results is suitable. An example of such a criterion is the modified G-criterion.

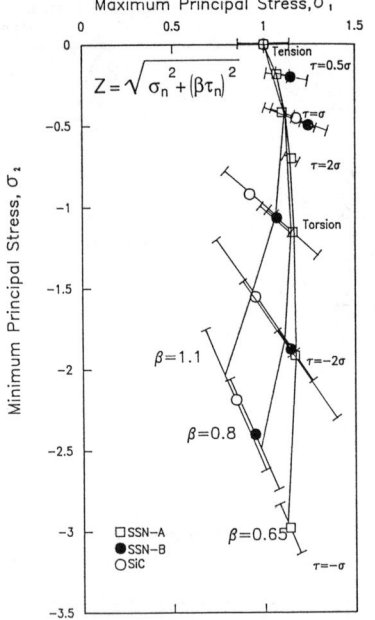

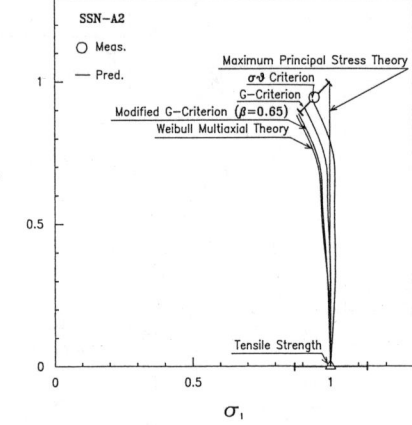

Fig.8. Comparison of the experimental results and predictions according to the modified G-criterion for SSN-A1, SSN-B, and SiC under tension-compression biaxial stresses. Plots represent experimental results, lines represent predictions.

Fig.9. Comparison of the experimental results and predictions according to various fracture Criterion for SSN-A2 under tension-tension biaxial stresses. Plots represent experimental results, lines represent predictions.

Fatigue Strength

This section describes the static fatigue strength and the cyclic fatigue strength of sintered Si_3N_4.

Static Fatigue Strength

The static fatigue lifetime of ceramics is generally predicted by the power law crack growth rate equation as in Equation (7) [11],[12].

$$\frac{da}{dt} = A K_I^n \quad (7)$$

where A and n are constants, and a is the flaw size. Stress intensity factor K_I is given by Equation (8).

$$K_I = Y \sigma a^{1/2} \quad (8)$$

where σ is the applied stress and Y is the geometrical factor. From Equations (7) and (8), lifetime t_s under constant stress in a static fatigue test is obtained as follows:

$$t_s = C_s \sigma^{-n} \quad (9)$$

$$C_s = \frac{2 \sigma_i^{n-2}}{(n-2) A Y^2 K_{IC}^{n-2}} \quad (10)$$

where σ_i is the inert strength for initial flaws. Apart from the static fatigue test, a dynamic fatigue test is also used to evaluate the crack growth behavior. The fracture strength σ_f under a constant stress rate $\dot{\sigma}$ in a dynamic fatigue test is obtained as follows, based on Equations (7) and (8):

$$\sigma_f^{n+1} = (n+1) C_s \dot{\sigma} \quad (11)$$

The time to failure t_f in a dynamic fatigue test can be converted into the static fatigue lifetime t_{eq} (hereinafter referred to as equivalent time) as follows, based on Equations (9) and (11):

$$t_{eq} = \frac{1}{(n+1)} t_f \quad (12)$$

Figures 10 shows the static fatigue characteristics of three types of sintered Si_3N_4, with different sintering additives, under a tensile stress [13]. Figure 10-(a) shows the results of the static and dynamic fatigue tests (the relationship between the applied stress and equivalent time) for SSN-A,

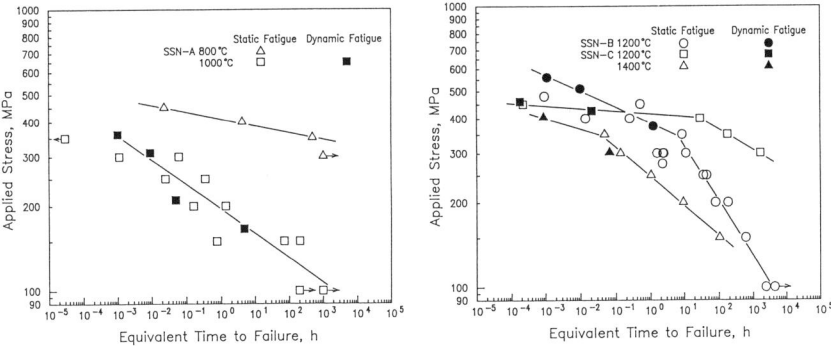

(a) SSN-A (b) SSN-B and SSN-C
Fig.10. Comparison of the static and dynamic fatigue life time.

and Figure 10-(b) shows those for SSN-B and SSN-C. At a high temperature, the three types of Si_3N_4 differ from each other in their static fatigue behavior. These differences are caused by the differences in the high-temperature characteristics of their grain boundary phases. While the grain boundary phases of SSN-A are glassy, those of SSN-B and SSN-C are crystalline. Glassy grain boundaries start to soften at a lower temperature than crystalline grain boundaries. Furthermore, the softening of the crystalline grain boundaries is suppressed at a higher temperature in SSN-C than in SSN-B. As a result, SSN-C has excellent static fatigue characteristics at high temperatures, followed by SSN-B then SSN-A.

For SSN-A, the equivalent lifetime of static and dynamic fatigue tests at 1000°C are in good agreement with each other. The strength degradation was caused by slow crack growth from existing flaws. However, remarkable creep deformation appeared in SSN-A at over 1100°C.

For SSN-B at 1200°C and SSN-C at 1200 and 1400°C, the results are divided into two stress regions. In the high stress region, it is possible to apply the power law crack growth formulation to static and dynamic fatigue data. In this region, the strength degradation was caused by slow crack growth from existing flaws. However, in the low stress region, it is not possible to apply Equation (7), based on the slow crack growth from existing flaws, to the static fatigue data. In this region, SSN-B and SSN-C showed creep deformation. Therefore, it is necessary to use another lifetime prediction procedure for this region.

The following Larson-Miller parameter P is generally used to predict the lifetime of metals before creep ruptures appear:

$$P = f(\sigma) = T (C + \log t_r) \qquad (13)$$

where T is absolute temperature, t_r is time to creep rupture, and C is a constant.

Figures 11-(a) and 11-(b) show the master Larson-Miller curve, where C=44 for SSN-B and C=30 for SSN-C are used [13]. The fatigue data of the high temperatures that cause creep deformation can be represented by a temperature-independent straight line. Thus, the Larson-Miller parameter can be applied to the prediction of lifetime in the creep deformation fracture region. C=20 is generally used for metals, however, C=20 is unsuitable for Si_3N_4. As the

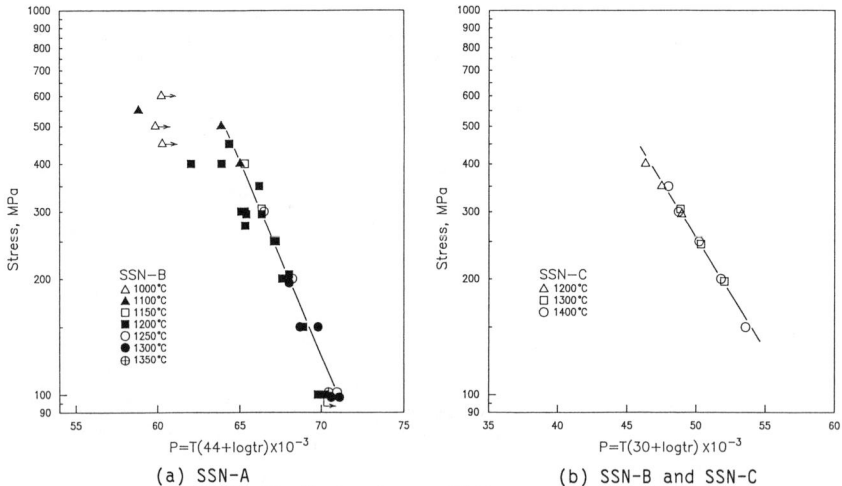

(a) SSN-A (b) SSN-B and SSN-C
Fig.11. Master Larson-Miller curve.

C value gets larger, the effect of temperature on the creep rupture lifetime increases.

The fracture surfaces of the Si_3N_4 can be classified into two distinctive types. The fracture surfaces that failed at low temperatures and high stress have pores and inclusions, and slow crack growth regions were observed around them. On the other hand, at high temperatures and low stress, relatively large regions with irregularities, which are creep deformation regions, were observed on the fracture surfaces. Such regions tended to increase with decreasing stress.

In summary, static fatigue in Si_3N_4 is generally caused by slow crack growth or creep deformation. Thus, the fracture mechanism must be identified to predict lifetimes appropriately.

Cyclic Fatigue Strength

This section describes examples of the cyclic fatigue characteristics of sintered Si_3N_4 at room and high temperatures.

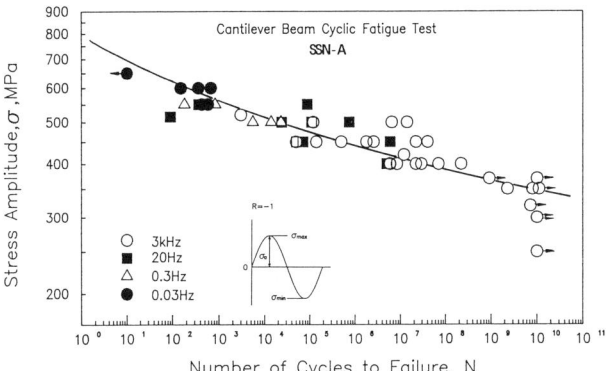

Fig.12. S-N curve of cantilever beam specimens of sintered Si_3N_4 under alternating load, at frequencies from 0.03Hz to 3kHz.

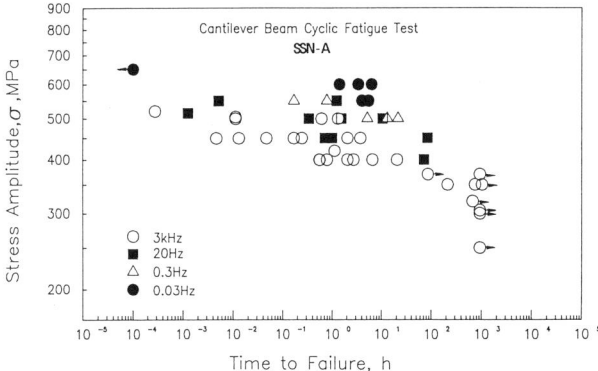

Fig.13. Time-to-failure data calculated from the S-N data for the tapered cantilever beam specimens in Figure 12.

Cyclic Fatigue Strength at Room Temperature

Figure 12 shows the stress-number (S-N) curve for tapered cantilever beam specimens of sintered Si_3N_4 under alternating loads, at frequencies from 0.03 Hz to 3 kHz [14]. Figure 13 shows the time-to-failure data calculated from the S-N data for the tapered cantilever beam specimens in Figure 12. The cyclic fatigue behavior of sintered Si_3N_4 appears to depend principally on the number of cycles rather than on time, because the fatigue strength at various frequencies did not show any discontinuities against the number of cycles (see Figure 12 and Figure 13). The solid curve in Figure 12 suggests a decrease in the fatigue limit to about 40% of its initial strength.

Figure 14 shows the mean stress/stress amplitude diagram of sintered Si_3N_4 [15]. σ_t on the horizontal axis represents the mean tensile strength of 6-mm diameter button-head type specimens. Open circles represent specimens surviving at 10^7 cycles, and closed circles represent the specimens that failed before 10^7 cycles. The broken line represents tensile strength. Alternating loads decreased the strength to a larger extent than pulsating loads. This diagram shows more clearly the safe stress state region. The solid line from σ_e to σ_t represents a modified Goodman line, which is expressed by Equation (14) and is used as a design criterion for metal components:

$$\sigma_a = \sigma_e (1 - (\sigma_m / \sigma_t)) \tag{14}$$

where σ_a is stress amplitude, σ_m mean stress, σ_t tensile strength, and σ_e alternating fatigue strength at stress ratio R=-1. The modified Goodman line agreed well with the results of the present fatigue test. The design methodology proposed for metal fatigue is applicable to the sintered Si_3N_4.

To clarify the effects of volume on cyclic fatigue, cyclic fatigue tests under alternating loads were carried out, using two kinds of specimens with different volumes. Figure 15 shows the results of cyclic fatigue tests at room temperature, in which tensile specimens and tapered cantilever beam specimens of sintered Si_3N_4 were used [16]. The fatigue limit of both specimens with different volumes seems to be the same.

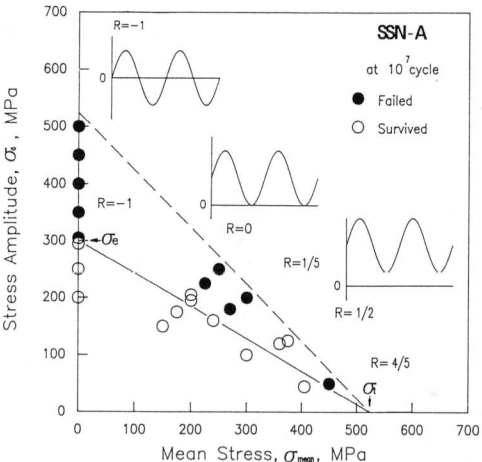

Fig.14. Mean stress versus stress amplitude for fatigue failure in sintered Si_3N_4 at 10^7 cycles measured at room temperature.

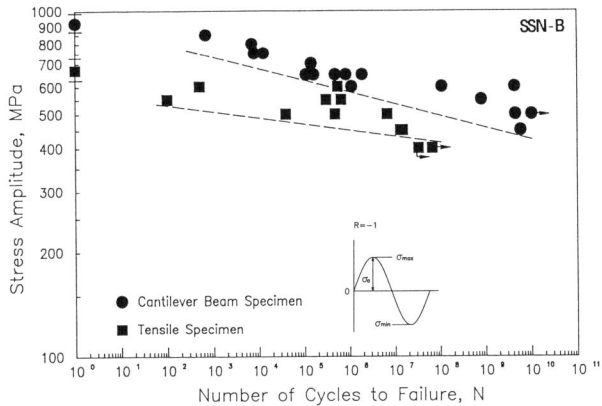

Fig.15. The effect of volume on cyclic fatigue at room temperature. Cantilever beam specimens and tensile specimens of sintered Si_3N_4 were used; the former has a smaller volume than the latter.

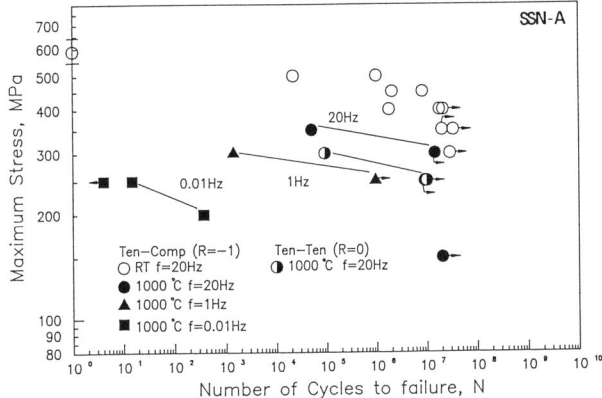

Fig.16. S-N diagram of sintered Si_3N_4 under tension-compression alternative loading at room temperature, 800 and 1000°C. The symbols on the vertical axis represent the tensile strength of the used materials.

Cyclic Fatigue Strength at Elevated Temperatures

Data for sintered Si_3N_4 with a glassy grain boundary phase at 800, 1000°C, and room temperature were plotted on a log-log plot of stress versus number of cycles to failure, as shown in Figure 16 [16]. The mean tensile strengths of 6-mm diameter button-head type specimens at room temperature, 800 and 1000°C were 585, 560, and 360 MPa, respectively. The arrow pointing to the left indicates the specimen failed at the onset of loading, and the ones pointing to the right indicate that the specimens survived at that cycle without failure. There was a large scatter, but the fatigue life of the specimens greatly increased as stress amplitude decreased.

At room temperature, the fatigue strength at 10^7 cycles was about 60% of

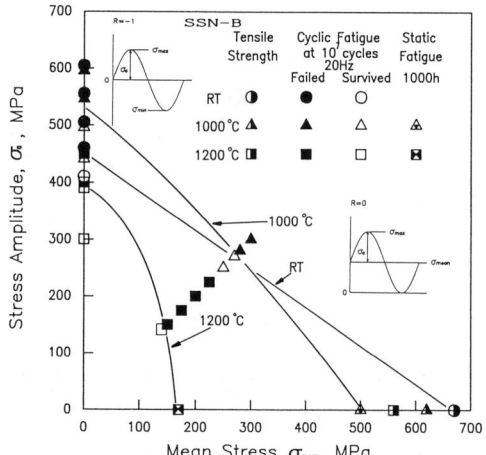

Fig.17. Tensile mean stress effects on cyclic fatigue of sintered Si_3N_4 at room temperature, 1000 and 1200°C.

the initial strength. Cyclic fatigue degradation at 1000°C with a stress ratio of -1 at a frequency of 20 Hz was smaller than at room temperature. In contrast to cyclic fatigue behavior at room temperature, the onset of the fatigue life at 1000°C depended on frequency, as shown in Figure 16. At elevated temperatures, the lower the frequency, the less the fatigue resistance. Low cycle fatigue at elevated temperatures were dominated by time-dependent fatigue.

The effects of mean tensile stress and static fatigue on the cyclic fatigue of sintered Si_3N_4 with a crystalline grain boundary phase at elevated temperature are shown in Figure 17 [16]. These results were obtained under load control at a constant frequency of 20 Hz. The vertical axis represents a fully reversed fatigue condition and the horizontal axis represents the static fatigue strength and tensile strength. As the temperature was elevated, static fatigue strength decreased. The boundary within which the safety stress region is present tends to become elliptical as the temperature elevates. An approximation for the mean stress effect is shown in Equation (15) for elevated temperatures.

$$\left(\frac{\sigma_a}{\sigma_e}\right)^n + \left(\frac{\sigma_m}{\sigma_s}\right)^n = 1 \qquad (15)$$

where σ_a is stress amplitude, σ_m mean stress, σ_e alternating fatigue strength and σ_s static fatigue strength. The n value increased with elevating temperatures. At elevated temperatures, fatigue strength can be predicted by a schematic diagram on which an elliptical curve connects the results of a alternating fatigue test to those of a static tensile fatigue test. This method is effective for examining the safety stress conditions of a Si_3N_4 on which cyclic stress is applied.

COMPONENT DESIGN

To design highly reliable ceramic components, it is necessary to precisely clarify the properties of the ceramics to be used and, secondly, to accurately predict the strength and lifetime of the components. To clarify the properties precisely, it is necessary to accurately evaluate the strength and its dispersion according to flexural and tensile tests and the generation and growth of flaws according to static and cyclic fatigue tests. It is also necessary to analyze the mechanism of each fracture mode by observing the fracture surfaces. To accurately predict the strength and lifetime of components, it is necessary to precisely analyze and obtain the temperature and stress distributions in the components using the finite element method (FEM) analysis on a computer. To precisely obtain temperature and stress distributions, it is necessary to precisely clarify the thermal and mechanical properties of the ceramics and the thermal and mechanical boundary conditions that would affect the components in the actual operating environment. This section outlines Fracture Map that has been developed for designing ceramic components.

Design Concept of Fracture Map

Figure 18 shows the mechanical design concept of ceramic components using

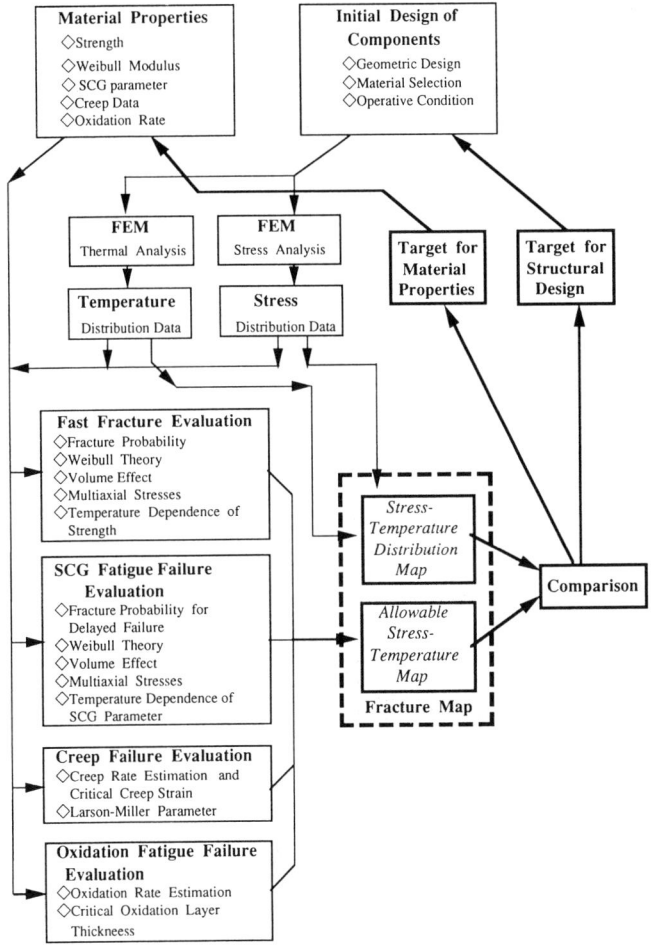

Fig.18. Mechanical design concept of ceramic components using Fracture Map.

Fracture Map. Fracture Map is used to evaluate whether a ceramic component can endure for the target lifetime in the operating environment. The evaluation covers the following fracture modes: fast fracture, SCG fatigue failure, creep failure, and oxidation fatigue failure. The reliability of a component can be guaranteed if the stress in the component is lower than the tolerable stress in all modes. The tolerable stress can be calculated according to the temperature and stress distributions in the component and the strength evaluation methods in the fracture modes mentioned above. For the concept of the tolerable stress, refer to [17] and [18] in the bibliography.

In the fast fracture and SCG fatigue failure modes, the fracture strength depends upon the distribution of existing flaws. Therefore, the volume effect of the components must be taken into consideration when evaluating the tolerable stress in terms of failure probability. In creep failure and oxidation fatigue failure modes, failures are mainly caused by newly formed flaws. Therefore, the volume effect is not considered when evaluating the tolerable stress. However, in either case, the tolerable stress can be calculated from the properties in each fracture mode, and component reliability can be evaluated. At present, calculation of the tolerable stress involves many assumptions; a more reliable and accurate method must be developed in the future. In addition, the fracture mechanism in the cyclic fatigue mode has not yet been clarified sufficiently. We are continuing to analyze the mechanism and develop a reliability evaluation method.

Application of Fracture Map

This section introduces an example of applying Fracture Map to a Si_3N_4 gas turbine rotor [19]. In a hot gas spin test where the turbine tip speed of the rotor is 750 m/sec and the turbine inlet temperature (TIT) is 1300°C, the temperature and stress distributions in the turbine rotor are obtained by three-dimensional FEM analysis as shown in Figures 19. Figure 20 shows the Fracture Map for the turbine rotor in the above test conditions. In Figure 20, the tolerable stresses are indicated by solid lines. They are calculated for critical creep strain of 0.3% and 1%, and for a 50% SCG fatigue failure probability in a five second test. The dots represent the temperatures and stresses at arbitrary points in the turbine rotor obtained by the FEM analysis. According to Figure 20, the maximum stress in the component (at portion A) is closest to the tolerable stress for the 50% SCG fatigue failure probability. This indicates that SCG is the most severe fracture mode in this test condition. The figure does not include the tolerable stress for oxidation fatigue failure, because the test time is so short (5 sec) that the tolerable stress becomes greater than that for SCG or creep. In the actual experiment,

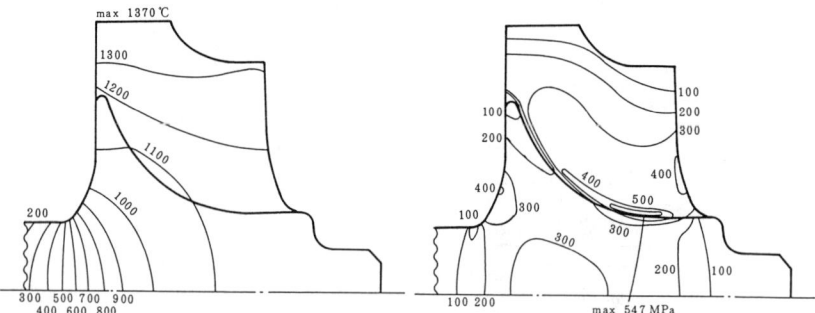

(a) Temperature distribution (b) Stress distribution
Fig.19. Temperature and stress distribution in a Si_3N_4 gas turbine rotor obtained by three-dimensional FEM analysis.

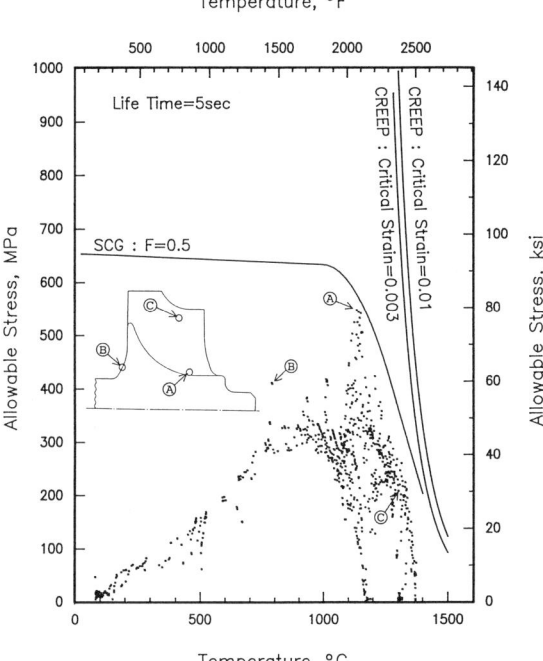

Fig.20. Fracture Map of a Si_3N_4 gas turbine rotor.

under the same conditions, the rotor fractured within a few seconds. The actual results were in good agreement with the predictions.

The above evaluation is only part of the reliability evaluation of a Si_3N_4 turbine rotor. When developing a turbine rotor, it is necessary to evaluate other properties including the strength of the joints between the ceramic rotor and the metal shaft and the FOD (foreign object damage) resistance of the ceramic blades.

SUMMARY

This paper has outlined the status of research and development on materials for ceramic gas turbine components. The contents are as follows.

(1) Strength properties of structural ceramics for gas turbine components, such as fast fracture strength and fatigue failure strength, were clarified.

(2) The applicability of Fracture Map to the material evaluation for gas turbine components was demonstrated by hot gas spin tests.

In the near future, all properties required for ceramics under practical use in the gas turbine should be made clear. Therefore, the long-term endurance tests using ceramic parts should be carried out under gas turbine operating conditions.

ACKNOWLEDGEMENT

A part of this work was performed under the management of the Engineering Research Association for High Performance Ceramics as a part of the R&D Project of Basic Technology for Future Industries supported by NEDO (New Energy and Industrial Technology Development Organization).

REFERENCE

1. H. Kawase, T. Matsuhisa, K. Kato, and T. Mizuno, ASME Paper No. 91-GT-270 (1991).
2. J.R. Smyth and R.E. Morey, ASME Paper No. 92-GT-381(1992).
3. M. Stute, H. Burger, M. Griguscheit, E. Holder, K.D. Morgenthaler, F. Neubrand, and M. Radloff, ASME Paper No. 90-GT-97(1991).
4. K. Honjo, R. Hashimoto, and H. Ogiyama, ASME Paper No. 92-GT-3(1992).
5. T. Itoh and H. Kimura, ASME Paper No. 92-GT-2(1992).
6. Y. Hara, Y. Furuse, T. Tsuchiya, F. Maeda, I. Tsuji, and K. Wada, 91-Yokohama-IGTC-101, pp. I-135-47 in Proceeding of Yokohama International Gas Turbine Congress.
7. Y. Kobayashi, E. Matsuo, T. Inagaki, and T. Ozawa, SAE Paper 910401(1991).
8. Y. Nakasuji, N. Yamada, H. Tsuruta, M. Masuda, and M. Matsui, in Fracture Mechanics of Ceramics, Vol.10, edited by R.C. Bradt, D.P.H. Hasselman, D. Munz, M. Sakai, and V.Ya. Shevchenko (Plenum Press, New York, 1992) pp. 211-226.
9. Y. Matsuo, Transactions of The Japanese Society of Mechanical Engineers-A, 46, 605-12(1980).
10. I. Oda, M. Matsui, T. Soma, M. Masuda, and N. Yamada, Nippon Seramikkusu Kyokai Gakujutsu Ronbunshi, 96 [5] 539-45(1988).
11. A.G. Evans, Int. J. Frac. 10, 251-59(1974).
12. T. Kawakubo and K. Komeya, J. Am. Ceram. Soc., 70[6], 400-405(1987).
13. M. Matsui, M. Masuda, and Y. Nakasuji, in Proceedings of the 1st International Symposium on the Science of Engineering Ceramics, edited by S. Kimura, K. Niihara (The Ceramic Society of Japan, Tokyo, 1991) pp. 177-182.
14. M. Masuda, N. Yamada, T. Soma, M. Matsui, and I. Oda, Nippon Seramikkusu Kyokai Gakujutsu Ronbunshi, 97 [5] 520-24(1989).
15. M. Masuda, T. Soma, M. Matsui, and I. Oda, Nippon Seramikkusu Kyokai Gakujutsu Ronbunshi, 96 [3] 277-83(1988).
16. M. Masuda, T. Soma ,and M. Matsui, Journal of the European Ceramic Society 6, 253-258(1990).
17. M. Matsui, T. Soma, Y. Ishida, and I. Oda, SAE Paper 860443(1986).
18. T. Soma, Y. Ishida, M. Matsui, and I. Oda, Advanced Ceramic Materials, Vol.2, No.4, pp.809-812(1987).
19. T. Ozawa, T. Matsuhisa, Y. Kobayashi, E. Matsuo, and T. Inagaki, SAE Paper 890247(1989).

DEVELOPMENT OF HIGH PERFORMANCE SILICON NITRIDE CERAMICS AND THEIR APPLICATIONS

YO TAJIMA
R&D Center, NTK Technical Ceramics, NGK Spark Plug Co., Ltd., 2808 Iwasaki, Komaki, Aichi, 485, Japan

ABSTRACT

Progress in sintering process and improvement of mechanical properties of silicon nitride ceramics are reviewed. Emphases are placed on contributions of advanced sintering techniques and better understanding of sintering additives and microstructure-properties relations. Current applications as engine components and cutting tools are described, and future prospect is considered.

INTRODUCTION

There has been a great deal of progress in the development of high performance silicon nitride ceramics owing to many factors including improvement of powder processing and better understanding of sintering phenomena and processing-microstructure-properties relations.[1] Based on the progress, it has become possible to produce various types of materials such as high toughness, high strength and heat resistant materials that have different chemistry and microstructures.

Since mid-1980's, applications of silicon nitride ceramics have considerably widened, ranging from engine components such as turbocharger rotors and glow plugs to industrial parts such as bearing materials and cutting tools. These applications have various requirements among which light weight, heat resistance, wear resistance and thermal shock resistance are important characteristics. In this paper, recent progress in the sintering process and the improvement of mechanical properties are reviewed, followed by descriptions of current status of applications and future prospect.

SINTERING PROCESS

Sintering process is a very important and characteristic process in producing silicon nitride ceramics. Gas pressure sintering (GPS) and hot isostatic pressing (HIP) have recently become common practice for achieving high reliability. Besides these advanced sintering techniques, improvement of sintering process is also related with development of new powders and better understanding of roles of sintering additives. Some of the advancement in these fields is described in the following.

Densification by GPS and HIP

GPS and HIP are phenomenologically somewhat similar but conceptionally totally different sintering processes.[2] The two processes are compared in Table 1. In the GPS, the major contribution for the densification is a high temperature, and a role of overpressure is to suppress the decomposition of silicon nitride. On the other hand, the major contribution for the densification is a high pressure in the HIP.

Because of these differences, characteristics of the resultant densified materials are also different. Due to the high sintering temperature, amounts of sintering aids can be reduced and grain growth is promoted by the GPS. The grain growth is effective in achieving high fracture toughness.[3,4] In the HIP, it has been shown that full densification is possible without adding sintering aids.[5] As the densification at lower temperatures is possible by the HIP, very fine microstructure can be produced by the HIP, which could result in high strength of the materials. The characteristic microstructures obtained by the GPS and HIP are shown in Fig. 1.

Table I. Comparsion of GPS and HIP

	GPS	HIP
Gas	N_2	N_2, Ar, etc.
Pressure	0.2 – 10 MPa	100 – 200 MPa
Temperature	High	Low – High
Pretreatment	Unnecessary	Encapsulation or Presintering
Major contribution for densification	High temperature	High pressure
Role of Gas Pressure	Suppress decomposition	Achieve densification
Characteristics – Amounts of sintering aids – Microstructure development	– Can be reduced – Grain growth promoted	– Can be reduced or eliminated – Grain growth can be minimized

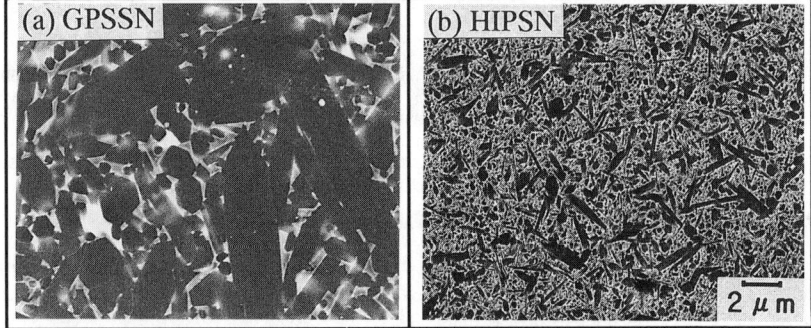

Fig. 1. Microstructures of (a)GPS and (b)HIP Si_3N_4. Both samples were polished and plasma-etched. GPSSN has a high fracture toughness (9 $MPam^{1/2}$), whereas HIPSN has a high flexural strength (1500 MPa).

As described so far, these two densification processes have their own advantages and disadvantages. Therefore, it is suggested that we should choose either process depending on what we want. For instance, if we want a high toughness material with coarse microstructure, the GPS would be a choice. If we want a high density, high strength material with fine microstructure, then the HIP would be a choice.

Sintering Additives

A fundamental role of sintering additives is promotion of densification, as the densification of silicon nitride is almost impossible without additives. In this respect, many oxide additives such as Al_2O_3, MgO and Y_2O_3 have been used. Broader views on additives would be possible in which their roles are regarded as not only densification aids but also key elements for the microstructural development, since the grain growth behavior and the nature of grain boundaries are, to some extent, influenced by the kinds and amounts of additives. Considering these viewpoints, two examples from the author's recent studies are described.

In the first example, roles of Cr_2O_3, V_2O_5 and SiO_2 were studied as the second additives for the Si_3N_4–Y_2O_3 system.[6,7] Additions of 4 wt% Y_2O_3 and 2 wt% of each second additive were examined. During the densification, these oxides were found to react with silicon nitride to supply excess oxygen resulting in better densification. The densification behavior was similar among the three compositions. In the cases of additions of Cr_2O_3 and V_2O_5, silicides were observed after the sintering. SEM observations revealed that the microstructures of the gas pressure sintered materials are quite different as shown in Fig. 2. Grain growth was most pronounced for the Cr_2O_3 addition followed by the V_2O_5 addition, whereas the SiO_2 addition did not promote the grain growth compared with the Si_3N_4–Y_2O_3 system.

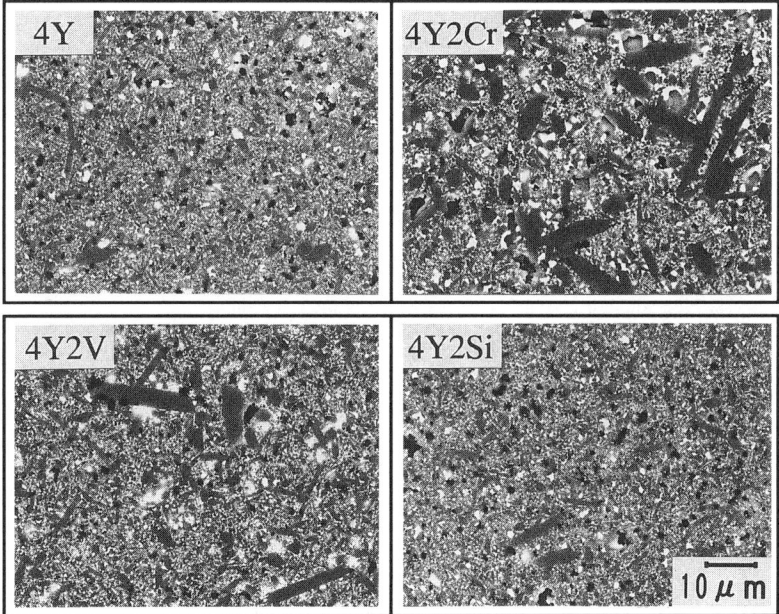

Fig. 2. Microstructures of 4 wt% Y_2O_3–doped GPSSN without or with the second additive (2 wt% Cr_2O_3, V_2O_5 or SiO_2). The materials were sintered at 1850°C for 4 h under 0.2 MPa N_2.

The second example is concerned with the modification of the grain boundary. The addition of MgO–ZrO$_2$ was studied to correlate the microstructures and properties.[8] It was found that the further addition of Al$_2$O$_3$ to the system markedly decreases the fracture toughness as shown in Fig. 3. Since the microstructures in terms of morphology and size of Si$_3$N$_4$ grains in these materials were not much different, it was considered that the difference in the toughness was due to the different nature of grain boundaries: bonding force becomes stronger as the amount of Al$_2$O$_3$ increases. The higher interfacial bonding force is thought to be responsible for the lower toughness analogous to fiber or whisker reinforced composites.

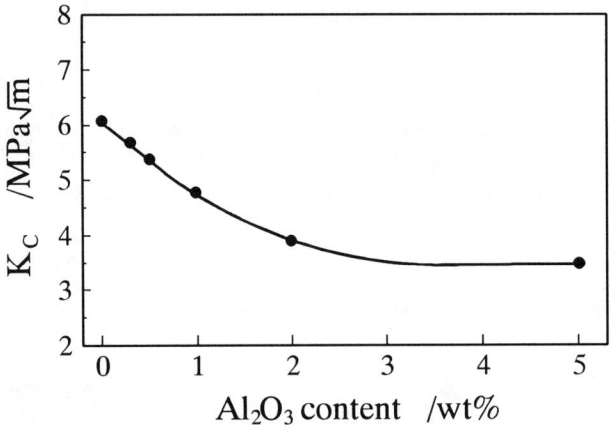

Fig. 3. Fracture toughness of GPSSN with additions of MgO, ZrO$_2$ and Al$_2$O$_3$ as a function of Al$_2$O$_3$ content.

PROPERTIES

Fracture toughness, strength and heat resistance are the three major properties that many researchers have tried to improve. As a result, significant improvements in these properties have been realized during the last decade. At the same time, it has become recognized that there is a trade-off in improving these properties; it is very difficult to achieve high fracture toughness, high strength and high heat resistance at the same time.

Fig. 4 shows the fracture toughness – strength relations for the silicon nitride ceramics that are commercially available and under development. The materials currently used for real applications have fracture toughness of 5–7 MPam$^{1/2}$ and a flexural strength of 700–1100 MPa. Attempts to improve these properties have resulted in two directions as shown in Fig. 4; higher strength and higher fracture toughness. It has been shown that the improvement of fracture toughness arises mainly from crack bridging by elongated silicon nitride grains resulting in rising R-curve behavior.[8,9] Thus, coarser microstructures are beneficial to the toughness improvement. On the other hand, the attempts to improve strength have been centered on achieving the fine microstructures with minimum sizes of defects while maintaining moderate fracture toughness. The examples of the microstructures derived from these two approaches are shown in Fig. 1.

The improvement of properties has been results of not only the improvement of processing but also the advancement of characterization techniques. Studies on fracture mechanisms have been quite active, and they have contributed to better materials design. Characterization of R-curve behavior has been conducted by several groups[8-10], and the relation between the R-curve and fatigue behaviors has been also studied.[8] More recently, the fracture mechanism was evaluated by using acoustic emission and back face compliance techniques,[11] and it was suggested that the frontal zone shielding in addition to the crack bridging also contributes to the toughening in in-situ toughened silicon nitride ceramics.

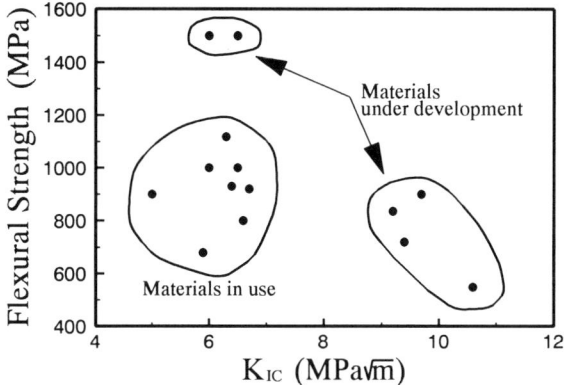

Fig. 4. Fracture toughness and strength of various silicon nitride ceramics.

APPLICATIONS

When considering applications, one of the most important characteristics of the silicon nitride ceramics is that their properties are well balanced compared with other ceramic materials. For instance, toughened zirconia can exhibit much higher strength and toughness, but its higher temperature properties and thermal shock resistance are poor. Or, silicon carbide has excellent heat resistance, but its toughness and thermal shock resistance are far behind. Another important aspect is that various types of materials can be produced by changing starting powders, kinds and amounts of additives and manufacturing processes. One could choose either high strength or heat resistant silicon nitride ceramics which have different chemistry and microstructures.

Silicon nitride ceramics are lighter and have better heat resistance than metals. These features make them the prime candidate materials for a gas turbine and other heat engines applications. Although the application to the gas turbine engine has not been commercialized yet, some engine components as well as some industrial parts including cutting tools are now in commercial production.

Engine Components

Table 2 summarizes engine components that are under commercial production.[12-25] Although purposes and requirements are somewhat different depending on the components, basic requirements are high strength, high thermal shock resistance and sufficient durability in order to assure that the components do not fail during the life of their services.

The first production ceramic turbocharger rotor using gas pressure sintered silicon nitride (GPSSN) was introduced in October 1985.[12,13] Prime objectives of the development of the ceramic rotor are improvement in acceleration response and heat resistance. The turbine rotor of a turbocharger operates under conditions of great stress, due to the high temperature and high speed of rotation. In addition, it must be reliably joined with a compressor rotor. Therefore, various development tasks including optimization of design, materials development, process engineering, joining technology and evaluation of reliability were conducted. Fig. 5 shows the result of durability tests during over-speed condition of the GPSSN rotors.[12] The failure stress, which was calculated by FEM as combined stress from centrifugal and thermal stress, distributed in the range much higher than 100% rated stress. The fatigue behavior may be estimated by the equation in the figure, and the fatigue parameter, n, in the equation seems to be sufficiently high. From these results, it was confirmed that the rotors made of GPSSN have high strength and high reliability.

In 1992, the larger size turbocharger rotor was introduced for the 460 hp on-highway truck diesel engine. The use of ceramics in this case has the effects of not only improving the acceleration response but also reducing particulate emissions.

Table II. Applications of silicon nitride ceramics to heat engines.

Component	Purposes	Requirements for materials
Turbocharger rotor	Response Power Clean exhaust gas	Light weight Heat resistance FOD resistance
Glow plug	Quick start Clean exhaust gas	Heat resistance Electrical insulation Thermal shock resistance
Swirl chamber	Power Clean exhaust gas	Heat resistance Thermal shock resistance
Rocker arm pad	Maintenance free	Wear resistance

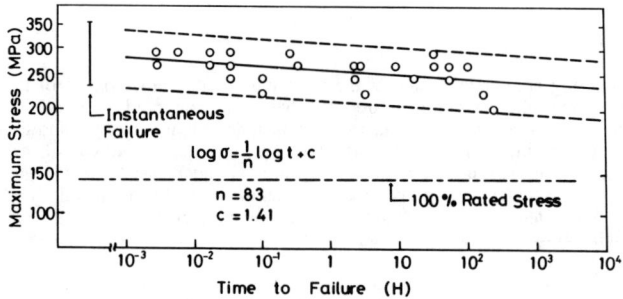

Fig. 5. Over-speed life test of GPSSN rotor at 900°C.

Ceramic glow plugs used in diesel engines consist of tungsten alloy heating element embedded in hot-pressed silicon nitride.[20] Due to the excellent heat resistance of the silicon nitride, quicker heating, longer after-glow and longer service life compared with conventional metal glow plugs are made possible with the ceramic glow plugs. Commercial production of the ceramic glow plugs started in 1981.

Ceramic swirl chambers are also used in diesel engines.[21-23] The lower half of the swirl chamber made of silicon nitride was developed for the improvement of heat resistance.[21,22] All ceramic swirl chambers were later developed in an attempt to reduce particulate emissions.[23]

In the liquid propane fueled taxis, the combination of the fuel-burning characteristics and the frequent starts and stops resulted in inadequate lubrication of the rocker arm wear pads. Metal pads failed rapidly, whereas silicon nitride provided greatly improved wear life.[24,25] Silicon nitride pads are either cast into or brazed to the metal rocker arm.

Cutting Tools

Silicon nitride is more heat resistant than tool-steel and tungsten carbide, and shows superior toughness and thermal shock resistance to alumina. These features made silicon nitride ceramics be widely applied as cutting tool materials. They have become popular because they allow vastly higher metal removal rates and subsequent productivity increase.[26]

An essential limitation, however, seems to exist in machinable workpiece materials. Gray cast iron with lamellar graphite and nickel-based superalloys can be machined successfully, whereas cast iron with globular graphite and steel generate rapid wear progress.[26-28] Reasons for good machinability of these workpiece materials are considered as follows.[28] Gray cast iron tends to be brittle, and so the chip/tool contact length is short. Therefore, the cutting temperature is relatively low resulting in the reduced wear. On the other hand, nickel-based superalloys have a relatively low chemical affinity for the tool materials.

The improvement of wear resistance has been attempted by several approaches.[29] CVD coatings of wear resistant materials such as alumina have been found to be effective. In this case, thickness of the coated layer is critical in obtaining good wear resistance. The improvement of abrasive wear was observed for the materials with smaller amounts of grain boundary phases. The improvement of the heat resistance was thought to be the cause. Modification of surface regions has been also suggested as the means of the improvement.

FUTURE PROSPECT

Recent studies on processing-microstructure-properties relations have enabled us, to some extent, to design the materials with better properties. Recognition of the importance of heterogeneous microstructures is one of the important outcomes. It should be emphasized, however, that our current knowledge is by no means complete and that there are still a lot of "black boxes" and unanswered questions. For instance, our knowledge on the important characteristics of powders that affect sintering behavior and resultant materials properties, effects and controllability of sintering atmospheres, precise characterizations of grain boundary phases and many other issues is quite limited. Many scientific contributions are needed.

Technological innovation along with the scientific contributions should promote the development of better materials. At present, the attainable upper limits of properties seem unclear. Although it seems that it is difficult to achieve high strength, high toughness and high heat resistance at the same time, there should be ways to overcome this problem. Strengthening by nanoscale particles[30] is one of the opportunities that should be further exploited. Also, better understanding and control of the grain boundaries would be indispensable for the further improvement.

Regarding the applications, although commercial production of silicon nitride ceramics has shown a sharp increase in its volume during the last decade, it is still on a much smaller scale compared with the production of alumina ceramics. One reason for this is undoubtedly the higher costs of silicon nitride ceramics. Therefore, reduction of costs by improving manufacturing processes is one of the most important tasks for the ceramic engineers working in this field. At the same time, further improvement of materials properties and innovation in design, evaluation, machining and joining among other technologies are essential to create new fields of applications because many mechanical engineers still believe that ceramic components are not as reliable as metal components. So the key words for the successful future will be, as they have always been, cost and reliability. Close interactions between users and manufactures are important in this context.

References

1. M. Mitomo and Y. Tajima, J. Ceram. Soc. Jpn. **99**, 1014 (1991).

2. Y. Matsuo, in Hot Isostatic Pressing: Theory and Applications, edited by K. Koizumi (Elsevier Science Publishers, New York, 1992), p.49.

3. Y. Tajima, K. Urashima, M. Watanabe and Y. Matsuo, in Ceramic Transactions, Vol. 1, edited by G.L. Messing, E.R. Fuller and H. Hausner (The American Ceramic Society, Inc., Ohio, 1988), p.1034.

4. T. Kawashima, H. Okamoto, H. Yamamoto and A. Kitamura, J. Ceram. Soc. Jpn. **99**, 320 (1991).

5. I. Tanaka, G. Pezzoti, T. Okamoto and Y. Miyamoto, Ceram. Eng. Sci. Proc. **10**, 817 (1989).

6. Y. Tajima, K. Mizuno, K. Matsubara and M. Watanabe, in Proc. of the 1st Int. Symp. on the Science of Engineering Ceramics, edited by S. Kimura and K. Niihara (The Ceramic Society of Japan, Tokyo, 1991), p.95.

7. K. Matsubara, K. Mizuno, Y. Tajima and M. Watanabe, presented at the Basic Science Division Meeting of the Ceramic Society of Japan, Osaka, Japan, 1992 (unpublished).

8. K. Urashima, Y. Tajima and M. Watanabe, in Fracture Mechnics of Ceramics, Vol. 9, edited by R.C. Bradt et al. (Plenum Press, New York, 1992), p.235.

9. Y. Maniette, M. Inagaki and M. Sakai, J. Europ. Ceram. Soc. **7**, 255 (1991).

10. C-W. Li and J. Yamanis, Ceram. Eng. Sci. Proc. **10**, 632 (1989).

11. K. Urashima, Y. Tajima, M. Enoki and T. Kishi, submitted to J. Am. Ceram. Soc.

12. Y. Hattori, Y. Tajima, K. Yabuta, Y. Matsuo, M. Kawamura and T. Watanabe, in Ceramic Materials and Components for Engines, edited by W. Bunk and H. Hausner (German Ceramic Society, 1986), p.165.

13. K. Katayama, T. Watanabe and K. Matoba, SAE Paper 861128 (1986).

14. K. Matoba, K. Katayama, M. Kawamura and T. Mizuno, SAE Paper 880702 (1988).

15. I. Matsuo and F. Nishiguchi, SAE Paper 880703 (1988).

16. M. Ito, N. Ishida and N. Kato, SAE Paper 880704 (1988).

17. B. Kitayama, Internal Combustion Engine **26**, 77 (1987).

18. C. Baker, R. Kobayashi and D. Baker, SAE Paper 890426 (1989).

19. T. Shimizu, K. Takama, H. Enokishima, K. Mikame, S. Tsuji and N. Kamiya, SAE Paper 900656 (1990).

20. O. Kamigaito, in Ceramics Engine, (Maruzen Ltd., Tokyo, 1987), p.92.

21. H. Matsuoka, H. Kawamura and S. Toeda, SAE Paper 840426 (1984).

22. S. Kamiya, M. Murachi, H. Kawamoto, S. Kato, S. Kawakami and Y. Suzuki, SAE Paper 850523 (1985).

23. Y. Tsukawaki, K. Shimono and M. Shigetsu, SAE Paper 861408 (1986).

24. K. Tashiro, I Tanaka, N. Miyamura and S. Nagano, J. Soc. Automotive Eng. Jpn. **39**, 1179 (1985).

25. M. Kano, J. Jpn. Soc. Tribologists **34**, 131 (1989).

26. L.R. Anderson, Cutting Tool Eng. [4], 65 (1990).

27. H.K. Tonshoff and S. Bartsch, Can. Metall. Quart. **28**, 353 (1989).

28. J. Aucote and S.R. Foster, Mat. Sci. Tech. [2], 700 (1986).

29. H. Tanaka, in Proc. of the 1st Int. Symp. on the Science of Engineering Ceramics, edited by S. Kimura and K. Niihara (The Ceramic Society of Japan, Tokyo, 1991), p.565.

30. K. Izaki, K. Hakkai, K. Ando, T. Kawakami and K. Niihara, in Ultrastructure Processing of Advanced Ceramics, edited by J.D. Mackenzie and D.R. Ulrich (John Wiley & Sons, New York, 1988), p.891.

APPLICATIONS OF SILICON NITRIDE BASED CERAMICS IN THE U.S.

R. NATHAN KATZ
Army Research Laboratory - Materials Directorate
Watertown, MA 02172

ABSTRACT

Silicon nitride based ceramics have unique combinations of mechanical, thermal-physical and electrical properties. Additionally, silicon nitride based materials can be fabricated into near net shape components by several processing routes. This unique combination of properties and processing versatility have encouraged the use of silicon nitride based ceramics in a variety of demanding commercial applications over the past decade. In the U.S., these applications currently include cutting tools, bearings, and selected components for diesel engines. Projected future applications include a broader range of reciprocating and gas turbine engine components, and electronic components. For each of these applications examples of use and the key properties of silicon nitride which enable enhanced systems performance will be discussed. Several materials developments which may broaden the range of applicability of silicon nitride based ceramics will be described.

INTRODUCTION

The past two decades have seen silicon nitride based structural ceramics emerge from the laboratory into the marketplace. Silicon nitride is typical of modern high performance ceramics which combine carefully controlled chemistry and deliberately designed microstructures to yield materials with unique combinations of properties capable of enhancing the performance of many engineering systems. Silicon nitride does not occur in nature. It was first prepared by Schutzenberger in 1879 [1], and the first process for its manufacture was patented by Mehner in 1896 [2]. The history of the development and commercialization of silicon nitride ceramics is a fascinating case study of international scientific cooperation and industrial technology transfer which has recently been summarized in references 3 and 4.

Silicon nitride is a strongly covalent material. Thus, pure silicon nitride can not be densified by conventional sintering processes [5]. Hence, the focus of much of the research on silicon nitride based ceramics during the 1950's through the 1970's was on developing processing techniques to yield fully dense, high strength and toughness materials. Much of this effort was driven with the goal of exploiting the low density, high temperature strength, and excellent thermal shock resistance of silicon nitride for gas turbine applications [6-8].

Silicon nitride based engineering ceramics are a family of materials processed by four major processing routes: reaction bonded silicon nitride (RBSN)[9], hot pressed silicon nitride (HPSN)[10], sintered silicon nitride (SSN)[11], and hot isostatically pressed silicon nitride (HIP-SN)[12]. With the exception of RBSN each processing route requires substantial amounts of oxide additives to promote sintering. These additives react with the silicon nitride starting powder and its native silica overlayer to create a grain boundary oxynitride phase whose characteristics generally control the high temperature behavior of the resultant silicon nitride product. An additional silicon nitride based ceramic is SiAlON, a solid solution alloy of silicon nitride and alumina [13]. RBSN, SSN, sintered SiAlON's, and HIP-SN can all be fabricated into near net shape components. This wide variety of processing routes and additive chemistries give the ceramic engineer the potential to custom tailor components for a large number of applications. Table I, lists typical properties for several representative silicon nitride based ceramics. (Note, the properties listed in Table I were not measured on any "real" material and are only meant to be illustrative.) It is these unique combinations of properties which have led to the selection of silicon nitride based ceramics in the applications discussed below.

TABLE I

REPRESENTATIVE PROPERTIES OF SILICON NITRIDE BASED CERAMICS

	RBSN	SSN	SiAlON	HIP-SN
DENSITY (Mg/m^3)	2.7	3.3	3.3	3.3
YOUNGS MODULUS (GPa)	200	300	300	300
STRENGTH (@ RT)	300a)	900a)	800a)	900b)
K_{IC} (MPa · m$^{1/2}$)	2	6	6.5	5.5-6.5
WEIBULL MODULUS (@ RT)	10	13	10	$\geq$20
THERM COND (w/m°C)	10	33	10-20	33
THERM EXPANSION (10^{-6}/°C)	3.1	3.1	3.2	3.1
WEAR IN ROLLING CONTACT FATIGUE	N/A	LOW	MODERATE	EXTREMELY LOW
DIELECTRIC CONSTANT @ 10^{10} Hz	6-7	6.5	-	6.5-7.5
$/NET SHAPE PART	LOWEST	MED	MED	HIGHEST

a) 4 pt BEND
b) DIRECT TENSION

CURRENT APPLICATIONS OF SILICON NITRIDE BASED CERAMICS

Cutting Tools

The hot hardness, fracture toughness, and thermal shock resistance of fully dense silicon nitride makes it well suited for use as a cutting tool material. Successful demonstrations of the effectiveness of dense silicon nitride in the single point turning and milling of grey cast iron, and of SiAlON in machining of Ni-based superalloys were made in the late 1970's.

Tool bits made of these materials entered commercial production in the 1980's, and have developed a significant niche in the market, by virtue of their ability to enhance metal removal productivity for selected applications.

Fully dense silicon nitride is the most cost effective cutting tool for single point turning of grey cast iron. One recent report compared the cost per part for O.D. turning of an 11 inch diameter grey cast iron brake rotor using three competitive tool bit materials [14]. The results were as follow: polycrystaline cubic BN = $0.024, SiAlON = $0.004, and hot-pressed silicon nitride = $0.002 per part. This study not only highlighted the cost effectiveness of HPSN in this application, but also illustrates the fact that ceramic cutting tool materials must now compete amongst themselves, as well as, with traditional materials.

Dense silicon nitride and SiAlON tools are also used in interrupted cut operations such as the milling of engine blocks and crankshafts [15]. In one reported case study of the turning of a grey cast iron crankshaft with a SiAlON tool, the metal removal rate increased by 150% and the tool life increased 10 fold [16]. This application was characterized by a very significant amount of interrupted cutting which produces a severe impact environment on the cutting tool. That silicon nitride based cutting tools perform well in such an environment is a consequence of their high toughness and thermal shock resistance.

Recent advances in silicon nitride based cutting tool technology include the commercial introduction of in-situ toughened monolithic silicon nitride inserts by Ceradyne and diamond coated silicon nitride tools by the Norton Company. The diamond coated tools are particularly effective in the machining of non-ferrous materials and composites. It has been reported that a diamond coated silicon nitride end mill yielded as much as 300 times the life of micrograin tungsten carbide when machining a SiC/ Al composite and 15 times the life in machining carbon/ carbon composites [17]. They have also demonstrated increased efficiency in machining composites containing Kevlar and Nomex [18]. The advantages of silicon nitride as the substrate for thin film CVD coated diamond tools is due to a combination of chemical compatibility and low mismatch between the CTE's of the two materials.

Bearings

Fully dense silicon nitride can significantly increase the performance of antifriction roller and ball bearings for a wide variety of applications. Hot pressed silicon nitride (NC-132) rolling elements have demonstrated increased fatigue life, increased speed capability, reduced heat generation, and increased corrosion resistance compared to high-performance M-50 steel [19]. More recently available grades of hot isostatically presessed (HIP) silicon nitride, such as Cerbec NBD 100 and NBD 200, have further increased the fatigue life and wear resistance of silicon nitride rolling elements [20]. Table II, lists the desired properties for a rolling element bearing material. With the exception of its relatively high elastic modulus, fully dense silicon nitride bearing materials meet or exceed all of these criteria. While one does not often list failure mode as a materials property, the fact that silicon nitride balls and rollers fail in the same non-catastrophic manner as steel elements is an important consideration for their commercial acceptability.

Because of the performance advantages enumerated in Table III, silicon nitride has entered commercial service as machine tool, turbomolecular pump, dental drill, and specialty

TABLE II

Desired properties for bearing materials.

Fracture toughness, K_{Ic}	High	>5 MN m$^{-3/2}$
Hardness	High	>1200 kg mm^{-2}
Elastic modulus	Low	<210 GPa
Density	Low	<4 Mg m^{-3}
Bend strength	High	>700 MPa
Corrosion resistance	High	
Upper use temperature	High	>800 °C
Failure mode	"Steel-like" Spallation	Small spalls

TABLE III

Performance benefits of ceramic and ceramic hydrid bearings.

Property / Behavior	Benefit (versus steel)
Wear life	Up to 10X greater (nongalling)
Fatigue life	3 to 10X, M50
Speed (DN)	50% greater (low density)
Nongalling	Marginal lubrication capability
Temperature capability	1000 °F
Heat generator	Substantial reduction
Corrosion resistance	Generally inert
Nonmagnetic	Accuracy for inst. bearings
Electrical insulator	Eliminate or reduce arcing across bearing

instrument bearings. Figure 1, highlights the system benefits and economic payoff provided by silicon nitride "hybrid bearings" (i.e., Silicon nitride balls and steel races) in one particular machine tool application. Turbomolecular pumps are a relatively recent market for silicon nitride balls. Currently, four manufacturers are selling pumps with silicon nitride balls which improve the performance and reliability of the pumps.

Due to the difficulties in attaching ceramic races to metallic shafts the vast majority of silicon nitride containing bearings are of the hybrid configuration. However, there are circumstances where corrosion, electrical or magnetic fields, or lubrication constraints mandate all ceramic bearings. An example of such an application is bearings for I-R seekers (figure 2) [21]. Another example is the use of all silicon nitride bearings utilized in tidal flow meters operating in sea water [21].

FIGURE 1

```
POTENTIAL COST BENEFITS OF
CERAMIC HYBRID BEARINGS
```

APPLICATION

- MACHINE TOOL SPINDLE BEARING
- Si_3N_4 BEARING ALLOWS GREASE (VERSUS OIL) LUBE AT DN OF 1×10^6 (VERSUS 0.6×10^6 FOR STEEL)
- DN OF 1×10^6 WAS REQUIRED

COST

- Si_3N_4 HYBRID BEARING, ADDED $800 PER SPINDLE
- ELIMINATION OF OIL LUBE SYSTEM SAVED $1,500 TO $3,000 PER SPINDLE

NET SAVINGS $700 TO $2,200 PER SPINDLE

Source: J. Hannoosh - Design News, November 23, 1988

FIGURE 2

IR Seeker Bearing (Courtesy of Peter Ward, Miniature Precision Bearing Div. of MPB Corp.)

Diesel Engine Components

Silicon nitride is presently utilized in a variety of automotive (both diesel and spark ignited) and truck engine and engine related components (principally turbochargers) in production in the U.S. and Japan as shown in Tables IV and V. In the U.S. the commercial application of silicon nitride components, to date, has been entirely in large truck engines. This is in marked contrast to Japan where the commercial use of silicon nitride has centered on automobile engines. Further, the first introduction of a silicon nitride component in a production engine in the U.S. was in 1989, whereas in Japan it was in 1981. The fact that truck engine production rates are typically only in the thousands to tens of thousands per year coupled with the recent introduction into production yields a substantially smaller market for silicon nitride engine components in the U.S. as compared to Japan at present. The rationale for the use of silicon nitride for various engine components has been discussed elsewhere by the author and others [22-24]. For the purposes of this paper it is important to discuss why the first commercialization of silicon nitride engine components in the U.S. has focused on large truck diesels.

TABLE IV

Ceramic components in production engines.

Component	Engine manufacturer	Engine type	Ceramic	Benefit
Turbo charger	Nissan	S-I	Sint. Si_3N_4	Lower inertia (Less lag)
Glo-plugs	Isuzu	Diesel	Sint. Si_3N_4	Faster start-up
Precombustion Chamber (Swirl chamber)	Isuzu Toyota Mazda	Diesel	Sint. Si_3N_4	Lower emissions Faster start-up Lower noise
Rocker-arm pads	Mitsubishi	S-I	Sint. Si_3N_4	Lower wear

TABLE V

Si_3N_4 COMPONENTS IN PRODUCTION DIESEL ENGINES - USA

PART	ENGINE MANUFACTURER	COMPONENT SUPPLIER	BENEFIT
TURBOCHARGER	DETROIT DIESEL SERIES 50 (250-300 HP)	NTK/GARRETT	↓ TURBO LAG ↓ EMISSIONS
	CATAPILLAR 3406C (~460 HP)	NTK/GARRETT	" "
CAM ROLLER FOLLOWER	DETROIT DIESEL SERIES 50	KYOCERA	↑ INJ PRESSURE ↓ EMISSIONS
FUEL INJECTOR LINK	CUMMINS	ENCERATECH	↓ WEAR

Early in the development of ceramics for diesel engines fuel economy as embodied in the so-called "Adiabatic", or more appropriately, low heat rejection engine concept was the principal objective. However, with the reduced criticality of the issue of oil price manipulation by cartels, the urgency of fuel economy as the driver for implementing ceramic technology diminished. However, over the past decade emissions control has become one of, if not the, technological driving force in engine development. Thus, the combination of wear resistance, low mass density, and excellent high temperature performance of silicon nitride has made it attractive for fuel injection and valve train components as well as for turbochargers. The turbocharger application is the most demanding. The truck turbocharger rotors listed in Table V are 4 inches in diameter, or approximately twice the diameter of the automobile turbochargers listed in Table IV. The use of the silicon nitride turbocharger rotor reduces the rotating moment of inertia thereby reducing turbocharger lag. Aside from improving performance and drivability the reduction in lag also reduces particulate emissions [25].

Wear Components

Non-automotive wear components represent a very small fraction of silicon nitride components used in the United States. Representative wear applications are fixtures for positioning and transferring metal parts during processes such as induction heating (Figure 3) or resistance welding where the combination of mechanical, thermal and electrical insulating properties of silicon nitride are critical. Key properties for these applications include; electrical insulation, low thermal conductivity, and good thermal shock resistance.

A particularly notable application is the use of HPSN as a pump seal in the first stage water circulation pumps in nuclear reactors. These components are large, up to 26 inches in diameter. They require both excellent wear resistance, and the utmost in mechanical reliability. Hundreds of these seals have been in use since the mid 1980's in reactors in both the U.S. and Europe [26].

FIGURE 3
HPSN Silicon Nitride Insulator Pads for an Induction Heating Process (Courtesy of John Schuldies, Industrial Ceramics Technology, Inc.)

Electronic Components

Vapor deposited silicon nitride has a long history of use as a dielectric layer in integrated circuits. Advantages of silicon nitride for electronic application include a high dielectric constant and a coefficient of thermal expansion that is a close match to silicon. However, until the past year bulk silicon nitride has not been commercially utilized as an electronic material. In 1991, a composite based on silicon nitride, zirconia, and zirconium oxynitride (LSZ) was introduced [27]. As shown in Figure 4, this material exactly matches the CTE of Si from RT to 400 C. This material is currently used as a substrate for Si integrated circuits where thermal mismatch can cause severe problems. One application is as a substrate for high pressure sensors.

FIGURE 4

Thermal Expansion of LZS Composite
in Comparison with Silicon and Other Ceramics

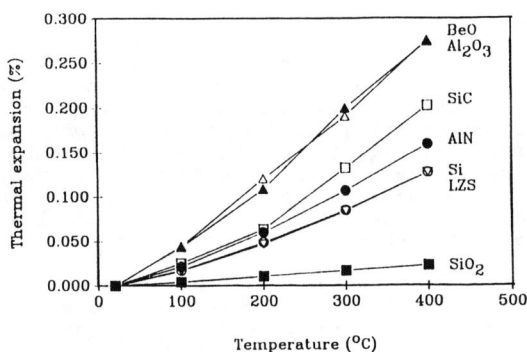

Courtesy of Ran-Rong Lee, Ceramics Processing Systems

FUTURE APPLICATIONS

In the years between now and the next century it is likely that there will be continued growth in the existing applications of silicon nitride based ceramics. The largest growth will be in the engine area. Valve train components including valves will see commercial application. An important development which may encourage the use of additional silicon nitride wear components in reciprocating engines is the potential use of natural gas engines to reduce pollution [28]. Additionally, as more engines use four valves per cylinder for increased performance, the use of lower density components in the valve train becomes an increasingly important means of reducing parasitic loads and thereby, increasing engine efficiency.

Limited introduction of silicon nitride components such as vanes and shrouds into non- man rated gas turbines (such as APU's) is a near certainty. Commercial exploitation of the wear properties, toughness and thermal shock resistance of silicon nitride materials for metal working dies will occur.

Demonstration of the potential of HPSN as a "blister punch" used in forming aluminum beverage cans was recently reported [29]. The use of SiAlON's, already employed in Europe and Japan for handling of molten non-ferrous metals, and as rolls and skid buttons in rolling mills in Japan is likely to occur to some extent in the U.S.

Electronic applications of bulk silicon nitride based ceramics are likely to see some growth in the area of specialty electronic substrates. The material's dielectric properties combined with outstanding rain and dust erosion behavior make it a candidate for radomes in advanced missile systems.

Oxynitride glasses posses unique mechanical and optical properties [30]. The high elastic modulus of oxynitride glass fibers makes them an interesting potential reinforcement for composites. Oxynitride fiber reinforced cements have recently been shown to have outstanding resistance to alkali attack [31].

CURRENT AND POTENTIAL FUTURE MARKET SIZE

The currant and future applications of silicon nitride based ceramics discussed above give strong reason to be optimistic with regard to an increasing demand for these materials in the U. S. marketplace. The obvious question is," how large is the present and likely future market likely to be"? Table VI presents the author's estimate of the current and projected year 2000 market for silicon nitride based ceramics in the U. S. The figures are based on interviews with many individuals within the U.S. silicon nitride community, and the author's own estimates and judgement.

TABLE VI
SILICON NITRIDE - U.S. MARKET ESTIMATE

APPLICATION	1992	2000 (1992$)
CUTTING TOOLS	$10-15 \times 10^6$	$20-25 \times 10^6$
BALLS/BEARINGS	10×10^6	$20-30 \times 10^6$
RECIPROCATING ENGINE COMPONENTS DIESEL	5×10^6	$30-90 \times 10^6$
GAS TURBINE COMPONENTS (INDUSTRIAL & APU)	--	9×10^6
MISC (WEAR PARTS, ETC.)	$1-2 \times 10^6$	$2-5 \times 10^6$
	$26-32 \times 10^6$	$81-159 \times 10^6$
RECIPROCATING ENGINE COMPONENTS AUTOMOTIVE	--	?

ESTIMATES BY R.N. KATZ

The estimation of current, let alone future, market sizes for an industry with as many players and niche markets as silicon nitride based ceramics, is at best an imprecise art. However, a study has recently been published which reports that the European market for silicon nitride was approximately $ 22 million in 1991 [32]. Thus, the values estimated above for the 1992 U.S. market would seem to be reasonable. The estimate for 2000 assume steady growth in the diesel, metalworking, and bearing areas. However, if the passenger car (automotive) market materializes, even for one or two components, the projected figures would be very conservative.

FUTURE DEVELOPMENTS

While there are many research developments of note being actively pursued in laboratories around the world, two in particular seem to hold the promise of increasing the applicability of silicon nitrides in the near term. These two developments are in-situ toughened silicon nitrides and electrically conductive silicon nitrides (and SiAlON's).

Researchers in the U.S. and Japan have developed in-situ toughened materials which combine high toughness (7 to 10 MPa*m1/2) and RT strengths of approximately 1 GPa. The microstructure is designed to contain elongated "whisker-like" self-reinforcing grains with smaller more nearly equiaxed grains filling in amongst them. These materials offer the promise of composites with the relative processing ease of monolithic ceramics and without some of the health risks associated with handling fine fibers. The Ceradyne cutting tools cited above are the first commercial exploitation of this type of microstructure and many others from a variety of firms are certain to follow.

Elimination of the need to machine silicon nitride with diamond tooling would help reduce the cost of many components. The fabrication of electrically conductive silicon nitride-TiN particulate composites opens the possibility of utilizing electro discharge machining (EDM) for silicon nitride parts.

SUMMARY

This paper has demonstrated that silicon nitride ceramics are presently contributing to the U. S. society in several ways, principally by enhancing metalworking productivity and reducing particulate emissions in several classes of diesel engine. Future growth in these areas, as well as wear, electronic and gas turbine applications is projected. This potential growth gives reason to be optimistic as to the future role of silicon nitride based ceramics in the U.S. economy.

REFERENCES

1. P. Schutzenberger,"On Silicon Nitride", C.R. Acad. Sci. (Paris), vol 89, 644-46 (1879)

2. H. Mehner, German Patent 88999, September 30, 1896

3. W.D. Kingery, Japanese / American Technological Inovation, (Elsevier Science Pub.,New York, 1991) pp 161-205

4. R. N. Katz,Interceram, 1992, 231-33

5. C. Greskovich, S. Prochazka, and J. H. Rosolowski, in Nitrogen Ceramics, ed., F. Riley (Noordhoff International Pub., Lyden, 1977) 351-57

6. E. M. Lenoe, R. N. Katz, and J. J. Burke, Ceramics for High Performance Applications III: Reliability, (Plenum Press, New York, 1983)

7. V. Tennery, Proceedings of thr 3rd International Symposium on Ceramic Materials and Components for Engines, (American Ceramic Soc., Columbus, OH, 1989)

8. J. R. Smyth, in Ceramics and Glasses, vol.4 Engineered Materials Handbook (ASM International 1991) pp 995-1002

9. J. A. Mangels,Progress in Nitrogen Ceramics,ed. F. Riley, (Martinus Nijhoff Publishers, Boston, 1983)pp 231-36

10. M. Torti, in ref 6, p 261

11. G. E. Gazza and R. N. Katz in Pressure Effects on Materials Processing and Design , eds. K. Ishizaki, E. Hodge, M. Concannon (Mater. Res. Soc. Proc. 251, Pittsburgh, PA, 1992) pp. 199-209

12. H. Larker, in ref 9, p 717

13. K. H. Jack, this volume;J. M<ater. Sci., $\underline{11}$, (1976) pp 1135-1158

14. L. R. Anderson, Cutting Tool Engineer, April (1990), pp.65-70

15. J. H. Adams, R. Anschutz, and G. Whitefield, in Ceramics and Glasses, vol 4 Engineered Materials Handbook (ASM International, 1991) pp 966-972

16. C. W. Beeghly and A. F. Shuster, presented at the Conf. on Adv. Tool Materials for High Speed Machining, Scottsdale,AZ 25-27 Feb. 1987, ASM preprint 8701-002

17. G. Geiger, Am. Ceram. Soc. Bulletin, $\underline{71}$, [10], (1992) p 1474

18. P. M. Stephan, Am. Cer. Soc. Bulletin, $\underline{71}$, [11],(1992) pp 1623 -29

19. R. N. Katz and J. G. Hannoosh, Int. J. High Technology Ceramics,$\underline{1}$,(1985) pp 69-79; R. N. Katz, in Friction and Wear of Advanced Ceramics, ed. S. Jahanmir (to be published by Marcel Dekker,1993)

20. J. W. Lucek, Rolling Wear of Silicon Nitride Bearing Materials, Am. Soc. of Mech. Engineers paper 90-GT-165 (1990)

21. R. A. Hanson and W. P. Ogden, in Split Ball Bearing Features (pub. by MPB Corp.) April 1991, pp 1 -7

22. R. N. Katz,Stuctural Ceramics, ed.J. Wachtman (Academic Press, Boston, 1989) pp 1 - 26

23. R. Kamo, in Ceramics and Glasses, vol 4 Engineered Materials Handbook, (ASM International, 1991) pp 987 - 994

24. L. C. Lindgren, R. L. Holtman, K. D. Kannmacher,J.D. Petty, and G.A. Costakis, in Ceramics and Glasses, vol 4 Engineered Materials Handbook, ibid , pp 716 -721

25. C.C. Baker and D. E. Baker, in Ceramics and Glasses, vol 4 Engineered Materials Handbook, ibid , pp 722 - 727

26. J Panzarino, Norton Co. (private communication)

27. R & D Magazine, October 1991, p 41

28. S. Nestlerode, Ceramic Industry July 1992, pp 40 -44

29. T. M. Sullivan, Ceramic Industry October 1992, pp 31 - 34

30. D. R. Messier, Revue de Chemie Minral, $\underline{22}$, 1985, pp518 -533

31. G. L. Leatherman, T. El-Korchi, T. Ho0lmes and R. N. Katz inFiber reinforced Cementitious Materials, eds. S. Mindess and J. Skalny (Mat. Res. Soc. Proc. $\underline{211}$, Pittsburg, PA, 1991) pp105 -110

32. High Tech Materials Alert, September 1992, p 3

SUPERPLASTIC SiAlON—A BIRD'S EYE VIEW OF SILICON NITRIDE CERAMICS

I-WEI CHEN and SHYH-LUNG HWANG[†]
Department of Materials Science & Engineering, University of Michigan, Ann Arbor, MI

ABSTRACT

Superplastic Sialons have very fine microstructures containing submicron grains and transient phases. Fabrication of these materials requires processing at relatively low temperatures. As a result, different stages of phase evolution, including oxide melt formation, nitride dissolution, α' and β'-SiAlON nucleation, and Sialon growth with and without concurrent deformation, can be captured in the development of these materials. In addition, only very low flow stresses are required for large strain deformation which, in turn, allows grain boundary and liquid phase processes to be manifested in the deformation behavior. Highlights of investigation of these aspects are reviewed here to shed light on the phase relationship, microstructural development, and grain boundary characteristics of silicon nitride.

INTRODUCTION

Superplastic Sialons have been developed only in the last few years.[1-4] (In addition, a superplastic β-Si_3N_4/SiC composite has been reported by Wakai et al. since 1990.[5-6]) Unlike conventional ceramics, these materials exhibit very high ductility in both uniaxial tension and biaxial tension at elevated temperatures. This novel feature is a direct result of the very fine microstructure found in these materials which have an initial grain size less than 0.3 µm. Ductility over 200% has been demonstrated between 1500°C and 1600°C with strain rate for deformation as high as 10^{-4}/s and flow stress typically less than 20 MPa.[1,3-4] These deformation conditions are sufficiently favorable for contemplating superplastic forming. Some initial experiments by punch stretching support this claim.[1-2]

The key to successful development of these fine-grained ceramics is the low hot-pressing temperature used (1550°C). In developing this process, a number of observations that revealed important details of the reaction pathways during densification of the complex Sialon systems were noted. It was found that the as-hot-pressed materials were far from equilibrium and underwent additional phase and compositional evolutions during the subsequent deformation and heat treatment steps. These evolutions have been carefully followed by TEM and x-ray analysis to shed light on the microstructural and microchemical aspects of the phase nucleation and growth processes. As expected, the deformation of these very fine-grained, liquid-containing ceramics is profoundly influenced by the kinetic and structural characteristics of the grain/grain and grain/liquid interfaces. Moreover, they were also found to exhibit novel rheological behavior. In all, these superplastic Sialons prove to be rather interesting materials which offer excellent opportunities to further understand silicon nitride ceramics. Therefore, consistent with the theme of this symposium, the intent of this review is to summarize these observations and highlight their implications on the more generic issues of phase relation and microstructural design.

COMPOSITION

The bulk composition investigated in our research lies on the (yttria) α'-SiAlON plane, i.e., on the Si_3N_4-Al_2O_3:AlN-YN:3AlN triangle. See Fig. 1. The general composition can be represented by the formula $Y_{m/3}Si_{12-(m+n)}O_{m+n}N_{16-n}$. Depending on the values of m and n,

---

[†] Now at Oak Ridge National Laboratory, Metals and Ceramics Div., Oak Ridge, TN

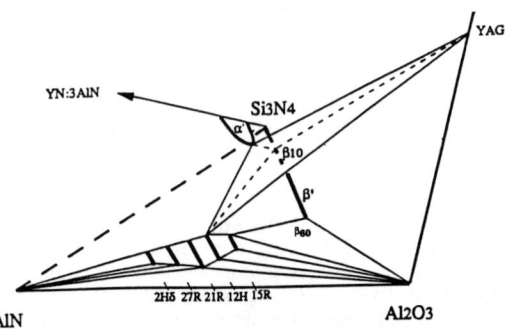

Fig. 1. Partial representation of Y-SiAlON system showing α' plane and α'-β_{10}-12H-YAG compatibility tetrahedron.[37]

major equilibrium phase assemblages of either α'-SiAlON phase, β'-SiAlON phase, or their mixture can be obtained. These phases, however, form only after equilibrium is attained. At earlier stages, phases of starting powders (α-Si_3N_4, β-Si_3N_4, Al_2O_3, AlN and Y_2O_3) and various transient substances can coexist. In addition, some AlN polytypes (12H/15R) form during the transient and, in cases where β'-SiAlON dominates, persist in the final equilibrium state. The amount of glassy phase after full densification at 1550°C is between 5 and 15 v/o.

PATHWAY OF REACTIVE HOT PRESSING/SINTERING

The reaction pathway of Sialon systems has received considerable attention in the literature.[7-10] For example, Tien and coworkers have examined the case of Si_3N_4/Al_2O_3/AlN reaction as well as the SiO_2/AlN reaction.[7] Since more than three powder reactants may be present in practice and the reactions may depend on both the phase relations and the physical/chemical characteristics of the powders, the kinetic pathway can be rather complicated. This issue is of considerable importance since the subsequent phase and microstructure evolutions are frequently influenced by the initial reaction pathway. Superplastic Sialons prepared by low temperature hot pressing has offered new information on this aspect. In the following we will summarize our findings on the reaction pathway of our Sialon system.

When Si_3N_4, AlN, Al_2O_3 and Y_2O_3 are used as the starting powder, a ternary eutectic liquid between oxides (Al_2O_3, Y_2O_3 and SiO_2, the latter from the oxidized surface of Si_3N_4) forms below 1350°C. This can be clearly seen in the hot pressing curves shown in Fig. 2 which all have a kink that lies between 1300 and 1350°C. The amount of the liquid formed, however, is quite small and is limited by the quantity of SiO_2. Therefore, no more than 10% densification is usually achieved at this stage. At the same time, the excess oxides also dissolve to form $Y_3Al_5O_{12}$ (YAG).

Subsequent densification requires the dissolution of Si_3N_4 and AlN into the oxide melt. The latter step was found to depend on whether AlN is present or not. For compacts which do not contain AlN, Si_3N_4 dissolution probably takes place between 1420 and 1550°C. The addition of AlN to the powder compact, however, causes a delay (50°C or more) in this process so that the densification curve is shifted to a higher temperature. This is shown in Fig. 2 by the curve labeled as 20 YAG + 15 AlN. The delaying effect of AlN is due to its tendency to trap oxide liquid, which is a result of the easier wetting between the melt and AlN than between the melt and Si_3N_4. This is evident from the wetting angle measurement performed for the eutectic oxide melt on either a polycrystalline Si_3N_4 or AlN substrate. Such data are shown in Fig. 3. Since the amount of ternary oxide liquid is initially limited, AlN powders can drain the liquid from the powder compact, as shown schematically in Fig. 4. Consequently, both densification and Si_3N_4 dissolution cease for a while. This continues until the temperature is sufficiently high for the melt to wet Si_3N_4.

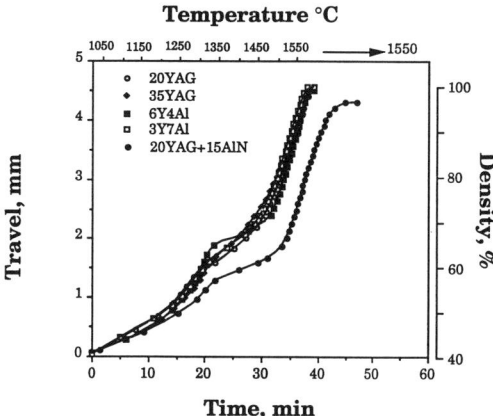

Fig. 2. Density/Ram travel versus time/temperature in hot pressing. Heating rate = 15°C/min to 1550°C then held at the same. A stress of 27 MPa was applied at all time. Only the curve labeled 20 YAG + 15 AlN contains AlN.

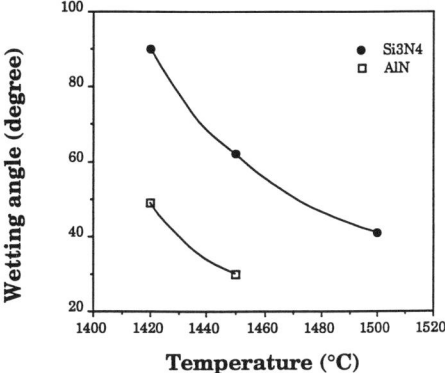

Fig. 3. Wetting angle between Y_2O_3-Al_2O_3-SiO_2 eutectic melt and substrates of Si_3N_4 and AlN.

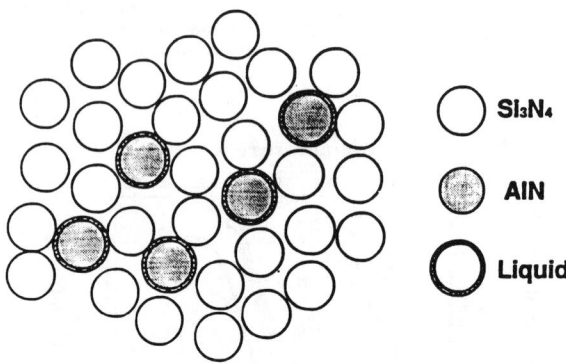

Fig. 4. Schematic showing liquid trapped by AlN particles.

The intervention of AlN has some profound influence on the subsequent phase development. Inasmuch as AlN is the predominant species that reacts with the ternary oxide melt between 1420 and 1550°C, an oxynitride melt which is rich in Al and N must form initially. This is supported by our finding that, in the reaction zone between AlN and the oxide melt, β_{60}-SiAlON, which is the richest solid solution of β'-SiAlON,[11-12] invariably formed. β_{60}-SiAlON, as will be described later, also tends to be the first Sialon phase precipitated from the oxynitride melt, again as a consequence of Al supersaturation. The intervention of AlN thus temporarily shifted the Sialon composition toward the Al-rich end.

Up to this stage, YAG appears to be compatible with β_{60}-SiAlON and the melt. This ends when, at above 1470°C, Si_3N_4 begins to dissolve and YAG starts to disappear. Their reaction, in the presence of the oxynitride melt, results in α' and β'-SiAlON of various compositions depending on the bulk composition selected. The typical composition of β'-SiAlON, though, is near β_{10}-SiAlON if α'-SiAlON is present.

The reactions discussed above can be summarized as:

Ternary Oxide Eutectic Reaction (1350°C)
$$SiO_2 + Al_2O_3 + Y_2O_3 \rightarrow (Si, Al, Y)_xO_y + YAG$$

AlN Reaction and Initial Precipitation (1420-1550°C)
$$(Si, Al, Y)_xO_y + AlN \rightarrow \beta_{60} + (Si, Al, Y)_x(O,N)_y$$

Si_3N_4/YAG Reaction & Secondary Precipitation (above 1470°C)
$$(Si, Al, Y)_x(O,N)_y + Si_3N_4 + YAG \rightarrow \beta_{10-60} + \alpha$$

This sequence of phase evolutions is consistent with the phase relations known for the Y-Al-Si-O-N system[13-15] provided the pivotal role of AlN intervention is granted. Assuming the liquid region extends from the Al_2O_3-SiO_2-Y_2O_3 oxide plane toward the Si_3N_4-AlN-Al_2O_3-SiO_2-plane as the temperature rises,[16] the pertinent composition line for phase reaction is initially connected between AlN and the ternary oxide eutectic. As the reaction proceeds and the temperature rises, this line is shifted toward that between Si_3N_4 and the oxynitride melt. At the same time, the actual composition of the reactive species originates from the ternary oxide eutectic point and travels progressively to points within the oxynitride melt-YAG-β_{60}-SiAlON compatibility prism, to outside the prism, and finally toward the α'-SiAlON plane. Such a kinetic pathway should be a rather general one applicable for reactive hot pressing and sintering of many Sialons.

NUCLEATION OF SiAlON PHASES

As we described above, β_{60}-SiAlON, β_{10}-SiAlON and an α'-SiAlON (at a composition in equilibrium with β_{10}-SiAlON) were the most common Sialon phases observed in our study. In theory, they could be nucleated heterogeneously on the initial Si_3N_4 powder, which had a high α content (95%) but also some β phase. They could also be nucleated homogeneously from the melt.

We have found abundant evidence of heterogeneous nucleation of both α'-SiAlON and β_{10}-SiAlON on α-Si_3N_4, and β_{10}-SiAlON on β-Si_3N_4. In addition, it was also common to find β_{10}-SiAlON growing from a seed of β_{60}-SiAlON, which itself could be homogeneously nucleated although it sometimes formed adjacent to an α-Si_3N_4 particle. The evidence was obtained from detailed TEM examination using various contrast and diffraction analysis, weak beam imaging, and STEM compositional mapping. One example of concurrent growth of α'-SiAlON and β_{10}-SiAlON on an α-Si_3N_4 core is shown in Fig. 5. Another that shows β_{10}-SiAlON nucleated on a β_{60}-SiAlON seed is given in Fig. 6.

Although there was always a nearly ideal alignment between the crystallographic orientation of the seed crystal and the outgrowth, the seed, whether it was α-Si_3N_4, β-Si_3N_4 or β_{60}-SiAlON, could be easily identified. This is because of the strain contrast between the seed and the outgrown crystal. (Note that there is a mismatch between α and β structures, as well as between α'-SiAlON or between β'-SiAlON of different compositions.) Microstructural growth features from different seeds do differ, however. For example, when an α-Si_3N_4 was the seed, the outgrowing α'-SiAlON and β'-SiAlON frequently assumed several different variants, as schematically illustrated in Figs. 7 and 8. Such features have been analyzed carefully in our study and, once understood, were used to identify other α-Si_3N_4 seed crystals. We found in our materials that the typical sizes of the α-Si_3N_4 and β-Si_3N_4 seeds were around 0.2 µm, comparable to the sizes of the starting powders. The β_{60}-SiAlON core, from which β_{10}-SiAlON grows, was also of a comparable size. In all, more than half of the grains we sampled contained a core of one of the above types.

The above direct observation of nucleation sites, including the transient nucleation of β_{60}-SiAlON, is the first of its kind. It has important implications on the sequence of phase nucleation and growth in Sialon systems. Considering the sectioning statistics from TEM thin crystals, we are convinced that essentially all Sialon grains must grow by a heterogeneous nucleation mechanism starting from some kind of identifiable seeds. The only possible exception is the nucleation of small β_{60}-SiAlON seeds themselves. The latter, if nucleated homogeneously, are most likely the product of large supersaturation (rich in Al and possibly N)

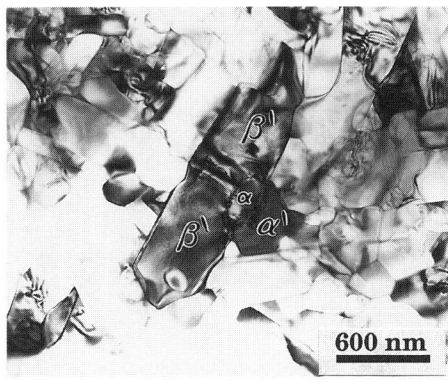

Fig. 5. Two β_{10}-SiAlON crystals and an α'-SiAlON crystal grown from the same α-Si_3N_4 seed.

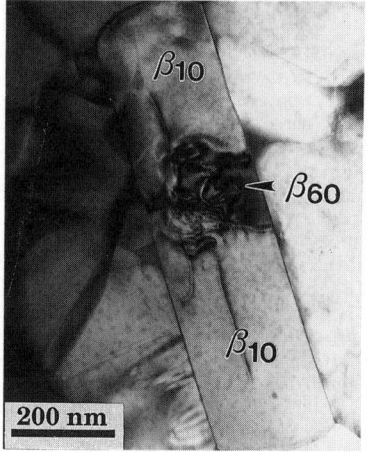

Fig. 6. A β_{10}-SiAlON rod grown from a β_{60}-SiAlON seed. The two ends have a slight misorientation (which can be seen by tilting not shown here) indicating independent growth events.

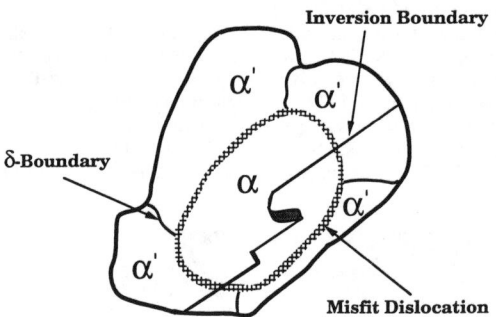

Fig. 7. Schematic growth morphology of α'-SiAlON from α-Si$_3$N$_4$ seed. δ-boundary originates at the interface and is due to a slight misorientation between two α'-Sialon variants. Inversion boundary is inherited from α-Si$_3$N$_4$ seed and represents impingement of two enantiomorphic forms of α-Si$_3$N$_4$.

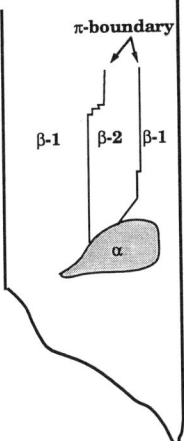

Fig. 8. Schematic growth morphology of β'-SiAlON from α-Si₃N₄ seed. π-boundary results from impingement of two growth habits which correspond to two equivalent ways of matching stacking sequences in α and β-Si₃N₄.

during the initial dissolution of AlN, as previously discussed. Later, as the supersaturation is dissipated following β₆₀-SiAlON precipitation and Si₃N₄ dissolution, the equilibrium composition should be restored so that the subsequent precipitation is reverted to α'-SiAlON and β'-SiAlON. We believe the supersaturation is now too small for homogeneous nucleation, and, in view of the large number of heterogeneous nucleation sites (α-Si₃N₄, β-Si₃N₄ and β₆₀-SiAlON) already available, growth can simply proceed at a progressively lower driving force until completion.

Direct observations of nucleation sites were made possible in our studies because of the relatively low hot pressing temperature (1550°C) and short time (0.5 hr) used which preserved the early transient evolutions. The high number of grains available in these very fine-grained materials have additionally provided favorable sampling statistics for TEM observations. In other studies of silicon nitride ceramics, higher processing temperatures were typically used.[17-18] Under such conditions, the kinetics were so fast that early nucleation events would have already been overtaken, and hence obliterated, by later growth and grain coarsening steps. Moreover, during grain coarsening, dissolution and reprecipitation occur simultaneously, which further reduces the number of grains containing first generation nuclei. From our observations, however, it should be clear that the operation of heterogeneous nucleation in general, with the possible exception of homogeneous nucleation of supersaturated β₆₀-SiAlON in transient, must be a quite common process in reaction hot pressing/sintering of all silicon nitride materials. Therefore, it should come as no surprise that subsequent microstructure development could be profoundly influenced by these nucleation events, and specifically by the size and composition of the starting powders.[19-22]

GROWTH OF SiAlON PHASES

Growth of Sialon phases in superplastic Sialons is always epitaxial. At larger sizes, the growth habit of β'-SiAlON is characteristically $(1\bar{1}00)$, which is flat, and the growth direction is [0001], which typically forms a rounded end. This is in agreement with that reported for other silicon nitride ceramics[23-24] and can be understood simply from structural considerations. Essentially, on the (0002) plane the Si-N chain follows a zig-zag pattern between two adjacent

(0002) planes, as shown in Fig. 9. Thus, this interface is naturally kinked and many attachment/growth sites are always provided for. It can also be shown from the Periodic Bond Chain (PBC) theory[25] that all planes between $(1\bar{1}01)$ and (0001) have similar characteristics. Thus, the [0001] end can grow normally by accepting atoms diffusing through the melt, and it maintains a rounded shape. In contrast, the $(1\bar{1}00)$ plane can be shown to be a "flat" face, according to the PBC theory, with major Si-N bonds continuous in two directions on the plane. Such a flat interface lacks attachment/growth sites to accept arriving atoms so that the growth is interface-controlled and generally sluggish. The above classification is also consistent with the notion of roughening transition proposed by Jackson:[26] $(1\bar{1}00)$, which has the strongest in-plane bonding, remains flat at all temperatures whereas planes between (0002) and $(1\bar{1}01)$, which have strong plane-to-plane bonding, should become rough above relatively low temperatures. A strong growth anisotropy and a hexagonal prismatic rod geometry are thus expected for β-Si_3N_4.

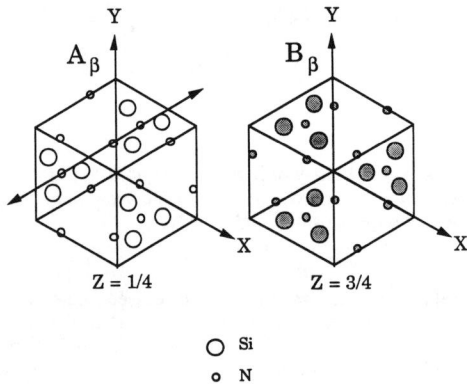

○ Si
o N

Fig. 9. Atomic arrangement of β-Si_3N_4 on (0002) planes. Note step rise from z = 1/4 to z = 3/4 at between of triangles.

In addition to the above general observations, we also found evidence that growth anisotropy tended to be stronger at a smaller driving force, and vice versa. A smaller driving force prevails during the later stage of evolution, when the grain size is large and Al content in β'-SiAlON is low (e.g., β_{10}-SiAlON). Conversely, a large driving force prevails during the early stage of evolution, when the grain size is small and Al content in β'-SiAlON is high (e.g., β_{60}-SiAlON). For example, we found that β-Si_3N_4 and β_{60}-SiAlON cores (the latter shown in Fig. 6), which seeded the later growth of β_{10}-SiAlON, always appeared nearly equiaxed. Likewise, as shown in Figs. 10(a) and (b), β'-SiAlON grains in similarly deformed (to 60% true strain) specimens are more elongated in the β_{10}-SiAlON composition (with a typical aspect ratio of 6 to 8) than in the β_{60}-SiAlON composition (with a ratio of only 4 to 5). We attribute this tendency of decreased growth anisotropy to progressive kinetic roughening of the interface at high driving forces. Theoretical considerations and computer simulations of growth in simple solids have found that attachment/growth sites become more populous when the driving force exceeds a few kT.[27] As this happens, the growth anisotropy between different planes which may or may not be rough according to PBC theory or Jackson's criterion diminishes, and the growth rate of an initially flat interface increases rapidly to approach that of a rough interface. Some examples from their computer simulations are reproduced in Fig. 11 to illustrate the point.

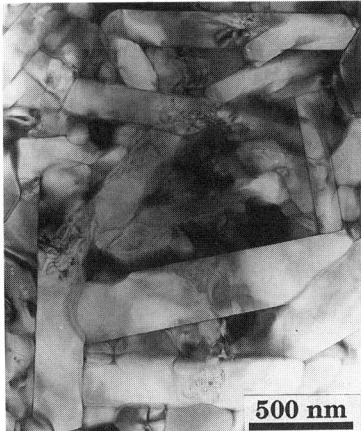

Fig. 10. β'-SiAlON grains deformed to 60% true strain in (a) β_{10}-SiAlON and (b) β_{60}-SiAlON.

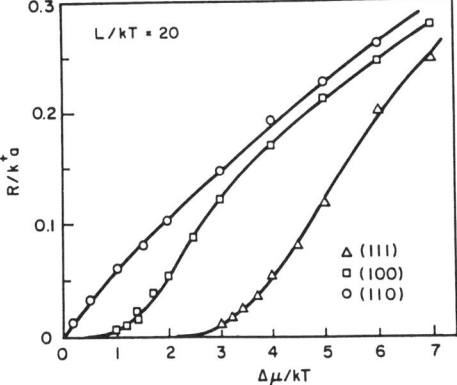

Fig. 11. Normalized growth rates of three FCC faces.[27] Although only (110) is above roughening transition at L/kT = 20 (L = heat of fusion, kinetic roughening at large driving forces brings the growth rate of three faces close to each other.

The above discussion on structural and kinetic roughening transitions for silicon nitride ceramics makes it clear once again that the investigation of very fine grained superplastic Sialons has afforded an excellent opportunity to examine the kinetic and microstructural processes. In this case, it is because of realizing the growth stages with very small grains and with transient compositions of very high Al content. These growth regimes were rarely investigated in the past since most of the grain growth studies of silicon nitride ceramics were focussed on later stage growth and, for practical reasons, lower Al content. Some interesting and essential information is thus missing because of such restrictions.

DEFORMATION AND INTERFACIAL PROCESSES

The main deformation characteristics are schematically indicated in Fig. 12. Since superplastic deformation was typically performed at a temperature comparable to, or slightly higher than, that of hot-pressing, concurrent phase evolution toward equilibrium was inevitable in these materials. Deformation, however, has an additional effect when the β'-SiAlON phase is present; it promotes the growth of β'-grains along the principal tensile stress direction. A highly oriented fibrous microstructure thus results which proves to be favorable for achieving tensile elongation due to fiber reinforcement. Indeed, a unique stress-strain curve, shown in Fig. 13, which continuously strain hardens (reminiscent of that of a semi-crystalline polymer such as polyethylene) has been recorded in uniaxial tension for such a material. It has been successfully modeled by taking into account grain growth and grain alignment along the tensile direction in a viscous matrix which is essentially Newtonian.[4] Superplastic forming can be most favorably carried out in this regime.

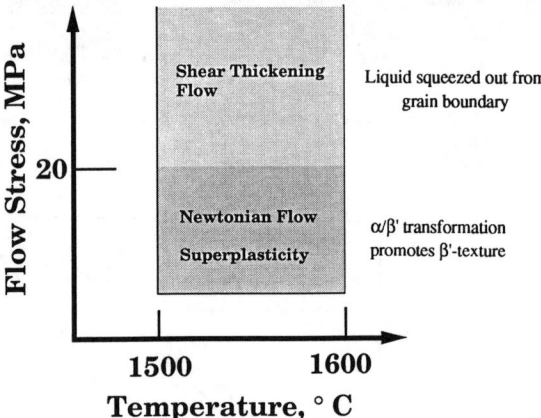

Fig. 12. Deformation regimes of superplastic Sialons. Concurrent microstructural processes also indicated.

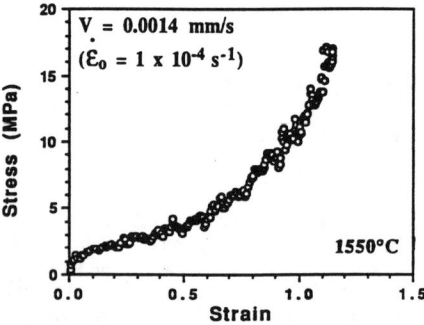

Fig. 13. Stress strain curve of a superplastic β'-Sialon at 1550°C showing unusual strain hardening.[4]

The Newtonian type of flow behavior has been confirmed, Figs. 14(a) and (b), in all of the superplastic Sialons deformed below 20 MPa or so. This is similar to that reported by Wakai et al. for a superplastic Si_3N_4-SiC-YAG composite[5-6] and by Raj et al. for a series of spodumene-based glass ceramics.[28] Thus, without exception, all the superplastic ceramics reported to date that contain a substantial liquid (greater than 3 v/o or so) apparently deform by Newtonian flow. In the case of superplastic Sialons, the flow stress seems to correlate with the viscosity of the melt, although the lack of data on the latter in most systems has thus far hampered a more definitive assessment of this correlation. In addition, shape distortions and coarsening of grains, as well as continued phase evolution toward equilibrium as described above, occur during deformation. These observations lead us to conclude that diffusion controlled solution-reprecipitation creep is the dominant deformation mechanism in these Sialons.

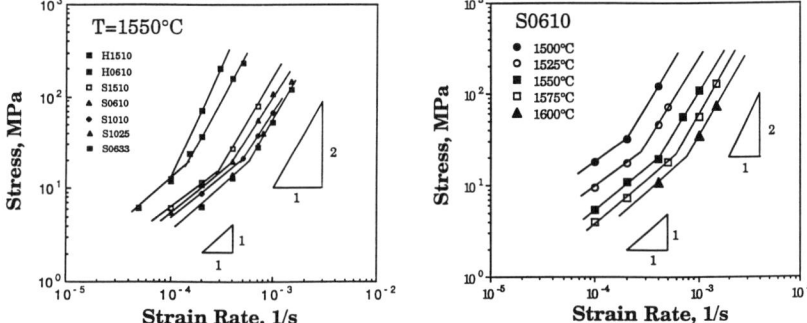

Fig. 14. Stress versus strain rate in compression showing Newtonian behavior at low stress and shear thickening at high stress. (a) various compositions at 1550°C; (b) various temperatures at one composition.[3]

It should be noted that the complex phase and microstructural evolution processes concurrent with deformation are highly sensitive to the composition, processing history, and deformation temperature. In particular, both the amount and the composition of the liquid phase can vary broadly. This, in turn, has a direct effect on the flow stress and formability of superplastic Sialons. In our experience, the superplastic forming temperature was narrowly confined to between 1550 and 1600°C. Below this temperature range, the flow stress increased sharply due to sluggish diffusion leading to premature failure at smaller strains. Above this temperature range, microstructural coarsening became too fast, and more seriously, melt evaporation increased, causing rapid deterioration of the surface quality of workpieces.

From a scientific point of view, the most interesting and surprising aspect of the deformation of these Sialons is the discovery of a novel shear-thickening transition.[3] This is noted in Fig. 14 by comparing the slope at below and above 20 MPa. The shear-thickening characteristics at above 20 MPa are manifested by a stress exponent less than unity for strain rate, or equivalently, a strain rate sensitivity of greater than unity for stress. This compares with the Newtonian behavior, characterized by the slope of unity, in Fig. 14 below 20 MPa. The transition stress was found ubiquitous in our study, independent of microstructure, composition, phase assemblage, and deformation conditions (i.e., temperature and strain rate). Moreover, the transition is reversible, albeit with a hysteresis in the stress-strain rate plot (i.e., when the strain rate is instantaneously decreased from above the shear-thickening transition, the decrease in flow stress is not instantaneous, and vice versa.)[3] This is despite the phase and microstructural evolution present. These features all point to a characteristic structural transition of the material that occurs at 20 MPa.

We have proposed that the structural transition at 20 MPa is due to a stress-induced "freezing" of the grain boundary liquid.[3] Referring to Fig. 15, we envision the liquid in-between two adjacent grains to contain a Stern layer (of a thickness Δ_s) on each side. In Sialon and Si_3N_4 ceramics, the grain boundary liquid is siliceous and may be modeled by a SiO_2 liquid. Thus, a reasonable estimate of the Stern layer thickness is a SiO_4 tetrahedron of a size of 0.3 nm. The liquid within the Stern layer is generally thought to have a more compact, almost solid-like structure and is much stiffer than an ordinary liquid. Therefore, when the liquid between the Stern layers is squeezed out, the remaining liquid within the Stern layer essentially "freezes" and the grain boundary is now in effect "dry." Using the modified DLVO theory by Clarke[29] to estimate the stress required to bring grains to a separation $2\Delta_s$, we have obtained a value of 35 MPa for our Sialons. This estimate seems to be able to account for the experimentally observed transition stress once the stress concentration factor (around 1.5) is taken into consideration.[3]

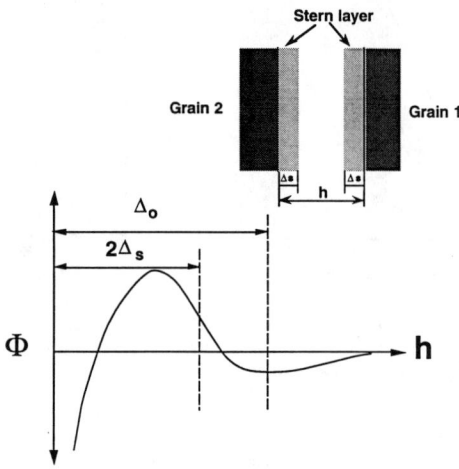

Fig. 15. Interaction Φ between grains at a separation h, with two Stern layers on the grain/liquid interfaces.[3]

To our knowledge, shear-thickening transition offers the only direct method to probe the interaction between grain, liquid and grain at a sub-nanometer scale. Although these interactions are thought to be general for silicon nitride ceramics, they are manifested in deformation for the first time in superplastic Sialons. This is because in all other silicon nitride ceramics studied in the past, the grain size was much larger and the temperature much lower which elevated the flow stress well beyond the level of shear thickening transition.[8,30-33] At such high stresses, other plastic deformation mechanisms, including cavitation,[32-33] generally occur and may have obscured the transition. In view of the continuing interest in understanding the liquid structure and grain-liquid interaction of the grain boundary phase in silicon nitride,[34] we hope the discovery of this novel phenomenon in superplastic Sialon will prove timely and valuable.

CONCLUSIONS

Superplastic Sialon is one of the new developments in silicon nitride ceramics in the last few years. Because of the very fine grain sizes and very low forming and deforming temperatures used, they offer a window of opportunity to capture new observations on silicon nitride and to shed light on the discussion of outstanding scientific issues in this field. These observations and their implications have been summarized in this review. Most notably, we have been able to elucidate the complex kinetics and pathways both for the reaction and dissolution of starting powders into a transient forming liquid and for the nucleation and growth of new phases from the liquid. A novel structural transition pertaining to the grain boundary liquid in a highly confined space under stress has also been noted. Together, they provide one glimpse into the rapid scientific progress and excitement that has been generated in the silicon nitride field in recent years.

Looking ahead, it seems clear that further research along a line similar to that plotted here is warranted. There are still outstanding issues, including the ones discussed in this review, which require much more attention. For example, while the growth anisotropy of β-Si_3N_4 and β'-SiAlON is now reasonably understood, the absence of a similar anisotropy for α'-SiAlON, despite some occasional observations of α-Si_3N_4 whiskers and rods,[35] remains a mystery. Clearly, the small structural difference, and, in the case of α'-SiAlON, the interstitial cations, must disrupt the bonding in a significant way to cause the loss of anisotropy. As another example, while heterogeneous nucleation of α' and β'-SiAlON is now well-established, the intriguing possibility of homogeneous nucleation under highly supersaturated conditions needs to be further explored.. In addition, it would be desirable to integrate this body of knowledge of nucleation and growth to specifically define the extent to which the size and composition of the initial powders can dictate the final microstructure (including the fibrous one that is known to be advantageous for toughness).[36] It would also be interesting to correlate the shear thickening transition, as a manifestation of grain/liquid/grain interactions, to high resolution TEM examinations in order to complete our picture of microscopic structure. Finally, it is important to emphasize that, for all these endeavors, the methodology followed by this investigation, which couples processing, microstructural characterization, and properties and which resorts to phase relations and kinetic theories, is likely to prove both profitable and indispensable.

ACKNOWLEDGEMENT

This research was supported by the National Science Foundation under Grant No. DDM 9024775. We are indebted to Professor T.Y. Tien for his interest and enlightening discussions during the course of this work.

REFERENCES

1. S.L. Hwang, PhD thesis, University of Michigan, 1992.

2. I-W. Chen and L.A. Xue, J. Am. Ceram. Soc., **73** (9) 2585 (1990).

3. I-W. Chen and S.L. Hwang, J. Am. Ceram. Soc., **75** (5) 1073 (1992).

4. X. Wu and I-W. Chen J. Am. Ceram. Soc., **75** (10) 2733 (1992).

5. F. Wakai, Y. Kodama, S. Sakaguchi, N. Murayama, K. Izaki and K. Niihara, Nature (London), **344** (6265) 421 (1990).

6. T. Rouxel, F. Wakai and K. Izaki, J. Am. Ceram. Soc., **75** (9) 2363 (1992).

7. S. Boskovic, L.J. Gauckler, G. Petzow and T.Y. Tien, Powder Metall. Int., **9** (4) 185 (1977); ibid., **10** (4) 184 (1978); ibid., **11** (4) 169 (1979).

8. M.H. Lewis, B.D. Powell, P. Drew, R.J. Lumby, B. North and A.J. Taylor, J. Mater. Sci., **12**, 61 (1977).

9. M.N. Rahaman, F.L. Riley and R.J. Brook, J. Am. Ceram. Soc., **63** (11-12) 648 (1980).

10. M. Kuwabara, M. Benn, F.L. Riley, J. Mater. Sci., **15**, 1407 (1980).
11. L.J. Gauckler, H. Lukas and G. Petzow, J. Am. Ceram. Soc., **58** (7-8) 346 (1975).
12. I.K. Naik and T.Y. Tien, J. Am. Ceram. Soc., **62**, 642 (1979).
13. W.-Y. Sun, T.Y. Tien and T.S. Yen, J. Am. Ceram. Soc., **74** (10) 2547 (1991).
14. W.-Y. Sun, T.Y. Tien and T.S. Yen, J. Am. Ceram. Soc., **74** (11) 2753 (1991).
15. S. Slasor and D.P. Thompson, in Non-Oxide Technical and Engineering Ceramics, edited by S. Hampshire (Elsevier Applied Sci., London, 1986), pp. 223-230.
16. R.A.L. Drew, S. Hampshire and K.H. Jack, in Progress in Nitrogen Ceramics, edited by F.L. Riley (Martinus Nijhoft, The Hague, 1984), pp. 323-330.
17. G. Ziegler, J. Heinrich and G. Wotting, J. Mater. Sci., **22**, 3041 (1987).
18. C. Chatfield, T. Ekström and M. Mikus, J. Mater. Sci., **21**, 2297 (1986).
19. F.F. Lange, J. Am. Ceram. Soc., **62** (7-8) 428 (1979).
20. M. Mitomo, M. Tsutsumi and H. Tanaka, J. Am. Ceram. Soc., **73** (8) 2441 (1990).
21. M. Mitomo and S. Uenosono, J. Am. Ceram. Soc., **75** (1) 103 (1992).
22. T.Y. Tien, this volume.
23. P. Drew and M.H. Lewis, J. Mater. Sci., **9**, 261 (1974).
24. C.M. Hwang, T.Y. Tien and I-W. Chen, in Sintering '87, edited by S. Somiya, M. Shimada, M. Yoshimura and R. Watanabe (Elsevier Applied Science, New York, 1988), pp. 1034-1039.
25. P. Hartman and W.G. Perdok, Acta Cryst., **8**, 49 (1955); ibid., **8**, 521 (1955); ibid., **8**, 525 (1955).
26. K. Jackson, in Liquid Metals and Solidification, (American Society for Metals, Cleveland, Ohio, 1958), pp. 174-186.
27. G.H. Gilmer and K.A. Jackson, in Crystal Growth and Materials, edited by E. Kaldis and H.J. Scheel (North-Holland, Amsterdam, Holland, 1977), pp. 80-114.
28. J.G. Wang and R. Raj, J. Am. Ceram. Soc., **67** (6) 385 (1984); ibid., **67** (6) 399 (1984).
29. D.R. Clarke, J. Am. Ceram. Soc., **70** (1) 15 (1987).
30. C.F. Chen and T.Y. Tien, Ceram. Eng. and Sci. Proc., **8** (7-8) 778 (1987).
31. R. Kossowsky, D.G. Miller and E.S. Diaz, J. Mater. Sci., **10**, 983 (1975).
32. P.J. Dixon-Stubbs and B. Wilshire, J. Mater. Sci. Lett., **14**, 2773 (1979).
33. F.F. Lange, B.J. Davis and D.R. Clarke, J. Mater. Sci., **15**, 601 (1980); ibid., **15**, 611 (1980).
34. H.J. Kleebe, M.K. Cinibulk, I. Tanaka, J. Bruley, R.M. Cannon, D.R. Clarke, M.J. Hoffmann and M. Rühle, this volume.
35. K. Niihara and T. Hirai, J. Mater. Sci., **14**, 1952 (1979).
36. P.F. Becher, H.T. Lin, S.L. Hwang, M.J. Hoffmann and I-W. Chen, this volume.
37. W.Y. Sun, T.Y. Tien and T.S. Yen, J. Am. Ceram. Soc., **74** (11) 2753 (1991).

PART II

Recent Progress

PART II A

Powder Synthesis and Processing Routes

SYNTHESIS OF Si$_3$N$_4$ POWDER
BY THERMAL DECOMPOSITION OF SI(NH)$_2$

SILVIA AMPUERO* PAUL BOWEN* AND TERRY A. RING**
*Powder Technology Laboratory, Materials Science Department, Swiss Federal Institute of Technology (EPFL), CH-1015 Lausanne.
**Chemical Engineering Department, University of Utah, Salt Lake City, Utah 84112.

ABSTRACT

A fine white α-Si$_3$N$_4$ powder has been produced by the thermal decomposition of the coprecipitation product of the reaction between SiCl$_4$ and NH$_3$. The Cl content, due to a reaction between Si(NH)$_2$ and the NH$_4$Cl by-product during the thermal treatment of the coprecipitate, has been reduced by using an isothermal step at ≈ 250°C in the heat treatment cycle.

INTRODUCTION

Although silicon nitride had already been isolated in 1844[1], it was not until 1960 that it was introduced as an engineering material by Parr et al[2]. Its outstanding combination of properties[3] -high strength, wear resistance, high decomposition temperature, creep and oxidation resistance in hostile atmospheres, good thermal shock resistance and low coefficient of friction- make it an ideal material for high temperature heat engine applications. The optimal performance and reliability of an advanced ceramic material is also determined by the densification process and the properties of the raw powder.

Several silicon nitride powder characteristics have been defined[4] for a good sinterability: Fineness to promote sintering[5], spherical shape for a good green compaction behavior, narrow size distribution[6] to inhibit grain growth. Chemical purity is also of major importance as an excessive impurity content can affect the final properties and the densification behavior. Rhodes et al[7] describe different types of impurities: Discrete inorganic or organic particles such as Fe. Anion impurities and organic compounds, that are often found in powders even after heating. In the case of silicon nitride, oxygen is the most common impurity, it forms films of SiO$_2$ or Si$_2$ON$_2$ on the surface of particles which may react with sintering aids such as Y$_2$O$_3$ to promote sintering, but an excess of oxygen can affect the elevated temperature strength and creep resistance, by changing the secondary phase composition. Another anion impurity is Cl which appears to inhibit the α to β transformation during sintering[8] and affect the final properties of the silicon nitride. Cation impurities, which usually concentrate as a second phase because of their poor solubility in the major phase, an example is the concentration of Ca[9] in the grain-boundary phase employed as a sintering aid for silicon nitride, this cation lowers the viscosity of the secondary phase and consequently the creep properties of silicon nitride.

Powder properties and also production cost[10] are determined by the silicon nitride powder synthesis route. Industrial processes for the production of silicon nitride powders are:
  a) Direct nitridation of silicon powder with nitrogen gas at elevated temperatures.
  b) Carbothermal reduction of silica powder with nitrogen in presence of carbon at high temperatures.
  c) Reactions between silicon halide, such as silicon tetrachloride and silane[11], and ammonia either in gas-phase (commercially available[6] since the 1970s) or in liquid-phase[12,13] sometimes followed by a thermal decomposition step.

Thermal decomposition of the reaction-product between silicon tetrachloride and ammonia is one of the more promising methods for producing high quality silicon nitride powder. The purity of the powder, other than C and Cl impurities, is controlled by reactant purity, commercially available in a highly pure state. Powder morphology and size distribution may be controlled by monitoring reactant supersaturation. Costs are low as little or no powder grinding/cleaning processes are needed, and the possibility of conducting the reaction at room temperature and as a continuous process make it more attractive.

The aim of the current work was to study the precipitation-thermal decomposition route with respect to powder quality. Particular emphasis has been placed on the thermal decomposition of the Si(NH)$_2$ / NH$_4$Cl coprecipitate and our results are compared with various mechanisms proposed in the literature.

EXPERIMENTAL

Silicon nitride powder was produced in a two-step process: first the precipitation reaction (1) of liquid silicon tetrachloride with gaseous ammonia at 0°C in hexane was conducted in a batch reactor, followed by the thermal decomposition of the products to obtain silicon nitride.

Precipitation

In a glove box under a nitrogen atmosphere, containing < 3 ppm O_2 and < 5 ppm H_2O, silicon tetrachloride (puriss. Fluka), without further purification, was introduced into a reactor containing hexane (pro analysi Merck), dried with molecular sieve. Ammonia (electronic-grade Carbagas) diluted with nitrogen (>99.995% Carbagas) was bubbled into the silicon tetrachloride solution, with continuous stirring at 0°C (using a water-ice bath), fig.1. The use of nitrogen has two purposes, to avoid moisture and oxygen contamination and to control the ammonia concentration in the gas flow.

Hexane was evaporated under a low vacuum with a small flux of nitrogen, at 45°C in a rotatory evaporator. Powders where handled and stored in a glove box. Typical reactant concentrations are: 7.6×10^{-2} M $SiCl_4$, 4×10^{-3} mol/min NH_3.

$$SiCl_4 \text{ (l)} + 6 NH_3 \text{ (g)} \xrightarrow{\text{hexane/0°C}} Si(NH)_2 \text{ (s)} + 4 NH_4Cl \text{ (s)} \quad (1)$$

Thermal Decomposition

The coprecipitate $Si(NH)_2$ and NH_4Cl is then heat treated in a long tube furnace, after purging with nitrogen, at temperatures around 1400°C for ≈4 hours under flowing nitrogen, to produce sub-micron α-phase silicon nitride powder.

Analysis

The thermal decomposition of the coprecipitate $Si(NH)_2$ and NH_4Cl has been studied by differential scanning calorimetry (DSC) and thermogravimetric analysis (TGA) (TA4000 Mettler system). All powder manipulations were carried out under a nitrogen atmosphere. Concurrent studies were made on $Si(NH)_2$ after removing the NH_4Cl by-product by washing the coprecipitate with liquid ammonia at -68°C. Diffuse reflectance measurements (Nicolet 700 FTIR), XRD (Siemens D500 difractometer, Kα ray of Cu), particle size distributions (Horiba Capa700), elemental analysis (Perkin Elmer 2400CHN Elemental Analyzer), Si content (gravimetry), and Cl content (titration with silver nitrate in the presence of potassium chromate) analyses were used to characterize the various powder samples.

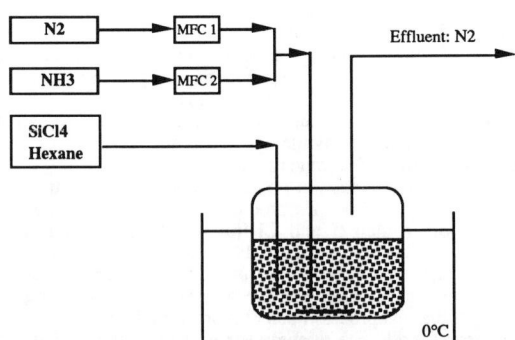

Fig.1 Schematic representation of the precipitation batch-reactor, for the reaction between $SiCl_4$ and NH_3. MFC 1 and MFC 2 are gas mass flow controllers.

Table I. Elemental analysis of different samples and theoretical compositions.
(+ a single data point, * two step heat treatment) (analysis precision ± 5%).

Sample	Formula	Si [wt%]	N [wt%]	H [wt%]	C [wt%]	Cl [wt%]	Si/N	N/H
theoretical	$Si(NH)_2$ + $4NH_4Cl$	10.3	30.9	6.7	-	52.1	0.3	4.6
coprecip.	$Si(NH)_2$ + $4NH_4Cl$	8.6	32.0	6.2	0.4	53.2	0.27	5.2
theoretical	$Si(NH)_2$	48.3	48.2	3.5	-	-	1.0	13.8
washed coprecip.		33.3	36.1	5.3	1.6	7+	0.9	6.8
LC Starck	Si_3N_4	60.1	39.9	-	0.2	-	1.5	
sample	Si_3N_4	not det.	39.5	0	0.18	1.8 < 0.01*		

RESULTS AND DISCUSSION

Precipitation

With the addition of ammonia to the silicon tetrachloride solution, precipitation takes place immediately, and a white slurry of silicon diimide and ammonium tetrachloride is formed in hexane. The reaction yield is ≈100% and a typical product concentration is 4.4 g/l of $Si(NH)_2$.

The dried precipitation products are fine white powders, readily charged electrostatically and therefore difficult to manipulate, they hydrolyze on contact with moisture decomposing to SiO_2 and evolving ammonia.

Elemental analyses of the coprecipitate, table I, show a molar ratio close to 1/4 for $Si(NH)_2/NH_4Cl$ (calculated from the Si/N weight ratio).

The infrared spectrum of the coprecipitate, Fig.2 a, shows characteristic bands of NH_4Cl: 3140, 3040, 2825 cm^{-1} (NH asymmetric elongation), 2010, 1760 cm^{-1} (NH vibration broad bands), 1400 cm^{-1} (NH bending vibration), and 1060 cm^{-1}. In the spectra of $Si(NH)_2$, washed coprecipitate, Fig.2 b, these peaks practically disappear, except the one at 1060 cm^{-1}, which also corresponds to SiN vibrations. The bands centered at 3335, 1545 and 1210[14] cm^{-1} (NH_2 stretching vibration and scissors bending vibration[15]) are more visible in trace b. SiN and Si characteristic bands are present in both traces: 970, 830 cm^{-1} (SiN stretching vibrations) and 430 cm^{-1} (Si breathing vibration). Vibration bands associated with SiH (2100, 2200 cm^{-1}) are not present in either trace. Absorption associated with SiO vibrations occur at 800 and 1100, because of the overlapping of these bands with SiN bands, the existance of SiO_2 cannot easily be determined by this method.

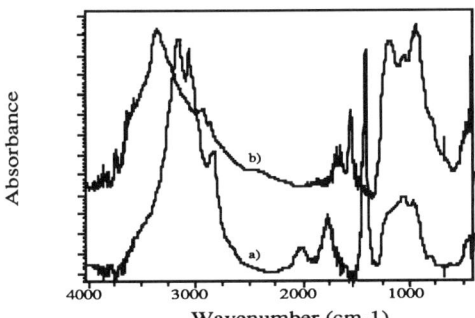

Fig.2 FTIR data: a) coprecipitate $Si(NH)_2$ + 4 NH_4Cl, b) washed coprecipitate.

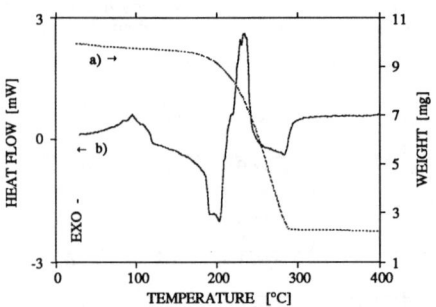

Fig.3 Thermal analysis of the coprecipitate (Si(NH)$_2$ + 4 NH$_4$Cl): a) TGA, b) DSC.

Thermal Decomposition

During the thermal decomposition of the coprecipitate Si(NH)$_2$ + 4 NH$_4$Cl a large weight loss (≈78%) takes place between 200-300°C, Fig.3 a, which mostly corresponds to NH$_4$Cl dissociation/evaporation. A smooth weight loss is also observed beginning almost at room temperature to ≈200°C and between 300-≈700°C attributed to the decomposition of Si(NH)$_2$ with the evolution of NH$_3$.

In the same fig. 3, the corresponding DSC data shows at first an exothermic peak around 100°C attributed to the decomposition of Si(NH)$_2$. Then in contradiction to the strongly endothermic peak expected for the sublimation of NH$_4$Cl around 300°C (see the DSC trace for NH$_4$Cl in the inset in Fig.4), a complicated trace appears between 200-300°C. The exothermic peak visible at ≈ 250°C could be due to further decomposition of an intermediate Si compound, or to a reaction between an intermediate Si compound and NH$_4$Cl. The thermal decomposition of NH$_4$Cl-free Si(NH)$_2$ (washed coprecipitate) shows no exothermic peak in this region, Fig.4 a), therefore a reaction between an Si compound and NH$_4$Cl around 250°C is proposed.

The thermal decomposition of a partially washed coprecipitate, Fig.4 b, shows the same trend as the coprecipitate in that an exothermic peak between 200-300°C is present, and because the amount of NH$_4$Cl available for sublimation is much less the heat effect at ≈250 °C

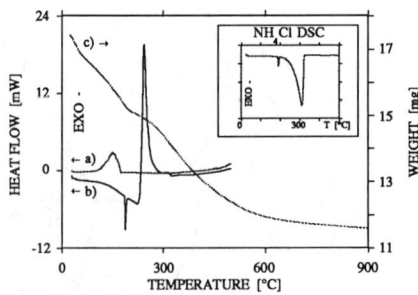

Fig.4 DSC data: a) washed coprecipitate (Si(NH)$_2$), b) partially washed coprecipitate (Si(NH)$_2$ + x NH$_4$Cl), c) TGA of partially washed coprecipitate (Si(NH)$_2$ + y NH$_4$Cl). NH$_4$Cl DSC included in the inset.

is more marked in this case, thus the hypothesis of a superposition of two phenomena taking place between 200-300°C (sublimation of NH4Cl and reaction of a Si compound and NH4Cl) is supported.

Several decomposition mechanisms have been proposed in the literature for the thermal decomposition of the Si(NH)2/NH4Cl coprecipitate into Si3N4. Based on thermogravimetric data some authors[13, 16 and 17] proposed a decomposition-mechanism for the Si compound assuming a complete sublimation of NH4Cl taking place simultaneously with the Si(NH)2 decomposition. Billy[18] followed by others[19] proposed a mechanism which is consistent with much of our data:

$$Si(NH)_2 \xrightarrow{60°-80°C} 1/2\ Si_2N_3H + 1/2\ NH_3 \quad (2)$$

$$1/2\ Si_2N_3H + 1/2\ NH_4Cl \xrightarrow{130°-150°C} 1/2\ Si_2N_3H_2Cl + 1/2\ NH_3 \quad (3)$$

$$1/2\ Si_2N_3H_2Cl \xrightarrow{300°-400°C} 1/3\ Si_3N_4HCl + 1/6\ NH_4Cl \quad (4)$$

$$1/3\ Si_3N_4HCl \xrightarrow{>400°C} 1/3\ Si_3N_4\ (amorphous) + 1/6\ HCl \quad (5)$$

Interpretation of our TGA/DSC data, Fig.3 and 4, is as follows. Reaction (2) takes place at 40-200°C, it is catalyzed by NH4Cl, this reaction seems to take place at 80-200°C in absence of NH4Cl. Between 200-350°C the reaction (3) between the Si compound and NH4Cl is completed. Each of these steps incorporates the loss of 1/2 mole of NH3 (10wt%). Further decomposition takes place with no apparent heat effect. The weight change observed between ≈40-200°C is ≈10 wt% (as expected for reaction (2)). In Fig.4 c (a partially washed coprecipitate) a continuous weight loss is observed from 200°C to > 700°C, the total amount of this weight loss is about 27.5 wt% which corresponds to the evolution of 1/2 mol NH3 (10wt%) during reaction (3), 1/6 mol NH4Cl (10.5 wt%) by reaction (4) and 1/6 mol HCl (7.2 wt%) by reaction (5). A steep weight loss is observed from room temperature to ≈40°C which might be due to the evolution of NH3 adsorbed on the surface of the particles, or to an initial Si compound such as $Si(NH_2)_4$[19] which transforms into $Si(NH)_2$, FTIR shows evidence for the existence of NH2 vibration modes, but they could also come from the Si(NH)2 molecule with the structural formula: N=Si-NH2.

Billy's work was carried out in vacuum, this could explain the differences in reaction temperatures when compared with our data.

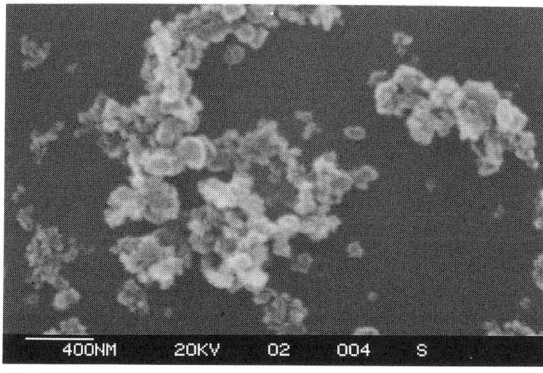

Fig.5 α-Silicon Nitride SEM image

Silicon Nitride Powder

A fine white silicon nitride powder was obtained. The proportions of the crystalline phases present depend on the heat treatment, typically heating at 1425°C for 4.25 hours gives 93 % α-silicon nitride.

The elemental analyses, table I, show a good silicon nitride powder stoichiometry, the oxygen content varies enormously if precautions are not taken to avoid oxidation during precursor handling and heat treatment. Carbon content was found to be 0.18 wt.%. Chlorine content was over 1.5 wt.% for some powders, but this was reduced when the heat treatment was made in two steps: the first at 200-250°C for 30 minutes to eliminate NH_4Cl and then at higher temperatures to complete the thermal treatment.

SEM images, Fig.5, show particulates of size under 200 nm, agglomerates as large as 2.4 μm were detected by particle size measurements.

CONCLUSIONS

We have shown that a reaction between an intermediate Si compound and NH_4Cl can take place in the thermal decomposition of an $Si(NH)_2/NH_4Cl$ coprecipitate leading to > 1% Cl content even after treatment at > 1400°C for up to 5 hours.

This Cl impurity level can be substantially reduced if the greater part of the NH_4Cl present is allowed to sublime between 200-250°C, before the reaction is complete (sublimation of NH_4Cl seems to take place at lower temperatures in the presence of $Si(NH)_2$). This has a great practical importance in that the coprecipitation of $Si(NH)_2$ + NH_4Cl can yield high quality powders, without the necessity of a tedious NH_4Cl washing step, or a high energy consuming reaction-process at liquid NH_3 temperatures or high pressure, and could be carried out in a continuos stirred tank reactor.

The exact mechanism for the reaction between a partially decomposed $Si(NH)_2$ and NH_4Cl is still unclear, further on-line elemental analysis using programmed heating ramps and GC-MS analysis of the decomposition products is hoped to shed further light on the mechanism.

This work was supported by the Swiss National Scientific Program FNP 19. We thank B.Senior for the SEM images.

REFERENCES

1. A. S. Berezhnoi, Silicon and its Binary Systems, (New York, 1960) p. 612.
2. N. L. Parr, G. F. Martin and E. R. W. May, Special Ceramics, ed. by P. Popper (Heywood & Company Ltd, London, 1960) p. 112.
3. C. C. Sorrell and E. R. McCartney, Materials Forum **9** (3), 148-161 (1986).
4. G. Wotting and G. Ziegler, Interceram, **35** (2), 32-35 (1986).
5. W. Symons and S. C. Danforth, Advances in Ceramics, **21**: Ceramic Powder Science, (The American Ceramic Society, Inc Copyright 1987) p 249-256.
6. K. Kendall, Powder Technology **58**, 151-161 (1989).
7. W. H. Rhodes and S. Natansohn, Ceramic Bulletin, **68** (10), 1804-1812 (1989).
8. D. R. Clarke, J. Am. Ceram. Soc., february, C21-C23 (1982).
9. W. Engel,Powder Metallurgy International 10 (3), 124-127 (1978).
10. J. M. Schoenung, Ceramic Bulletin **70** (1), 112- 116 (1991).
11. S. Prochazka and C. Greskivich, Ceramic Bulletin 57 (6), 579-586 (1978).
12. G. M. Crosbie, R. L. Predmesky, J. M. Nicholson and E. D. Stiles, Ceramic Bulletin **68** (5), 1010-1014 (1989).
13. K. S. Mazdiyasni, C. M. Cooke, Dayton, U.S. Patent No. 3 959 446 (25 May 1976).
14. E. A. Pugar and P. E. D. Morgan, Better Ceramics through Chemistry III, ed. by C. J. Brinker, D. E. Clark and D. R. Ulrich (Materials Research Soc., Pittsburgh, 1988).
15 D. V. Tsu, G. Lucovsky, and M. J. Mantini, Physical Review B, **33** (19), 7069-7076 (1986).
16 K. S. Mazdiyasni, and C. M. Cooke, J. Am. Ceram. Soc., **56** (12), 628-633 (1973).
17 M. C. Sneed and R. C. Brasted, Comprehensive Inorganic Chemistry **5** (D. Van Nostrand Co., New York, 1966) p. 173.
18 M. Billy, Ann. Chem., 4, 818-851 (1959).
19 V. O. Glemser and P. Naumann, Z. Anorg. Allg. Chemie, 298, 134-141 (1959).

STRUCTURAL CHARACTERIZATION OF SILICON CARBONITRIDE CERAMICS FROM POLYMERIC PRECURSORS USING NUCLEAR MAGNETIC RESONANCE

CORINE M. GERARDIN*, F. TAULELLE AND J. LIVAGE
Chimie de la Matiere Condensee, Universite P. M. Curie, 4, place Jussieu. 75252 Paris, France.
*present address : Department of Geological and Geophysical Sciences, Guyot Hall, Princeton, NJ08540, USA.

ABSTRACT

The pyrolytic conversion of a polyvinylsilazane precursor to Si-C-N ceramics is studied using solid-state NMR. ^{13}C and ^{29}Si magic angle spinning and ^1H static NMR experiments allow us to investigate the structure of the disordered multiphased intermediates formed during the pyrolysis. The Si environments are quantitatively analyzed by ^{29}Si MAS NMR spectra simulations. The relative abundance of $\underline{Si}HCN_2$, $\underline{Si}C_2N_2$, $\underline{Si}N_3C$ and $\underline{Si}N_4$ sites at different temperatures are used to calculate elemental compositions that characterize the Si-containing phase. The comparison of these compositions with those obtained from chemical analyses shows evidence of the presence of a free carbon phase. We have also determined the H and C contents in the free carbon phase.

INTRODUCTION

Silicon carbide and silicon nitride ceramics are known for their good mechanical properties and high thermal and chemical stability[1,2]. In recent years, many procedures for the synthesis of new materials have been developed. Silicon carbonitride[3,4] materials have been prepared by pyrolysis of organometallic polymers such as polysilazanes. In order to improve the properties of the ceramics it is important to be able to characterize not only the resulting ceramic powder, but also all the intermediates formed during the preparation of the materials. Various structural evolutions during the heat-treatment of the polymers are not completely understood : crosslinking of the starting polymer, ceramization of the precursor, phase separation with formation of free carbon and silicon carbonitride phases. The objective of our study is to show how nuclear magnetic resonance studies are useful in quantitatively describing the local sites and the different phases in amorphous multiphase Si-C-N materials.

To follow the structural transformations from the polymer to the final material, ^{13}C and ^{29}Si magic angle spinning (MAS) NMR experiments are performed with and without cross-polarization (CP). The hydrogen content in the materials is precisely measured using solid-state static ^1H NMR spectroscopy. Quantitative analysis of the ^{29}Si MAS NMR spectra allow us to determine the proportions of silicon sites and then to deduce the relative amounts of each phase, silicon nitride, silicon carbonitride and free carbon.

These NMR methods are applied to the study of the pyrolysis of a polyvinylsilazane[5] preceramic polymer. Its idealized formula is -(HSiVi-NH)$_n$-, where Vi is a vinyl group. The synthesis of this precursor has been previously described[6, 7]. Oligosilazanes are prepared by ammonolysis of vinyldichlorosilane :

n SiHCl$_2$Vi + 3n NH$_3$ → -(ViHSi-NH)$_n$- + 2n NH$_4$Cl.

The oligomers, which present an unusually high functionality, are cured in order to lead to a high ceramic yield. The possible cross-linking[6] reactions are :
- hydrosilylation of vinyl groups, promoted in hexane by hexachloroplatinic acid :

$\equiv$SiH + $\equiv$Si-CH=CH$_2$ → $\equiv$Si-CH$_2$-CH$_2$-Si$\equiv$ (α-addition)

or $\equiv$Si-CH(CH$_3$)-Si$\equiv$ (β-addition)

- transamination which leads to trisubstituted N sites :

2 $\equiv$Si-NH-Si$\equiv$ → $\equiv$Si-N-Si$\equiv$ + $\equiv$Si-NH$_2$; $\equiv$Si-NH$_2$ + $\equiv$Si-NH-Si$\equiv$ → $\equiv$Si-N-Si$\equiv$ + NH$_3$
                          | |
                          Si$\equiv$ Si$\equiv$

- polymerization of vinyl groups :

$\equiv$Si-CH=CH$_2$ + $\equiv$Si-CH=CH$_2$ →   -CH-CH$_2$-CH-CH$_2$-
                                              | |
                                              $\equiv$Si $\equiv$Si

The polyhydrogenovinylsilazane, PVSZ, has been extensively characterized by ^{1}H, ^{13}C, ^{29}Si, ^{15}N and ^{14}N liquid NMR[5]. Carbosilane bridges, Si-C$_x$-Si, are formed during hydrosilylation reactions, consuming some vinyl groups and Si-H bonds. A few NSi$_3$ nitrogen sites have appeared during transamination reactions.

EXPERIMENTAL SECTION

Elemental analyses were performed by Service Central de Microanalyses (CNRS Vernaison, France). The NMR spectra were recorded on a Bruker MSL400 spectrometer, operating at 400.13 MHz, 100.62 MHz and 79.5 MHz, respectively for ^{1}H, ^{13}C and ^{29}Si nuclei. Magic angle spinning NMR spectra were acquired and the samples were spun at 4 kHz. A contact time of 1.5 ms and a repetition time of 10 s were used for the ^{13}C cross-polarization (CP) MAS NMR spectra. The ^{29}Si one-pulse experiments required a repetition delay of 20 to 60 s. Static ^{1}H NMR spectra were recorded to obtain the hydrogen content[5]. A one-pulse sequence with a recycle time of 4 to 10 s and a specific probe, with no background proton signal, were used.

RESULTS AND DISCUSSION

(1) Qualitative analysis of the pyrolytic conversion of the PVSZ precursor

The ^{13}C CPMAS NMR spectrum of PVSZ-500 (Fig. 1) shows a unique broad peak centered at 9 ppm which is characteristic of C$_{sp3}$ carbon sites. Below 500°C, all vinyl groups (C$_{sp2}$ sites) have been transformed to carbosilane bridges Si-C$_x$-Si by hydrosilylation and/or by polymerization. At 600°C, some weak signals (Fig. 1) appear from 50 to 150 ppm which we assigne to protonated C_{sp2} sites. At 900°C a few C_{sp3} sites (δ=6ppm) are still present, but most of the carbon atoms are in unsaturated C_{sp2} protonated environments.

The ^{29}Si MAS NMR spectra of the pyrolytic residues are shown in Fig. 2. According to a previous study[8, 9] based on a correlation established between ^{29}Si chemical shifts and Si partial charges, we can identify three main signals at 500°C : $\underline{Si}N_4$ sites at -45.8 ppm, $\underline{Si}HCN_2$ and $\underline{Si}N_3C$ sites at -23.8 ppm and $\underline{Si}C_2N_2$ sites at 2.2 ppm. The $\underline{Si}HCN_2$ and $\underline{Si}C_2N_2$ sites have been formed by hydrosilylation and/or polymerization reactions. Above 500°C, some N-rich sites, ie $\underline{Si}N_3C$ and $\underline{Si}N_4$, appear and become more numerous at the expense of the $\underline{Si}C_2N_2$ sites up to 900°C. At this temperature, infrared spectra show that residual functional groups (Si-H, N-H and C-H bonds) are no longer present. The observation of C=C bonds above 600°C agrees with the formation of an aromatic carbon network, as detected by transmission electronic microscopy[10]. From 900°C to 1400°C, carbon and nitrogen atoms are bonded to Si in a covalent structure : $\underline{Si}N_3C$, $\underline{Si}N_4$ and $\underline{Si}C_2N_2$ sites form the silicon carbonitride phase.

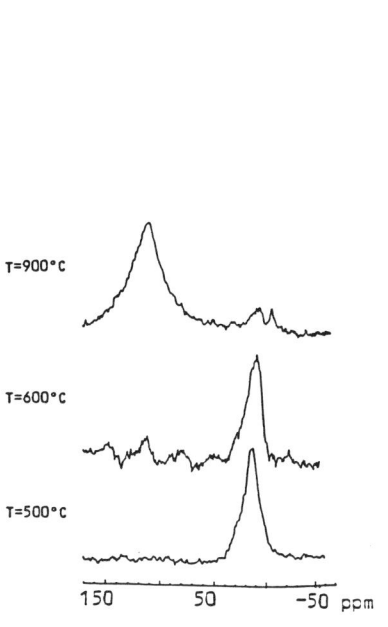

Fig. 1 : ^{13}C CPMAS NMR spectra of the pyrolysis products

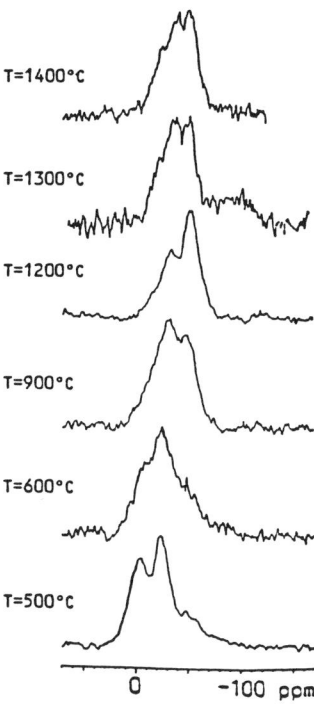

Fig. 2 : ^{29}Si MAS NMR spectra of the pyrolysis products.

(2) Quantitative analysis of the pyrolytic conversion of the PVSZ precursor

The samples have been characterized by static solid-state ^{1}H NMR in order to determine hydrogen contents[5] from spectrum integration. This measurement of the H content is more reliable than that obtained from chemical analyses[5]. The Si, C and N elemental compositions have been determined by chemical analyses. The complete atomic compositions are presented in table I.

The ^{29}Si MAS NMR spectra have been simulated (Fig. 3) to estimate the relative intensities of the different components. The $\underline{Si}N_3C$, $\underline{Si}N_4$, $\underline{Si}C_2N_2$ and $\underline{Si}HCN_2$ sites (with the $\underline{Si}HCN_2$ sites present only until 600°C) accounted for.

Figure 4 shows the changes in the Si environments during the pyrolysis. The number of N-rich silicon sites increases and become dominant as the temperature increases. The peak intensity corresponding to $\underline{Si}N_3C$ sites is maximum at 900°C. From 1200°C to 1400°C, the Si-C-N phase contains a majority of $\underline{Si}N_4$ sites, this is in agreement with TEM[8] observations of numerous large Si$_3$N$_4$ crystals at 1400°C.

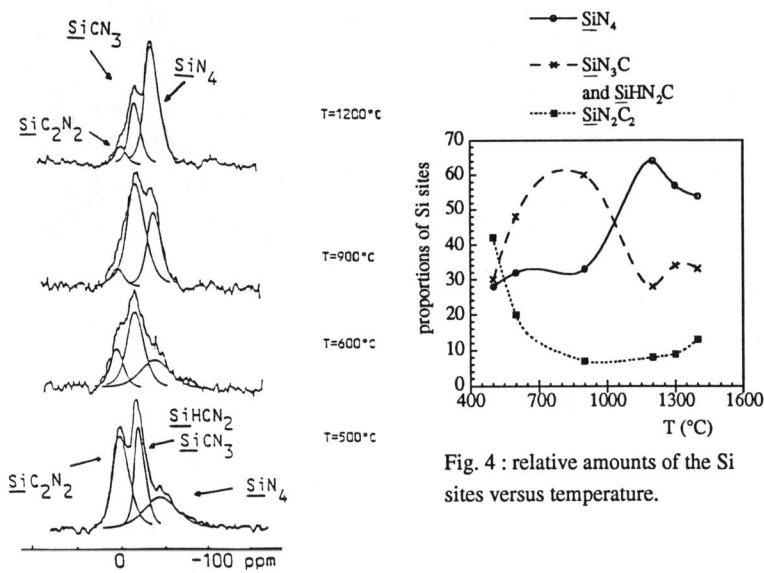

Fig. 3 : ^{29}Si MAS NMR simulated spectra

Fig. 4 : relative amounts of the Si sites versus temperature.

Elemental compositions of the Si-C-N phase (characterized by $\underline{Si}C_xN_{4-x}$ species) are calculated from the percentages of the various sites. The following assumptions were made : below 900°C, the N atoms have the two environments $\underline{N}HSi_2$ and $\underline{N}Si_3$ in the same amounts. Above 900°C, all N sites are $\underline{N}Si_3$. Carbon atoms in the Si-C-N phase, i. e. directly bonded to Si, are of two types up to 600°C : Si-CH$_2$-CH$_2$-Si or Si-CH(CH$_3$)-Si ; at 900°C, the C atoms are essentially $\underline{C}HSi_3$ sites and at 1200°C, the C atoms are in $\underline{C}Si_4$

environments. Table II presents the atomic compositions calculated from simulated NMR spectra.

Table I : atomic compositions of the pyrolyzed samples from chemical analyses and ^{1}H NMR.

T (°C)	% Si	% C	% N	% H
500	10.6	22.3	12.7	54.4
600	14.3	25.1	15.2	45.4
900	22.1	38.2	21.4	18.4
1200	27.7	37.3	25.1	9.9
1300	26.8	43.8	25.6	3.8
1400	27.1	45.4	25.7	1.8

Table II : atomic compositions of the Si-C-N phase in the fired materials from ^{29}Si MAS NMR spectra.

T (°C)	% Si	% C	% N	% H
500	14	16	14	56
600	23	20	25	32
900	37	9	32	22
1200	43	5	52	0
1300	44	5	51	0
1400	44	6	50	0

A comparison between tables I and II shows that C and H contents in the Si-C-N phase are smaller than the total C and H contents. This demonstrates the presence of a free carbon phase containing C and H atoms. Above 900°C, most of the carbon is present as a separate phase. The percentages of C, %C$_{(freeC)}$, and H, %H$_{(freeC)}$, in the free C phase are determined using the values in tables I and II : they vary from 10% at 500°C to 42% at 1400°C and 20% at 500°C to 2% at 1400°C, respectively.

The H/C$_{(freeC)}$ atomic ratio characterizing the free carbon phase decreases considerably with temperature from approximately 2 at 500°C to 0.04 at 1400°C (Fig. 5). This is in agreement with the TEM observations[8] : from 900°C to 1400°C, aromatic units join together to form larger structures of aromatic cycles. The progressive association of polyaromatic layers occurs through elimination of hydrogen atoms around the units.

The proportions of the phases present from 1200°C to 1400°C can be determined from the percentages of silicon sites and the content of free carbon in the materials. If it is assumed that all SiN$_4$ sites form a silicon nitride phase at 1400°C, the following atomic proportions are obtained : 30% Si$_3$N$_4$, 26% SiN$_x$C$_y$ phase and 44% free carbon. Usually, a phase composition is extracted only from elemental analyses based on the presence of Si$_3$N$_4$, SiC and free C. This is incorrect since no silicon carbide phase actually exists. An accurate description must be based on the presence of a silicon carbonitride phase which is mainly composed of SiCN$_3$ sites. The composition of the Si-C-N phase is thus calculated to be SiN$_{0.9}$C$_{0.3}$.

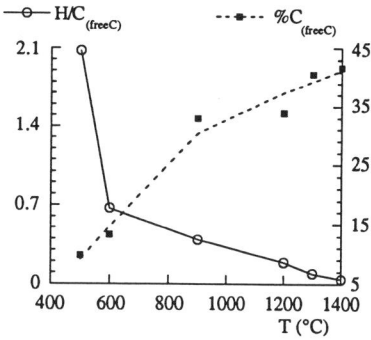

Fig. 5 : Evolution of %C$_{(freeC)}$ and H/C$_{(freeC)}$ during the pyrolytic conversion.

CONCLUSION

We have studied the pyrolytic conversion of a polyvinylsilazane preceramic polymer using solid-state NMR. The hydrogen content in the pyrolysis products was determined using ^{1}H static NMR spectra. The H contents were combined with Si, C and N elemental compositions obtained from chemical analyses to determine atomic compositions of the different samples. The atomic compositions show that the N/Si ratio at 1400°C equals 0.95 instead of 1 in the precursor, indicating that very few transamination reactions occured during the pyrolysis. The C/Si ratio only drops from 2 in the precursor to 1.7 at 1400°C, leading to a high carbon content in the final ceramic.

^{13}C CPMAS NMR spectra clearly show the transformation of vinyl groups to C_{sp3} sites at temperature lower than 500°C, due to hydrosilylation and polymerization reactions. Between 500°C and 900°C, the majority of carbon atoms progressively transform to unsaturated C atoms agreeing with TEM detection of the formation of an aromatic carbon network.

Simulation of ^{29}Si MAS NMR spectra show the changes in the relative amounts of the SiC_xN_{4-x} sites with increasing temperature. The N atoms content in the silicon carbonitride phase, which is quantitatively described by ^{29}Si MAS NMR spectra, increases as the C atoms content decreases. Elemental compositions characterizing the Si-C-N phase are calculated from the fractions of Si sites. A comparison between these compositions and the compositions of the whole materials indicates that all carbon and hydrogen atoms are not incorporated into the Si-C-N phase. The amount of free carbon is determined at each temperature of pyrolysis, we found that 87% of C atoms are in the free carbon phase at 1400°C. The H/C$_{(free\ C)}$ ratio decreases from about 2 at 500°C to 0.04 at 1400°C. The polyvinylsilazane-derived silicon carbonitride ceramic is composed at 1400°C of the following phases : 30% Si_3N_4, 26% silicon carbonitride, $SiN_{0.9}C_{0.3}$, and 44% free carbon.

ACKNOWLEDGMENT : The authors thank N. S. Choong Kwet Yive, R. J. P. Corriu, D Leclercq, H. Mutin and A. Vioux for providing the polymeric precursor and A. Lavedrine, D Balhoul and P. Goursat for the heat-treatments of the precursor.

REFERENCES

[1] G. Ziegler, J. Heinrich, G. Wotting, J. Mater. Sci. 22, 3041 (1987).
[2] T. Iseki, T. Hase, In Fine Ceramics; S. Saito, ED.; Elsevier Applied Science Publishers Ltd. : Essex, England. 1985; p 188.
[3] D. Seyferth. In Silicon-based polymer science. A comprehensive Resource; Adv. Chem. Ser. 224; J.M. Zeigler, F. W. Fearon, Eds.;American Chemical Society: Washington DC, 1990, p 565.
[4] K. J. Wynne and R. W. Rice, Ann. Rev. Mater. Sci., 14, 297 (1980).
[5] Corine Gerardin, PhD dissertation, Université P. et M. Curie, Paris, France (1991).
[6] N. S. Choong Kwet Yive, R. J. P. Corriu, D. Leclercq, H. Mutin, A. Vioux, New. J. Chem., 15, 85 (1991).
[7] idem., Chem. Mater. 4, 141 (1992).
[8] C. Gerardin, M. Henry, F. Taulelle, Mat. Res. Soc. Symp. Proc. 271, 777 (1992)
[9] M. Henry, C. Gerardin, , F. Taulelle, Mat. Res. Soc. Symp. Proc. 271, 243 (1992)
[10] O. Delverdier, PhD dissertation, Université de Pau et des Pays de l'Adour (1991).

SPECTROSCOPIC CHARACTERIZATION OF THE NITRIDATION PROCESS OF POLYMERIC PRECURSORS TO SI-M-N-O SYSTEMS (M=Ti, Zr, Al).

F. BABONNEAU* AND G. D. SORARU**
*Chimie de la Matière Condensée, Université Paris 6, 4 place Jussieu, 75252 Paris (France)
**Dipartimento di Ingegneria dei Materiali, Universita di Trento, 38050 Mesiano-Trento (Italy)

ABSTRACT

A commercial polycarbosilane was modified with various metallic alkoxides to get mixed ceramic precursors in the Si-M-C-O systems (M=Ti, Zr Al). Pyrolysis under ammonia leads to the formation of various ceramic materials depending on the nature of M : Si_3N_4/TiN, Si_3N_4/ZrO_2 or β'-SiAlON phases.

The polymer-to-ceramics conversion have been studied by MAS-NMR (^{29}Si, ^{27}Al) and X-ray absorption (Ti K-edge). This paper will show how these two complementary spectroscopic techniques can be used to follow the nitridation process by probing local environments of different elements. In addition, it will be pointed out how X-ray absorption can be a powerful tool for the detection of the crystallization of nitride cubic phases.

INTRODUCTION

A large variety of ceramics can now be produced via the pyrolysis of polymers [1]. This new synthetic route to ceramics is particularly suitable for the production of fibers [2], coatings [3] or reactive amorphous powders for sintering [4]. A real progress in this field requires a better understanding of the structural evolution during the polymer-to-ceramics conversion, and thus involves the characterization of amorphous intermediates. Magic Angle Spinning Nuclear Magnetic Resonance (MAS-NMR) is a very powerful technique in this field, especially in the case of silicon-containing systems [5, 6]. For other elements such as Ti or Zr, which are not easily accessible by NMR, X-ray absorption techniques can bring very interesting information [7, 8]. Polymetallocarbosilanes, ceramic precursors in the Si-M-C-O systems (M=Ti, Zr, Al), were prepared from a chemical modification of a commercial polycarbosilane. Their pyrolysis under ammonia leads to the formation of various ceramic materials depending on the nature of M : Si_3N_4/TiN, Si_3N_4/ZrO_2 or β'-SiAlON phases. This paper will show how the polymer-to-ceramics conversion can be followed by MAS-NMR (^{29}Si, ^{27}Al) and X-ray absorption (Ti K-edge).

EXPERIMENTAL

The different polymetallocarbosilanes, PTC, PZC and PALC, were synthesized from polycarbosilane, PCS (Dow Corning, X9-6348), and respectively titanium n-butoxide Ti(OBu^n)$_4$, zirconium n-propoxide Zr(OPr^n)$_4$ and aluminum sec-butoxide modified with ethylacetoacetate, Al(OBu^s)$_2$(etac) according to a published procedure [9]. The structure of commercial PCS can be ideally represented by $[Si(CH_3)_2$-CH_2-$SiHCH_3$-$CH_2]_n$ [10]. The amounts of the two reagents are chosen to give an M/Si molar ratio of 0.25 (M=Ti, Zr, Al). The polymers were fired in NH$_3$ flow at various temperatures up to 1000°C. For higher temperatures up to 1600°C, the heat treatments were done in N$_2$ flow.

^{27}Al and ^{29}Si MAS-NMR spectra were recorded on a MSL 400 Bruker equipment. For the ^{29}Si NMR experiments (79.6 MHz), a 7-mm Bruker probe was used with 5 kHz spinning rate. The pulsewidth was 2 μs ($\Theta \cong 30°$) and the delay between pulses 1 min. For the ^{27}Al NMR spectra (104.2 MHz), a 4-mm Bruker probe was used with high spinning rates (15 kHz).

X-ray absorption experiments have been performed at LURE (Orsay - France) using the EXAFS III line (1.85 GeV) equipped with a two crystals Si 311 monochromator. XANES spectra were recorded with a 0.3 eV step while EXAFS spectra were recorded with a 2 eV step. Energy calibration was done with a metallic titanium foil taking the first maximum at 4964.2 eV. Data were analyzed according to the usual EXAFS formalism using MacKale amplitude and phase functions [11].

RESULTS
. *Pyrolysis of PALC under ammonia.*

PALC was already characterized [12]. The ^{29}Si MAS-NMR spectra (Figure 1a) is quite similar to the starting PCS with two main peaks due to SiC$_4$ (0 ppm) and SiC$_3$H (-17 ppm) units. The ^{27}Al MAS-NMR response corresponds to the spectrum of amorphous alumina obtained from an aluminum alkoxide (Figure 1b) with a main peak around 3 ppm due to octahedral Al sites. A structural model proposed for PALC is based on the dispersion of amorphous alumina-based particles in PCS chains.

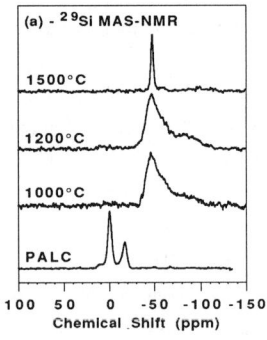 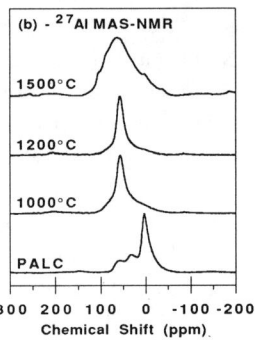

Figure 1 : ^{29}MAS-NMR (a) and ^{27}MAS-NMR (b) spectra of PALC and pyrolyzed products.

XRD study performed on pyrolyzed samples shows that the system is amorphous up to 1400°C, when a β'-SiAlON phase is formed. From the cell parameters, the phase was identified to β'-Si$_{2.5}$Al$_{0.5}$O$_{0.5}$N$_{3.5}$ [13].

At 1000°C, the ^{29}Si MAS-NMR spectrum (Figure 1a) shows a main peak centered at -46.6 ppm, typical of a silicon nitride environment. The components at low field reveal the presence of Si-O bonds. The ^{27}Al MAS-NMR spectrum (Figure 1b) presents a peak at 56.2 ppm, due to tetrahedral Al sites. At 1200°C, the spectra are similar showing a structural stability of the amorphous phase formed at 1000°C. At 1500°C, the spectra change drastically : the ^{29}Si MAS-NMR signal sharpens and corresponds to a crystalline silicon nitride environment [14]. The ^{27}Al MAS-NMR signal becomes broader, centered at -62 ppm with components present at ≈5, ≈85 and ≈105 ppm. This spectrum is similar to that reported for the crystalline β'-$Si_{2.5}Al_{0.5}O_{0.5}N_{3.5}$ phase [15].

Pyrolysis of PTC under ammonia.

PTC was already characterized in a previous paper [16]. The main difference between the ^{29}Si MAS-NMR spectra of PTC (Figure 2) and PALC comes from the presence of an additional peak at 10 ppm, assigned to the formation of Si-OBu^n groups from Si-H bonds. The XANES and EXAFS data (Figure 3) show that the local environment is close to titania [8]. The proposed structural model is thus similar to that proposed for PALC with oxide-based particles dispersed in PCS chains.

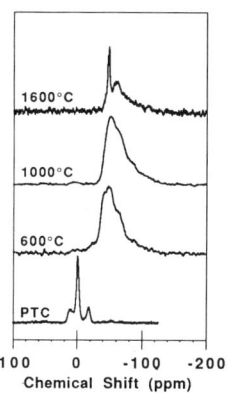

Figure 2 : ^{29}Si MAS-NMR spectra of PTC and pyrolyzed products

X-ray diffraction patterns on pyrolyzed samples are amorphous up to 1500°C. At 1600°C, diffraction peaks are present, but their broadness reveals the presence of an amorphous phase. Several crystalline phases have been identified : TiN and α and β–Si_3N_4.

The ^{29}Si MAS-NMR spectrum of the sample pyrolyzed at 600°C (Figure 2) presents a main peak at ≈-50 ppm corresponding to $SiN_{4-x}O_x$ sites. The maximum should correspond to x=0, but the lowfield components show the presence of sites similar to those expected for a silicon oxynitride phase. This result shows that the Si-C bonds are all cleaved at this temperature, in agreement with published data [6]. The spectrum does not change at 1000°C. At 1600°C, two sharp signals appear at -46.7 and -48.6 ppm due to crystalline silicon nitride phases. The reported values are -48.7 ppm for the β-Si_3N_4 and -46.8 and -48.9 ppm for the α-Si_3N_4 [14]. However, most of the Si sites seem still present in an amorphous phase characterized by a broad NMR signal centered at ≈-58 ppm, typical of silicon oxynitride phases.

The Ti K-edge XANES spectra are presented in Figure 3a. The spectrum of PTC presents a pre-edge peak at 4967.0 eV with a smaller peak at 4971.0 eV, and two maxima at 4986.7 and 4998.9 eV, typical values for an oxide environment [8]. The XANES spectrum of crystalline TiN was recorded as a reference (Figure 3a) : it shows a shoulder at 4969.0 eV for the pre-edge peak, and two maxima at 4982.3 and 4995.5 eV. The position of the peaks are thus quite sensitive to a difference of environment. During the pyrolysis, the position of the peaks regularly shifts towards values characteristic of crystalline TiN (Figure 4). It seems that progressively, the Ti-O bonds are replaced by Ti-N bonds. A comparison between the spectra of PTC pyrolyzed at 1600°C and TiN shows however some differences. It suggests that the average environment of the titanium atoms is not exactly that expected for pure TiN. A careful analysis of the XRD pattern shows that the peak positions do not correspond exactly to TiN : this point is under investigation.

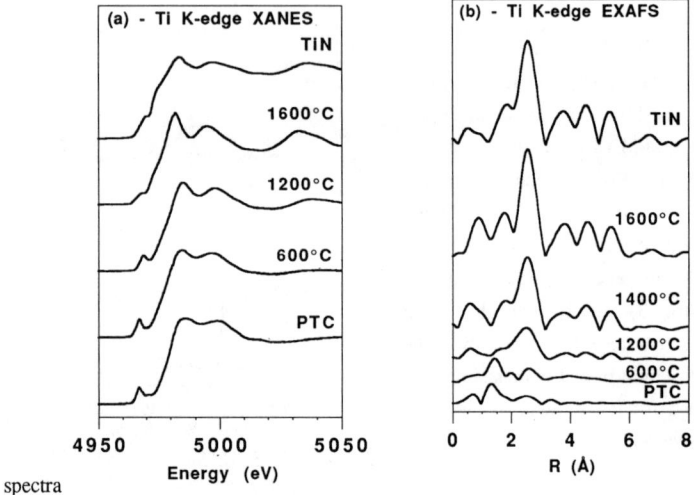

Figure 3 : Ti K-edge XANES spectra (a) and $k^3.\chi(k)$ Fourier Transform of Ti K-edge EXAFS spectra (b) of PTC and pyrolyzed products.

$k^3.\chi(k)$ Fourier transforms of the EXAFS spectra are presented in Figure 3b. For pyrolysis temperatures below 1000°C, the signal intensities are quite low, characteristic of a disordered environment. At 1200°C, a peak at ≈2.5 Å appears clearly with some components at higher distances, between 4 and 6 Å. The intensity of these peaks greatly increases with the pyrolysis temperature. At 1600°C, the radial distribution function is quite similar to that of crystalline TiN.

The Fourier transform of the EXAFS spectrum of crystalline TiN presents an intense peak due to 12 Ti atoms at 3.00 Å. The presence of rather intense peaks at long distances, up to 6 Å is quite unusual. They are characteristic of the cubic structure of TiN : alignments of atoms in a structure lead to an enhancement of the peak intensity, and is responsible for a focussing effect. This effect is very interesting to detect the onset of formation of such cubic phases. At 1200°C, the presence of signals at long distances shows that titanium atoms are already in a crystalline TiN environment. Nevertheless, a simulation of the main peak corresponding to Ti-Ti distances at ≈3 Å, leads to an average number of 4.9 neighboring atoms per Ti atom. This value is far from the expected value for crystalline TiN (12), and suggests that only a part of the Ti atoms are in an ordered environment. The number of Ti neighbors at ≈3 Å was extracted from a simulation of the EXAFS signal. The evolution versus pyrolysis temperature is presented in Figure 5. The regular increase of this number of Ti neighbors is characteristic of the growth of the crystalline phase. At 1600°C, this number reaches the expected value of 12.

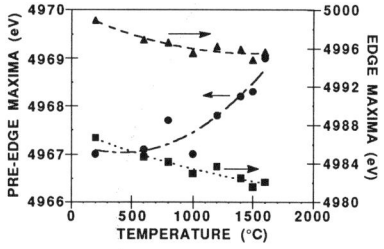 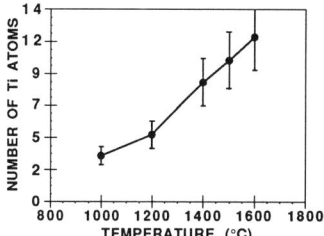

Figure 4 : Evolution of the pre-edge peak and of the edge maxima during the pyrolysis of PTC.

Figure 5 : Evolution of the number of Ti neighboring atoms during the pyrolysis of PTC.

DISCUSSION

Three polymetallocarbosilanes were chosen, prepared from a common synthetic procedure. They can all be described as a dispersion of oxide-based particles within PCS chains. The main difference is the nature of the oxide, titania, zirconia, alumina. Study of the zirconium containing system (PZC) has already been published [17]. X-ray diffraction studies report for the three systems different crystallization behaviors at 1500°C : the PALC system crystallizes in a single β'-SiAlON phase, while the PZC leads to a clear phase separation between tetragonal zirconia and β-silicon nitride (with a small amount of α phase). At the same temperature, the PTC system is poorly crystalline, and reveals at 1600°C, the presence of crystalline titanium and silicon nitrides (α and β), as well as an amorphous phase.

The purpose of this paper is to show how the use of complementary spectroscopic techniques permits a detailed structural characterization of the amorphous intermediates formed during the polymer-to-ceramics conversion. The simplest behavior is that of PZC : the precursor can be described as a nanocomposite between oxide particles and PCS chains. The nitridation process only affects the PCS chains, and both phases transform with no interaction with each other, to give crystalline zirconia and silicon nitrides. The opposite behavior is that of PALC system : the nitridation process affects not only the silicon environment, but also the aluminum one. At 1000°C, all the aluminum sites that were in an octahedral environment in the precursor are now in a tetrahedral symmetry. From the ^{27}Al NMR spectrum, it is difficult to give a precise description of the sites. The chemical shift value (56 ppm) corresponds more to AlO_4 than to AlN_4 sites (110 ppm). However, the change in symmetry shows a strong interaction between the two systems: the aluminum sites adopt a symmetry compatible with that of silicon atoms. This strongly suggests the formation of a single amorphous SiAlON phase at this temperature, that will then crystallize into pure β'-SiAlON at 1400°C.

The behavior of the PTC system is more complicated. At 600°C, the silicon environment changed to a nitride environment like for the previous systems. However, even at 1600°C, an important amorphous silicon oxynitride phase is present, together with crystalline silicon nitrides (α and β phases). The pyrolysis under ammonia also affects the titanium environment which progressively changes from oxide to nitride. This process seems continuous even above 1000°C, when the pyrolysis is done under nitrogen. X-ray absorption clearly shows at 1200°C the onset of formation of the cubic TiN phase, with a progressive growth of crystallites up to 1600°C.

CONCLUSION

The polymer-to-ceramics conversion under ammonia of three polymetallocarbosilanes in the Si-M-C-O system (M=Ti, Zr, Al) has been studied by MAS-NMR and X-ray absorption. The importance of such spectroscopic techniques to probe the evolution of the local environments of different elements (Si, Al or Ti) is clearly demonstrated. Structural models are proposed for the amorphous intermediates formed during the pyrolysis. This study also shows the importance of X-ray absorption techniques to detect clearly the formation of crystalline cubic phase such as TiN, in samples which present poor diffraction patterns.

ACKNOWLEDGEMENTS

The authors would like to greatly acknowledge F. Ribot and M. Verdaguer for the EXAFS experiments. This study was supported by NATO and CNR (Italy).

REFERENCES
1. M. Peuckert, T. Vaahs, M. Brück, Adv. Mater., 2, 398 (1990)
2. S. Yajima, Handbook of Composites, Vol.1 - Strong Fibers, W; Watt and B.V. Perov Eds, 1985, Elsevier, pp. 201.
3. H.E. Fischer, D.J. Larkin, L.V. Interrante, Mater. Res. Soc. Bull., XVI, 59 (1191)
4. R. Eiedel, M. Seher, G. Becker, J. Europ. Ceram. Soc., 5, 113 (1989).
5. G.D. Sorarù, F. Babonneau, J.D. Mackenzie, J. Mater. Sci., 25, 3886 (1990).
6. T. Taki, M. Inui, K. Okamura, M. Sato, J. Mater. Sci.Letters, 8, 1119 (1989).
7. C. Laffon, A.M. Flank, P. Lagarde, M. Laridjani, R. Hagege, P. Olry, J. Cotteret, J. Dixmier, J.L. Miquel, H. Hommel, A.P. Legrand, J. Mater. Sci., 24, 1503 (1989).
8. F. Babonneau, P. Barré, J. Livage, M. Verdaguer, Mater. Res. Soc. Symp. Ser. 180, 1035 (1990).
9. S. Yajima, T. Iwai, T. Yamamura, K. Okamura, Y. Hasegawa, J. Mater. Sci., 16, 1349 (1981).
10. S. Yajima, Y. Hasegawa, J. Hayashi, M. Iimura, J. Mater. Sci., 13, 2569 (1978).
11. B.K. Teo, in EXAFS: Basic Principles and Data analysis, Inorganic Chemistry Concepts, Vol. 6, Springer Verlag, Berlin, 1986.
12. F. Babonneau, G.D. Sorarù, K.J. Thorne, J.D. Mackenzie, J. Am. Ceram. Soc. 74, 1725 (1991).
13. G. D. Sorarù, M. Mercadini, R. Dal Maschio, F. Taulcllc, F. Babonncau (submitted to J. Am. Ceram. Soc.)
14. K.R. Carduner, R.O.Carter III, M.E. Milberg, G.M. Crosbie, Anal. Chem., 59, 2794 (1987)
15. M.E. Smith, J. Phys. Chem. 96, 1444 (1992).
16. F. Babonneau, J. Livage, G.D. Sorarù, G. Carturan, J.D. Mackenzie, New J. Chem. 14, 539 (1990)
17. G.D. Sorarù, A. Ravagni, R. Dal Maschio, G. Carturan, F. Babonneau, J. Mater. Res. 7, 1266 (1992)

FABRICATION AND CHARACTERIZATION OF β-SiAlON COMPONENTS FROM POLYMERIC PRECURSORS

G.D. SORARU*, V. M. SGLAVO*, F. VULCAN*, F. BABONNEAU**
* Dipartimento di Ingegneria dei Materiali, Università di Trento, 38050, Mesiano-Trento, Italy.
** Chimie de la Matière Condensée, Université Paris 6, 4 place Jussieu, 75252 Paris, France.

ABSTRACT

This paper reports on the fabrication of microcrystalline β-SiAlON components by direct nitridation of polymeric green compacts. The preceramic material was obtained from a polycarbosilane modified with an aluminum alkoxide. The microstructural transformation from the polymeric bodies to the final dense ceramic components was monitored by several techniques such as BET, XRD and dilatometry. Mechanical characterization, performed by indentation at various stages of the conversion process, will also be presented.

INTRODUCTION

Chemical routes to advanced ceramics allow for the production of materials with advanced properties as well as the manufacture of materials which are traditionally difficult to shape, such as fibers or coatings [1-3]. Indeed, this route to non-oxide ceramics such as SiC or Si_3N_4, has been developed after the pioneering work of Yajima who demonstrated the feasibility of producing high strength, high modulus silicon carbide fibers via pyrolysis of polycarbosilane (PCS) [4]. Among the various ceramic systems already prepared via the polymer route, it has been recently shown that microcrystalline β-SiAlON powders can be prepared after pyrolysis under ammonia of a polyaluminocarbosilane (PALC), obtained by reacting a PCS with a modified aluminum alkoxide [5,6].

The use of preceramic polymers in the processing of dense ceramic bodies has been attempted as well. In fact, extra-fine grained silicon carbide- and silicon nitride-based ceramics have been prepared from polymer-derived reactive amorphous powders [7,8]. More recently, bulk ceramics have been obtained by *direct pyrolysis of polymeric green compacts* [9]. A single phase Si-C-N ceramic component, 90% dense, has been successfully prepared from polysilazane compacts at temperature as low as 1000°C [10]. This approach offers many advantages compared to the traditional powder processing including low sintering temperatures and low impurity levels. Moreover, it extends the shaping technology typical of the polymeric materials to the ceramic processing. The key point of this innovative fabrication route is the use of a *partially crosslinked polymer* in order to avoid, in the early stages of pyrolysis, the melting and the subsequent bloating of the samples. This paper will show how, according to this new processing route, dense, microcrystalline β-SiAlON components can be fabricated via nitridation of polymeric green compacts obtained using PALC as polymeric precursor.

EXPERIMENTAL

Polyaluminocarbosilane (PALC) was produced from commercial PCS (Dow Corning X9-6348) and aluminum sec-butoxide modified with ethylacetoacetate, $Al(OBu^s)_2(etac)$ following a procedure already proposed in the literature [11]. After the refluxing stage, the solvent was removed and the temperature slowly raised up to 300°C to achieve crosslinking. The infusible polymer was then ground and the powders with a size less than 60 μm were used to fabricate green compacts by uniaxial pressing at 75 MPa for 30 min. Disks 13 to 20 mm in diameter and 3 to 5 mm thick were produced. The polymeric samples were nitridated in NH_3 flow (100 cc/min) in a quartz tube at different temperatures up to 1000°C for 1 hour with an heating rate of 1°C/min. In order to follow the evolution at higher temperatures, some of the pellets nitridated at 1000°C were subsequently fired in N_2 flow (100 cc/min) in a graphite furnace at various temperature up to 1500°C. In this case an heating rate of 2.5°C/min (1 hour at the maximum temperature) was applied. At the end of the pyrolysis treatments the weight loss and the specific surface area (as measured by the nitrogen absorption method - BET -) was recorded. Apparent density was evaluated from the geometrical volume and, on selected samples, bulk density was measured on powdered samples either using a helium picnometer or with toluene as fluid. Porosity was measured by mercury pressure porosimetry.

In order to follow the sintering process, dilatometric tests and differential thermal analysis experiments were performed in flowing nitrogen at 5°C/min up to 1550°C on pellets already pyrolized under ammonia at 1000°C. Structural characterization of the final β-SiAlON ceramic was accomplished by X-ray diffraction analysis on powdered samples. Grain size was estimated from the width of the diffraction peaks using the Scherrer equation. Microstructural investigation of the ceramic phase was performed by SEM on polished or fracture surfaces. Mechanical characterization, namely, hardness (H_V), elastic modulus (E) and fracture toughness (K_{IC}) was achieved by means of Vickers and Knoop indentation tests performed at 40-60 and 10 N, respectively [12-13].

RESULTS AND DISCUSSION

A detailed characterization of PALC and of its structural transformation during the nitridation process has been already reported and will be briefly summarized here. PALC polymer can be described as a dispersion of amorphous alumina-based particles into PCS chains [11]. ^{29}Si and ^{27}Al MAS-NMR analysis performed at various stages of the pyrolysis process showed that the nitridation of PALC starts at 400°C and it is complete at 600°C, when all the silicon atoms are present in an amorphous silicon nitride environment [5,6]. The same structure is still present at 1000°C and can be described as a homogeneous amorphous SiAlON phase. By increasing the pyrolysis temperature, the disordered phase converts, at 1400°C, into a microcrystalline β-SiAlON ceramic.

PALC pellets fabricated in this study exhibit an apparent density of 1.09±0.02 g/cc. By comparing this result with the density of the polymer (1.20 g/cc), measured by helium picnometer on powders, the total amount of porosity in the green bodies can be estimated to be

≈10 %. Indeed, results of mercury porosimeter analysis agree well, indicating an open porosity in the range 5-15%. The high density of the green compacts (≥ 85% th. density), compared to the usual density of green ceramic bodies (≈ 60 % th. density), could be related with a plastic deformation process of the polymeric powders. Moreover, mercury porosimeter investigations clearly indicate a bimodal pore size distribution at 100 nm and 10 nm. The larger voids seems to be due to interparticle porosity while the finer ones could be related to intrinsic porosity of the polymeric powders.

The nitridation under ammonia of the yellowish PALC compacts leads to the formation, at 1000°C, of white, crack-free amorphous SiAlON pellets with a linear shrinkage of 20%. After an additional pyrolysis step under flowing nitrogen at 1500°C the amorphous samples convert into a microcrystalline β-SiAlON monolithic components with a further linear shrinkage of 20 % (see Figure 1).

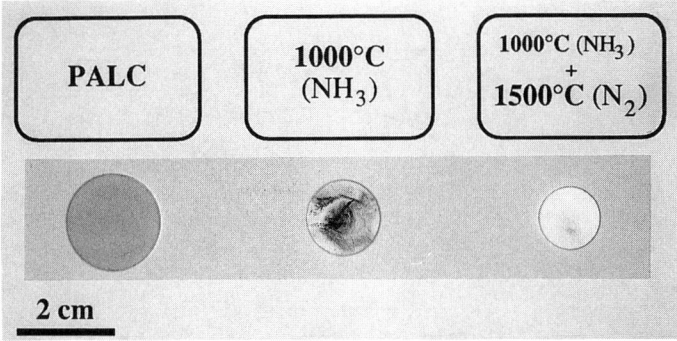

Figure 1. PALC compact and corresponding derived SiAlON pellets at various stages of the pyrolysis process.

The evolution of weight loss, density and specific surface area of polymeric compacts during the conversion process are reported in Figures 2 and 3.

A weight loss of 25% has been observed from room temperature up to 400°C. It is due mainly to the decomposition of the organic component of the modified aluminum alkoxide, since PCS chains are stable up to this temperature [5]. This process leads to a corresponding increase of the specific surface area up to 350 m^2/g without any appreciable variation in the apparent density. In the temperature range 400-600°C the nitridation of PCS occurs with a further weight loss of up to ≈ 35% and an additional increase of surface area to 500 m^2/g. In this case, however, the formation of a tridimensional silicon nitride network leads to an increase in density to 1.4 g/cc, indicating that a sintering mechanism is active in this temperature range. Above 600°C the weight of the samples does not change up to the maximum temperature reached in this study (1500°C). However, its apparent density remains constant (1.4 g/cc) up to 1200°C then increase dramatically above 1300°C, approaching a value 2.5 g/cc at 1500°C. The evolution of

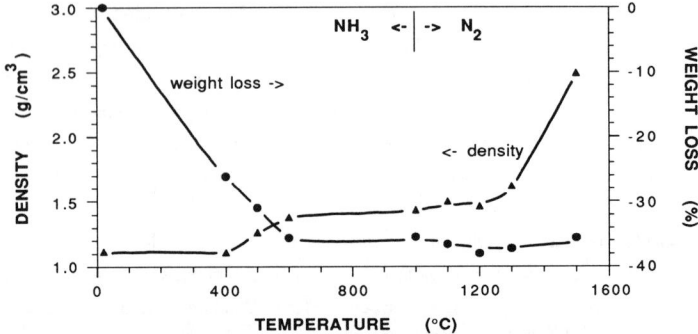

Figure 2: Evolution of apparent density and weight loss during the pyrolysis process of PALC compacts.

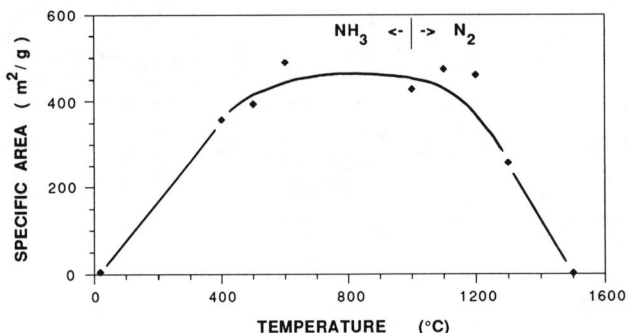

Figure 3: Evolution of specific surface area during the pyrolysis process of PALC compacts.

surface area with the pyrolysis temperature is similar to the density, i.e., it remains constant (500 m^2/g) up to 1200°C and decreases rapidly above 1300°C, being close to zero at 1500°C. In order to get a better insight into the sintering mechanism that is active above 1300°C, dilatometric and DTA tests have been performed on samples previously nitridated at 1000°C. As can be seen in Figure 4, the dilatometric curve clearly shows that the shrinkage of the amorphous SiAlON pellets starts around 1300°C and it is complete before 1500°C. DTA analysis indicates that, in the same temperature interval, a strong exothermic effect, related to the crystallization of β-SiAlON phase [5,6], is present. Indeed, the crystallization of the sample has been checked by XRD analysis and, from the position and the width of the diffraction peaks, the phase was identified as $\beta\text{-}Si_{2.5}Al_{0.5}O_{0.5}N_{3.5}$ with a mean particle size of 400 Å. Thus, the observed sintering process has to be related with the amorphous to crystalline transformation that occurs *via* the formation of a liquid phase. Indeed, TEM observations of the microcrystalline phase obtained at 1500°C indicate the presence of a thin intergranular glassy

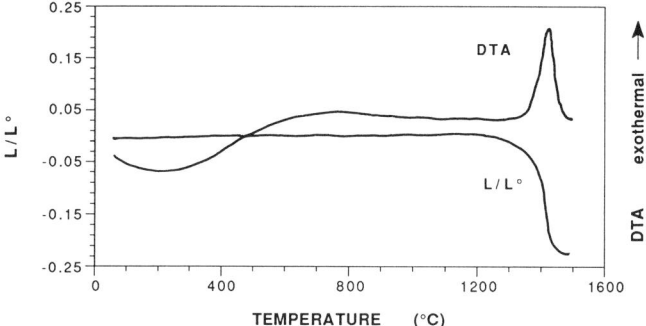

Figure 4: Dilatometric and DTA curves for amorphous SiAlON ceramics nitridated at 1000°C.

phase [5]. Therefore, the sintering process seems to be due to the formation of a liquid phase from which the crystalline β-SiAlON phase precipitates.

Monolithic β-SiAlON ceramics obtained after sintering at 1500°C exhibit a density of 85% as can be estimated by comparing the values of apparent density and bulk density, 2.98 g/cc (measured by liquid picnometer on powders). A typical polished section is shown in Figure 5 and shows the presence of residual porosity in the range 1 to 15 μm. Mechanical characterization has been performed on samples pyrolyzed at 1300°C and 1500°C, i.e., before

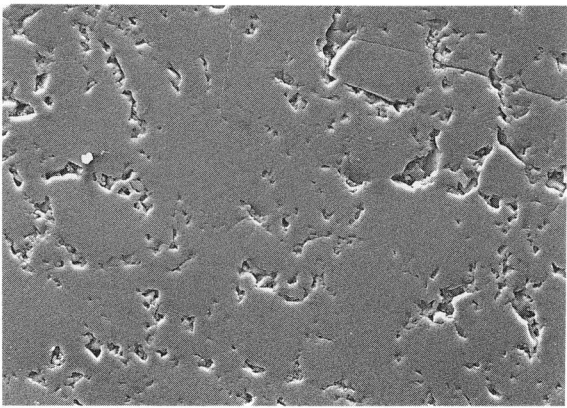

Figure 5. SEM micrograph of a polished surface of β-$Si_{2.5}Al_{0.5}O_{0.5}N_{3.5}$

and after the sintering process that is active in this temperature interval, and the results are reported in Table 1. At 1500°C the observed properties compare well with the characteristic values reported in the literature for SiAlON materials processed at higher temperatures [14].

Table 1. Measured values of hardness (H_V), elastic modulus (E) and fracture toughness (K_{IC}) for amorphous (1300°C) and microcrystalline (1500°C) β-SiAlON pellets.

Temperature (°C)	H_v (GPa)	E (GPa)	K_{IC} (MPa m$^{1/2}$)
1300	2.0±0.3	28±12	1.2±0.5
1500	11.0±1.4	198±28	4.7±0.8

CONCLUSION

Microcrystalline β-SiAlON materials have been fabricated by direct nitridation of polyaluminocarbosilane green compacts. Pyrolysis under ammonia at 1000°C results in the formation of amorphous SiAlON ceramics with a very high value of surface area. BET, dilatometric, density and XRD measurements showed that the amorphous material densifies at ≈ 1400°C leading to the formation of a 85% dense β-$Si_{2.5}Al_{0.5}O_{0.5}N_{3.5}$ ceramic.

ACKNOWLEDGEMENTS

CNR (Italy) and NATO are gratefully acknowledged for their financial support.

REFERENCES
1. W. E. Rhine, H. K. Bowen, Ceram. Intern., 17, 143 (1991).
2. J. Lipowitz, Bull. Am. Ceram. Soc. 70, 1888 (1991).
3. M. Peuckert, T. Vaahs, M. Bruck, Adv. Mater., 2, 398 (1990).
4. S. Yajima, J. Hayashi, M. Omori, Chem. Lett., 931 (1975).
5. G. D. Soraru, A. Ravagni, R. Campostrini, F. Babonneau, J. Am. Ceram. Soc., 74, 2220 (1991).
6. F. Babonneau, G. D. Soraru, Mat. Res. Soc. Symp. Proc., this volume
7. H. Kodama, T. Miyoshi, Adv. Ceram. Mater., 3, 177 (1988).
8. V. M. Sglavo, R. Dal Maschio, G. D. Soraru, A. Bellosi, J. Mat. Sci., Sub. for pub.
9. P. Greil, M. Seibold, T. Erny, B. Ahlers, presented at the 2nd ECerS Meeting, Ausburg, Germany, 1991 (unpublished).
10. G. Passing, R. Riedel, H. Schonfelder, R. J. Brook, presented at the 2nd ECerS Meeting, Ausburg, Germany, 1991 (unpublished).
11. F. Babonneau, G. D. Soraru, K. J. Thorne, J. D. Mackenzie, J. Am. Ceram. Soc., 74, 1725 (1991).
12. K. Niihara, R. Morena, D. P. H. Hasselmann, J. Mat. Sci. Lett., 1, 13 (1982).
13. D. B. Marshall, T. Noma, A. G. Evans, J. Am. Ceram. Soc., 65, C175 (1982).
14. T. Ekstrom, M. Nygren, J. Am. Ceram. Soc., 75 259 (1992).

Mg-Si-Al-O-N Glasses Prepared By Nitriding Cordierite Powders Derived from Polymer Precursors

Richard M. Laine, Clint R. Bickmore, Kurt F. Waldner, Brian L. Mueller and Hal W. Estry
Departments of Materials Science and Engineering, and Chemistry
University of Michigan, Ann Arbor, MI 48109-2136

ABSTRACT

We have previously reported the synthesis of cordierite polymer precursors directly from MgO, SiO_2 and alumina.[1] Removal of organics at 400°C in air and/or in oxygen gave amorphous, atomically mixed, oxide powders with surface areas of 250-400 m^2/g. It is possible to convert these high surface area oxide powders to Mg-Si-Al-O-N powders by heating in a flowing NH_3 atmosphere. The effects of selected processing conditions on nitrogen incorporation, in a set of well defined oxide powders, were examined. The powders are nitrided in flowing NH_3, at temperatures of 700-1200°C, to obtain N contents of up to 5 atomic percent.

INTRODUCTION

Silicon aluminum oxynitride glasses and ceramics, M-Si-Al-O-N, (M = Li, Ca, Mg, Y) are typically synthesized by a high temperature melt process from powder mixtures of metal oxide M_xO_y, AlN and SiO_2, or Al_2O_3 and Si_3N_4. This approach, while effective, has several drawbacks that include the need for several high temperature processing steps (i.e. synthesis of AlN and/or Si_3N_4, plus the final melt processing step), and the difficulty of achieving good quality, high N content (> 4-5 at. %) materials. The latter problem derives from the fact that higher N content glasses and ceramics require higher processing temperatures to obtain uniform materials. This is simply a consequence of the fact that viscosity and melting temperature go up as N content goes up.

Higher melt processing temperatures mean that processing is done closer to the point where reaction (1) can affect the quality of the resultant glasses. Attempts to process high nitrogen

$$SiO_2 + Si_3N_4 \xrightarrow{>1600°C} 2SiO + Si + 2N_2 \qquad (1)$$

content M-Si-Al-O-N glasses and ceramics (M = Li, Ca, Y, Mg) can be plagued by the presence of nitrogen gas derived reboil pores and silicon metal inclusions.[2-5]

One potential solution to these problems is to develop a low-temperature, chemical processing approach to M-Si-Al-O-N glasses using of N-containing organometallic polymers.[6-10] A second approach, and the one chosen here, is to nitride amorphous, M-Si-Al-O powders in an effort to introduce nitrogen only in the final step.

Our choice of this approach is predicated on our discovery of several novel routes to atomically mixed multimetallic precursor polymers that can be processed to oxide fibers, coatings or high surface area powders with retention of atomic mixing as previously demonstrated.[1,11,12] The cordierite ($2MgO \cdot 2Al_2O_3 \cdot 5SiO_2$) precursors used in the current studies were prepared either as described previously[1] or according to the following schematic which will be described in detail elsewhere.[13] The current work attempts to identify conditions whereby the high surface area powders can be effectively nitrided.

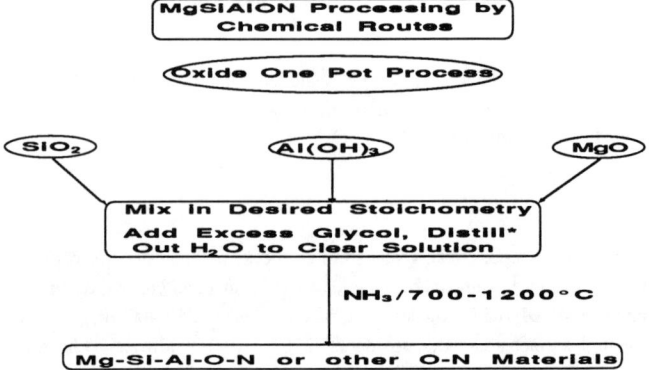

The choice of cordierite as the target oxide ceramic was based on our previous work[1] and the fact that Mg-Si-Al-O-Ns with the similar metals stoichiometry have been used previously to form high N-content glasses.[14-17]

RESULTS AND DISCUSSION

Figure 1, the TGA for the polymer from the above process (20°C/min in air), indicates a

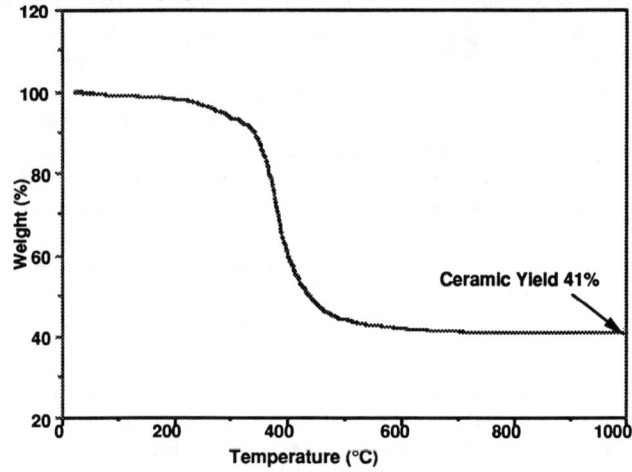

Figure 1. TGA of Cordierite Precursor Made via One Pot Method.

ceramic yield of 41%. Because most of the organic material is lost following heating to temperatures of ≈ 400°C, as found in the earlier studies,[1] our efforts to prepare useful oxide powders focused on pyrolysis of bulk precursor polymer at 400°C, first for 2 h in air and then either an additional 2 h in air or in O_2.

The oxide powders produced at 400°C are amorphous and offer quite high surface areas. Air/O_2 pyrolysis of 1-5 g precursor samples typically gives (by BET) surface areas of up to 450 m^2/g. Scale-up to 10-20 g precursor sample sizes provides powders with somewhat reduced surface areas of ≈ 250 m^2/g. Given that the nitridation process is a gas/solid reaction, it can be anticipated that the higher the surface area, the more efficient the nitridation process, as has been verified in previous nitridation studies.[10,14,15] Thus, the 400°C powders offer potential for incorporation of considerable nitrogen providing that atomic mixing was maintained during pyrolytic removal of the organics, and remains during nitridation.

Samples of the 400°C oxide powders were heated at 10°C/min in flowing air to either 700, 800, 900, 1000, 1100 or 1200°C and held at temperature for 4 h. The x-ray powder diffraction patterns of powders produced by the previous method[1] and those produced as illustrated in the above schematic, were recorded on a Rigaku θ–θ instrument. The first generation powders,[1] begin to crystallize at 900°C with the formation of μ-cordierite, which continues to crystallize up to approximately 1100°C. On heating to 1300°C, the only phase observed in the XRD is β-cordierite. However, at intermediate temperatures (1000-1200°C) some segregation occurs as evidenced by the appearance of small amounts of extraneous phases (e.g. sapphrine). The new powders also crystallize at 900°C to form μ-cordierite. Crystallization continues to 1000°C. Above 1000°C, the μ-cordierite converts smoothly to β-cordierite, without formation of other phases, which suggests that atomic mixing has been retained.[13]

Nitridation Studies

Because our goal is to make Mg-Si-Al-O-N glasses, nitridation of the ≈ 450 m^2/g cordierite powders[1] was run at 700°C to avoid: (1) sintering with loss of surface area; (2) crystallization, and (3) segregation. Tables 1 and 2 list the N contents obtained following nitridation.

Heat Treatment	No Combustion Aid			Combustion Aid (Tin)		
	C	H	N	C	H	N
700°C-4hr 1 g sample (stagnant air)	0.34, 0.30	0.52, 0.58		0.46, 0.32	0.37, 0.43	
700°C-4hr 1 g sample (stagnant air)	0.14, 0.14	0.45, 0.35	0.54, 0.56	0.38, 0.35	0.28, 0.52	0.41, 0.47
700°C-4hr 1 g sample (0.5 m/s LFR NH_3)	1.05, 1.17	1.19, 1.33	1.34, 1.69	1.79, 1.63	1.23, 1.20	1.92, 1.90
700°C-4hr 1 g sample (0.5 m/s LFR NH_3)	1.99, 1.64	1.17, 1.18	2.10, 1.86	2.18, 2.44	1.20, 1.18	2.02, 2.22
700°C-4hr 2 g sample (0.5 m/s LFR NH_3)	1.75, 1.68	1.02, 1.04	1.95, 1.92	2.18, 2.33	1.20, 1.06	2.00, 1.99

Table 1. C,H,N Analyses. LFR = Linear flow rate in ms^{-1} NH_3 (based on tube dia./total vol. of flow/sec).

The use of tin as a combustion aid provides more realistic N contents. Comparison with standard LECO N-analyses (Table 2), suggests that nitrogen uptakes are of the order of 1.5-

Heat Treatment	Sample/Conditions	N Content (Sn/Ni combustion aid)		
700°C-4hr	1 g sample stagnant air	0.27	0.25	0.22
700°C-4hr	1 g sample 0.5 m/s LFR NH_3	1.42	1.24	1.39
700°C-4hr	2 g sample 0.5 m/s LFR NH_3	1.53	1.55	1.64

Table 2. LECO Analyses for Three Heat Treated Powders.

2.0 atom equivalents following nitridation at 700°C for 4 h even with larger sample sizes. Still higher N contents (> 5 at % by LECO) were obtained upon heating under identical conditions at 1200°C for 4 h. The fact that the nitrogen contents increase only by a factor of 2-3, is a consequence of significant changes in powder surface area. The 250 m^2/g 400°C powder, produced using the newer route, undergo significant reductions in surface after nitriding for 4 h at 700°C in NH_3. Measured surface areas were ≈ 150 m^2/g. Nitriding under identical conditions, at 1000°C, gave surface areas of only ≈ 1 m^2/g.

ESCA Analyses

If the nitridation process is effective, then uniform nitridation must occur. It is necessary to answer the question, where does the nitrogen go? Figures 2 and 3, which compare core electrons for the Si-X and Al-X (X = O or N) bonds in the 1200°C oxide powder with those found for a 1200°C, nitrided powder, clearly show that considerable replacement of O with N has occured adjacent to Si and Al atoms. The 1200°C nitrided sample Si peak is centered at approximately 101 eV, which is typical for pure silicon nitride (101.4 eV) and quite different from the 1200°C oxide peak at ≈ 103.3 eV which correlates with Si in silica (103.3-103.7 eV). The 1200°C nitrided sample Al peak centered at 74 eV is near that of AlN (73.9 and 74.4 eV) as compared with the 1200°C oxide peak at ≈ 75.3 eV. The 700°C ESCA data show similar progressions to the 1200°C ESCA data; although the shifts are smaller as expected based on chemical analyses. Kamiya et al have used the shift data to estimate the extent of nitriding and compare it to the chemical analyses.[19]

The ESCA results suggest considerable N uptake; but, ESCA analyses are only valid to depths of 2.5 nm below the surface. When considered in light of the above chemical analyses, it is clear that the nitridation process at 1200°C, where surface areas are low, occurs strictly at the surface. The results are supported by the observation of crystallization of β-cordierite, mullite and sapphrine in the 1200°C powders.

Given that we seek to prepare glasses, nitridation of high surface area powders, <u>where much of the material is at the surface</u>, offers a reasonable approach to a more uniformly nitrided powder. Furthermore, the need to maintain a glassy state suggests the study of extended nitri-

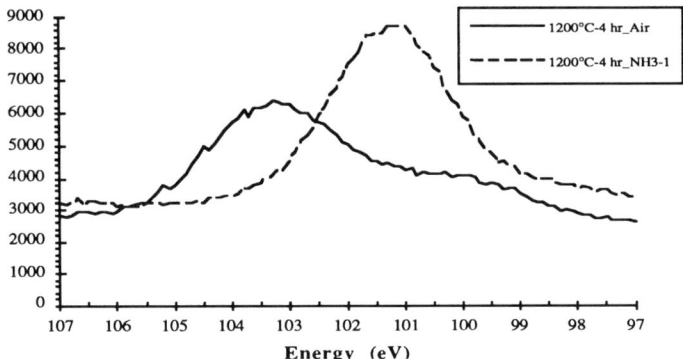

Figure 2. ESCA of the Silicon 2p Peak for 1200°C Powders.

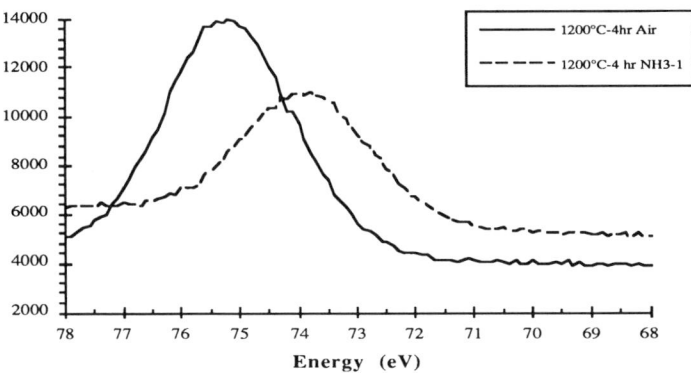

Figure 3. ESCA of the Aluminum 2p Peak for 1200°C Powders.

nitridation at 700 or 800°C, where surface areas are high and no crystallization is observed.

Although, we will discuss the validity of this approach in detail at a later date, some brief results are as follows. XRD studies indicate that nitridation at 700°C for 4, 12, or even 100 h does not promote the formation of any crystalline phases.[13b] These data are supported by infrared studies.[13b] Furthermore, surface areas drop only to 60 m^2/g at 100 h. Thus, it may be possible to get uniform incorporation of nitrogen into the powders at lower temperatures.

ACKNOWLEDGEMENTS

We would like to thank the U. S. Army Advanced Materials and Technology Laboratories, Watertown, Mass. for support of this work through Contract No. DAAL-0491-C--0068.

REFERENCES
1. Z.-F. Zhang, M. L. Hoppe, J. A. Rahn, S.-M. Koo and R. M. Laine, in "Synthesis and Processing of Ceramics: Scientific Issues," edited by W. E. Rhine, T. M. Shaw, R. J. Gottschall and Y. Chen, (Mat. Res. Soc. Symp. Proc. **249**, 1992) p. 81-86 .
2. D. R. Messier, "Army Oxynitride Glass Research," Ceram. Bull. **68**, 1931 (1989).
3. D. R. Messier, R. P. Gleisner and R. E. Rich, "Yttrium-Silicon-Aluminum Oxynitride Glass Fibers," J. Am. Ceram. Soc. **72**, 2183 (1989).
4. S. Sakka, K. Kamiya and T. Yoko, "Preparation and Properties of Ca-Al-Si-O-N Oxynitride Glasses," J. Noncryst. Sol. **56**, 147 (1983).
5. R. E. Loehman, "Oxynitride Glasses," J. Noncryst. Solids (1980) **42**, 433.
6. Y.-F. Yu, T.-I. Mah, "Si-O-N Ceramics from Organosilicon Polymers," in Better Ceramics Through Chemistry II, edited by C. J. Brinker, D. E. Clark, D. R.Ulrich.(Mat. Res. Soc. Symp. Proc. **73**, Pittsburgh, PA, 1986) p. 559 and references therein.
7. R. M. Laine, Y. D. Blum, R. D. Hamlin, A. Chow, "Organometallic Polymers as Precursors to Ceramic Materials: Silicon Nitride and Silicon Oxynitride" in Ultrastructure Processing of Ceramics, Glasses and Composites II, edited by D. J Mackenzie and D. R. Ulrich, J. Wiley & Sons, N. Y. (1988) p. 756.
8. M. L. J. Hackney, L. V. Interrante, G. A. Slack, P. J. Schields, "Organometallic Precursors to $Al_wSi_xN_yC_z$ Ceramics," in Ultrastructure Processing of Ceramics, Glasses and Composites, edited by D. J Mackenzie, D. R. Ulrich, J. Wiley & Sons, N. Y. (1988) p. 99.
9. G. D. Soraru, A. Ravagni, R. Campostrini. "Synthesis and Characterization of ß'-SiAlON Ceramics from Organosilicon Polymers", J. Am. Ceram. Soc. **74**, 2220 (1989).
10. Preceding paper in this symposium.
11. R. Laine, K. Blohowiak, T. Robinson, M. Hoppe, P. Nardi, J. Kampf, J.Uhm, "Synthesis of Novel, Pentacoordinate Silicon Complexes from SiO_2," Nature **353**, 642-644 (1991) .
12. R. M. Laine, K. A. Youngdahl and P. Nardi, U. S. Patent 5,099,052 (March 24, 1992).
13. a. R. M. Laine, C. Bickmore, B. Mueller and T. Hinklin, unpublished work.
14. N. S. Jameel and D. P. Thompson, "The Preparation of Nitrogen Glass-Ceramics in the Magnesium Sialon System," in Special Ceramics 8; Taylor and Popper Eds., Proc. of the Brit. Ceram. Soc., (1986) pp 95-108.
15. K. H. Jack, "Sialon Glasses," in Nitrogen Ceramics, edited by R. L. Riley, Noordhoff International, Reading, MA (1978) pp 257-262.
16. T.-Y. Tien, G. Petzow, L. J. Glackler, J. Weiss "Phase Equilibrium Studies in Si_3N_4-Metal Oxide Systems," in Progress in Nitrogen Ceramics, Martinus Nihoff Publ. (1983) p. 89.
17. T. Hayashi and T.-Y. Tien, "Formation and Crystallization of Oxynitride Glasses in the System Si, Al, Mg/O, N," Yogyo-Kyokai-Shi, **94**, 54 (1986).
18. J. Sjoberg and R. Pompe, "Nitridation of Amorphous Silica with Ammonia", J. Am. Ceram. Soc. **75**, 2189 (1992).
19. K. Kamiya, M. Ohya, T. Yoko, "Nitrogen-Containing SiO_2 Glass Fibers Prepared by Ammonolysis of Gels Made from Silicon Alkoxides", J. Noncryst. Sol., **83**, 208 (1986).

EFFECT OF POWDER PROCESSING ON SLIP CASTING PARAMETERS AND SINTERING OF Si_3N_4-SiO_2-$Y2O3$

GARY GILDE AND GEORGE GAZZA
Army Research Labs, Watertown Site, AMSRL-MA-CA, Watertown, MA 02172-0001

ABSTRACT

A procedure was developed for slip casting Si_3N_4-Y_2O_3-SiO_2 compositions. The silicon nitride was oxidized to supply the silica. The procedure resulted in 65 w/o slips with viscosities of 35 mPa·s which when cast, yielded compacts which were 56 % of theoretical density. Slips with 70 w/o solids loading were also prepared having viscosities of 120 mPa.s. The properties of the slips were influenced by the starting silicon nitride powder, dispersant, and milling media. The resulting compacts could be gas pressure sintered to full density, at temperatures as low as 1850 C, using nitrogen pressures as low as 4 MPa. Strengths as high as 750 MPa were achieved and the initial results show good creep resistance at 1350 C.

INTRODUCTION

Slip casting of Si_3N_4-Y_2O_3-SiO_2 compositions is difficult because at a pH less than 8 Y_2O_3 forms positive counterions of 3+ by the dissolution of it's hydroxide[1] and at a pH greater than 9 the solubility of silica increases rapidly[2]. Trivalent yttria ions can flocculate the slip at low concentrations. In order to keep the concentration of yttria ions as low as possible, it is necessary to keep the pH above 8. When the silica becomes soluble it is difficult to produce a homogeneous distribution of the silica and also the viscosity of the slip increases. Maintaining the pH between 8 and 9 is difficult. When Si_3N_4 is milled or aged in water, the pH usually rises above 9 due to the dissociation of Si_3N_4 to form ammonia. Typical strategies for dispersing silicon nitride often include raising the pH above 9 to take advantage of its high negative surface charge at high pH. When slip casting compositions with high silica content this is not advantagous.

The goal of this study was to determine if Si_3N_4-Y_2O_3-SiO_2 compositions can be formulated and slip cast by oxidizing the surface of the silicon nitride to uniformly distribute the SiO_2 and increase the negative zeta potential of the Si_3N_4, while using commercial polyelectrolyte dispersants to adjust the pH of the suspension to between 8.3 and 8.8. It was believed that oxidizing the silicon nitride surface would aide in dispersing it and would yield a high quality "green" body with improved distribution of silica on the surface of the powder, higher "green" densities, and a more uniform pore distribution.

EXPERIMENTAL

Two silicon nitride powders were used in this study, UBE-E-10 from UBE Industries (Japan) and Hermann Starck's LC-12-SX(Germany). The silicon nitride was oxidized at 1000 C in 1/2 inch high alumina crucibles for the selected time. The oxygen content, after oxidation, was determined by Inert Gas Fusion. For the standard composition, a three hour oxidation of the UBE-E-10 was selected to adjust the silica content and the yttria was varied from 5.5 to 6.5 w/o. The yttria/silica composition was controlled to provide a molar ratio of approximately 1:2.

Five anionic polyelectrolytes, listed in Table 1 and an organic amine (3-amino-1-propanol) were tested as dispersants. Viscosity vs. dispersant level curves (demand curves) were then established for the five polyelectrolytes. A 1w/o addition of 3-amino-1-propanol was used, as suggested by Starck. Slips were milled for 24 hour with various dispersant levels and the viscosity was measured using a shear rate of 15.84 s^{-1}. Based on initial demand curves, Nopcosperse 644A was selected for further investigation. Viscosity vs. dispersant level curves for 60, 65, and 70 w/o slips were developed using UBE and Starck silicon nitride powders in both oxidized and the as-received condition. Either yttria and silica or yttria and alumina were used as sintering aids. Silicon nitride, alumina or nylon covered steel milling media were used with either distilled or distilled, deionized water.

Name	Distributor	% Active Ingredient	Molecular Weight	Active Ingredient	PH
Darvan C	R.T. Vanderbilt, Inc.	25 %	10-16000	Ammonia Polymethacrylate	7.5-9.0
Darvan 821A	R.T. Vanderbilt, Inc.	40 %	≈ 6000	Ammonia Polyacrylate	7.0
Gradol 250A	Grader Chemical Co.	25 %	unknown	Ammonia Polymethacrylate	1% solution 9.5 - 10.5
Alcosperse 244	AICO Chemical	40 %	3200	Ammonia Polyacrylate	7.0 - 9.0
Nopcosperse 644A	Henkel Corporation	35 %	unknown	Proprietary Blend Ammonium polyelectrolyte	8.0 - 8.5

Table 1: Material Specifications of Dispersants (Manufacturer Specifications)

RESULTS AND DISCUSSION

1. Oxidation of starting powder

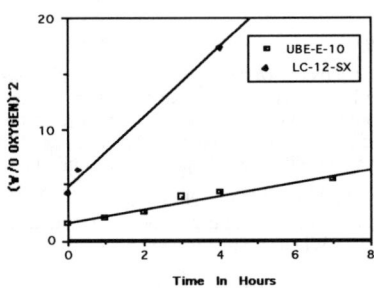

Figure 1. Weight percent oxygen vs. oxidation time at 1000C.

Figure 1 shows the oxygen content of the two silicon nitride powders after oxidation at 1000C for various times. The Starck powder oxidizes much more rapidly. This would be expected because of its greater surface area. Both powders exhibit parabolic oxidation behavior shown when an exponential plot of the weight percent of oxygen gained is graphed vs. oxidation time in Figure 1. When the plots are normalized for surface area, the slopes of each plot do not change significantly. This suggests that the kinetics of oxidation are somewhat different for each powder.

Another factor that may be influencing the oxidation rates is the ratio of surface to bulk oxygen. The LC-12-SX, which is made by the direct nitridation of silicon, has an oxygen content of 2.0 % approximately 45 % of which is surface oxygen distributed over a surface area of 20 m2/g. UBE-E-10 has a total oxygen content of 1.3 % and a surface area of 13 m2/g. UBE docs not specify the amount of surface oxygen vs. the total oxygen. It is suspected that the oxygen is principally distributed on the surface. A thicker layer of surface silica, a greater amount of substitutional oxygen, and a lower surface area could explain the different oxidation rates.

2. Effect of oxidation of starting powders on sintering

During the sintering experiments, the as-received LC-12-SX would not densify with 6.0 w/o yttria where as the UBE-E-10, oxidized for three hours, with 6.0 w/o yttria, was easily sintered to full density. The total oxygen content for the as-received LC-12-SX and the UBE-E-10 with a three hour oxidation was 2.0 %. When the LC-12-SX was oxidized for 15 minutes and 6 w/o yttria was added, it could be sintered to 92 % of theoretical density. It appears that there is a critical amount of surface silica needed to activate the sintering for this composition. When UBE-E-10 (as-received plus 6 w/o yttria) was fired it would not sinter to full density. When it was oxidized for three hours, it was easily sintered to full density with yttria contents ranging from 5.5 to 6.5 w/o. Compacts that were pressure cast using oxidized UBE with 6 w/o yttria could be sintered to full density at 1850C for 180 minutes with a nitrogen overpressure of 4 MPa using a two step sintering cycle. This is a low firing temperature for a refractory composition, fired with a 4 MPa over pressure of nitrogen.

By oxidizing the silicon nitride to provide the silica for this composition it is expected that the silica will be more homogeneously distributed than if it is added as a powder. Since the silica is uniformly distributed on the surface of the silicon nitride powder it should promote a uniform distribution of the liquid phase to enhance sintering at lower temperatures. In comparing the sintering behavior of the slip cast compacts made with oxidized silicon nitride to the sintering behavior of compacts formed from cold pressing and CIPing as-received silicon nitride with silica additions from previous studies[3], it appeared the oxidized silicon nitride sintered at lower temperatures. It is recognized that the slip cast bodies also had higher densities and were probably less agglomerated.

3. Selection of dispersant

After generating the oxidation curves for the powders, a standard slip casting procedure was developed. The first step was to select a dispersant. Five anionic polyelectrolytes and 3-amino-1-propanol were tested. A 1.0 w/o addition of 3-amino-1-propanol, as suggested by Starck, produced a good slip and appeared to be promising. However, sintered pieces made from compacts using this dispersant had a layered appearance. The lower portion of the compact was white, the top portion gray. The pH of slips made with the 3-amino-1-propanol was 10.3. At this high pH, it is believed that the silica became soluble and was concentrated on the bottom portion of the compact as the piece was dewatered in the pressure caster. This area became the white area on firing. For this reason, it was decided to use a anionic polyelectrolyte and work at pH levels between 8.3 and 8.9.

Figure 2 compares the demand curves for the five dispersants tested using 50 w/o slips of LC-12-SX with 5.0 w/o yttria and 5.0 w/o alumina. The slips were milled in 500 ml Nalgene bottles with 250 grams of alumina milling media. Table 1 shows the characteristics of each dispersant. As can be seen in Figure 2 the two polyacrylates behave very similarly

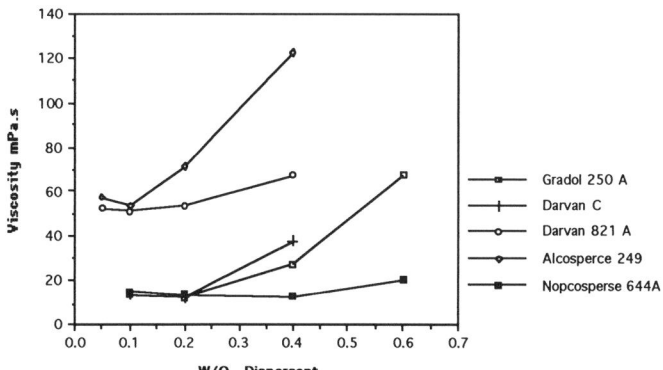

Figure 2. Demand curves for the five dispersants tested with 50 w/o slips of LC 12 SX with yttria and alumina sintering aids dispersed in distilled water and milled with alumina milling media.

giving identical minimum viscosities at the the same w/o dispersant. The two polymethacrylates were more effective dispersants for the silicon nitrides used in this study, producing significantly lower viscosities than the polyacrylates. Again, both polymethacrylates appeared to behave similarly. The Nopcosperse, although not producing a lower viscosity than either of the polymethacrylates tested, was more resistant to over deflocculation. For this reason, the Nopcosperse was selected over the other dispersants.

4. Effect of powder oxidation on the properties the slip

A comparison of Figure 2 and 3 shows that as the solids loading of the slip is increased from 50 to 65 w/o, the slips become more sensitive to over deflocculation. Figure 3 shows that oxidizing the UBE-E-10 did not result in a lower viscosity slip than the as-received powder. This is in conflict with the results found by Greil[4]. However, in his study, Greil used Starck's LC-12 silicon nitride. When LC-12-SX powder was oxidized in this study, it also resulted in a lower viscosity slip. Starck's LC-12 and LC-12-SX are made by the direct nitridation of silicon while the UBE powder is synthesized by a wet chemical process. Differences in behavior may be related to the different powder synthesis methods. Greil[5] proposed that there is a minimum silica layer thickness that is needed to shield the silicon nitride surface. It is plausible that if most of the oxygen is on the surface, the surface layer of silica may exceed this minimum thickness even for a silicon nitride with low oxygen content. This does not explain the difference in the pH of slips produced with oxidized silicon nitride and as-received silicon nitride. Slips made from oxidized silicon nitride typically had a pH of 8.3 while the as received silicon nitride resulted in slips with a pH above 9. If the as-received silicon nitride surface is shielded with a silica layer, it would be expected that the pH of the slip should be similar to that of the slip produced with oxidized silicon nitride, unless adsorbed ammonia is responsible for the rise in pH. UBE-E-10 silicon nitride was heated to 1000 C in nitrogen to remove any surface impurities. The behavior of the UBE-E-10 held at 1000 C for three hours

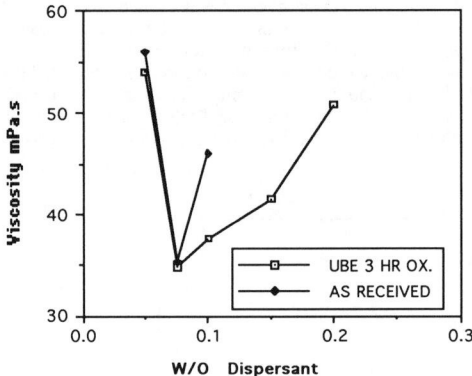

Figure 3. Viscosity vs dispersant level for two 65 w/o slips prepared with either as received UBE-E-10 plus silica additions or oxidized UBE-E-10. Both were dispersed in distilled deionized water and milled with nylon milling media.

was the same as the as-received silicon nitride, therefore the difference in pH is not believed to be the result of adsorbed surface impurities.

It can be seen in Figure 4 that oxidizing the LC-12-SX significantly changes its behavior. The oxidized slip at 60 w/o is much more resistant to over deflocculation, with the minimum viscosity at 0.1 w/o dispersant. The as-received LC-12-SX is more sensitive to over deflocculation, with a minimum viscosity at 0.3 w/o dispersant. The minimum viscosity for the as-received powder slip is also significantly higher than that of the oxidized LC-12-SX.

Oxidized UBE-E-10 acts as a strong acid lowering the initial pH of a slurry of silicon nitride and distilled deionized water to 4.5 When the nylon media was used for milling, it took 50 hours for the pH to rise to 7. The dramatic lowering of the pH when oxidized silicon nitride is used is believed to be associated with the high silica surface. It is hypothesized that at 1000C, where the silicon nitride is oxidized, a dehydroxylated pyrogenic silica surface is produced. This would explain why the oxidized silicon nitride acts as an acid[6]. The nature of this dehydroxylated pyrogenic silica surface and its behavior in aqueous suspensions merits

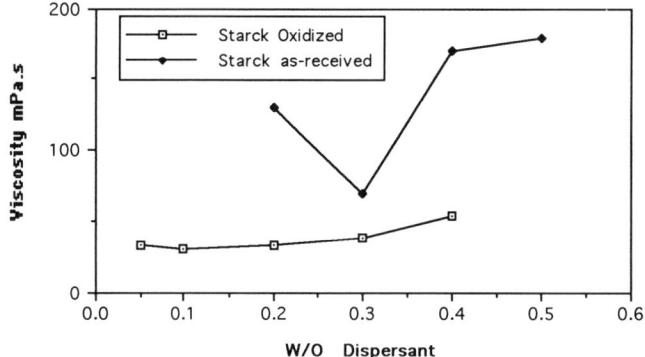

Figure 4. Slip viscosity vs. w/o dispersant for 60 w/o slips with either as-recieved Starck LC-12-SX or Starck LC-12-SX oxidized for 15 minutes with 5 w/o yttria and 5 w/o alumina as sintering aids. The slips were prepared with distilled, deionized water using alumina milling media.

further investigation. It is known that the charge density of a dehydroxylated pyrogenic silica are higher than for a rehydrated surface. Below a pH of 9-10 a dehydroxylated surface acts opposite to that of a hydroxylated surface[6].

For a slip of the standard composition, the pH is 8.3 after 24 hours of milling. The pH of the water and dispersant alone is also 8.3. When the oxidized silicon nitride is put in water containing the dispersant, the pH is lowered. After 72 hours of mixing on a shaker with no milling media, the pH is 7. When the 6 w/o yttria is added, the pH rises to 8.4 in minutes. This mix was then milled for 24 hours using nylon media. The pH after 24 hours remained at 8.4. Both the oxidized silicon nitride and the yttria have a major effect on the pH. When slips with 70 w/o solids or higher are prepared, it was found that it was necessary to add at least some of the yttria before adding the silicon nitride or it was not possible to get an initial solids loading 65 w/o.

5. Influence of milling media

During the course of this investigation, it became apparent that the use of different milling media were affecting the viscosity. Originally, the nylon covered steel media were used because it was reasoned that they would eliminate chipping of the milling media. It was also observed that the nylon coated steel milling media produced slips with lower viscosities. This can be seen by comparing the 65 w/o slips in Figure 5. Demand curves for 65 w/o slips prepared with UBE-E-10 oxidized for three hours with 6 w/o yttria using distilled, deionized water were milled with either alumina or nylon covered steel milling media. The nylon milled slips gave consistently lower viscosity. The pH of the slips milled with the nylon media ranged between 8.3 and 8.5, whereas the pH of the slips milled with alumina media ranged from 8.7 to 8.9. It is believed that the nylon covered steel media does not remove the surface layer of silica and the slip pH remains stable. The alumina media fractures the silicon nitride particles or abrades the surface exposing the Si_3N_4 to water where it reacts to form ammonia. The hydrolysis reactions are discussed by Griel[1]. The increase in viscosity could be due to the increase in pH, the increase of ions in the slip, or the increase in surface area due to milling.

Figure 6 shows that the pH vs. milling time for UBE silicon nitride powder as-received and oxidized (for three hours), is dependent on the milling media. The slips were milled with either silicon nitride milling media or nylon covered media. Additionally, pH vs. milling time is also shown for as-received Starck silicon nitride milled with nylon media. The graph illustrates that for slips milled with the nylon covered media, the pH does not rise above 9, even after 200 hours of milling. When the silicon nitride milling media is used, it causes the pH to rise above 10 in under 150 hours. The milling media plays an important role in determining the final pH of the slip. Unfortunately, after extended use, the polymer coating on the steel ball

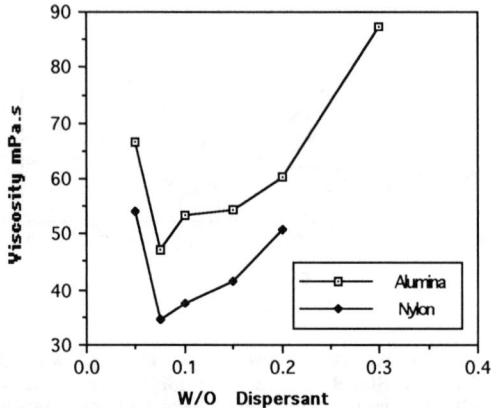

Figure 5. Viscosity vs dispersant level for 65 w/o slips, with UBE-E-10 oxidized for three hours dispersed in distilled deionized water. Milled with nylon or alumina milling media.

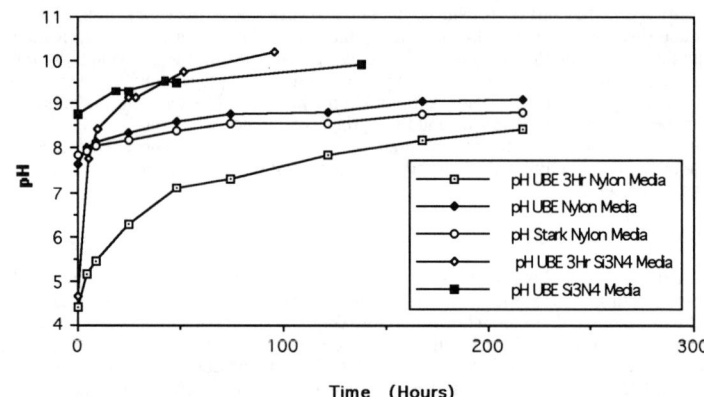

Figure 6. Milling time vs. pH for as received UBE-E-10, LC 12 SX, or UBE-E-10 oxidized for three hours milled with nylon and silicon nitride milling media.

begins to wear off of the substrate and the slurry begins to pick up iron. Future work is planned to find a way of coating the milling media that is more wear resistant

CONCLUSIONS

(1) Under the slip conditions studied, the Nopcosperse 644A polyelectrolyte was found to be

the most effective dispersant for the compositions investigated.
(2) Oxidizing both the UBE and Starck silicon nitride starting powders enhanced the sinterability of the compositions studied.
(3) Reductions in slip viscosity were observed when oxidized Starck LC-12-SX powder was used instead of as-received powder. This effect was not observed with oxidized UBE powder.
(4) The use of nylon coated milling media produced lower slip viscosities than ceramic milling media when used to prepare slips with the oxidized silicon nitride.

REFERENCES

1. P. Greil, Review: Colloidal Processing of Silicon Nitride Ceramics, Ceramic Material and Components for Engines, V.J. Tennery Editor, Amer. Ceramic Soc., pp 319-329
2. R. K. Iler, The Chemistry of Silica, (John Wiley and Sons, New York, 1979), p. 42
3. G.E.Gazza, B. Robinson, Sintering of Si3N4-Y2O3-SiO2 compositions with Mo2C Additions, Ceramic Material and Components for Engines, V.J. Tennery Editor, Amer. Ceramic Soc., pp 390-399
4. H. Stadelmann, G. Petzow, P. Greil. Effects of Surface Purification on Properties of Aqueous Silicon Nitride Ceramic Suspensions. J. Eur. Ceram. Soc., $\underline{5}$ 155-163, (1989)
5. P. Griel, R Nitzche, H. Friedrich, and W. Hermel, Evaluation of Oxygen Content on Silicon Nitride Powder Surface from the Measurement of the Isoelectric Point. J. Eur. Ceram. Soc., $\underline{7}$, 353-359, (1991)
6. R. K. Iler, The Chemistry of Silica. (John Wiley and Son, New York, 1972), pp. 658-663

TAPE CASTING OF SILICON NITRIDE

Young-Wook Kim and June-Gunn Lee
Structural Ceramics Lab., Korea Institute of Science and Technology
P.O.Box 131, Cheongryang, Seoul, Korea.

ABSTRACT

Tape casting behavior of submicron silicon nitride powder(UBE SN-E10) with Al_2O_3 and Y_2O_3 as sintering additives was studied. It was found that Hypermer KD1(polymeric dispersant) is effective dispersant for submicron silicon nitride in organic solvent system. Optimization of the rheological properties of the slurries allows the homogeneous green tapes with green densities of 45-50% theoretical. The cast tapes could be sintered to closed porosity with densities of higher than 96% theoretical.

INTRODUCTION

Tape casting is an important process for forming large-area, thin, flat ceramic parts. The chief advantage of the tape-casting process over other forming techniques is that it is the best method of forming flat articles with thickness in the $25\mu m$ to $1200\mu m$ range [1]. Tapes produced by this method were used widely in making multilayer capacitors, substrates for ceramic packages and heaters, and other passive electronic components, e.g. thermistors, inductors, piezoelectrics. There are numerous other references and patents relating to tape casting in the ceramic field [2-7].

The aim of this study has been to investigate tape casting as a forming technique of thick film for submicron silicon nitride powder in terms of slurry system selection, tape casting conditions, and its sintering behavior.

EXPERIMENTAL PROCEDURE

High purity submicron silicon nitride powder* was used for this study. It is manufactured by an imide decomposition process, with properties as; (a) surface area, 11.4 m^2/g, (b) α-phase >95%, (c) Oxygen 1.49 wt.%. High purity yttria** and alumina*** were used as sintering additives. Reagent grade methyl ethyl ketone (MEK), methyl iso-buthyl ketone (MIBK), and cyclohexanone were used as solvents.

Gravity sedimentation experiments were performed by adding 5 vol.% of solids to various dispersant/solvent slurry systems. The dispersions were agitated for 5 min using homogenizer(3000 rpm), and then quickly transferred to a 25 mℓ cylinder. The cylinders were allowed to stand at ambient temperature without disturbance for at least 120 h before the final sedimentation volume was recorded.

The solvent, dispersant, and ceramic powder were premilled with high purity Al_2O_3 grinding balls for 8 h. Binder and plasticizer were subsequently added to the slurry and milled for 24 h and the slurries were allowed to age for 24 h for complete absorbance of polymers. Before casting, the slurry was deaired in a vacuum for 1 h.

Viscosity was measured at a shear rate of 292 s^{-1} using the cone-plate type microviscometer(Brookfield Engineering Labs, Inc., Model HBTDV-IICP).

Tapes were cast onto silicon coated Mylar film using a laboratory scale caster. A two blade casting head was used to give uniform thickness [1].

---

*   SN-E10, Si_3N_4, Ube Industries, Ltd., Ube, Japan.
**   99.99%, Y_2O_3, Aldrich Chemicals Co., Milwaukee, U.S.A.
***   99.99%, Al_2O_3, Sumitomo Chemicals Co., Osaka, Japan.

Casting was performed at a speed of 0.25 m/min, at ambient temperature of 22°C ± 4°C, and an ambient humidity of 70% ± 10%.

The cast tapes were allowed to dry for 72 h, and the tapes were stripped from the Mylar film and were cut into appropriate shapes for property measurement and subsequent processing. Green density was determined from geometrical measurements and sample weights.

The tapes were sintered using the following heating schedule; Slow heating (1°C/min) to 360°C followed by an 1 h hold in air was employed to facilitate burnout of organics, slow heating(3°C/min) to 1300°C and rapid heating to sintering temperature in N_2 was followed by 4 h soak and cooling to room temperature. The sintering temperature ranged from 1760 to 1780°C. Sintered density and water absorption were measured using the Archimedes principle.

RESULTS AND DISCUSSION

Slurry system selection

In choosing the various organic components for the tape casting slurry, the binder is often chosen first. Other components then are found that are compatible with the binder. Some of factors that must be considered in the selection of binder are: (1) thickness of tape to be made; (2) casting surface; (3) solvent type desired.[1] By considering above factors, polyvinylbutyral (PVB) was chosen as binder.

In choosing the solvent, low volatility is desired for thick tape (>1000 μm) to minimize crack during drying. Table I shows the properties of various solvents, and the solvents having evaporation rate less than 4000 x 10^{-8} g/cm^2/sec at 25°C were selected for this study.

Inially, 25 commercial dispersants and 30 solvents or solvent blends were screened. A solvent blend is usually desirable because the solubility of a polymer in a mixed solvent is commonly greater than in any individual pure solvent, and it also provides a controlled evaporation rate,[10] and a controlled solubility of the dispersant.[5] Each solvent blend has a different evaporation rate. Three types of sedimentation behavior were observed in this experiments. (1) In the most stable dispersions a cloudy supernatant persisted throughout the experiment and a dense sediment cake grew slowly from the bottom of cylinder. (2) In the least stable dispersions the entire solid mass sedimented rapidly to its

Table I. Properties of various solvents listed in order of their evaporation rate[8,9]

Solvent	Evaporation rate (g/cm^2/sec x 10^8, 25°C)	Solubility parameter	Hydrogen bonding characteristic
Acetone	5830	10.0	Moderate
n-Hexane	5560	7.3	Poor
Cyclohexane	4487	8.2	Poor
Ethyl acetate	4440	9.1	Moderate
MEK	3620	9.3	Moderate
Methanol	3152	14.5	High
Iso-propyl acetate	2250	8.4	Moderate
Ethanol	1950	12.7	High
Toluene	1900	8.9	Poor
Iso-prophyl alcohol	1305	11.5	High
MIBK	1094	8.4	Moderate
Xylene	650	8.7	Poor
Buthanol	338	11.4	High
Cyclohexanone	290	9.9	Moderate

final volume and left a clear supernatant, usually within 1 h. (3) Weakly flocculated particles settled rapidly to yield a clear supernatant and then continued to rearrange and pack into a relatively dense cake.

The representative sedimentation volume of 5 vol.% silicon nitride slurries in various solvents is shown in Table II for a variety of dispersants. A loading of 1-2.5% dispersant per dry weight of ceramic powder was found to be sufficient to obtain the minimum sedimentation volume for each dispersant. The most effective dispersants were considered to be those which allowed the highest particle packing or lowest sedimentation volume. From the sedimentation test, Hypermer KD1 as dispersant and MEK-MIBK-cyclohexanone as solvent blend achieved the minimum sedimentation volume, and the system was chosen for further study.

Rheological studies of the Hypermer KD1-MEK-MIBK-cyclohexanone system

Slurries were made by preparing separate samples at each KD1 concentration. In this initial experiments, binder and plasticizers were omitted in order to keep the number of components to minimum. Silicon nitride and sintering additives(8 wt.% Y_2O_3, 2 wt.% Al_2O_3) were mixed by ball milling for 24 h. Each suspension contained 60 wt.% solids. The viscosity of the slurries at the highest shear rate vs concentration plot as shown in Fig.1 (cone-plate viscosity at 292 s^{-1} and 25°C) indicates the optimum state of dispersion at 1.75 wt.% of the Hypermer KD1. A viscosity of 300 mPa·s was seen at the maximum dispersion point. To examine the above results, cast surface was observed by scanning electron microscopy(SEM), as shown in Fig. 2. The dense and most uniform cast occurs at where viscosity was a minimum with 1.5-2.0 wt.% KD1, indicating that this concentration is optimum for dispersion.

Tape casting

Table III gives the formulation of a slurry with a high ceramic to nonceramic ratio, low viscosity, good stability, and homogeneity. Its ceramic content is 31.4 vol% (61.0 vol% after evaporation of the solvents) and its viscosity is about 850 mPa·s at a shear rate of 292 s^{-1}. Homogeneous green tapes with relative green densities of 45-50% are

Table II. Sedimentation data for silicon nitride (5 vol% solid in 25 mℓ solvent)

Dispersant			Solvent	Sedimentation volume (mℓ)
Trade name	Maker	Identity		
Zonyl FSN100	Du Pont	Fluorosurfactant	MEK-Ethanol	7.2
Zonyl FSN	Du Pont	Fluorosurfactant	MEK-Ethanol	7.0
Witcamide 5138	Witco Chem. Co.	Lipophilic alkanolamide	Toluene	6.1
Witcamine PA78B	Witco Chem. Co.	Salt of fatty imidazoline	Iso-prophylalcohol -Toluene	5.8
Emcol CC-36	Witco Chem. Co.	Polypropoxytrialkyl ammonium chloride	Iso-prophylalcohol -Toluene	5.0
Tergitol NP-4	Union Carbide Co.	Nonyl phenol polyethylene glycol ether	Toluene	4.6
Glyoxal	Kanto Chem.Co.	Oxalaldehyde	Iso-prophylalcohol	3.5
7347C	San Nopco	Proprietary	Methanol-Butanol	3.3
Hypermer KD1	ICI Chem.	Proprietary	MEK-Cyclohexanone	2.8
Hypermer KD1	ICI Chem.	Proprietary	MIBK-Cyclohexanone	2.7
Hypermer KD1	ICI Chem.	Proprietary	MEK-MIBK-Cyclohexanone	2.5

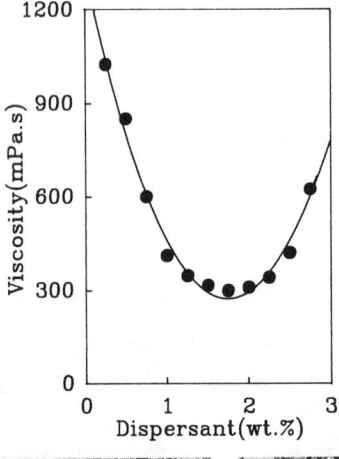

Fig. 1. Viscosities of Si_3N_4 slurries at 292 s^{-1} as a function of Hypermer KD1 concentration.

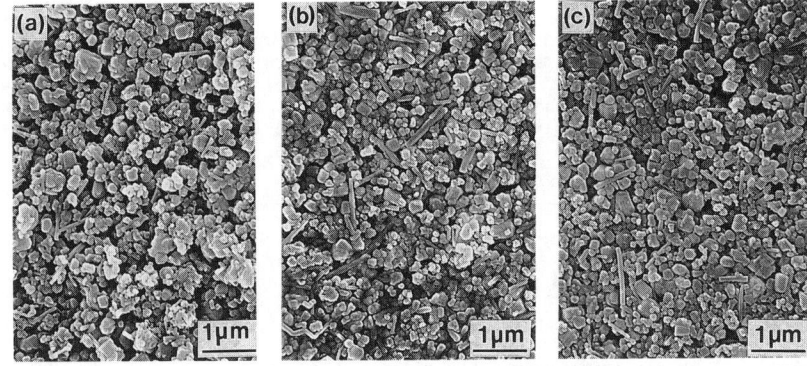

Fig. 2. Surface of the cast cakes as a function of Hypermer KD1 concentration: (a) 1.0%, (b) 1.75%, and (c) 2.5%.

obtained with this composition.

In order to know the effect of binder to plasticizer ratio, the binder to plasticizer ratio was varied keeping the ceramic to polymer ratio constant (ceramic/(ceramic+polymer)=0.787). An increase in the binder to plasticizer ratio resulted in an increase in slurry viscosity (Fig. 3).

Table III. Optimum composition of tape-casting slurry for silicon nitride

Component	Function	Composition wt.%	vol%
Si_3N_4 + 8wt.% Y_2O_3 + 2wt.% Al_2O_3	Ceramic powder	62.0	31.4
MEK	Solvent	7.3	15.2
MIBK	Solvent	9.0	18.9
Cyclohexanone	Solvent	8.2	14.5
Hypermer KD1	Dispersant	1.1	1.7
Polyvinyl butyral	Binder	7.4	10.8
Polyethylene glycol 400	Plasticizer	2.5	3.7
Benzyl butyl phthalate	Plasticizer	2.5	3.8

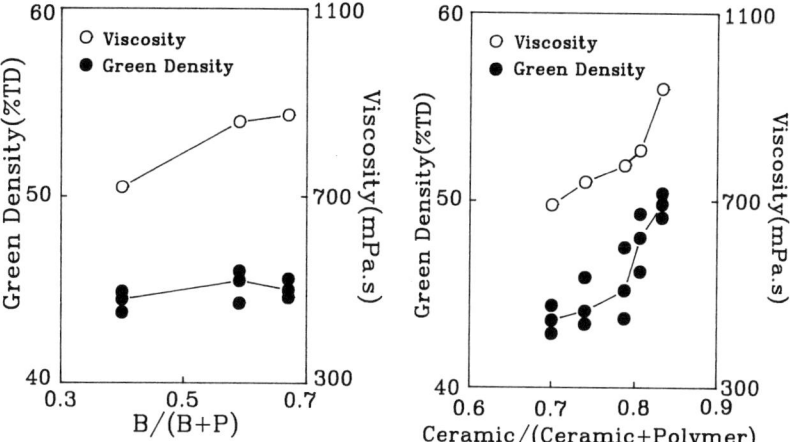

Fig. 3. Effect of the binder to plasticizer ratio on the viscosity and green density (B = binder, P = plasticizer).

Fig. 4. Effect of binder plus plasticizer concentration on the viscosity and green density.

However, it did not change the green density, as expected [11]. The increase in slurry viscosity is expected since the binder is a long-chain polymer (molecular weight (MW)=66500) whereas the plasticizers are short-chain polymers (MW=400).

In order to know the effect of ceramic to polymer ratio, the ceramic powder to binder plus plasticizer ratio was varied keeping the binder to plasticizer ratio constant (B/(B+P)=0.59). Since ceramic powder was added at the expense of binder plus plasticizer, this procedure is approximately equivalent to changing the ceramic powder loading of the slip. Increasing the ceramic powder loading resulted in an increase in viscosity and in green density (Fig. 4).

Sintering

Homogeneous green tapes with relative densities of 38-50% were sintered according to the schedule described in experimental procedure. Table IV

Table IV. Physical properties of debinded and sintered tapes*

Batch	Composition		Relative green density (%)	Relative sintered density (%)	Water absorption (%)
	B/(B+P)	Ceramic/(ceramic+polymer)			
1	0.40	0.670	38.1	81.5	12.8
2	0.40	0.687	40.0	84.2	11.9
3	0.59	0.700	43.6	96.0	0.09
4	0.59	0.740	45.9	96.9	0.05
5	0.59	0.787	47.5	96.8	0.07
6	0.59	0.806	49.3	96.8	0.07
7	0.59	0.833	50.4	97.2	0.05

*Sintered at 1770°C for 4 h under 0.1 Mpa of nitrogen.

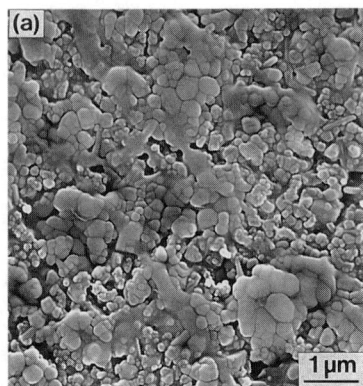 

Fig. 5. Microstructural changes of batch 7 (Table IV) at each fabrication step: (a) green tape and (b) sintered tape.

shows physical properties of sintered tapes for various batches. Each batch has a different binder to plasticizer ratio and binder plus plasticizer concentration. It shows that the higher green density leads to the higher sintered density. Also green density of >43% is a necessary condition for the sintering to closed porosity.

Fig. 5 shows microstructural changes of batch 7 at each fabrication step. It shows homogeneous pore distribution in green tape (Fig.5(a)) and high density sintered tape (Fig.5(b)).

CONCLUSIONS

Tape casting is a viable process for green forming method of submicron Si_3N_4. This method enables the forming of homogeneous, thin, flat tapes with green densities of 45-50 % theoretical. Hypermer KD1 and mixtures of MEK, MIBK, and cyclohexanone were highly effective dispersant and solvent system for Ube SN-E10 Si_3N_4 slurries.

Green density of >43% was a necessary condition for sintering to closed porosity (>96% theoretical density).

REFERENCES

1. R.E.Mistler, D.J.Shanefield, and R.B.Runk in Ceramic Processing Before Firing, edited by G.Y.Onoda and L.L.Hench (Wiley Sons, New York, 1978) pp. 411-448.
2. G.N.Howatt, R.G.Breckenridge, and J.M.Brownlow, J.Am.Ceram.Soc. 30 (8), 237-242 (1947).
3. G.N.Howatt, U.S.Patent No. 2 582 993 (22 January 1952).
4. H.W.Stetson and W.J.Gyurk, U.S.Patent No. 3 698 923 (17 October 1972).
5. D.J.Shanefield and R.E.Mistler, Ceramic Bull. 53 (5), 416-420 (1974).
6. J.C.Williams in Ceramic Fabrication Processes, edited by F.Y.Wang (Academic Press, New York, 1976) pp.173-198.
7. R.E.Becker and W.R.Cannon, J.Am.Ceram.Soc. 73 (5), 1312-1317 (1990).
8. E.C.Larson in Technology of Paints, Varnishes and Lacquers, edited by C.R.Martens (Chapman-Reinhold, Inc., New York, 1968) pp. 281-326.
9. T.C.Patton, Paint Flow and Pigment Dispersion, 2nd ed. (John Wiley & Sons, Inc., New York, 1979), pp. 335-354.
10. L.W.Tai and P.A.Lessing, J.Am.Ceram.Soc. 74 (1), 155-160 (1991).
11. A.Karas, T.Kumagai, and W.R.Cannon, Adv.Ceram.Mater. 3 (4), 374-377 (1988).

RHEOLOGICAL BEHAVIOR OF SILICON NITRIDE WHISKERS

PATRICIO TORO* and OSCAR WITTKE**
*Depto. Química and **Depto. Física,
F.C.F.M. Universidad de Chile
P.O.Box. 2777. FAX: 56-2-6994119. Santiago. Chile

ABSTRACT

Si_3N_4 whiskers are found through SEM analysis of powder obtained by carbothermal reduction of calcined Chilean rice husks. X-ray and FTIR analysis showed that the ceramic material was mainly α-Si_3N_4. A stable aqueous suspension with 30 or 40 wt% Si_3N_4 was obtained at pH = 8.5. A maximum viscosity at pH = 4.0 and low rpm was also observed and this rheological behavior is consistent with the zeta potential data. An irregular flow behavior near pH 4.0 shows a decrease in viscosity that is related to active impurities as detected by potentiometric titrations.

These rheological studies conducted allowed production of a Si_3N_4 green ceramic body measuring 170 x 25 x 5 mm formed by slip casting. The optimal aqueous suspension of powder contained the following: 17.0 g Si_3N_4, 1.0 g Cu_2O, 1.3 g bentonite, and 20.0 g water.

INTRODUCTION

Silicon nitride is an attractive candidate as a high strength material at high temperatures. Si_3N_4 whiskers, also referred to as crystalline whiskers of α-Si_3N_4, produced through carbothermal reduction, are highly appreciated because their mechanical strength approaches the theoretical value.[1] Much research on whisker synthesis has been conducted already.

The work presented here discusses a method of manufacturing α-Si_3N_4 whiskers using carbothermal reduction. The rheological behavior of an aqueous suspension thereof and optimal conditions for producing a ceramic body by slip casting are also reported.

EXPERIMENTAL PROCEDURE

Rice husk supplied by a Chilean producer gave the following analysis : 37.63 % C , 4.81 % H. 19.30 % SiO_2. This rice husk was first washed with deionized water and hydrochloric acid (50 vol.%), then milled under 200 mesh Tyler (approx. < 63 microns) before any thermal treatment. Thereafter, calcination at 500 °C and 1000 °C yielded powders containing 96% silicon dioxide. The fraction of soluble potassiun salts amounted to less than 5%.

The procedure adopted to obtain α-Si_3N_4 powder by thermal transformations (at 1400°C) of rice husk pretreated as above was carried out as described before [2,3].

Viscosity determinations were made with a Brookfield Viscometer, Model RVT-CP. Zeta potential (ZP) measurements were made by means of a Zeta-Meter Inc. New York Model Cramer 1626 using the microelectrophoretic technique. The potentiometric titrations (PT) of powder suspensions were performed employing a Radiometer Automatic Titrations Assembly. The potentiometric titration of the suspension with the pH adjusted to 3.0, by means of hydrochloric acid, was made using a standard solution of potasium hydroxide up to pH 10.5. ZP and PT measurements were undertaken at constant ionic strengths maintained using potasium chloride, 1.0×10^{-3} M.

RESULTS AND DISCUSSION

Rice husk calcination at 1400°C provided Si_3N_4 powder with a 67% yield based on initial material. The powder consisted of a whisker-like structure of Si_3N_4 monocrystals visible in the most white portion located on the surface. The remaining part of the powder forming the inner layers was darker. This shows that the process is controlled by the diffusion of gaseous nitrogen and thus prevails in the inner layers. SEM analysis of powder produced in this fashinon clearly shows the Si_3N_4 whiskers in the white portion thereof. On the other hand, the darker portions also contained some Si_3N_4 as detected by X-ray diffraction. The corresponding needles appeared in a more incipient state of growth.

The powder produced by carbothermal reduction of rice husk was crystalline and showed a haloed XRD pattern at $2\Theta(Cu$-$K\alpha) = 4 - 90°$ expected for α-Si_3N_4 (Figure 1).

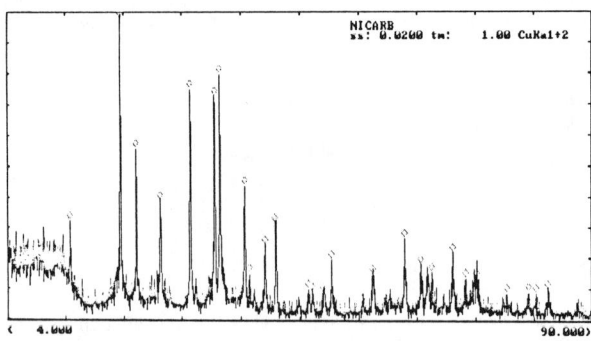

Figure 1. X-ray diffraction analysis of Si3N4

SEM micrographs show that most of the whiskers were broken during post-crystallization handling; however, a few of them carried drops at their tips (Figure 2). The analysis of similar whiskers by other authors indicated the presence of considerable amounts of iron in such drops.[4] Hence, it can be inferred that crystal-growth, as observed here, is also a result of an interaction between vapor, liquid and solid phases.

Figure 2. SEM micrographs of whiskers Si3N4 :
A) 12 micrometer/cm and B) 5 micrometer/cm

Furthermore, FTIR analysis shows the production of Si_3N_4 in view of the presence of a broad absorption band near 1000 cm^{-1} and of a sharp absorption band in the region below 800 cm^{-1} corresponding to Si-N bonding. Figure 3 shows a FTIR spectrum in the range of 1000 cm^{-1} to 400 cm^{-1} for the produced and commercial Si_3N_4 powders.

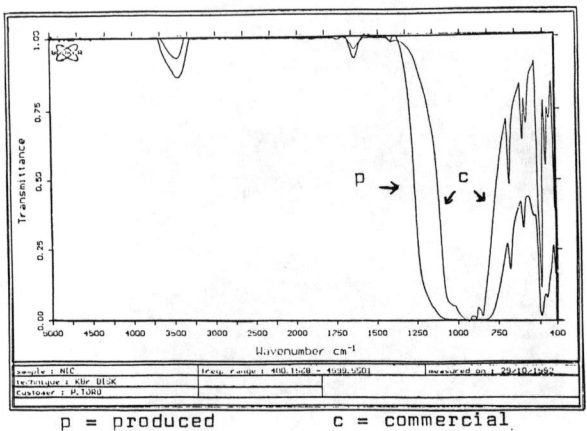

Figure 3. FTIR spectrum of the Si_3N_4 powder produced by the carbothermal processing of rice husk.

The viscosity and pH measurements for 30 and 40 wt% aqueous suspensions show shear thinning behaviour. The viscosity curve for 40 wt% Si_3N_4 content, Figure 4, shows maximun viscosity at a pH around 4.5, and a minimum value which is required for a stable suspension at pH above 8.5. At pH values below 4.5 the viscosity decreases again, but only at pH 3.0 is a deflocculated suspension obtained. The latter suspension is not similar to the one observed at pH 8.5; the viscosity at pH 3.0 is almost twice as high as that at pH 9.0 for 10 and 20 rpm (Figure 4). This viscosity behaviour of the Si_3N_4 whisker suspension is related to the ZP measurements of the same whiskers suspension, which shows an isoelectric point (IP) of 5.0 and extreme ZP values of 35 mV at pH 3.0 and -50 mV at pH 9.0.

To describe the potential determining surface groups of Si_3N_4 powder, we may assume that these groups are amphoteric silanol (Si-OH) and/or basic secondary amine (Si_2-NH) groups, depending on the pH of their suspensions. When no preferential adsorption of the supporting electrolyte (i.e., potassium chloride) on powder surfaces exists, the isoelectric point (IP) is closely related to the charge of superficial groups. That is to say, if the surface contains ony amino groups, then the IP is at pH = 9.0, and if it contains only silanol groups then the IP is at pH = 2.0, according to determinations carried out on powders that contained these groups [5]. In this instance, our powder exhibits a large Si/N ratio due to the predominance of silanol groups on the surface. Finally, the titration curve of the powder produced is related to this surface model. The point of zero charge determined by titration is in the same acid pH range as the isoelectric point. This correlation is expected, as a matter of fact, especially in the absence of any additive on powders, as in the present instance.[6]

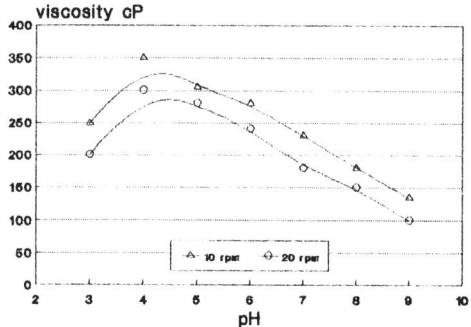

Figure 4.- 40 wt% Si3N4

The amounts of the binder (bentonite) and of the sintering aid (copper oxide) as well as the pH value were adjusted to obtain the optimal rheological properties, that is, a low viscosity or a high PZ, to produce a slip cast green body. The gravimetric proportions of 17.0 g Si_3N_4, 1.0 g Cu_2O, 12.3 g bentonite and 20.0 g water were used for slip casting, providing an α-Si_3N_4 whisker ceramic body measuring 170 x 25 x 5 mm (Figure 5).

(i)   (ii)   (iii)

Figure 5. A α-Si_3N_4 ceramic body of: (i) whiskers, (ii) commercial, (iii) particles.

CONCLUSIONS

Si_3N_4 whiskers were obtained by carbothermal reduction of calcine Chilean rice husks. Viscosity, zeta potential and potentiometric measurements as functions of pH have been made for a Si_3N_4 whiskers suspension with the purpose of identifying the pH at which the system is deflocculated and/or flocculated.

A more stable suspension of Si_3N_4 whiskers may be obtained at pH 8.5 than at pH 3.0.

It is possible to obtain a green compact ceramic body by slip casting of these Si_3N_4 whiskers.

ACKNOWLEDGEMENTS

The present work is only a part of the research and development activities now being conducted on advanced ceramics by the Interdisciplinary Team on Materials, of the Faculty of Physical and Mathematical Sciences of the University of Chile. The authors are grateful to the Fondo Nacional de Ciencia y Tecnologia (FONDECYT) for the grant of Project Contract Nr 1099/90, to the Departamento Técnico Investigación of the Universidad de Chile and to the Dirección Investigación of said Faculty for the financial assistance suplied to this work and to its presentation to the Materials Research Society at the Fall 1992 Symposium to be held in Boston (MA), USA.

REFERENCES

1.- K.Niwano, Silicon Nitride-1, edited by Sh.Somiya, M.Mitomo and M.Yoshimura, Ceramic Research and Development in Japan - Vol. 1, Elsevier App. Sci., 1990, pp.117.

2.- I.A.Rahman and F.L.Riley, J.European Ceramic Soc. **5**, 11 (1989).

3.- a) P.Toro et.al., Proc. XXXV Congreso Brasilero Ceramica, Belo Horizonte, Brasil 1991.
b) C.Mancilla. Químico thesis, IPS Santiago, Chile. 1992.

4.- Y.Kohtoku, Silicon Nitride-1, edited by Sh.Somiya, M.Mitomo and M.Yoshimura, Ceramic Research and Delopment in Japan - Vol. 1 (Elsevier Science Publishers, New York, 1990) pp. 71-80.

5.- L.Bergström et.al., Ceramic Today-Tomorrow's Ceramics, edited by P.Vincenzini (Elsevier Science Publishers, New York, 1991) pp. 1005-1014.

6.- W.Rudzinski et.al., Langmuir, **8**, 1154 (1992)

MICROWAVE NITRIDATION OF SILICON COMPACTS UTILIZING A TEMPERATURE GRADIENT

JEFFREY J. THOMAS, HAMLIN M. JENNINGS, AND D. LYNN JOHNSON
Depts. of Materials Science and Engineering and Civil Engineering
Northwestern University, Evanston, IL 60208-3108

ABSTRACT

Silicon compacts nitrided utilizing the temperature gradient inherent to microwave heating were more fully converted to silicon nitride than was possible with similar compacts nitrided isothermally. Although nitrogen depletion prevented the reaction rate in the center from exceeding that at the surface, the temperature gradient partially counteracted the effect of nitrogen depletion. Thus the microwave-heated specimens could be nitrided fully before the reduction in porosity that accompanies the reaction eliminated the diffusion of nitrogen into the compact.

INTRODUCTION

Silicon nitride is a potentially useful material for structural applications such as gas turbine engines because of its excellent high-temperature properties, such as thermal shock resistance. Reaction-bonded silicon nitride (RBSN), which is formed by heating a silicon powder compact in a nitrogen atmosphere, has certain advantages over fully dense sintered silicon nitride that make it attractive despite its lower strength. RBSN maintains its bulk dimensions during nitriding, which makes it easier to fabricate complex shapes. It also does not require the addition of the densification aids which adversely affect the high-temperature properties of the sintered material [1].

The nitridation mechanisms and kinetics have been thoroughly reviewed in the literature [2-4]. The reaction rate is increased both at higher nitrogen pressures and at higher temperatures. One of the key drawbacks to isothermal RBSN is that the size and density of the final component are limited by the requirement that nitrogen diffuse through the porosity of the compact. As the reaction proceeds, the pore size and pore volume decrease, causing the outside of the compact to react faster than the inside [5]. If the starting compact is too dense, unreacted silicon will remain in the interior of the final product, degrading the mechanical properties.

Microwave processing of ceramic materials is an area of much current interest for several reasons, most notably the acceleration of diffusional processes such as sintering [6,7]. Another unique aspect of microwave heating is the formation of temperature gradients in the specimens. This non-isothermal temperature distribution is a significant advantage for the processing of RBSN because it tends to cause the specimen to react preferentially in the hotter interior, thus counteracting the effects of restricted nitrogen diffusion.

The ability of microwave heating to cause an inside-out reaction already has been documented using 19 mm diameter by 7 mm thick disks of varying green densities [8,9]. Here there was ready access to nitrogen at the faces of the disks, so nitrogen depletion effects were minimal.

The present study addressed the nitridation of rod-shaped silicon compacts which were large enough so that nitrogen depletion was a significant factor in the reaction process. Comparisons with conventional (isothermal) heating demonstrated that reacting under a temperature gradient increases the size of compact which can be fully converted to RBSN in the center.

EXPERIMENTAL

The silicon powder used (HQ/10 grade, Elkem Metals Co., Pittsburgh, PA) had a maximum particle size < 10 μm, mean particle size of 2-3 μm and was 99.6% pure. Compacts were formed by isostatic pressing at 275 MPa to form rods 250 mm long and 15 mm in diameter, with a green density of 64%. These rods were then cut into 45 mm lengths for the nitriding experiments. A reactant gas mixture of 99% nitrogen and 1% hydrogen was used for all the experiments. The gas was passed over copper filings at 750°C to remove oxygen and then through molecular sieves to remove water before entering the reactor. The microwave and the conventional nitriding apparatus both had gas delivery systems which could be evacuated and backfilled to remove the air.

The microwave experiments were done in a vertical quartz tube which passed through a tunable cylindrical cavity (Wavemat Inc., Plymouth, MI) using 2.45 GHz microwave energy. Initially, specimens were placed inside an open alumina fiberboard basket (Zircar AL-30, Zircar Products, Florida, NY) which exposed the surface of the reacting rod. This maximized the temperature gradient but resulted in uneven heating patterns and an opaque condensate on the quartz tube. These problems were alleviated by placing the specimens into a thin alumina fiberboard cup and then pouring pure silicon nitride powder (SN-9FW, Denka, Tokyo, Japan) around the rod to form a powder bed (see Fig. 1). This insulation reduced but did not eliminate the temperature gradients.

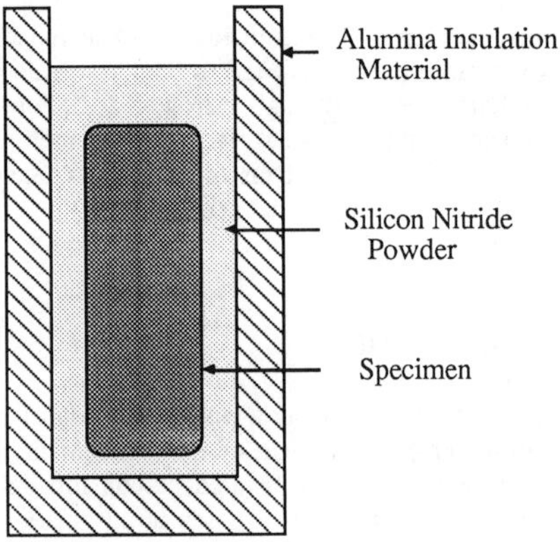

Figure 1. Schematic cross-section of Si rod specimen in Si_3N_4 powder and alumina fiberboard cup.

The insulation cup was suspended inside the cavity from an analytical balance to provide weight gain measurements, and an optical pyrometer (Model 100, Accufiber, Beaverton, OR) was focussed on the outside of the cup to provide relative temperature measurements, as shown in Fig. 2. Note that no direct measurements of the specimen temperature were possible during the microwave experiments due to the insulation. The processing was controlled by adjusting the microwave power so as to maintain the preprogrammed weight gain rate.

Conventionally nitrided RBSN specimens were also made, using an alumina muffle tube furnace with a SiC heating element. A silicon nitride powder bed was used to duplicate the conditions of the microwave experiments. The only significant difference between the microwave and conventionally heated rods was the presence of temperature gradients in the former.

To determine the percent conversion from silicon to RBSN, the final weight gain was compared to the theoretical maximum of 66.5% of the green weight. To measure composition profiles, rods were sectioned and polished (away from the ends) and microhardness tests were taken across the diameter using a Vickers indenter with a 3 kg load. The hardness of partially reacted compacts was previously found to increase rapidly with the amount of RBSN present [8,9]. Because of the color change from black to light grey as the reaction proceeds, composition profiles were easily observed after cutting the specimens.

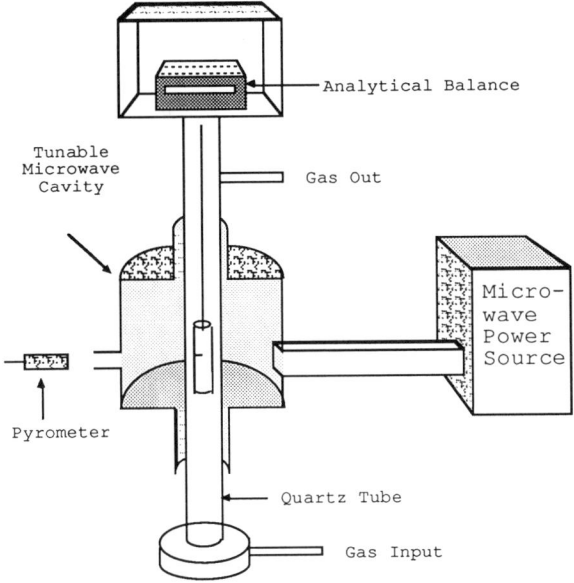

Figure 2. Schematic diagram of the microwave RBSN system.

RESULTS AND DISCUSSION

Green specimens were nitrided to full conversion in the microwave apparatus using a constant weight gain rate of 5% of the original weight per hour for most of the run. Temperature and conversion data from a typical experiment appear in Fig. 3. The temperature of the specimen could not be measured, but did not exceed the melting point of Si. After reaching 75% conversion, the surface temperature was held at 800°C until no further weight gain occurred. The final weight gain of the specimen indicated 98% conversion overall. The actual conversion percentage was probably higher, because of a small amount of weight loss during nitriding.

The hardness profile across the diameter of the rod indicated that the center was as fully reacted as the outside (see Fig. 4). This was corroborated by the uniformly light grey appearance of the inside of the specimen. X-ray diffraction scans taken at both the center of the rod and near the surface revealed no residual silicon peaks. The detectability limit for this experiment was about 1 weight percent of unreacted silicon.

Conventional nitridation below the melting point of silicon (1410°C) did not result in full conversion to RBSN. A two stage heating schedule was used: a four hour hold at 1250°C which resulted in 35-40% conversion, followed by a 24 hour hold at 1375°C which increased the conversion to a maximum of 88%. Longer hold periods at 1375°C did not result in increased conversion. Figure 4 shows a hardness profile for a conventionally reacted rod. The lower hardness in the center is a result of poor nitrogen penetration into the compact caused by the closing of the pore channels near the surface. This composition gradient was clearly visible as a transition from light grey near the surface to dark grey in the center. X-ray diffraction scans taken in the darker interior verified the presence of unreacted silicon.

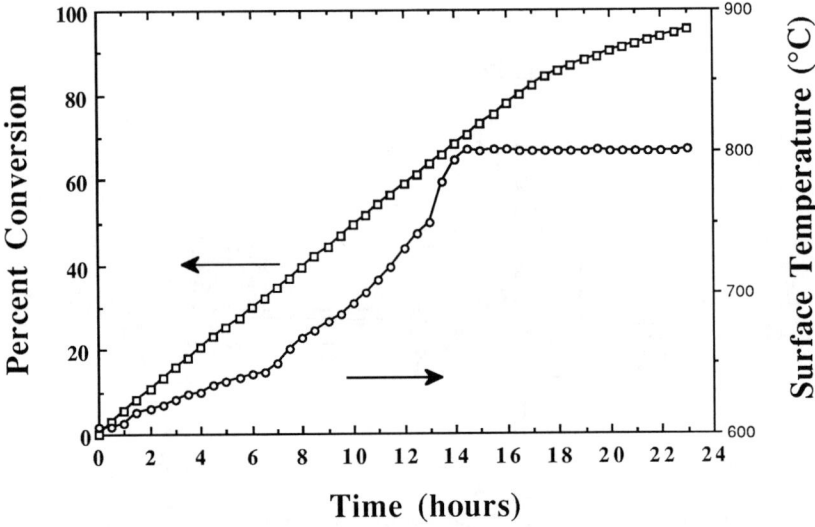

Figure 3. Conversion and surface temperature data for a typical nitriding experiment. The initial weight gain rate was 5%/hr.

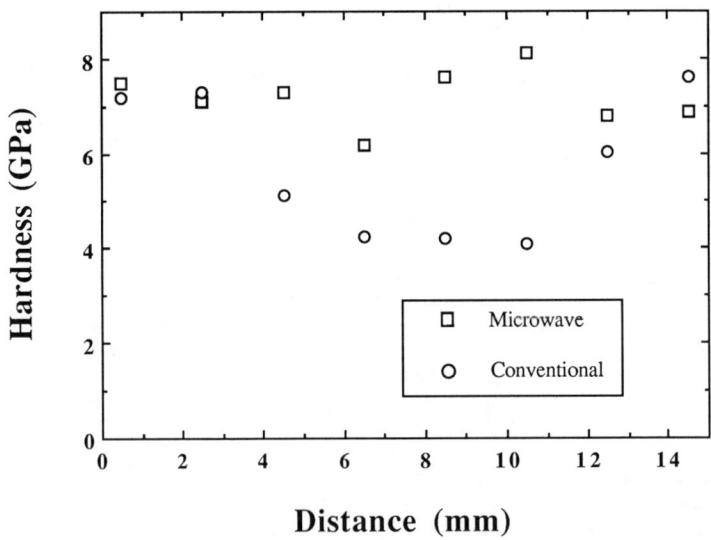

Figure 4. Hardness profiles for microwave and conventionally nitrided rods.

By raising the temperature to 1425°C for eight hours after holding at 1375°C, the conversion was increased to 96.5% as measured by the final weight gain. However, a region of melted silicon formed in the center of these specimens which was visible as a shiny metallic area. It is probable that these rods could be nitrided fully without melting using a heating schedule that raised the temperature from 1375°C to 1425°C over a period of a few days. By contrast, the microwave-heated rods were fully reacted without melting in less than 24 hours.

Microwave-heated rods which were partially nitrided were less reacted in the center than near the surface, although the difference was much smaller than with conventional heating. This indicates that the temperature gradient in specimens heated in the powder bed is not sufficiently large to fully counteract the effects of reduced nitrogen availability inside the compact. To fully nitride larger specimens in the microwave apparatus, the temperature gradient could be increased by reducing the amount of insulation. For a given specimen, the magnitude of the temperature gradient is approximately proportional to the amount of heat lost from the surface [10]. Increasing the surface temperature from 800°C to 1000°C would double the amount of radiative heat loss and would thus double the magnitude of the temperature gradient.

SUMMARY

The experiments performed in this study demonstrate that the size and green density of silicon compact that can be fully nitrided are increased by the temperature gradients associated with microwave heating.

Direct comparisons between microwave and conventional (isothermal) RBSN processing demonstrated that the microwave-heated specimens could be fully reacted, while the conventionally-heated specimens could not. The difference is attributed to the increased reaction rate in the interior of the microwave-heated specimens caused by the temperature gradient.

ACKNOWLEDGEMENTS

This work is supported by the National Institutes of Standards and Technology under Contract SC91209.

REFERENCES

1. M. K. Ferber, T. N. Tiegs, and M. G. Jenkins, "Effect of Post-Sintering Microwave Treatments on the Mechanical Performance of Silicon Nitride," *Ceram. Eng. Soc. Proc.* **12** [9-10], pp. 1993-2004 (1991).

2. A. J. Moulson, "Reaction-Bonded Silicon Nitride: Its Formation and Properties," *J. Mater. Sci.* **14**, pp. 1017-51 (1979).

3. H. M. Jennings, B. J. Dalgleish, and P. L. Pratt, "Reactions Between Silicon and Nitrogen, Part 1: Mechanisms," *J. Mater. Sci.* **18**, pp. 951-67 (1983).

4. M. Barsoum, P. Kangutkar, and M. J. Koczak, "Nitridation Kinetics and Thermodynamics of Silicon Powder Compacts," *J. Am. Ceram. Soc.* **74** [6], pp. 1248-53 (1991).

5. I. Amato, D. Martorana, and M. Rossi, "The Nitriding of Silicon Powder Compacts," *Powder Met.* **18** [36], pp. 339-48 (1975).

6. W. H. Sutton, "Microwave Processing of Ceramic Materials," *Ceram. Bull.* **68** [2], pp. 376-86 (1989).

7. M. A. Janney and H. D. Kimrey Jr, "Microwave Sintering of Alumina at 28 GHz," pp. 919-24 in Ceramic Transactions, Ceramic Powder Science II, B, eds G. L. Messing, E. R. Fuller Jr, and H. Hausner, American Ceramic Society, Westerville, OH (1988).

8. J. J. Thomas, R. R. Jesse, D. L. Johnson and H. M. Jennings, "Nitridation of Non-Isothermal Silicon Compacts," pp. 277-83 in Microwave Processing of Materials III, Vol. 269, Eds R. L. Beatty, W. H. Sutton and M. F. Iskander, Materials Research Society, Pittsburgh, PA (1992).

9. J. J. Thomas, R. J. Christensen, D. L. Johnson, and H. M. Jennings, "Non-Isothermal Microwave Processing of Reaction-Bonded Silicon Nitride," *J. Am. Ceram. Soc.* (submitted), (1991).

10. D. G. Watters, "Advanced Study of Microwave Sintering," PhD thesis, Northwestern University, 1989.

SINTERED REACTION-BONDED SILICON NITRIDE BY MICROWAVE HEATING

TERRY N. TIEGS, JAMES O. KIGGANS, JR., AND KRISTIN L. PLOETZ
Oak Ridge National Laboratory, P.O. Box 2008, Oak Ridge, Tennessee 37831-6087

ABSTRACT

Sintered silicon nitride has many desired properties, however, for most applications these materials are too expensive to compete with metal parts. Sintered reaction-bonded silicon nitride (SRBSN) is more economical, with raw material costs <27% those of comparable high-purity materials, making it competitive with metal parts. Conventional processing of SRBSN requires long nitridation times and a two-step firing process. Microwave (MW) heating reduces the reaction times and is performed in a one-step process, thereby simplifying the operation. The flexural strength of the MW-SRBSN is equivalent to the strength of some materials made from higher-cost powders. Thus, these materials may be appropriate for a number of applications.

INTRODUCTION

Silicon nitride-based materials have the best overall combination of properties and thus are the leading candidates for use as high-temperature structural ceramics.[1] Significant progress has been made in recent years in producing materials with superior strength, fracture toughness, wear behavior, and creep resistance. However, this progress was driven by property optimization with the cost a secondary consideration. Consequently, these materials tend to be very expensive and are not competitive with metal parts on a replacement basis in many applications. Reduction of costs has been recognized as a major factor for the introduction of advanced ceramics into the marketplace, especially for silicon nitride.[2-4] A recent study concluded that raw material costs for silicon nitride powders constitute a significant portion of the total cost of a part.[5] For example, with a cam roller follower, raw materials account for ~37% of the total cost. Other major costs include labor and machining (~44%) and capital charges (~11%). Other projects are currently addressing cost reductions for the machining and near-net-shape fabrication areas.[6]

Sintered reaction-bonded silicon nitride (SRBSN) has been studied for about 15 years, so there is a considerable database available.[7,8] The process is an attractive alternative to the expensive high-purity powders for a number of reasons. Most importantly, silicon is economical compared to high purity silicon nitride powders. A comparison shows high purity silicon nitride powders typically cost >$25/lb ($55/Kg), whereas silicon powders are <$5/lb ($11/Kg).[a] An added benefit for the RBSN is that the silicon gains about 60 wt.% during nitridation, further reducing the overall raw material cost per pound of the final product. Another advantage of SRBSN is that during sintering there is less shrinkage than for compacts starting with silicon nitride powders, thereby improving control over the dimensional tolerances of the parts produced.[9]

Normally, SRBSN is produced in a two-step process.[9-11] The first step, referred to as nitridation, is the exothermic reaction of the silicon powders with a N_2 containing gas at temperatures between 1200 and 1400°C. It is done with the parts exposed (non-insulated) to prevent the exothermic reaction from producing a temperature rise above the melting point of silicon, thus ruining the parts. The second step is the densification of the RBSN at temperatures from 1750 to 1800°C to effect liquid-phase sintering. This is typically done with the parts packed in Si_3N_4 powder to minimize weight losses.

Traditionally, the production of RBSN and SRBSN has been accomplished using resistance-heated furnaces, so what advantage does microwave heating provide? In recent years, microwave heating has been introduced as an alternative process for producing RBSN materials. This earlier research showed that nitridation occurs at a faster rate and initiates at slightly lower temperatures (~50°C) in the microwave as compared to conventional heating.[12,13] Thus, microwave heating offers a way to reduce total fabrication times of components. Another time-saving advantage of the microwave heated materials is that the parts are nitrided in an insulation

---

[a] Costs are based on purchases by Oak Ridge National Laboratory of 20 lb (10 Kg) of materials from commercial sources. Large scale purchases could reduce the costs by 30-50%.

package consisting of silicon nitride powders. Consequently, the RBSN parts do not have to be repackaged and can be heated directly up to the sintering temperatures. Thus, the microwave heated materials are reacted and densified in just a one-step process, which may affect overall economics of the fabrication costs by eliminating a handling operation.[14,15] It has also been shown that a reverse thermal gradient exists in microwave heated parts.[15,16] In the present case, this will result in nitridation of internal material first with the reaction moving to the external surfaces. Thus, pore closure by the silicon nitride reaction product growing into the pore space is minimized. This situation is in contrast to conventional heating where the reaction starts at the external surfaces and pore closure is a problem, especially for very large parts.

For these reasons, fabrication of microwave heated SRBSN using cost-effective materials was undertaken to estimate the economics of the raw materials in the process and to determine the properties that can be obtained.[15,17] In these previous studies, it was shown that the mechanical properties of the MW-SRBSN were not equivalent to ceramics made from the higher-cost powders, but were appropriate for a number of applications requiring lower temperatures and stress levels. At ambient temperature, typical strength, toughness (K_{Ic}) and hardness values of those initial materials in that study were 480 MPa, 4.8 MPa$\sqrt{m}$, and 14 GPa, respectively. It was desirable to improve both the strength and toughness of these materials since these were the properties that were lower than the higher cost materials. Consequently, in the present study, samples were fabricated using different processing procedures and compositions to improve on these properties. The impact on the cost was also assessed.

EXPERIMENTAL PROCEDURES

The starting materials for the SRBSN and sintered silicon nitride (SSN) consisted of appropriate amounts of silicon, α-Si_3N_4, Al_2O_3, La_2O_3, and Y_2O_3 as shown in Table 1. The final composition for the SRBSN samples after nitriding and sintering was estimated to be either Si_3N_4 - 11.5 wt. % La_2O_3 - 3 wt. % Al_2O_3 or Si_3N_4 - 9 wt. % Y_2O_3 - 3 wt. % Al_2O_3. The powders were turbomilled in either water or isopropanol (IPA) with 0.7 wt.% PVP K-15[b] and 1 wt.% Darvan 821A[c] added as dispersants. To form samples, the powder mixtures were either (1) aged and slip-cast into tiles approximately 0.8 cm X 8.5 cm X 8.5 cm or (2) dried and cold isostatically pressed (CIP) at 207 MPa into discs approximately 7 cm in diameter and 1 cm thick. All the SRBSN parts were pre-sintered in argon at 1200°C for 1 hour and then green machined to produce flat surfaces. The densities at this point were 60-65% T. D. The SSN tiles had a binder burnout heat-treatment to 600°C in air prior to sintering.

Microwave processing of the pre-sintered silicon tiles was conducted in a 500 L cylindrical microwave cavity with a 2.45 GHz power generator. The tiles were packed in Si_3N_4 powder containing 4% Y_2O_3 and 4% SiC, inside of a 15 cm X 15 cm X 12.5 cm alumina fiberboard box. A molybdenum sheathed thermocouple with a boron nitride sleeve was inserted between the tiles to measure the sample temperature. Nitridation was performed with N_2 -4 % H_2-5 % He at ~0.1 MPa (16 psi) with additional N_2 added as the reaction proceeded. After nitridation, the materials were heated to the sintering temperature and maintained for the appropriate time. An entire heating cycle to 1800°C required approximately 27 h. The SSN tiles were fired in a similar manner in the microwave, but bypassed the nitridation step and were heated directly to the sintering temperature in nitrogen.

The conventional processing for the SRBSN samples was carried out in a graphite-resistance heated furnace in a two-step process. In step one, the pre-sintered tiles were placed on a bed of Si_3N_4 powder in a graphite crucible with a N_2- 4 % H_2 gas flow and heated to 1480°C. The samples were then completely covered with the Si_3N_4 packing powder and heated in N_2 gas at 5°C/min. to 1800°C. Note that, in this conventional process, a two step process was required, because tiles completely packed in powder in step one experienced partial melting during the exothermic nitridation reaction stage. Further details of the experimental procedure can be found in previous publications.[13-15] The SSN tiles were fired in a similar manner in the graphite furnace, but bypassed the nitridation step and were heated directly to the sintering temperature in nitrogen.

Densities were determined by the Archimedes method. Tiles from the 1750 and 1800°C runs were machined into bend bar specimens with nominal dimensions of 3 mm x 4 mm x 50

---

[b] GAF Chemicals, Wayne, NJ: Polyvinylpyrrolidone K-15

[c] R. T. Vanderbilt, Norwalk, CT

mm. Flexural strength testing was done in four point bending with inner and outer spans of 20 mm and 40 mm, respectively.

Table 1 - Compositions of sintered reaction-bonded silicon nitride materials.

Sample No.	Sample Type	Silicon Impurity (wt. %)	SiO$_2$ Additive (wt.%)	Sintering Aid Content* (wt.%)	Initial α-Si$_3$N$_4$ Content (wt. %)	Milling Liquid/ Forming Method	Material Cost - ($/lb) As-Sintered†
TM-137	SRBSN	<0.05 [1]	None	11.5 % La$_2$O$_3$ 3% Al$_2$O$_3$ [6,7]	10 [3]	Water/ Slip-cast	6.42
TM-139	SRBSN	< 0.05 [1]	None	11.5.% La$_2$O$_3$ 3% Al$_2$O$_3$ [6,7]	5 [4]	Water/ Slip-cast	5.17
TM-141	SRBSN	< 0.05 [1]	None	11.5 % La$_2$O$_3$ 3% Al$_2$O$_3$ [6,7]	10 [3]	Water/CIP	6.42
TM-142	SRBSN	<0.05 [1]	1.6	9% Y$_2$O$_3$ 3% Al$_2$O$_3$ [7,10]	10 [3]	IPA/CIP	9.89
TM-145	SRBSN	<0.5 [2]	1.6	9% Y$_2$O$_3$ 3% Al$_2$O$_3$ [7,10]	10 [3]	IPA/CIP	9.30
TM-150	SRBSN	<0.05 [1]	None	11.5% La$_2$O$_3$ 3% Al$_2$O$_3$ [6,7]	10 [3]	IPA/CIP	6.19
TM-152	SRBSN	<0.05 [1]	1.6	11.5% La$_2$O$_3$ 3% Al$_2$O$_3$ [6,7]	10 [3]	IPA/CIP	6.19
TM-133	SSN	N. A.	None	11.5 % La$_2$O$_3$ 3% Al$_2$O$_3$ [8,9]	>98 [5]	Water/ Slip-cast	37.26
TM-132	SSN	N. A.	None	9% Y$_2$O$_3$ 3% Al$_2$O$_3$ [9,10]	>98 [5]	Water/ Slip-cast	41.76

*All compositions contain the same molar content of sintering additives.
#Material cost based on purchase price in 10 kg lots. Cost will decease at larger quantities.
†Assumes a yield from nitridation of silicon of 58% for water milled and 62% for IPA milled materials.
[1]Elkem Metals Co., Buffalo, NY; Grade Si-HQ; 4.2 μm mean particle size.
[2]Elkem Metals Co., Buffalo, NY; Grade Metallurgical Si; 3.4 μm mean particle size.
[3]Starck, Berlin, Germany; Grade LC-10N Si$_3$N$_4$
[4]Starck, Berlin, Germany; Grade S1 Si$_3$N$_4$
[5]Ube Industries, Japan; Grade E-10 Si$_3$N$_4$
[6]La$_2$O$_3$-Molycorp, White Plains, NY; Grade 5205, >99.9%.
[7]Al$_2$O$_3$-Reynolds, Malakoff, TX; Grade RC-HP DBM.
[8]La$_2$O$_3$-Molycorp, White Plains, NY; Grade 5200, >99.99% .
[9]Al$_2$O$_3$-Ceralox, Tucson, AZ; Grade HP(0.5μm).
[10]Y$_2$O$_3$-Molycorp, White Plains, NY; Grade 5600, >99.99%.

Table 2 - Summary of results on the densification and strength of the SRBSN materials.

Sample	Heating Type	Sintering Conditions (°C/h)	Sintered Density (g/cm^3, % T. D.)	Flexural Strength at 25°C (MPa)
TM-137 (SRBSN)	Conventional	1800/1	3.09, 91.3	N. D.
TM-137 (SRBSN)	Microwave	1800/1	3.37, 99.9	480±86
TM-139 (SRBSN)	Conventional	1800/4	3.20, 95.4	N. D.
TM-139 (SRBSN)	Microwave	1800/2	3.23, 96.4	415±54
TM-141 (SRBSN)	Microwave	1800/1	3.35, 99.4	310±54
TM-142 (SRBSN)	Microwave	1800/2	3.25, 97.7	677±45
TM-145 (SRBSN)	Microwave	1800/1	3.12, 93.7	N. D.
TM-145 (SRBSN)	Microwave	1800/2	3.23, 97.0	785±98
TM-150 (SRBSN)	Microwave	1800/1	3.16, 93.6	N. D.
TM-152 (SRBSN)	Microwave	1800/1	3.34, 99.1	526±17
TM-133 (SSN)	Microwave	1700/1	3.32, 98.3	654±16
TM-132 (SSN)	Microwave	1700/1	3.27, 98.5	N. D.

RESULTS

A comparison of the costs for the different sample types is shown in Table 1. The cost advantage of the SRBSN materials is readily evident. Purchases of large quantities of these materials would be expected to decrease the costs by 40 % from those presented here. Thus, a commercial scale operation for either SRBSN composition, assuming a 40% reduction, would be expected to have raw materials costs on the order of $6/lb or less. Comparison of the raw material costs reveals that all of the SRBSN compositions in the present study are <27% of those for either of the SSN materials.

Densification behavior of the SRBSN and SSN materials is summarized in Table 2. As shown, high densities were obtained with samples fired at temperatures of 1800°C for 1-4 h. Microwave sintering of the SRBSN was more effective than conventional sintering for similar conditions as indicated by the results from samples TM-137 and TM-139. This behavior has been observed in previous comparisons of microwave and conventional sintering of Si_3N_4 powder compacts and SRBSN.[16-18] As a result, comparative sintering runs were not made for all of the sample types and the microwave sintering was emphasized. The specimens fabricated with the high purity powders (TM-133 and -132) sintered to densities >98% at temperatures of 1700°C with microwave heating.

Densification was also dependent on the powder processing conditions and the initial α-Si_3N_4 content. The materials milled in water generally sintered to high densities easily as exhibited by TM-137 and TM-141. It is a well known fact that significant oxidation of the Si occurs during milling in water and it is this additional SiO_2 in these samples that improved densification. When the same composition was processed in IPA (TM-150), the SiO_2 content was diminished and densification was lower. To compensate for the low SiO_2 content, a small amount was added to the same composition (TM-152) and high densities were again easily obtained. Consequently, for the other compositions processed in IPA (TM-142 and TM-145), additional SiO_2 was added at a level to simulate the SiO_2 content in a typical Si_3N_4 powder. The effect of the initial α-Si_3N_4 content is shown in a comparison of TM-137 and TM-139, where a lower α-Si_3N_4 level resulted in lower densities. The cause for this behavior is presently being examined in more detail to further understand the processes involved. From an economic standpoint, it would be desirable to lower the α-Si_3N_4 content since it represents a significant portion of the total cost of these materials.

Mechanical testing was performed on selected materials that achieved high density as shown in Table 2 and several factors were attributed to the strength variations. Processing in water was detrimental to the strength of the SRBSN materials. This is illustrated by the CIPed samples, where the strength was significantly improved by using IPA during the milling step (TM-141 vs TM-152). Other research has shown agglomeration of the Si during water processing and this is evidently affecting the strength in the water milled materials.[20] By far the most important factor affecting the strength was the rare earth additive. By substituting Y_2O_3 for La_2O_3 (TM-142 vs TM-152), the strength increased from ~525 MPa to ~675 MPa even though the density was lower.

The highest mean ambient temperature strength was 785 MPa. This strength is higher than for similar SRBSN materials processed at temperatures ≤1800°C by conventional heating.[18] The highest strength materials contained Y_2O_3-Al_2O_3 and used the lower purity silicon. It is known that the lower purity results in a higher α-Si_3N_4 content in the samples after nitridation, but before the sintering step. Generally, the samples with the low purity silicon had ~70% α-phase content after nitridation as compared to <50% for the nitridation product from the high purity silicon. It is believed this higher α-content results in more acicular grain growth during the α-to-ß transformation and improved properties. However, the lower purity powders are expected to have lower high temperature properties due to the less refractory nature of the intergranular phases. Flexural testing at elevated temperatures showed appreciable strength decrease at temperatures above 1000°C (Fig. 1). The sample containing Y_2O_3-Al_2O_3 and using the high purity Si (TM-142) showed good strength retention up to 1000°C.

While significant strength improvements were made over previous materials, the fracture toughness (K_{Ic}) values ranged from 4.5 to 5.1 MPa√m for the SRBSN materials in this study. This is the same as earlier results and is lower than the values obtained for the samples starting with the high purity Si_3N_4 powders (6.8-7.0 MPa√m).

CONCLUSIONS

SRBSN is a cost-effective method to fabricate silicon nitride ceramics. Raw materials costs are less than 27% of those for high-purity silicon nitride materials, which improves the cost-competitiveness of these materials with metal parts. Conventional SRBSN requires long nitridation times and two-step firing. By using microwave heating, nitridation times are reduced and all firing is performed in a one-step continuous process, simplifying the operation. The current materials produced by this fabrication method have strength appropriate for numerous applications at temperatures up to approximately 1000°C.

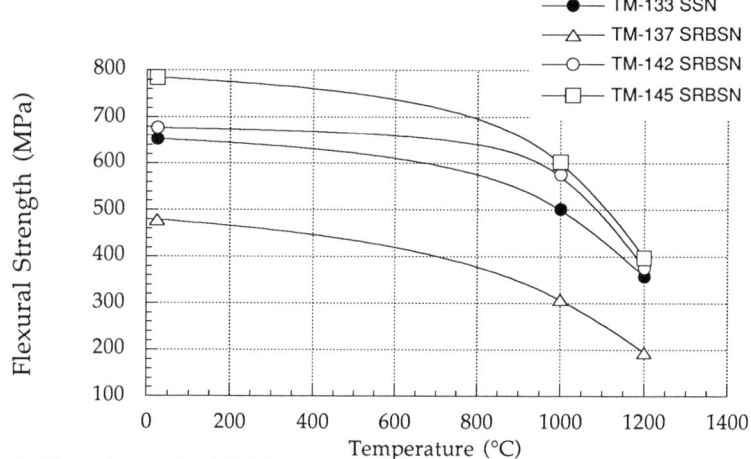

Fig. 1 - Flexural strength of SRBSN and SSN materials as a function of temperature. Samples TM-133 and TM-137 were nominally Si_3N_4-11.5% La_2O_3-3% Al_2O_3. Samples TM-142 and TM-145 were nominally Si_3N_4-9% Y_2O_3-3% Al_2O_3. TM-137, TM-142 and TM-145 were reaction-bonded and sintered at 1800°C for 1-2 h. TM-133 was fabricated from high-purity materials and sintered at 1700°C for 1 h. All samples were heated in the microwave.

REFERENCES

1. R. N. Katz, Nitrogen Ceramics 1976-1981, pp. 3-20 in Progress in Nitrogen Ceramics, ed. F. L. Riley, Martinus Nijhoff Pub., The Hague, Netherlands (1983).
2. L. M. Sheppard, "Cost-Effective Manufacturing of Advanced Ceramics," Am. Ceram. Soc. Bull., 70[4]692-707(1991)
3. T. Quadir, R. W. Rice, J. C. Chakraverty, J. A. Breindel, and C. C. Wu, "Development of Lower Cost Si_3N_4," Ceram. Eng. Sci. Proc., 12[9-10]1952-1957(1991)
4. J. M. Schoenung, "Analysis of the Economics of Silicon Nitride Powder Production," Am. Ceram. Soc. Bull., 70 [1] 112-116 (1991).
5. S. Das and T. R. Curlee, "The Cost of Silicon Nitride Powder and the Economic Viability of Advanced Ceramics," Am. Ceram. Soc. Bull., 71 [7] 1103-1111 (1992).
6. R. B. Schulz and D. R. Johnson, "Transportation, Energy and Ceramics," Ceram. Eng. Sci. Proc., 12[9-10]947-956(1991).
7. A. J. Moulson, "Reaction-Bonded Silicon Nitride: Its Formation and Properties," J. Mater. Sci., Vol. 14, pp. 1017-1051 (1979)
8. F. L. Riley, "Nitridation and Reaction Bonding," pp. 265-288 in Nitrogen Ceramics, F. L. Riley (ed.), Noordhoff, Netherlands (1977)
9. J. A. Mangels and G. J. Tennenhouse, "Densification of Reaction-Bonded Silicon Nitride," Am. Ceram. Soc. Bull., 59 [12] 1216-1222 (1980).
10. R. M. Williams and A. Ezis, "Slip Casting of Silicon Shapes and Their Nitriding," Am. Ceram. Soc. Bull., 62 [5] 607-619 (1983).
11. A. Ezis, "The Fabrication and Properties of Slip-Cast Silicon Nitride," in Ceramics for High Performance Applications, Brook Hill Publishing Co., 207-222 (1974).
12. T. N. Tiegs, J. O. Kiggans, and H. D. Kimrey, "Microwave Processing of Silicon Nitride," pp. 267-272 in Microwave Processing of Materials-II, Vol. 189, Materials Research Soc., Pittsburgh, PA 1991.
13. J. O. Kiggans, C. R. Hubbard, R. R. Steele, H. D. Kimrey, C. E. Holcombe, and T. N. Tiegs, "Characterization of Silicon Nitride Synthesized By Microwave Heating," in Ceramic Transactions, Microwaves, Theory, and Applications in Materials Processing, Am. Cer. Soc., Westerville, Ohio, 267-272 (1991).
14. J. O. Kiggans and T. N. Tiegs, "Characterization of Sintered Reaction Bonded Silicon Nitride Processed By Microwave Heating," pp. 285-290 in Microwave Processing of Materials-III, Vol. 269, Materials Research Soc., Pittsburgh, PA 1992.
15. T. N. Tiegs, J. O. Kiggans, and K. L. Ploetz, , "Cost-Effective Sintered Reaction-Bonded Silicon Nitride for Structural Ceramics" to be published in Ceramic Engineering and Science Proceedings, American Ceramic Society
16. J. G. P. Binner, "Microwave Processing of Ceramics," Brit. Ceram. Proc., 45, 97- 108 (1990).
17. T. N. Tiegs and J. O. Kiggans, "Fabrication of Silicon Nitride Ceramics by Microwave Heating," pp. 665-671 in Proc. 4th International Symp. Ceram. Mater. & Compon. for Engines, Elsevier Applied Sci., New York (1992).
18. T. N. Tiegs and J. O. Kiggans, and H. D. Kimrey, "Microwave Sintering of Silicon Nitride," Ceram. Eng. Sci. Proc., 12[9-10] 1981-1992 (1991).
19. J. A. Mangels, "Sintered Reaction-Bonded Silicon Nitride," Ceram. Eng. Sci. Proc., 2[7-8] 589-603 (1981).
20. R. G. Stephen and F. L. Riley, "The Influence of Pre-Oxidation on the Processing of Silicon Powder," pp. 307-314 in 4th Internat. Symp. on Ceram. Mater. Components for Engines, Elsevier Applied Science, NY (1992)

ACKNOWLEDGMENTS

Research sponsored by the U.S. Department of Energy, Assistant Secretary for Conservation and Renewable Energy, Office of Transportation Systems, as part of the Ceramic Technology Project of the Materials Development Program, under contract DE-AC05-84OR21400 with Martin Marietta Energy Systems, Inc.

MICROSTRUCTURAL CHARACTERISATION OF MICROWAVE SINTERED SILICON NITRIDE CERAMICS

Kevin P. Plucknett and David S. Wilkinson

Dept. of Materials Science and Engineering,
McMaster University, Hamilton, Ontario, CANADA.

The microstructure of a series of microwave sintered silicon nitride based ceramics have been assessed using a combination of optical microscopy/image analysis and analytical electron microscopy. Materials were studied as-received and after post-sinter hot-isostatic pressing. The grain size of microwave sintered materials was appreciably finer than conventionally processed ceramics of similar composition, although the mechanism involved is not clear. The as-received ceramics exhibited a reverse porosity gradient (with the highest porosity level at the surface) due to heat dissipation to the cooler surroundings during sintering. This also resulted in a small increase in the β' grain aspect ratio close to the surface arising from an increase in the glass phase viscosity as the temperature decreases. Post-sinter HIPing of microwave sintered samples resulted in the elimination of most of the bulk porosity, but not near the surface. This is due to the reverse porosity gradient previously described, which leads to a transition from closed to open porosity with decreasing density near to the surface.

Introduction

High density silicon nitride (Si_3N_4) ceramics can be prepared using several conventional processing routes, including uniaxial hot-pressing and pressureless-sintering. Hot-pressed Si_3N_4 is limited to the fabrication of relatively simple shapes, although a high degree of composition control is possible. Pressureless-sintering of Si_3N_4 allows the production of complex shaped articles, however multiple sintering additives are required in higher volumes than for hot-pressing. Glass encapsulated hot-isostatic pressing offers the benefits of both of these routes, albeit at considerably increased processing complexity and cost.
A further alternative which has been reported recently, involves pressureless-sintering using microwave heating [1-5]. Microwave heating has previously been utilized to sinter various ceramics including Al_2O_3 and related composites [6-8], B_4C [9] and ZrO_2 [10]. Tiegs et al [1] demonstrated enhanced densification of microwave sintered Si_3N_4, prepared with Y_2O_3 and Al_2O_3, when compared to conventionally heated materials. It was noted that higher microwave sintered densities could be obtained by either increasing the additive content or by adding secondary particulate phases that couple well during microwave heating. It was also demonstrated that improved densification could be obtained when sintering at a higher frequency microwave power.
Further development of microwave sintering techniques for the densification of Si_3N_4 based ceramics have been performed by Alcan Int. [3-5]. This work has concentrated upon ensuring adequate coupling to, and therefore heating of, the silicon nitride via the use of a 'powder-bed' around the sample that acts as a susceptor during heating [11]. Refinement of this process route has allowed the preparation of high density (>95% of theoretical) silicon nitride based ceramics for cutting tool applications [4,5]. In addition relatively large batches (up to 1Kg) of material can be densified in one sintering cycle [5]. Several potential benefits were apparent when microwave sintering these cutting tools, notably significant energy savings (up to ~78%) and a considerably shorter processing cycle time [5]. The present work has involved the microstructural study of several Si_3N_4 based ceramics densified using the Alcan microwave sintering technique.

Experimental Procedure

The microwave sintered silicon nitride ceramics used in the current study have been supplied by Alcan International Ltd. (Kingston, Ontario), with the compositions summarised in table 1. All compositions were prepared using Ube SN E-10 α-Si_3N_4 powder. Samples were sintered in air, with a typical sintering time of 90 minutes (~30 minutes at temperature), although longer cycle times are used for larger batches (up to 1 kg). The actual sintering temperatures used were not specified. Further details of the processing route followed can be found in the relevant literature [3,4,11]. Post-sinter hot-isostatic pressing (HIPing) was performed at a temperature of 1800°C and pressure of 200 MPa for one hour. An argon/nitrogen mixture (ratio 2:1) was used for HIPing.

Table 1. Summary of the compositions of Alcan microwave sintered silicon nitride based ceramics (* Hot-isostatically pressed).

Composition	Sintering Additives (weight %)	Density (g.cm^{-3})	% Theoretical Density
SN55	Y_2O_3 (5), Al_2O_3 (5)	3.10-3.13	93.4-94.3
SN75H*	Y_2O_3 (7), Al_2O_3 (5)	3.20	95.2
SN55H*	Y_2O_3 (5), Al_2O_3 (5)	3.25	97.9

Compositions SN55 and SN55H were provided in the form of rectangular samples (15x15x8mm), while the SN75H samples were in the form of long cylinders (26mm length and 10mm dia.). In addition to study of the as-received and HIPed materials sections of composition SN55 were subjected to a post-sinter annealing heat-treatment in argon at 1200°C for 100 hours.

Densities of microwave sintered samples were determined by immersion in distilled water, with porosity distributions obtained using a Leco 2001 Image Analyser attached to a Zeiss optical microscope. Scanning electron microscopy (SEM) of plasma etched samples was performed using a Philips 515. Thin foils were examined using conventional transmission and scanning transmission electron microscopy (CTEM and STEM respectively) using a Philips CM12 (operating at 120 kV), with energy dispersive X-ray (EDX) microanalysis conducted in STEM mode using a 7.5 nm probe size to minimize beam broadening. 'k'-factors for quantification of EDX data were calculated from spectra obtained from binary oxide standards. Identification of crystalline phases was obtained using X-ray diffraction (XRD).

Results and Discussion

Optical Microscopy Observations and Image Processing

Reflected light optical micrographs of the near-surface and bulk regions of materials SN55, SN55H and SN75H are shown in Fig.1. Analysis of the porosity distribution in the as-sintered material reveals the highest porosity volume occurs at the surface and then gradually decreases into the bulk material (Fig.2a). A similar trend is apparent in the HIPed material, although the overall porosity volume is lower and decreases much more rapidly with distance from the surface until the bulk material is fully dense (Fig.2b). The reverse density gradient observed in the as-sintered ceramic (SN55) indicates a temperature gradient during sintering with the lowest temperature occurring at the specimen surface. This behaviour may be expected during microwave heating as heat is dissipated from the sample surface to the surrounding atmosphere.

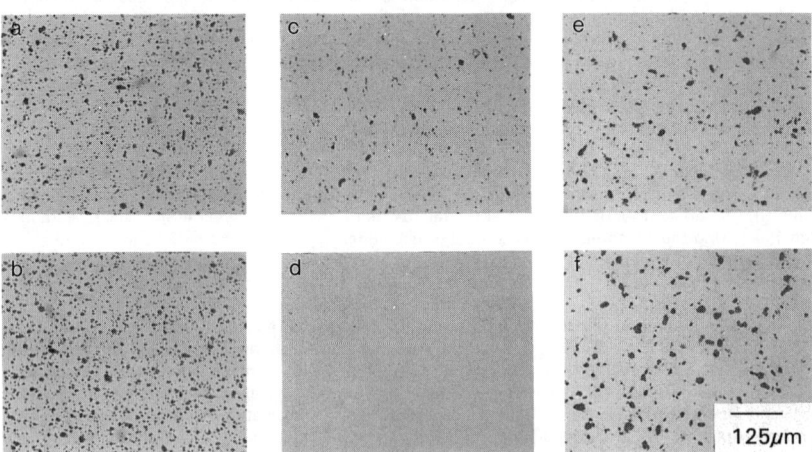

Figure 1. Reflected light optical micrographs of a,b) SN55, c,d) SN55H and e,f) SN75H (a,c,e near surface and b,d,f bulk microstructure).

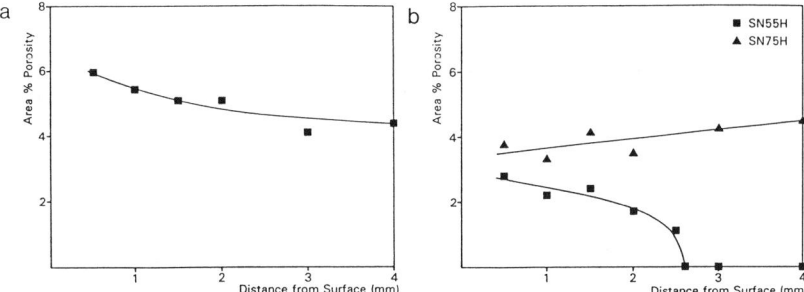

Figure 2. Porosity area distributions for a) as-received (SN55) and b) HIPed microwave sintered ceramics (SN55H and SN75H).

The presence of a reverse density gradient results in a transition from open porosity near the sample surface to closed porosity nearer the centre. This is consistent with the data presented in Fig.2b for the HIPed ceramic SN55H as the material can reach full density only in the region of closed porosity. The high gas pressure acting within the surface connected open porosity will inhibit further densification, resulting in the retention of porosity in this region. Previous studies on post-sinter HIPing of silicon nitride have indicated that complete densification does not occur, even when HIPing samples that do not contain open porosity [12,13]. The enhanced bulk densification in the present example is believed to arise from the much finer grain size of the microwave sintered ceramics when compared to conventionally processed materials (discussed in detail in the next section), as it has recently been demonstrated that very fine grained silicon nitride based materials can be superplastically deformed at temperatures greater than 1600°C [14,15]. Significant density gradients were not observed in composition SN75H, which showed an approximately uniform porosity level throughout after HIPing. This is believed to be due to lower green densities in the SN75H samples, arising from the pressing method.

Microstructural Assessment
XRD typically demonstrated complete transformation of α- to β-Si_3N_4 in the as- sintered material (SN55), although some samples contained up to ~15% residual α-Si_3N_4. Examination of the high angle peaks of the XRD traces revealed a small shift to lower diffraction angles, indicating that the matrix phase is actually a low substitution β'-Si-Al-O-N [16].
SEM micrographs of the microwave sintered, and microwave sinter plus HIP, ceramics are shown in Fig. 3. It is clear that microwave sintering results in a considerable refinement in the scale of the microstructure (Fig. 3a), with grain sizes significantly less than 1μm, when compared to conventionally sintered silicon nitride ceramics. In addition the grain size distributions are much narrower than generally observed, despite a relatively short ball milling time during powder processing (a total of 20 hours) [5]. Typically these grain sizes are an order of magnitude lower than conventionally processed materials of similar composition [17]. The attainment of such fine and uniform grain sizes indicates a much higher β' nucleation rate during liquid phase sintering with negligible Ostwald ripening. The exact mechanism behind these observations is difficult to determine without further in-depth study, however there are two main possibilities;

i) the actual temperature during liquid phase sintering is significantly higher than conventional sintering temperatures, which coupled with the short processing time results in the observed microstructure, with minimal dissociation of Si_3N_4 (observed weight losses are less than 0.5% [18]), or alternatively,
ii) selective coupling of the microwave energy to the oxide sintering additives leads to preferential heating of the oxides and glass formation, with the subsequent possibility of a temperature gradient driven α-Si_3N_4 dissolution process resulting in higher nitrogen contents in the glass phase and consequently a higher β'-Si_3N_4 nucleation rate.

In reality it is probable that a combination of these processes are active when microwave sintering these materials. Post-sinter HIPing of the microwave sintered materials results in notable grain growth (Fig. 3b,c), particularly in the higher additive content ceramic (SN75H).

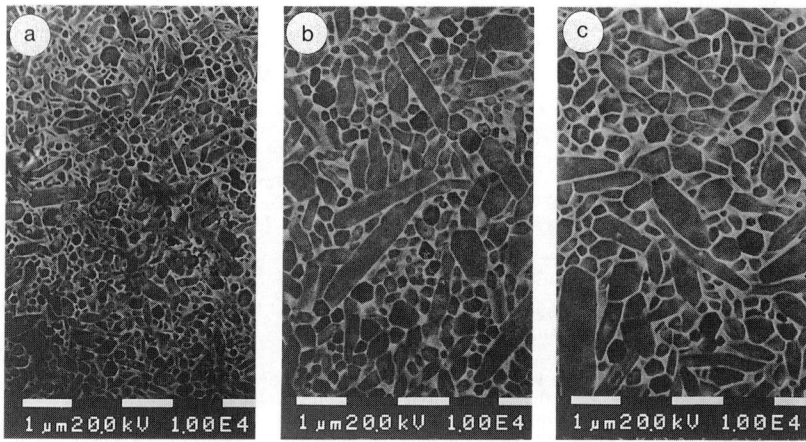

Figure 3. SEM micrographs of the bulk microstructure of each of the microwave sintered ceramics after plasma etching, a) SN55, b) SN55H and c) SN75H.

Measurements of the average grain length ($\bar{L}$) and width ($\bar{d}$) as a function of distance from the surface for material SN55 are shown in Fig. 4a. Both material parameters show a high degree of consistency, as has been previously noted for these materials [5]. Fig. 4b presents similar data for the average aspect ratio ($\bar{a}$) and the average value of the highest 10% of recorded aspect ratios ($\bar{a}_{95}$). The $\bar{a}_{95}$ aspect ratio concept was originally suggested by Wötting et al as a more accurate measure of the actual aspect ratio than the overall average [19]. This is due to the fact that imaging of the materials is performed on a planar surface which may be sectioned through any given grain at a position that does not reveal the true size and aspect ratio of that particular grain. This fact should also be noted when measuring the 'average' grain lengths and widths. The same general consistency in the materials microstructural parameters observed in Fig. 4a are also shown in Fig. 4b with the exception that the apparent aspect ratio ($\bar{a}_{95}$) increases from ~4:1 to ~5:1 at the sample surface (measured ~20μm from the surface). The observations of an increase in β' grain aspect ratio near to the sample surface of composition SN55 was confirmed by TEM (Fig. 5a,b). These observations are again indicative of a lower temperature at the sample surface during sintering, resulting in an increase in the viscosity of the intergranular glass phase and the development of higher aspect ratio β' grains, as previously observed by Wötting et al [19].

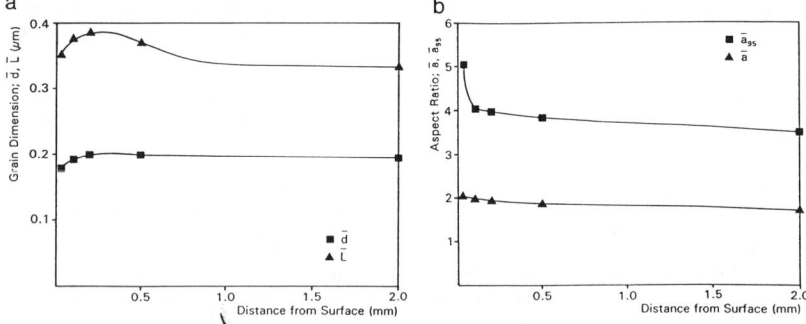

Figure 4. Measurements of a) the average grain length ($\bar{L}$) and width ($\bar{d}$) and b) the average aspect ratio ($\bar{a}$) and the average value of the highest 10% of recorded aspect ratios ($\bar{a}_{95}$) as a function of distance from the surface for material SN55.

The analysed intergranular glass compositions were generally consistent for each thin foil, although isolated regions deviating significantly from these mean values were observed. Regions of varied glass phase composition were attributed to inhomogeneous additive distribution during milling, and were occasionally associated with microstructural defects, for example entrapped liquid inclusions within β' grains. The intergranular glass phase is more uniformly distributed than is typical of conventionally sintered silicon nitride, with relatively few glass pockets. This is a consequence of the fine grain size, resulting in a large area of grain boundary capable of absorbing glass onto nanometre scale equilibrium surface layers. For example, reducing the grain length by a factor of 10 (i.e. from 4 to 0.4μm, for an aspect ratio of 2:1) results in an increase in β' surface area by a factor of 10. Therefore, approximately 3 vol.% of glass would be accounted for as a 1nm thick layer for 0.4μm grains whereas only ~0.3 vol.% glass would be contained in the same thickness layer for 4μm grains.

Figure 5.   TEM micrographs of the microstructure of microwave sintered SN55, a) ~15μm from surface and b) the bulk ceramic.

The application of a post-sinter annealing heat-treatment to the microwave sintered ceramic SN55 results in partial crystallisation of the yttrium aluminosilicate glass to β-$Y_2Si_2O_7$ (Fig. 6a). Bright/dark field imaging in the TEM indicates that each β-$Y_2Si_2O_7$ grain envelopes several β' grains (Fig. 6b), as previously noted by other researchers [20].

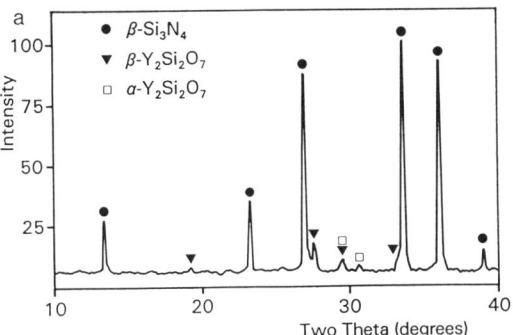 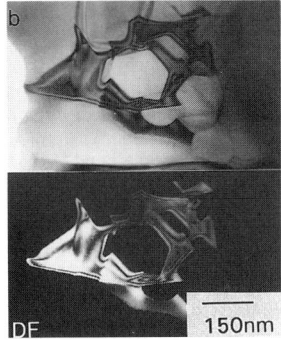

Figure 6.   a) XRD demonstrates partial devitrification of the intergranular glass to β-$Y_2Si_2O_7$. b) Bright field/dark field imaging in the TEM shows each β-$Y_2Si_2O_7$ crystal envelopes several β'-Si-Al-O-N grains.

Summary

- Si_3N_4 based ceramics, densified using a novel microwave sintering process developed by Alcan Int., have been microstructurally characterised using optical and electron microscopy.
- Microwave sintering results in the preparation of high density β'-Si-Al-O-N materials, of low substitution level with a very fine and uniform grain size, significantly finer than conventionally processed commercial silicon nitride based materials of similar composition.
- Typically these as-sintered materials exhibit a 'reverse' density gradient, with the lowest density at the sample surface due to radiant heat losses during sintering.

- Significant heterogeneous grain growth occurred in the microwave sintered ceramics during post-sinter HIPing at 1800°C.
- All ceramics studied possessed an amorphous intergranular aluminosilicate phase after sintering, which could be partially devitrified via the use of a post-sinter annealing treatment.

Acknowledgements

The authors would like to thank Drs. Mark Patterson and Prasad Apte (now at Sherritt Gordon, Alberta) of Alcan International, Kingston, Ontario, for providing the microwave sintered samples and Dr. Pankaj Mehrotra and Mr. William Huston, of Kennametal Inc., Latrobe, PA, for plasma etching.

References

1. T.N. Tiegs, J.O. Kiggans and H.D. Kimrey, *Ceram. Eng. Sci. Proc.*, **12** 1981-1992 [9-10] (1991).
2. T.N. Tiegs, J.O. Kiggans and H.D. Kimrey, pp. 267-272 in *Microwave Processing of Materials II*, Materials Research Society, Pittsburgh, PA, 1990.
3. M.C.L. Patterson, P.S. Apte, R.M. Kimber and R. Roy, pp. 33-36 in *Proc. Int. Conf. on Microwaves and High Frequency*, Nice, France, 1991.
4. M.C.L. Patterson, P.S. Apte, R.M. Kimber and R. Roy, pp. 301-310 in *Microwave Processing of Materials III*, Materials Research Society, Pittsburgh, PA, 1993.
5. M.C.L. Patterson, P.S. Apte, R.M. Kimber and R. Roy, pp. 291-300 in *Microwave Processing of Materials III*, Materials Research Society, Pittsburgh, PA, 1993.
6. T.T. Meek, R.D. Blake and J.J. Petrovic, *Ceram. Eng. Sci. Proc.*, 8 [7-8] 861-871 (1987).
7. M.A. Janney and H.D. Kimrey, pp. 919-924 in *Ceramic Powder Science II*, American Ceramic Society, Westerville, OH, 1988.
8. P.S. Apte, R.M. Kimber and M.C.L. Patterson, pp.167-174 in *Structural Ceramics Processing, Microstructure and Properties*, Riso National Laboratory, Roskilde, Denmark, 1990.
9. J.D. Katz, R.D. Blake, J.J. Petrovic and H. Sheinberg, pp. 219-226 in *Microwave Processing of Materials I*, Materials Research Society, Pittsburgh, PA, 1988.
10. C.E. Holcombe, T.T. Meek and N.L. Dykes, pp.227-234 in *Microwave Processing of Materials I*, Materials Research Society, Pittsburgh, PA, 1988.
11. P.S. Apte, R.M. Kimber, M.C.L. Patterson, R.Y. Roy, D.N. Mitchell, *International Patent Application No. WO 91/05747* (2nd May 1991).
12. G. Zeigler and G. Wötting, *Int. J. High Tech. Ceram.*, 1 31- (1985).
13. K.P. Plucknett and M.H. Lewis, *Ceram. Eng, Sci. Proc.*, 12 [7-8] 1327-1344 (1991).
14. I-W. Chen and S.-L. Hwang, *J. Am. Ceram. Soc.*, 75 [5] 1073-1079 (1992).
15. X. Wu and I-W. Chen, *J. Am. Ceram. Soc.*, 75 [10] 2733-2741 (1992).
16. Y. Oyama and O. Kamigaito, *Japan. J. Appl. Phys.*, 10 [11] 1637 (1971).
17. K.P. Plucknett and D.S. Wilkinson, Presentation No. 13-SX-92 at the 94th American Ceramic Society Annual Meeting, Minneapolis, MN, April 1992.
18. M.C.L. Patterson, Personal Communication.
19. G. Wötting, B. Kanka and G. Ziegler, pp. 83-96 in *Non-Oxide Technical and Engineering Ceramics*, Elsevier, London, England, 1986.
20. D.A. Bonnell, pp. 877-882 in *Ceramic Microstructures '86: Role of Interfaces*, Materials Science Research, Vol. 21, Plenum Press, New York, 1987.

Strength Optimization Through Powder Modification*

ARVID E. PASTO[†], F. AVELLA[††], S. NATANSOHN[††], AND W. J. ROURKE[†††]
[†]Oak Ridge National Laboratory, Oak Ridge, TN 37831
[††]GTE Laboratories Incorporated, MA 02254
[†††]Duracell Worldwide Technology Center, Needham, MA 02194

ABSTRACT

The effect of oxygen content of silicon nitride powders on the properties of resulting ceramics was studied by physically and chemically treating the powder to modify its surface oxygen content. These powders were compounded with yttria and hot-pressed into dense ceramics. Strength and oxidation resistance of these ceramics were measured and correlated with the powder and ceramic compositions as well as the resulting intergranular phases. Results showed that the phases varied with slight differences in the initial powder oxygen content as predicted, and that strength could be correlated to initial oxygen concentration. Best results were obtained when the oxygen content was increased by thermal oxidation. A Taguchi Methods experimental study designed to optimize the thermal treatment resulted in silicon nitride ceramics with strength improvements of 22 and 37% at ambient temperature and 1370°C, respectively. Oxidation resistance was also improved.

INTRODUCTION

Recent studies[1-3] have shown that the surface chemical and physical characteristics of silicon nitride powder can be readily modified, and that silicon nitride ceramics fabricated from these powders have significantly different properties. A study [2] of means to alter surface oxygen content showed conclusively that thermal oxidation of the surface of the powder was superior to chemical means of attaining the same end, as strength at both ambient and elevated temperature of a silicon nitride - 6 w/o yttria (PY6) ceramic was improved. This work was extended [3] to provide an explanation of the strength increase, through the development of specific phases in the silicon nitride-silica-yttria-yttrium nitride pseudoquaternary phase diagram. A significant portion of the latter work was the performance of a large Taguchi Methods experiment on the thermal oxidation of silicon nitride powder. This paper describes the experiment in detail.

The Taguchi Method [4,5] for parameter design was utilized for this study. This method of process optimization relies on determination of process control factors such that the process yields a product whose quality characteristic (response factor) is optimized and has minimal sensitivity to "noise". Noise is defined as undesirable effects on the process or product caused by process factors which one cannot or does not want to control. In the present case, the authors attempted to maximize the following response factors: density achieved by hot-pressing, strength of the silicon nitride ceramic at ambient and elevated temperature, and oxidation resistance. Noise factors must be controlled during the experimentation, and for this case, only one was chosen: lot-to-lot powder variation. The Taguchi method utilizes fractional factorial arrays for assigning experimental values to the control and noise factors. Control factors included process parameters such as temperature of the oxidation furnace, time of oxidation, etc. A set of experiments was defined by the matrices, and the resulting data were analyzed by analysis of variance techniques. A statistical measure of performance called "signal-to-noise ratio" was calculated for each control factor to provide optimal control factor settings for each response factor.

EXPERIMENTAL PROCEDURE AND RESULTS

CHARACTERIZATION OF SILICON NITRIDE POWDERS:
The silicon nitride powder was a high purity, high alpha-phase content material (manufactured by Ube Industries, Ltd.,Tokyo, Japan; designated as SN-ESP) selected for its ability to yield high strength ceramics. Detailed characterization has been described elsewhere [1,2]. Some of the data

---

*Research sponsored by the U. S. Department of Energy, Assistant Secretary for Conservation and Renewable Energy, Office of Transportation Technologies, as part of the Ceramic Technology Project of the Materials Development Program under contract DE-AC05-84OR21400 with Martin-Marietta Energy Systems, Inc.

Table 1. Properties of silicon nitride powders used in this investigation and of the ceramics made therefrom.

Lot No.	1	2
POWDER PROPERTIES:		
Phases by x-ray diffraction	Major alpha phase; minor other(likely beta).	
Surface area (sq.m/g)	7.54±0.25	7.49±0.23
Mean particle size [d50 (μm)]	0.58±0.01	0.52±0.02
Total oxygen content (w/o)	1.14±0.03	1.04±0.02
CERAMIC PROPERTIES:		
MOR (MPa) at 25°C:		
Billet 1	1020±100	845±160
Billet 2	1050±60	
MOR (MPa) at 1370°C:		
Billet 1	550±30	500±10
Billet 2	560±50	
Oxidation resistance (Δw/A in g/m2) after 500 hr at		
1200°C	1.4	1.5
1370°C	4.0	3.5

are listed in Table 1 for the two lots used in the experiment described here. The oxygen content of these two lots differs a small but significant amount.

Relevant results of the chemical characterization are:

1. the powders contain very low levels of cationic impurities ($\leq$ 100 ppm total),
2. anionic impurities are principally oxygen (1.1 w/o), carbon (0.2 w/o), chlorine (<100 ppm), and fluorine (4 ppm), and
3. 30 - 36% of the oxygen, and virtually all of the carbon and fluorine are located on the silicon nitride particles' surfaces.

The flexural strength of the hot-pressed PY6 ceramics was measured in an attempt to establish correlations between powder characteristics and ceramic properties. Modulus-of-rupture (MOR) values resulting from four-point bend testing are summarized in Table 1. The oxidation resistance of these materials was also examined (see Table 1, or Reference 1 for more details).

The yttria used was Molycorp Grade 1600, and its characterization results have been described previously [1,2]. The material was of high purity (<200 ppm impurity cations) and had a surface area of about 11 sq. m/g.

POWDER OXIDATION:

The silicon nitride or milled silicon nitride/yttria powders were loaded into silica trays to a given bed depth, and subjected to heat-treatment in a tube furnace under a specific flow rate of high purity air or oxygen. The powder was heated at 300°C/h to the final selected temperature, maintained at this temperature ±10°C for the selected time, and then furnace cooled at the inertial rate.

CERAMIC FABRICATION:

The powder was dry-milled in a polyethylene container with silicon nitride milling media using 0.5 w/o stearic acid as a milling aid. Three different procedures for the mixing of the yttria with the silicon nitride, and the milling of these materials, were pursued. These procedures are described in the section describing the Taguchi experiment variables below. The resulting powders were heated in air for 4 h at 600°C to effect removal of the stearic acid, screened, and then hot-pressed under static argon at atmospheric pressure. Hot-pressing was accomplished in a graphite die internally coated with BN powder to minimize carbon contamination. A pressure of 34 MPa (5000 psi) was applied for 4 h at 1725°C.

CERAMIC CHARACTERIZATION:
Flexural test specimens were sliced from the densified billet and ground with 325 grit diamond wheels to yield specimens of 1.27 x2.54 x 25.4 mm dimension. The four lengthwise edges were chamfered 0.25 mm at 45°. The specimens were broken in a universal test machine using four-point flexure fixtures with 10.16 mm inner and 22.86 mm outer spans. The fixtures were fabricated from graphite, with SiC load pins, allowing testing at elevated temperatures under argon atmosphere. The flexure strength was determined at room temperature and 1370°C at a crosshead speed of 0.51 mm/min.

Oxidation resistance tests were conducted at 1200 and 1370°C on the polished specimens prepared from bars broken during flexural strength testing. The specimens were placed on thin platinum wires to allow air access to the total surface area of the samples. They were heated in ambient air atmosphere and their weight was monitored at regular time intervals to an accuracy of ± 2µg throughout an exposure period of 500 h.

TAGUCHI METHODS EXPERIMENT ON THERMAL OXIDATION:
Matrix, Factors, and Levels
An L_{18} ($2^1 \times 3^7$) orthogonal array was chosen for assigning a range of control factor values (levels) to a matrix of eighteen experiments. Originally capable of accommodating one factor at two levels and seven factors at 3 levels, this array was modified to accept one factor at six levels, one factor at two levels and five factors at three levels with orthogonality maintained. Figure 1 illustrates the matrix of eighteen experiments that was performed with each of two powder lots acting as noise factor levels. A description of the factors and levels along with designations used in later discussions and figures is given as follows:

A. Temperature of powder treatment (levels #1 - 6): 800, 900, 950, 1000, 1050 and 1100°C.
B. Time at temperature (3 levels): A sliding scale was used, with 2, 6 and 12 h for the 800 and 900°C experiments; 2, 4 and 6 h for the 950, 1000 and 1050°C experiments; and 1, 2, and 4 h for the 1100°C treatment.
C. Powder bed depth (levels #1 - 3): 12, 3 and 24 mm.
D. Three different powder processing sequences were used: level #1 in which the powders were first milled with the yttria sintering aid and then treated thermally (designated MO); level #2 in which the powders were first milled for half of the intended milling time without yttria, then treated thermally and finally milled with yttria for the remainder of the milling schedule (MOM); and level #3 in which the powder was first heated and then milled with yttria (OM).
E. The total milling times had levels #1 - 3 set at 18, 36 and 72 h; milling times were split into halves for the middle processing sequence (MOM) discussed in factor D above.
F. The atmosphere was either air (level #1) or oxygen (level #2).
G. Gas flow (levels # 1 - 3): 0 (static atmosphere), 10 and 80 ml/min were used.
N. The noise factor was the silicon nitride powder and had two levels: Lot #1 and Lot #2.

RESULTS:
The billets produced in the 36 experiments were evaluated for density, strength, and oxidation resistance, and the raw data for the eighteen billets of Lot #1 are listed in Table 2. The density values are those for the hot-pressed billets, while ambient strength is an average of 10 test values and elevated temperature strength is the average of 5 tests per billet. Oxidation resistance is an average result from two specimens per billet.

These data were subjected to analysis of variance; level averages for the raw data and for signal-to-noise ratios for strength are plotted in Fig. 2. This figure depicts the main effects of the control parameters on the quality characteristics of the silicon nitride ceramics. Signal-to-noise ratios logarithmically combine the mean effect of a factor level with its variability to produce a figure of merit for the level. They were calculated to indicate a *larger-the-better* condition for density and strength and a *smaller-the better* condition for oxidative weight gain [4]. Each data point in Figure 2 represents an average of several strength values, because each parameter level is included in multiple experiments with two different powders. In the case of heat-treatment temperature, for example, each point in the Figure represents the average of 60 data points at room temperature and 30 at 1370°C.

Exp. No.	A Temp. (°C)	B Time (hr)	C Bed Dep. (mm)	D Process	E Mill Time (hr)	F Atmos.	G Flow Rate (ml/min)	Powder Lot 1	Powder Lot 2
1	800	2	12	1	18	1	0		
2	800	6	3	2	36	2	10		
3	800	12	24	3	72	1	80		
4	900	2	12	2	36	1	80		
5	900	6	3	3	72	1	0		
6	900	12	24	1	18	2	10		
7	950	2	24	2	72	1	10		
8	950	4	12	3	18	2	80		
9	950	6	3	1	36	1	0		
10	1000	2	3	1	72	1	80		
11	1000	4	24	2	18	1	0		
12	1000	6	12	3	36	1	10		
13	1050	2	24	3	36	2	0		
14	1050	4	12	1	72	1	10		
15	1050	6	3	2	18	1	80		
16	1100	1	3	3	18	1	10		
17	1100	2	24	1	36	1	80		
18	1100	4	12	2	72	2	0		

Figure 1. The Taguchi Methods experimental matrix for powder oxidation.

Table 2. Results from the Taguchi Methods experiments for powder Lot #1.

Exp.	Density (% Theo.)	MOR(MPa) at 1200°C	MOR(MPa) at 1370°C	500 Hour Weight Gain in air(g/sq. m) at 25°C	500 Hour Weight Gain in air(g/sq. m) at 1370°C
1	99.5	1021	675	1.12	1.98
2	99.2	1000	692	0.86	1.67
3	99.7	999	735	0.62	1.48
4	99.3	987	727	0.81	1.67
5	99.7	1145	739	0.7	1.4
6	99.3	1102	745	0.76	1.4
7	99.8	1157	734	0.72	1.73
8	99.7	1068	691	0.73	1.7
9	99.0	1134	700	0.53	1.52
10	97.7	1088	730	0.97	1.29
11	99.6	1015	768	0.51	1.15
12	99.5	1138	700	0.66	1.38
13	99.4	1125	722	0.57	1.3
14	92.8	431	338	0.44	0.98
15	98.8	1064	734	0.59	1.42
16	99.0	1129	769	0.56	1.22
17	92.4	608	524	0.49	1.24
18	97.5	1136	715	0.57	1.51

Density

The data indicate that powders heat-treated at temperatures above 1000°C do not consolidate to high densities. Highest densities are realized when the powder is treated in the shallowest bed, a plausible effect because the gas can more easily and thoroughly access the powder particles. Heat treatment followed by milling with the sintering aid is the powder processing route most effective in producing dense ceramics. Short milling times are beneficial as is the use of static oxygen. There is no discernible effect on the hot-pressed density related to the two powder lots.

Flexural Strength

The main effects of control and noise factors on flexural srength at room temperature and 1370°C are shown in the upper half of Fig. 2, while signal-to-noise ratios (S/N) are plotted in the lower half. The observed trends are similar, but not identical, to the effects on density. The best room temperature flexural strength is found in ceramics made from powders heat-treated at 950°C. Heat-treatment at lower temperatures, particularly at 800°C, results in ceramics with lower strength even though their density is high. The effects of treatment time and bed depth on flexural strength are identical to those observed for density, where the poorest result was obtained for the middle time level. The second and third type of powder processing routes (MOM and OM) give equivalent strength and are superior to the first type (MO). Powders processed with the shortest milling times and heated in static oxygen result in the strongest ceramics as well as the densest ones. For strength, the difference between the two powder lots is also negligible.

The factor effects on flexural strength at 1370°C are essentially the same as those at ambient temperature except for the effect of the treatment temperature factor. Here the best high temperature flexural strength is realized in ceramics made from powders heated at 1000°C, not at 950°C which was best for room temperature strength.

Oxidation Resistance

The effects of the control factors on the oxidative weight gain of the PY6 ceramics at 1200 and 1370°C are small except for temperature of powder heat-treatment. The data show that oxidation resistance generally improves with increasing heat treatment temperature, being the highest at 1000 to 1100°C, as well as with increasing treatment time and bed depth. Other factor effects are insignificant.

Predicted Optimization

The Taguchi Method allows one to estimate the value of a quality characteristic that could be obtained using the optimum factor levels determined in the experimental study [4]. The optimum factor levels for strength extracted from this study were A_4, B_3, C_2, D_3, E_1, F_2, and G_1, a combination not evaluated in the experimental series (see above discussion for factor and level designations.) The predictive calculation assumes that control factor effects are additive and utilizes only those factors having the strongest impact, in this case A_4, B_3, and D_3. The choice of level A_4 (1000°C) was a compromise between it and level A_3 (950°C) in considering the highest MOR values for low and high temperature in Figure 3. The results of the prediction were MOR values of 1250 MPa at ambient temperature and 860 MPa at 1370°C (Table 3.)

Confirmation Tests

Confirmation experiments were performed using powder Lot #2. In reviewing the optimum factor levels for strength to be used in the confirmation tests, temperature levels A_3 and A_4 (950 and 1000°C) were similar as were time levels B_1 and B_3 (2 and 6 hrs), and it was decided to run an experiment with each of those levels, as well as the intermediate time level B_2 (4 h).

The results of these tests, given in Table 3, confirmed the beneficial effects of powder treatment, in that a significant improvement over baseline materials was obtained in these hot-pressed ceramics. The optimal materials exhibited a flexural strength of about 1100 MPa at room temperature and 770 MPa at 1370°C. The specimens also exhibited a better oxidation resistance (data not shown): they had about half the weight gain at either 1200 or 1370°C of that observed on baseline materials.

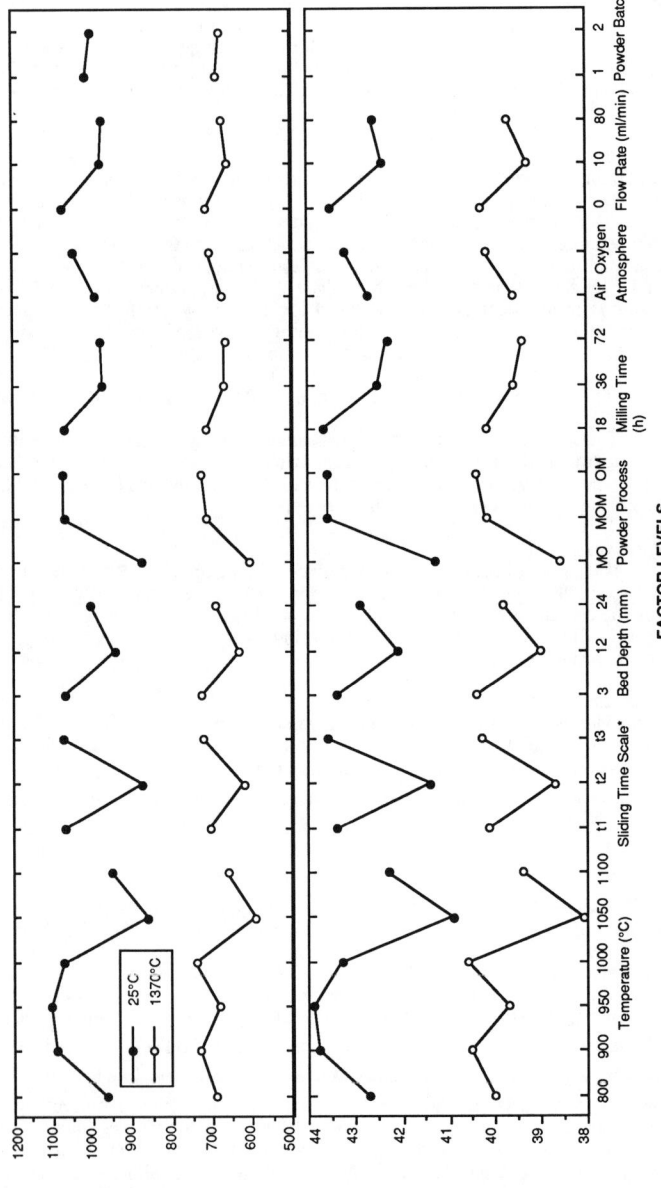

Figure 2. Raw data (top) and signal-to-noise analysis (bottom) of strength results.

Table 3. Comparison of results obtained in the Taguchi Methods experiment with predicted MOR values and the results of confirmation experiments

Description	Run No.	Control Factor Settings A B C D E F G	MOR Values (MPa)(2) 25°C	MOR Values (MPa)(2) 1370°C	Oxygen Content(3)
Baseline [Lot #1]	--	------------------------	1035	555	11.4
Maximum [Lot #1]	7	3 1 3 2 3 1 2	1157	734	18.3
	16	6 1 2 3 1 1 2	1129	769	28.4
Maximum [Lot #2]	7	3 1 3 2 3 1 2	1186	714	17.2
	5	2 3 2 3 3 1 1	1070	772	16.1
Prediction(4)	----	4 3 2 3 1 2 1	1250	860	-----
Confirmation tests [Lot #2]	CR1	4 1 2 3 1 2 1	1090	770	21.0
	CR2	4 2 2 3 1 2 1	1090	735	22.9
	CR3	4 3 2 3 1 2 1	1060	685	28.9
	CR4	3 1 2 3 1 2 1	940	700	15.6
	CR5	3 2 2 3 1 2 1	995	690	18.1
	CR6	3 3 2 3 1 2 1	1060	720	19.2

(1) Letters and numbers refer to factor designations and levels in Figure 1 and text.
(2) Maximum values are underlined.
(3) Expressed as mg/g in the powder.
(4) Factor levels A4, B3, and D3 used in calculation.

However, in none of these confirmation tests were the maximum predicted ambient strength values attained, or even approached. This implies that the underlying assumptions of the analysis, such as additivity and linearity of the parametric factor effects, may not be entirely valid.

DISCUSSION OF RESULTS

The Taguchi Methods experiment showed strong correlation between strength at ambient and elevated temperatures and oxygen content, with strength increasing up to a powder oxygen level of about 18 mg/g (1.8 w/o). The results in Table 3 illustrate this strength improvement. Fig. 3 plots the strength data as a function of the oxygen content (expressed now as equivalent percent) over the entire range of oxygen contents examined. The tie line between Si_3N_4-$Y_2Si_2O_7$ is taken from the relevant Si_3N_4-SiO_2-Y_2O_3-YN phase diagram [3]. It can be seen that optimum strengths are attained in PY6 both at ambient and at elevated temperatures when the oxygen content of the silicon nitride powder is manipulated to attain a value consistent with development of *only* the two phases represented on the tie line. An Y/Si ratio near unity was measured [3] for the intergranular phase of the highest strength materials produced in the Taguchi Methods experiments. Higher and lower oxygen levels caused development of phases other than silicon nitride and yttrium silicate, with the result that strength diminished. The slope of the strength-oxygen content curve is such that it is far better to err on the high oxygen content side, if strength is the main objective.

SUMMARY AND CONCLUSIONS

Ambient and elevated temperature strength as well as oxidation resistance of PY6 silicon nitride ceramics are maximized when the oxygen content, including the oxygen present on the

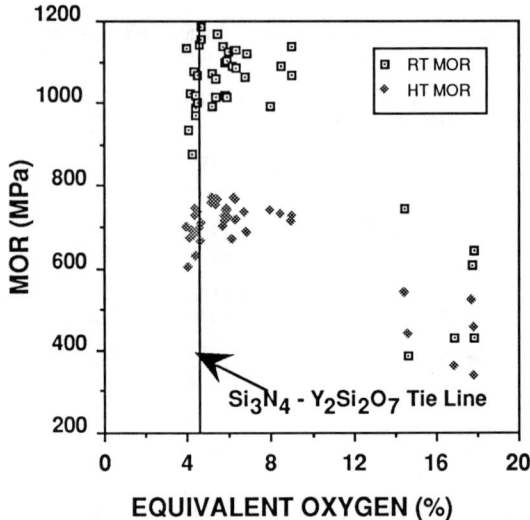

Figure 3. Ambient and elevated temperature strength as a function of composition in the system silicon nitride-6 w/o yttria.

surface of the Si_3N_4 powder, is oxidatively adjusted to result in a product whose composition lies on the tie-line between Si_3N_4 and $Y_2Si_2O_7$. A Taguchi Methods experiment was utilized to optimize the oxidative treatment process, with the result that ambient temperature strength was increased by 22%, and elevated temperature strength by 37%. Use of Taguchi Methods allowed relatively quick investigation of a large number of process control factors through utilization of a designed experiment in a fractional factorial array.

ACKNOWLEDGEMENT

The authors are indebted to L. Fitzpatrick and L. Schlafer (GTE Laboratories) for their painstaking assistance in sample preparation and testing.

REFERENCES

1. A. E. Pasto and S. Natansohn, "Development of Improved Processing Methods for High Reliability Structural Ceramics for Advanced Heat Engines", ORNL/Sub/89-SD548/1, July, 1992.

2. S. Natansohn and A. E. Pasto, "Improved Processing Methods for Silicon Nitride Ceramics", Presented at the Int'l. Gas Turbine and Aeroengine Congress and Exposition, Orlando, FL, June 3-6, 1991, ASME Paper 91-GT-316.

3. S. Natansohn, A. E. Pasto, and W. J. Rourke, "Effect of Surface Modifications on the Properties of Silicon Nitride Ceramics," presented at the Ann. Mtg., Am. Ceram. Soc., Minneapolis, MN 4/13-16/92; to be published in the J. Am. Ceram. Soc., 1993.

4. D. M. Byrne and S. Taguchi," The Taguchi Approach to Parameter Design", 40th Annual Quality Congress Transactions, American Society for Quality Control, 1987.

5. P. T. Jessup, "The Value of Continuing Improvement", Proceedings of the International Communications Conference, "ICC 85", Institute of Electrical and Electronics Engineers, 1985.

INFILTRATION/PYROLYSIS PROCESSING OF FIBER-REINFORCED SILICON NITRIDE

STUART T. SCHWAB, RENEE C. GRAEF, CHERYL R. BLANCHARD, YI-MING PAN, AND DAVID L. DAVIDSON
Engineering & Materials Sciences Division, Southwest Research Institute, 6220 Culebra Road, San Antonio, Texas 78238-5166

ABSTRACT

While its high-temperature strength, resistance to oxidation, and other properties make silicon nitride an attractive candidate for many advanced structural applications, its propensity for brittle failure has hindered its widespread adoption. One approach to avoiding brittle failure is through incorporation of continuous fiber-reinforcement; however, conventional (powder-based) methods of silicon nitride fabrication can degrade fibers and are not amenable to the production of complex shapes. The Southwest Research Institute has developed a number of polymeric precursors to silicon nitride which are available as thermosetting liquids, and we have shown that these materials can be used in combination with near net-shape manufacturing techniques to produce fiber-reinforced silicon nitride composites. Mechanical property tests conducted at room temperature suggest that these polymer-derived composites exhibit fracture behavior comparable to those produced through conventional techniques; micromechanical investigations conducted at 800°C indicate that non-brittle failure is maintained at elevated temperature.

INTRODUCTION

Chemical methods of producing advanced ceramics are of considerable interest because traditional, powder-based processing techniques appear incapable of producing components that can satisfy advanced performance requirements in a cost efficient manner. Because of its strength-to-weight ratio, oxidation resistance, low thermal expansion, and thermal shock resistance, silicon nitride (Si_3N_4) is a particularly attractive material for high-temperature applications, such as in the hot zone of jet or internal combustion engines. Despite its high potential to improve the performance of power systems, Si_3N_4 brittle fracture behavior has hampered its utilization. One approach to alleviating brittle fracture behavior is through the incorporation of high-strength fibers; however, Si_3N_4 powders are abrasive, and so traditional techniques degrade fiber properties during processing. In addition, traditional methods are not amenable to the production of complex parts in near net-shape.

Polysilazanes have been shown to be effective precursors to silicon ceramics [1-3]. Although currently used in the production of Si-C-N [4] and Si_3N_4 [5] fibers, many other potential applications of preceramic polysilazanes remain largely unexplored. Expanding on chemistry reported by Stock and Somieski in the early part of this century [6], Southwest Research Institute (SwRI) has developed a particularly useful family of polymeric precursors to Si_3N_4. These preceramic polysilazanes are obtained as waxes or low-viscosity, thermosetting liquids which exhibit very high ceramic yields ($\geq 85\%$ by weight) and produce carbon-free Si_3N_4. Although originally developed as binders for Si_3N_4 powder processing [7], these materials have also demonstrated their utility as coating precursors for refractory composites [8]. Their low viscosity and high ceramic yield indicated that they might prove to be useful matrix precursors for the manufacture of fiber-reinforced Si_3N_4 through techniques similar to those currently used to manufacture carbon-carbon composites [9]. We report here the results of our preliminary evaluations of the utility of polysilazanes in the manufacture of fiber-reinforced Si_3N_4.

EXPERIMENTAL

General

Unless otherwise noted, all manipulations of uncured polysilazanes were carried out under anhydrous and anaerobic conditions using common synthetic techniques in combination with an inert atmosphere/vacuum manifold system or an argon-filled drybox (Vacuum Atmospheres HE-43-2 with HE-493 Dri-train) [10]. All NMR experiments were conducted at the NMR center at Colorado State University. Samples were handled under strictly anhydrous and anaerobic conditions. ^{29}Si and ^{1}H chemical shifts are reported relative to TMS; ^{15}N chemical shifts are reported relative to ammonia.

The polymer synthesis [11] and composite fabrication [12] have been described elsewhere. Briefly, composite fabrication was accomplished by saturating Nicalon fabric (8-HS weave, 3K tow) with neat polysilazane, stacking the pre-cut plies, enclosing them in a double-sided vacuum bag, and curing the assembly under applied pressure in an autoclave. The cured article was removed from the vacuum bag and pyrolyzed under flowing anhydrous nitrogen. As Nicalon fiber is known to degrade at elevated temperature, the maximum pyrolysis temperature was limited to 1000°C. Reinfiltration was accomplished by processing in a vacuum bag with additional matrix precursor surrounding the composite. Curing and pyrolysis procedures were the same as those described for the initial lay-up.

Results & Discussion

To gain insight into the nature of the polysilazane-derived Si_3N_4 and to develop an appropriate firing procedure for the composite, the polymer-to-ceramic conversion process was examined by infra red (DRIFT), ^{1}H (CRAMPS), and ^{29}Si MAS-NMR spectroscopy. The results of the DRIFT experiments are presented as Figure 1, where it can be seen that, despite weight loss as measured by thermogravimetric analysis (TGA) being complete by 800°C, substantial amounts of Si-H and N-H functions are detected (at ca 3380 and ca 2180 cm^{-1}, respectively) in the char products obtained as high as 1200°C. Both the 1400°C and 1700°C char spectra appear devoid of these "organic" functions and are essentially identical to spectra obtained from commercial Si_3N_4 powders.

The results of the CRAMPS investigation are presented in Figure 2, where substantial Si-H and N-H functions (δ ~5.0 and δ ~0.5, respectively) are again observed in the char products obtained at temperatures as high as 1200°C. The char material produced at 1000°C exhibits a substantially smaller N-H resonance than the char material produced at 800°C, and shows two Si-H magnetic environments. While we believe that the bifurcation of the resonance at δ 5.0 reflects an increase in the SiH/SiH$_2$ ratio, the presence of a substantial radical population in the char materials, as determined by electron spin resonance (ESR), complicates peak assignment. Regardless of the exact peak assignments, difficulty was encountered in obtaining a spectrum from the 1200°C char material because of the very small proton population, and a spectrum was not obtained from the 1400°C char material despite extended acquisition times.

The ^{29}Si MAS-NMR spectra of the polysilazane char products (Figure 3) display a major resonance at ca δ -48, which arises as a composite of the signal from the α- and β-Si_3N_4 phases present [13]. In the spectra of the 800°C and 1000°C char products, this resonance is broadened by contributions at ca δ -33 and ca δ -21, which we believe arise from residual SiH$_2$ and SiH functions. The intensity of these signals has decreased substantially in the char product obtained at 1200°C, which also exhibits what could be a very weak resonance at ca δ -80. In the 1400°C char product, a new resonance is observed at δ -80.6, which is consistent with the presence of elemental silicon. Elemental silicon was not detected in the 1700°C char product, which displays only a single, sharp resonance at δ -48.2. This spectrum is consistent with that expected from a material with a phase composition of approximately 60% α-Si_3N_4 and 40% β-Si_3N_4.[13]

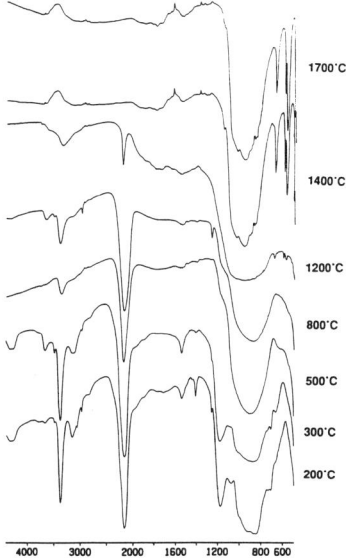

Figure 1. DRIFT spectra of polysilazane fired at indicated temperatures.

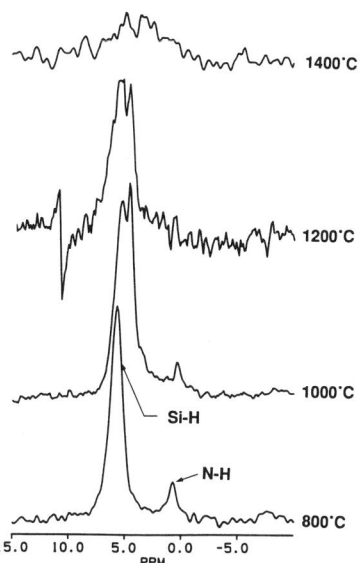

Figure 2. Solid State ^{1}H (CRAMPS) spectra of polysilazane fired at indicated temperatures.

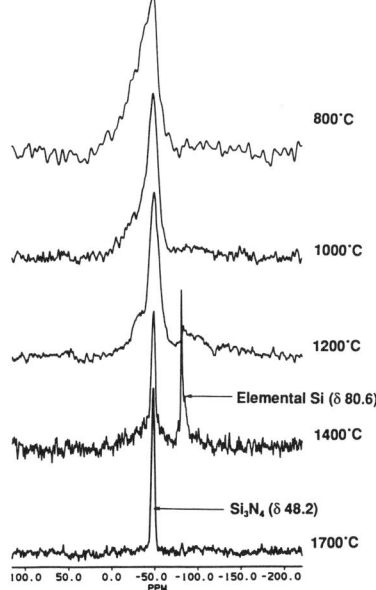

Figure 3. Phase composition of polysilazane-derived ceramics.

Powder X-ray Diffraction (XRD) analysis was used to further characterize the phase composition of the polymer-derived ceramic. As seen in Figure 4, the chars produced at temperatures below 1400°C are essentially amorphous, with some elemental silicon detected in the 1200°C product. Although the ^{29}Si NMR analysis suggests that the majority of the silicon present in the char products obtained at 800°C and above is in the magnetic environment expected for Si_3N_4, it appears that sufficient chemical functions (mainly Si-H) remain in the products obtained at temperatures below 1400°C for the crystal lattice to be sufficiently distorted for Bragg's Law not to be satisfied. At temperatures above 1200°C, sufficient energy is present for the incipient crystallites to shed the residual chemical functionality, and the polymer-derived material appears to be fully crystalline by 1400°C. Elemental silicon is a substantial constituent of the material produced at 1400°C; however, as indicated by the ^{29}Si NMR analysis and confirmed by the XRD analysis, the 1700°C char product is nearly devoid of free silicon.

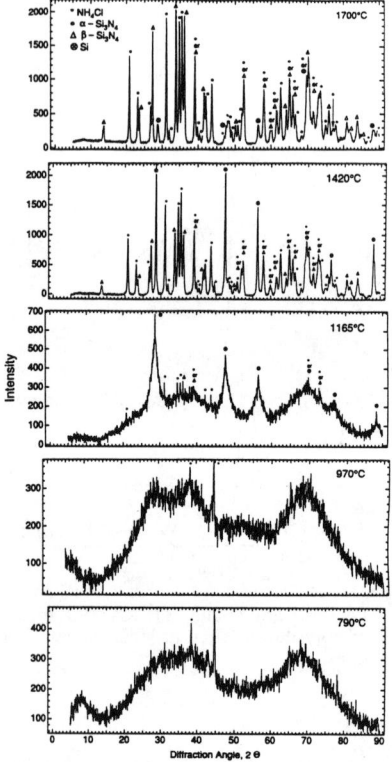

Figure 4. X-Ray Diffraction Patterns Obtained from Polysilazanes Fired to Indicated Temperatures.

Detailed analysis of the XRD data [14] reveals the phase composition of the char material varies with firing temperature and atmosphere, as depicted in Figure 5. This analysis indicates that the amount of $\beta\text{-Si}_3\text{N}_4$ increases with decreasing free silicon, while the percentage of $\alpha\text{-Si}_3\text{N}_4$ remains essentially unchanged. When the pyrolysis is carried out under an ammonia atmosphere, the consumption of free silicon is accelerated. The char material produced at 1400°C exhibits nearly the same phase composition as that produced at 1700°C under nitrogen. While extended firing times at 1400°C under ammonia could yield a stoichiometric, fully crystalline Si_3N_4 matrix, the Nicalon fiber is known to degrade at temperatures greater than roughly 1000°C, thus composite firing temperatures were limited to 1000°C.

Composite Fabrication & Properties

Initial attempts at composite fabrication were hampered by excessive loss of the matrix precursor to the surrounding bleeder cloth while under pressure in the autoclave. In an effort to improve the process conditions for optimum fabric infiltration, several rheological studies were conducted. If one defines the "cross-linking" temperature as that at which the viscosity and elastic modulus intersect, the heating rate was found to have an effect on the cross-linking temperature. A heating rate of 2°C min^{-1} yields a cross-linked article at roughly 134°C, while a heating rate of 10°C min^{-1} yields a cross-linked article at roughly

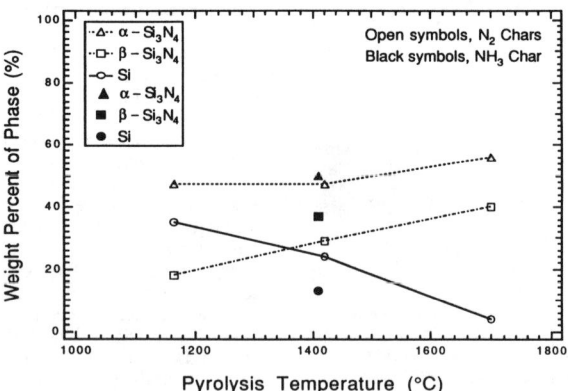

Figure 5. Phase composition of polysilazane-derived ceramics.

107°C. Incorporation of this and other rheological data into the process cycle yielded a procedure in which typically 90% or more of the applied polysilazane resin was incorporated into the composite. A composite measuring ca 2.5 in x 6 in was fabricated using graphite-coated Nicalon fabric (3K tow, 8-harness satin weave reinforcement).

After a total of eight infiltration/pyrolysis cycles, the composite was found to have a density of approximately 2.2 gcm^{-3}, an apparent porosity (ASTM C-373) of 7.3%, and a fiber volume (by SEM cross section map) of 59%. Examination of a polished composite cross section revealed reasonably dense matrix within the fiber bundles (Figure 6), while examination of a fracture surface (Figure 7) revealed substantial "fiber pull-out," which is believed to impart fracture toughness to composite materials [15].

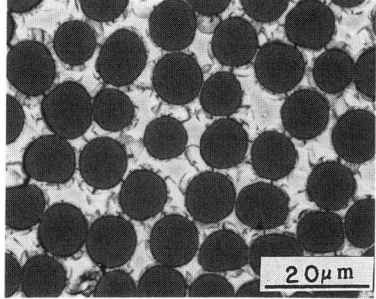 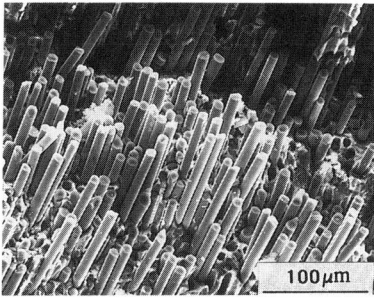

Figure 6. Polished cross section of composite fiber bundle.

Figure 7. Composite fracture surface.

Three-point bend tests were conducted to determine mechanical properties and to verify the non-brittle failure mode suggested by the fiber pull-out observed in the fracture surface. The fracture properties of the composite, as well as the strength and modulus, were determined through tests of notched and unnotched specimens at room temperature. Details of the mechanical test procedure are provided elsewhere [12]. These tests show the composite to exhibit a modulus (E) and strength (modulus of rupture) values of ca 100 Gpa, and ca 275 MPa, respectively. The chevron notch tests reveal K_{Ic} and Work of Fracture values of ca 33 MPam$^{-1/2}$ and ca 20 KJm^{-2}, respectively. Despite the amorphous nature of the matrix, these values fall in the high end of the range of fracture toughness values, determined using the same procedures, reported for hot-pressed SiC monofilament reinforced Si_3N_4 [16]. Tensile tests conducted at 800°C indicate that strength and non-brittle failure is maintained at elevated temperature.

CONCLUSION

Polymer infiltration/pyrolysis processing with polysilazanes provides a convenient, net-shape method of fabricating fiber-reinforced Si_3N_4 composites. The fracture behavior of polysilazane-derived composites is comparable to Si_3N_4 composites prepared by hot-pressing. If the polymer infiltration/pyrolysis processing technique can dependably produce Si_3N_4 components with a reproducible and high degree of reliability, these materials should soon find increased utilization in advanced power systems.

ACKNOWLEDGEMENTS

This work was supported by the Southwest Research Institute Internal Research Program and the Air Force Office of Scientific Research (Contract F49620-91-C-0045). The authors acknowledge with gratitude the SwRI technical support of Mr. S. Salazar, Mr. D. Weed, Mr. J.

Fey, Mr. A.E. Nicholls, and Ms. L.L. Ramon. The authors are also grateful for the assistance of the following personnel of the NMR Center at Colorado State University: B.L. Hawkins, S.F. Dec, M.F. Davis, R.Lewis, and G.E. Maciel.

REFERENCES

1. a) K.J. Wynne, "Ceramics via Polymer Pyrolysis," in *Transformation of Organometallics into Common and Exotic Materials: Design and Activation*, NATO ASI Series E - No. 141, R.M. Laine, Ed., Martinus Nijhoff: Dordrecht, 1988, pp 89-96; b) K.J. Wynne, R.W. Rice, *Ann. Rev. Mat. Sci.*, **1984**, *14*, pp 297-334; c) R.W. Rice, *Am. Ceram. Soc. Bull.*, **1983**, *62*, pp 889-892.

2. Laine, R.M.; Blum, Y.D.; Tse, D.; Glaser, R.D., "Synthetic Routes to Oligosilazanes and Polysilazanes: Polysilazane Precursors to Silicon Nitride," in *Inorganic and Organometallic Polymers*; pp 124-142.

3. a) Seyferth, D.; Wiseman, G.H.; Schwark, J.M.; Yu,Y.-F.; Poutasse, C.A., "Organosilicon Polymers as Precursors for Silicon-Containing Ceramics," in *Inorganic and Organometallic Polymers*; M. Zeldin, K.J. Wynne, and H.R. Allcock, Eds., ACS Symposium Series 360; American Chemical Society: Washington, DC, 1988, pp 143-155; b) Seyferth, D.; Wiseman, G.H., *J. Am. Cer. Soc.*, **1984**, *67*, C-132ff; c).Seyferth, D.; Wiseman, G.H., in *Ultrastructure Processing of Ceramics, Glasses and Composites*, L.L. Hench and D.R. Ulrich, Wiley, New York, 1986, Chapter 38

4. a) Lipowitz, J.; Rabe, J.A.; Carr, T.M. *Polymer Preprints*, **1987**, *28*, pp 411-414; b) Lipowitz, J.; Freeman, H.A.; Goldberg, H.A.; Chen, R.T.; Prack, E.R. *Mat. Res. Soc. Symp. Proc.*, **1986**, *73*, pp 489-494; c) Legrow, G.E.; Lim, T.F.; Lipowitz, J.; Reaoch, R.S. *ibid*, pp 553-558

5. O. Funayama, T. Isoda, H. Kaya, T. Suzuki, and Y. Tahsiro, *Polymer Preprints*, **1991**, *32*, pp. 542-454.

6. a) A. Stock and K. Somieski, *Ber. dt. Chem. Ges.*, **1919**, *52*, pp. 695-725.

7. a) S.T. Schwab, R.C. Graef, D.L. Davidson, Y.-M. Pan, *Polymer Preprints*, **1991**, *32* (Aug), pp. 556-558; b) S.T. Schwab and C.R. Blanchard-Ardid, "The Use of Organometallic Precursors to Silicon Nitride as Binders," *Mat. Res. Soc. Symp. Proc.*, **1988**, *121*, pp. 581-587

8. S.T. Schwab and R.C. Graef "Repair of Oxidation Protection Coatings on Carbon-Carbon Using Preceramic Polymers" NASA CP 3133 (1991) pp. 781-798.

9. P.R. Becker, *Cer. Bull.*, **1981**, *60*, pp. 1210-1214.

10. a) D.F. Shriver, M.A. Drezdzon, *The Manipulation of Air-Sensitive Compounds*, 2nd Edition, John Wiley: New York, 1986; b) A.L. Wayda, M.Y. Darensbourg, *Experimental Organometallic Chemistry*, ACS Symposium Series 357; American Chemical Society: Washington, DC, 1987.

11. S.T. Schwab, "Polysilazane Precursors for Silicon Nitride and Resultant Products" (patent pending).

12. S.T. Schwab, R.C. Graef, and D.L. Davidson, "Infiltration/Pyrolysis Processing of SiC Fiber-Reinforced Si_3N_4 Composites," *Proceedings of the 16th Annual Conference on Composites, Materials, and Structures* (Cocoa Beach, FL, Jan 12-15, 1992).

13. K.R. Carduner, R.O. Carter, M.E. Milberg, and G.M. Crosbie, *Anal. Chem.*, **1987**, *59*, pp. 2794-2797.

14. C.P. Gazzara and D.P. Messier, *J. Am. Cer. Soc.*, **1977**, *56*, pp. 777-780

15. a) M.D. Thouless and A. G. Evans, *Acta Metall*, **1988**, *36*, pp. 517-522; b) A. G. Evans, "High Toughness Ceramics and Ceramic Composites," in *Materials Architecture: Proceedings*, edited by J.B. Bilde-Sorensen, N. Hansen, D. J. Jensen, T. Ieffers, H. Linholt, and O.B. Peterson (Roskilde: RISO National lab, 1989), pp. 51-91.

16. Y.-M Yang, S. T. J. Chen, S. M. Jeng, R. B. Thayer, J.-F. LeCoustaouec, *J. Mater. Res.*, **1991**, *6*, pp. 1926-1936.

A NOVEL PROCESSING ROUTE FOR THE FABRICATION OF MONOLITHIC AND COMPOSITE SILICON NITRIDE

R. V. Raman and S. V. Rele,
Ceracon Inc., 1101 N. Market Blvd. Suite 9, Sacramento, CA 95834.

ABSTRACT

Current hot isostatic consolidation methodology used for the fabrication of complex-shaped Si_3N_4-based components requires the use of an expensive glass encapsulation technique and extended thermal exposure (in hours) of the specimen. An alternative consolidation approach involving the use of solid pressure transmitting media under high pressure, has enabled the consolidation of Si_3N_4 alloys without the need for glass encapsulation.

Characterization of microstructures and mechanical properties of this (MOR, fracture toughness) material has been carried out and will be presented. It has been noted that in $Si_3N_4/8\%Y_2O_3$-$4\%Al_2O_3$ composition, consolidated using this approach, a significantly larger volume fraction of α phase has been retained compared with typically observed conversion in $\alpha \Rightarrow \beta$ in hot isostatically pressed material or sintered material.

Key issues for addressing densification and microstructure control using this process are presented. This rapid consolidation approach appears to be a promising alternative to hot isostatic pressing for the fabrication of complex-shaped Si_3N_4 components.

INTRODUCTION

Because of the good combination of mechanical, thermal, and thermo-mechanical properties, silicon nitride (Si_3N_4) and composites based on this material are some of the most promising materials for high temperature applications. In particular, Si_3N_4 has high strength at high temperatures, good thermal stress resistance due to the low coefficient of thermal expansion, and relatively good resistance to oxidation compared to other high-temperature structural materials. This combination of properties can be used to increase operating temperatures of the engine, thus increasing its efficiency. Potential applications for which Si_3N_4 is under consideration are the all-ceramic gas turbine engine or in the replacement of metallic components in internal combustion engines. Moreover, other engineering applications are also under consideration, such as energy conversion systems, industrial heat exchangers, wear-resistant materials in metals processing, and as material for ball and roller bearings [1, 2, 3, 4]. The planned application in the vehicle gas diesel engines requires complex-shaped dense Si_3N_4, which is currently produced by hot-pressing, or hot-isostatic pressing of injection-molded specimens.

During hot-isostatic pressing, high pressure is applied via a gas to consolidate a powder compact or to remove residual porosity from pre-sintered materials. All starting materials that exhibit open porosity have to be encapsulated before HIPping to prevent the penetration of the highly pressurized gas into the compact. Up to now, refractory metals, such as tantalum, molybdenum, tungsten ceramics and different types of glasses have been used. Some of the problems associated with encapsulation include the fabrication of the container material, the decapsulation of complex-shaped components without causing any damage, and canning of complex-shaped components. Special encapsulation techniques have been developed, such as the application of a glass-particle envelope and subsequent hot-evacuation through the still porous and permeable glass envelope and sealing of the glass envelope, or embedding turbine blades into boron nitride powder by cold isostatic pressing. After this, the compact is encapsulated in a silica glass tube. In this case, the boron nitride powder bed acts as a pressure transmitter in avoiding stress and reaction zones [5,6].

Up to now it seems that the results of HIP densification of pre-sintered Si_3N_4 are strongly dependent on various microstructural characteristics of the sintered starting materials, e.g., density, grain size, grain morphology, and characteristics of the grain boundary. Density increase was only observed if the density of the pre-densified Si_3N_4 prior to HIPping was higher than 93% th.d.[7]. Second, the strength data after HIPping is dependent on the type of pre-sintered Si_3N_4[8]. The α/β-Si_3N_4 phase composition and the grain morphology of the pre-densified material are thought to be decisive parameters for controlling strength values. The use of nitrogen as the pressurizing gas led to higher densities and higher strength values than those obtained using an argon atmosphere because high-pressure nitrogen gas suppressed the decomposition of Si_3N_4 and gave density and strength values as a function of pressure for different temperatures. The major drawbacks of the HIP process include the need for encapsulating specimens, and inadequate control over microstructure due to exposure of the specimen to extended time at temperature. This is especially critical for the case of composite specimens, where formation of extended interfacial reaction zones needs to be avoided.

A consolidation process which has capabilities that allows for better control of microstructure and that can use preforms without encapsulation should be of significant interest. The Ceracon Process has shown such capabilities in the consolidation achieving near-net shape, full density from powdered metals. It is a simple consolidation technique that utilizes conventional forging equipment and set-up. The process is a quasi-isostatic, hot consolidation technique that utilizes a ceramic particulate material as a pressure transmitting medium instead of a gas media as used in HIPping. Pressures up to 1.24 GPa can be used and part temperatures of up to 2600 C, have been measured. The process as detailed in **Figure 1**, consists of fabrication of a green preform, part heating and grain heating, transfer to the Ceracon die, consolidation, part removal and grain recycling.

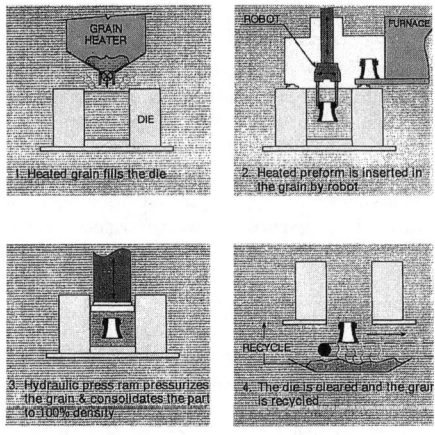

Figure 1: The Ceracon Process

The short time consolidation capability of this quasi-isostatic, hot consolidation, high pressure process, has been applied to a broad range of material systems including high temperature ceramic superconductors. The approach of using this quasi-isostatic process for consolidating Si_3N_4 based material had not been explored before. This approach, if successful, may have significant advantages over hot isostatic pressing consolidation approaches, including the use of preforms without glass encapsulation, as well as better control over matrix and interfacial microstructure.

The overall objective of this work was to demonstrate the feasibility of densifying Si_3N_4 based monolithic and composite materials using a quasi-isostatic hot consolidation process and to demonstrate the use of tailored interface to achieve preferred debonding at the Si_3N_4 to SiC interface, thus leading to novel microstructures having unique properties.

EXPERIMENTAL WORK:

In this work, initially a model matrix α Si_3N_4 composition, containing liquid phase sintering additives 4%Al_2O_3 and 8%Y_2O_3, was evaluated to obtain experimental validation of the process' expected capabilities. The alpha silicon nitride powder used in this study was procured from Ube Industries, Inc. and is of designation E-10. The binder phases Al_2O_3 and yttria were procured from Rhone Poulenc. The cold isostatic pressing technique was exclusively used for the fabrication of both monolithic and composite preform specimens.

CERACON CONSOLIDATION OF MONOLITHIC AND COMPOSITE Si_3N_4

Consolidation experiments were carried out on monolithic Si_3N_4-4% Al_2O_3-8 %Y_2O_3. Presintering of each part was carried out in a grain bed of boron nitride.

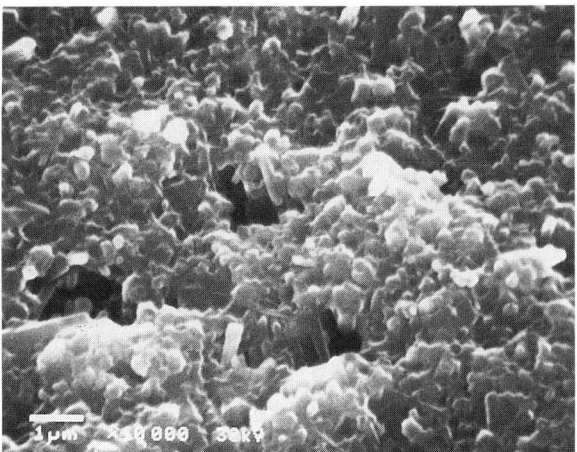

Figure 2: Ceracon Processed Si_3N_4 Showing Submicron Grain Size

The pressure transmitting media used was also varied from pure carbonaceous media to a mixture of carbonaceous and Al_2O_3 grain. The effect of pressure, consolidation temperature, and presintering mode on density achieved was evaluated. A maximum density of 98.1% was achieved using an optimum combination of part heating approach, highest pressure (1.24 GPa), and starting part temperature (1800°C).

A typical microstructure of a consolidated specimen is shown in **Figure 2**. This material consists of extremely fine submicron (< 1μm) grain size that is of the same size as the starting powder size with porosities of sizes less than the critical flaw size of 1 micron. A X-ray diffraction pattern (**Figure 3**) shows the presence of mostly alpha phase with some beta present.

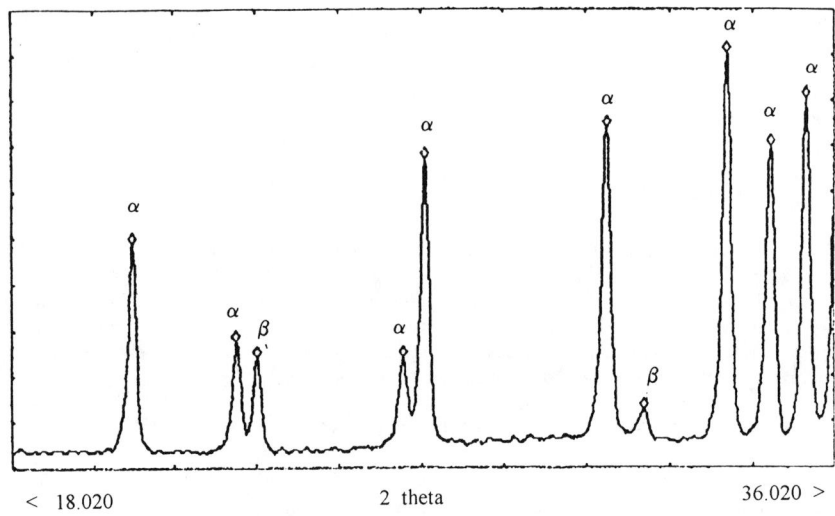

Figure 3: X-Ray Diffraction Pattern of the Ceracon Consolidated Si_3N_4

4.0 MECHANICAL PROPERTIES:

The Ceracon consolidated specimens were cut into modulus of rupture bars of size 25 mm long x 1.27 mm wide x 2.54 mm. The MOR strength measurements were carried out using a four point bending mode at the High Temperature Materials Laboratory (HTML) section of Oak Ridge National Laboratory. Fracture toughness (KIc) values were measured by making hardness indentations and measuring the crack size and calculating the toughness values. The indentation load used in fracture toughness determination was 10 kg. The MOR data for monolithic and composite specimens as well as toughness and hardness values are shown in **Table 1**.

	RT MOR Strength MPa	1200°C Strength MPa	Fracture Toughness MPa sqrt (m)	Hardness GPa
Range	324-540	287-515	6	14
Average Value	433	401	6	14

Table I: Mechanical Properties of Ceracon Consolidated Material

In monolithic specimens, MOR strengths ranging from 324 MPa to 540 MPa were obtained at room temperature with an average strength of 433 MPa. At 1200°C, values ranged from 287 MPa to 515 MPa and the average strength was 401 MPa. The analysis of the mechanical test data shows that monolithic material possesses good toughness and moderate strength. The high toughness is due to the presence of very fine pores in the material. The high temperature strength of 515 MPa at 1200°C must result from the ultrafine grain size of this material. The hardness and toughness values again parallel the strength values observed. For a monolithic material, a hardness value of 14 GPa and toughness value of 6 GPa was obtained.

CONCLUSIONS:

- The Ceracon process can use preforms having interconnected porosity and there is no need for expensive glass encapsulation of preforms because solid pressure transmitting media is used.

- The silicon nitride material fabricated by the Ceracon Process has a unique ultrafine grained (0.2 µm) microstructure of predominantly alpha phase, same as the starting powder

- The unique combination of ultrafine grain size and presence of mostly alpha phase leads to a high hardness, good toughness, and good high temperature strength material, which means that the process can be evaluated to fabricate more complex components for insertion into heat engines and other high temperature components.

REFERENCES:

1. F. L. Riley, (ed.), "Nitrogen Ceramics" (Noordhoff-Leyden, 1977).

2. W. Bunk, M. Bohmer and H. Kissler, (eds), "Keramische Komponenten fur Fahrzeug-Gasturbinen", Vols I to III (Springer-Verlag, Berlin, 1984).

3. E. M. Lenoe, R. N. Katz and J. J. Burke (eds), "Ceramics for High-Performance Applications". Vol. III (Plenum, New York, London, 1979).

4. F. L. Riley (ed.), "Progress in Nitrogen Ceramics" (Martinus Nijhoff, 1983).

5. F. F. Lange, ibid 56 (1973) 518.

6. J. Heinrich and M. Bohmer, Sci. Ceram. 11 (1981) 439.

7. Arvid Pasto, private communucation with R. Raman, October 1991.

8. Communication between Jeff Neil-GTE Labs, and Dr. Ramas V. Raman, 1992.

NUCLEATION AND GROWTH OF CVD POLYCRYSTALLINE Si_3N_4 FILMS AT LOW TEMPERATURES

Frederick S. Lauten, Janet Rankin and Brian W. Sheldon
Division of Engineering, Brown University, Providence, RI 02912.

ABSTRACT

Silicon nitride films were deposited from silane and ammonia onto single crystal silicon substrates at a total pressure of 4.7 kPa. At temperatures below 1200° C, continuous films of polycrystalline Si_3N_4 deposited within a narrow region of low SiH_4 partial pressures but at relatively high growth rates, for example, > 10 μm/hr at 1170° C. The early stages of crystalline film growth were studied with a combination of analytical techniques: FT-IR spectroscopy, x-ray and electron diffraction and electron microscopy. During deposition faceted Si_3N_4 grains nucleate on a growing nanocrystalline/amorphous interlayer.

INTRODUCTION

Past studies have demonstrated that the microstructure of polycrystalline Si_3N_4 films deposited by chemical vapor deposition (CVD) varies appreciably with processing conditions. Furthermore, specific mechanical and chemical properties are strongly dependent on film microstructure. While an empirical understanding of the relationship between processing conditions and microstructure has been developed in some systems, the nucleation and growth mechanisms that determine these microstructures are not well understood.

Continuous crystalline films are typically deposited at 1275° to 1500° C [1-3]. At these higher temperatures, Si/(Si+N) molar ratios of the input gases are usually on the order of 0.01 to 0.1. While crystalline deposits of Si_3N_4 have been reported at temperatures as low as 1025 to 1100° C, the deposits were not continuous films but rather whiskers or crystallites within an amorphous film [4,5]. The research presented here explores the early stages of deposition of Si_3N_4 onto Si substrates at low temperatures and demonstrates that relatively high deposition rates can be achieved.

A common difficulty encountered when studying a CVD process is insufficient knowledge concerning mass transfer within the reactor. A variety of interrelated reaction conditions such as: temperature gradients, upstream mass depletion, mass flow rate, substrate position or geometry, chamber dimensions, or distance of gas inlet from substrate, are often cited as severely affecting the deposition rate and the nature of the deposits within a given CVD reactor. Mass transfer in an impinging jet is well defined in many systems [6,7]. For this reason an impinging jet CVD reactor was used for the resarch presented here.

EXPERIMENTAL PROCEDURE

Deposition occurred within a cold wall reactor chamber fashioned from fused silica quartz glass. The substrate was placed within a silicon/silicon nitride coated graphite impinging jet cell which was heated with a 450 kHz RF induction heater. A schematic representation of the impinging jet cell is shown in Figure 1. The process temperature was recorded by a single beam IR thermometer which was focused on a black body hole located directly below the substrate.

In order to simplify the chemical system, SiH_4, NH_3 and Ar were the only reactant gases used in this study. In the experiments presented in this paper, deposition occurred at temperatures between 1070 and 1250° C with Ar and NH_3 flow rates of 1100 and 400 sccm and SiH_4 flow rates varying from 0.4 to 4.0 sccm. Silicon substrates were used because they did not contaminate the chamber, were transparent to FT-IR transmission measurements, and were readily obtained in relatively pure form. The substrate was ultrasonically cleaned in trichloro-ethylene, etched in 0.48 molar HF acid, rinsed in distilled water, and dried with ethanol. Prior to heating the chamber was evacuated to < 1.33 Pa, backfilled with Ar and evacuated twice. After the second cycle the pressure was stabilized to the deposition pressure of 4.67 kPa. The CVD process consisted of three stages, a temperature ramp with an H_2 etch, nitridation, and deposition. During the ramp stage, the sample was heated at a constant rate for two minutes in flowing H_2 to 1000° C. During the nitridation stage, NH_3 at 400 sccm was added to the H_2 gas flow and the sample was heated to the deposition temperature. Within the last 15 seconds of the nitridation, the H_2 was replaced with the equivalent Ar mass flow. The H_2 etch occurring during the ramp stage removed most of the remaining native oxide from the surface. The nitridation stage accomplished two goals. First it inhibited the development of etch pits. Second it formed a

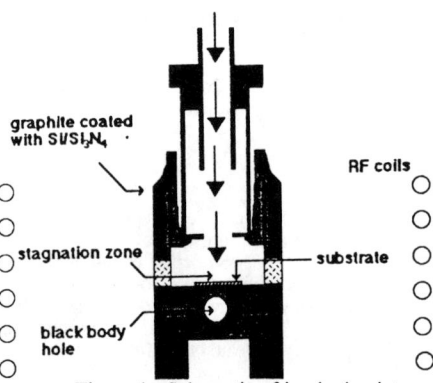

Figure 1. Schematic of impinging jet configuration.

passivating nitride layer which reduces the amount of Si diffusing from the substrate into the growing deposit layer. The deposition stage began with the onset of SiH_4 flow. When deposition was complete, the RF generator, pressure control circuit, and NH_3, SiH_4 and Ar flows were turned off and the sample cooled to room temperature under flowing H_2.

A number of techniques were used to analyze the films. A Jeol 840F SEM was used to image the surface characteristics of the deposited material and to determine the film thickness by cleaving the substrate and measuring the thickness of the deposited layer. The accuracy of this method has been confirmed by comparison to similar measurements made on a TEM. Specimens with both cross-section and plane-view geometries were investigated in a Phillips 420T TEM. Both selected area and micro-electron diffraction were used to determine the crystallinity of particular regions of the samples.

Fourier Transform Infrared and X-ray diffraction spectra were obtained from many deposited films. In both techniques the films were analyzed as deposited on the substrate. The films were deposited on silicon wafers with rough back surfaces. In order to collect acceptable FT-IR spectra, the unpolished side of the silicon was polished with 1 μm diamond particles. Absorbance peaks due to the polishing compounds have been subtracted from the FT-IR spectra presented here.

Certain experimental conditions must be satisfied in order to ensure that the largest possible area of the substrate is within the stagnation zone. In impinging jet configurations, the stagnation zone is defined as a region which is a certain distance from the jet inlet in which mass transfer is a maximum. When an area on the substrate is within this zone, it is uniformly accessible to the diffusing species in the jet. In particular, there is an optimum separation between substrate and jet inlet where the stagnation zone covers the largest area on the substrate. This distance is dependent on the Reynolds number, Re. In addition the gas flow rate, deposition temperature, and jet inlet substrate separation must be such that recirculation effects and free convection as characterized by the Grashoff number, Gr, are minimal when compared to the forced mass flow of the jet [8]. For the experimental conditions presented here Re=112 and Gr/Re^2 <0.01. Additionally at these conditions Si depletion upstream of the jet inlet was less than 10%. This was determined by the mass change of the quartz glass jet inlet sleeve before and after deposition.

RESULTS AND DISCUSSION

Figure 2 shows experimental results which have defined the narrow range of SiH_4 partial pressures in which continuous crystalline films deposit at low temperatures. The typical morphologies of the deposits grown above 1140° C in the three growth regimes shown in Figure 2 are grossly different. In the two amorphous regimes at both low and high partial pressures the amorphous films are smooth and continuous. The continuous

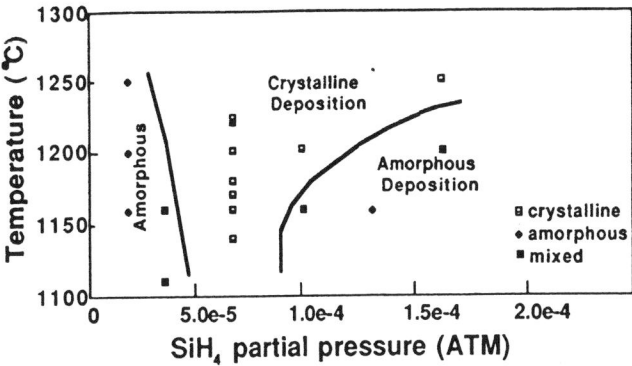

Figure 2. Silicon nitride film microstructures as a function of temperature and SiH_4 concentration. The flow rates of NH_3 and Ar are 400 sccm and 1100 sccm respectively. (P_{Tot} = 4.666 kPa)

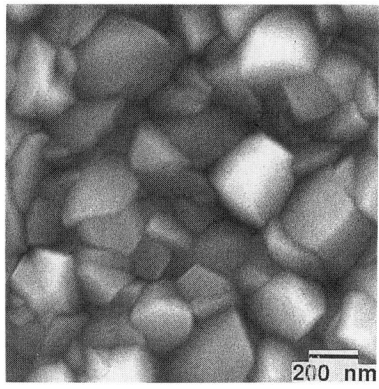

Figure 3. Continuous crystalline film typical of deposition at low temperatures. This sample was deposited at 1195° C for 7 minutes.

crystalline films, however, are composed of sharply faceted micron sized grains as seen in Figure 3. In the narrow range of partial pressures where continuous crystalline films deposit at low temperature growth rates are relatively high. For example at 1170° C the crystalline Si_3N_4 film deposited at a rate slightly greater than 10 μm/hr.

As can be seen in Figure 4, the continuous crystalline films do not deposit directly onto the Si substrate itself, but onto a granular interlayer. The surface of the interlayer is apparently composed of rounded connected particles which have a diameter on the order of 100 nm. The Si_3N_4 layer consists of particles having faceted morphology. Transmission electron microscopy analysis gives a more specific view of the film structure. Figure 5 shows a sample viewed in cross section. A 0.4 μm thick interlayer grows from the substrate surface. This is the granular layer which is seen on the SEM micrograph. In order to verify that the interlayer resulted from the reaction of SiH_4 and NH_3, a number of Si substrates were nitrided with NH_3. No deposits were resolved using SEM. In Figure 5 a 10 to 20 nm thick region exists directly on the substrate. While possibly an effect due to specimen thickness, this region may be a result of the two minute pre-deposition nitridation stage. Selected area electron diffraction patterns revealed that the interlayer is composed of nanocrystalline regions dispersed within amorphous material. The diffraction patterns do not contain d-spacings which correspond to Si. Due to the wide range of d-spacings in the phases of Si_3N_4 or Si, N, H compounds, the chemical composition of the nanocrystalline material can not be verified; however, it is likely that the crystalline material detected by electron diffraction is Si_3N_4. From the TEM micrograph one can see that the granular character of the interlayer changes through its thickness. At the earliest stages of deposition, near the substrate surface, the interlayer has a fine granular structure. As the interlayer grows, the granular structure becomes coarser. The crystallites of Si_3N_4 nucleate abruptly on this coarse granular interlayer when the granularity is on the order of tens of nm.

X-ray diffraction peaks are not generated by the samples which have not yet grown faceted particles. We attribute this to the fact that the interlayer does not contain enough crystalline material to be detected by the diffractometer. We do, however, believe that it is the formation of the nanocrystalline interlayer and its subsequent control of nucleation of Si_3N_4 which is one important factor in the growth of continuous crystalline films at this low temperature.

FT-IR spectroscopy provides additional information on the film microstructure. Figure 6a is typical of the spectrum generated by smooth continuous amorphous silicon nitride films [9]. As expected it consists of a peak centered around 860 cm^{-1} corresponding to amorphous silicon nitride. Figure 6b is the spectrum generated by a sample grown at the same conditions as Figure 4, but for one and a half minutes instead of five. Only a small

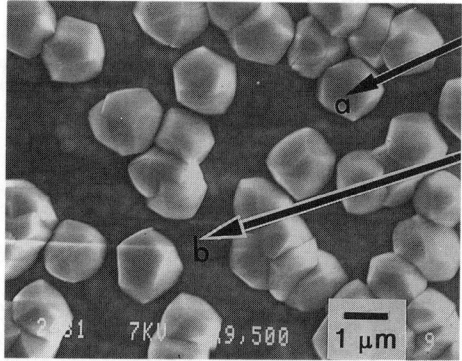

Figure 4. SEM micrograph of sample grown for five minutes at deposition conditions of 1170° C, 35 Torr and Ar/NH$_3$/SiH$_4$ flow of 1100/400/1.6 sccm (SiH$_4$/[SiH$_4$ + NH$_3$] = 0.004). (a) Si_3N_4 crystal. (b) Nanocrystalline interlayer.

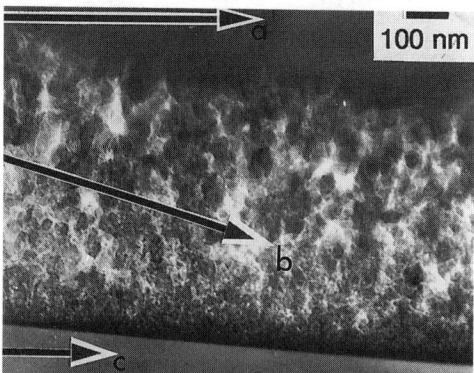

Figure 5. TEM cross section of sample shown in Figure 4. (a) Si_3N_4 crystal. (b) Nanocrystalline interlayer. (c) Si substrate.

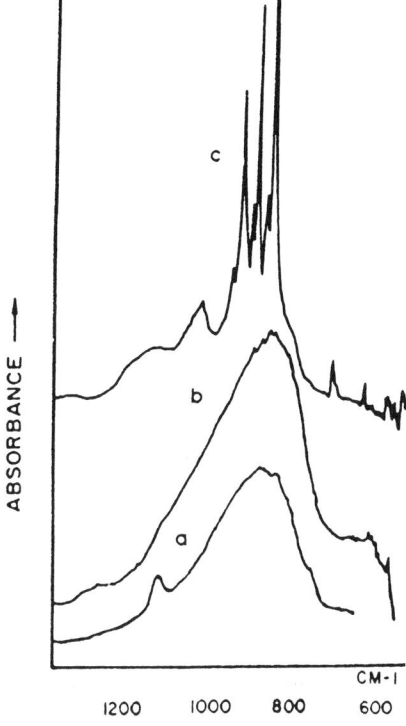

Figure 6. FT-IR spectra from three samples. (a), a typical amorphous film of silicon nitride. (b), the amorphous/nanocrystalline interlayer grown for 1.5 minutes at 1170° C. (c), the continuous polycrystalline Si_3N_4 film shown in Figure 3.

number of faceted Si_3N_4 grains have nucleated on top of the interlayer. Again the spectra consists of a peak centered around 860 cm^{-1}. In contrast a continuous crystalline film (for example that shown in Figure 3) generates peaks centered at 1040 cm^{-1} and near 860 cm^{-1} as expected for crystalline Si_3N_4 [10]. This spectrum is Figure 6c. Any peaks between 1150 and 1200 cm^{-1} are attributed to Si-O bonds which formed on the film surface during post deposition handling. Results from FT-IR are in good agreement with TEM results and indicate that the interlayer is composed of amorphous silicon nitride with some nanocrystalline regions.

An explanation of the experimental observations requires some extension of conventional nucleation theory. One possibility is that the interlayer reduces the surface free energy to some critically low value such that large faceted Si_3N_4 grains are able to nucleate. This, however, would not explain why the crystalline material which was detected in the interlayer by electron diffraction does not act as nucleation sites and lead to the formation of faceted grains earlier in the process. Under the reaction conditions used here, standard nucleation theory predicts a critical nucleus of less than ten atoms. However, this value is highly approximate. If the critical nucleus size were somewhat larger, the crystalline material in the interlayer would be too small to act as nucleation sites. With this model, the nanocrystalline regions in the interlayer are essentially embryos (subcritical nuclei). Thus, nucleation events which lead to the formation of large Si_3N_4 grains occur only after those regions have reached critical size.

The observed deposition mechanism and the experimental results shown in Figure 2 provide insight into the relatively fast growth of continuous crystalline Si_3N_4 films at low temperature. The transition from crystalline to amorphous deposition as partial pressure is increased at constant temperature is well documented for silicon nitride and other material systems [11]. This phenomenon is commonly explained by the idea that the impingement rate of adatoms is large enough to impede the diffusion of deposited adatoms to crystalline sites. Thus crystalline deposition occurs when the SiH_4 concentration is low enough to permit nucleation and the subsequent growth of a continuous microcrystalline film. Relatively high growth rates were possible because mass transfer to the substrate in the impinging jet is much more efficient than in other CVD configurations. This combined with a low depletion of Si upstream of the substrate allows relatively high crystalline growth rates even at these low Si species supersaturations, e.g., $>10.2\ \mu mhr^{-1}$ at 1160° C. The transition from crystalline to amorphous deposition when the SiH_4 concentration was further reduced has not often been discussed. One possible explanation is that when the rate of adatom impingement becomes too low, the nucleation of crystalline material is suppressed because the Si species supersaturation is too low to create a critical sized nucleus.

SUMMARY

A narrow region of low SiH_4 partial pressures has been experimentally defined in which continuous Si_3N_4 polycrystalline films deposit at technologically low temperatures and relatively fast growth rates. At both greater and lower SiH_4 partial pressures amorphous films deposit at the same temperatures. During crystalline film growth sharply faceted grains of Si_3N_4 nucleate abruptly on an nanocrystalline/amorphous interlayer. It has been proposed that nucleation sites form as the interlayer grows. Furthermore the high growth rates observed when SiH_4 concentrations are one to two orders of magnitude lower than those commonly used is attributed to the high efficiency and low upstream depletion of the impinging jet gas delivery system which was used.

ACKNOWLEDGEMENTS

This research was supported by the U.S. Department of the Air Force, Office of Scientific Research. under contract AFOSR-91-0357 with Brown University. The authors gratefully acknowledge Theodore Kirst for assisting in FT-IR analysis.

1   H. Doi, N. Kikuchi, Y. Oosawa, *Mats. Sci. Eng.* **A105/106**, 465 (1988).
2   A.C. Airey, S. Clarke and P. Popper, *Proc. Brit. Ceram. Soc.* **22**, 305 (1973).
3   K. Niihara and T. Hirai, *J. Mat. Sci.* **11**, 593 (1976).
4   K.E. Bean *et al.*, *J. Electrochem. Soc.* **114**, 733 (1967).
5   K. Kijima, N. Setaka and H. Tanaka, *J. Cryst. Grow.* **24/25**, 183 (1974).
6   V. G. Levich, *Physiochemical Hydrodynamics*, Prentice Hall, Englewood Cliffs, NJ (1962).
7   H. Rebenne and R. Pollard, *J. Am. Ceram. Soc.* **70** [12], 907 (1987).
8   L.Vandenbulcke and G.Vuillard, *J.Electrochem. Soc.:Solid State Science and Technology* **124** [12], 1931 (1977).
9   K. Okada, H. Sakane and Y. Sugioka, *J. Phys. Soc. Japan* **23**, 655 (1967).
10 Yu.N. Volgin and Yu.I. Ukhanov, *Opt. Spektrosk.* **38**, 727 (1975).
11 W.Y. Lee, J.R. Strife and R.D. Veltri, *J. Am. Ceram. Soc.* **75** [19], 2803 (1992).

PART II B

Grain Boundary Phases and Glasses

INTERGRANULAR MICROSTRUCTURE AND OXIDATION BEHAVIOUR OF Si_3N_4 CERAMICS FORMED WITH Y_2O_3, Al_2O_3 AND ZrO_2

L.K.L. FALK*, E.U. ENGSTRÖM* AND K. RUNDGREN**
*Department of Physics, Chalmers University of Technology, S-412 96 Göteborg, Sweden
**Swedish Ceramic Institute, Box 5403, S-402 29 Göteborg, Sweden.

ABSTRACT

Analytical electron microscopy, X-ray diffractometry and secondary ion mass spectrometry (SIMS) have been used to characterize the intergranular microstructure of nitrided pressureless sintered (NPS) Si_3N_4 ceramics and the phase transformations which occur in the intergranular regions during oxidation. The phase transformations, which take place after crystallisation of the residual intergranular glass, are caused by an inward transport of oxygen and result in the development of sub-scalar phase gradients. Cubic Y_2O_3 stablized ZrO_2 partitions from the oxynitride liquid phase sintering medium when a small amount of $ZrO_2(+ 3$ mol% $Y_2O_3)$ is added together with the Y_2O_3 and Al_2O_3 sintering additives. The latter material has a reduced oxidation resistance, presumeably due to the high oxygen conductivity of the cubic ZrO_2 structure.

INTRODUCTION

Si_3N_4 ceramic materials formed by nitridation and pressureless sintering of Si:Si_3N_4 powder compacts were first developed using Y_2O_3 and Al_2O_3 as metal oxide sintering additives [1,2]. The comparatively large amount of metal oxides which was required for full densification resulted in a microstructure consisting of high aspect ratio β-Si_3N_4 grains separated by a Si, Y and Al rich residual glass phase present as thin intergranular films merging into larger pockets at multi grain junctions. A substantial part of this glass crystallised during a post densification heat treatment at temperatures around 1100 °C and above [2,3].

In order to influence the intergranular microstructure of nitrided pressureless sintered (NPS) Si_3N_4 ceramics, e.g. volume fraction of residual glass and glass composition, small amounts of partially stabilized ZrO_2 have been added to the initial composition. The presence of a ZrO_2 phase in the microstructure of the Si_3N_4 ceramic may also have a toughening effect as indicated by work on Si_3N_4/ZrO_2 composite ceramics [4].

This paper is concerned with the intergranular microstructure and the oxidation behaviour of NPS Si_3N_4 materials. The paper will also discuss some of the work which has been concerned with phase transformations in the intergranular regions during heat treatment and oxidation of NPS Si_3N_4.

EXPERIMENTAL PROCEDURES

The starting powder compacts had a Si to Si_3N_4 mass ratio of 6:4, and the added metal oxides equals to 6 wt% Y_2O_3 and 2 wt% Al_2O_3 or 6 wt% Y_2O_3, 2 wt% Al_2O_3 and 2 wt% $ZrO_2(+ 3$ mol% $Y_2O_3)$ respectively, after nitridation. Nitridation was carried out at 1350 °C which resulted in full conversion of the Si to α- and β-Si_3N_4. Pressureless sintering to near theoretical density was carried out at 1780 to 1820 °C in an N_2 atmosphere with the nitrided compacts embedded in a Si_3N_4 protective powder bed. Detailed descriptions of the formation process and the reactions occuring during nitridation and pressureless sintering are given elsewhere [1,2].

The weight gain during oxidation of the two compositions at temperatures between 1200 and 1400 °C was registered in a thermobalance, and the material formed with Y_2O_3 and Al_2O_3 was also heat treated in air at temperatures in this interval. The microstructures before and after heat treatment were characterized by analytical electron microscopy and X-ray diffractometry. In-depth profiling by secondary ion mass spectrometry (SIMS) was used to detect elemental concentration profiles which build up in the sub-scalar region during oxidation. Analysis of Si_3N_4 materials by SIMS is discussed in detail in reference [5].

RESULTS AND DISCUSSION

The Microstructure of Sintered Materials

The α-Si_3N_4 was fully converted to the β structure during pressureless sintering of the two NPS Si_3N_4 compositions. TEM studies indicated that the β-Si_3N_4 grains were faceted and elongated, Figure 1. This prismatic morphology is obtained when the β grains can grow in an isotropic liquid phase environment [6].

The addition of Y_2O_3 partially stabilized ZrO_2 to the starting powder mixture resulted in two types of intergranular regions, Figure 1. In areas, the ZrO_2 had formed crystalline pockets, and selected area electron diffraction together with centered dark field imaging in the TEM showed that these pockets had the same crystallographic orientation within such areas. Electron diffraction implied that the ZrO_2 phase had the cubic fluorite structure [7]. The diffraction patterns also showed the diffuse electron scattering which has been attributed to the presence of

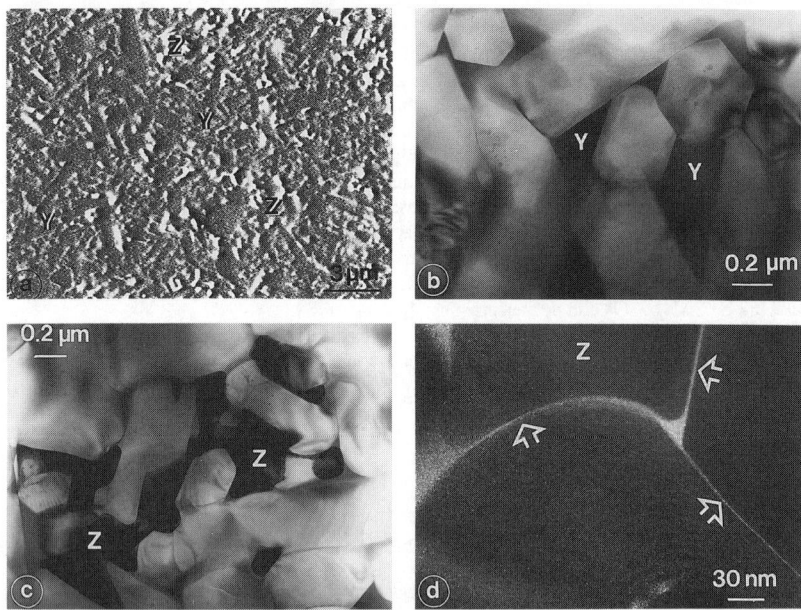

Figure 1: The microstructure of the NPS material formed with 6 wt% Y_2O_3, 2 wt% Al_2O_3 and 2 wt% ZrO_2(+ 3 mol% Y_2O_3) imaged (a) by backscattered electrons in the SEM. Y rich glass (Y) and ZrO_2 (Z) containing regions and are imaged by TEM in (b) and (c) respectively. Thin intergranular glass films, arrowed in (d), were separating adjacent grains.

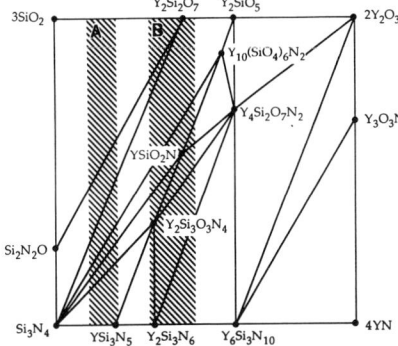

Figure 2: The Si_3N_4-SiO_2-Y_2O_3-YN behaviour diagram [10]. Y/Si ratios of glass pockets in materials formed with (A) and without (B) ZrO_2(+ 3 mol% Y_2O_3) were determined by EDX in the STEM.

the disordered vacancy structure in stabilized ZrO_2 structures [8]. The ZrO_2 grains contained a significantly higher concentration of Y_2O_3 than the partially stabilized ZrO_2 starting powder; the concentration of Y_2O_3 in the ZrO_2 pockets varied between 13 and 16 mol%. These concentrations support the observation of a retained high temperature cubic structure [9]. An amorphous phase, in most cases present as a thin grain boundary film, was separating the ZrO_2 pockets from adjacent, prismatic, β-Si_3N_4 grains, Figure 1d.

The areas containing the cubic ZrO_2 were separated by a microstructure similar to that of NPS Si_3N_4 formed without ZrO_2, Figure 1b; comparatively large Y, Si and Al rich glassy pockets were present at multi grain junctions, and adjacent β-Si_3N_4 grains were separated by a thin amorphous film. These glass pockets had a significantly higher Si/Y ratio than the glass in NPS Si_3N_4 formed without the ZrO_2 addition (Figure 2) which is consistent with the observed Y uptake of the ZrO_2 during densification. The glass in the ZrO_2 containing material did, in general, not contain any detectable amounts of Zr. In both materials, the composition of the intergranular glass showed a pronounced variation between different pockets in the microstructure (Figure 2) which indicates an inhomogeneous composition of the oxynitride liquid phase sintering medium.

The intergranular microstructure of the ZrO_2 containing areas implies that the Y_2O_3 stabilized cubic ZrO_2 partitioned from the oxynitride liquid phase present during pressureless sintering of the nitrided compacts. The crystallographic relationship between adjacent ZrO_2 pockets indicates that the crystallisation involved comparatively few nucleation sites, and that the ZrO_2 grains grew in the form of a three dimensional intergranular network. Hence, the Y, Si and Al rich intergranular glass is the residue of the liquid phase sintering medium remaining after growth of prismatic β-Si_3N_4 grains and partitioning of the ZrO_2 phase. The significantly increased Y_2O_3 concentration in the ZrO_2 after densification shows that Zr, Y and O from ZrO_2 and Y_2O_3 additions may react via an oxynitride liquid phase sintering medium whereby Y will be incorporated into the cation lattice of the ZrO_2 structure.

Phase Transformations During Heat Treatment

ZrO_2 has previously been reported as an effective crystallisation agent when added to these Si_3N_4 based systems [11]. However, a fully crystalline and homogeneous intergranular microstructure was not obtained when the metal oxides added to the NPS Si_3N_4 had a ZrO_2/Y_2O_3 weight ratio of 0.31.

The residual glass in NPS Si_3N_4 formed with Y_2O_3 and Al_2O_3 can, however, be substantially crystallised by a post-densification heat treatment. Heat treatments in air carried out for short times at 1100 and 1200 °C showed that it is possible to crystallise a significant amount of the glass prior to any appreciable oxidation of the Si_3N_4 grains [2,3]. The crystallised pockets were generally poly crystalline which implies that the crystallisation

Table I: Secondary crystalline phase contents at different depths below the oxidised surface on heat treated NPS Si_3N_4 materials [3,5]. The phase compositions are given in approximate order of predominance.

Temperature (°C)	Time (h)	10-20 μm	400 μm	1.5 mm
1200	1, 6, 24	A, α, W, (YS)		A, α, W, (YS)
1300	100	α, W		α, W, A
1350	25	α, β, A	A, α	
1350	125	β	β, α, A	
1400	7	β		α, W, A

Legend: A = Y,N-apatite $(Y_{10}(SiO_4)_6N_2)$; W = Y,N-wollastonite $(YSiO_2N)$; α = α-$Y_2Si_2O_7$; β = β-$Y_2Si_2O_7$; YS = $YSiO_5$

involved a high density of nucleation sites. X-ray diffractometry and electron diffraction in the TEM showed that different crystallisation products formed, Table I, which could be expected from the inhomogeneous composition of the residual glass in the NPS material, Figure 2. Strain contours in TEM images showed that crystallisation was associated with an introduction of internal stresses, Figure 3a.

A prolonged time at temperatures between 1200 and 1400 °C resulted in a more homogeneous intergranular microstructure containing single grain pockets. Full crystallisation was however never obtained, Figure 3b; thin intergranular films, both in Si_3N_4/Si_3N_4 grain boundaries and in boundaries between Si_3N_4 and crystalline pockets, as well as smaller glass pockets remained undevitrified [3]. This behaviour may be due to hydrostatic stresses supported by the constrained residual glass present after densification [12].

During oxidation at temperatures in the range 1300 to 1400 °C, phase transformations occured in the intergranular regions and this resulted in a reduced number of crystalline phases and also in phase gradients below the oxide scales, Table I [3,5]. It can be noted that there was a predominance of phases with a higher O/N ratio and/or lower Y/Si ratio close to the scales. A comparatively slow heating rate (20 °C/min) was used in order to allow crystallisation of the residual glass before the oxidation temperature was reached. The SIMS profiles in Figure 4 show that the N content of the sub-scalar material was virtually unaffected by the oxidation. Together with the higher in O/N ratio of the crystalline intergranular phases close to the oxide scale, this implies that an inward diffusion of oxygen caused an internal oxidation of the

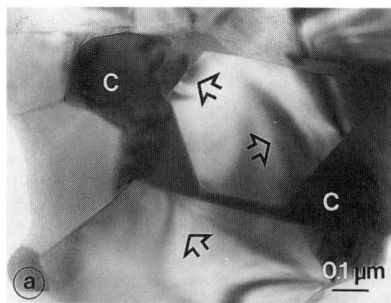 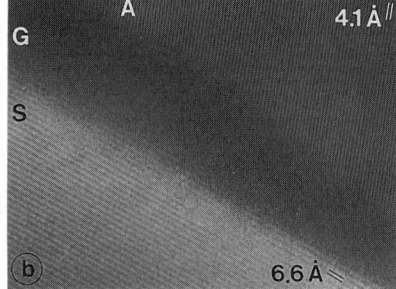

Figure 3: (a) Partially crystallised pocket (C) in material heat treated 10 min at 1100 °C. The TEM image shows strain contours (arrowed). (b) Residual glass film (G) between grains of β-Si_3N_4 (S) and Y,N-apatite (A) in material heat treated 24 h at 1200 °C.

intergranular microstructure [3,5]. As discussed elsewhere, the instrument used for these analyses did not allow the detection of O [5].

In-depth profiling by SIMS also showed that elemental concentration profiles built up below the oxidised surface during oxidation at 1200 °C and temperatures above [5]. Figure 4 shows in-depth SIMS profiles after oxidation at 1350 °C for 25 h. It has been suggested that the transition from α- to β-$Y_2Si_2O_7$ which occurs during oxidation (Table I), and at higher temperatures than reported previously [13], is affected by the elemental contents of the intergranular regions [3,5]. The oxide scales were enriched in cations originating from the intergranular regions, while a zone below the scales was depleted in cations. This is particularly clear for the impurity elements Fe, Ca and Ti which originate from the starting powders, Figure 4. The narrow depletion zone closer to the scale observed for Y after oxidation at these temperatures could possibly contribute to the predominance of phases with a lower Y/Si ratio just below the oxide scales. The elemental profiles are the result of a formation of a reaction couple between the silica rich oxide scale and the intergranular glass in the Si_3N_4 material [14].

Oxidation Behaviour

During prolonged oxidation, the initially formed silica layer will become enriched in cations originating from the intergranular regions (Figure 4), and this will promote oxygen transport through the oxide scale [15]. This may explain the increase in weight gain which was registered by the thermobalande in the temperature interval 1240 to 1270 °C during continuous heating from 1200 to 1400 °C (heating rate 0.2 °C/min).

The NPS material formed with Y_2O_3 and Al_2O_3 showed a significant increase in oxygen uptake at 1350 °C. As discussed in the previous section, this increased oxygen uptake did not only result in a continuous oxidation of the Si_3N_4 but also in an oxidation of the intergranular microstructure; after oxidation at higher temperatures, there was a predominance of more O-rich intergranular phases below the oxide scale, see Table I. An increased oxygen uptake at around 1350 °C could possibly be promoted by local liquid formation; the lowest reported eutectic temperature in the SiO_2-Al_2O_3-Y_2O_3 system is 1345 °C [16]. TEM has previously shown that larger glass pockets were present in materials oxidised at 1350 and 1400 °C [3]. A slight reduction in oxidation rate was observed with time at 1400 °C, presumeably due to a reduced oxygen transport through a thicker and mainly crystalline oxide scale [5,15].

Also the NPS Si_3N_4 formed with Y_2O_3 partially stabilized ZrO_2 showed an increase in oxygen uptake between 1340 and 1370 °C. At higher temperatures, this material had a significantly reduced oxidation resistance. The increased oxidation rate caused by the ZrO_2

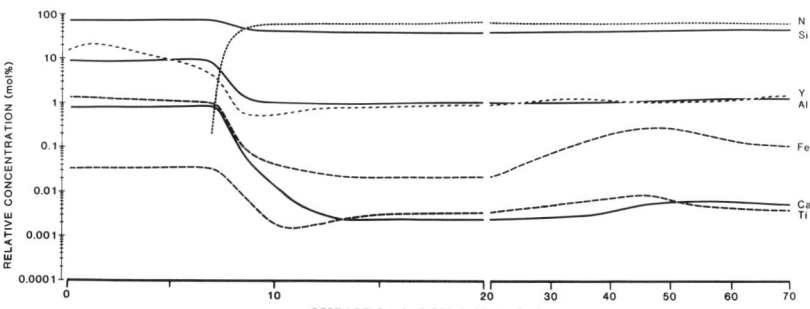

Figure 4: In-depth SIMS profiles through the oxide scale into the bulk material after oxidation at 1350 °C for 25 h. The intersection between the Si and N profiles has been taken as a measure of oxide scale thickness. This material was NPS Si_3N_4 formed with Y_2O_3 and Al_2O_3.

addition may be explained by the high oxygen conductivity of the cubic ZrO_2 which would promote the inward transport of oxygen [17]. The oxidation resistance would also be reduced if larger volumes of residual glass were retained during oxidation.

X-ray diffractometry showed that the oxide scale on the material formed with Y_2O_3 and Al_2O_3 contained β-$Y_2Si_2O_7$ and α-SiO_2 after oxidation at 1400 °C. Al_2SiO_5 and $ZrSiO_4$ were present in the oxide scale after oxidation of the ZrO_2 containing NPS material at this temperature.

CONCLUDING REMARKS

A post-densification heat treatment of NPS Si_3N_4 formed with Y_2O_3 and Al_2O_3 makes it possible to substantially crystallise the residual intergranular glass. Prolonged oxidation at high temperatures results in the development of elemental and phase gradients below the oxidised surface, and the intergranular microstructure will be dependent upon time and temperature of heat treatment/oxidation. Addition of Y_2O_3 partially stabilized ZrO_2 together with the Y_2O_3 and Al_2O_3 results in a decreased oxidation resistance. This may be due to the presence of secondary phases with a higher oxygen conductivity; Y_2O_3 stabilized cubic ZrO_2 partitions from the oxynitride liquid phase sintering medium during densification.

ACKNOWLEDGEMENT

This work was supported by the Swedish National Board for Technical Development (STU).

REFERENCES

1. L.K.L. Falk, G.L. Dunlop and R. Pompe, J. Mater. Sci. **20**, 3545 (1985).
2. L.K.L. Falk, G.L. Dunlop and R. Pompe, Mater. Sci. Eng. **71**, 123 (1985).
3. L.K.L. Falk and G.L. Dunlop, J. Mater. Sci. **22**, 4369 (1987).
4. F.F. Lange, L.K.L. Falk and B.I. Davis, J. Mater. Res. **2**, 66 (1987).
5. L.K.L. Falk and E.U. Engström, J. Am. Ceram. Soc. **74** (9), 2286-92 (1991).
6. M.H. Lewis, B.D. Powell, P. Drew, R.J. Lumby, B. North and A.J. Taylor, J. Mater. Sci. **12**, 61 (1977).
7. V. Lanteri, R. Chaim and A.H. Heuer, J. Am. Ceram. Soc. **69** (10), C-258-261, (1986).
8. B. Hudson and P.T. Moseley, J. Solid State Chem. **19**, 383 (1976).
9. H.G. Scott, J. Mater. Sci. **10**, 1527 (1975).
10. D.P. Thompson, in Tailoring Multiphase and Composite Ceramics, edited by R.E. Tessler, G.L. Messing, C.G. Pantano and R.E. Newnham (Plenum Publishing Corp., New York, 1986), pp. 79-91.
11. W. Braue, G. Wötting and G. Ziegler, in Ceramic Materials and Components for Engines, edited by W. Bunk and H. Hausner (Verlag Deutsche Keramische Gesellschaft, Bad Honnef, 1986), pp. 503-510.
12. R. Raj and F.F. Lange, Acta Met. **29**, 1993 (1981).
13. J. Ito and H. Johnson, Am. Mineral. **53**, 1940 (1968).
14. D.R. Clarke and F.F. Lange, J. Am. Ceram. Soc. **63** (9-10) 586-593 (1980).
15. J. Schlichting, in Energy and Ceramics, edited by P. Vincenzini (Elsevier, Amsterdam, 1983), pp.390-398.
16. I.A. Bondar and F.J. Galakhov, Izv. Akad. Nauk. SSSR, Ser. Khim. **7**, 1325 (1964).
17. J.F. Baumard and P. Abelard, in Advances in Ceramics, Vol. 12, Science and Technology of Zirconia II, edited by N. Claussen, M. Rühle and A.H. Heuer (American Ceramic Society, Columbus, OH, 1984) pp. 555-571.

ELEMENTAL ANALYSIS OF MATRIX GRAIN BOUNDARIES IN SiC WHISKER REINFORCED Si_3N_4 BASED COMPOSITES

J. LIU,* K. DAS CHOWDHURY,* R. W. CARPENTER* AND W. BRAUE,**
*Center for Solid State Science, Arizona State University, Tempe, AZ
**German Aerospace Research Establishment, D5000, Cologne, Germany.

ABSTRACT

The structures of SiC/Si_3N_4 interfaces and Si_3N_4 matrix grain boundaries in Ceramic Matrix Composites (CMC) were investigated by high resolution electron microscopy. The light element chemistry of the interfaces was analyzed by high spatial resolution (~3 nm) position resolved EELS in a field emission TEM and by high spatial resolution EDS in a dedicated scanning transmission electron microscope (STEM). High-angle annular dark-field (HAADF) imaging (resolution < 1 nm) technique was used to determine the distribution of yttrium atoms at matrix grain boundaries and at SiC/Si_3N_4 interfaces. HAADF images suggest that yttrium might diffuse into Si_3N_4 crystals bounding the interfacial and grain boundary regions.

INTRODUCTION

The structure and toughness of ceramic matrix composites are critically dependent on the deforming and sliding characteristics of their reinforcement/matrix interfaces and matrix grain boundaries [1]. These interfacial/grain boundary characteristics are in turn dependent on local chemical bonding, which depends on local composition. Local interface and boundary composition in CMC's is generally influenced by sintering aid additives during processing which accelerate densification by liquid phase sintering. High spatial resolution analysis is required to determine elemental distributions in interface/boundary regions of these nonequilibrium materials. We showed earlier using EELS that oxygen and nitrogen gradients at these interfaces/boundaries defined the interface/boundary chemical width about an order of magnitude larger than the structural width [2]. Here, for the first time, we apply HAADF imaging to measurements of heavy element composition gradients in the same regions and correlate the results with EDS. The results showed that heavy metal distributions from the sintering aids follow the oxygen distribution in both the dissimilar interfaces and grain boundaries. The results showed that HAADF imaging can be used with a spatial resolution < 1 nm to measure composition gradients in interface/boundary regions that will change local lattice constants, elastic moduli, flow stress and other important interfacial parameters.

EXPERIMENTAL METHODS

The specimens were prepared from Toyo-Soda α-Silicon nitride (TS10) powder ball milled, with sintering aids namely, 5.5 wt% yttria and 1.1 wt% alumina. 20 vol% Huber β-SiC(w) and silicon nitride mixture were then presintered at 1500°C in 0.1 MPa Argon. Complete densification was achieved by hot-isostatic-pressing the compact at 1780°C in 190 MPa Argon.

The structure of interfaces was examined in an Akashi 002B microscope, with a point resolution of 0.18 nm. The chemistry at and near the interfaces was analyzed by position-resolved EELS (PREELS) technique developed in this laboratory. A Philips 400ST TEM fitted with a field emission gun, coupled to a Gatan 666 parallel EELS detector was used for this purpose. An approximately 3 nm diameter probe with a current density of the order of 10^8 A/m^2 at the specimen and a beam convergence half angle of 5 mrad were used. The acceptance half angle for the parallel EELS detector was 10 mrad. The PREELS technique was described in detail elsewhere [3]. These electron optical conditions will broaden an atomically sharp chemical interface by about 6 nm [4] which is much smaller than the chemical interface widths we observed. The EELS and EDS spectra reported here were taken from specimen regions thin enough for quantitative HREM so that probe broadening did not yield significant loss of resolution.

HAADF images are formed by collecting high-angle scattered electrons with an annular dark-field detector in a dedicated STEM VG HB501 with an image resolution < 1 nm. HAADF images give high resolution imaging contrast which is sensitive to the atomic numbers of the scatterers. HAADF Images are also free of the complications of dynamical scattering and coherent interference effects which are limitations of bright-field electron microscopy. By increasing the inner collection angle of the annular dark-field detector to very high angles (~ 70-100 mrad) HAADF images can be interpreted as a two-dimensional chemical map of the examined sample with high atomic number contrast provided that the sample thickness does not vary abruptly and that electron channeling effects are avoided [5, 6]. The effects of beam broadening are also reduced by collecting high-angle scattered signals since the cross-section of high-angle scattering events is smaller. These high-angle scattered signals, however, are not spectroscopic. Therefore, the interfacial chemistry of the CMC was also analyzed by a windowless energy dispersive spectrometer with a probe size < 1 nm. The combination of the more efficient HAADF signal with the conventional analytical signals (electron energy loss and characteristic X-ray signals) has proven very powerful for characterizing the chemistry of grain boundaries with heavy element segregation. HAADF intensity line scans across Si_3N_4 matrix grain boundaries and SiC/Si_3N_4 interfaces were obtained by off-line processing of the digitally acquired HAADF images. Special care was taken when aligning the whisker/matrix crystals in order to obtain edge-on interfaces and also to avoid electron beam channeling effects. This was accomplished by examining the convergent beam nanodiffraction patterns from the crystals and by observing the contrast variations of the grains in HAADF images.

EXPERIMENTAL RESULTS

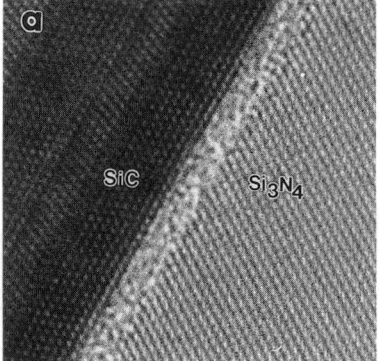 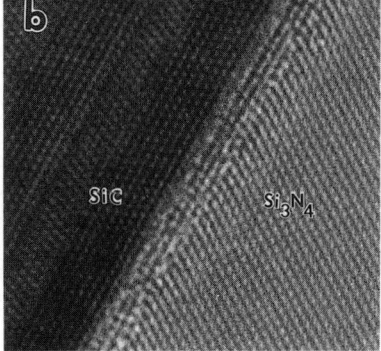

Fig. 1. HREM images of $SiC(w)/Si_3N_4$ interfaces revealing (a) the presence of an amorphous interfacial film with a width of ~ 2 nm and (b) direct bonding between the SiC whisker and the Si_3N_4 matrix grain.

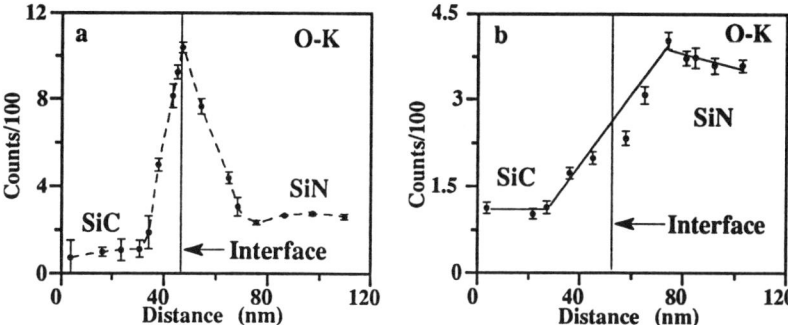

Fig. 2. Variations of oxygen concentration as a function of distance from the whisker/matrix interface. A sharp increase of oxygen (a) is detected at the interface shown in Fig. 1a and no sharp rise of oxygen concentration (b) is detected at the interface shown in Fig. 1b.

Figures 1a and 1b show, respectively, HREM images of different areas of a SiC whisker/matrix interface in the CMC. The detailed discussions of the physical nature of the interfacial layer was given elsewhere [2]. Figure 1a reveals a 1nm-wide bright amorphous band while Figure 1b shows an almost crystalline interface. Both these images were obtained from the same interface but the interfacial regions chosen were around 20 nm apart. The bright contrast seen on the matrix grain side of the interface (Figure 1b) does not imply non-edge-on orientation of the interface with respect to the beam direction. Upon close inspection of the interface, no moire fringes are visible at all. Moreover, the lattice fringes from the matrix grains abut very closely with one of the two sets of <111> fringes from the SiC(w). The curvature of fringes making a large angle with the interface and the change in contrast of fringes parallel to the interface result from local changes in sample thickness at the interface, caused by non-uniform ion-milling of the specimen. Position-resolved EELS results shown in Figure 2 indicate that there is a sharp rise in oxygen concentration (Figure 2a) at the SiC whisker/matrix interfacial region shown in Figure 1a but there is no such sharp increase in oxygen concentration (Figure 2b) at the SiC whisker/matrix interface shown in Figure 1b. Therefore, it is concluded that the oxygen rich interfacial layer is discontinuous along the SiC(w)/Si$_3$N$_4$ interfaces. It should be noted that the structural width of the oxygen rich amorphous film at the SiC(w)/Si$_3$N$_4$ interface shown in Figure 1a is only about 1 nm but the chemical width of the interfacial region is much broader (~ 40 nm).

Figures 3a and 3b show, respectively, a typical HAADF image and 3-D intensity line scans of a Si$_3$N$_4$ matrix grain boundary. The bright band in the HAADF image suggests that this interfacial region contains elements that give higher intensity of the high-angle scattered electrons. Since HAADF images are not spectroscopic nanometer resolution EDS point analysis was used to determine the specific elements present at the grain boundaries. Figure 3c shows EDS spectra obtained at the grain boundary (dotted line) and inside the Si$_3$N$_4$ matrix grain (solid line). It can be seen from these EDS spectra that the interfacial region contains yttrium in addition to silicon and oxygen and a small amount of aluminum. It should be noted that a significant amount of nitrogen is also detected in this Si$_3$N$_4$ grain boundary. No yttrium, however, is present inside Si$_3$N$_4$ grains. The presence of yttrium atoms at the grain boundaries gives the high bright contrast in the HAADF image of Figure 3a. From the analyses of many EDS spectra and high resolution X-ray mapping of the

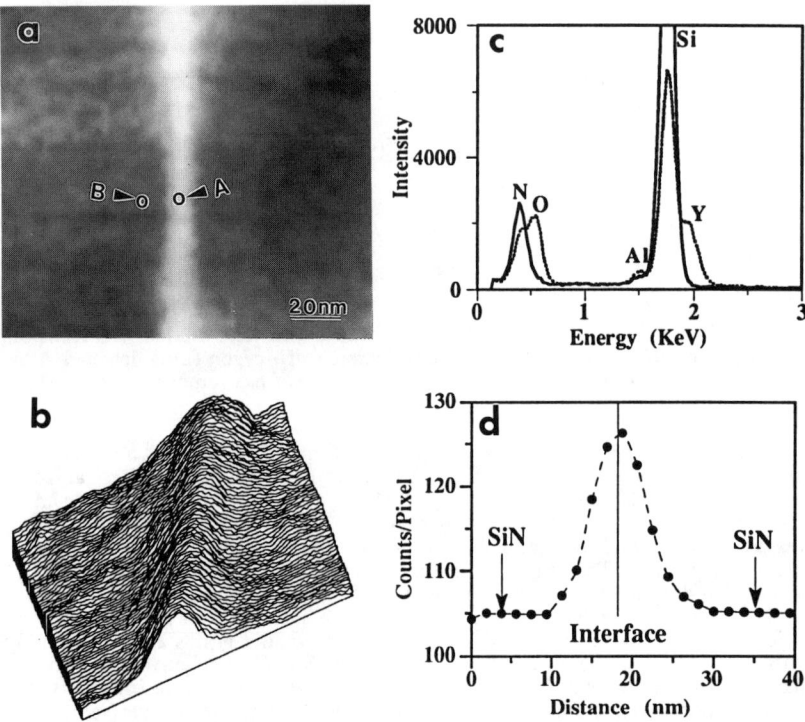

Fig. 3. (a) HAADF image and (b) 3-D intensity line scans of a Si_3N_4/Si_3N_4 grain boundary in CMC. Fig. 3c shows EDS spectra obtained at the interface (dashed line) indicated by A in Fig. 3a and from within Si_3N_4 grain (solid line) indicated by B in Fig. 3a. Fig. 3d is a HAADF intensity line scan across the grain boundary showing the symmetric distribution of yttrium atoms at the Si_3N_4 matrix grain boundary with a width of about 20 nm.

CMC we concluded that the bright contrast of the interfacial regions in HAADF images of CMC always represent the presence of yttrium atoms. Furthermore, it is found that oxygen is always associated with the distribution of yttrium atoms at Si_3N_4 matrix grain boundaries and at SiC(w)/Si_3N_4 interfaces.

Quantitative information about the distribution of yttrium atoms in the interfacial region can be extracted from HAADF intensity line scans across the interfaces. Figure 3d shows such an intensity line scan averaged over about 40 nm laterally along the interface. The yttrium has a symmetric distribution about the boundary with a total width about 20 nm. Both symmetric and asymmetric grain boundary elemental distributions have been observed and the chemical widths of these grain boundary regions range from about 20 nm to 60 nm. These are significantly different from the structural widths of about 1~2 nm as revealed by HREM images. HAADF images showed that the yttrium distribution at Si_3N_4 grain boundary regions is continuous in the CMC.

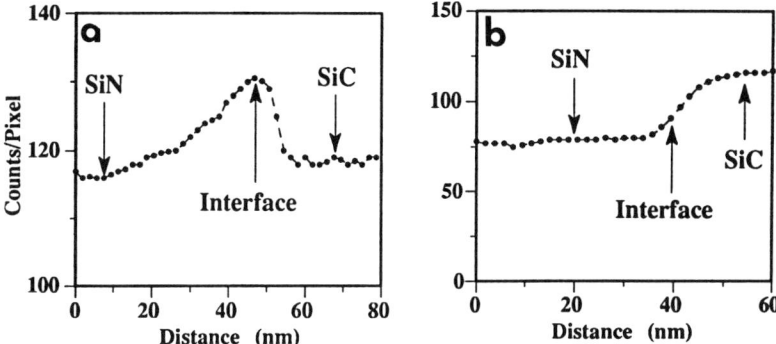

Fig. 4. HAADF intensity line scans across SiC(w)/Si3N4 interfaces revealing the presence (a) and absence (b) of yttrium atoms at the whisker/matrix interfaces.

The distribution of yttrium atoms at SiC(w)/Si3N4 interfaces was investigated by HAADF imaging technique. Figure 4a shows a HAADF intensity line scan across a SiC(w)/Si3N4 interface showing the distribution of yttrium at the SiC whisker/matrix interfacial region. The yttrium atoms diffused much deeper into the Si3N4 grains than into the SiC whisker, yielding an asymmetric compositional profile across the SiC whisker/matrix interface. The total width of the yttrium distribution at this interface is about 40 nm. Unlike that for Si3N4/Si3N4 grain boundaries the interfacial region of yttrium rich materials present at the SiC(w)/Si3N4 interfaces is discontinuous. Figure 4b is a HAADF intensity line scan across another region of the SiC(w)/Si3N4 interface. No sharp increase in the HAADF intensity at this interface is observed, suggesting that no yttrium atoms are present at this interfacial region. Nanometer resolution EDS point analysis also indicates that there is no sharp rise of oxygen concentration either at this interfacial region. The higher intensity of the SiC grain shown in Figure 4b is probably caused by a thickness effect.

DISCUSSION

The structural widths of the SiC whisker/Si3N4 matrix and Si3N4/Si3N4 matrix grain boundary interfaces are typically 0.5-2 nm. The chemical widths of these interfaces, however, are an order of magnitude broader than the structural widths. Position resolved PEELS results showed that a discontinuous oxygen rich region is present at the SiC whisker/matrix interfaces and a continuous oxygen rich region exists at the Si3N4 matrix grain boundaries. The presence of the oxygen rich region near the interfaces was previously attributed to the substitution reaction during synthesis between aluminum and silicon ions present in the sintering aids and the matrix powder [2, 7]. Diffraction analyses and HREM images suggested that the CMC matrix studied here was fully transformed into β-phase during the synthesis process. In α-Si3N4 the reaction between alumina and Si3N4 leads to more substitution of silicon by aluminum than nitrogen by oxygen and additional cations are needed to balance charge. Therefore, large modifying yttrium cations can be incorporated into large interstices in the structure. For β-Si3N4, however, the substitution is complete and no additional cations are needed to keep the charge balance.

HAADF images, however, showed that yttrium distribution is similar to oxygen at whisker/matrix and matrix grain boundary interfaces. This suggests that yttrium dissolve into β-Si$_3$N$_4$ grains during the synthesis of the CMC. Results from EDS point analysis and elemental mapping indicate that yttrium and oxygen have the same spatial distribution at the interfaces and grain boundaries in the CMC. It should be noted that aluminum diffused into the β-Si$_3$N$_4$ matrix (Figure 3c). Our data indicate that yttrium, aluminum and oxygen diffuse into the β-Si$_3$N$_4$ matrix for distances much larger than the structural width of the interfaces/boundaries, but still small on an absolute scale. The diffusion range in SiC is even smaller (Figure 4a). In this case, for β-Si$_3$N$_4$, it appears that yttrium substitutes for some of the aluminum, probably on aluminum sites, during cation-oxygen diffusion into the silicon nitride. It is also possible that the Y-Al diffusion had occurred early during processing of α-Si$_3$N$_4$ if the α to β transformation is slow relative to Y-Al diffusion. However, no rejection of yttrium as precipitates in the β-Si$_3$N$_4$ grains has been observed by HREM, so we believe the former mechanism is the most likely explanation for the presence of yttrium in β-Si$_3$N$_4$.

CONCLUSION

Whisker/matrix interfaces in SiC(w)/Si$_3$N$_4$ and matrix grain boundaries in Si$_3$N$_4$ were investigated by HREM, position resolved EELS and HAADF imaging techniques. HREM images show that a discontinuous layer of amorphous material with a width about 1 nm is present at SiC(w)/Si$_3$N$_4$ interfaces. Position resolved EELS, nanometer resolution EDS and HAADF results reveal, however, a much broader chemical width of the oxygen and yttrium rich interfacial regions.

We demonstrated that the combination of high resolution structural imaging with high resolution HAADF imaging and high spatial resolution microanalysis is both necessary and powerful for analyzing interfacial structures in composite materials.

ACKNOWLEDGMENTS

This research was supported by Shell Development Company (J.Liu) and by a grant from US DOE Basic Energy Sciences, DE-FG02-87ER-45305 and conducted at the Center for High Resolution Electron Microscopy, which is supported by the National Science Foundation under grant No. DMR-9115680.

REFERENCES

[1] D. B. Marshal, B. B. Cox, W. L. Moms and M. C. Shaw, in : Advanced Composite Materials, Ed., M. Sacks, (Published by American Ceramic Society, 1991) 459.
[2] K. Das Chowdhury, R. W. Carpenter and W. Braue, Ultramicroscopy **40**, 229 (1992).
[3] J. K. Weiss, P. Rez, and A. A. Higgs, Ultramicroscopy **41**, 291(1992).
[4] M. Catalano, M. J. Kim, R. W. Carpenter, K. Das Chowdhury and W. Braue, submitted to J. Mater. Res.
[5] S. J. Pennycook, S. D. Berger and R. G. Culbertson, J. Microscopy **144**, 229 (1986); S. J. Pennycook, Ultramicroscopy **30**, 58 (1989).
[6] J. Liu and J. M. Cowley, Ultramicroscopy **34**, 119 (1990); **40**, 352 (1992).
[7] K. H. Jack, in: Progress in Nitrogen Ceramics , Ed., F. L. Riley, (Martinus Nijhoff Publishers, 1983) 45.

INTERFACE STRUCTURE OF Si_3N_4 MATRIX COMPOSITE WITH NANO-METER SCALE SiC PARTICLES

GEN SASAKI,* KATSUAKI SUGANUMA,** TERUAKI FUJITA,** KENJI HIRAGA,*** AND KOICHI NIIHARA****
*Research Center for Advanced Science and Technology, The University of Tokyo, Komaba 4-6-1, Meguro-ku, Tokyo 153, Japan.
**National Defense Academy, 1-10-20 Hashirimizu, Yokosuka, 239, Japan.
***Institute for Materials Research, Tohoku University, 2-1-1 Katahira, Sendai 980, Japan.
****Institute of Scientific and Industrial Research, Osaka University, 8-1 Mihogaoka, Ibaraki 567, Japan.

ABSTRACT

β-Si_3N_4 matrix composites reinforced with nano-meter scale SiC particles were fabricated by hot-pressing the mixture of SiC and Si_3N_4 fine powders. Interface microstructure and crack propagation in the composite were observed by high resolution electron microscopy. The interface between SiC and Si_3N_4 had good coherence, but no preferred orientation was observed. At the interface between SiC and Si_3N_4, the (111) face of SiC particles was flat, and sometimes a disturbed lattice structure thinner than 1 nm was observed. Near the main cracks, the microcracking at fine SiC particles/Si_3N_4 matrix interface was frequently observed. The cracks propagated exactly along this interface. The interface microcracking seems to contribute to the improvement of fracture toughness.

INTRODUCTION

Recent investigators[1-4] attempted to improve the mechanical and thermomechanical properties of Si_3N_4 by adding ceramic fiber or particles to the monolithic Si_3N_4. The dispersoid SiC seems to be superior because of its high hardness and strength at high temperatures. In previous reports,[5-6] a Si_3N_4 matrix composite dispersed with nano-meter SiC particles was fabricated by hot-pressing the mixture of fine powders. The mechanical properties of the composite improved over those of the monolithic Si_3N_4. Figure 1[5] represents the influence of SiC volume fraction on bending strength and fracture toughness of the composites at room temperature. As SiC content increases to 5 or 10 vol%, bending strength and fracture toughness increases until a maximum value is reached. This dependence is due to strengthening caused by the dispersed nano-meter scale SiC particles in the Si_3N_4 grains and the rod-like grain shape of Si_3N_4. With increasing SiC content, the mechanical properties were finally degraded. This is caused by the fine and equiaxed grain

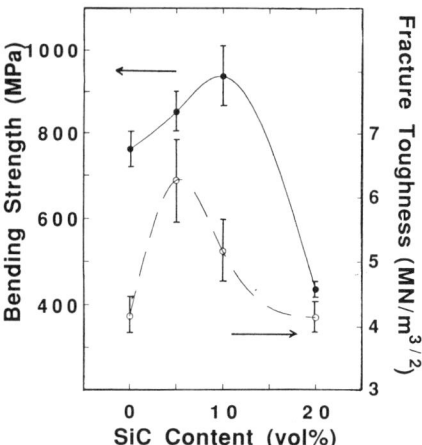

Fig. 1. Bending strength and fracture toughness as a function of SiC content.[5]

structure of Si_3N_4 and the reduction in the relative density. In addition, with increasing SiC content, the fracture morphology changed from transgranular to intergranular. The interface structure has not been investigated, although it strongly affects the mechanical properties. Some of the present authors have investigated[7] the interface structure in the Si_3N_4-SiC nano-composite fabricated by hot-pressing the CVD Si-C-N amorphous powder. Good matching was found on the straight interfaces between SiC and Si_3N_4. The preparation of the composite was done by
in-situ precipitation, and is different from the case of the composite investigated in this present study.

The purpose of the present study is to investigate the interface structure between SiC and Si_3N_4 in the mechanically mixed composite, to consider the relationship between the crack propagation mechanism and the microstructures, and finally to discuss the relationship between interface structure and mechanical properties in this composite.

EXPERIMENTAL PROCEDURE

The Si_3N_4 used as the matrix was an α-phase powder with an average particle size of 0.2 μm (Ube Industries Co., Ltd., E10 grade). The SiC used as the dispersoid was a β-phase powder with an average particle size of 0.3 μm (Ibiden Co., Ltd., UF grade). Volume fraction of the added SiC powder varied from 0 to 0.20. Y_2O_3 powder (Nippon Yttrium Co., Ltd.) was added as the sintering aid with a weight fraction of 0.08 of Si_3N_4 and SiC powder mixture. These powders were mixed by ball-milling with an Al_2O_3 ball using an ethanol solvent for 2.8×10^5s in a polyethylene vessel. Hot-pressing was performed at 2073 K for 3.6×10^3 s in N_2 atmosphere under a pressure of 30 MPa.

The microstructure of the interface and the crack propagation were observed by transmission electron microscopy (TEM) using a JEOL JEM-200CX operated at 200 kV. Cracks were introduced by the indentation method using a Vickers hardness tester with 1 kg load for 10 s.

RESULTS AND DISCUSSION

Figure 2 shows the typical microstructures of the composite. Inside Si_3N_4 grains, many nano-meter scale SiC particles were dispersed uniformly. The sizes of dispersed SiC particles were from 10 nm to 200 nm. Micrometer scale SiC particles were also seen at grain boundaries.

Figure 3 shows a high-resolution image of the interface between SiC and Si_3N_4. The SiC is a dispersed nano-meter scale particle inside the Si3N4 grains. Only the SiC lattice is visible in this image. The interface has some curvature and shows good bonding. No void, no dislocation and no intergranular phase were observed near the interface. Flat interfaces were frequently observed along the (111) plane of SiC. Flat {111} twin planes can also be seen.

Figure 4 shows another high-resolution image of the interface between SiC and Si_3N_4. The flat interface along the (111) plane of SiC was observed. The observed lattice image of Si_3N_4 was a (100) plane. However, no preferred orientation between SiC and Si_3N_4 was found. There was a disturbed lattice layer thinner than 1 nm at the interface between SiC and Si_3N_4. The disturbed layer seems to form by the relaxation of high interface energy caused by the anisotropy of the interface between SiC and Si_3N_4. This layer was not observed in the composite fabricated by hot-pressing from amorphous Si-C-N powder.[7] Recall that in the case of amorphous powder, SiC particles were formed by precipitation during hot-pressing. In such a case, the interface tended to have a preferred orientation corresponding to a low energy configuration.

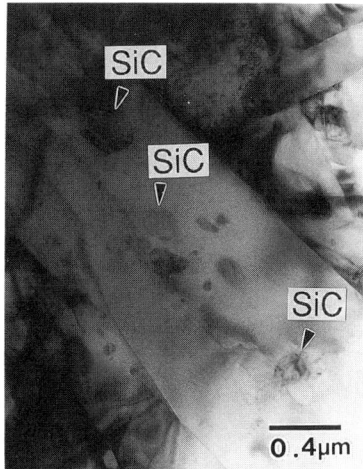

Figure 2. TEM micrograph of the nano-meter scale SiC particles dispersed in β-Si₃N₄ grains.

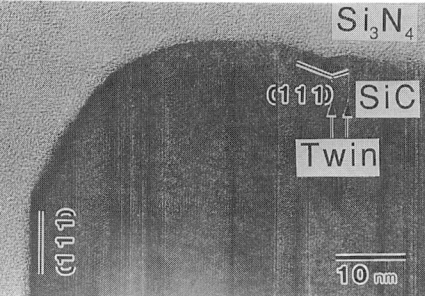

Figure 3. High-resolution electron micrograph of interface structure between SiC and Si₃N₄ at SiC particle dispersed in β-Si₃N₄ grain.

Figure 4. High-resolution electron micrograph of interface between β-Si₃N₄ and (111) face of SiC at SiC particle dispersed in β-Si₃N₄ grain.

Figure 5 shows the microstructure of starting SiC powder. The diameter of the powder particle in figure 5(A) ranged from 10 nm to 2000 nm. SiC powder particles were equiaxed and frequently had flat faces. Figure 5(b) shows the high resolution image of flat faces of the particle surface. Flat faces consist of {111} planes of SiC. This indicates that a flat interface on the (111) plane observed in the composite did not form during hot-pressing. It is likely that, as the (111) plane of SiC is a cleavage plane,[8] ball-milled SiC powder developed many flat {111} surfaces during processing.

Figure 6 shows a high-resolution image of the interface between SiC and Si₃N₄ without the disturbed layer. Imaged planes of SiC and Si₃N₄ are (111) and (110) planes, respectively. The interface has a gentle curvature and was observed edge-on. There were no interface dislocations. The lattices of SiC and Si₃N₄ had good coherence. The angle between the (111) plane of SiC and the interface is about 24.5°. On the other hand, the angle between the (110) plane of Si₃N₄ and the interface is about 31.5°. These show that the lattice distances of SiC and Si₃N₄, being perpendicular to the interface, are 0.553 nm and 0.550 nm, respectively. These lattice distances are nearly the same, and the misfit of these lattice distances is less than 2%.

Figure 7 shows a cracked microstructure with the crack propagating inside the Si₃N₄ grain. The crack was always deflected around the dispersed SiC particles and never

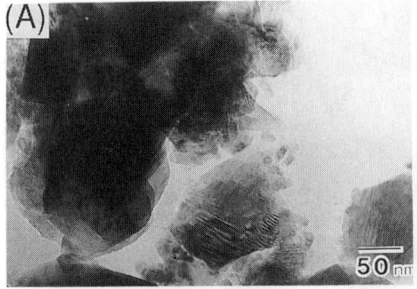

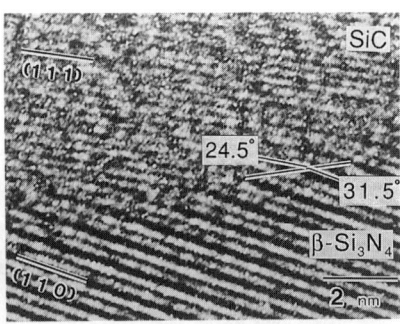

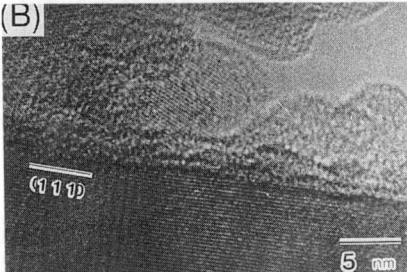

Figure 6. High-resolution electron micrograph of interface between SiC and β-Si_3N_4 without disturbed layer.

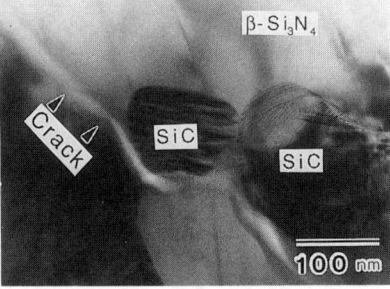

Figure 5. TEM micrographs of SiC powder as starting materials. Figure (A) shows the powder morphology, and figure (B) shows high-resolution electron micrograph at the surface of powder along (111) plane.

Figure 7. TEM micrograph of the deflected crack by SiC particle in β-Si_3N_4 grain.

propagated through a SiC particle. Thus, the strength of the interface between SiC and Si_3N_4 seems to be weak and such crack deflection may lead to an increase in toughness.

Figure 8 shows the microstructure at crack termination in a Si_3N_4 grain. The general position for the crack termination near a nano-meter scale SiC particle in a Si_3N_4 grain is seen in Fig. 8(A). Interface decohesion between SiC and Si_3N_4 near the main crack was observed in Figure 8(B). The width of interface decohesion is about 5 nm and such decohesion is exactly on the interface between SiC and Si_3N_4. No dislocation generation or movement, normally accompanying fracture, could be observed around the interface. Such decohesion which causes stress relaxation could suppress further crack propagation and increase fracture toughness. Yet the nano-meter scale SiC particles are small enough, as compared with the critical crack length for fracture, so that the weak bond between SiC and Si_3N_4 need not affect the decrease in strength. Figure 8(C) shows the high-resolution image of crack termination in Si_3N_4 grains, which shows crack termination to be atomistically sharp. In front of the crack termination, broad black and white contrasts caused by the strain field were also observed. There is no defect, such as a pinning particle, that is associated with crack termination.

Figure 9 shows the microstructure of intergranular crack propagation. The crack propagated exactly at the interface between SiC-Si_3N_4 and SiC-intergranular phase. The intergranular phase consisted of a sintering aid and Si_3N_4, and existed at triple points of the grain boundaries. This transgranular fracture was observed frequently with increasing SiC

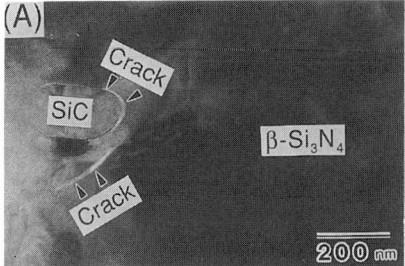
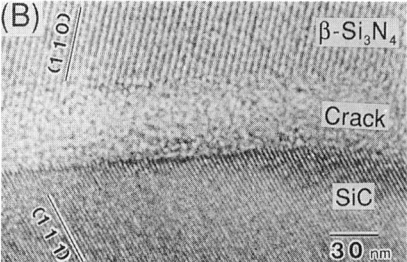
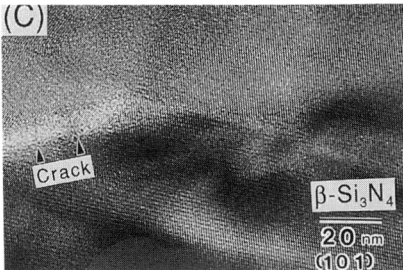

Figure 8. TEM micrograph of crack termination in β-Si₃N₄ grain. Figure (A) shows crack termination near nano-meter scale SiC particles, figure (B) shows the high resolution electron micrograph of interface decohesion between SiC and Si₃N₄, and figure (C) shows the high-resolution electron micrograph of crack termination in β-Si₃N₄ grain.

contents or shorter sintering time. These samples had SiC particles segregated at the grain boundaries. Such segregation seem to form the fracture origin.

CONCLUSION

The interface structure of a nano-meter scale SiC-particle-dispersed Si_3N_4 composite fabricated by hot-pressing the mixture of fine SiC and Si_3N_4 powders was investigated by conventional and high-resolution transmission electron microscopy. The relationship between interface structure and mechanical properties were discussed. The results are summarized as follows:

Figure 9. TEM micrograph of crack propagation at the interface between SiC-Si₃N₄ and SiC-intergranular phase in intergranular fracture.

(1) The interface between SiC and Si_3N_4 was curved in general, but flat faces were frequently observed. The flat part in the interface was along the (111) plane in SiC. Any preferred orientation between SiC and Si_3N_4 was not observed.

(2) The interface between SiC and Si_3N_4 had good coherency. Disturbed lattice layers, thinner than 1 nm, were frequently observed. The misfit of SiC and Si_3N_4 interface without disturbed layer was less than 2%, and no boundary dislocation was observed in this interface.

(3) Around the main cracks, the interface decohesion between Si_3N_4 and nano-meter scale SiC particles was frequently observed. This may induce stress relaxation by the debonding, which will suppress crack propagation.
(4) In intergranular fracture, the crack propagated along the SiC-Si_3N_4 and SiC-intergranular phase interface. Large SiC particles could be the fracture origin in this case.

ACKNOWLEDGEMENTS

The authors thank Dr. A. Nakahira and Associate Professor T. Suga for their help and discussion during this work.

REFERENCES

1. F.F. Lange, J. Am. Ceram. Soc., **56**, 518-22 (1973).

2. C. Greskovich and J.A. Palm, J. Am. Ceram. Soc., **63**, 597-99 (1980).

3. H. Komada, T. Suzuki, H. Sakamoto and T. Miyoshi, J. Am. Ceram. Soc., **73**, 678-83 (1990).

4. R. Lundberg, L. Kahlman, R. Pompe, R. Carlsson and R. Warren, Am. Ceram. Soc. Bull., **66**, 330-33 (1978).

5. G. Sasaki, H. Nakase, K. Suganuma, T. Fujita and K. Niihara, J. Ceram. Soc. Jpn., **100**, (4), 536-40 (1992).

6. G. Sasaki, H. Nakase, K. Suganuma, T. Fujita and H. Niihara, (Proc. 1st Intl. Symp. on the Science of Engineering Ceramics, 1991) pp. 291-96.

7. K. Niihara, K. Suganuma, A. Nakahira and K. Isaki, J. Mater. Sci. Lett., **9**, 598-99 (1990).

8. G. Sasaki, K. Hiraga, M. Hirabayashi, K. Niihara and T. Hirai, Adv. Ceram. Mat., **3** (4), 378-81 (1988).

CHARACTERIZATION OF HIP'ED, HIGH PURITY Si_3N_4 GRAIN BOUNDARIES

PING LU*, AND S. C. DANFORTH** AND W. T. SYMONS***
*Department of Materials Science, **Department of Ceramic Engineering, Rutgers, The State University, Piscataway, NJ 08855-0909
***A. C. Rochester, Div. of General Motors, 1300 N. Dort Hwy, Flint, MI 48556

ABSTRACT

Laser synthesized, ultra-fine, amorphous, Si_3N_4 powders were densified via HIP'ing without any oxide sintering aids. *Exposed* samples were made from powder that had been exposed to the atmosphere, thereby picking up an oxide surface layer, and *unexposed* samples were made from powders processed entirely under glove box conditions, i.e. without oxygen contamination. TEM (and sintering) studies indicate that the *exposed* samples HIP'ed at temperatures in excess of the melting point of SiO_2, densified via a solution-reprecipitation mechanism, with a resultant intergranular glassy phase of high purity SiO_2. In contrast, *unexposed* samples had to be HIP'ed to 2050°C to achieve a density of ~70 %ρ Th. These samples consisted of equiaxed β-Si_3N_4 grains, with localized high density regions where no inter-granular phase (crystalline or glassy) was detected to within 0.66 nm.

INTRODUCTION

A number of investigators have densified Si_3N_4 without the use of extrinsic sintering aids such as Al_2O_3, Y_2O_3, etc., utilizing the native SiO_2 layer on the starting Si_3N_4 powders to promote liquid phase sintering. Most investigations, such as Larker et al.[1], and Honma et al.[2], and more recently, Miyamoto et al.[3], and Tanaka et al.[4-5] have used crystalline α-Si_3N_4 starting powders. Symons et al.[6-7] have used ultra-fine, amorphous powders (with and without oxygen contamination) in an effort to produce Si_3N_4 with properties not dominated by any intergranular phase. Recently, Zeng et. al.[8] have shown that for Si_3N_4 with up to 20% added SiO_2, creep rates at 1400°C were nearly independent of SiO_2 content, provided that: 1) the SiO_2 was very pure, and 2) there was not enough SiO_2 present in the grain boundaries for easy cavitation to occur, i.e. ≤ 10 - 20 wt% SiO_2. This paper focuses on the characterization of the grain boundaries in Si_3N_4 HIP'ed from laser synthesized amorphous Si_3N_4 powders with and without significant levels of oxygen in the starting powders.

EXPERIMENTAL PROCEDURE

Si_3N_4 powders were prepared by a laser driven gas phase reaction.[6-7, 9] The Si_3N_4 powders were processed both with (*exposed*) and entirely without (*unexposed*) atmospheric exposure, see Table I. Samples were die pressed, CIP'ed at 350 MPa, outgassed at 600 °C for 12 hrs in a 10^{-5} torr vacuum, and subsequently HIP'ed in BN lined Pyrex™ ampules for 1-3 hr at temperatures from 1650°C to 2050°C at 193 MPa. TEM samples were prepared by mechanical polishing, dimpling, and finally, Ar ion milling at 5keV and 1 mA until perforation. An ISI-002B high resolution transmission electron microscope, which has a resolution limit of 0.18 nm, was used for observations.

RESULTS AND DISCUSSION

Samples HIP'ed From *Exposed* Si_3N_4 Powder

Earlier results [6-7] showed that densities of samples HIP'ed from *exposed* powder increased dramatically as the HIP'ing temperature exceeded 1800°C (Fig. 1), due to the melting of a SiO_2 glass phase and concurrent densification via a liquid solution-reprecipitation process. The amorphous *exposed* Si_3N_4 powders crystallized to 100% α–phase by 1650°C, i.e. crystallized fully prior to densification. Conversion of the α– to β–phase occurred above 1800°C, by solution-reprecipitation, concurrent with densification, and was complete by ~1950°C.

Table I.   Characteristics of Si_3N_4 Powders

Characteristics	*Exposed*	*Unexposed*
TEM (nm)	16.2 ± 2.1	---
BET (ESD) (nm)	16.9	16.7
Shape	Spherical	Spherical
BET Surface Area (m^2/g)	122	126
He Pycn. Density (g/cm^3)	2.91	2.85
Oxygen Content		
Neutron Act. An. (wt%)	4.0@	0.2
XPS (at.%)	11@	5
XPS N/Si (at. Ratio)	0.577@	0.608

@ Dependent on atmospheric exposure time.

Samples HIP'ed below 1750°C consisted of equiaxed α-Si_3N_4 grains, with no evidence of an intergranular glassy phase. Once the melting point of SiO_2 is exceeded, the samples densified rapidly, and consisted of elongated, prismatic β-Si_3N_4 grains. This grain morphology is consistent with the well known solution-reprecipitation mechanism for Si_3N_4. STEM and EDX work by Symons [6-7] showed that the grain boundary phase contained only Si and O, (with N below the detection limit of ~3 wt%). It was therefore concluded that the grain boundary phase was high purity SiO_2. The presence of the liquid phase can clearly be observed in the dark field image of Fig. 2.[10] The secondary phase is bright and the crystalline β-Si_3N_4 grains are dark. The intersections of three (triple junctions) and four grain junctions are filled with the secondary phase. A thin layer of the secondary phase is also present at the majority of grain boundaries (two grain junctions), i.e. Fig. 3.

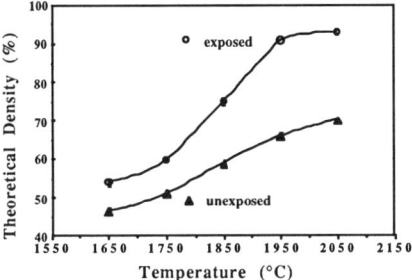

Fig. 1. Density vs. temperature. for samples HIP'ed at 193 MPa, 1 hr.

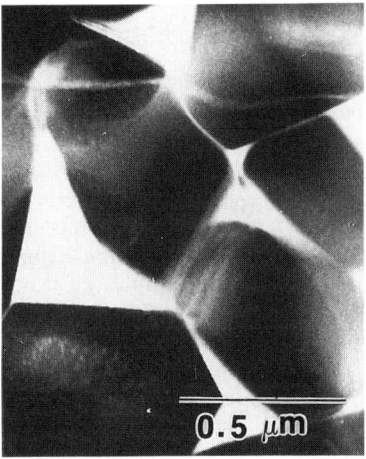

Fig. 2. Dark field TEM image of *exposed* Si_3N_4 HIP'ed at 1950°C, 1 hr.

The thickness of this SiO_2 grain boundary phase is 1.5 nm, in fairly good agreement with the value of 1.68 nm calculated by Clarke for an equilibrium Si_3N_4-SiO_2-Si_3N_4 interface.[11]

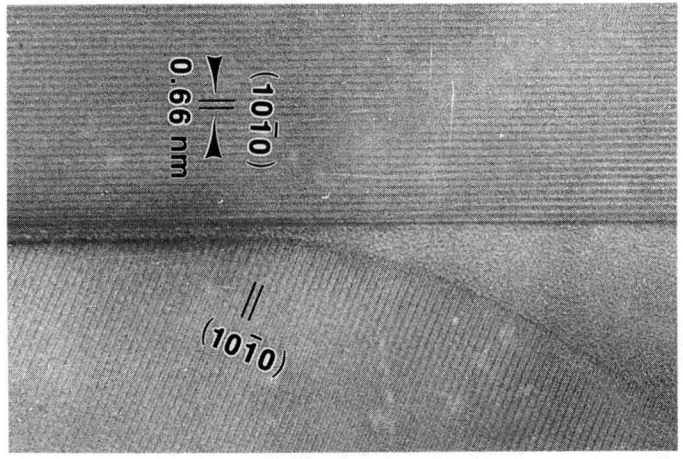

Fig. 3. HRTEM of *exposed* sample HIP'ed at 193 MPa for 1 hr at 1950°C, showing the 1.5 nm grain boundary glassy SiO_2 phase.

Samples HIP'ed from *Unexposed* Si_3N_4 Powder

HIP'ed *unexposed* samples crystallized to 80% α-Si_3N_4 and 20% β-Si_3N_4 at 1650°C, and converted to 100% β-phase above 1850°C. After HIP'ing for 60 minutes at 193 MPa and 1950°C and 2050°C, samples had densities of 65 %ρ Th. and 70 %ρ Th. respectively, Fig. 1, and had equiaxed β-Si_3N_4 grain morphologies. *Unexposed* samples HIP'ed at 1950°C possessed a relatively low average density, however, locally dense regions (as high as that in the *exposed* samples HIP'ed at 1950°C) were observed throughout the sample. The sample HIP'ed from *unexposed* powder at 2050°C showed more high density regions than the *unexposed* sample HIP'ed at 1950°C, with a 0.8 µm average grain size. Densification is presumed to occur via a combination of: rearrangement, enhanced atomic mobility during the amorphous to α- to β-Si_3N_4 transformations, and grain boundary diffusion in local high density regions between 1950°C and 2050°C.

In contrast to HIP'ed *exposed* samples, β-grains in *unexposed* HIP'ed samples are not surrounded by any detectable secondary phase. Dark field TEM, EDX-STEM, and HRTEM images were not able to reveal the presence of <u>any</u> secondary phases at the triple junctions or at two grain boundaries. Fig. 4 is a HRTEM image showing a typical two grain boundary for the *unexposed* Si_3N_4 sample HIP'ed at 1950°C for 1 hr.

This interface (Fig. 4a)) is aligned parallel to the electron beam, as supported by the appearance of overlap fringes, shown in Fig. 4b, when the interface is tilted by 2°. Despite the boundary's curvature, the central region of the interface is without overlap and shows no grain boundary phase to at least the 0.66 nm lattice fringe spacing. No grain boundary glassy phase was observed in triple junctions either.[10]

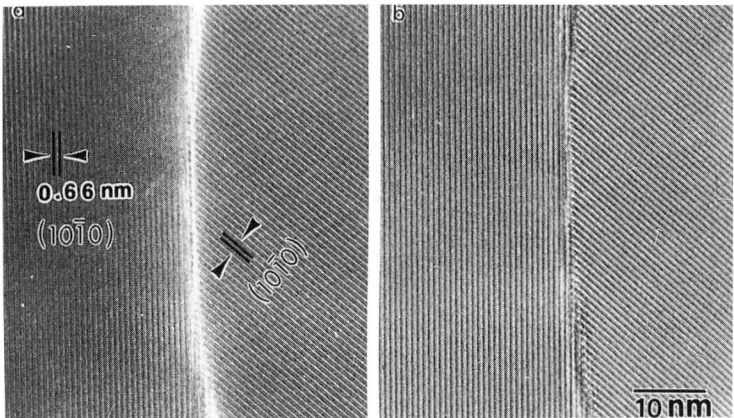

Fig. 4. HRTEM of *unexposed* Si3N4 HIP'ed at 193 MPa for 1 hr at 1950°C, showing a typical "clean" grain boundary, a) boundary nearly parallel to, and b) boundary tilted ~2° perpendicular to interface.

The literature contains additional published work [12,13] examining the interfacial characteristics in Si3N4 ceramics with various grain boundary chemistries. Low energy grain boundaries of the following types: (0001)//(0001), (10$\bar{1}$0)//(10$\bar{1}$0) and (0001)//(10$\bar{1}$0) are reported to be free of any glass phase even with a large volume fraction of glassy phase present in the material. The two (i.e. Fig. 4) and three grain junctions of *unexposed* samples studied in this work, do not correspond to the low energy types that characteristically do not contain any glassy phase. These results indicate that the grain boundaries in the *unexposed* HIP'ed samples have very little (if any) glassy grain boundary phase.

SUMMARY

Transmission electron microscopy of samples HIP'ed from *exposed* Si_3N_4 powder, at temperatures in excess of the melting point of SiO_2, shows microstructural evidence that is consistent with densification via a solution-reprecipitation mechanism, i.e. formation of a continuous SiO_2 grain boundary phase and elongated β-Si_3N_4 grains. In contrast, samples of *unexposed* Si_3N_4 powder must be HIP'ed to 2050°C to achieve a density of 70 % ρ Th. In this state, the sample consists of equiaxed β-Si_3N_4 grains, with both low and (locally) high density regions. No inter-granular phase, crystalline or glassy, was detected in these *unexposed* samples.

ACKNOWLEDGMENTS

We thank Dr. D. Niesz for his thoughtful discussions, and the Rutgers Center for Ceramic Research and the New Jersey Commission on Science and Technology for supporting this work.

REFERENCES

1. H. Larker, J. Adlerborn, and H. Bohman, Soc. Autom. Eng. Report 770335, (1977).
2. K. Honma, H. Okada, T. Fujikawa, and T. Tatuno, Yogyo-Kyokai-Shi **95**, 91 (1987).
3. Y. Miyamoto, K. Tanaka, M. Shimada, and M. Koizumi, "Survey of HIP Sintering Condition and Characterization of Dense Silicon Nitride Without Additives", <u>Ceramic Materials and Components for Engines</u>, Ed. W. Bunk, and H. Hausner (Deutsche Keramische Gesellschaft, Germany, 1986), p. 271.
4. I. Tanaka, G. Pezzotti, T. Okamoto, Y. Miyamoto and M. Koizumi, Ceram. Eng. Sci. Proc. **10** (7-8) 817-822 (1989).
5. I. Tanaka, G. Pezzoti, T. Okamoto, Y. Miyamoto and M. Koizumi, J. Am. Ceram. Soc., **72** (9) 1656-60 (1989).
6. W. Symons, and S. C. Danforth, in <u>Ceramic Materials and Components for Engines</u>, edited by V. J. Tennery (Am. Cer. Soc., OH, 1989), p.67.
7. W. Symons, PhD Thesis, Rutgers University, October, 1989.
8. J. Zeng, I. Tanaka, Y. Miyamoto, O. Yamada, and K. Niihara, J. Amer. Cer. Soc., **75** (1), 195-200 (1992).
9. A. Bhanap, MS Thesis, Rutgers University, 1991.
10. P. Lu, W. Symons, S. C. Danforth, In press, *J. Mater. Sci.*, 1993.
11. D. R. Clarke, in Ann. Rev. Mater. Sci., edited by R. A. Huggins, J. A. Giordmaine, J. B. Wachtman, Jr., (Annual Reviews Inc., **17**, CA, 1987) pp. 57-74.
12. D. A. Bonnell, *Mat. Sci. Forum*, **47**, 132 (1989).
13. J. Y. Leval, A. Thorel, *Mat. Sci. Forum*, **47**, 143 (1989).

PARTIAL DEVITRIFICATION OF SINTERED SILICON NITRIDE DURING STATIC FATIGUE TESTING

W. BRAUE* AND G. D. QUINN**

*German Aerospace Research Establishment (DLR), D-5000 Cologne 90, Germany

**National Institute of Standards and Technology (NIST), Ceramics Division, Gaithersburg, MD 20899, USA

Abstract

The static fatigue behavior of sintered Y_2O_3/Al_2O_3-fluxed Si_3N_4 in air is controlled by slow crack growth or creep fracture. Partial devitrification of the amorphous grain boundary phase at $1000°C$ and $1100°C$ improves the static fatigue resistance with specimens surviving up to 1500 hrs. during stress rupture experiments. In this study the early stages of partial devitrification during static fatigue testing at $1000°C$ are investigated by conventional and analytical transmission electron microscopy with emphasis on nucleation and growth of $\delta\text{-}Y_2Si_2O_7$ and $X_1\text{-}Y_2SiO_5$ and possible constraints from different stress states. The results show that the stress state does not affect the nature of the secondary phase assemblage. However, the amount of crystallization is higher within the tensile region of the flexural specimens than in areas which experienced compressive stresses.

Introduction

Si_3N_4-based polyphase ceramics are densified via liquid-phase sintering and contain amorphous phases at triple grain junctions and grain boundaries which may be described as metal-Al-Si-O-N oxynitride glasses depending on the sintering additive composition. This intergranular non-crystalline phase is prone to partial devitrifivation if post-sintering annealing treatments, similar to the processing of glass-ceramic materials, are applied [1]. The controlled crystallization of refractory crystalline secondary phases from the transient oxynitride liquid has been particularly explored in the Y-Al-Si-O-N system [2–4] and has resulted in a significant improvement in the high-temperature resistance of sintered Si_3N_4.

It has been shown [5] that during static fatigue testing of Y_2O_3/Al_2O_3-fluxed Si_3N_4 in air, partial devitrification can occur at a sufficient rate at $1000°C$ and $1100°C$ to retard or completely offset time-dependent failure which otherwise is controlled by slow crack growth or creep fracture. The fracture map (Fig. 1a) depicts the stress-temperature regimes where each mechanism is dominant. The principal devitrification products in the high tensile stress portion of the flexural specimens are $\delta\text{-}Y_2Si_2O_7$ and minor N-apatite $Y_{10}(SiO_4)_6N_2$. The oxynitride preferentially crystallizes at higher temperatures and/or longer annealing times.

We were interested in the early stages of partial devitrification and give a comprehensive TEM report on nucleation and growth of $\delta\text{-}Y_2Si_2O_7$ and $X_1\text{-}Y_2SiO_5$ during static fatigue testing at $1000°C$. This approach includes possible effects from the state of stress on both the amount of crystallization and the nature of secondary phases formed.

Experimental Details

A commercially sintered Si_3N_4 grade (SNW-1000 from GTE WESCO, Belmont, California) containing 13 wt.% Y_2O_3 and 3 wt.% Al_2O_3 was employed in this study.

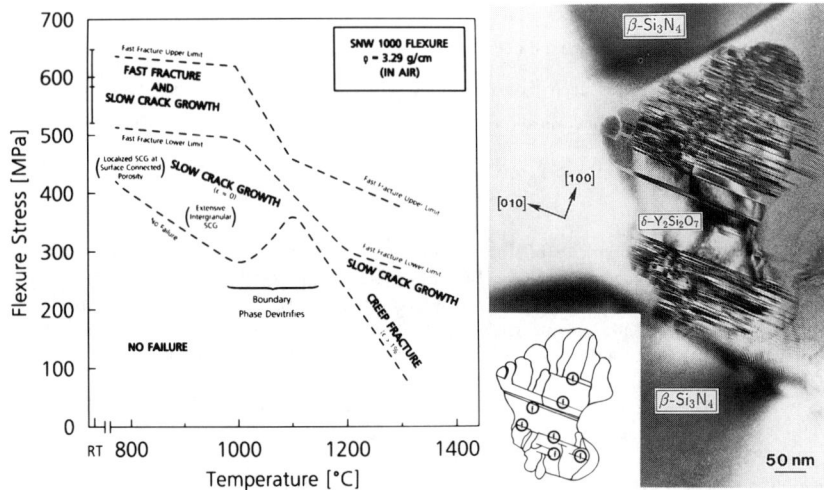

Fig. 1: (a) Fracture mechanism map for SNW-1000 Si_3N_4 in air [5] (b) δ-$Y_2Si_2O_7$ at triple grain junction (1000°C, no-stress)

The bulk composition of the batch (neglecting the Al-content for first approximation) is controlled compositionally to plot into the Si_3N_4-$Y_2Si_2O_7$-N-apatite $Y_{10}(SiO_4)_6N_2$ compatibility triangle of the Si_3N_4-Y_2O_3-SiO_2 ternary system. As-received SNW-1000 exhibits an amorphous intergranular phase.

The static fatigue tests at room and high temperatures were performed in four-point flexure as described elsewhere [6]. Specimens from stress rupture experiments at 1000°C, 250 MPa/5 hrs. and 300 MPa/5 min. respectively, were selected for the purpose of this study. The *no-stress* reference sample was prepared from an unloaded fast-fracture stress rupture experiment at 1000°C.

The TEM work was performed with a Philips EM 430T operating at 300 kV. Particular emphasis was on convergent beam electron diffraction (CBED) for correct identification of the fine-grained Y_2SiO_5 and/or $Y_2Si_2O_7$ polymorphs. Thin foils were prepared by standard cermographic techniques involving dimpling and argon ion beam thinning. To study the different stress states in the microstructure, the near-surface regions of a bend bar were sampled with the major foil plane parallel to the plane of tensile or compressive stress, respectively. Cross-section specimens with the foil plane perpendicular to the bend bar axis were prepared to monitor the microstructure in the proximity of the neutral axis.

Results

As a general finding of this study, the nature of the secondary phases derived from partial devitrification is independent of stress. The amount of crystallization however was higher on the tensile side of the flexural specimens than in compressive areas. The orthorhombic δ-$Y_2Si_2O_7$ polymorph was identified as the principal devitrifcation product during static fatigue testing at 1000°C. At triple grain junctions the δ-phase typically consists of an intergrowth of small dendrites (Fig. 1b) which coalesce into larger aggregates. The

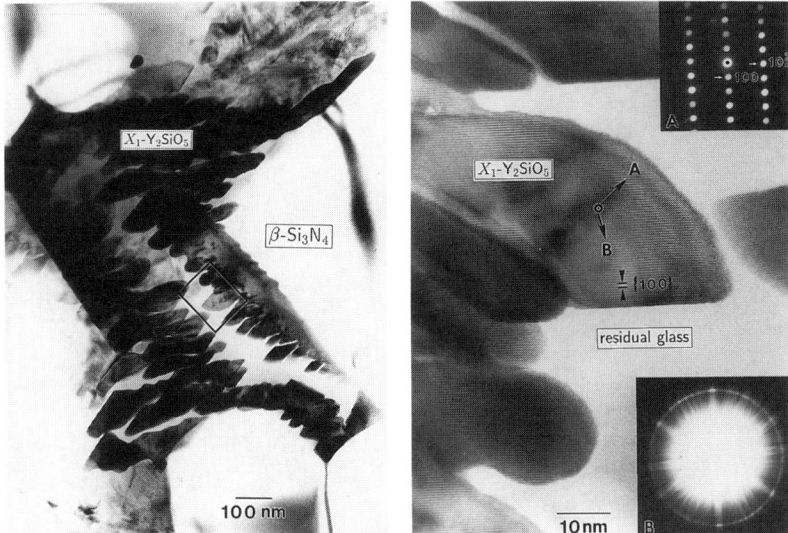

Fig. 2: (a) X_1-Y_2SiO_5 dendrites (orientation $\parallel$ [010]) growing into glassy pocket, (from 250 MPa/5 hrs., compression), for details of boxed region see (b).

dendrites share a common growth axis, but are slightly misoriented with respect to each other, giving rise to discontinuities in the stacking faults (thus revealing the peripheries of individual dendrites, as shown schematically in the inset of Fig. 1b). Matrix grains and the δ-phase are separated by a thin amorphous layer. The second devitrification product was unambiguously identified as X_1-Y_2SiO_5 by applying a combination of HREM, CBED and small-probe microanalysis. Similar to the δ-disilicate, the X_1 orthosilicate is characterized by a dendritic grain morphology (Figs. 2a/b). Some dendrites are bound by macroscopic {100} planes while others appear more cellular exhibiting smooth surfaces. During crystallization of the di- and orthosilicates, the retained glassy phase is depleted with respect to yttrium, but acts as major sink for aluminium and nitrogen. The average Y/Si ratio in the amorphous phase is approximately 1:1. Data scatter, however, is substantial and it is difficult to find homogeneous areas of glassy phase which are clearly connected to the surface of the foil and do not interfere with the fine-grained secondary phase particles.

Occasionally X_1-Y_2SiO_5 dendrites lie flat on β-Si_3N_4 matrix grains causing moire fringe contrast. In Fig. 3a the β-Si_3N_4 {101} and X_1-Y_2SiO_5 {121} lattice spacings give rise to a rotation moire with a fringe spacing of 6.2 nm. Aligning the boundary plane of this particular β-Si_3N_4/X_1-Y_2SiO_5 interface parallel to the electron beam was beyond the tilting range of the microscope. However, HREM images of similar phase boundaries oriented edge-on clearly reveal an amorphous layer at the interface with no evidence of heterogeneous nucleation of the secondary phase directy on the matrix grains. In Fig. 3b it is an orthosilicate dendrite (orientation $\parallel$ [032]) growing inside a triple grain junction which is separated from the matrix (orientation $\parallel$ [001]) by an amorphous layer of approximately 3 nm in thickness.

Individual "embryo" dendrites of the X_1 phase are frequently observed during the early stages of partial devitrification while the δ-phase mostly crystallizes as more extended aggregates (Fig. 1b). From this observation it was hypothesized in the beginning of

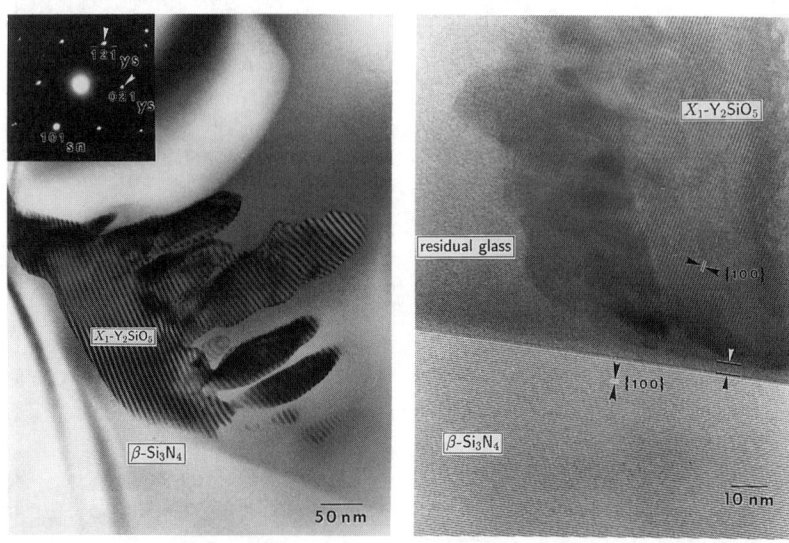

Fig. 3: X_1-Y_2SiO_5/β-Si_3N_4 boundaries as imaged perpendicular (a) and parallel (b) to the β-Si_3N_4 matrix grains.

this research, that the more Y-rich orthosilicate might act as a precursor for the disilicate following a reaction scheme $X_1\text{-}phase + liquid\,I \rightarrow \delta\text{-}phase + liquid\,II$ supported by diffusional interaction with the residual liquid phase. This transition stage, however, could not be identified in-situ. The di- and orthosilicates do frequently coexist within the same triple grain junction, as displayed in Fig. 4a. This BF image was taken from a cross-section thin foil of a 300 MPa/5 min. stress rupture experiment, emphasizing the very dynamic nature of partial devitrification during static fatigue testing at $1000°C$.

Discussion

Due to fundamental arguments related to energetic constraints during nucleation and growth of crystalline secondary phase particles along intergranular boundaries [7], devitrification in liquid-phase sintered Si_3N_4 is restricted to triple grain junctions only, leaving a thin film of amorphous phase at most grain-and phase boundaries. Therefore, the creep and static fatigue resistance of SNW-1000 after devitrifcation is still controlled by the intrinsic deformation behavior of the residual glassy layer via viscous flow or dissolution-reprecipitation processes [8], although the deformation characteristics of the di- and orthosilicates are of significance also. A higher degree of crystallization on the tensile side of flexural specimens has also been reported from devitrification of other vitreous-bonded ceramic materials [9]. As predicted by theoretical considerations [10], more glass is squeezed out of interfacial boundaries parallel to the applied stress in tension than in compression. Such geometry effects may account for the different degree of crystallization observed. They may also explain differences in the tension/compression creep rates if the level of microstructural homogeneity [9] is properly addressed.

The δ-phase is the high-temperature polymorph of the $Y_2Si_2O_7$ system characterized by a stability field above $1535°C$ [11]. The metastable crystallization of δ-$Y_2Si_2O_7$ during

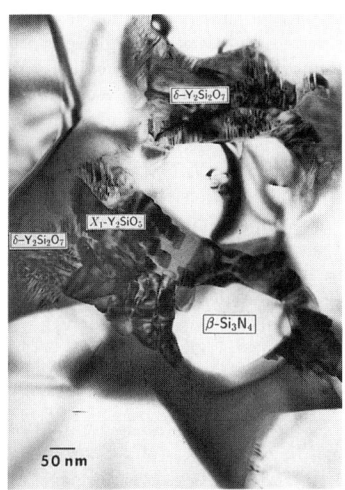 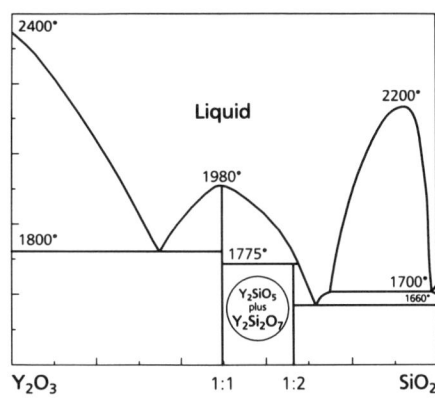

Fig. 4: (a) Crystallization of δ-$Y_2Si_2O_7$ and X_1-Y_2SiO_5 within the same triple grain junction.
(b) Schematic version of the Y_2O_3-SiO_2 binary system, after [4].

partial devitrification of SNW-1000 at temperatures as low as 1000°C is an interesting finding which may be rationalized in terms of kinetic constraints following Ostwald's step rule. This finding is in agreement with other reports on the stability of $Y_2Si_2O_7$ polymorphs [2–4], although a comparison is not easily achieved because of differences in the impurity content, the annealing atmospheres, as well as the heating and the quenching rates involved which even account for Si_3N_4 batches of similar additive composition. In a previous study on devitrification of SNW-1000 [5] it was suggested that the δ-phase may preferentially crystallize in stressed, as opposed to unstressed specimens, annealed under similar conditions because it's highly faulted structure may be able to accommodate deformation better than the other $Y_2Si_2O_7$ polymorphs. The present results however clearly provide evidence that the formation of the δ-phase is totally independent of stress. The dominance of the δ-polymorph, however, remains a remarkable result which requires further attention.

It is interesting to compare the crystallization of δ-$Y_2Si_2O_7$ and X_1-Y_2SiO_5 in SNW-1000 with the general phase relationships in the Y_2O_3-SiO_2 binary system [4]. The $Y_2SiO_5 + Y_2Si_2O_7$ two-phase stability field of the phase diagram (Fig. 4b), which indeed is reflected in the microstructure of SNW-1000 (Fig. 4a), is limited at 1775°C by the peritectic reaction $Y_2SiO_5 + liquid \rightarrow Y_2Si_2O_7$. No evidence was obtained that this diffusional interaction of the di- and orthosilicates with the residual liquid could survive metastably in the Si_3N_4 materials investigated. Although the Y_2O_3-SiO_2 phase diagram system predicts Y_2SiO_5 as a potential liquidus phase candidate, neither the X_1 nor the X_2 polymorph was reported in the literature at high temperatures and prolonged soaking times. The δ-phase, however, was again the most stable $Y_2Si_2O_7$ polymorph at 1600°C [4].

Both Y-bearing silicates share two characteristic features: i) the dendritic grain morphology, which is due to the rejection of Al and N to the residual glassy phase thus creating instabilities in the solid/liquid interface [2] and ii) their fine-grained nature, which obviously involves multiple nucleation sites. Nucleation and growth of the δ-phase and the X_1 orthosilicate clearly contrast with the crystallization of N-apatite during devitrification of SNW-1000 at $1100°C$ and above [5]. Starting from fewer nucleation sites, growth of N-apatite results in interconnected skeletons of uniform orientation spreading throughout the microstructure.

Because of the good lattice match, particularly between β-Si_3N_4 and X_1-Y_2SiO_5, epitaxial growth of the secondary phase on matrix grains was anticipated. The existence of an amorphous film at such phase boundaries does not imply that direct nucleation of X_1-Y_2SiO_5 on matrix grains might not be generally possible. In a highly dynamic microstructure a phase boundary may locally develop different wetting/dewetting characteristics [12], however, additional research is required to explore the detailed nucleation mechanism of crystalline secondary phases within glassy triple grain pockets of sintered Si_3N_4.

Conclusions

The static fatigue behavior of SNW-1000 at $1000°C$ and $1100°C$ in air is improved during partial devitrifcation of the intergranular phase. The early stages of this process are characterized by dendritic growth of δ-$Y_2Si_2O_7$ and X_1-Y_2SiO_5 at triple grain junctions. The state of stress only affects the amount of crystallization, not the nature of secondary phases. The temporary benefit of partial devitrification to the static fatigue resistance of SNW-1000 is balanced by nucleation and growth of creep cavitation from residual amorphous pockets as well as from grain and phase boundaries. Time dependent failure resumes at $1200°C$ and above.

Acknowledgement

The stimulating discussions with R. Pleger, DLR, Cologne, Germany, and H. J. Kleebe, MPI für Metallforschung, Stuttgart, Germany, are gratefully acknowledged.

References

[1] G. Leng-Ward and M. H. Lewis, in: *Glasses and Glass-Ceramics*, edited by M. H. Lewis (Chapman & Hall, New York 1989), p.106
[2] T. R. Dinger, R. S. Rai and G. Thomas, J. Am. Ceram. Soc. **71**, 236 (1988)
[3] W. E. Lee, C. H. Drummond, G. E. Hilmas, S. Kumar, J. Am. Ceram. Soc. **73**, 3575 (1990)
[4] S. Kumar and C. H. Drummond, J. Mater. Res. **7**, 997 (1992)
[5] G. D. Quinn and W. Braue , J. Mat. Sci. **25**, 4377 (1990)
[6] G. D. Quinn, J. Mat. Sci. **25**, 4361 (1990)
[7] R. Raj and F. F. Lange, Acta metall. **29**, 1993 (1981)
[8] J. E. Marion, A. G. Evans, M. D. Drory and D. R. Clarke, Acta metall. **31**, 1445 (1983)
[9] S. M. Wiederhorn, B. J. Hockey, R. F. Krause and K. Jakus, J. Mat. Sci. **21**, 810 (1986)
[10] J. R. Dryden, D. Kucerovsky, D. S. Wilkinson and D. F. Watt, Acta metall. **37**, 2007 (1989)
[11] K. Liddell and D. P. Thompson, Br. Ceram. Trans. J. **85**, 17 (1986)
[12] D. R. Clarke, Journal de Physique, **C4**, 51 (1985)

FABRICATION AND PROPERTIES OF Si_3N_4 WITH RARE EARTH APATITE GRAIN BOUNDARY PHASES

TERRY N. TIEGS, STEPHEN D. NUNN, KRISTIN L. PLOETZ, PAUL A. MENCHOFFER, AND CLAUDIA A. WALLS, Oak Ridge National Laboratory, P.O. Box 2008, Oak Ridge, Tennessee 37831-6087

ABSTRACT

The rare earth-oxide and nitride apatites were examined as grain boundary phases in silicon nitride to assess their potential for developing high toughness materials using gas-pressure-sintering. Densification was dependent on the quantity of additives used with high densities achieved at equivalent oxygen contents of ~8%. Fracture toughnesses (K_{Ic}) up to 8-10 MPa√m were obtained for some compositions. Ambient temperature flexural strengths were in the range of 400-720 MPa; however, the strengths at elevated temperatures (1200°C) were reduced from these values.

INTRODUCTION

The mechanical properties of Si_3N_4 ceramics have been recognized as being superior to most other ceramic systems. The explanation for much of the property improvements is the acicular grain growth of the ß-phase and the interlocking morphology of these grains in the densified materials.[1] Recent studies have shown that the grain growth can be manipulated to produce materials with very high fracture toughness values (K_{Ic}) in the range of 8-11 MPa√m.[2-5]

These high toughness materials are made by sintering at elevated temperatures to promote grain growth. At temperatures > 1825°C, Si_3N_4 readily decomposes under 0.1 MPa of nitrogen pressure (1 atmosphere) and consequently, nitrogen gas over-pressures are used to suppress the decomposition. Pressures of 1-10 MPa (10-100 atmospheres) are usually required to accomplish this and the technique is referred to as gas-pressure sintering (GPS).[6,7] Densification occurs by both particle rearrangement and solution-reprecipitation processes where the additives react with the surface SiO_2 on the starting α-Si_3N_4 particles to form a liquid phase.[8] This liquid phase, in turn, reacts with the α-Si_3N_4. Precipitation of ß-Si_3N_4 grains occurs from the liquid phase and, because of differences in growth kinetics on different crystallographic planes, the ß-Si_3N_4 grains grow with acicular or elongated morphologies. Diffusion through the liquid phase determines the densification behavior, the resulting grain sizes, the aspect ratio of the ß-Si_3N_4 grains and the extent of the α-to-ß conversion.[9]

A multitude of sintering additives has been used to densify silicon nitride. The desired qualities of the additives are that they form a liquid phase at the sintering temperature, to aid in densification, but that will be of a transient nature so they crystallize as a refractory grain boundary phase. Some possible alternate grain boundary phases include the rare earth-oxide and nitride apatites as discussed by Thompson.[10] These include a wide range of stable apatites of composition $A_2B_8(SiO_4)_6O_2$ where A is a divalent cation (typically Mg, Mn, Pb, Ca, Sr, Ba) and B is a trivalent cation (typically Y, La, Ce, Nd). An advantage of the nitrogen apatite is that it can oxidize without a large volume expansion. In this study, a range of silicon nitride compositions was prepared using various combinations of Y_2O_3, La_2O_3, Nd_2O_3, SrO, and BaO to develop these grain boundary phases. The microstructure development and properties of the resulting materials were examined to assess their potential for developing high toughness materials.

EXPERIMENTAL PROCEDURES

The starting materials consisted of appropriate amounts of Si_3N_4, Al_2O_3, La_2O_3, Y_2O_3, Nd_2O_3, SrO, and BaO as shown in Table 1. A composition containing Y_2O_3-Al_2O_3 was also included as a standard material since a large body of data already exists on similar compositions. The powders were turbomilled for ~2 h in deionized water (1.3 wt.% PVP K-15[b] and 1 wt.% Darvan 821A[c] were added as dispersants) to fully deagglomerate and mix the

---

[b] GAF Chemicals, Wayne, NJ: Polyvinylpyrrolidone K-15
[c] R. T. Vanderbilt, Norwalk, CT

various constituents.[11] Samples were formed by two methods. The mixtures for the TRSN series were first dried, screened to -100 mesh and isopressed at 207 MPa into discs approximately 7 cm in diameter and 1 cm thick. The SC series were slip-cast into tiles approximately 0.8 cm x 8.5 cm x 8.5 cm and then dried. The densities of the specimens at this point were ~60% T. D. for both methods. Binder burnout consisted of a heat-treatment to 600°C in air prior to sintering.

Sintering was performed in a graphite element furnace with gas overpressure capability. During sintering the samples were covered with a mixture of Si_3N_4-25% BN-5% Y_2O_3. Two GPS sintering conditions were examined in the current study: (1) 1850°C for 2 h at 50 psi followed by 1900°C for 2 h at 300 psi, and (2) 1900°C for 2 h at 100 psi followed by 1950°C for 2 h at 300 psi. A two-step firing procedure was used to minimize any density gradients across the cross-section of the samples, as this has been known to occur with one-step firing processes.[12] Both GPS runs had a 1 h hold at 1400°C after sintering with no applied overpressure (0.1 MPa) for crystallization of the grain boundary phases.

Densities were determined by the Archimedes method. Selected samples of high density were machined into bend bar specimens with nominal dimensions of 3 mm x 4 mm x 50 mm. Flexural strength testing was done in four point bending with inner and outer spans of 20 mm and 40 mm, respectively. Fracture toughness was determined by both indentation and indentation/fracture methods.[13,14]

RESULTS

A summary of the densities obtained during GPS is given in Table 2. Significant densification was observed with the apatite containing samples having high equivalent oxygen contents, whereas at the low oxygen contents, densification was difficult even at the higher temperature sintering conditions. Differences in densities among the three rare earth additives were related to their refractory nature. For example, the samples with the La_2O_3 addition exhibited better densification than the Y_2O_3 containing materials because of the lower eutectic liquid formation temperature for the La_2O_3 bearing materials. The samples containing Nd_2O_3 were intermediate in their densification behavior between the La_2O_3 and the Y_2O_3 containing samples. By the same reasoning, the sample with BaO (SC-291) showed less densification than a comparable sample using SrO (SC-293).

X-ray diffraction of the samples after sintering showed complete α-to-ß transformation and the formation of a crystalline second phase in all of the samples. While the peaks for the second phase were similar to those from known apatite phases, some peak shifting suggests that considerable solid solution between the oxygen and nitrogen apatites occurred. Consequently, no positive identification of the phases present could be made.

Mechanical property testing showed moderate room temperature flexural strengths for the TRSN samples in the range from ~400-750 MPa (Fig. 1). These strengths are similar to other GPS samples that have been reported.[2] One factor affecting the strength was exaggerated grain growth experienced by some of the compositions. For example, Fig. 2 shows the large grains observed in TRSN-3-2 (Si_3N_4-8% Y_2O_3-5.7 % SiO_2-1.8 % SrO sintered at 1900°C/2 h-1950°C/2h) which had a room temperature strength of 475 MPa. Such large grains are well known to have detrimental effects on the strength of silicon nitride materials. Similar grain growth was observed in other compositions with low strength. Modification of the sintering cycle temperature, times and pressures will be used in the future to control the grain growth behavior.

While the strengths at room temperature were reasonable, the strengths obtained at elevated temperature were significantly reduced (Fig. 1). The drop in strength at 1200°C for the compositions containing the rare earth apatite additives is comparable to the strength reduction for the sample containing Y_2O_3-Al_2O_3. This would indicate that these materials are not suitable for high temperature applications under stress.

Another reason for moderate strengths of the TRSN series appears to be the green forming technique used. The TRSN-series of samples were dried, screened and CIPed after milling. Apparently, hard agglomerates were formed in the drying steps and were carried over into the parts. Examination of fracture surfaces showed differential shrinkage around these hard agglomerates and the formation of large defects. This behavior is illustrated with a comparison of CIP-ed materials with the same composition that was slip-cast (TRSN-3-1 vs. SC-293). It shows a consistently higher strength for the slip-cast materials under identical sintering conditions.

Table 1. Summary of silicon nitride compositions with rare earth apatite grain boundary phases.

Sample ID	Additive Content (wt.%)					Equivalent % Oxygen[a]	Forming Method
	SiO_2[b]	Y_2O_3[c]	La_2O_3[c]	Nd_2O_3[c]	SrO[d]		
TRSN-1	---	6[e]	---	---	---	5.4	CIP
TRSN-2	5.5	---	11.2	---	1.8	8.0	CIP
TRSN-3	5.7	8	---	---	1.8	8.0	CIP
TRSN-4	5.5	---	---	11.5	1.8	8.0	CIP
TRSN-11	2.7	5.6	---	---	1.3	4.4	CIP
TRSN-12	2.7	---	7.9	---	1.3	4.4	CIP
TRSN-13	2.7	---	---	8.1	1.2	4.4	CIP
TRSN-17	3.5	8.2	---	---	1.9	6.0	CIP
TRSN-18	3.3	---	11.4	---	1.8	6.0	CIP
TRSN-19	3.3	---	---	11.7	1.8	6.0	CIP
SC-290	5.5	---	11.2	---	1.8	8.0	Slip-Cast
SC-291	5.6	7.9	---	---	2.7[f]	8.0	Slip-Cast
SC-293	5.6	8.0	---	---	1.8	8.0	Slip-Cast

[a] SiO_2 content of silicon nitride powder included. Ube Industries, Japan; Grade E-10 Si_3N_4
[b] Cabot Corp., MA; Cabosil M-5,
[c] Molycorp, White Plains, NY; >99.99%
[d] Mallinckrodt, St. Louis, MO; Reagent Grade
[e] 2% Al_2O_3 also added. Reynolds, Malakoff, TX; Grade RC-HP DBM.
[f] BaO, Mallinckrodt, St. Louis, MO; Reagent Grade

Table 2. Summary of results on densification of silicon nitride with rare earth apatite grain boundary phases.

Sample ID	Intergranular Phase Composition[a] (Wt.% Oxides)	GPS Cycle (°C/h)	Sintered Density (% T. D.)
TRSN-1-1	6Y-2Al	1850/2-1900/2	99.3
TRSN-2-1	11.2La-5.5Si-1.8Sr	1850/2-1900/2	99.8
TRSN-3-1	8Y-5.7Si-1.8Sr	1850/2-1900/2	99.0
TRSN-4-1	11.5Nd-5.5Si-1.8Sr	1850/2-1900/2	99.8
TRSN-11-1	5.6Y-2.7Si-1.3Sr	1850/2-1900/2	72.7
TRSN-12-1	7.9La-2.7Si-1.3Sr	1850/2-1900/2	84.3
TRSN-13-1	8.1Nd-2.7Si-1.2Sr	1850/2-1900/2	82.7
TRSN-17-1	8.2Y-3.5Si-1.9Sr	1850/2-1900/2	80.1
TRSN-18-1	11.4La-3.3Si-1.8Sr	1850/2-1900/2	95.6
TRSN-19-1	11.7Nd 3.3Si 1.8Sr	1850/2-1900/2	90.0
TRSN-1-2	6Y-2Al	1900/2-1950/2	99.2
TRSN-2-2	11.2La-5.5Si-1.8Sr	1900/2-1950/2	99.7
TRSN-3-2	8Y-5.7Si-1.8Sr	1900/2-1950/2	99.9
TRSN-4-2	11.5Nd-5.5Si-1.8Sr	1900/2-1950/2	99.9
TRSN-11-2	5.6Y-2.7Si-1.3Sr	1900/2-1950/2	75.9
TRSN-12-2	7.9La-2.7Si-1.3Sr	1900/2-1950/2	88.5
TRSN-13-2	8.1Nd-2.7Si-1.2Sr	1900/2-1950/2	88.7
TRSN-17-2	8.2Y-3.5Si-1.9Sr	1900/2-1950/2	84.3
TRSN-18-2	11.4La-3.3Si-1.8Sr	1900/2-1950/2	99.1
TRSN-19-2	11.7Nd 3.3Si 1.8Sr	1900/2-1950/2	96.2
SC-290	11.2La-5.5Si-1.8Sr	1850/2-1900/2	99.1
SC-291	7.9Y-5.6Si-2.7Ba	1850/2-1900/2	91.9
SC-293	8Y-5.5Si-1.8Sr	1850/2-1900/2	99.6

[a] Y=Y_2O_3; Al=Al_2O_3; La=La_2O_3; Nd=Nd_2O_3; Si=SiO_2; Sr=SrO; Ba=BaO.

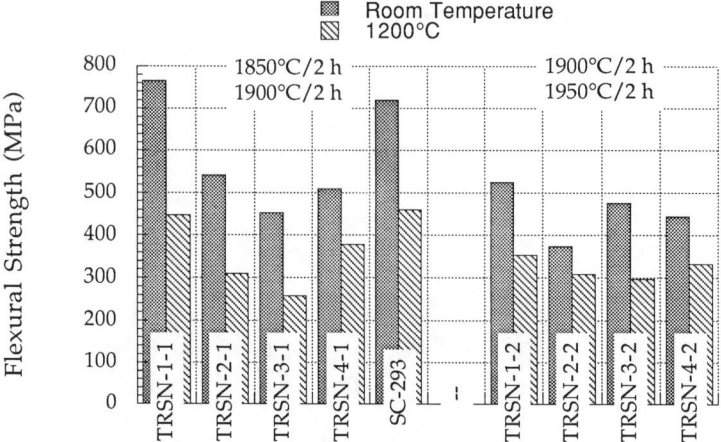

Fig. 1. Summary of results on flexural strength of GPS silicon nitride compositions at room temperature and 1200°C.

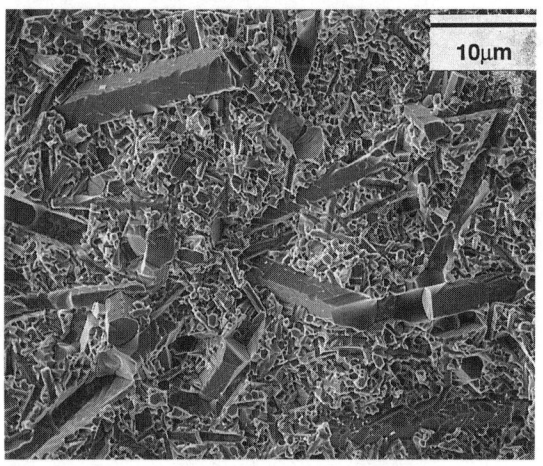

Fig. 2. Fracture surface of GPS silicon nitride showing exaggerated grain growth. Sample is TRSN-3-2 (Si_3N_4-8% Y_2O_3-5.7 % SiO_2-1.8 % SrO sintered at 1900°C/2 h-1950°C/2h).

During GPS, the high gas over-pressures result in dissolution of nitrogen into the intergranular phases and the formation of nitrogen-rich phases. These phases are metastable at one atmosphere. There was some evidence that during the crystallization step used in the present samples (1 h hold at 1400°C and 0.1 MPa nitrogen), excess nitrogen exsolved and formed gas bubbles at the grain boundaries. Future work will include modification of the crystallization cycle temperature, times and pressures to eliminate these types of defects.

A summary of the results on the fracture toughness (K_{Ic}) is shown in Fig. 3. Measurements were made by two different methods which showed only rough agreement between them. Generally, the highest toughnesses were observed with the Y_2O_3-SrO containing materials (TRSN-3) while the lowest toughness were associated with the samples

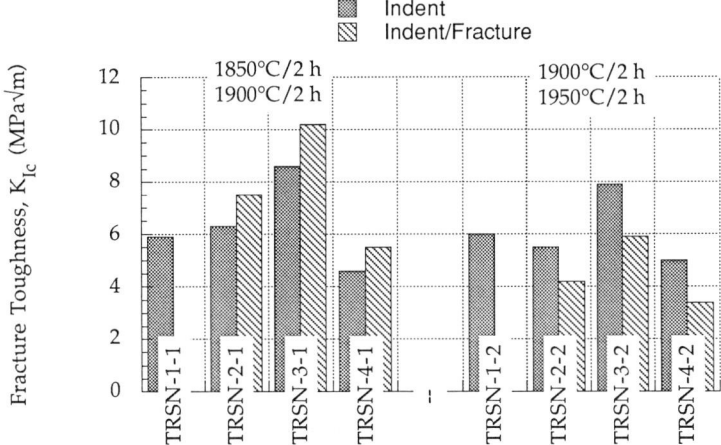

Fig. 3. Summary of results on fracture toughness of GPS silicon nitride compositions by indentation and indentation/fracture techniques.

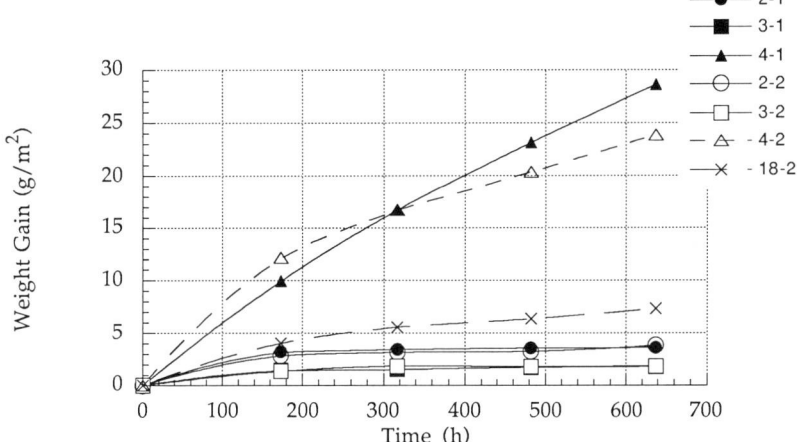

Fig. 4. Summary of results on oxidation of TRSN series samples at 1000°C.

containing Nd_2O_3 (TRSN-4). The highest fracture toughness were for the Y_2O_3-SrO containing sample sintered at 1850°C/2 h-1900°C/2 h (TRSN-3-1) which had values from 8.6-10.1 MPa√m. These values are comparable to toughnesses obtained in previous studies.[2-5]

One of the advantages mentioned for the nitrogen apatites in silicon nitride was that they do not experience volume increases upon oxidation. In the past, volume expansion during intermediate temperature oxidation, especially for melilite phases, has been observed to cause severe degradation of the materials.[15] Weight gain measurements at 1000°C showed parabolic kinetics for most of the compositions (Fig. 4). This indicates the formation of a protective layer and little to no volume expansion of the reaction product during oxidation. However, the samples containing Nd_2O_3 (TRSN-4) exhibited nearly linear weight gain kinetics indicating the formation of a non-protective layer.

CONCLUSIONS

Gas-pressure-sintering can be used to densify silicon nitride with rare earth-oxide and nitride apatites as the grain boundary phases. Densification was dependent on the quantity of additives used with high densities achieved at equivalent oxygen contents of ~8%. Fracture toughnesses (K_{Ic}) up to 8-10 MPa$\sqrt{m}$ and room temperature flexural strengths in the range of 400-720 MPa were obtained for most compositions. However, the strengths at elevated temperatures (1200°C) are reduced.

ACKNOWLEDGMENTS

Research sponsored by the U.S. Department of Energy, Assistant Secretary for Conservation and Renewable Energy, Office of Transportation Systems, as part of the Ceramic Technology Project of the Materials Development Program, under contract DE-AC05-84OR21400 with Martin Marietta Energy Systems, Inc.

REFERENCES

1. F. F. Lange, "Relation Between Strength, Fracture Energy and Microstructure of Hot-Pressed Si_3N_4," *J. Am. Ceram. Soc.*, 56 [10] 518-522 (1973)
2. E. Tani, S. Umebayashi, K. Kishi, K. Kobayashi, and M. Nishijima, "Gas-Pressure Sintering of Si_3N_4 With Concurrent Addition of Al_2O_3 and 5 wt % Rare Earth Oxide: High Fracture Toughness with Fiber-Like Structure," *Am. Ceram. Soc. Bull.*, 65[9]1311-1315(1986)
3. C. Li and J. Yamanis, "Super-Tough Silicon Nitride With R-Curve Behavior," Ceram. Eng. Sci. Proc., 10[7-8]632-645(1989)
4. K. Matsuhiro and T. Takahashi, "The Effect of Grain Size on the Toughness of Sintered Si_3N_4," Ceram. Eng. Sci. Proc., 10[7-8]807-816(1989)
5. T. Kawashima, H. Okamoto, H. Yamamoto, and A. Kitamura, "Grain Size Dependence of the Fracture Toughness of Silicon Nitride Ceramics," J. Ceram. Soc. Japan (1991)
6. M. Mitomo, "Pressure Sintering of Si_3N_4," *J. Mater. Sci.*, 11, 1103-1107 (1976)
7. H. F. Priest, G. L. Priest, and G. E. Gazza, "Sintering of Si_3N_4 Under High Nitrogen Pressure," *J. Am. Ceram. Soc.*, 60 [1-2] 81-85 (1977)
8. P. Drew and M. H. Lewis, "The Microstructures of Silicon Nitride Ceramics During Hot-Pressing Transformations," *J. Mater. Sci.*, 9, 261-269 (1974)
9. M. H. Lewis, G. Leng-Ward, and C. Jasper, "Sintering Additive Chemistry in Controlling Properties of Nitride Ceramics," pp. 1019-1033 in *Ceramic Transactions, Ceramic Powder Science, II*, B., Edited by G. L. Messing, E.R. Fuller, Jr., and H. Hausner. American Ceramic Society, Westerville, OH (1988)
10. D. P. Thompson, "Alternative Grain-Boundary Phases for Heat-Treated Si_3N_4 and ß'-Sialon Ceramics," Brit. Ceram. Proc.,45, 1-13 (1990)
11. D. E. Wittmer, "Alternative Processing Through Turbomilling," *Am. Ceram. Soc. Bull.*, 67[10]1670-1672 (1988)
12. Y. Hattori, Y. Tajima, K. Yabuta, J. Matsuo, and T. Watanabe, "Gas Pressure Sintered Silicon Nitride Ceramics for Turbocharger Applications," pp. 165-172 in *Proc. 2nd Internat. Symp. on Ceramic Materials and Components for Engines*, Edited by W. Bunk and H. Hausner, Verlag Deutsche Keramische Gesellschaft, Bad Honnef, FRG (1986).
13. G. R. Anstis, et al, "A Critical Evaluation of Indentation Techniques for Measuring Fracture Toughness: I, Direct Crack Measurements," J. Am. Ceram. Soc., 64[9]533-538(1981).
14. P. Chantikul, et al, "A Critical Evaluation of Indentation Techniques for Measuring Fracture Toughness: II, Strength Method," J. Am. Ceram. Soc., 64[9]539-543(1981).
15. 24. F. F. Lange, S. C. Singhal, and R. C. Kuznicki, "Phase Relationships and Stability Studies in the Si_3N_4-SiO_2-Y_2O_3 Pseudoternary," J. Amer. Ceram. Soc., 60 (5-6), 249-252 (1977)

SILICON NITRIDE CONTAINING RARE EARTH SILICATE INTERGRANULAR PHASES

STEPHEN D. NUNN, TERRY N. TIEGS, KRISTIN L. PLOETZ, CLAUDIA A. WALLS, AND NELSON BELL, Oak Ridge National Laboratory, Oak Ridge, TN 37831-6087

ABSTRACT

Si_3N_4 ceramics prepared with refractory grain boundary phases to improve high temperature properties are difficult to densify by conventional sintering methods. Gas-pressure sintering may be used to promote densification and development of acicular grains for improved fracture toughness. The current study examined rare earth silicate sintering aids with the composition $M_2Si_2O_7$, where M is a trivalent cation (Y, La, Nd). M_2O_3 and SiO_2 additions were varied to develop a number of compositions in the Si_3N_4–Si_2N_2O–$M_2Si_2O_7$ ternary phase field. Pressureless sintering and gas-pressure sintering were used to densify the samples. Densification, microstructure development, oxidation resistance, and mechanical properties were evaluated and compared with respect to compositional variations and processing conditions.

INTRODUCTION

The utilization of silicon nitride ceramics in high temperature applications, such as engine components, requires a material with oxidation resistance, high retained strength at elevated temperatures, and high fracture toughness.[1] While high density, high strength silicon nitride can now be produced routinely, some of the sintering aids typically used to promote densification, particularly Al_2O_3 and MgO, result in the formation of a relatively low melting temperature intergranular glassy phase which causes severe degradation of strength at high temperatures.[2-4]

The use of reduced amounts of sintering aid or of a more refractory sintering aid composition can greatly improve the high temperature strength of silicon nitride, however, it also increases the difficulty of obtaining high density ceramics.[5] Hot pressing or hot isostatic pressing (HIPing) is often required to process these more refractory compounds. Both of these procedures have limitations. Only simple geometries can be formed by hot pressing, and HIPing is a complex and expensive process.

More recently, gas-pressure sintering has been used to densify refractory silicon nitride compositions at high processing temperatures.[6-8] In gas-pressure sintering, an over-pressure of nitrogen (3 to 100 atm.) is used to suppress the decomposition of silicon nitride which would normally occur at temperatures above about 1750°C. Under these conditions, silicon nitride can be fired to temperatures over 2000°C, allowing the use of a much lower amount of sintering aid and/or a more refractory sintering aid.

In the present study, silicon nitride compositions containing highly refractory rare earth silicate phases were processed by conventional sintering (1 atm. N_2) and by gas-pressure sintering. The densification, oxidation resistance, and mechanical properties were evaluated.

EXPERIMENTAL PROCEDURE

The compositions which were examined are shown in Table I. These consist of silicon nitride with additions of SiO_2 and M_2O_3 (M = Y, La, or Nd) to form the disilicate phase $M_2Si_2O_7$. The amounts of the additions were varied to result in oxygen contents from 3.3 to 8.0 eq. %. The oxygen content of the silicon nitride powder was taken into consideration in calculating the batch additions. Sample compositions were prepared by mixing aqueous suspensions of the required constituents in a turbomill containing ZrO_2 media. Dispersing aids (GAF PVP K-15 and Darvan 821A) were added to the mill in an amount between 1 and 2 wt. % of the solids. Some of the slurries were slip cast in plaster-of-paris tile molds. For the majority of the samples, the milled powder was dried and sieved before being pressed into billets. The billets were formed by uniaxial pressing at 1000 psi in a cylindrical die, followed by isostatic pressing at 50 ksi.

Table I. Silicon nitride/rare earth disilicate compositions.

Sample	Weight %					Equivalent %		
	Si3N4	SiO2	Y2O3	La2O3	Nd2O3	Oxygen	Silicon	R E*
Y3	93.49	2.52	3.98			3.31	98.72	1.28
La3	91.86	2.48		5.66		3.31	98.72	1.28
Nd3	91.69	2.48			5.83	3.32	98.72	1.28
Y4	91.44	3.29	5.26			4.39	98.28	1.71
La4	89.35	3.23		7.41		4.40	98.29	1.71
Nd4	89.13	3.22			7.64	4.40	98.29	1.71
Y6	87.95	4.31	7.75			6.15	97.43	2.57
La6	85.03	4.17		10.81		6.15	97.43	2.57
Nd6	84.72	4.15			11.13	6.15	97.43	2.57
Y8	85.96	6.47	7.57			7.91	97.48	2.52
La8	83.11	6.33		10.56		7.97	97.48	2.52
YLa8	84.51	6.40	3.80	5.29		7.95	97.47	2.53

Si3N4 - UBE E-10, UBE Industries. * RE= rare earth.
SiO2 - Cab-O-Sil, Cabot Corp.
Y2O3, La2O3, and Nd2O3 - 99.99%, Molycorp.

Table II. Firing conditions for the two-step gas-pressure sintering runs.

Gas-Pressure Run	Step 1			Step 2		
	Temp.	Time	Pressure	Temp.	Time	Pressure
1	1850°C	2 hr	50 psi	1900°C	2 hr	300 psi
2	1900°C	2 hr	100 psi	1950°C	2 hr	300 psi

The organics were removed from the samples by firing in air to a maximum temperature of 600°C. The samples were then fired in a graphite resistance furnace under 1 atm. N_2 at temperatures of 1700, 1750, or 1800°C for 5 hr. or in a gas-pressure sintering furnace using a two-step firing schedule. The gas-pressure sintering conditions are shown in Table II. The two-step firing schedule has been found to improve the uniformity of densification and to promote the formation of a bimodal grain size distribution which is believed to enhance rising R-curve fracture toughness behavior.[9-11] All of the firing schedules included a one-hour hold at 1400°C during cool down to promote crystallization of the intergranular phase.

The densities of the sintered materials were measured by the liquid displacement method using high purity ethyl alcohol. The fired billets were machined to produce test bars measuring 3 x 4 x 50 mm for oxidation, modulus of rupture, and fracture toughness testing. The edges of the test bars were beveled to reduce stress concentrations during bending.

The oxidation resistance was evaluated by exposing the samples to 1000°C in air for 800 hours. Oxidation was monitored by interrupting the test periodically to measure the weight gain of the samples.

The fracture toughness was calculated either by direct measurement of indentation crack lengths using Anstis' equation[12] or by the indentation strength method using Chantikul's equation.[13]

The room temperature modulus of rupture was measured in four-point flexure using a bend test fixture having an outer span of 38 mm and in inner span of 19 mm. The fracture strength was also measured at 1200°C using a SiC test fixture having spans of 40 and 20 mm. The fracture surfaces were examined by optical and scanning electron microscopy (SEM).

Samples were also examined by x-ray diffraction analysis to determine the phases which were present in the sintered materials.

RESULTS AND DISCUSSION

A comparison of the densities obtained by conventional and gas-pressure sintering is shown in Fig. 1. The conventionally sintered samples had relatively low fired densities. Only the

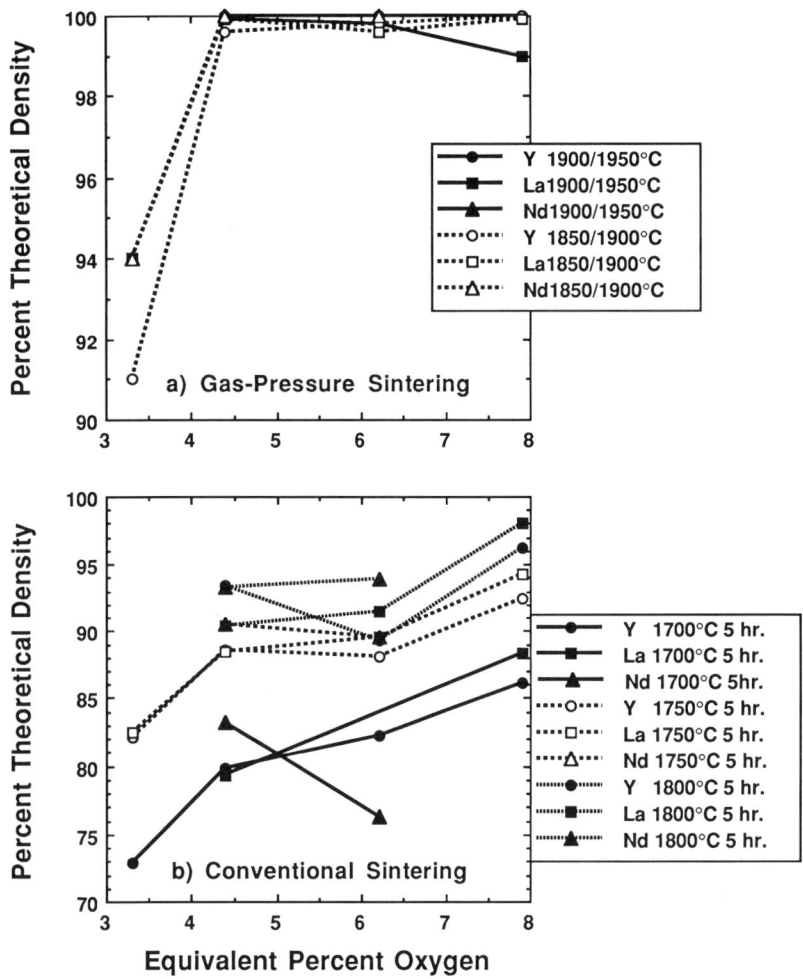

Fig. 1. A comparison of the sintered densities as a function of the sintering aid content (eq. % oxygen) for samples containing different rare earth sintering aids which were densified by a) gas-pressure sintering under two different temperature/pressure conditions and b) conventionally fired at three different temperatures.

samples which were fired at 1800°C and which contained the highest amount of sintering aid (8 eq. % oxygen) approached reasonably high densities (96-98% theoretical). In contrast, most of the samples which were gas-pressure sintered were nearly 100% dense. The only exceptions were those samples containing the minimum amount of additive (3.3 eq. % oxygen). The densities for these samples were less than 95%. Gas-pressure sintering clearly enhanced the densification of these refractory silicon nitride compositions, probably due to increased liquid phase formation and higher sintering rates at the higher firing temperatures which are possible using this process.[6]

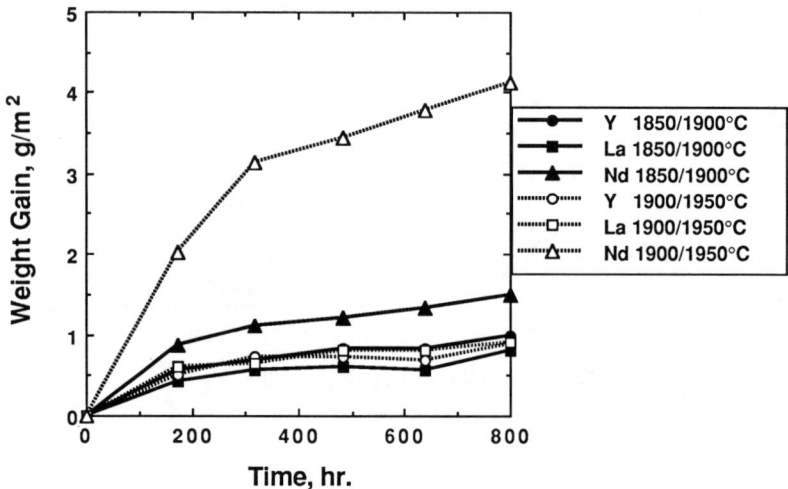

Fig. 2. A plot of the weight gain of the gas-pressure sintered samples containing 4.4 eq. % oxygen after long-term exposure to oxidizing conditions at 1000°C.

The 1000°C oxidation resistance of the gas-pressure sintered samples containing 4.4 eq. % oxygen is shown in Fig. 2. After a total exposure time of 800 hrs., the samples containing yttrium and lanthanum show a weight gain of only about 1 g/m^2. This is a very low weight gain for silicon nitride at this temperature[14] and shows the highly refractory nature of the intergranular phase formed by the sintering aids. The samples containing neodymium show less resistance to oxidation, especially the sample fired at the higher temperatures. This may be due to the formation of less refractory oxynitride phases. The secondary phases which were formed could not be identified from the x-ray diffraction peaks for the neodymium samples. The patterns did not correspond to any JCPDS file patterns.

The fracture strength of selected gas-pressure sintered samples is shown in Fig. 3. The strength of many of the dry pressed samples was anomalously low and was traced to processing-related defects in the green bodies. The slip cast samples showed generally higher room temperature strengths, primarily due to the higher quality of the green ceramics. These samples contained the highest levels of sintering aid additives, however, and showed the greatest loss in strength at high temperature.

The most interesting observation from the MOR tests is that the elevated temperature strength of many of the samples was nearly the same as the room temperature strength. This suggests a flat strength vs. temperature curve for these materials. Improvements in the processing procedure and optimization of the gas-pressure sintering conditions may allow an increase in the magnitude of the strengths at both room and elevated temperatures.

Examination of the fracture surfaces of the samples showed that the materials developed highly acicular grains and appeared to have a bimodal grain size distribution. An example of a typical microstructure is shown in the SEM micrograph in Fig. 4. The microstructure consists of mostly fine grains (<1 μm in diameter) intermixed with numerous large grains (2-4 μm in diameter). Both the large and small grains have a high aspect ratio. Such a microstructure should result in a material with high fracture toughness.[10,15,16] However, the measured fracture toughness values were in the range of 3 to 6 MPa√m which is a low to moderate toughness for silicon nitride. The reason for the relatively low toughness is not completely understood at this time, but may be due to the development of strong interfacial bonding between the grains which would limit the extent of crack deflection, debonding, and pull-out toughening mechanisms.

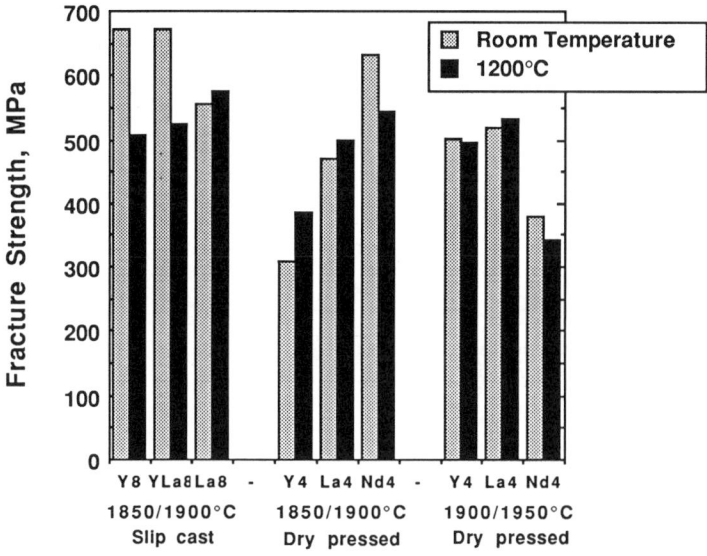

Fig. 3. Room temperature and 1200°C fracture strength results for gas-pressure sintered samples having different compositions, preparation methods, and firing conditions.

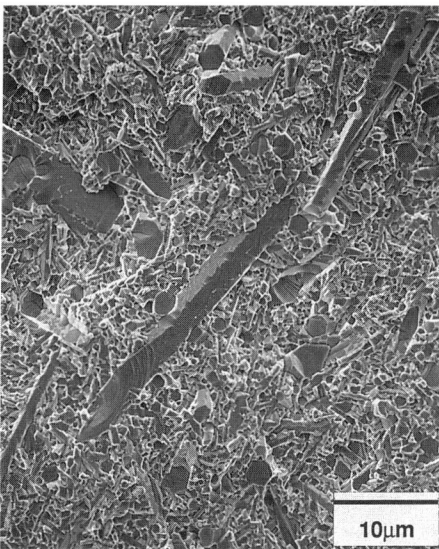

Fig. 4. Fracture surface of a gas-pressure sintered sample (La6 Run 2) showing a typical microstructure containing highly acicular grains and a bimodal grain size distribution.

CONCLUSIONS

Silicon nitride containing sintering aids having the rare earth disilicate composition, $M_2Si_2O_7$ where M = Y, La, or Nd, were fired to high densities by a two-step gas-pressure sintering process. The fired samples containing Y and La showed excellent oxidation resistance after 800 hrs. exposure at 1000°C. The Nd-containing samples were less resistant to oxidation. The room temperature fracture strengths of the samples prepared by slip casting were moderate for silicon nitride, 550-700 MPa. A number of the samples showed excellent strength retention at 1200°C. Fracture toughness values were typical for silicon nitride despite the development of highly acicular grains in the microstructure. Further experiments will be required to improve processing and to optimize the composition and firing conditions to obtain materials with high strength which can be retained at elevated temperatures and which show good fracture toughness.

ACKNOWLEDGEMENT

Research sponsored by the Ceramic Technology Project, DOE Office of Transportation Technologies, under contract DE-AC05-84OR21400 with Martin Marietta Energy Systems, Inc.

REFERENCES

1. D. Carruthers and L. Lindburg, in Ceramic Materials and Components for Engines, edited by V. J. Tennery (American Ceramic Society, 1988) pp. 1258-1272.
2. R. L. Tsai and R. Raj, *J. Am. Ceram. Soc.*, **63** [9-10] 513-17 (1980).
3. S. H. Knickerbocker, A. Zangvil, and S. D. Brown, *J. Am. Ceram. Soc.*, **68** [4] C99-101 (1985).
4. N. Hirosaki, A. Okada, and M. Mitomo, *J. Mater. Sci.*, **25** (1990) 1872-76.
5. N. Hirosaki, A. Okada, and Y. Akimune, *J. Mater. Sci. Letters*, **9** (1990) 1322-23.
6. C. Greskovich, *J. Am. Ceram. Soc.*, **64** [12] 725-30 (1981).
7. A. Okada and N. Hirosaki, *J. Mater. Sci.*, **25** (1990) 1656-61.
8. M. Mitomo and S. Uenosono, *J. Am. Ceram. Soc.*, **75** [1] 103-108 (1992).
9. Y. Hattori, Y. Tajima, K. Yabuta, J. Matsuo, M. Kawamura, and T. Watanabe, in Ceramic Materials and Components for Engines, edited by W. Bunk and H. Hausner (Verlag, FRG, 1986) pp. 165-172.
10. E. Tani, S. Umebayashi, K. Kishi, K. Kobayashi, and M. Nishijima, *Am. Ceram. Soc. Bull.*, **65** [9] 1311-15 (1986).
11. C.-W. Li and J. Yamanis, *Ceram. Eng. Sci. Proc.*, **10** [7-8] 632-45 (1989).
12. G. R. Anstis, P. Chantikul, B. R. Lawn, and D. B. Marshall, *J. Am. Ceram. Soc.*, **64** [9] 533-38 (1981).
13. P. Chantikul, G. R. Anstis, B. R. Lawn, and D. B. Marshall, *J. Am. Ceram. Soc.*, **64** [9] 539-43 (1981).
14. D. M. Mieskowski and W. A. Sanders, *J. Am. Ceram. Soc.*, **68** [7] C160-C163 (1985).
15. K. Kishi, S. Umebayashi, and E. Tani, *J. Mater. Sci.*, **25** (1990) 2780-84.
16. P. F. Becher, *J. Am. Ceram. Soc.*, **74** [2] 255-69 (1991).

CHEMICALLY INDUCED DEFECTS IN OXYNITRIDE GLASSES

DONALD R. MESSIER AND PARIMAL J. PATEL
U.S. Army Research Laboratory Materials Directorate,
Watertown, MA 02172-0001.

ABSTRACT

The ultimate usefulness of oxynitride glasses and fibers depends upon the minimization or elimination of metallic defects that arise during processing. Despite this, the origins and chemistry of such defects in oxynitride glasses have received scant attention in the literature. The defects reduce glass transparency and cause oxynitride glass fibers to fail at relatively low stress levels. The same types of defects undoubtedly occur in the grain boundary glass phase of sintered Si_3N_4 with unknown effects on material properties.
Examples are shown of Si-rich metallic defects in oxynitride glasses, and their effects on glass and fiber properties are discussed. Chemical reactions that produce the defects are considered, as are chemical analysis results supporting the proposed reaction mechanisms.

INTRODUCTION

It has been apparent for some time that many glass properties are improved by the substitution of N for O in the glass structure.[1-5] Of particular interest for structural applications are the increases in hardness, elastic modulus, and, ultimately, strength offered by oxynitride glasses as compared to oxide glasses. At present, however, the usefulness of oxynitride glass is limited by colloidal, Si-rich metallic inclusions; the inclusions limit the transparency of the bulk glass and also prevent the attainment of the extremely high tensile strengths expected for oxynitride glass fibers. Additionally, such inclusions may have unrecognized, but possibly significant, effects on the behavior of the glassy grain boundary phase in sintered silicon nitride.
The objectives of this paper are to demonstrate the deleterious effects of metallic inclusions on oxynitride glasses and fibers, to discuss high temperature chemical reactions that produce the inclusions, and to consider methods of minimizing or eliminating metallic inclusions in the glasses.

DEFECTS IN OXYNITRIDE GLASSES AND FIBERS

Figure 1 shows highly transparent, but nevertheless grey-colored Mg-Si-Al-O-N glass disks. The grey color is typical of most oxynitride glasses whatever the composition.[6] Experience with glasses in several oxynitride glass systems over an extended length of time [7] has shown that the darkening is attributable to metallic precipitates similar to those shown in Figure 2 in a Y-Si-Al-O-N glass. In this particular instance, the precipitates contain some Fe (a common impurity in commercial Si_3N_4 powder) and are much larger than in current glasses prepared with higher purity batch materials and improved melting procedures. Glass preparation has now been improved to the point where such precipitates are undetectable via optical microscopy.

Our experience has also been that, unless distinguishable morphologically (sometimes they occur as spherical inclusions on fracture surfaces), the precipitates are undetectable via scanning or transmission electron microscopy using the usual techniques. Considering also that the inclusions in good glasses are sub-micron and probably irregularly shaped, identifying and characterizing the metallic precipitates is a challenging task.

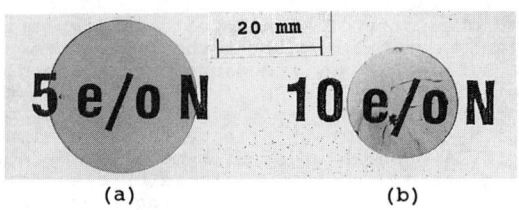

Figure 1. Mg-Si-Al-O-N Glass Disks Containing 4 Atom % N (a), and 8 Atom % N (b). Each Disk is 5 mm thick.

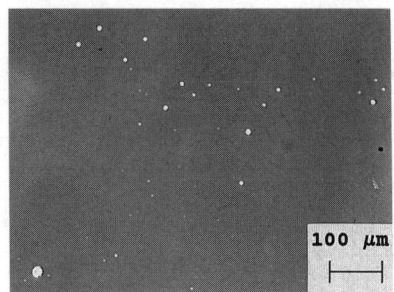

Figure 2. Metallic Inclusions Visible in Polished Section of Y-Si-Al-O-N Glass.

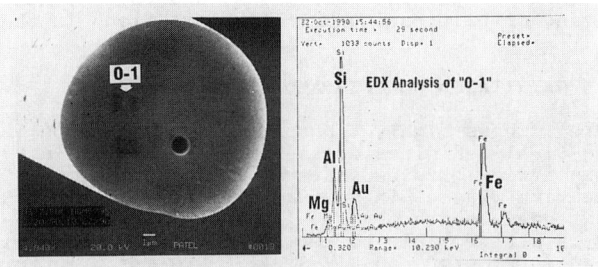

Figure 3. Mg-Si-Al-O-N glass fiber fracture surface showing a critical defect rich in Fe. Occasional axial voids such as the one shown are not strength limiting.

Figure 3 shows a fracture surface of a Mg-Si-Al-O-N glass fiber that failed from an internal Fe silicide defect. As mentioned above, this particular defect was identifiable because of its spherical morphology, and because it was the fracture origin. Observations such as this were used as feedback for processing improvements that have increased fiber tensile strength to the point that fracture origins are no longer identifiable. The best tensile strengths (2800 MPa) obtained at ARL to date for oxynitride glass fibers however are still less than for "s" glass (4500 MPa) suggesting that very small colloidal metallic inclusions may still be strength-controlling.

CHEMICAL ORIGINS OF DEFECTS

Important, if uninteresting scientifically, sources of defects in oxynitride glasses and fibers are impurities originating from impure batch materials. In addition to the above mentioned Fe impurities, we have also observed fiber failure from Al-rich defects traceable to free Al in the AlN used as a batch component. Moreover, we have even found metallic Si contamination in the high purity silica used as one batch component. The highly reducing N_2 atmosphere used in glass melting converts easily reducible oxides such as those of Fe into metals, enhancing their reactivity with Si, and also maintains elemental impurities such as Al and Si in their metallic forms preventing the latter from dissolving in the glass. These sources of metallic defects can be minimized however by using high purity batch materials, and by pre-melting the oxide part of the batch in air before adding the nitrogen source, usually Si_3N_4, and re-melting in N_2.

A key issue in oxynitride glass technology that also relates to Si_3N_4 sintering is understanding thermal decomposition processes. Particularly relevant to the glass are the chemical reactions that produce the colloidal Si in that material. Although it has often been proposed, for example by Baik and Raj [7] that such Si results from simple thermal decomposition of Si_3N_4, the following analysis shows why that process is irrelevant for oxynitride glasses.

The thermal decomposition of Si_3N_4 is represented by:

$$Si_3N_4(s) = 3\ Si(l) + 2\ N_2(g) \qquad (1)$$

and that of Fe-Contaminated Si_3N_4 by:

$$Si_3N_4(s) + x\ Fe(s,l) = Si_3Fe_x(l) + 2\ N_2(g) \qquad (2)$$

Typically, the glasses are melted in N_2 at 100 kPa; thermal decomposition can therefore occur only when the N_2 pressure from Reaction (1) or (2) exceeds that value. In the first case, the activity of Si (in its standard state) is one, and in the second, representing the decomposition of Fe-contaminated Si_3N_4, the activity of Si is less than one, the exact value depending upon the amount of combined Fe. For Equation (1) the decomposition pressure reaches 100 kPa at 2152 K.[8] In the second case, if we assume that the Si activity is as low as 0.5 (more likely it is higher), the N_2 pressure reaches 100 kPa at 2052 K.[8] Since none of the oxynitride glasses that we have dealt with reach temperatures higher than 1950 K at any stage of processing, this particular decomposition mechanism is clearly ruled out.

The most likely mechanism for the formation of colloidal Si in oxynitride glass melts is illustrated in Figure 4. According to this mechanism, proposed by Zintl [9] for SiO_2 and later by Geld and Esin [10] for siliceous slags, oxygen loss during melting produces divalent Si that on cooling disproportionates to tetravalent and elemental Si, the latter appearing as colloidal precipitates in the glass. Mulfinger [11,12] observed a similar phenomenon attributed to the same mechanism in oxynitride glass, and Schrimpf and Frischat [13-15] discussed similar findings in their work on oxynitride glasses. It is interesting to note that the mechanism shown in Figure 4 has nothing to do with the presence of N in the glass; it is merely a consequence of heating in a reducing atmosphere.

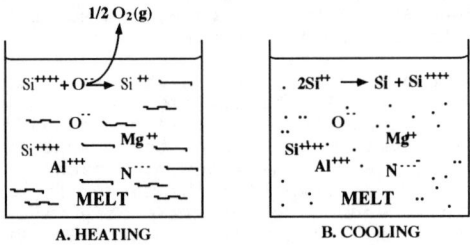

Figure 4. Mechanism for Formation of Metallic Precipitates in Oxynitride Glass.[9-15]

Although evidence for the mechanism shown in Figure 4 is persuasive, it is entirely circumstantial with the exception of the work of Geld and Esin [10] who based their conclusions on chemical analysis data that showed their slags to be oxygen deficient. If the mechanism holds for oxynitride glasses, one would likewise expect those glasses to be oxygen deficient as well. In order to test the latter premise, data from chemical analyses on various oxynitride glasses from a number of sources [2,10,16-20] were used to calculate O/Si ratios as summarized in Figure 5. While even more analyses are available in the literature, valid calculations were only possible where none of the elements was done by difference, i.e., the only data used were for cases in which each cation and anion was analyzed for separately. Although they show more scatter than may be desirable, the data in Figure 5 support the idea of oxygen deficiency in oxynitride glasses.

MINIMIZATION/ELIMINATION OF DEFECTS

An obvious first step in the elimination of metallic defects is the elimination of the impurities that cause or exacerbate the defects. Batch materials should be free of metallic impurities such as Si or Al that will remain as elements during processing. Also, Fe in any form should be eliminated as far as possible.

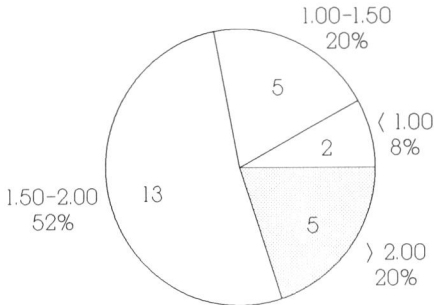

Figure 5. O/Si Ratios from 25 Chemical Analyses of
Various Oxynitride Glasses. (Stoichiometric O/Si = 2.00)

Elimination of the defects that arise from divalent Si in
the melt presents formidable difficulties, partly because of
our incomplete understanding of the processes from which they
arise, and also because of a total lack of thermodynamic data
on oxynitride glass melts.

If Si^{++} does indeed originate from the process
illustrated in Figure 4, it could in principle be eliminated
by controlling the partial pressure of O_2 over the melt. As
stated above, however, the O_2 pressure is unknown. Even if
the data were available, it is often difficult to equilibrate
glass melts with the melting atmosphere short of bubbling the
gas through the melt, something not attempted in any of the
work discussed herein.

Another approach to the minimization of Si^{++} in the melt
is to add oxidizers.[21] A problem, however, is finding an
oxidizing agent that is not reduced to its metallic state by
the extreme reducing conditions used to make oxynitride
glasses. One additive that has shown promise to date is CeO_2
which appears to oxidize the Si^{++} as follows:

$$2 Ce^{+4} (melt) + Si^{++} (melt) = Si^{+4} (melt) + 2 Ce^{+3} (melt) \quad (3)$$

Much uncertainty remains concerning the thermodynamics and
kinetics of oxygen loss; more work is required to establish
how effective the additive is and to ascertain the optimum
amount to add to the glass batch.

SUMMARY

The strength and optical quality of oxynitride glasses
and fibers are degraded by metallic defects. The defects
result from oxygen loss owing to the reducing atmosphere
required for glass melting and are exacerbated by certain
impurities. Glass and fiber quality have been improved
considerably through the use of high purity batch materials,
by melting the oxide part of the glass in air to oxidize

metallic particles, and by using CeO_2 to internally oxidize divalent Si in the glass. The tensile strengths of Mg-Si-Al-O-N glass fibers have been substantially increased by processing changes guided by the foregoing.[22] Further improvements will require better understanding and control of the glass making process. Particularly useful would be a better understanding of the minimum size below which the colloidal inclusions no longer degrade glass or fiber properties.

REFERENCES

1. K.H. Jack, in Nitrogen Ceramics, Edited by F.L. Riley (Noordhoff, Leyden, 1977), pp. 257-261.
2. K. R. Shillito, R.R. Wills, and R.B. Bennett, J. Am. Ceram. Soc., **61** [11-12] 537 (1978).
3. R.E. Loehman, J. Am. Ceram. Soc., **62** [9-10] 491-494 (1979); J. Non-Cryst. Solids, **42** 433-446 (1980); ibid., **56** 123-134 (1983).
4. R.A.L. Drew, S. Hampshire, K.H. Jack, in Special Ceramics 7. edited by D. Taylor and P. Popper. Proc. Brit. Ceram. Soc., [31] 119-132 (1981).
5. D.R. Messier, Rev. Chim. Min., **22** 518-533 (1985).
6. D.R. Messier and R.P. Gleisner, U.S. Army MTL TR 92-6 (1992).
7. S. Baik and R. Raj, J. Am. Ceram. Soc., **68** [5] C-124-C-126 (1985).
8. JANAF Thermochemical Tables, 2nd ed. (U.S. Govt. Printing Office, Washington, D.C., 1971).
9. E. Zintl, Z. anorg. allgem. Chem., **245** [1] 1-7 (1940).
10. P.V. Geld and O.A. Esin, J. Appl. Chem. USSR, 23 1277-1283 (1950).
11. H.-O. Mulfinger, J. Am. Ceram. Soc., **49** [9] 462-467 (1966).
12. T. Kelen and H.-O. Mulfinger, Glastechn. Ber., **41** [6] 230-242 (1968).
13. C. Schrimpf, Doctoral Dissertation, Tech. U. Clausthal, 1982.
14. C. Schrimpf and G.H. Frischat, J. Non-Cryst. Solids, **52** 479-485 (1982).
15. C. Schrimpf, and G.H. Frischat, Glastech. Ber., **57** [5] 97-111 (1984).
16. J. Homeny and D.L. McGarry, J. Am. Ceram. Soc., **67** [11] C-225-C-227 (1984).
17. P.E. Jankowski and S.H. Risbud, J. Am. Ceram. Soc., **63** [5-6] 350-352 (1980).
18. D.R. Messier, Ceram. Eng. Sci. Proc., **3** [9-10] 565-575 (1982).
19. D.R. Messier and E.J. DeGuire, J. Amer. Ceram. Soc., **67** (9) 602-605 (1984).
20. D.R. Messier (unpublished data).
21. A. Paul, Chemistry of Glasses. (Chapman and Hall, London, 1982), p. 148.
22. P.J. Patel, D.R. Messier, R.E. Rich, in Technology 2001, (NASA Conference Publication 3136, Vol. 2, 1991) pp. 258-264.

PART II C

Microstructure–Design, Development and Characterization

REACTION SINTERING OF β-Si₃N₄/α'-SIALON CERAMICS

S. BOSKOVIC*, K.J. LEE**, AND T.Y. TIEN**
*Institute of Nuclear Sciences, Vinca 11001 Belgrade, Yugoslavia
**Materials Science and Engineering, The University of Michigan, Ann Arbor, MI 48109

ABSTRACT

Compositions in the α'-SiAlON-β-Si₃N₄ solid solution region in the system Si,Al,Y/N,O which contained a third phase as a sintering aid were prepared. Mixtures of starting materials were reaction sintered to full densities without applied pressure. Phases were identified and lattice parameters of α' and β phases were measured and compared with standards. The results were used to construct α'-SiAlON-β-Si₃N₄ solid solution tie lines in the two phase region. As expected, the specimens with a higher α'-SiAlON content showed higher hardness and lower toughness values.

INTRODUCTION

Mechanical properties of monolithic silicon nitride ceramics depend on the morphology of β-Si₃N₄ grains and the nature of the grain boundary phase. Since liquid was needed during densification and for the morphology development of the β-Si₃N₄ grains in silicon nitride ceramics, the existence of the glassy grain boundary phase could not be avoided, and is thus, the cause for the inferior high temperature mechanical property.

Sun et al.[1] have shown that two phase fields "β-Si₃N₄-α'-SiAlON" and three-phase fields "β-Si₃N₄-α'-SiAlON-AlN polytypoids" exist in the system Si,Al,Y/N,O under sub-solidus temperatures. Melting of these phases had neven been observed since they vaporize at very high temperatures. Ceramics containing any combination of these three phases should therefore have superior high temperature mechanical properties if the glassy grain boundary phase is eliminated.

Sheu and Tien[2] synthesized ceramics containing elongated β-Si₃N₄ grains and dispersed fine α'-SiAlON particles. These two-phase ceramics have excellent mechanical properties at room temperature as well as at high temperatures. Sheu et al. prepared specimens by hot pressing those containing various ratios of these two phases and found that the composition containing

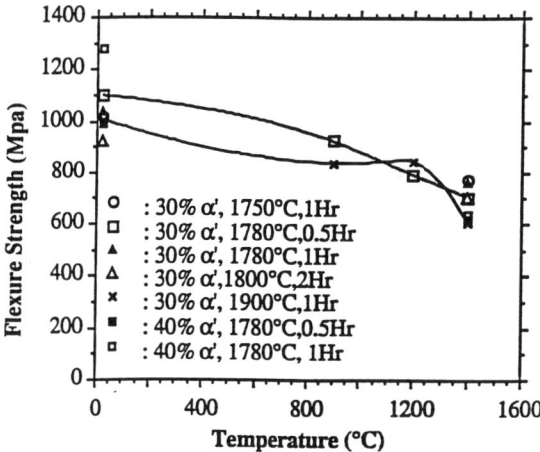

Fig. 1 Flexural strength vs. testing temperature of samples with 30% and 40% α'-SiAlON.

Fig. 2　SEM miscrograph of sample with 30% α'-SiAlON. Sample was hot-pressed at 1780 °C for 1 hour.

70 vol% β-Si_3N_4 and 30 vol% α'-SiAlON showed the highest flexural strength both at room temperature and elevated temperatures. The temperature dependence of flexural strength of these compositions are reproduced in Fig. 1. A microstructure of the hot pressed two-phase specimen containing 70% β-Si_3N_4 grains and 30% α'-SiAlON grains is shown in Fig. 2. This microstructure shows the elongated β-Si_3N_4 grains and finely dispersed α'-SiAlON particles.

The sub-solidus phase relationship in the system Si,Al,Y/N,O[1] is shown in Fig. 3. The single phase α'-SiAlON region is located in the triangle Si_3N_4-AlN:Al_2O_3-YN:3AlN. The β-Si_3N_4 and α'-SiAlON two-phase field is delineated as a polygon (Fig. 4). As shown in Fig. 3, the line connecting Si_3N_4 and Y_2O_3:9AlN is the intercept of the two compatibility triangles Si_3N_4-AlN:Al_2O_3-YN:3AlN and Si_3N_4-AlN-Y_2O_3. This line, cut through the β-Si_3N_4 and α'-SiAlON two-phase field, is indicated in Fig. 4. Compositions studied by Sheu et al. were selected on this composition line.

Sheu et al.[3] have also reported results on the mechanical properties of hot pressed ceramics containing β-Si_3N_4, α'-SiAlON and AlN polytypoids. These ceramics also show desirable properties. Fig. 5 shows a series of compatibility tetrahedra containing β-Si_3N_4-α'-SiAlON-AlN polytypoids. These two-phase and three-phase ceramics were formulated using Si_3N_4, AlN, Al_2O_3 and Y_2O_3 as starting materials and were densified to full density. The densification of these ceramics can be explained by transient liquid phase sintering. In the beginning of the sintering process, eutectic liquids were formed at the contact points of the starting material particles. The amount of liquid was increased as temperature increased during sintering. When enough liquid was formed, densification proceeded according to liquid phase sintering. When sintering was prolonged, the reaction took place between the transient liquid and the solid which formed the equilibrium phases, i.e., β-Si_3N_4 and α'-SiAlON.

Both of these two-phase and three-phase ceramics demonstrated good flexural strength at room temperature as well as at high temperatures. However, their fracture toughness was lower than those of monolithic silicon nitride. Lai et al.[4] have shown that the flexural strength and fracture toughness of monolithic silicon nitride depends on the grain size and aspect ratio of β-Si_3N_4 grains. Samples studied by Sheu et al.[3] were hot pressed without control of the microstructural development. It is the intention of this investigation to prepare samples of these ceramics by reaction sintering in order to develop desirable microstructures having optimum

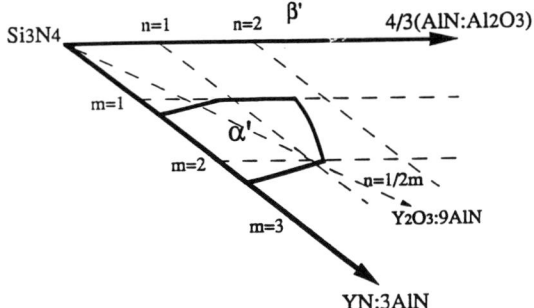

Fig. 3 Enlarged corner of α'- sialon plane in Y-sialon system.

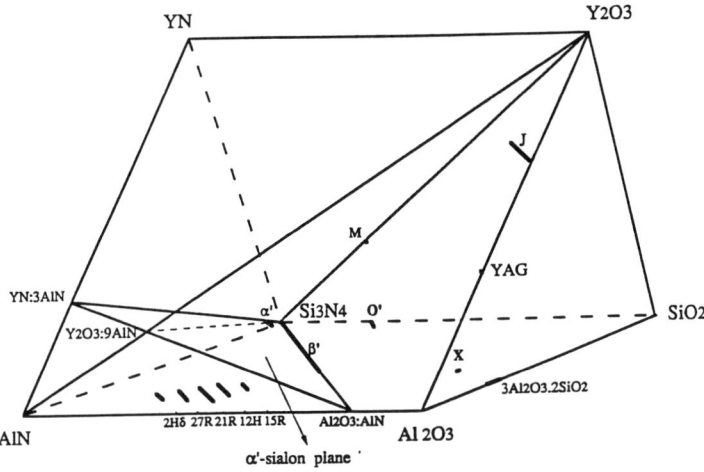

Fig. 4 Representation of α'- sialon plane in Y-sialon system.

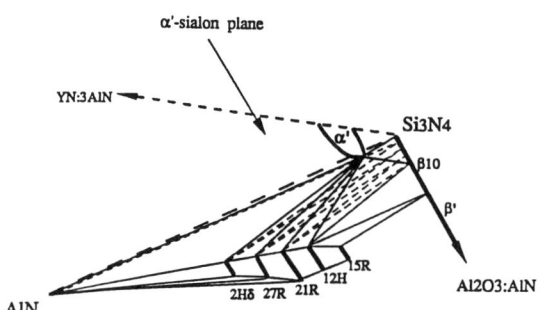

Fig. 5 α'–SiAlON is compatible with polytypoids (from 2Hδ to12H), AlN and β' forming eight compatibility tetrahedra: α'--12H--21R--β10; α'--21R--β10--β8; α'--21R--β8--27R; α'--β8--27R--β5; α'--27R--β5--2Hδ; α'--β5--2Hδ--β2; α'--2Hδ--β2--AlN and α'--β2--AlN--Si3N4

properties and to avoid the expensive hot pressing process. We intentionally added a small amount of a third phase to form a residual liquid which facilitates densification. The compositions chosen were in the three-phase region: α', β plus garnet.

Sun et al.[5] delineated the α'-SiAlON single phase region and β-Si$_3$N$_4$ plus α'-SiAlON two-phase region. However, β-Si$_3$N$_4$-α'-SiAlON tie lines had not been determined. It is also the intent of this paper to establish some of the tie lines in the two-phase region.

EXPERIMENTAL RESULTS

Densification studies

Compositions forming single phase α'-SiAlON and two-phase materials β-Si$_3$N$_4$-α'-SiAlON in the two-phase region β-Si$_3$N$_4$-α'-SiAlON plus a small amount of garnet as a sintering aid were formulated using Si$_3$N$_4$, AlN, Al$_2$O$_3$ and Y$_2$O$_3$ as starting materials. Both Si$_3$N$_4$ and AlN were obtained from H. Starck. Other chemicals used were all chemical pure grade. Compositions studied are shown in Fig. 6. Mixtures of starting materials were weighed,

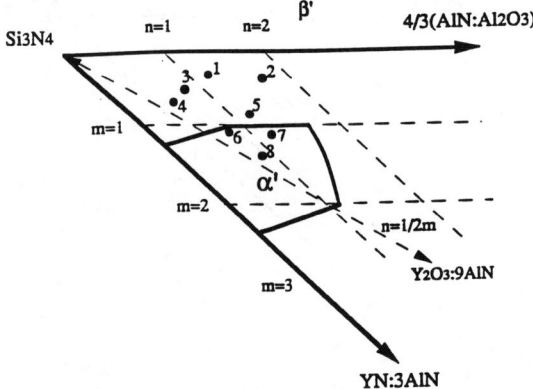

Fig. 6 Composition of samples prepared in Y-sialon system.

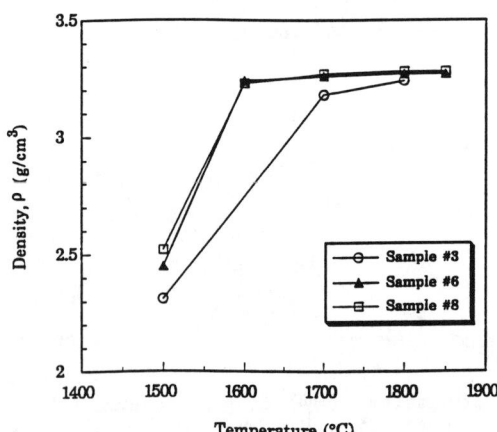

Fig.7 Sintered density as a function of temperature.

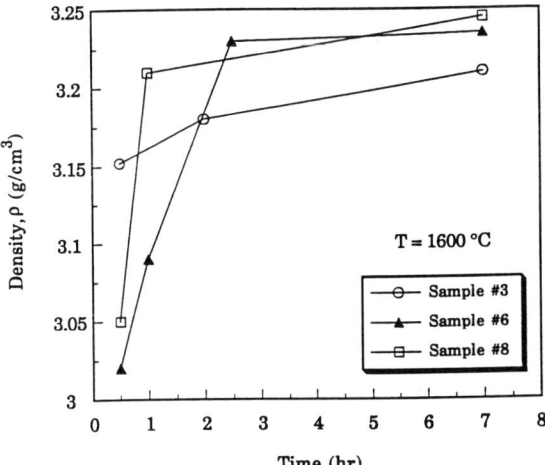

Fig.8 Sintered density as a function of sintering time.

ball-milled under isopropanol for two hours and then dried. Compacts were formed under an isostatic pressure of 25 MPa and sintered at 1600 to 1800°C for two hours under one atmosphere of nitrogen in a furnace with graphite as the heating element. Sintered densities were measured and the results are given in Fig. 7. Full density was reached for specimens sintered at temperatures above 1700°C. Density change with time for the same specimens sintered at 1600°C are given in Fig. 8. Compositions containing a higher α'-SiAlON content show an abrupt shrinkage with sintering time. This phenomenon can be understood using the classic liquid phase sintering theory; as the amount of liquid increases at higher yttria composition, a shorter time will be needed for the particle rearrangement to be completed. For a composition containing 3% yttria (sample #3), not enough liquid is formed during sintering for densification by particle rearrangement to occur.

Table I shows the amount of phase change during reaction for different compositions. In the beginning of the reaction, melilite formed and slowly reacted with α-Si$_3$N$_4$ forming

Table I. Phase compositions of sintered sample (2hr).

Sample	1500 °C					1600 °C				1700 °C			1800 °C		
	α	α'	β'	G*	M*	α	α'	β'	M*	α'	β'	Polytype*	α'	β'	Polytype*
1													14	81	5
2						1	2	85	12	4	83	13	5	84	15
3	61		39							16	80	4			
4						18	46	86		40	58	2	46	54	
5						10	45	45		43	53	4	49	48	3
6	38	41	17		4	14	70	16		71	20	4	76	13	7
7						12	72	16		77	21	2	87	11	2
8	24	26	6	20	24	9	89			99		1	99	1	

* G : YAG
  M : Melilite,
  Polytype of AlN : 15R or 12H.

α'-SiAlON at 1600°C. At the same time α-Si$_3$N$_4$ transformed to β-Si$_3$N$_4$ by solution-precipitation. At temperatures above 1700°C, liquid formed immediately which aided densification and α'-SiAlON formation. These results indicate that a transient liquid was present during the early stage of sintering and reacted in the later stage forming α'-SiAlON.

Phase Relationships

X-ray diffraction shows that the sintered specimens contained either single phase α'-SiAlON or mixtures of β-Si$_3$N$_4$ and α'-SiAlON phases. α'-SiAlON content in samples with different yttria content are given in Fig. 9. Lattice parameters of both α'-SiAlON and β-Si$_3$N$_4$ solid solutions were measured. These data were used to construct the tie lines in the α'-SiAlON plus β-Si$_3$N$_4$ solid solution region. These tie lines are given in Fig. 10. The standard lattice parameters used to compare with these samples were from Sun[5] and Gauckler.[6]

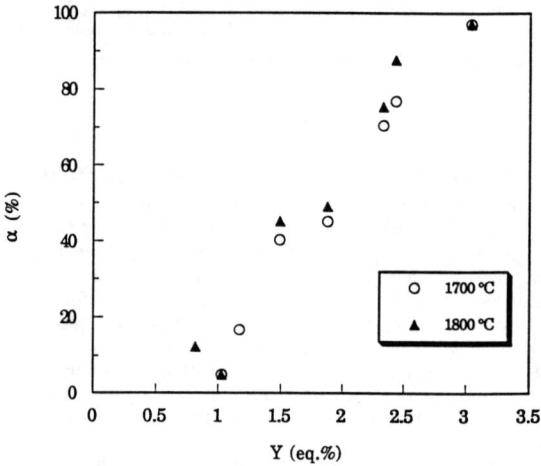

Fig.9 Dependence of α-phase content on Y$_2$O$_3$ concentration.

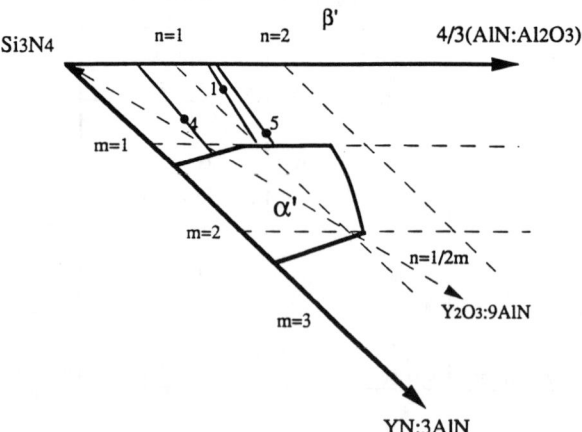

Fig. 10 Construction of tie line in α' + β two phase region in Y-sialon system.

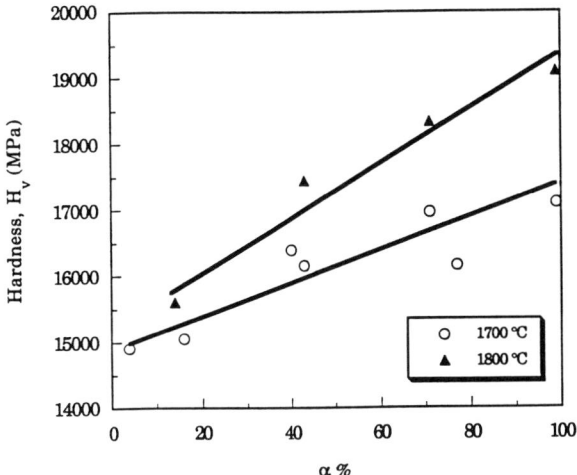

Fig.11 Hardness vs. total α-phase content of samples sintered for 2 hrs.

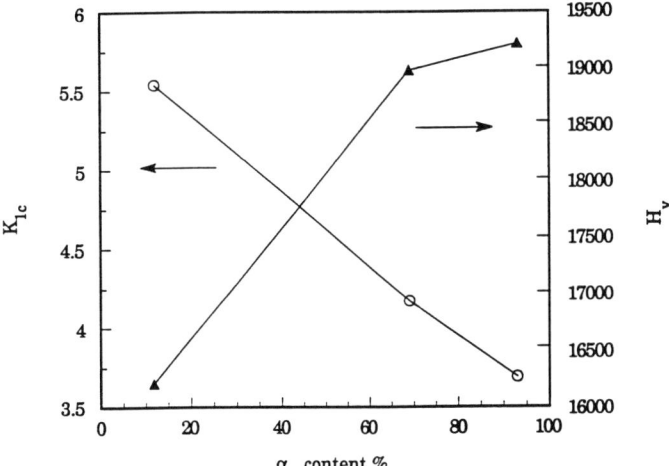

Fig. 12 H_v and K_{1c} of sintered specimen as a function of α-content.

Hardness

Hardness of these samples was also measured and the results are given in Fig. 11. As expected, compositions with a high α'-SiAlON content showed a higher hardness. Fracture toughness was also measured by the indentation method and the results are given in Fig. 12.

SUMMARY

Compositions in the two-phase region α'-SiAlON plus β-Si$_3$N$_4$ solid solution with a small amount of third phase as a sintering aid were prepared. Mixtures of starting materials were reaction sintered to full densities without applied pressure. The phases identified and lattice parameters measured were compared with standards and were used to construct α'-SiAlON-β-Si$_3$N$_4$ solid solution tie lines. As expected, the specimens containing α'-SiAlON showed a higher hardness and lower toughness.

ACKNOWLEDGMENTS

This research was partially supported by the U.S.-Yugoslavia Joint Board for Scientific and Technology Cooperation, administered by the U. S. National Science Foundation and SIZ for Science of SR Serbia.

REFERENCES

1. W-Y. Sun, T-Y. Tien and T-S. Yen, J. Am. Ceram. Soc. **74** (11), 2753-58 (1991).

2. T.S. Sheu and T.Y. Tien, Bimonthly Report, Ceramic Technology for Advanced Heat Engine Project, Oak Ridge National Laboratory, April 30, 1991, p. 69.

3. T.S. Sheu and T.Y. Tien, Bimonthly Report, Ceramic Technology for Advanced Heat Engine Project, Oak Ridge National Laboratory, August 28, 1992, p. 70.

4. K.R. Lai and T.Y. Tien, Bimonthly Report, Ceramic Technology for Advanced Heat Engine Project, Oak Ridge National Laboratory, April 30, 1991, p. 69.

5. W-Y. Sun, T-Y. Tien and T-S. Yen, J. Am. Ceram. Soc. **74** (10), 2547-50 (1991).

6. L.J. Gauckler, J. Weiss, G. Petzow, and T.Y. Tien, J. Am. Ceram. Soc. **61** (9-10), 397-8 (1978).

TRANSIENT VISCOUS PHASE REACTION SINTERED (TVPRS) SILICON OXYNITRIDE CERAMICS.

Kevin P. Plucknett and David S. Wilkinson,

Dept. of Materials Science and Engineering,
McMaster University, Hamilton, Ontario, CANADA.

High density silicon oxynitride ceramics were fabricated by reaction-sintering a mixture of silicon nitride and silica without the use of sintering aids. Precursor silicon nitride compacts were prepared by conventional means after which they were subjected to a low temperature oxidation heat-treatment ($\sim 1000°C$) producing a composite silicon nitride/silica compact. Oxidized compacts were then reaction-sintered in a nitrogen atmosphere at temperatures between 1400 and 1800°C using a range of protective 'powder-bed' compositions. A 'powder-bed' comprising a mixture of boron nitride, silicon nitride and silica was found to be most effective in preventing decomposition and subsequent weight loss. Nearly complete reactive transformation to silicon oxynitride was observed under optimised sintering conditions.

Introduction

Silicon oxynitride (Si_2N_2O) is a promising structural ceramic material for high temperature applications, exhibiting many properties comparable to the more widely studied silicon nitride (Si_3N_4) ceramics and some that are superior (notably oxidation resistance). Several conventional processing routes exist for the fabrication of high density Si_2N_2O that are generally identical to those already derived for the densification of Si_3N_4, for example hot-pressing or pressureless-sintering. Previous researchers have utilized additions of single metal oxides to aid densification via the presence of a liquid phase during hot-pressing, including Al_2O_3 [1,2], Y_2O_3 [3] and CeO_2 [4,5]. Alternatively Si_2N_2O can be densified without applied pressure when a combination of sintering aids are used, for example Y_2O_3 and Al_2O_3 [6,7] or Y_2O_3 and MgO [8].

Both of these processing routes suffer from the same limitations that have previously been outlined for Si_3N_4 based ceramics, namely the shaping restrictions and increased cost imposed by hot-pressing and the degradation of high temperature properties that occurs with the use of multiple sintering aids. To a certain extent these drawbacks can be overcome by the use of glass encapsulated hot-isostatic pressing (HIPing), which allows the fabrication of dense ceramics to near net shape with either the minimization or total elimination of sintering aids. This approach has been taken to fabricate dense, additive-free Si_2N_2O ceramics from a stoichiometric mixture of SiO_2 and Si_3N_4 [9,10] as well as Si_3N_4/Si_2N_2O composites with Y_2O_3 additions [11,12]. However HIPing is an inherently expensive process and it is desirable to find alternatives to this processing route for the fabrication of dense silicon oxynitride ceramics.

This communication will report upon the initial investigation of a unique processing route that allows the fabrication of high density Si_2N_2O ceramics and composites without either sintering additives or applied pressure [13]. The approach taken in the current work involves the creation of a composite ceramic green body comprised of fine, SiO_2 coated Si_3N_4 powder, which can then be reaction-sintered to high density via the presence of a transient viscous phase (the amorphous SiO_2 coating), leading to the terminology transient viscous phase reaction sintering (TVPRS).

Experimental Procedure

A commercially available Si_3N_4 powder (Ube SN E-10) was used during the current study as a precursor for silicon oxynitride formation. Green compacts were prepared by uniaxial compaction, followed by cold isostatic pressing.

To incorporate silica into the green compacts they were subjected to an oxidizing heat-treatment in a static air atmosphere. Optimal conditions, to obtain a uniform level of oxidation throughout the Si_3N_4 compact, were determined by measuring the weight gain during oxidation at various temperatures between 950 and 1200°C for a duration of 200 hours. After oxidation selected compacts were pressureless-sintered in a graphite resistance furnace at a temperatures between 1400 and 1800°C for a period of two hours. Sintering was conducted in either a high purity nitrogen or argon atmosphere with the oxidized green compact situated within a protective 'powder-bed'. Three 'powder-bed' compositions were studied; 'pure' boron nitride (BN), an equal weight percentage mixture of BN and α-Si_3N_4 and a mixture of 50 wt.% BN, 40 wt.% α-Si_3N_4 and 10 wt.% SiO_2. In addition samples were also sintered without the use of a protective 'powder-bed'. Both graphite and alumina crucibles were used for specimen containment during sintering. Further samples were also pressureless-sintered in air at 1600°C for a period of two hours.

After sintering, samples were characterised by measurement of weight loss and density (via immersion in distilled water). Phase composition was determined by X-ray diffraction (XRD). The microstructure of sintered samples was studied by scanning electron microscopy (SEM).

Results and Discussion

Oxidation of Si_3N_4 green compacts
Passive oxidation of Si_3N_4 typically occurs under combined conditions of high temperature and high oxygen partial pressure, and SiO_2 forms following the reaction;

$$Si_3N_{4(s)} + 3O_{2(g)} \rightarrow 3SiO_{2(s,l)} + 2N_{2(g)} \tag{1}$$

The formation of silica during the passive oxidation of Si_3N_4 is accompanied by a significant weight gain which can be either continuously monitored during oxidation or determined by intermittent sample removal from the heat-treatment furnace. In the present example oxidation kinetics were obtained by the second method due to the lengthy nature of the oxidation heat-treatment (up to 200 hours). The oxidation weight gain of the silicon nitride compacts used in the current study is presented in figure 1 for temperatures between 950 and 1200°C. At temperatures below 1050°C an approximately linear increase in mass is observed with increasing heat-treatment duration. At intermediate temperatures (i.e. 1050°C) the oxidation kinetics approximate to parabolic, while at higher temperatures asymptotic oxidation kinetics are observed.

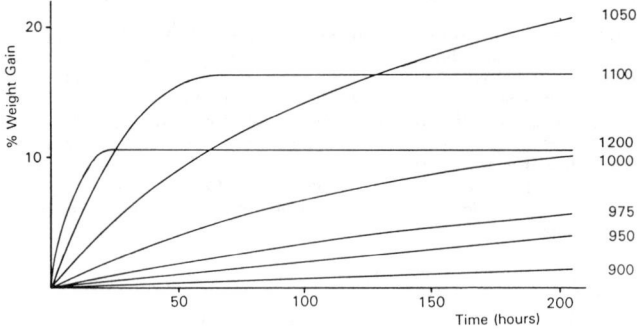

Figure 1. Measurement of the oxidation weight gain as a function of time for ClPed pellets of Ube SN E-10 α-Si_3N_4 powder.

The change in oxidation behaviour with temperature indicates a general change in the oxidation mechanism of the porous silicon nitride compact and is essentially identical to that observed during the oxidation of reaction bonded silicon nitride (RBSN) [14]. The uniform oxidation of a porous silicon nitride body requires an even oxidation rate throughout the green body. However, two concurrent processes are active during oxidation, namely the diffusive flow of oxygen into the interconnecting pores and the reaction of oxygen with silicon nitride. It has been demonstrated that at low oxidation temperatures (~1000°C) a uniform level of oxidation is observed, with an approximately constant volume of SiO_2 present in both the bulk ceramic as well as at the surface [15]. At higher temperatures the SiO_2 concentration is highest at the surface and then decreases gradually into the bulk region of the specimen. This transition in oxidation behaviour with temperature is also reflected in a change in oxidation kinetics from approximately linear at the lower temperatures to parabolic and finally asymptotic as the temperature is increased, similar in nature to the results presented in figure 1. This effect is shown schematically in figure 2.

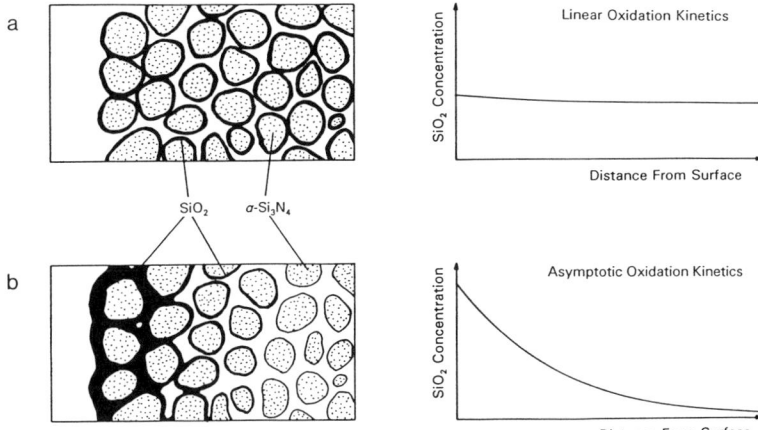

Figure 2. Schematic representation of the oxidation mechanism occurring in porous pellets of Si_3N_4 for a) linear oxidation kinetics and b) asymptotic oxidation kinetics.

It should be noted that in previous studies on the internal oxidation of RBSN the determination of SiO_2 concentration profiles was simplified by the presence of SiO_2 in a crystalline form (i.e. as cristobalite), however in the current study it has been found via XRD that the SiO_2 formed does not crystallise during oxidation but is retained in an amorphous state (figure 3), with α-Si_3N_4 being the only detectable crystalline phase. It is probable that the formation of an amorphous oxidation product in the current case may be favoured due to the high purity nature of the α-Si_3N_4 powder used. Previous studies on the (internal) oxidation of RBSN ceramics have been performed on materials derived from silicon that typically contain a high concentration of metallic impurities and often added catalysts (such as Fe or metal oxides) to promote nitridation of the precursor silicon. It has been demonstrated that the presence of low concentrations of impurities can have a significant effect upon the devitrification kinetics of amorphous silica [16-18]. The work of Laczka has shown that, depending upon the glass processing route, the addition of small amounts of metallic oxides can either increase or decrease the devitrification rate [16]. The presence of trace amounts of 'free' silicon in silica has been found to reduce the rate of devitrification [18]. It is therefore difficult to predict what effects the additional constituents of RBSN will have, although it is reasonable to attribute the lack of devitrification in the present case to the high purity of the α-Si_3N_4 powder used together with the relatively low oxidation temperature.

Sintering of oxidized Si_3N_4 compacts
After determination of suitable oxidizing conditions to produce a uniform level of SiO_2 formation a series of α-Si_3N_4 compacts were oxidized at 1000°C for a period of approximately 70 hours and prepared for sintering. The formation of Si_2N_2O from a mixture of Si_3N_4 and amorphous SiO_2 occurs following the reaction;

$$Si_3N_{4(s)} + SiO_{2(l)} \rightarrow 2Si_2N_2O \quad (2)$$

It can therefore be seen that a 1:1 molar ratio of Si_3N_4 to SiO_2 is required to form stoichiometric Si_2N_2O. The oxidation period of 70 hours was chosen such that the molar ratio of Si_3N_4 to SiO_2 was slightly greater than 1:1 and therefore after sintering a small volume fraction of Si_3N_4 will be retained. An excess of Si_3N_4 may be preferred over an excess of SiO_2 (i.e. a Si_3N_4:SiO_2 ratio less than 1:1) as the presence of residual SiO_2 may enhance creep cavitation and flow at elevated temperatures and, if the SiO_2 devitrifies, the volume change during the polymorphic α to β cristobalite phase transformation would lead to microcrack formation. However it has recently been noted that the creep resistance of HIPed Si_3N_4 ceramics with SiO_2 additions is not significantly reduced until the SiO_2 content exceeds 20 vol.% [19].

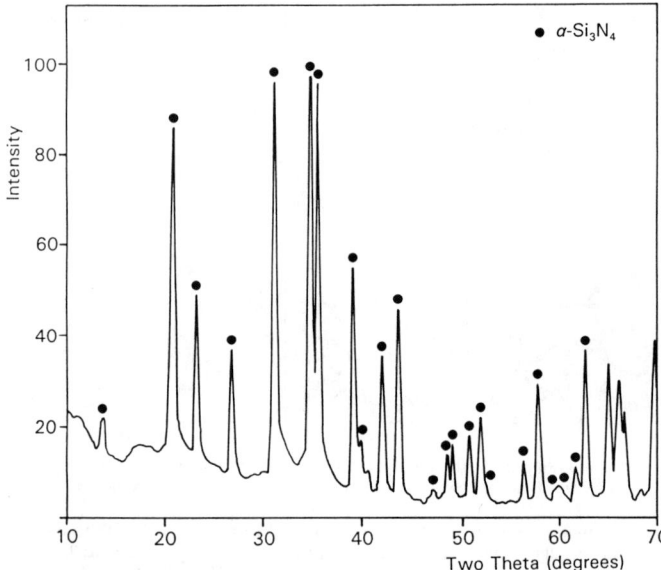

Figure 3. XRD trace of an oxidized Ube SN E-10 α-Si_3N_4 pellet.

Oxidized Si_3N_4 compacts were sintered using various 'powder-bed' compositions at 1600°C for a period of two hours. Sintering was typically performed in either a high purity nitrogen or argon atmosphere using either an Al_2O_3 or graphite crucible. The effects of both 'powder-bed' composition and crucible material upon sample weight loss and density after sintering are shown in table I. Samples sintered without the use of a protective 'powder-bed' exhibit significant weight loss during sintering (>40%). However when a mixed Si_3N_4/BN or Si_3N_4/BN/SiO_2 composition is used for the 'powder-bed' only minimal weight loss is recorded, particularly with the SiO_2 additions, indicating negligible decomposition during sintering. As decomposition of either SiO_2 or Si_2N_2O, once formed, are most probable it is apparent that the presence of SiO_2 in the 'powder-bed' is required to inhibit decomposition. This is most likely to occur via the formation of a partial pressure of SiO from the 'powder-bed' and hence, particularly when sintering in nitrogen, a mixed nitrogen/SiO atmosphere around the sample during sintering. The formation of a porous scale, and gas bloating, at the surface of pressureless-sintered Si_2N_2O ceramics has been previously noted [6]. This may be attributed to Si_2N_2O decomposition following the reaction;

$$2Si_2N_2O_{(s)} \rightarrow Si_3N_{4(s)} + SiO_{(g)} + \tfrac{1}{2}O_{2(g)} \tag{3}$$

with subsequent weight loss due to the formation of volatile species. Such a reaction scale was not observed in the current study when using the three component powder bed, however signs of surface decomposition during the later stages of sintering were apparent in samples sintered in the two component 'powder-bed'. The decomposition of Si_2N_2O, to form β-Si_3N_4, is also apparent from the surface X.R.D. traces of sample 3 when compared to that processed with the BN/Si_3N_4/SiO_2 'powder-bed', as demonstrated by table I and figure 4. Under optimised processing conditions high densities are obtained (> 96% of theoretical), i.e. samples 3 and 4 in table I. Sintering studies at lower temperatures (1400°C) have shown that the viscous phase sintering mechanism utilized in the present case is particularly efficient, with comparable sintered densities to those observed at the higher sintering temperatures (i.e. 1600°C), although transformation to Si_2N_2O does not occur.

Sintering of an oxidized Si_3N_4 compact in air at 1600°C resulted in the formation of a surface oxide scale ~0.5mm thick, which was primarily comprised of cristobalite. XRD of the densified bulk ceramic material, after removal of the surface scale, revealed that no transformation to Si_2N_2O had occurred with the retention of α-Si$_3$N$_4$. From a thermodynamic viewpoint it can be expected that SiO$_2$ will be retained at high oxygen partial pressures, and hence the formation of Si$_2$N$_2$O will be unfavourable [20].

Study of the fracture surface of sample 4 in the SEM indicates an equiaxed grain morphology with grain sizes in the sintered materials of ~0.4-0.5μm.

Table I. Summary of the results obtained when sintering oxidized α-Si$_3$N$_4$ compacts in a nitrogen atmosphere at 1600°C for two hours (+ 10-30%, * <10%, † not assessed).

No.	Powder-bed Comp. (wt.%)	Crucible	Weight Loss (%)	Density (g.cm^{-3})	Phase Content	
					Surface	Bulk
1	None	Al$_2$O$_3$	43.2	†	β-Si$_3$N$_4$ α-Si$_3$N$_4$*	Si$_2$N$_2$O β-Si$_3$N$_4$+
2	BN (100)	Al$_2$O$_3$	27.8	2.40-2.45	β-Si$_3$N$_4$ α-Si$_3$N$_4$+	Si$_2$N$_2$O, β-Si$_3$N$_4$+ α-Si$_3$N$_4$*
3	BN/Si$_3$N$_4$ (50/50)	Al$_2$O$_3$	1.37	2.83	Si$_2$N$_2$O β-Si$_3$N$_4$*	Si$_2$N$_2$O, β-Si$_3$N$_4$+ α-Si$_3$N$_4$*
4	BN/Si$_3$N$_4$/SiO$_2$ (50/40/10)	Al$_2$O$_3$	0.37	2.87	Si$_2$N$_2$O α-Si$_3$N$_4$*	Si$_2$N$_2$O, α-Si$_3$N$_4$* β-Si$_3$N$_4$*
5	BN/Si$_3$N$_4$/SiO$_2$ (50/40/10)	Graphite	0.64	2.36	α-Si$_3$N$_4$ β-Si$_3$N$_4$+	Not Assessed

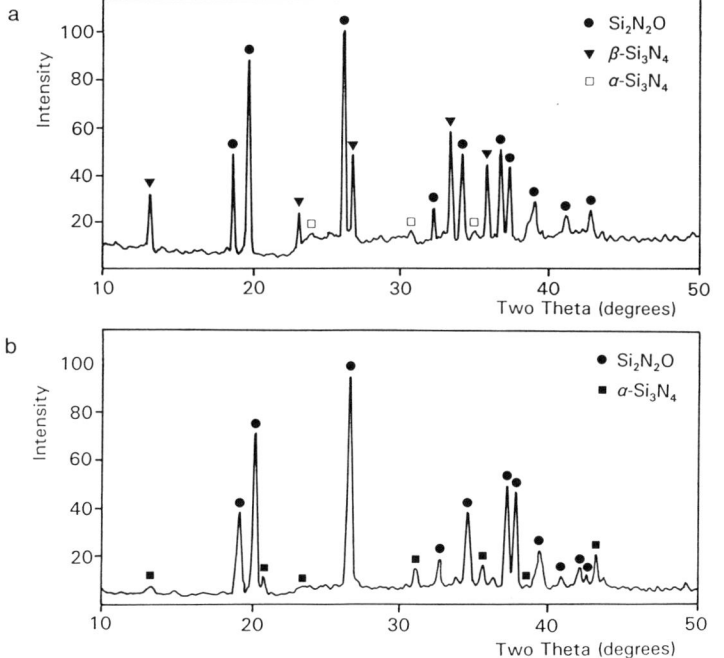

Figure 4. XRD traces of the surface of α-Si$_3$N$_4$/SiO$_2$ compacts pressureless-sintered at 1600°C in a) a mixed BN/Si$_3$N$_4$ powder bed and b) a mixed BN/Si$_3$N$_4$/SiO$_2$ powder bed.

In addition to the preparation of samples without sintering additives further Si_2N_2O have been prepared with the addition of 5 wt.% Y_2O_3. Similar results to sample 4 were obtained, with minimal weight loss and a high proportion of Si_2N_2O formed when sintering at 1600°C. Residual α-Si_3N_4 was transformed to β-Si_3N_4 in these samples.

Conclusions

A simple two stage densification route has been devised for the fabrication of high density Si_2N_2O based ceramics. The first stage involves the creation of a composite α-Si_3N_4/SiO_2 green body via low temperature oxidation (~1000°C) of an α-Si_3N_4 preform. Determination of the oxidation kinetics of the preforms at various temperatures between 900 and 1200°C was used to assess the predominant oxidation mechanism at each temperature. Approximately linear oxidation kinetics were observed at temperatures below 1050°C, with the subsequent formation of a uniform SiO_2 concentration throughout the green body. The duration of oxidation was then controlled to yield an approximately equimolar mixture of Si_3N_4 and SiO_2. A range of sintering conditions were studied and it was found that optimisation of the sintering process (i.e. minimal weight loss, complete Si_2N_2O formation etc.) occurred when using an Al_2O_3 crucible in a nitrogen atmosphere. In addition the selection of a suitable protective 'powder-bed' composition was also of extreme importance. The initial study has demonstrated that a mixed boron nitride/silicon nitride/silica combination leads to minimal weight loss, maximum Si_2N_2O formation and the prevention of surface decomposition during sintering.

References

1. Z.K. Haung, P. Greil and G. Petzow, *Ceram. Int.*, **10** [1] 14-17 (1984).
2. M. Mitomo, S. Ono, T. Asami and S.-J.L. Kang, *Ceram. Int.*, **15** [6] 345-50 (1989).
3. M. Billy, P. Boch, C. Dumazeau, J.C. Glandus and P. Goursat, *Ceram. Int.*, **7** [1] 13-18 (1981).
4. M. Ohashi, H. Tabata and S. Kamazaki, *J. Mat. Sci. Lett.*, **7** [4] 339-40 (1988).
5. M. Ohashi, S. Kamazaki and H. Tabata, *J. Am. Ceram. Soc.*, **74** [1] 109-14 (1991).
6. M.H. Lewis, N.D. Butler and C.J. Reed, *Mat. Sci. Eng.*, **71** 87-94 (1985).
7. C. O'Meara and J. Sjöberg, pp. 647-663 in *Sintering of Advanced Ceramics*, Ceramic Transactions 7, American Ceramic Society, Westerville, OH, 1990.
8. K.P. Plucknett and M.H. Lewis, *Unpublished Research*.
9. R. Larker and L. Hermansson, pp. 375-381 in *Proceedings of the International Conference on Hot-Isostatic Pressing*, Lulea, Sweden, (1988).
10. R. Larker, *J. Am. Ceram. Soc.*, **75** [1] 62-66 (1992).
11. I.P. Tuersley, G. Leng-Ward and M.H. Lewis, pp. 231-246 in *Advanced Engineering with Ceramics*, British Ceramic Society Proc. No. 46, The Institute of Ceramics, Stoke on Trent, UK, 1990.
12. K.P. Plucknett and M.H. Lewis, *Ceram. Eng. Sci. Proc.*, **13** [9-10] 991-999 (1992).
13. K.P. Plucknett and D.S. Wilkinson, *U.S. Patent Application*, (July 1st 1992).
14. F. Porz and F. Thummler, *J. Mat. Sci.*, **19** [4] 1283-95 (1984).
15. F. Thummler and G. Grathwohl, pp. 547-555 in *Progress in Nitrogen Ceramics*, Martinus-Nijhoff (1983).
16. M. Laczka, *J. Am. Ceram. Soc.*, **74** [8] 1916-1921 (1991).
17. P.P. Bihuniak, *J. Am. Ceram. Soc.*, **66** [10] C.188-189 (1983).
18. F.E. Wagstaff and K.J. Richards, *J. Am. Ceram. Soc.*, **48** [7] 382-383 (1965).
19. J. Zeng, I. Tanaka, Y. Miyamoto, O. Yamada and K. Niihara, *J. Am. Ceram. Soc.*, **75** [1] 195-200 (1992).
20. H. Wada, M.-J. Wang and T.Y. Tien, *J. Am. Ceram. Soc.*, **71** [10] 837-840 (1988).

FORMATION AND DENSIFICATION OF R-α'-SiAlONS
(R = Nd, Sm, Gd, Dy, Er and Yb)

P.L. WANG, W.Y. SUN and T.S. YEN
Shanghai Institute of Ceramics, Chinese Academy of Sciences, Shanghai, P.R. China.

ABSTRACT

The densification behavior and formation characteristics of R-α'-Sialons (R = Nd, Sm, Gd, Dy, Er and Yb) hot-pressed in the temperature range of 1500°C-1750°C have been studied. The temperatures required for full densification decrease and the content of the α'-Sialon phase increases with an increasing atomic number of R.

INTRODUCTION

Si_3N_4-based ceramics are attractive for high temperature and wear resistant applications. However, they are difficult to densify without sintering additives because of the covalent nature of the Si-N bonding and the very low self-diffusion coefficient of the atoms. Oxide additives are generally used to promote densification by forming a liquid-phase with the SiO_2 on the surface of Si_3N_4. The liquid phase, however, remains as a grain-boundary glassy phase after cooling and usually deteriorates high temperature properties. One effective approach is to reduce the amount of intergranular phase by forming a transient liquid phase which can react with Si_3N_4 forming α'-Sialon. The α'-Sialon phase,[1] isostructural with α-Si_3N_4, contains two large isolated interstices where some large metal ions can be accommodated. The general formula of α'-Sialon is represented as $M_xSi_{12(m+n)}Al_{m+n}O_nN_{16-n}$, where m(Si-N) are substituted by m(Al-N), n(Si-N) are substituted by n(Al-O) and the valency discrepancy introduced by the substitutions is compensated by metal ion M. The elements M, which have been reported to enter an α'-Sialon structure, are Li, Ca, Mg, Y and R (Nd, Sm, Gd, Dy, Er and Yb).[2,3]

The present work reports the densification behavior, reaction sequences, microstructural features and mechanical properties of R-α'-Sialon (R = Nd, Sm, Gd, Dy, Er and Yb) compositions obtained by hot-pressing, with the aim of understanding the effect atomic numbers of rare earth elements have on these characteristics. Y-α'-Sialon, with the same corresponding composition, was used for comparison.

EXPERIMENTAL PROCEDURES

The nominal compositions studied in the present work are $R_{0.36}Si_{10.38}Al_{1.62}O_{0.54}N_{15.46}$, which are located on the join between Si_3N_4-R_2O_3:9AlN. The starting powders used were α-Si_3N_4 (LC12, H.C. Stark, Berlin, 1.8 wt% O), AlN (laboratory made, 2 wt% O), R_2O_3 (R = Nd, Sm, Gd, Dy, Er and Yb) and Y_2O_3. R_2O_3 and Y_2O_3 (99.9% pure) were the products of Yaolung Chemical Works, China. The starting materials were mixed in absolute alcohol and milled in an agate mortar for 1.5 h. To study densification behavior and reaction sequence hot-pressing was carried out in a graphite-resistance furnace under a pressure of 20 MPa in N_2 at temperatures from 1500 to 1750°C for 1 h. The specimens used for determination of mechanical properties and for TEM observation were hot pressed at 1750°C for 2 h.

Fracture toughness was measured by the indentation method and phases were identified by the X-ray diffraction technique. Intensities of reflection lines (210) of α-Si_3N_4, (200) of β-Si_3N_4, (211) of R(Y)-melilites and (100) of AlN were used to make calibration curves for the semi-quantitative estimation of crystalline phases.

RESULT AND DISCUSSION

Densification Behavior

During hot-pressing, shrinkage of the specimens can be observed. All the compositions start to shrink at about 1550°C with the exception of Yb-α'-Sialon composition, which shrinks at about 1490°C. The temperatures required to reach full densification vary with the rare-earth elements, as shown in Fig. 1. The Yb-α'-Sialon composition can be fully densified at 1550°C. The Dy-, Er- and Yb-α'-Sialon compositions reach the same values at 1600°C. Nevertheless, 1700°C is necessary for the other R-α'-Sialon compositions.

Huang et al.[2] reported that the solubilities of rare earth ions in α'-Sialon with the compositions along the Si_3N_4-R_2O_3:9AlN line increase while the ionic radii of rare earth ions decreases (i.e., the atomic number gets higher), and, for the Yb ion, the top limit of solubility has a value of $x_{max} = 1.0$. The tendency toward densification of R-α'-Sialon compositions seems to follow the same order. However, the main factor affecting the densification temperature is closely related to the temperature at which the liquid phase appears. In general, the liquid phase is first formed by the reaction between oxide additives and native oxides on the surface of nitrides. The dissolution of nitrogen into the oxide liquid phases would increase their viscosity. The solubility of nitrogen in Nd-sialon glass can reach 25 a/o[4], whereas in the Y-Si-Al-O-N system the highest solubility is only about 15 a/o[5]. The behavior of heavy rare earth oxides are, in many aspects, similar to Y_2O_3. The higher firing temperature required for densification of light R-α'-Sialon compositions might imply that the concentration of N in light rare earth Sialon glasses is higher than that in heavy R-Sialon glasses. On the other hand, the formation of melilite in the light rare-earth compositions at a very early stage, as mentioned below, might also hinder the densification.

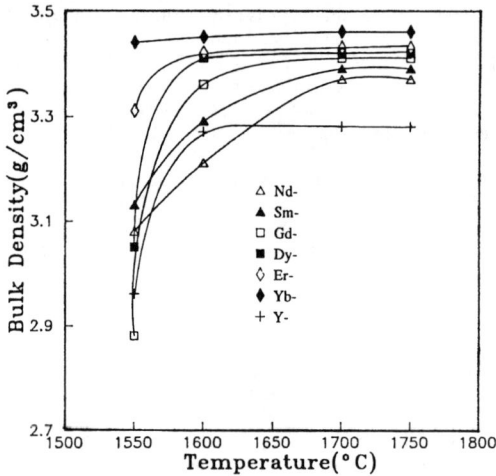

Fig.1 Densification behaviour of R(Y)-α'-Sialon compositions

Reaction Sequences

Figure 2 shows the reaction sequences of R(Y)-α'-Sialon compositions, and Table 1 gives the phase compositions of the specimens hot pressed at 1500°C for 1 h. At 1500°C R(Y)-containing oxynitride phases (i.e., melilite, K-phase and J-phase) are already formed. With increasing temperature, more α'-Sialon, in company with more or less β'-Sialon, are formed at the expense of α-Si_3N_4, β-Si_3N_4, AlN and the previously formed R(Y)-melilite

Table 1. Phase present in the R-α'-Sialon compositions hot-pressed at 1500°C for 1h

Comp.	Nd-α'	Sm-α'	Gd-α'	Dy-α'	Y-α'	Er-α'	Yb-α'
Strong	α	M, α	α, M	α, M	α	α	α
Medium	K	--	--	--	J	J, α'	α', J
Weak	AlN, β	AlN, β, α'	AlN, α', β	AlN, α', β	α', AlN, β	AlN, β	Y, AlN, β

*M(Melilite) = $R_2Si_3O_3N_4$, K(Wollastonite) = $RSiO_2N$, J(Wohlerite) = $R_4Si_2O_7N$, Y = YAG

or other oxynitride phases. Small amounts of unreacted α-Si_3N_4 and AlN remain until hot-pressed at 1750°C for 2 h. The amount of α'-Sialon phase and the ratio of α'/(α' + β') in the R(Y)-α'-Sialon compositions obviously increase as the atomic number of the rare earth element increases, and Yb-α'-Sialon composition has the highest ratio of α'/α' + β' and contains Jss phase instead of melilite as a minor phase. However, the formation of R-melilite seems to have the opposite tendency.

Sun et al.[5] have studied the subsolidus phase relationships in the system Si,Al,Y/N, O and found that α-Sialon (boundary facing β'), β-Si_3N_4 and melilite establish a compatibility tetrahedron. Considering the oxygen content inherently associated with Si_3N_4 and AlN, the composition used in this work is located just inside the α-Sialon (boundary facing β')-β-Si_3N_4-melilite compatibility tetrahedron. This may be the reason for the existence of a small amount of β'-Sialon and melilite phases in the Y-α'-Sialon composition. Although phase relationships in the R-Si-Al-O-N systems have not been thoroughly studied, the appearance of a small amount of β'-Sialon and melilite in the R-α'-Sialon compositions studied, especially in the heavy rare earth-containing α'-Sialon compositions, can also be attributed to the same reason. However, the fact that relatively large amounts of β'-Sialon and melilite phase remain in the Nd- and Sm-α'-Sialon compositions indicates that the formation of light rare earth-containing melilites at lower temperatures might hinder the conversion of α-Si_3N_4 to α'-Sialon thus promoting the formation of β'-Sialon. The formation of the β'-Sialon phase decreases slowly with the increase in atomic number of rare earth elements in R-α'-Sialon compositions. However, the unit cell dimensions are almost the same (corresponding to a z value of 0.5). The kinetic priority of the occurrence of light rare earth-melilite probably reduces the liquid phase thus retarding the densification of R-α'-Sialon compositions, as mentioned above. The effect of atomic number of rare earth elements on the formation of R-α'-Sialon is similar to the result obtained by Ekström et al.[7]

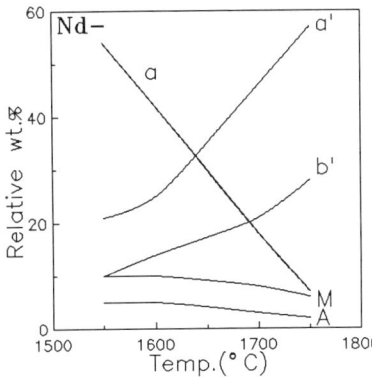
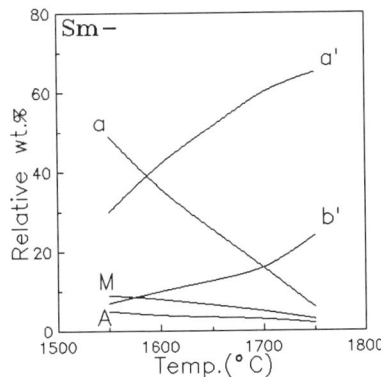

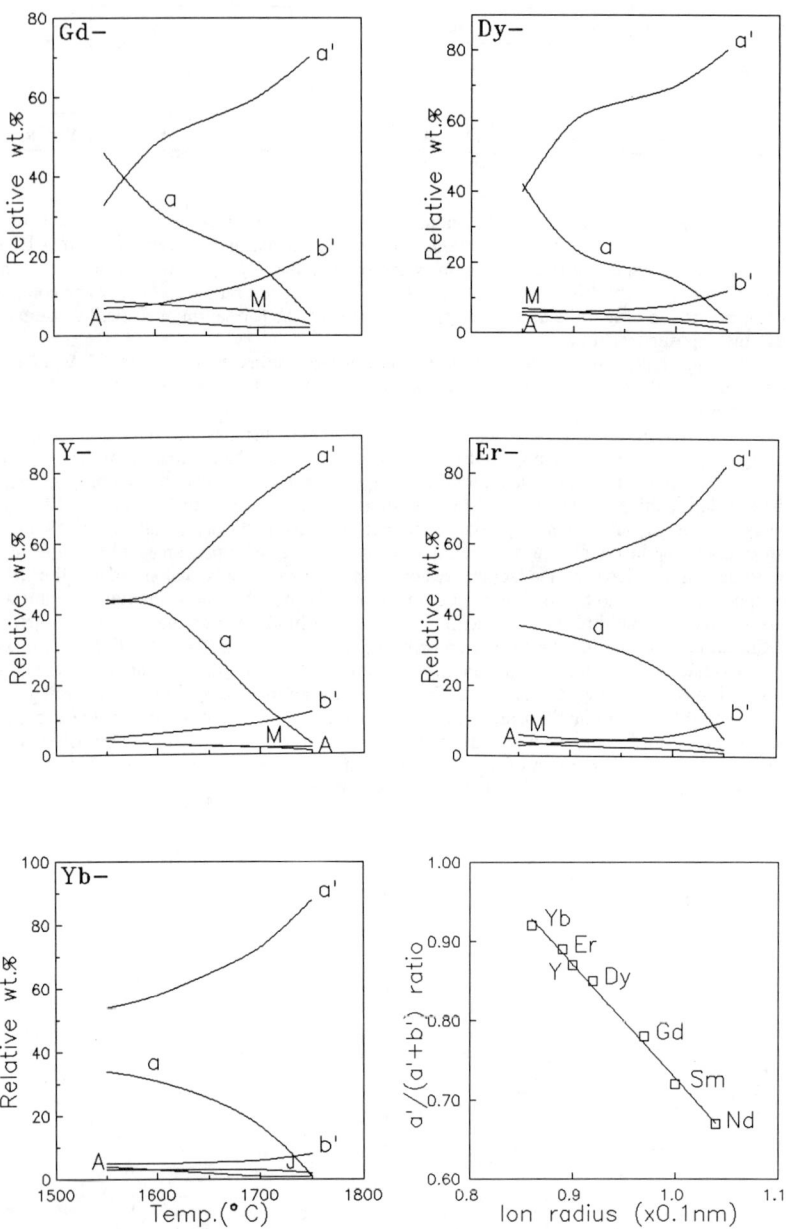

Fig.2 The reaction sequence and the ratio of $\alpha'/(\alpha'+\beta')$ for all R-α'-Sialon compositions. (a=α-Si_3N_4, a'=α'-Sialon, b'=β'-Sialon, M=$R_2Si_3O_3N_4$, A=AlN, J_{ss}=$Yb_4Si_2O_7N_2$-$Yb_4Al_2O_9$).

Microstructure Features

R(Y)-α'-Sialon compositions have similar morphologies. Figures 3a and b are the TEM photographs for Gd- and Yb-α'-Sialon compositions. The largest α'-Sialon grains with equiaxed shape are about 2μ. The metal elements in α'-Sialon grains were detected by EDAX to be Si, Al and rare earth elements (or Y) in decreasing sequence of the peak height. As shown in Fig. 3, the grain boundary thickness is very thin and has a mean thickness of about 0.2μ-0.3μ. The clean grain boundary can be attributed to the transient liquid-phase sintering mechanism. The crystalline grain boundary phase which was presumed to be melilite phase was also analyzed by EDAX and the results show that besides Si and R(Y), Al does occur, as Jack[8] and Slasor[9] reported. The existence of melilite phase at the grain boundaries was further confirmed by electron diffraction. Figure 4 shows a big grain boundary region in Nd-α'-Sialon compositions. The electron diffraction pattern along the direction of [001] corresponds to a tetragonal lattice with unit cell parameters close to melilite. The peak heights of the intergranular phase in Nd-, Gd-, and Y-α'-Sialon compositions are in the sequence of Si > Nd(Gd,Y) > Al. However, in the Yb-α'-Sialon specimen, the Al peak is higher than the Yb peak which indicates that the intergranular phase in the Yb-α'-Sialon composition is not a Yb-melilite phase, as was also indicated by X-ray diffraction results.

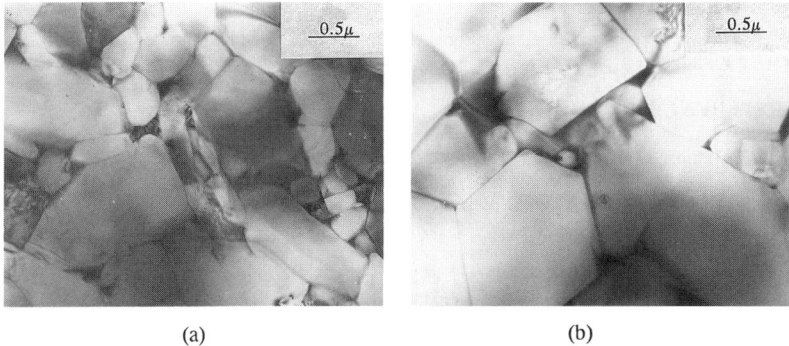

(a)   (b)

Fig.3 TEM micrographs of (a) Gd- and (b) Yb-α'-Sialons compositions

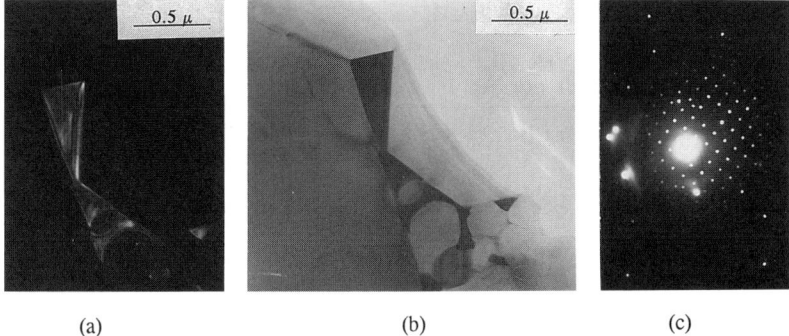

(a)   (b)   (c)

Fig.4 TEM micrographs of grain boundary in Nd-α'-Sialon composition: (a) dark-field, (b) bright-field, (c) electron diffraction along [001]

Mechanical Properties

The hardness (HRA) of all R-α'-Sialon compositions and the fracture toughness (K_{1c}) for Nd-, Gd-, Y-, Yb-α'-Sialon compositions are listed in Table 2. These hardness values are nearly the same with the exception of Nd-α'-Sialon. The relatively lower hardness and higher fracture toughness of Nd-α'-Sialon composition can be attributed to the lower content of α'-Sialon, which is known to have a higher hardness and lower fracture toughness than β'-Sialon.

Table 2. Hardness(HRA) and fracture toughness(K_{1c}) of R(Y)-α'-Sialon compositions

Composition	Nd-α'	Sm-α'	Gd-α'	Dy-α'	Y-α'	Er-α'	Yb-α'
HRa	94.1	95.1	95.3	95.2	95.1	95.3	95.6
K_{1c}(MPa·m$^{1/2}$)	6.78		5.43		5.68		5.06

ACKNOWLEDGEMENT

This work is supported by the National Natural Science Foundation of China under Contract No. 59082420.

REFERENCES

1. S. Hampshire, H.K. Park, D.P. Thompson and K.H. Jack, Nature, **274**, 880 (1978).

2. Z.K. Huang, P. Greil and G. Petzow, J. Am. Ceram. Soc., **66**, C-96 (1983); Z.K. Huang, T.Y. Tien and T.S. Yen, ibid., **69**, C-241 (1986).

3. Z.K. Huang, W.Y. Sun and D.S. Yan, J. Mater. Sci. Letters, **4**, 255 (1985); S.F. Kuang, Z.K. Huang, W.Y. Sun and T.S. Yen, ibid., **9**, 69 (1990); **9**, 72 (1990).

4. R.A.L. Drew, Ph.D. Thesis, University of Newcastle upon Tyne, 1980.

5. D.P. Thompson (private communication).

6. W.Y. Sun, T.Y. Tien and T.S. Yen, J. Am. Ceram. Soc., **74**, 2547 (1991); Sci. in China Series A, **35** (7), 877 (1992).

7. T. Ekström and M. Nygren, J. Am. Ceram. Soc., **75**, 259 (1992).

8. K.H. Jack, in Non-Oxide Technical and Engineering Ceramics, edited by S. Hampshire (Proc. Inter. Conf. Limerick, Ireland, 1985) pp. 1-30.

9. S. Slasor, K. Liddell and D.P. Thompson, Br. Ceram. Proc., **37**, 51 (1986).

PARAMETERS AFFECTING PRESSURELESS SINTERING OF α'-SIALONS WITH LANTHANIDE MODIFYING CATIONS

K.P.J. O'REILLY* M. REDINGTON*, S. HAMPSHIRE* M. LEIGH**
*Materials Research Centre, University of Limerick, Ireland
**De Beers Industrial Diamond Division, Shannon, Ireland

ABSTRACT

α-silicon nitride forms a range of solid solution $M_x(Si,Al)_{12}(O,N)_{16}$ where x = 2 and M is Li^+, Ca^{2+}, Y^{3+} or certain lanthanide cations. This paper reports the formation of Ln-α'-sialons by the reaction of Si_3N_4 with AlN and Ln_2O_3 where Ln=Nd, Sm, Eu, Gd, Tb, Dy, Ho, Er, Tm, Yb, Lu and Y. Transformation reactions of Nd, Sm, Gd, Ho, Er and Y cations occurring during liquid phase pressureless sintering are reported. The effects of the volume and viscosity of the liquid phase and the amount of secondary N-melilite on sintering are discussed.

1. INTRODUCTION

α'-sialon ceramics have received considerable interest over the last 15 years because of their thermal and chemical stability combined with high hardness which gives rise to potentially useful engineering ceramics[1-7]. Densification of silicon nitride based ceramics occurs via liquid phase sintering which requires an additive in order to form an oxynitride liquid phase that promotes shrinkage and phase transformation by a solution-diffusion-precipitation process[8]. In the case of α'-sialons, the additive itself can be accommodated in solid solution offering the possibility of producing a single-phase material[1,2,3,7]. α'-sialons have the general composition of $M_x(Si,Al)_{12}(O,N)_{16}$, where $0 \leq x \leq 2$ and M is a modifying cation such as Li^+, Ca^{2+}, Y^{3+} and most of the lanthanide (Ln^{3+}) cations[5,7] but may also be a mixture of cations[9] (e.g. Li^+/Ca^{2+}, Li^+/Nd^{3+}, Ca^{2+}/Nd^{3+}).

The structure of α'-sialons is derived from α-silicon nitride, which has a trigonal space group P31c[10], by partial replacement of Si by Al and N by O. The modifying cation maintains electrical neutrality by being "stuffed" into the structure. Crystallographic[1,4,7,11,12] and EXAFS[13] studies suggest that these modifying cations occupy large closed interstices in the (Si,Al)-(O,N) network positioned at (0.333, 0.667, z) and (0.667, 0.333, 0.500+z). Previous studies[14] have reported practical difficulties associated with the formation and densification of pure, single-phase α'-sialons. As α'-sialon forms during reaction-sintering, the volume of liquid decreases as the modifying cation is taken into solid solution. A number of investigations have concentrated on α'-formation in RE-Si-Al-O-N systems[14-17]. Reaction sequences go through an intermediate stage in which a N-melilite phase[14,15] is observed which appears to inhibit densification and α' yield. The maximum amount of melilite and the temperature at which this occurs varies depending on the modifying cation.

This paper describes an investigation of lanthanide oxides as additives to establish which Ln^{3+} cations stabilise α'-sialon of a particular composition. Transformation reactions are outlined for some of the Ln^{3+} modifying cations. The kinetics of densification during liquid phase sintering allow the volume of liquid phase formed to be assessed. The effects of the volume and viscosity of the liquid phase and the amount of N-melilite formed on sintering are discussed.

2. EXPERIMENTAL PROCEDURE

The starting compositions were prepared from mixtures of Ln_2O_3, Si_3N_4 and AlN (supplied by H.C. Starck, Germany). These were wet-mixed in isopropanol for 30 minutes, followed by dry-mixing to overcome the sedimentary effects caused by evaporating off the alcohol. The powder mixtures were then first preformed into a cylinder (10mm diameter x 5mm height), embedded in boron nitride in an alumina crucible and then fired at the appropriate temperature and time under one atmosphere of nitrogen. Weight and bulk density measurements were taken before and after firing, before the sample was pulverised and submitted for powder X-ray diffraction to determine the phases present in the sintered sample.

3. RESULTS AND DISCUSSION

Following work done by Hampshire et al.[1], an initial study was carried out using yttrium and the lanthanides as modifying cations to stabilise the α'-sialon structure. The composition $M_{0.55}Si_{8.35}Al_{3.65}O_{0.8}N_{15.2}$ was fired at 1750°C for 30 minutes. α'-sialon was formed in addition to other phases and these are shown in Table I. A more detailed investigation was carried out on the more promising systems by varying the ratios of the starting powders M_2O_3, Si_3N_4 and AlN. Phase assemblages, bulk densities and weight losses were recorded for each system.

Table I

Lanthanides				Phases %	
	α	β	α'	AlN	M-N-Melilite
La	45	45	-	10	-
Ce	40	55	-	5	-
Pr	45	45	-	10	-
Nd	35	-	35	10	20
Pm		Not Investigated			
Sm	-	-	75	5	20
Eu	35	-	35	10	20
Gd	5	-	75	5	15
Tb	35	-	35	10	20
Dy	20	-	50	10	20
Ho	-	-	70	5	25
Er	-	-	70	10	20
Tm	40	-	35	10	15
Yb	25	-	55	5	15
Lu	10	-	65	5	20
Y	-	-	90	5	5

To understand more fully the behaviour and formation of α'-sialon during sintering, the kinetics of densification were studied at various temperatures for sintering times of up to 2 hours.

Arising from this, a variety of parameters governing the behaviour of α'-sialon formation were investigated. The more pertinent of these are as follows. From the initial study of the optimum composition, the overall α'-sialon yield was established with respect to the cations used. From Table I, it can be seen that La, Ce and Pr do not not promote α'-sialon formation. α→β transformation occurs with no N-melilite formation. All the other lanthanides result in at least 35% α'-sialon yields and up to 90% α'-sialon in the case of Y as the modifying cation. From Table I, N-melilite is seen to form in varying amounts, in conjunction with α'-sialon. N-melilite is seen to compete with α'-sialon for the stabilising cation due to its cation-rich nature, $M_2Si_3O_3N_4$. After 30 minutes at 1750°C, some untransformed starting phases remain. These were unreacted α-Si_3N_4 and AlN. A coarse AlN grain size is assumed to be partly responsible for this.

The more detailed study supported the findings of the initial study. Bulk densities and weight losses were recorded. Qualitative interpretations of the α'-sialon formation tendencies were deduced.

The kinetic studies were based on the Kingery model[18]. From plots of log shrinkage against sintering time, quantitative estimates of the volume of liquid present during sintering were based on the volume shrinkage found from the end of the rearrangement stage. The rates of α → α' transformation varied over time, and an average transformation rate in terms of % α' formed per minute was calculated. Finally estimates of viscosity were based on the metal oxide heat of formation. Viscosity, as liquid flow, is assumed to take place by bond breaking and reforming. A strong metal-oxygen bond will be difficult to break leading to a high viscosity liquid. This assumption is supported by studies of Tg for oxynitride glasses:

Table II. Log Viscosities (Pa.s) at 1000°C

%N→ Cation	10	17	25
Y	10.6	11.0	11.5
Nd	9.8	10.8	11.4
La	9.8	10.1	11.0
Sm	8.7	9.7	10.7

From this, the important parameters affecting the pressureless sintering of α'-sialons using lanthanide modifying cations, where an optimum sintering temperature of 1750°C is used, are: 1) α'-sialon yield, 2) % volume of liquid, 3) % N-melilite, 4) liquid viscosity. Other dependent parameters are: 5) % untransformed phases, 6) bulk density, 7) α'-sialon transformation rate. These are qualitatively expressed in Table III. To investigate the parameters that affect the pressureless sintering an overlay approach will be used. It is desirable in sintering to approach 100% α'-sialon yield with full density.

Figure 1 shows a plot of α'-sialon yield against liquid volume. It is clear from the density contours that bulk density increases as the liquid volume increases. An allowance has to be made for the atomic weight of the modifying cation. Yttrium is a substantially lighter cation than the lanthanides. However the bulk density increase is non linear. Up to 15%, liquid volume has little effect on densification whereas liquid volumes in excess of 25% lead to practically full densification. Figure 1 also shows that there is a non-linear relationship between α'-sialon yield and liquid volume. Low volumes of liquid clearly promote high α'-sialon yields, but at the expense of low bulk densities. As the % liquid

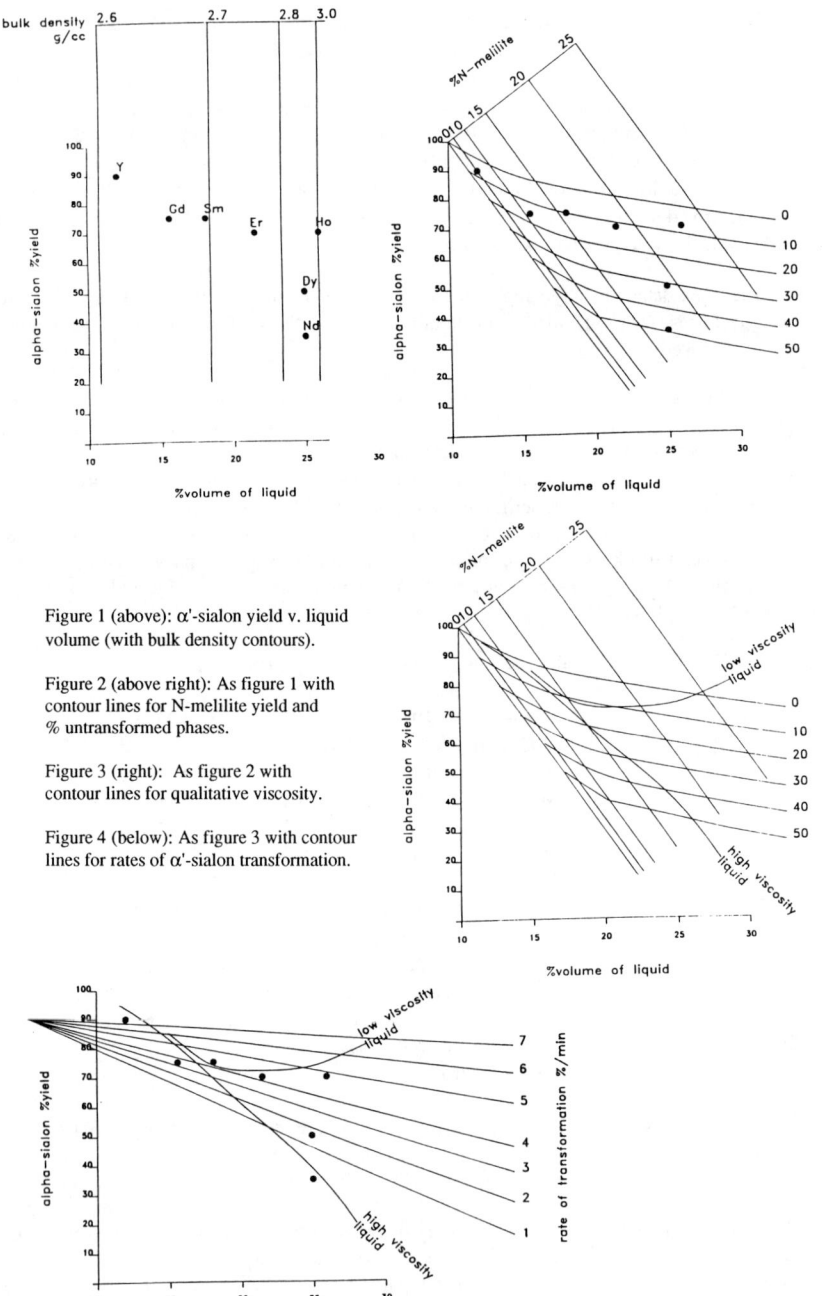

Figure 1 (above): α'-sialon yield v. liquid volume (with bulk density contours).

Figure 2 (above right): As figure 1 with contour lines for N-melilite yield and % untransformed phases.

Figure 3 (right): As figure 2 with contour lines for qualitative viscosity.

Figure 4 (below): As figure 3 with contour lines for rates of α'-sialon transformation.

volume increases, the situation is not as clear cut. Higher liquid volumes result in a spread of α'-sialon yields.

Table III

Cation	Y	Nd	Sm	Gd	Dy	Ho	Er
Parameter							
%α'-sialon yield	90	35	75	75	50	70	70
% volume of liquid	12	25	18.1	15.6	24.9	26	21.5
% N-melilite	5	20	20	15	20	25	20
Ht.formation (KCals/mol)	171	169	137	171	146	148	146
% untransformed phases	5	45	5	10	30	5	10
Bulk density (g/cc)	2.82	2.80	2.71	2.62	2.88	2.98	2.75
α'-transformation rate %/min	7.65	0.84	4.38	3.45	1.04	4.26	4.54

Figure 2 indicates the situation when other phases are taken into account. N-melilite competes with α'-sialon for the modifying cation. The amounts of unreacted phases are also plotted as a series of contour lines. The spread of α'-sialon yields at high liquid volumes is seen in the context of N-melilite formation. Where larger, i.e. up to 20%, quantities of N-melilite are formed, there is direct competition with the yield of α'-sialon. When neither yields are high, this is reflected by the high fraction of untransformed starting phases. At high liquid volumes there is a governing feature of the liquid which inhibits solution. If solubility is sufficiently high at high liquid volumes, then N-melilite is increased (along with increased α'-sialon yields). This is controlled by both the volume and the viscosity of the liquid. These qualitative contour lines are plotted in figure 3. At low liquid volumes, variations in liquid viscosity have little effect on α'-sialon yield. As the liquid volume increases, the governing feature of the liquid is its viscosity. A low viscosity liquid allows rapid transport and widespread melilite formation. Higher viscosity liquids affect and inhibit solution and transport of the starting powders. Nevertheless, it still must be pointed out that the lower viscosity liquids promote α'-sialon yield just as much as N-melilite yield.

The issue of solution and transport in liquids of different quantities and viscosities can be further explored by assessing the average rate of α'-sialon transformation in terms of % α'-sialon formed per minute. This is shown in figure 4. At low liquid volumes high α'-sialon yields are achieved at a high conversion rate. As the volume of liquid increases, the rate of α'-sialon formation shows more variation with respect to overall α'-sialon yield. A large volume of high viscosity liquid results in low α'-sialon formation rate. As the viscosity increases the faster α'-sialon formation rate results in greater α'-sialon yields in addition to somewhat increased N-melilite yields. The volume of liquid will affect the rate of build up to supersaturation and the smaller the liquid volume, the more rapid this will be, resulting in faster α'-sialon formation and consequently, a reduction in the liquid volume as this reaction proceeds and the M^{3+} is taken into solid solution from the liquid phase.

4. CONCLUSIONS

Where pressureless sintering of α'-sialon is carried out at 1750°C, all the lanthanide cations, with the exception of La, Ce and Pr, are capable of stabilising the α'-sialon structure. Differences in liquid volume fraction during sintering occur depending on the lanthanide metal oxide used. The greater the quantity of liquid formed, the greater the resultant bulk density. At low liquid volumes there is a greater α'-sialon yield, regardless of liquid characteristics such as viscosity. At high volumes of liquid a lower α'-sialon transformation rate is associated with a higher liquid viscosity. This results in a low α'-sialon yield. At high volumes of liquid, high α'-sialon transformation rates are associated with lower viscosity liquids. In these lower viscosity liquids, although α'-sialon yields are increased there is greater competition for the cation from the N-melilite. Thus high liquid volumes, desirable for good densification, can at worst give low α'-sialon yields with much untransformed silicon nitride and, at best, give moderate α'-sialon yields with increased N-melilite formation.

REFERENCES

1. S. Hampshire, H.K. Park, D.P. Thompson and K.H. Jack, Nature, **274**, 880 (1978).
2. G. Grand, J. Demit, J. Ruste and J.P. Torre, J. Mater. Sci., **14**, 1749 (1979).
3. M. Mitomo, H. Tanaka, K. Muramatsu, N.II and Y. Fujii, J. Mater. Sci., **15**, 2661 (1980).
4. H.K. Park, D.P. Thompson and K.H. Jack, in Science of Ceramics **10**, edited by H. Hausner (Deutsche Keramische Gesellschaft, Germany,1980) p. 251.
5. M. Mitomo and Y. Uemura, J. Mater. Sci., **16**, 552 (1981).
6. Z-K. Huang, P. Greil and G. Petzow, J. Am. Ceram. Soc. 66, C-96 (1983).
7. K. H. Jack, in Progress in Nitrogen Ceramics, edited by F.L. Riley (Martinus Nijhoff, The Hague,1983) pp.45-60.
8. K.H. Jack, in Non-oxide Technical and Engineering Ceramics, edited by S. Hampshire (Elsevier - Applied Science, Barking, England, 1986) pp. 1-30.
9. M. Redington, K.P.J. O'Reilly and S. Hampshire, in Ceramic Materials and Components for Engines, Proceedings of the 3rd International Symposium, edited by V.J. Tennery, (American Ceramic Society, Westerville, OH, 1989) pp. 127-138.
10. R. Marchand, Y. Laurent and J. Lang, Acta. Crystallogr., B**25**, 2157 (1969).
11. F. Izumi, M. Mitomo and J. Suzuki, J. Mater. Sci. Lett., **1**, 533 (1982).
12. F. Izumi, M. Mitomo and Y. Bando, J. Mater. Sci. **19**, 3115 (1984).
13. M. Cole, K.P.J. O'Reilly, M. Redington and S. Hampshire, J. Mater. Sci. **26,** 5143 (1991).
14. S. Slasor and D.P. Thompson, in Non-oxide Technical and Engineering Ceramics, edited by S. Hampshire (Elsevier-Applied Science, Barking, England, 1986) pp.223-230.
15. S. Hampshire, K.P.J. O'Reilly, M. Leigh and M. Redington, in High-Tech Ceramics, edited by P. Vincenzini (Elsevier, Amsterdam, 1987) p. 933.
16. S. Slasor and D.P. Thompson, J. Mater. Sci. Lett. **6**, 315 (1987).
17. Z.K. Huang, T.Y. Tien and D.S. Yan, J. Am. Ceram. Soc. **69**, C-241 (1986).
18. S. Hampshire and K.H. Jack, in Special Ceramics 7, edited by D. Taylor and P. Popper, (Proc. Brit. Ceram. Soc., **31,**1981) pp. 37-49.

NON-OXIDE ADDITIVES AS SINTERING AIDS FOR Si$_3$N$_4$-BASED CERAMICS

C. GE, Y. XIA AND L. CHEN
Laboratory of Special Ceramics and Powder Metallurgy, University of Science and Technology Beijing, Beijing 100083, China

ABSTRACT

This article is a review of the selected activities since 1979 in using non-oxide additives as sintering aids for silicon nitride. Emphasis was placed on the work done in the authors' laboratory, and it is shown that ZrN and AlN in combined use can result in producing silicon nitride with good properties.

INTRODUCTION

Silicon nitride is recognized as one of the most important high temperature structural ceramics for the development of ceramic engines. Generally, oxide additives (such as MgO, Al$_2$O$_3$, Y$_2$O$_3$, etc.) have to be used to promote liquid phase sintering and densification. However, the glassy phase formed at the grain boundaries after sintering often causes serious degradation of high temperature properties. A great deal of work has been done on improving high temperature properties of Si$_3$N$_4$ with oxide additives. Utilization of non-oxide additives instead can significantly reduce the amount of glassy phase and raise its softening temperature; however, they make Si$_3$N$_4$ difficult to sinter. C. Greskovich[1] used a small amount of Be$_3$N$_2$ (or BeSiN$_2$) as a sintering aid and prepared a single phase hot-pressed Si$_3$N$_4$ ceramic with a relative density of > 95% with equiaxed grain structures from high purity Si$_3$N$_4$ powders. By further control of the chemistry and hot-pressing parameters, single-phase β-Si$_3$N$_4$ solid solutions with an approximate composition of Si$_{2.9}$Be$_{0.1}$N$_{3.8}$O$_{0.2}$ were prepared with a relative density of > 98% and an average grain size of 1 μm. Later in 1981[1] he also reported that Si$_3$N$_4$ compacts could be reproducibly sintered to a relative density of > 99% by a gas-pressure sintering process. Good oxidation resistance, creep resistance, and high temperature bending strength were obtained, although the toughness was relatively low (K_{IC}-4MPam$^{1/2}$).

In this paper the authors describe the follow-up research in using non-oxide additives to silicon nitride, with special emphasis on their own work.

NITRIDE ADDITIVES OTHER THAN Be$_3$N$_2$

To avoid the toxicity of Be compounds, work has been directed to other non-toxic nitride additives. For example, N. Uchida, M. Shimada and M. Koizumi[2] reported that highly dense Si$_3$N$_4$ ceramic bodies with a single nitride additive such as TN, AlN, ZrN, VN, NbN, YN, Mg$_3$N$_2$, TaN, or HfN, could be fabricated by high pressure hot-pressing. They also found that AlN, VN, YN and Mg$_3$N$_2$ were quite effective additives for consolidating Si$_3$N$_4$; their relative densities after hot isostatic pressing were > 96%.[3] More recently, Osami Abe[4] prepared Si$_3$N$_4$ containing 1.24-5.96% oxygen by adding alkaline-earth nitride as a sintering aid. Apart from Be$_3$N$_2$, other additives such as Mg$_3$N$_2$, Ca$_3$N$_2$, or Sr$_3$N$_2$ were used. In this report, the density of Be$_3$N$_2$-doped Si$_3$N$_4$ was only ~90% of the theoretical value, and the ceramic had a relatively low room temperature strength (~400 MPa). However, hot isostatically pressed Si$_3$N$_4$ with Mg$_3$N$_2$ achieved flexural strength values of 1070 MPa at room temperature, 800 MPa at 1200°C and 620 MPa at 1300°C.

DUPLEX ADDITIVES

Ge and Xia[5] found that duplex ZrN-AlN additives could promote densification of Si_3N_4 to > 95% of the theoretical value with gas pressure sintering. Similar density could not be achieved with single AlN or ZrN additive alone under the same processing conditions. It was found[6] that fully dense Si_3N_4 with these additives (specimens ZAN) has a 3-point bending strength (σ_{RT}) of 686 MPa at room temperature and indentation fracture toughness (K_{IC}) of 6.5 MPam$^{1/2}$. The strength ratios at 1250°C and room temperature was 90% and at 1400°C and room temperature was 70%. For comparison, Si_3N_4 with $MgAl_2O_4$ as a sintering aid (specimens MAO), was also studied and found to have much lower strength ratios (~60% and ~40%, respectively). Figure 1, which shows SEM fractographs of specimens tested at 1400°C, indicates the different characteristics of these specimens. Many small "spheres" appeared at the fracture surface of oxide-added Si_3N_4, presumably due to liquid phase formations under the action of surface tension. This phenomenon was not observed in nitride-added Si_3N_4. A lower oxidation rate constant, $3.5 \times 10^{-11} kg^2 m^{-4} sec^{-1}$, at 1400°C, for the nitride-added Si_3N_4 was also measured. Such value was two orders of magnitude lower than that of $MgAl_2O_4$-added material.

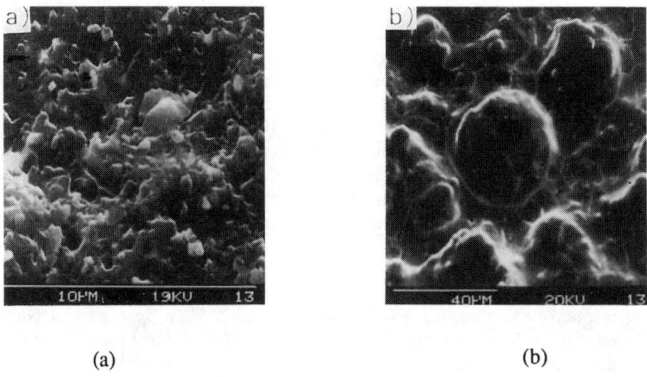

(a)　　　　　　　　　　　　(b)

Fig. 1. SEM fractographs after bending strength test at 1400°C: (a) ZAN; (b) MAO

Ge et al. later examined the microstructural changes in these nitride-added materials and found that transformation was complete at 1750°C.[7] Densification, on the other hand, was achieved in the range of 1750-1800°C. This fact can be explained with the liquid phase sintering mechanism and is related to the small amount of oxide that existed on the nitride surfaces (Al_2O_3 and ZrO_2). The more effective densification of Si_3N_4 with two nitride additives than single nitride additives can be simply interpreted with the behavior diagram of the Si_3N_4-SiO_2-AlN-Al_2O_3-ZrN-ZrO_2 system,[8] namely, the solidus temperature is lowered and the composition range of the liquid phase field is expanded when more nitrides (and oxides from powder surface contamination) are present.

Nevertheless, it is suggested that much better high temperature properties still obtained due to a significant reduction in liquid phase volume and increase of its softening temperature and viscosity in comparison with Si_3N_4 sintered with oxide additives. This is supported by TEM observations.[9] Some minor intergranular glassy phase was visible at a few triple regions, but not in most of the two-grain boundaries (Figs. 2, 3).

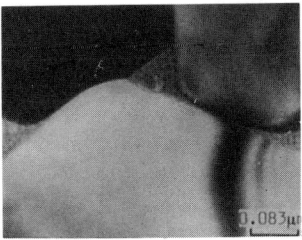

Fig. 2. Minor intergranular glassy phase in triple-point regions of β'-Si$_3$N$_4$ grains.

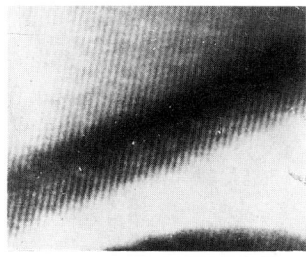 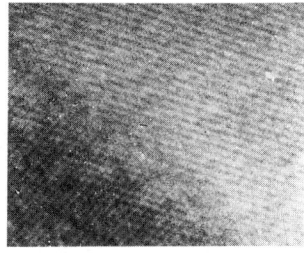
(a)   (b)

Fig. 3. Lattice images of ZAN: (a) intergranular glassy phase at triple point of β'-Si$_3$N$_4$ grains; (b) no apparent glassy phase between two adjacent grains.

An additional interesting side observation is the superlattice structures found in α'-solid solution in hot-pressed materials of similar composition. It is suggested that Ca^{2+} was located interstitially in the lattice and some Ca$_y$(Si$_{1-x}$Al$_x$)$_3$N$_4$-type compound might have formed.[10] The distribution of the latter structure may follow certain periodicity and lead to the formation of superlattice structures (Fig. 4).

Fig. 4. Superlattice structure of α'-Si$_3$N$_4$ of ZAN.

APPLICATIONS AS CUTTING TOOLS

Developments in Si_3N_4 cutting tool inserts with AlN/ZrN additives have been carried out in the authors' laboratory. These inserts exhibit a good combination of high wear resistance, hot-hardness, thermal shock resistance, toughness, chemical inertness and impact resistance. They have been found suitable for turning and milling those materials which are otherwise very difficult to machine using conventional cemented carbide inserts. Rough-machining of high-hardness (HRc 59-64), wear-resistant alloyed cast iron, chilled iron, quenched alloyed steels (HRc 60-65), Ni alloys, hard-facing cermet coatings, etc. have been attempted (Fig. 5).

(a)

(b)

Fig. 5. New Si_3N_4-based cutting inserts with AlN/ZrN-type non-oxide additives: (a) turning Cr-Mo alloyed cast iron parts (hardness: HRc 61-64); (b) turning Ni-Cr alloyed cast iron roll (hardness: Hs 77-85).

They can also be used for high-rate machining of grey cast iron, heat-treated steels and Al-Si alloys. They show good versatility and, generally speaking, 3-20 times better cutting efficiency, as compared to cemented carbide cutting inserts. This can result in a 30-50% saving in electricity, labor, tools and machinery. A pilot plant has been installed and production of Si_3N_4-based cutting inserts with AlN/ZrN additives is now on-going at the authors' laboratory. This kind of Si_3N_4 cutting insert is also being used in some branches of Chinese industry.

REFERENCES

1. C. Greskovich, J. Mater. Sci., **14**, 2427 (1979); J. Am. Ceram. Soc., **64**, (12) 725 (1981).

2. N. Uchida, M. Shimada and M. Koizumi, (Proc. Inter. Symposium on Ceramic Components for Engines, Hakone, Japan, 1983) pp. 404-11.

3. N. Uchida, M. Shimada and M. Koizumi, J. Am. Ceram. Soc., **68**, (2) C-38 (1985).

4. O. Abe, Ceram. Int., **16**, 53 (1990).

5. G. Changchun and X. Yuanluo, in Horizons of Powder Metallurgy, Part 2, edited by W. A. Kaysser and W.J. Huppman (Verlag Schmid GmbH, Freiburg, 1986), pp. 1163-66.

6. G. Changchun, X. Yuanluo and C. Limin, in Modern Developments in Powder Metallurgy, Vol. 21, compiled by U. Gummeson and D.A. Gustafson (MPIF & APMI, GmbH, Freiburg, New Jersey, USA, 1988), pp. 619-33.

7. G. Changchun, X. Yuanluo and C. Limin, Euro-Ceramics, Vol. 1, edited by G.de With, R.A. Terpstra and R. Metselaar (Elsevier Applied Science, London and New York, 1989), pp. 1.426-1.430.

8. G. Changchun, X. Yuanluo, C. Limin and Y. Yi, edited by R. Carlsson, T. Johansson and L. Kahlman, (4th Inter. Symposium on Ceramic Materials and Components for Engines, Elsevier Applied Science, London and New York, 1991) pp. 221-28.

9. J. Weiss, L.J. Gauckler, H.L. Lukas, G. Petzow and T.Y. Tien, J. Mater. Sci., **16**, 2997-3005 (1981).

10. W. Suling, J. Inorganic Mater., Shanghai, **3**, 199-206 (1987).

MICROSTRUCTURAL DESIGN BY SELECTIVE GRAIN GROWTH OF β-Si_3N_4

Naoto Hirosaki*, Yoshio Akimune*, Mamoru Mitomo**
*Nissan Motor Co., Ltd., 1, Natsushima-cho, Yokosuka, 237 Japan
**National Institute for Research in Inorganic Materials, 1-1, Namiki, Tsukuba-shi, 305 Japan.

ABSTRACT

Raw β-Si_3N_4 powder was gas-pressure sintered with Y_2O_3-Nd_2O_3 additives at > 1700°C. Grain growth behavior was investigated in relation to sintering conditions. Selective growth of large grains was accomplished by sintering the powder at high temperatures with small amounts of additives. As a result, in-situ composites were obtained from β-powder.

The desired material properties have been attained by controlling the microstructural design using large grains. Materials with high reliability, having a Weibull modulus of about 50, were fabricated by maintaining a uniform size and distribution of elongated grains. Tough materials, having fracture toughness of $10 MPa\sqrt{m}$, were developed by increasing the diameter of elongated grains. This method was applied to the sintering of refractory grade powder with the aim of lowering sintered material cost. Fairly good mechanical properties have been obtained even with impure powders.

INTRODUCTION

Dense silicon nitride ceramics have been fabricated by liquid phase sintering. Homogeneous microstructures developed when β-Si_3N_4 powder was hot-pressed [1]. On the other hand, heterogeneous microstructures with the rod-like grains developed during sintering of α-Si_3N_4. Rod-like grains increase the toughness of the material [1]. Thus, α-Si_3N_4 powder has been usually used for engineering structural applications.

Abnormal grain growth of Si_3N_4 further increases fracture toughness. When α-powder was fired at 1800-2000°C using a gas-pressure sintering (GPS) method, some large elongated grains with high aspect ratios developed in smaller more equiaxial grains [2]. These materials with duplex microstructures are known as "in-situ composites" or "self-reinforced materials", and have fracture toughness as high as 8-11 $MPa\sqrt{m}$ [3-5].

The microstructure of an in-situ composite is also developed when β-Si_3N_4 powder is sintered at higher temperatures using a GPS method [6, 7]. A high α content is not always necessary for the development of in-situ composites. The size distribution of β-Si_3N_4 grains is the main driving force for selective growth of elongated grains [8]. In conventional sintering of β raw powder, the firing temperature has been too low to develop a rod-like morphology.

In this work, selective growth of large β-Si_3N_4 grains was controlled by the firing time, and the effect of abnormal grain size and number on mechanical properties was investigated to obtain the desired duplex microstructural design.

EXPERIMENTAL PROCEDURE

The raw β-Si_3N_4 powder (Grade SN-P21FC, Denkikagaku, Tokyo, Japan) was pure as indicated by the properties in Table 1. The properties of a low-purity, refractory grade powder (Grade SN-B1), are also indicated in the table. The particle size distribution of SN-P21FC was rather wide, as it contained some larger particles as shown in Figure 1. Larger particles would act as nuclei for grain growth. Si_3N_4 powder and 1 to 10 mol% of an equimolar ratio of Y_2O_3 and Nd_2O_3 [99.9% pure, Shin-etsu Chemical, Tokyo, Japan] were ball-milled in ethanol for 94 h. After being dried, the powder mixture was die-pressed under 20 MPa and isostatically pressed under 200 MPa; the resultant pressed specimens

Table 1 Properties of high-purity powder (SN-P21FC) and low-purity powder (SN-B).

		SN-P21FC	SN-B
α-phase content (wt%)		5	<1
Specific surface area (m^2/g)		9.6	2.9
Impurities (wt%)	Fe	0.0016	0.99
	Al	0.011	0.34
	Ca	0.0061	0.32
	Mg	0.0005	0.011
	O	0.73	2.0

Figure 1 Scanning electron micrograph of raw β-Si₃N₄ powder (SN-P21FC).

were about 6 by 6 by 50 mm. Gas-pressure sintering was performed in a graphite resistance furnace.

Specimens for bending tests were ground to 3 by 4 by 40 mm using an 800-grit diamond wheel. Bending tests were performed on 15 specimens for each condition using a three-point method with a 30-mm span at a crosshead speed of 0.5 mm/s. The microstructure was observed using scanning electron microscopy (SEM). Fracture toughness was measured by the single edge precracked beam (SEPB) method [9].

RESULTS AND DISCUSSION
Effect of Temperature and Additive Amount on Grain Growth

Sintering conditions for grain growth of β-Si$_3$N$_4$ were investigated. β-Si$_3$N$_4$ powder (SN-P21FC) containing 1 to 10 mol% Y$_2$O$_3$-Nd$_2$O$_3$ was fired at 1700° to 1900°C for 4 h in 10 MPa N$_2$ gas. Figure 2 shows the relative density of the sintered material. The compacts for all compositions fired at 1900°C densified up to >98% of the theoretical. The β raw powder showed better sinterability than the α powder with similar purity and specific surface area.

Figures 3 shows the microstructures of Si$_3$N$_4$ (a) with 1 mol% oxides fired at 1800°C, (b) with 1 mol% oxides fired at 1900°C, and (c) with 10 mol% oxides fired at 1900°C. Abnormal grain growth of β-Si$_3$N$_4$ was observed at 1900°C and was enhanced with a smaller amount

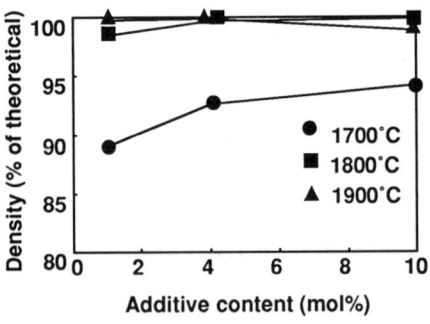

Figure 2 Density of Si₃N₄ containing 1-10 mol% of Y₂O₃-Nd₂O₃ fired at 1700-1900°C

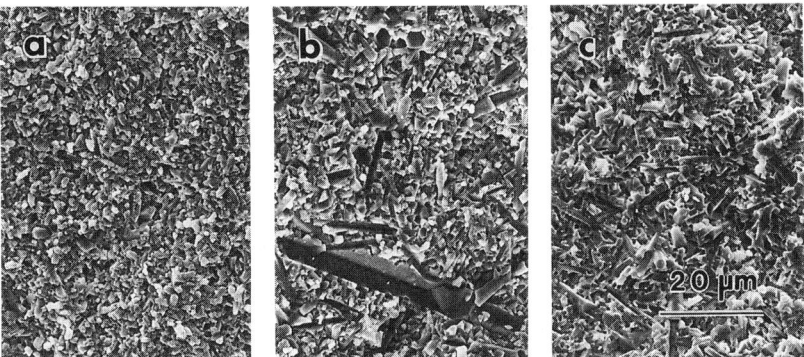

Figure 3 Microstructure of Si3N4 (a)with 1 mol% oxides fired at 1800°C, (b)with 1 mol% oxides fired at 1900°C, (c) with 10 mol% oxides fired at 1900°C.

of oxide additives. The rate of grain growth in a diffusion controlled mechanism is given by

$$-\frac{dr}{dt} = \frac{2\gamma\Omega^2}{RT}DC_0(\frac{1}{r_1} - \frac{1}{r_0})\frac{1}{\bar{x}} \quad (1)$$

where γ, Ω, D, C_0, and $\bar{x}$ are the interface free energy, molar volume of solute, diffusion constant, solubility on a flat plane, and mean separation of two grains, respectively[10]. Sintering at high temperature enhanced diffusion (D), thereby promoting grain growth. This accounts for the difference in microstructures seen in Figure 3(a) and (b). The mean distance between two grains ($\bar{x}$) increased with the amount of liquid although $\bar{x}$ is not always the distance between adjacent particles. This indicates that a smaller amount of liquid enhances grain growth, which explains the difference between Figure 3(b) and (c).

The grain size distribution of the raw powder also affects the grain growth of β-Si$_3$N$_4$ as shown by r_1 and r_0 in eq. (1). Finer β-Si$_3$N$_4$ (SN-P21F grade) results in a homogeneous microstructure even sintered in GPS [8]. However, nuclei (seeds) introduced artificially into small grains (SN-P21F) develop a rod-like morphology [8]. Important conditions to be controlled to develop in-situ composites from β-powder are (1) grain size distribution, (2) amount of oxide additive, and (3) firing temperature.

Microstructural Control by Selective Grain Growth

Selective growth of β-Si$_3$N$_4$ grains was controlled by the heating time, and the effect of the microstructure on mechanical properties was investigated. β-Si$_3$N$_4$ powder (SN-P21FC) containing 1 mol% Y$_2$O$_3$-Nd$_2$O$_3$ was fired at 1900°C for 2 h in 10 MPa N$_2$ gas. The density of the materials was 99% of the theoretical. The materials were further heated at 2000°C for 2 h (SN2), 4 h (SN4), and 8 h (SN8) in 30 MPa N$_2$ gas to enhance abnormal grain growth.

The relative density of sintered SN2, SN4, and SN8 was 98.9 %, 99.1 %, and 98.7 %, respectively. Figure 4 shows Weibull plots of the three-point bending strength. SN2 had bending strength of 689 MPa and a Weibull modulus of 53. When α-powder was sintered under similar conditions, the Weibull modulus of the material was about 9 [11]. Both the average strength and Weibull modulus decreased as the firing time increased. The

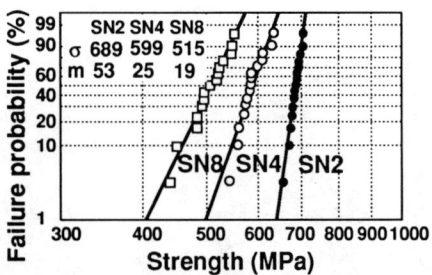

Figure 4 Weibull plots of Si3N4 containing 1 mol% oxides fired at 2000°C for 2h (SN2), 4h (SN4), and 8h (SN8).

fracture toughness of SN2, SN4, and SN8 was 8.5MPa$\sqrt{m}$, 10.3MPa$\sqrt{m}$, and 8.8MPa$\sqrt{m}$, respectively.

Figure 5 shows the microstructure of SN2, SN4, and SN8. The specimens had a duplex microstructure composed of small matrix grains and large fibrous grains; the matrix grains were 0.2 to 2 μm in diameter and 1 to 5 μm in length, and the large grains were 2 to 20 μm in diameter and 10 to 300 μm in length. The large grains developed with increasing sintering time; the grain diameter and length increased and the number of grains per unit volume increased. When the large grains were relatively small in number (SN2), they were dispersed uniformly and did not come in contact. However, when the large grains grew substantially in size and in number as a result of further heating (SN4 and SN8), they came in contact and tended to cluster. When α-powder was used as the raw powder, contact between large grains was also observed [11].

Large grains act as fracture flaws. A large grain or a cluster of large grains **was** observed at the fracture origin of all the SN2, SN4, and SN8 specimens. The reduction in strength observed with a longer sintering time was attributed to the increased length of large grains. One large grain of a uniform length was observed at the fracture origin of SN2 specimens. Since the flaw size of the fracture origin was uniform, the Weibull modulus increased up to 53. SN4 and SN8 specimens fractured from one large grain or multiple ones of various sizes. As a result, the Weibull modulus decreased.

The grain growth of β raw powder during GPS is simpler than that of α powder. In the sintering of α-powder, the nuclei for grain growth were formed through an α-to-β phase transformation [12]. It is quite difficult to control the size and distribution of large β grains, so that the resultant microstructure is not reproducible. In the sintering of β-powder, the nuclei for grain growth were relatively larger grains in the raw powder. The number and distribution of nuclei in materials made from β-powder can be easily controlled during powder processing. Thus, the degree of the grain growth can be readily manipulated by controlling the size distribution of raw β-Si$_3$N$_4$. As a result, the heterogeneities of the duplex microstructure (size and distribution of large grains) are more easily controlled in the sintering of β-powder, which makes the microstructure development more reproducible. In this study, the ball-milling process provided an adequate grain size distribution for developing a microstructure of uniformly large grains. This would explain the high Weibull modulus of materials from β-Si$_3$N$_4$.

In the case of self-reinforced silicon nitride from α-powder, fracture toughness has been reported to increase proportional to the square root of the diameter of large grains, indicating that the bridging of large grains contributes to greater toughness [4, 5, 12]. In

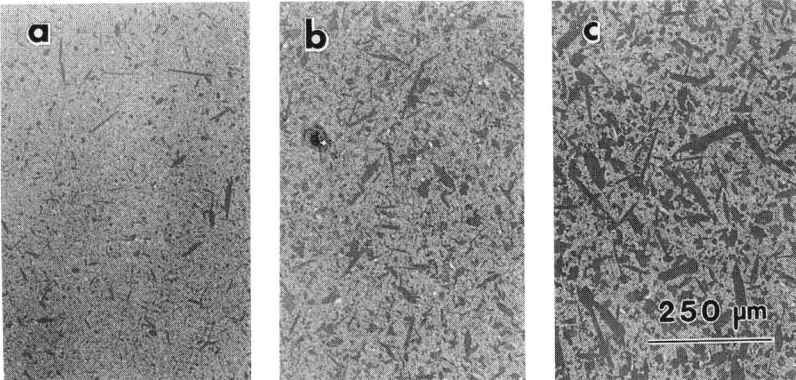

Figure 5 Microstructure of Si₃N₄ containing 1 mol% oxides fired at 2000°C for (a)2 h, (b)4 h, (c)8 h.

this study, SN4 showed fracture toughness as high as 10.3 MPa$\sqrt{m}$, which is attributed to the crack bridging by large grains. The fracture toughness of SN8, however, was lower than that of SN4, although the diameter of large grains still increased. Thus, the decrease in toughness of SN8 cannot be explained only by the bridging mechanism. Microcracking also contributes to the toughening of silicon nitride [13, 14]. Microcracking toughening is affected by the grain size [15]. When the grain size is relatively small, the extent of microcracking increases with increasing grain size, which toughens the material. However, when larger grains produce overly extensive microcracking, the linking of microcracks can cause macrocrack, thereby reducing fracture toughness [15]. This suggests that the elongated grains of SN4 increased fracture toughness as a result of bridging, whereas excessively large grains of SN8 reduced fracture toughness owing to microcracking.

Sintering of Refractory Grade β-Si₃N₄ Raw Powder

Refractory grade β-Si₃N₄ powder (Grade SN-B, Denkikagaku, Tokyo, Japan) was tested as a raw powder for engineering structural materials with the aim of lowering the sintered material cost. This powder contained coarse grains and Fe, Al, Ca, Mg, and oxygen impurities (Table 1). Si₃N₄ containing 2 mol% of an equimolar ratio of Y₂O₃ and Nd₂O₃ was sintered at 1900°C for 4 h in 10 MPa N₂. As shown in Figure 6, small rod-like grains were observed in this material although a microstructure of duplex composites was not obtained. This material had a relative density of 99 %, an average strength of 600 MPa with a Weibull modulus of 11, and fracture toughness of 5.3 MPa$\sqrt{m}$. Using the present method, fairly good mechanical properties have been attained even from impure powders.

CONCLUSION

An in-situ composite microstructure was obtained using β-Si₃N₄ raw powder by (1) controlling the grain size distribution, (2) adding small amount of oxide, and (3) firing the powder at >1900°C.

Designed microstructures were attained by selective grain growth induced by heat treatment at 2000°C for 2 h (SN2), 4 h (SN4), and 8 h (SN8) following sintering at 1900°C for 2 h. SN2 had a Weibull modulus as high as 53 because uniform-size large grains were

Figure 6 Microstructure of the material using low-purity Si3N4 powder (SN-B) containing 1 mol% oxides fired at 1900°C for 4 h.

distributed uniformly in the microstructure. Elongated grains of SN4 increased fracture toughness, whereas excessively large grains of SN8 reduced it because of microcracking.

When refractory grade β-Si$_3$N$_4$ powder was sintered using the present method, a microstructure with small rod-like grains was obtained. This material displayed strength of 600 MPa and fracture toughness of 5.3 MPa$\sqrt{\text{m}}$.

REFERENCES

1. F. F. Lange, *J. Am. Ceram. Soc.* **56**, [10] 518-522 (1973).

2. M. Mitomo and K. Mizuno, *Yogyo-Kyokai-Shi*, **94**, [1] 106-11 (1986).

3. E. Tani, S. Umebayashi, K. Kishi, K. Kobayashi, and M. Nishijima, *Am. Ceram. Soc. Bull.* **65**, [9] 1311-15 (1986).

4. C-W. Li and J Yamanis, *Ceram. Eng. Sci. Proc.* **10**, [7-8] 632-645 (1989).

5. T. Kawashima, H. Okamoto, H. Yamamoto, and A. Kitamura, *J. Ceram. Soc. Japan* **99** [4] 320-323 (1991).

6. M. Mitomo, M. Tsutsumi, H. Tanaka, S. Uenosono, and F. Saito, *J. Am. Ceram. Soc.* **73**, [8] 2441-45 (1990).

7. N. Hirosaki, M. Ando, Y. Akimune, and M. Mitomo, *J. Ceram. Soc. Japan* **100** [6] 826-829 (1992).

8. M. Mitomo, in Proceeding of the 1st International Symposium of the Science of Engineering Ceramics, edited by S. Kimura and K. Niihara, (The Ceramic Society of Japan 1991), p.101-107.

9. T.Nose and T. Fujii, *J. Am. Ceram. Soc.*, **71**, [5] 328-33 (1988).

10. S. Sarian and H. W. Weart, *J. Appl. Phys.*, **37** [4] 1675-81 (1966).

11. N. Hirosaki and A. Okada, *J. Ceram. Soc. Jpn.*, **97** [6] 673-675 (1989).

12. M. Mitomo and S. Uenosono, *J. Am. Ceram. Soc.* **75**, [1] 103-108 (1992).

13. R. W. Rice, K.R.McKinney, C. C. Wu, S. W. Freiman and W. J. M. Donough, *J.Mater. Sci.*, **20** [4] 1392-1406 (1985).

14. A. Okada and N. Hirosaki, *J. Mater. Sci.*, **25** 1656-61 (1990).

15. R. W. Rice, S. W. Freiman, and P. F. Becher, *J. Am. Ceram. Soc.* **64** [6] 345-350 (1981).

THE EFFECT OF GLASS CHEMISTRY ON THE MICROSTRUCTURE AND PROPERTIES OF SELF REINFORCED SILICON NITRIDE

ALEKSANDER J. PYZIK, DANIEL F. CARROLL AND C. JAMES HWANG
The Dow Chemical Company, Central Research & Development, Advanced Ceramics Laboratory, Midland, MI 48674

ABSTRACT

The advantage of self-reinforced silicon nitride is the in-situ control of the microstructure. This control is provided in large degree by the chemistry of glassy phase which can be adjusted to tailor the morphology of silicon nitride grains as well as the matrix - reinforcement interface. The presence of high aspect ratio silicon nitride grains is necessary but not sufficient condition to produce materials with optimum properties. For maximum flexure strength and fracture toughness an optimized glass matrix is required.

INTRODUCTION

The benefits of whisker reinforcement for improved fracture toughness have been clearly demonstrated in many ceramic systems [1,2]. However, the use of whiskers in commercial applications is generally limited due to potential health hazards. Recently, the concept of whisker reinforcement has been applied to silicon nitride materials since the high temperature beta phase has a tendency to form elongated needle-like grains. Even though this tendency has been known for a long time, the controlled formation of elongated beta grains to produce materials with maximum fracture toughness has only been demonstrated within the last few years[3-5]. The in-situ formation of elongated beta silicon nitride grains provides an opportunity to easily reinforce a material without the difficulties associated with handling and processing whisker reinforcements. Most of the reported work in the past has been done on Si_3N_4-Y_2O_3-Al_2O_3[5-6], Si_3N_4-MgO[7] and Si_3N_4-Y_2O_3-MgO [8-9] systems. This paper will describe a self-reinforced silicon nitride material based upon a new glass system and discuss how the glass chemistry affects microstructural development and mechanical properties.

EXPERIMENTAL

The self-reinforced silicon nitride materials were made using a silicon nitride powder that was composed of 95.3 % of the alpha phase with a surface area of 10.5 m^2/g, 1.18 wt. % oxygen, < 30 ppm Fe, < 100 ppm Cl and < 50 ppm Ca and Al. The average particle size was 0.11 µm. The elemental composition (by XPS) was: Si 37 %, N 34 %, O 28 % and C 1 % (all atomic %). The additives were selected base upon their purity and particle size. All compositions were attrited for 1 hour in methanol, dried, passed through a 60 mesh screen and hot pressed at 1800°C for 0.5 to 6 hours or pressureless sintered at 1700°C to 1750°C for 6 to 12 hours. Microstructures were examined by scanning electron microscopy. Glass content and grain size were measured by line intercept method from back scattered electron images. Glass chemistry and elemental analysis were carried out by analytical electron microscopy and auger electron microscopy.
Selected billets were sectioned for fracture toughness and flexure strength evaluation. Approximately 10-12 specimens (3 x 4 x 45 mm) were broken in four point bending to determine the room temperature flexure strength. The flexure

strengths measured at elevated temperatures were also tested in four point bending in an air environment. The fracture toughness was measured by fracturing five chevron notched bend beam (CNB) specimens in three point-bending. The dimensions of the CNB specimens were b=3.0 mm, w=4.0 mm, a_0/w=0.42, a_1/w=1.0 and s/w=10. The fracture toughness was calculated using the method developed by Xian[10].

RESULTS AND DISCUSSION

The fabrication of self-reinforced silicon nitride is based upon the ability to control chemistry of the starting Si_3N_4 powder and the glass components during densification under controlled pressure and temperature conditions. The source and characteristics of the silicon nitride powders affects the general shape and number of β-Si_3N_4 grains in the microstructure. The glass chemistry and glass content determine the kinetics of microstructural evolution and grain morphology. The size and aspect ratio of the β-Si_3N_4 grains can be tailored by adjusting the ratio of densification aid (containing elements such as Mg, Sr, Be, Ba or Ce) to transformation aid (e.g. Y, Sc, La and Ac) in presence of whisker growth agent (e.g. Ca, Mn, Sc, Nb, Al and La). Depending on the selected glass composition, the particular ratio of densification aid/transformation aid/whisker growth agent required to form the self-reinforced microstructure will change. Figure 1 shows the general ranges of glass compositions resulting in silicon nitride ceramics with high fracture toughness. The diagram is based upon SiO_2, densification aid and transformation aid assuming that small quantity of whisker growth agent is always present. The small field represents a range of glass compositions for the Si_3N_4-SiO_2-Y_2O_3-MgO-CaO system. The large field represents a range of compositions where Y_2O_3, MgO and CaO were replaced by other elements. Variations in microstructures were observed in both fields depending on the glass composition. Changes in the weight ratio of transformation aid to densification aid (with constant amount of whisker growth agent) for several levels of SiO_2 lead to a conclusion that best properties can be obtained between ratios 3:1 and 1:2. Experiments with different glass contents have shown that the silicon nitride grains tend to have higher aspect ratio in materials with lower glass contents. This result is not due to a faster growth rate in the c-direction of the silicon nitride grains but rather to a very slow growth rate in the a-direction. The optimum content of whisker growth agent can vary with glass composition, but is typically within the range of 0.1 and 1.0 wt. %.

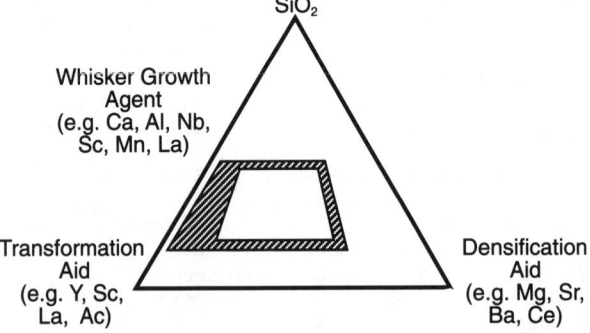

Fig.1. Ranges of glass compositions resulting in high toughness materials

The growth of elongated grains in silicon nitride ceramics is a complex phenomenon in which the mechanism is still not fully understood. Experimental observations indicate that the grain morphology can be affected by (i) the crystallographic modification, size and surface chemistry of the silicon nitride powder, (ii) the processing conditions, and (iii) the chemistry of oxynitride glass. While the first two factors are important, the third is critical. This is because the glass composition not only affects the morphology of silicon nitride grains, but also the glass-silicon nitride interfacial bond and the crystallinity of the glass matrix. Since it is well established that materials with the same reinforcement can have very different properties depending on the whisker-matrix interface[1-2], control of the glass chemistry is the key to enhanced properties of self-reinforced silicon nitrides.

Most silicon nitride ceramics have metastable microstructures "frozen in" at some stage of development after processing. Once the α to β transformation is completed, the beta grains grow through a solution-reprecipitation mechanism. Depending on the silicon nitride-glass system, the rate of grain growth along c-axis and a-axis often vary. In the Si_3N_4- MgO system, the maximum values of flexure strength and fracture toughness occur just after α to β transformation[7]. Longer processing times tend to decrease the grains' aspect ratios, degrading the mechanical properties of the material. A similar behavior is also observed in the Si_3N_4-Y_2O_3-SrO-Al_2O_3 system. The materials with highest toughness and strength are formed under short densification cycles which correlated well with the completion of the α to β transformation. Prolonged heat treatments reduced the mechanical properties. Fig. 2 illustrates this behavior in two self-reinforced silicon nitride materials. Curve A represents the fracture toughness and flexure strength as a function of processing time for the Si_3N_4-Y_2O_3-SrO-Al_2O_3 system. Curve B represents the mechanical properties for the Si_3N_4-Y_2O_3-MgO-CaO system. This system is different from the prior self-reinforced silicon nitride in that the grains exhibit a continuous increase in the aspect ratio long after the α to β transformation has been completed.

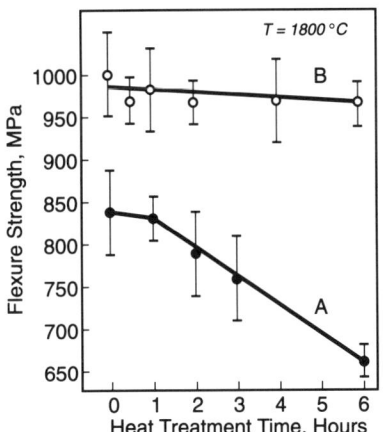
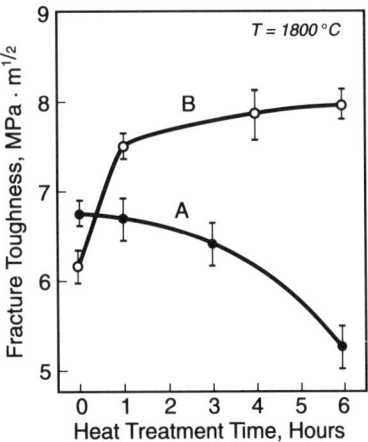

Fig. 2. Flexure strength and fracture toughness of two self-reinforced silicon nitrides, (A) Si_3N_4-Y_2O_3-SrO-Al_2O_3 and (B) Si_3N_4-Y_2O_3-MgO-CaO.

Depending on the glass chemistry, the microstructural evolution differs in both systems. In the Si3N4-Y2O3-SrO-Al2O3 system, elongated grains develop quickly prior to the completion of the α to β transformation. When the transformation is completed, solution-reprecipitation processes deposit more material along the a-axis than the c-axis, decreasing the aspect ratios of the silicon nitride grains. In the Si3N4-Y2O3-MgO-CaO system, grain growth occurs mostly in the c-direction throughout the densification process. For example, in a silicon nitride powder containing 3 % of sintering additives (based on Y2O3-MgO-CaO glass), the transformation starts at about 1400°C with elongated growth starting above 1650°C. After 12 hours of heat-treatment, the elongated silicon nitride grains have diameters ranging from 0.1 to 0.7 μm and lengths approaching 50 μm (figure 3A). The resultant morphology is strongly controlled by the glass chemistry. If the same silicon nitride powder is heat-treated under identical conditions in the presence of excess SiO2 only, the grains would have the morphology shown in Figure 3B.

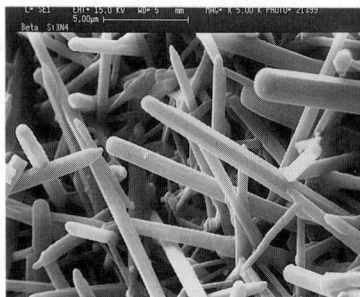 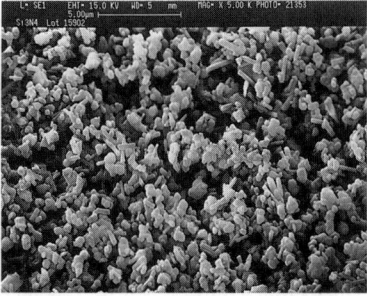

Fig. 3. Morphology of silicon nitride grains grown from different glass; (A) Y2O3-MgO-CaO and (B) SiO2.

It has been shown for several ceramic systems that the increase in diameter of the reinforcing phase increases fracture toughness, but reduces the overall flexure strength of the material[1]. This same trend is observed in self-reinforced silicon nitride. Fracture toughness of 9 MPa m$^{1/2}$ can be achieved in Si3N4-Y2O3-Al2O3 when the average grain diameter is about 1.7 μm [1]. Similar fracture toughness have been achieved in the Si3N4-Y2O3-MgO-CaO system but at a much smaller average grain diameter of 0.3 to 0.4 μm [11]. This result suggests that the formation of elongated grains is a necessary, but not a sufficient, condition to produce high fracture toughness in ceramics. Besides the reinforcement phase (i.e. elongated Si3N4 grains), properties of the glassy phase and the character of interfacial bonding are critical. For example, the change in aspect ratio of Y2O3 to MgO from 1:2 to 1:4 does not result in a significant change of microstructure. The average aspect ratio is approximately 3.5 and the average grain diameter ranges from 0.31 to 0.33 μm. At the same time, however, this change in the Y2O3/MgO ratio decreases the fracture toughness from 8.1 MPa m$^{1/2}$ to 6.2 MPa m$^{1/2}$. This difference indicates that even with elongated grains present, the fracture characteristics depend upon the glass matrix chemistry. Our experimental data on other glass systems confirm that identical morphology of reinforcements (silicon nitride) do not result in the same mechanical properties.

The crack path in the Y_2O_3-MgO-CaO glass is very tortuous. Since the strength of the interface is difficult to control, the fracture behavior of the glassy matrix itself becomes a significant variable in the fracture process. Materials that contain 14.7 wt % glass have approximately 60 % glass on the fracture surfaces. Auger analysis of in-situ fractured surfaces shows that the catalyst content (in this case Ca) on the fracture surface is 2.2 to 4 times higher than the average Ca content of the glass phase. This result suggests a preferential crack propagation through Ca-rich regions. The whisker growth agent not only promotes the formation of high aspect ratio grains, it also influences the matrix-reinforcement interface, enhancing tortuous crack growth. Therefore, in the formation of self-reinforced silicon nitrides, it is important to select the glass chemistry to provide conditions for both elongated grain growth and a weak interfaces for maximum fracture toughness.

The broad range of chemistries available to form self-reinforced silicon nitrides offer the flexibility to tailor a material's grain structure and glass chemistries to meet the mechanical property requirements for specific applications. Table I summarizes how the mechanical properties can be tailored for four different self-reinforced silicon nitride systems. In each one of these materials, the glass phase chemistry has been optimized to meet specific mechanical properties or processing requirements. Material A, for example, is optimized to have a maximum strength and fracture toughness at low to intermediate temperatures (<1100°C). The fracture toughness and hardness of material B is optimized to provide a material with superior wear resistance in such applications as cutting tools. Material C is a self-reinforced silicon nitride where the glass composition has been adjusted so the material can be processed by pressureless sintering methods. This material has good mechanical strength and high fracture toughness at low to intermediate temperatures. The self-reinforced composition in material D is tailored to provide a glass phase which can be crystallized after processing for improved high temperature mechanical properties. This material exhibits an excellent high temperature strength of 550 MPa at 1375°C.

Table I. Mechanical Properties of Selected Self-Reinforced Silicon Nitrides

Property	Material A	Material B	Material C	Material D
Density (g/cc)	3.25	3.23	3.23	3.22
Flexure Strength (MPa)	1000-1200 (25°C)	950-1100 (25°C)	750-850 (25°C)	850-900 (25°C)
	890 (1100°C)	625 (1100°C)	510 (1100°C)	760 (1200°C)
	270 (1300°C)	400 (1300°C)	365 (1265°C)	550 (1375°C)
Fracture Toughness (MPa-m$^{1/2}$)	8.5-9.0	7.8-8.5	7.6-8.5	6.2-6.9
Hardness (kg/mm^2)	1450-1500	1600-1650	1500-1550	1750-1850
Ave. Grain Diameter	0.3	0.4	0.6	0.5
Densification Method	HP	HP	Sintering	HP/HIP

CONCLUSIONS

1. A family of self-reinforced silicon nitrides has been developed based upon a three-component glass system and a whisker growth agent. The whisker growth agent plays a major role in determining the aspect ratio of the silicon nitride grains as well as the fracture characteristics of the glass phase.

2. The advantage of self-reinforced ceramics is the in-situ control of the microstructure. This control is provided to a large degree by chemistry of the glassy phase which can be adjusted to tailor the matrix-reinforcement interface.

3. The growth of elongated Si_3N_4 grains is a complex phenomenon in which the mechanism responsible is still not fully understood. However, experimental observations show that (i) not every Si_3N_4-glass system can produce microstructures with elongated grains, (ii) not every Si_3N_4-glass system which can produce microstructures with elongated grains can also be controllable to yield an optimum grain size required by application, and (iii) not every Si_3N_4 with elongated grains and desired grain size will result in a ceramic material with improved mechanical properties.

ACKNOWLEDGEMENTS

The authors wish to thank Don Beaman, Dave Susnitzky and Harold Klassen for analytical characterization; Harold Rossow, Tim Allen and Sherry Fuller for sample preparation. Special thanks are due to Alan Hart for supporting this research. The authors also appreciate the discussions with Don Beaman, Barbara Pyzik and Art Prunier.

REFERENCES

1. P. F. Becher, J. Am. Ceram. Soc., 74, 2, 255 (1991)
2. G.H. Campbell, M. Ruhle, B. J. Dalgleish and A. G. Evans, J. Am. Ceram. Soc., 73, 3, 521 (1990)
3. K. Matsuhiro and T. Takahashi, Ceram. Eng. Sci. Proc., 10, 7-8, 807 (1989)
4. C. W. Li and J. Yamanis, Ceram. Eng. Sci. Proc., 10,7-8, 632 (1989)
5. G. Wotting, B. Kanka and G. Ziegler, in Non-Oxide Technical and Engineering Ceramics, edited by S. Hampshire (Elsevier, London and New York, 1986), p. 83
6. M. Mitomo, M. Tsutsumi, H. Tanaka, S. Uenosono and F. Saito, J. Am. Ceram. Soc., 73, 8, 2441 (1990)
7. G. Wotting and G. Ziegler, Ceramics International, 10, 1, 18 (1984)
8. S. Hampshire, M. J. Pomeroy and B. Saruhan, In Technical Ceramics, ed. by P. Vincenzini (Elsevier, Amsterdam, 1987), p. 941
9. F. M. Mahoney, M.J. Hoffman, G. Petzow and C. Boberski, in Ceramic Materials and Components For Engines, ed. by R. Carlson, T. Johansson and L. Kahlman (Esevier Applied Science, London and New York, 1992), p.649
10. W. Shang-Xian, in Chevron-Notched Specimens: Testing and Stress Analysis, ASTM STP 855, ed. by S. W. Frieman and F. I. Baratta (ASTM, Philadelphia, 1984), p.177
11. A. J. Pyzik, D. F. Carroll, C. J. Hwang and A. R. Prunier, in Ceramic Materials and Components For Engines, ed. by R. Carlson, T. Johansson and L. Kahlman (Elsevier Applied Science, London and New York, 1992), p. 584.

ROLE OF SINTERING PARAMETERS ON MICROSTRUCTURE DEVELOPMENT AND MECHANICAL PROPERTIES OF SINTER/HIP SILICON NITRIDE

Arnd Kühne, Rainer Oberacker and Georg Grathwohl
University Karlsruhe, IKM, Haid- und Neustr. 7, D-7500 Germany

ABSTRACT

Sinter/HIP consolidation with systematic variation of all relevant process parameters was performed for low additive (Y_2O_3-Al_2O_3) silicon nitride material. With SEM-investigations as well as bending and creep tests the relevance of process parameters on microstructure development and mechanical properties was studied.

INTRODUCTION

Dense silicon nitride with low amount of additives is a favourable material for mechanical applications at elevated temperatures. Consolidation of such materials, however, requires a high temperature/ pressure sintering process. By capsule-HIP full density can be achieved even with very low amounts of additives but a major disadvantage of this route are high costs due to the cladding. For successful application of alternative containerless methods like sinter/HIP and postHIP, the content of liquid phase forming sintering aids must be high enough to achieve pore closure in a conventional sintering step prior to the high pressure step. Due to the higher amount of glassy grain boundary phases in the sintered body the maximum service temperature of such material usually should be assumed to be not as high as with capsule-HIP material. Besides a highly refractory nature, best strength and toughness properties are required. In the case of silicon nitride this can only be obtained if the formation of typical needle-like interlocking grains takes place during sintering [1]. Independent of the consolidation method a certain amount of liquid phase is always necessary for this microstructural evolution. With the commonly used sintering additive yttria about 2 wt% are favourable for capsule-HIP consolidation while a combination of yttria/ alumina of at least 6 wt% is normally required in the case of sinter/HIP or postHIP processing. Especially for the sinter/HIP procedure a significant reduction in the additive content seems to be possible if all processing stages of the complex consolidation operation are consequently optimized. This optimization with respect to full densification and adequate microstructures leading to high strength even at high temperatures is the main objective of the present research work.

A sinter/HIP process for silicon nitride can be divided in a sintering and a pressure step being controlled by different key parameters which can be optimized on the background of some simple theoretical considerations [2]. Determining a temperature course, the problems of thermal stress, silicon nitride dissociation, differential sintering and grain growth must be taken into account. Additionally, nitrogen overpressure is required to avoid dissociation but also results in pore entrapped gas hindering further compaction [3,4]. These theoretical considerations were verified in own studies on the densification behaviour [5] using a special HIP-dilatometer furnace [6].

Regarding the pressure step, a favourable microstructure at the moment of pressurization will result in higher densities [7-9]. As known from own experiments [10], fine and non-interlocking grains provide the basis for effective pressure assisted densification. Therefore pressurization ideally begins as soon as pore closure has been completed. For further optimization, additional variations in the pressure step including the pressure level and the soaking time are of interest.

EXPERIMENTAL

For the experiments a typical sinter/HIP composition [11] was used, consisting of silicon nitride (Ube E-10) with 4,2 wt% yttria (HCStarck feinst) and 2 wt% alumina (Alcoa A49) as sintering aids. The powder-mixture was wet-milled in a ball mill with silicon nitride lining and -balls for 20 hours in ethanol. After drying and sieving, the powder was cold isostatically compacted at 600 MPa into plates (58 * 40 * 6,5 mm^3) with a green density of 55 %td. Sinter/HIP experiments were performed in an ASEA QIH-6 HIP. Microstructure of salt-melt etched fracture surfaces was investigated in a SEM. With a 4 pt device (40/ 20 span) RT bending strength of 4,5 * 3,5 * 45 mm^3 bending bars was determined. Fracture toughness was discovered with naturally sharp precracked bars in 3 pt loading. Creep behaviour was examined at 1350°C/ 100 MPa using the same geometries as for the room temperature bending tests.

RESULTS AND DISCUSSION

A first set of experiments was realized to determine the influence of process parameters on the occurrence of pore closure and the related microstructure. Besides nitrogen pressure, the heating rates and temperatures were varied over a wide range. As can be seen from fig. 1, the amount of closed porosity is independent of such variations and is only a function of density. Pore closure always occurs at about 91 %td with the microstructure developed at this density being nearly identical. Fig. 2 illustrates this for materials sintered at significantly different temperatures.

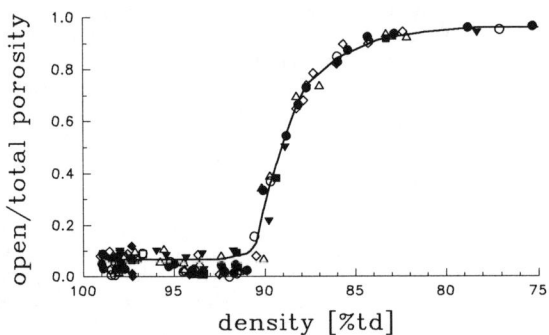

Fig. 1: Relation between density and open porosity for a broad variation of processing parameters in the sintering step ($\Delta T/dt$ = 2 - 100°C/min, T_{sinter} = 1750 - 1950°C, P_{max} = 0.5 - 10 MPa)

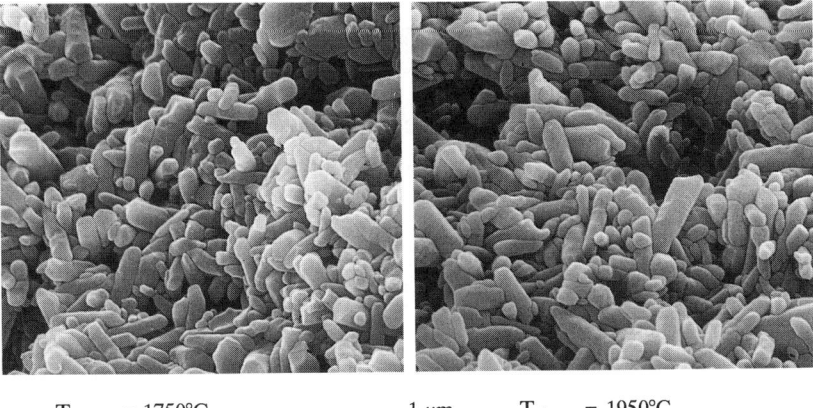

$T_{sinter} = 1750°C$ —— 1 μm $T_{sinter} = 1950°C$

Fig. 2: Microstructure at pore closure (≈ 91 %td) of materials sintered at different temperatures

The results from a quantitative microstructural analysis are given in fig. 3, showing a very slight tendency for increasing grain size and decreasing aspect ratio with higher sintering temperature. Thus, with regard to the efficiency of pressurization the parameters of the sintering course up to pore closure can be chosen from a wide range. Nevertheless, the above mentioned specific problems related to the pressure level must be taken into account.

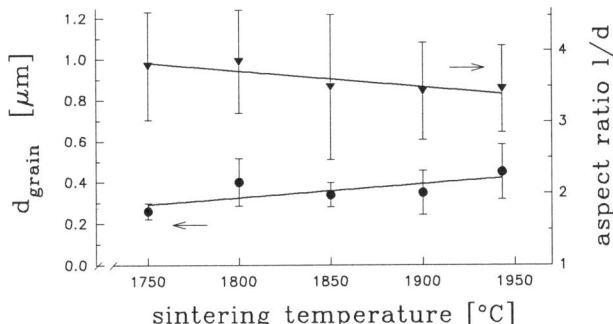

Fig. 3: Results from quantitative microstructural analysis of material sintered to pore closure at different temperatures (density ≈ 91 %td)

In the high pressure step variations of temperature, soaking time and maximum pressure were carried out. Besides faster densification with higher pressurization rates, no clear differences became evident from dilatometer curves [5]. The resulting microstructure, however, was somewhat coarser having a larger aspect ratio in the case of longer soaking times and higher sintering temperatures (fig. 4).

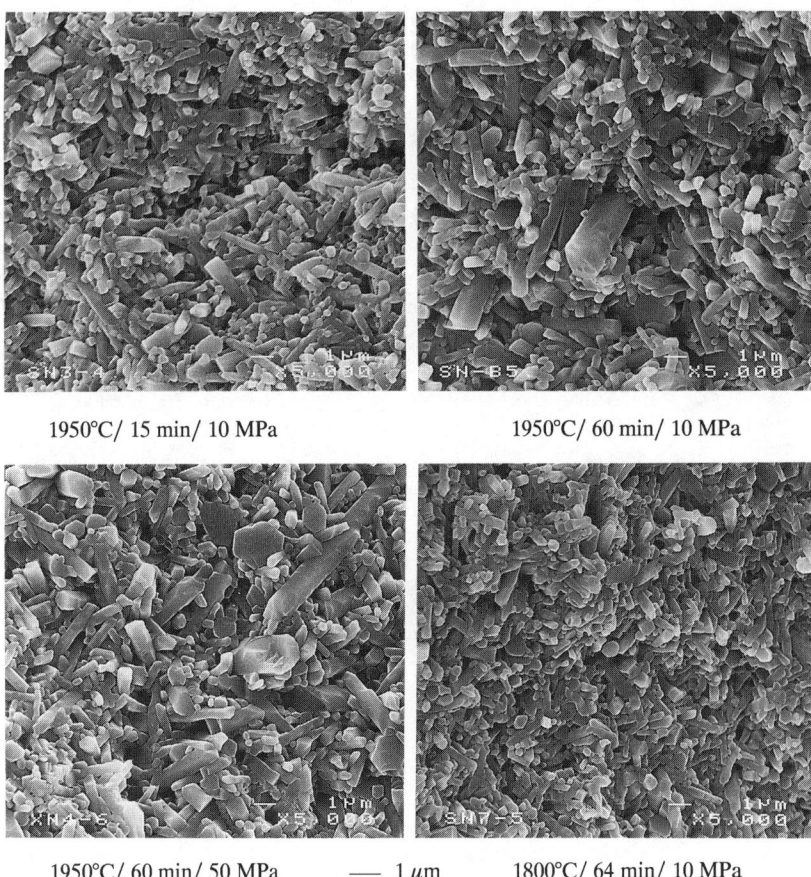

| 1950°C/ 15 min/ 10 MPa | 1950°C/ 60 min/ 10 MPa |
| 1950°C/ 60 min/ 50 MPa — 1 μm | 1800°C/ 64 min/ 10 MPa |

Fig. 4: Microstructure of dense silicon nitride after sinter/HIP with variations of the processing parameters T_{sinter}/ t_{soak}/ P_{max} in the pressure step

Additionally, small differences in density were found, with longer soaking time and higher maximum pressure giving the best values. This is affecting the mechanical properties, where higher bending strength was found to be correlated to density, and fracture toughness increased with aspect ratio and density (fig. 5). For comparison, values from a postHIP fabrication route are given in fig. 5, too. These samples were sintered to closed porosity using the same sintering parameters as with the sinter/HIP samples. After cooling to room temperature a typical postHIP cycle with a high maximum pressure at a relatively low temperature followed. Despite of the pressure of 100 MPa, only about 99.7 %td were achieved resulting in poor mechanical properties. This is similar to findings from other authors [12], and a possible reason could be the high initial gas pressure during postHIP processing. Gas may be transported into sealed pores or dissolved in the matrix via diffusion through the grain boundary phase, being compressed and then hindering full densification. The weight gain found in the postHIP step seems to support this assumption.

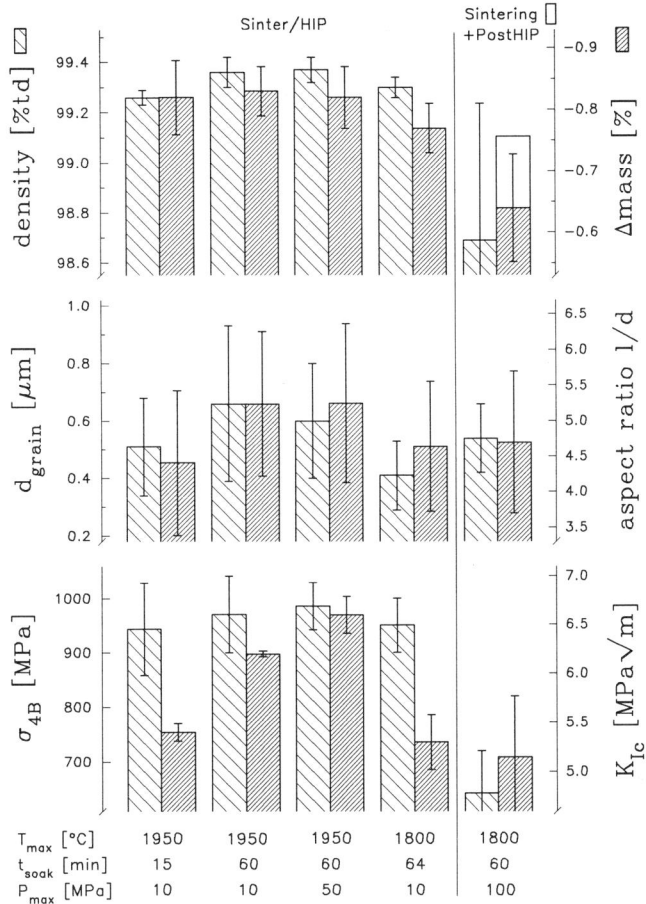

Fig. 5: Properties after sinter/HIP variations compared to post/HIP-processed material (mean values and standard deviation of 10 - 15 samples)

Evaluation of creep curves revealed similar behaviour of the materials prepared using different process parameters. For comparison another composition with only 2 wt% yttria and 1 wt% alumina was also tested. This material was used to verify process optimization findings and could also be fully densified by sinter/HIP. As to be expected, the creep rate of this material was significantly lower (fig. 6). These results confirm that creep behaviour is insensitive to slight differences in density and microstructure, but is primarily influenced by the grain boundary state.

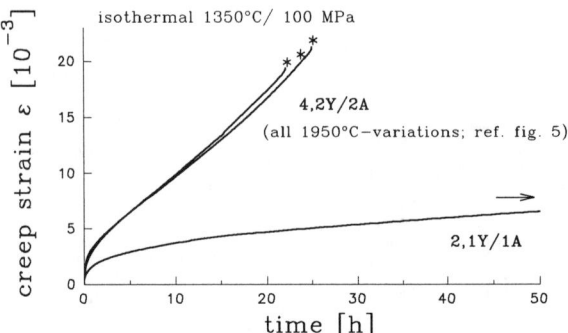

Fig. 6: Creep behaviour of dense sinter/HIP silicon nitride with parameter variations and with reduced amount of sintering additives, respectively

SUMMARY AND CONCLUSION

Broad variations in the different stages of a sinter/HIP process for low additive silicon nitride were realized. No significant influence of process parameters on the occurrence of pore closure could be found, and the microstructures developed up to this point is always similar. Therefore, other factors as weight loss or process duration can be used for deciding on the parameters for the sintering step. In the pressure step an increase in maximum pressure, longer soaking times and higher temperatures give higher density and larger aspect ratios resulting in better mechanical properties. PostHIPed material showed rather poor properties due to significantly lower densities. This is thought to be caused by effects arising from the high initial gas pressure. Creep curves were identical for materials from several parameter variations. However, a clear improvement can be achieved with reduced amounts of additives, as the creep behaviour is mainly controlled by the grain boundary phase.

REFERENCES

1. F.F. Lange, J. Am. Ceram. Soc., **62**, 428-430 (1979)
2. R. Oberacker, A. Kühne, M. Komac, M. and F. Thümmler, Interceram, **40**, 400-410 (1991)
3. C. Greskovich, J. Am. Ceram. Soc., **64**, 725-730 (1981)
4. S.-J.L. Kang and K.J. Yoon, J. Europ. Ceram. Soc., **1**, 135-139 (1989)
5. A. Kühne, R. Oberacker and F. Thümmler, in *Euro-Ceramics II*, (1991) in press
6. A. Kühne, R. Oberacker and F. Thümmler, pmi, **23**, 113-119 (1991)
7. C. Greskovich and J.H. Rosolowski, J. Am. Ceram. Soc., **59**, 336-343 (1976)
8. G. Wötting and G. Ziegler, in *Ceramic Materials and Components for Engines*, edited by W. Bunk and H. Hausner (DKG, Köln, 1986), pp. 235-242
9. A. Bellosi, V. Biasini and S. Guicciardi in *Euro-Ceramics*, edited by G. de With, R.A. Terpstra and R. Metselaar (Elsevier Sci. Publ., London, 1989), pp. 3.216-3.220
10. A. Kühne, R. Oberacker and F. Thümmler in *Proceedings ISO4* (MPR Publ. Services, 1991), pp. 36.1-36.17
11. J. Heinrich and M. Böhmer, in [8], pp. 243-253
12. T. Kito, K. Yabuta, M. Watanabe and Y. Matsuo, Y. in *Hot Isostatic Pressing: Theory and Applications*, (ASM Int., 1990) pp. 155-158

GRAIN MORPHOLOGY AND INTERGRANULAR STRUCTURE OF Si_3N_4 BASED CERAMICS FORMED BY HIP

H. BJÖRKLUND*, L. K. L. FALK*, J. WASÉN**, J. E. ADLERBORN+ AND H. T. LARKER+
* Department of Physics and **Department of Engineering Metals, Chalmers University of Technology, S-412 96 Göteborg, Sweden
+ ABB Cerama AB, S-915 00 Robertsfors, Sweden

ABSTRACT

The microstructures of unreinforced and reinforced Si_3N_4 ceramics have been characterized by analytical electron microscopy in combination with quantitative microscopy. Special attention was paid to the effect which different additives, viz. metal oxides and SiC- and β-Si_3N_4-whiskers, have upon matrix morphology and intergranular microstructure. It was demonstrated that SiC-whiskers may suppress Si_3N_4 grain growth while an addition of β-Si_3N_4-whiskers results in a coarser Si_3N_4 microstructure. The different microstructures have been related to the mechanical properties of the ceramics.

INTRODUCTION

A number of investigations have shown that room temperature fracture toughness and strength of Si_3N_4 ceramics can be increased by additions of whiskers [1-3]. Crack-deflection by the whiskers, crack bridging and whisker pullout have been proposed as major toughening mechanisms [4]. Whiskers, as well as metal oxide sintering additives, may also have an influence upon matrix microstructure [5,6]. A Si_3N_4 microstructure consisting of high aspect ratio β grains promotes high room temperature toughness and strength and can be formed by a careful choice of metal oxide additives [6,7].

The grain size and shape distribution of a material can be characterized by stereological methods. The grain areas in sections through the material are measured by quantitative microscopy and the grain size distribution may be calculated from the grain area distribution provided that the grain shape is known [8]. These methods have previously been employed to characterize systems such as WC-Co [8].

This paper concerns the effect which metal oxide and whisker additions have upon Si_3N_4 grain morphology and intergranular microstructure. Grain area distributions have been determined by quantitative microscopy. The Si_3N_4 ceramics were densified by hot isostatic pressing (HIP) which enables formation of an isotropic material. Also, HIP makes it possible to reduce the amount of metal oxide additives required to reach full density, and some Si_3N_4 starting powders may be densified without addition of metal oxides [3,5].

EXPERIMENTAL

The experimental materials in this investigation are shown in Table I. The samples were densified by HIP using the ABB Cerama proprietary glass encapsulation process. The four compositions in Table I resulted in Si_3N_4

materials with bulk densities above 99.5 % of the theoretical value calculated from the starting powder compositions. The metal oxide containing material could be densified using a lower pressure at a lower temperature during a shorter period of time. Detailed descriptions of processing, resulting microstructures and mechanical properties are given elsewhere [3,5].

The general microstructures of the materials in Table I were studied by scanning electron microscopy (SEM) of polished surfaces etched in molten sodium hydroxide. The material containing SiC-whiskers was plasma etched prior to etching in sodium hydroxide in order to distinguish the SiC whiskers from the Si_3N_4 grains.

The Si_3N_4 grain area distributions in sections through the materials were measured using a semi-automatic method [8]. Enlarged SEM micrographs of polished and etched sections were placed on a digitalizer and the corners of the Si_3N_4 grains were marked manually in order to outline the grains. In each material, 200-400 grains were measured.

RESULTS AND DISCUSSION

Matrix morphology

Retained α-Si_3N_4 could not be detected in any of the materials which shows that a complete transformation from α- to β–Si_3N_4 occured during densification also when metal oxides were not added. The presence of Si_2N_2O in the materials formed with the different additions can be explained by an increased oxygen content originating from the metal oxides or from impurities in the whisker materials [9,10].

The material formed without additives showed mainly equi-axed grain sections in the SEM. Only few sections had an elongated shape, Figure 1a. The addition of metal oxides resulted in a larger number of smaller grain sections and a more fibrous appearance, Figure 1b. When SiC whiskers were added, there was no obvious difference in matrix morphology compared to the material formed without whisker and metal oxide additions, Figure 1c. The material formed with β–Si_3N_4-whiskers had a much coarser microstructure compared to the other materials, Figure 1d.

These differences in the microstructures of the materials were clearly reflected in the grain area distributions, see Figure 2. The metal oxide containing material

Table I. Powder compositions, process conditions and phase compositions of the examined specimens [5].

Specimen	Densification			Phase composition
	T (°C)	t (h)	P (MPa)	
Si_3N_4*	1950	2	250	β-Si_3N_4
Si_3N_4 +MO	1775	1	200	β-Si_3N_4, Si_2N_2O
Si_3N_4 +25 wt% β-SiC**	1950	2	250	β-Si_3N_4, Si_2N_2O, β-SiC
Si_3N_4 +25 wt% β-Si_3N_4***	1950	2	250	β-Si_3N_4, Si_2N_2O

Legend: *UBE SNE 10, **Tateho SCW#1, ***UBE-SN-WB
MO = 2.5 wt% Y_2O_3 and 0.2 wt% Fe_2O_3

had the lowest mean grain area, 0.05 $(\mu m)^2$, with 76 % of the observed grain areas smaller than 0.06 $(\mu m)^2$. This could possibly be explained by an increased nucleation in the larger liquid volume present during densification of this material. Also, reduced grain growth could be expected because of the lower densification temperature.

The mean grain area measured for the SiC whisker reinforced material was lower than that measured for the material formed without additives, 0.09 compared to 0.16 $(\mu m)^2$. SEM, as well as TEM [5] did not indicate any significant difference in Si_3N_4 grain shape. It may therefore be concluded that the SiC-whisker addition resulted in a reduced Si_3N_4 matrix grain size. The lower frequency of larger grain areas does also imply that the SiC-whiskers will suppress matrix grain growth. Observations indicating that SiC-whiskers hinder Si_3N_4 grain growth have been reported earlier [11,12].

When $\beta-Si_3N_4$ whiskers had been added to the Si_3N_4 ceramic, the mean grain area was 0.56 $(\mu m)^2$ which implies that exaggerated grain growth took place during densification. This was promoted by the $\beta-Si_3N_4$ whiskers which are significantly larger than the Si_3N_4 starting powder particles, and also larger than the grain size of the unreinforced material.

Intergranular microstructure

The three materials formed without metal oxide additives contained thin amorphous films separating adjacent $\beta-Si_3N_4$ grains. Only very small glass

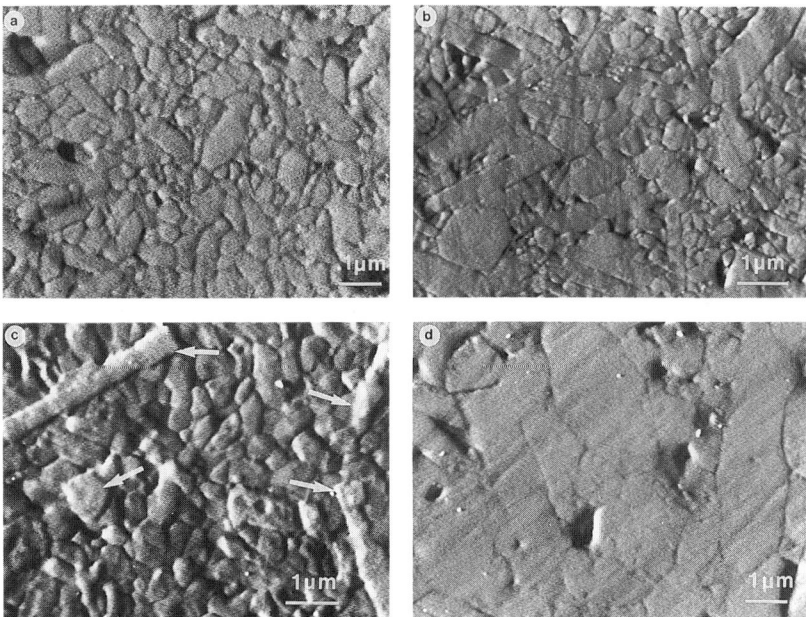

Figure 1. Backscattered SEM images of etched sections of the materials formed (a) without additives and with (b) metal oxides, (c) SiC-whiskers (arrowed) and (d) β–Si_3N_4-whiskers.

pockets could be observed at multi grain junctions in these materials. The formation of this amorphous phase was promoted by the surface silica present on the Si3N4 starting powder particles. In the SiC whisker containing material, extremly thin intergranular glass films were present also between whisker and matrix.

Addition of metal oxides resulted in an increased amount of residual glass separating the β-Si3N4 grains. Amorphous intergranular films merged into larger Y and Si rich glass pockets at multi grain junctions.

Mechanical properties

According to Becher *et al.*, both whiskers and matrix contribute to the fracture toughness of whisker reinforced noncubic ceramics [13]. Hence, matrix microstructure will have an influence upon mechanical properties of reinforced Si3N4 ceramics. This is also indicated by the Si3N4 materials in this investigation.

The material formed without additives had comparatively low room temperature strength and toughness, Figure 4. The toughness of these materials were determined by indentation technique using the equation proposed by

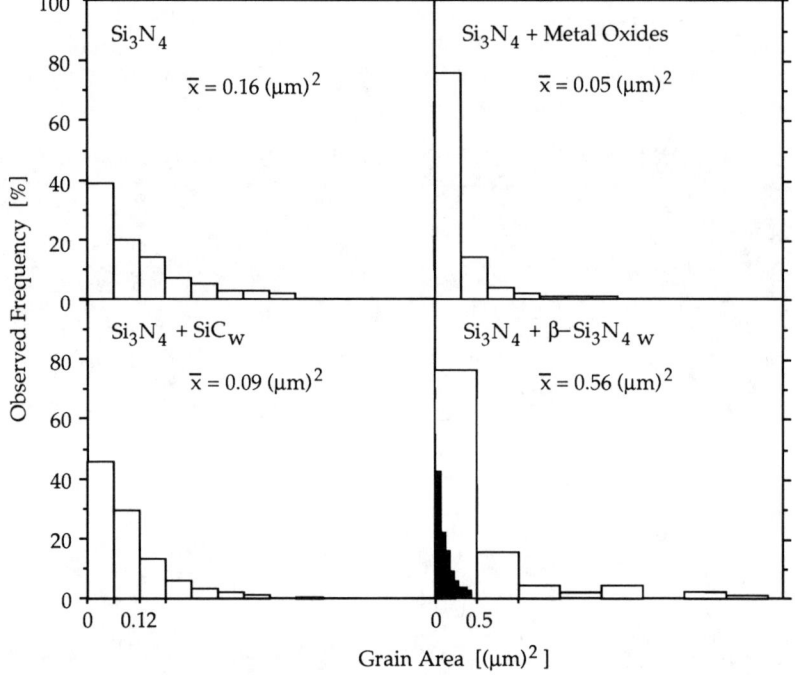

Figure 2. Observed matrix grain area frequency for the examined materials. Each bar represents an interval of 0.06 $(\mu m)^2$, except in the diagram for the material formed with an addition of β–Si3N4 whiskers where a bar represents an interval of 0.5 $(\mu m)^2$. The grain area distribution for the material formed without additives is superimposed (black bars) on the grain area distribution for the β–Si3N4 whisker reinforced material.

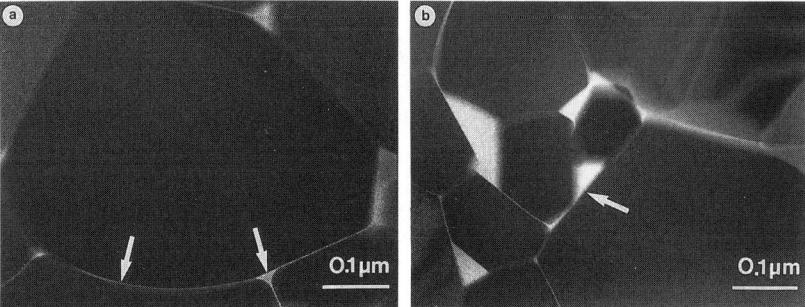

Figure 3. TEM centered dark field images formed using diffuse scattered electrons. The intergranular glassy phase appears bright (arrowed). Material formed (a) without additives and (b) with metal oxides.

Niihara [14]. An addition of 25 wt% SiC or β–Si$_3$N$_4$ whiskers had a clear strengthening and toughening effect. The material formed with β–Si$_3$N$_4$ whiskers had a higher toughness than the SiC whisker reinforced material, but there was no pronounced differnce in measured strength values [3]. This suggests that the larger grains in the coarser microstructure of the material formed with β–Si$_3$N$_4$ whiskers promote an increased toughness, but also that large Si$_3$N$_4$ grains may act as flaws.

The metal oxide sintering additives increased toughness and strength significantly, Figure 4. It has been demonstrated previously that high room temperature strength and toughness of Si$_3$N$_4$ ceramics are associated with a fibrous β–Si$_3$N$_4$ microstructure [6,7]. A determination of Si$_3$N$_4$ grain shape has not yet been done, but the larger liquid volume provided by the metal oxides

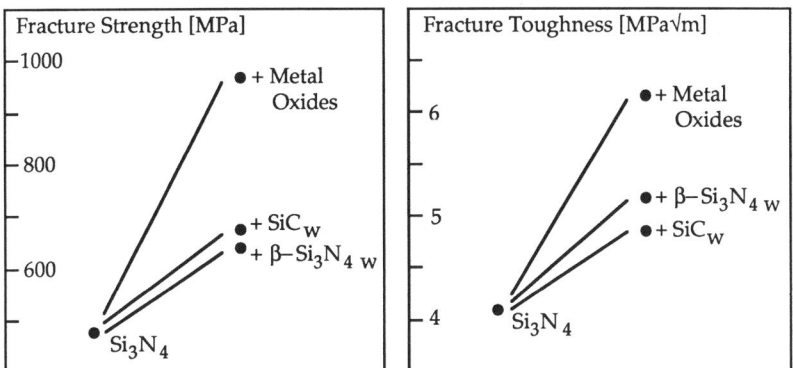

Figure 4. Mean values of 3-point flexural strength and indentation fracture toughness at room temperature. After reference [3].

would promote a development of prismatic, high aspect ratio β–Si_3N_4 grains [15]. This is also indicated by the SEM micrographs in Figure 1. The amount and chemistry of the residual intergranular glass may affect the contribution from different toughening and strengthening mechanisms in Si_3N_4 ceramics [2].

CONCLUDING REMARKS

The effect which different additives have upon Si_3N_4 grain size and shape can be estimated by SEM in combination with quantitative microscopy. Additions of metal oxides and whiskers will control Si_3N_4 grain morphology and intergranular microstructure, and this will have an influence upon mechanical properties of reinforced Si_3N_4 ceramics.

ACKNOWLEDGEMENT

The project was supported by the Swedish National Board for Technical Development (STU).

REFERENCES

1. S.T. Buljan, J.G. Baldoni and M. L. Huckabee, Am. Cer. Soc. Bull. **66**, 347 (1987).
2. G.H. Campbell, M. Rühle, B.J. Dalgliesh and A.G. Evans, J. Am. Ceram. Soc. **73** (3), 521 (1990).
3. H.T. Larker and J.E. Adlerborn , in Ceramic Materials and Components for Engines, edited by V.J. Tennery (American Ceramic Society Inc., Westerville, Ohio, 1989) pp. 227-236.
4. P.F. Becher, H. Chun-Hway, P. Angelini and T.N. Tiegs, J. Am. Ceram. Soc. **71** (12), 1050 (1988).
5. L.K.L. Falk, H. Björklund, J.E. Adlerborn and H.T. Larker, in Ceramic Materials and Components for Engines, edited by R. Carlsson, T. Johansson and L. Kahlman (Elseviers Science Publishers, London, 1992) pp. 699-706.
6. E.M. Knutson-Wedel, L.K.L. Falk, H. Björklund and T. Ekström, J. Mater. Sci. **26**, 5575 (1991).
7. F.F. Lange, Int. Met. Rev., No. 1, 1 (1980).
8. J. Wasén and R. Warren, Mater. Sci. Tech., **5**, 222 (1989).
9. J. Homeny, L.J. Neergard, K.R. Karasek, J.T. Donner and S.A. Bradley, J. Am. Ceram. Soc. **73** (8), 102 (1990).
10. K.R. Karasek, S.A. Bradley, J.T. Donner, M.R. Martin, K.L. Haynes and H.C. Yeh, J. Mater. Sci. **24**, 1617 (1989).
11. B.J. Hockey, S.M. Wiederhorn, W. Liu, J.G. Baldoni and S.-T. Buljan, J. Mater. Sci. **26**, 3931 (1991).
12. C. Olagnon, E. Bullock and G. Fantozzi, Ceramics International **17**, 53 (1991).
13. P.F. Becher, E.R. Fuller Jr. and P. Angelini, J. Am. Ceram. Soc. **74** (9), 2131 (1991).
14. K. Niihara, R. Morena, D.P.H and Hasselman, in Fracture Mechanics of Ceramics 5, edited by R.C. Bradt, D.P.H. Hasselman and F.F. Lange (Plenum Press, New York, 1983) pp. 97-105.
15. M.H. Lewis, B.D. Powell, P. Drew, R.J. Lumby, B. North and A.J. Taylor, J. Mater. Sci. **12**, 61 (1977).

CHARACTERIZATION OF HOT-PRESSED SILICON NITRIDE CERAMICS WITH ALKOXIDE-DERIVED OXIDE MIXTURES AS THE SINTERING AID

Y. SATO*, C. SAKURAI*, M. UEKI* AND K. SUGITA**
* Advanced Mat. & Tech. Res. Lab., Nippon Steel Corp., Kawasaki 211, Japan
**Technical Development Bureau, Nippon Steel Corp., Futtsu 299-12, Japan

ABSTRACT

A homogeneous mixture of Y_2O_3, CeO_2 and MgO with a final weight ratio of 3: 1: 2 was prepared by the alkoxide method. The powder mixture was then added into Si_3N_4 powder in amounts ranging from 4 to 12 wt%, and consolidated by hot-pressing. Microstructure and mechanical properties of the sintered bodies were determined and compared to those of materials prepared by the conventional route of mixing the oxide powders as sintering aids individually in essentially same composition. The β-fraction (modification ratio) in same composition was higher in the sintered bodies made through the alkoxide method than those made through the conventional one. The room temperature flexural strength was maximized with 6wt% addition of the alkoxide derived oxide, whereas, 12wt% addition of the total oxide was required to maximize the strength by conventional processing.

INTRODUCTION

Silicon nitride(Si_3N_4) based ceramics have various excellent properties such as toughness, strength and wear resistance that are superior to other ceramics. In the fabrication of silicon nitride ceramics, densification is achieved through liquid phase sintering using oxides as sintering aids. Oxides such as yttria(Y_2O_3), alumina(Al_2O_3), magnesia(MgO) and ceria(CeO_2) [1-3] are commonly used for this purpose. In the course of sintering of the silicon nitride ceramics, densification proceeds with an accompanying α→β crystalline phase transformation.[4-6] The rate of the transformation is influenced by the state of the starting powders[4] and the sintering conditions.[5] The addition of oxides also affects the phase transformation.[6] According to the aforementioned findings, it is apparent that the properties of silicon nitride ceramics are strongly affected by the amount and kind of oxide additives.

In general, the room temperature strength of silicon nitride ceramics is determined by the acicularity of the Si_3N_4 grains and their adhesive strength with grain boundary phases formed mainly by the oxide addititives. When the grain boundary phase is amorphous, the strength of the interface between Si_3N_4 grain and boundary phases is quite high and contributes greatly to the flexural strength of the sintered body at room to intermediate (≤800°C) temperatures. However, at much higher temperatures, the strength decreases due to softening of the glassy phase. Therefore, to increase the high temperature (≥1000°C) strength of the silicon nitride ceramics, crystallization of the grain boundary phases has commonly been practiced. Due to poor coherency between different structures of crystals, the strength of the interface in the silicon nitride grains with the crystallized grain boundary phases is not so high compared to materials having the interface between crystal and glass.

In the present study, we intended to develop the silicon nitride ceramics with high strength at room temperature by hot-pressing. Therefore, our concept for the material design is acicular shaped grains with a certain preferred orientation bound by a glassy phase. In such a program, we aimed to decrease the total amount of oxides as the sintering aids without lowering the room temperature strength. As an attractive method to achieve this, we tried to prepare the oxide mixture of sintering aids through sol-gel processing using metal alkoxides as the starting material. Applying this process we can prepare the multicomponent glasses and ceramics at low temperature since chemical homogeniety on a molecular level can be easily achieved in solution.[7] Such sol-gel processing of monolithic gels,[8] fibers[9] and films[10] from metal alkoxides has produced many scientific and technological advances.

Through the evaluation of the mechanical properties and microstructure of the hot-pressed silicon nitride ceramics with various amounts of oxide mixture, and by comparing them with the conventional materials, we have investigated the effectiveness of the application of the sol-gel processing to the fabrication of silicon nitride ceramics.

EXPERIMENTAL PROCEDURE

Preparation of Oxide Mixture from Alkoxide

The metal alkoxides used were solutions of $Y(OC_4H_9)_3$ in xylene, $Ce(OC_4H_9)_4$ in toluene and $Mg(OC_2H_5OC_2H_5)_2$ in ethoxyethanol. These alkoxide solutions were mixed to achieve the weight ratio of Y_2O_3: CeO_2: MgO = 3: 1: 2 in the final powder mixture. During mixing the solution strongly, it was hydrolyzed completely by the addition of excess water to obtain a mixture sol. The mixture obtained was then dried using a rotary evaporator at 60°C and then rinsed with water. Rinsing and drying were repeated twice, and then the dried gel powder was finally obtained. The thermal decomposition process of the dried powder was traced up to 1100°C by both differential (DTA) and gravimetric thermal (TG) analyses. Crystalline phases in the powder were identified by X-ray diffractometry (XRD) in the range from $2\theta=10$ to $70°(CuK\alpha)$ before and after calcining at 300° and 600°C. The calcination temperature was determined through the above mentioned analyses. Applying this optimum condition, the dried gel powder was calcined to get the desired oxide powder mixture.

Fabrication and Evaluation of Sintered Body

The alkoxide derived oxide mixture was added in amounts of 4, 6, 8 and 12wt%, to the raw powder of silicon nitride (Ube SN-E10) and ball-milled for 24h with acetone. The ball-milled powder was then dried and ground with a mortar. The powder was hot-pressed at 40MPa in flowing N_2 gas. In the course of the hot-pressing operation, the temperature was held for 1h at both 1600° and 1700°C.

To evaluate the alkoxide derived mixture oxide as a sintering aid, materials with commercial oxides used as sintering aids were also prepared by the conventional method. The oxide powders used were Y_2O_3 (Nippon yttrium, purity >99.9%, particle size 0.4μm), CeO2 (Shinetsu kagaku, >99.9%, 0.8μm) and MgO (Ube 1000A, >99.98%, 0.1μm). For the comparison, the same ratio and amount of oxide additions were adopted as those used in the alkoxide method, and the oxide was mixed and hot-pressed by same procedure described previously.

Densities of the hot-pressed silicon nitride ceramics were measured by the water displacement (Archimedes) method. After machining the sample to a size of 3×4×40mm, 3-point flexural strength was measured, using a span of 30mm, at room temperature (JIS R 1601). Fracture toughness, K_{IC} was also measured by the single edge pre-cracked beam(SEPB) method (JIS R 1607) using the 3×4×20mm samples. The microstructure of the ceramics was observed using a scanning electron microscope (SEM) (Hitachi S-4000) after plasma etching, and also using a transmission electron microscope (TEM) (Hitachi H-800).

RESULTS AND DISCUSSION

Thermal Analysis and X-ray Diffractometry of Powder Mixture

Figure 1 shows results of the thermal analysis of the powder mixture derived from alkoxide. The weight reduction of the powder occurred continuously with elevation of temperature and a steep reduction in the weight accompanying an endothermic reaction occurred at around 300°C, then the weight change ceased at around 600°C. Figure 2 shows the X-ray diffraction (XRD) patterns of the powder mixture after calcination at both 300° (b) and 600°C (c) as well as before calcination (a). In the XRD pattern taken from the sample without calcination (a), peaks representing $Mg(OH)_2$ crystal were observed. Although the peaks remained in Fig. 2 (b), which is from the sample calcined at 300°C, they disappeared completely for the sample calcined at 600°C (c).

As apparent from the results shown in Fig. 1, the reduction in weight ceased at around 600°C. This may be relevant to the disappearence of the $Mg(OH)_2$ phase in the XRD pattern. In the sample calcined at 600°C, the peaks from Y_2O_3 crystals appeared in the XRD pattern. Through these analyses, the optimum condition for the calcination was determined as 600°C for 2h in air because the gel powder decomposed completely into the oxides at 600°C.

Figure 3 shows the transmission electron micrograph of the oxide powder mixture after calcination. The mixture was composed of fine particles less than 0.05μm in diameter. MgO particles having a diameter of 0.04μm (shown by A in Fig. 3) and clusters of Y_2O_3 and CeO_2 approximately 0.015μm in size (shown by B in Fig. 3) were mixed homogeneously.

β-fraction of the Silicon Nitride Ceramics after Hot-pressing

Figure 4 shows the comparison of the XRD patterns of hot-pressed silicon nitride ceramics. At

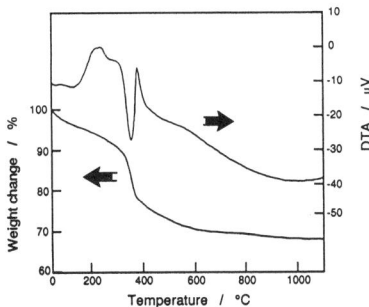
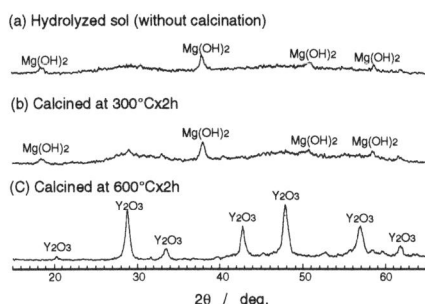

Fig. 1 Thermal decomposition behavior of the hydrolysis product of powder mixture.

Fig.2 Transition of X-ray diffraction (XRD) pattern of hydrolyzed sol due to calcination; (a) without calcination, (b) calcined at 300°C for 2h and (c) calcined at 600°C for 2h.

a given hot-pressing condition, described previously, the α→β transformation was entirely completed in the materials containing 12wt% oxides as sintering aid with no dependence on whether alkoxide or conventional processing methods were used. Namely, the fraction of β-Si3N4 phase (β-fraction) was almost 100% in such cases. With 8wt% or less oxide addition, however, α-phase remained in the hot-pressed bodies without transforming. Measuring the peak height (X-ray intensity) in the XRD pattern for both α and β crystals,[11] the β-fraction in the hot-pressed bodies was calculated and plotted against the total amount of oxides to be added into the silicon nitride for both methods as shown in Fig. 5. Comparing the materials with same level of oxide addition, it is apparent that the samples using the alkoxide derived oxide have a higher β-fraction than those from the conventional method. The oxide powder mixture derived from the alkoxide consists of very fine particles with higher chemical homogeniety on a molecular level compared to those from the conventional method, so the liquid phase is formed at a lower temperature in the alkoxide derived oxide. Therefore it can be considered that the α→β transformation was enhanced through activated microscopic mass trasportation.

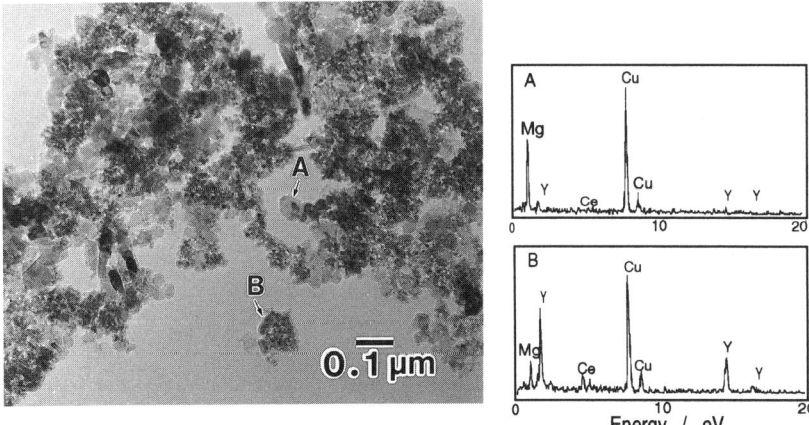

Fig. 3 TEM micrographs and spectra obtained by energy dispersed analysis of X-ray (EDX) of oxide powder mixture derived from metal alkoxide. (Peaks representing Cu are from sample holder)

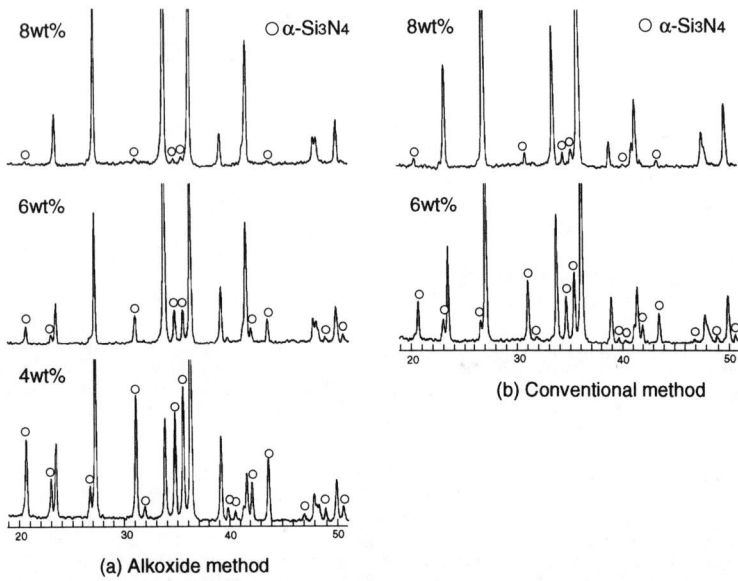

Fig. 4 Comparison of XRD patterns of hot-pressed silicon nitride ceramics prepared by using alkoxide derived sintering aid and by the conventional method; comparing 6 and 8wt% additions. The peaks representing α-Si3N4 are lower in the patterns taken from alkoxide method (a) compared to the conventional method (b).

Mechanical Properties of Hot-pressed Silicon Nitride Ceramics

Figure 6 shows the comparison of the relations between 3-point flexural strength at room temperature and the amount of oxide additives for the materials prepared by both alkoxide and conventional methods. Although the strength of silicon nitride ceramics prepared by the conventional method increased monotonically with the amount of oxide addition and the maximum was 1450MPa at 12wt%, the behavior of materials with alkoxide derived oxide differed largely. The maximum strength was achieved at 6wt% and was higher than that exhibited in the 12wt% oxide addition in the conventional method. The strength then decreased with higher oxide addition.

As shown in Fig. 7, the fracture toughness, K_{IC} increased with the oxide additions from 6.3MPam$^{1/2}$ at 6wt% to 7.3MPam$^{1/2}$ at 12wt% for the materials prepared by the conventional method. This increase in K_{IC} is due mainly to grain growth. On the other hand, K_{IC} in the case of materials prepared by the alkoxide method was almost constant at 6.8MPam$^{1/2}$ for 6wt% or more addition.

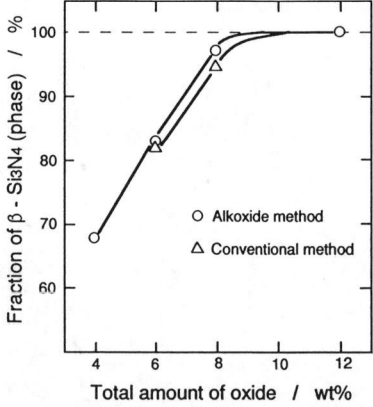

Fig. 5 Effect of total amount of oxide added on the β-fraction of hot-pressed silicon nitride ceramics prepared by both alkoxide and conventional methods.

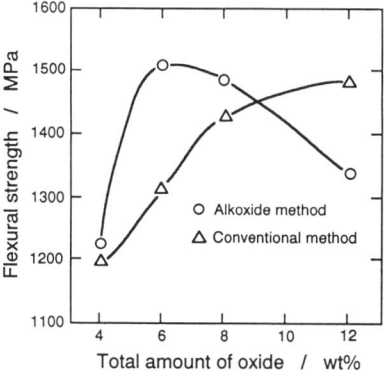

Fig. 6 Effect of total amount of oxide added on the flexural strength of hot-pressed silicon nitride ceramics prepared by both alkoxide and conventional methods.

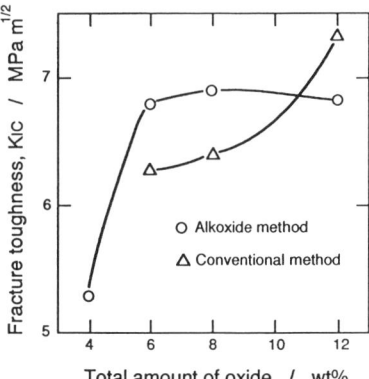

Fig. 7 Effect of total amount of oxide added on fracture toughness of hot-pressed silicon nitride ceramics prepared by both alkoxide and conventional methods.

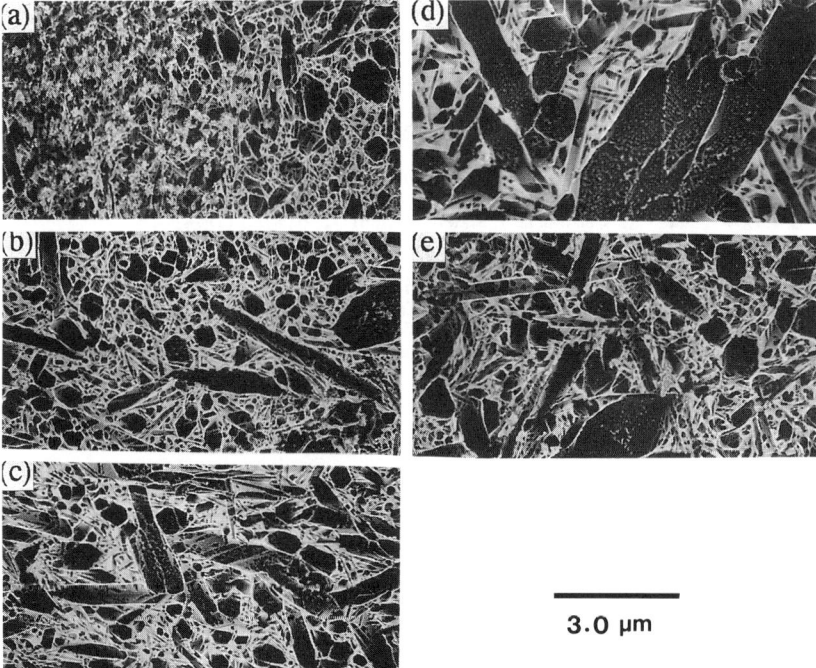

Fig. 8 Scanning electron micrographs of hot-pressed silicon nitride after plasma etching. (a) 4, (b) 6, (c) 8 and (d) 12wt% of alkoxide derived oxide mixture, and (e) 12wt% of oxide addition by the conventional method.

Microstructure

Figure 8 shows the scanning electron micrographs of the hot-pressed silicon nitride with surfaces perpendicular to the hot-pressing axis after plasma etching. In the sample with 4wt% additives produced from the alkoxide method, acicularity of the grains was not so high and some equiaxed grains remained (a). In the sample with 6wt% oxide addition, acicular grains developed uniformly (b). Comparing the micrographs within the materials prepared by the alkoxide method as a function of the amount of oxide addition, exaggerated grain growth and clustering of the β-Si_3N_4 grains became prominent at additions of 12wt% oxide (d) due to activated mass transfer in the alkoxide derived sintering aid. On the other hand, in the micrograph of the material prepared by the conventional method with the addition of 12wt% oxide (e), the β-Si_3N_4 grain size and morphology were similar to those in materials using the alkoxide derived oxides with 8wt% oxide addition. From microstructural point of view, it became apparent that we can reduce the amount of sintering aid from 12 to 8wt% by applying the alkoxide derived oxide without deteriorating the mechanical properties.

CONCLUSIONS

A fine homogeneous oxide powder mixture of Y_2O_3-CeO_2-MgO was successfully prepared by the alkoxide method. Using the powder as the sintering aid for silicon nitride, the α→β transformation was promoted at lower temperatures due to an activated mass transfer, and high strength silicon nitride ceramics were obtained with smaller amounts of oxide additions compared to the conventional method. The flexural strength was maximized at 6wt% addition of the alkoxide derived oxide mixture, in contrast with this, 12wt% of oxide addition was required for the conventional method. Although fracture toughness was almost constant at 6.8MPam$^{1/2}$ for the alkoxide method, it increased with oxide addition from 6.3MPam$^{1/2}$ for 6wt% to 7.3MPam$^{1/2}$ for 12wt% addition in the conventional method.

REFERENCES

1. G. R. Terwilliger and F. F. Lange, J. Am. Ceram. Soc. **57** (1), 25-29 (1974)
2. A. Tsuge and K. Nishida, Am. Ceram. Soc. Bull. **57** (4), 424-426,431 (1978)
3. M. Mitomo and K. Mizuno, Yogyo-Kyokai-Shi **94** (1), 106-111 (1986)
4. S. Tanaka, K. Nakano and S. Horikiri, in Proceedings of the IISS Symposium, edited by S. Somiya, M. Shimada, M. Yoshimura and R. Watanabe (Elsevier Science Publishers, New York, (1987), pp. 980-985,
5. L. J. Bowen and T. G. Carruthers, J. Mater. Sci. Lett. **13**, 684-687 (1978)
6. D. R. Messier, F. L. Riley and R. J. Brook, J. Mater. Sci. **13**, 1199-1205 (1978)
7. H. Dislich, J. Non-Cryst. Solids, **80**, 115-121 (1986)
8. T. Fukui, C. Sakurai and M. Okuyama, ibid., **139**, 205-214 (1992)
9. C. Sakurai, T. Fukui and M. Okuyama, Am. Ceram. Soc. Bull. **70** (4), 673-674 (1991)
10. S. Sakka, J. Iron and Steel Inst. Japan **77** (3), 326-335 (1991)
11. C. P. Gazzara and D. R. Messier, Am. Ceram. Soc. Bull. **56** (9), 777-780 (1977)

DIRECT OBSERVATION OF MICROSTRUCTURE CHANGE DURING DENSIFICATION OF SILICON NITRIDE CERAMICS WITH A NOVEL CHARACTERIZATION METHOD

Y. IWAMOTO, H. NOMURA, I. SUGIURA, J. TSUBAKI, H. TAKAHASHI,* K. ISHIKAWA,* N. SHINOHARA,* M. OKUMIYA,* Y. YAMADA,** H. KAMIYA,*** K. UEMATSU****
JFCC, Mutsuno, Atutaku, Nagoya, Japan
*Asahi Glass Co., Hazawa, Kanagawaku, Yokohama, Japan
**Ube Industries, Ube, Japan
***Department of Chemical Engineering, Nagoya University, Nagoya, Japan
****Department of Chemistry, Nagaoka University of Technology, Nagaoka, Japan

ABSTRACT

A novel characterization method is applied to study the evolution of microstructures during densification of silicon nitride ceramics. This characterization method involves an immersion liquid for making green and sintered bodies transparent, and a subsequent direct optical microscopic examination. Granules were prepared with the spray drying process and formed into green bodies by CIP. After sintering at various temperatures, the specimens were examined for microstructural evolution. Large pores were located at the center and boundary regions of granules left in the green bodies; they were not removed by densification and resulted in large pores in the sintered body, possibly forming fracture origin in ceramics.

INTRODUCTION

Defects are present in virtually all ceramics and govern various properties such as fracture strength.[1] Reduction in the concentration and size of defects is very important for the improvement of strength and reliability of high performance ceramics. To achieve this, it is necessary to fully understand the behavior of microstructural evolution during densification with particular focus on the behavior of large processing defects.

Behavior of large processing pores during densification has been a subject of both theoretical and experimental studies. Thermodynamic argument has shown that pores much larger than the surrounding grain are stable and grow during densification.[2] This argument was recently confirmed in alumina ceramics which were densified through the solid-state sintering process.[3] It is very important to understand the behavior of processing pores in a liquid phase sintering process; since all important high strength ceramics, as well as functional ceramics, densify through the liquid phase sintering process.

This paper presents the direct observation on the behavior of processing pores during liquid phase sintering in silicon nitride. This is a representative high performance ceramic, and the formation mechanism of fracture origin is of great practical interest. Following the previous study,[4] we have adopted the liquid immersion method for characterizing pores in green and sintered bodies. In this method, specimens were made transparent with an immersion liquid and were examined with a transmission optical microscope. This microscope provides clear, semi-three dimensional macro-and microstructural information otherwise difficult to obtain by any other method. Its characteristically large examination volume provides accurate information on features of very low concentration, such as fracture origins. The information obtained is also very objective. Mercury porosimetry and SEM observation were also applied to provide supplemental characteristics of micro-pores and high resolution images.

EXPERIMENT

Materials used in this study were all commercial grade. Silicon nitride (Ube, SNE-10) powder containing 5 wt% alumina and 5 wt% yttria as sintering additives (5 kg) and deflocculant (150 g) were mixed with pure water (3.5 kg) by an attrition mill. After adding a binder (equivalent solid content: 0.375%) and passing through a sieve (320 mesh) to remove foreign objects, the slurry mixture was spray dried to form granules. The granules were mold pressed into pellets at 20 MPa and subsequently cold isostatically pressed at 150 MPa. The pellets were then heated (10°C/min) to various temperatures and quenched. For structural examination, the granules were placed on a slide glass of the optical microscope and made transparent with a drop of immersion liquid (refractive index: 2.05, Cargille, New Jersey, USA). These transparent granules were then examined under an optical microscope in the transmission mode. In addition, the partially sintered specimens were thinned to a few tenths of a millimeter with sandpaper, then made transparent with the immersion liquid, and subjected to the same optical microscopic examination as above. Phases present in the specimens were examined by the powder x-ray diffraction method. The microstructure was examined by the conventional SEM technique. Micro-pores were examined by mercury porosimetry.

RESULTS

Figure 1 shows the structure of powder granules prepared in this study. They have a basically spherical shape and wide size distribution. A majority of granules contain deep dimples. A dimple seen from its top appears as a light small round feature in the center of a granule. The side view shows that the dimple is very deep and penetrates almost entirely through the granule. The dark circumferential region outlining each granule corresponds to the binder segregated at the surface of the granule. The binder can only be seen when viewed from the tangential direction. Small dark spots in the granules correspond to the particles of sintering additives which have a refractive index different from that of silicon nitride.

According to x-ray diffraction, the major phase in the specimen was alpha-silicon nitride at temperatures up to 1650°C and was beta-silicon nitride at 1750°C. Only beta-silicon nitride was found at 1800°C. Figure 2 shows the change in relative densities with sintering temperature. The densification started at about 1350°C. It became very rapid in the temperature range of 1450-1650°C and slowed down at higher temperatures. Note that the specimens were quenched immediately after reaching these temperatures. The maximum relative density of 97% was achieved by sintering for 1 h at 1800°C.

Figure 3 shows the change in microstructures with densification, which was examined by the liquid immersion technique. Pores can be observed as either light or dark parts, depending on whether air was trapped during liquid immersion; pores with air trapped appear dark, and those without air light. Except for the trapped air and/or immersion liquid, the pores were empty; they were not filled with liquid phase which is formed by the reaction between sintering additives and silicon dioxide. In the temperature range of zero shrinkage, the granules retain their spherical shape. Pores in the central region of the granules also remained in tact. With densification, the boundaries between granules became less clear and

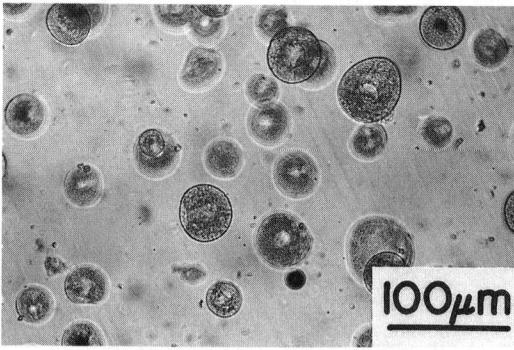

Fig. 1 Structure of powder granules examined by the liquid immersion technique.

pores at the boundaries of the granules became rounded. The pore in each granule was not removed and its shape remained almost the same during the densification process. The specimen resulted in a unique structure which is characterized by regularly arranged pores of similar size.

Figure 4 shows a SEM micrograph of a specimen sintered at 1800°C for 1 h. The structure in Fig. 3 is again present in this micrograph. Pores originated by the pores in the center of granules are arranged in the same regular manner as in the green and partially sintered bodies. The spacing between them was again approximately several tens of microns. Smaller pores were also found at the boundaries of the granules. The pore volume estimated with this micrograph was approximately a few percent, suggesting that the matrix of the specimen was fully dense except for these large pores.

Figure 5 shows the change in pore size distribution with densification. There are two groups of pores. One is centered in the submicron region and constitutes the majority of the total pore volume of the specimen. The other is large pores in the ten to a hundred micron region. The volume of large pores was so small that it was difficult to characterize them accurately in this figure. With densification, the pore volume of the first group decreased without noticeably changing the pore size while the pore volume of the second group was virtually unaffected.

DISCUSSION

Clearly the processing pores formed in the central and boundary regions of the granules cannot be removed by the compaction process nor the densification process in liquid phase sintering. The large pores at the center of the granules do not change their shape appreciably in densification. Medium-sized pores located between granules changed their shape and were rounded after densification. Small pores between primary particles can be eliminated according to mercury porosimetry. These small pores constitute the majority of pore volume in the green body, and their removal gives rise to the density increase in sintering.

These behaviors of large- and medium-sized pores are similar to what was found for alumina in solid state sintering[3] where large pores were also stable and could not be removed in densification. The monotonous decrease in the volume of small pores with sintering was different from what was found in solid state sintering; in alumina, small pores grew with densification before final removal. The behavior was explained by the presence of an agglomerate. The thermodynamic argument presented in past papers also holds true for the stability of large pores; the curved solid-gas interface provides a driving force for pore growth in large pores. Likewise, a curved interface for small pores provides a driving force for pore shrinkage.[2] The slight difference in the behaviors of pores in alumina vs. silicon nitride can be attributed to the sintering mechanism.

The large pores were clearly empty during densification. The liquid phase formed by the reaction between the sintering additives and silicon dioxide is absent in the large pore. This

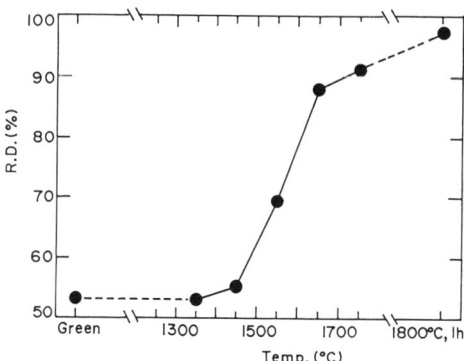

Fig. 2 Change in relative density with sintering.

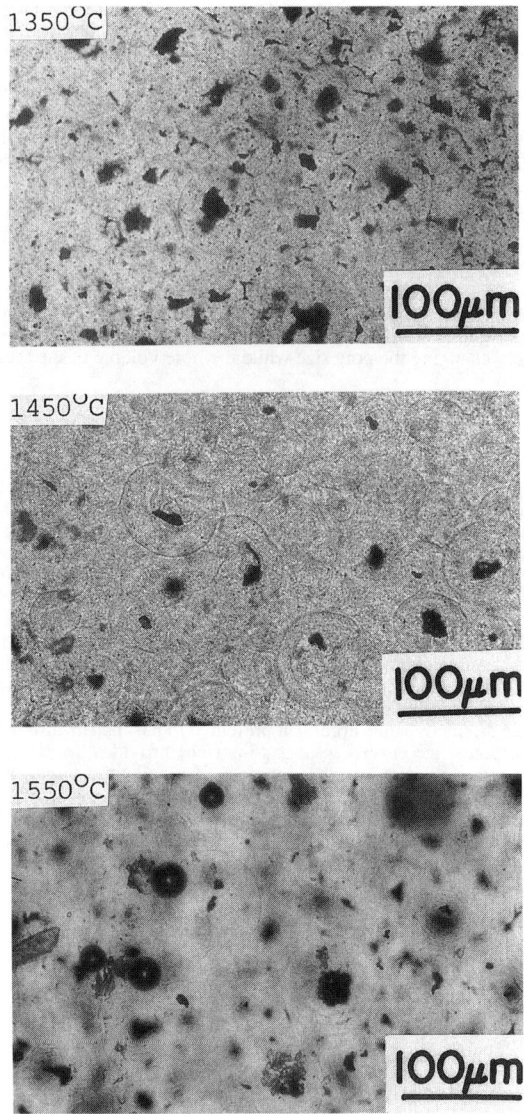

Fig. 3 Change in microstructure with sintering.

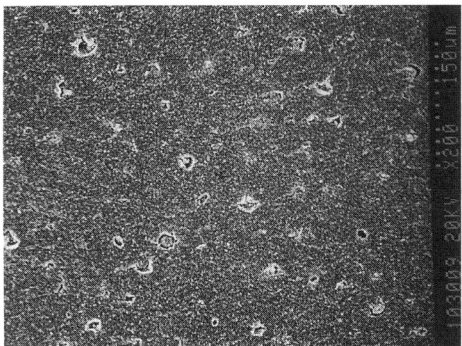

Fig. 4 Microstructure of specimen after sintering.

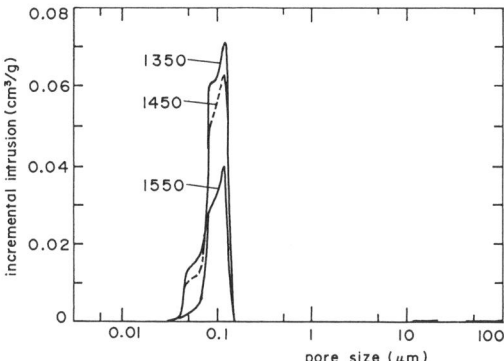

Fig. 5 Change in pore size distribution with sintering.

result is understandable when we consider the capillary force acting on the liquid; liquid is pulled into small pores since its thermodynamic stability is increased with this process if it wets the surface of the silicon nitride particle.

In the final microstructure, large scale non-uniformity originally present in the green body is preserved. Large pores in the sintered body are inherited from the large pores in the green body. The main origins of these large pores are dimples in the granules and pores formed at the boundaries of the granules. Large pores in ceramics behave as strength-limiting flaws. In the present specimen, a rather uniform strength is expected since fracture origins of similar size are uniformly distributed. The specimen is expected to have a low to medium level of strength with a high m-value in the Weibull plot. This result provides a guideline for producing high-strength ceramics. However, special efforts are needed to achieve a green body free of large scale non-uniformity.

CONCLUSION

The following conclusions were reached from examination of the microstructural evolution in silicon nitride ceramics:
(1) Large processing pores were present at the central and boundary regions of granules in the green body, which was formed from granules containing pores.
(2) These processing pores were stable and not removed by sintering. They remain as large pores in the sintered body.
(3) These large pores are distributed with regular spacing in the sintered body and reduce the strength of the ceramic.
(4) Large pores are empty; they are not filled with the liquid phase formed by the reaction between the sintering additives and silicon dioxide.

REFERENCES

1. A.G. Evans and G. Tappin, Br. Ceram. Soc., **20**, 275-97 (1972).

2. B.J. Kellett and F.F. Lange, J. Am. Ceram. Soc., **72**, 725-34 (1989).

3. K. Uematsu, M. Miyashita, J.-Y. Kim and N. Uchida, J. Am. Ceram. Soc., **75**, 1016-18 (1992).

4. K. Uematsu, M. Miyashita, J.-Y. Kim, Z. Kato and N. Uchida, J. Am. Ceram. Soc., **74**, 2170-74 (1991).

ACOUSTIC EMISSION STUDY OF Si_3N_4

E. C. SUBBARAO* and V. SRIKANTH
Materials Research Laboratory, The Pennsylvania State University, University Park, PA 16802.

J. C. WALCK and C. A. TARRY
GTE Products Corporation, Towanda, PA 18848

ABSTRACT

Sintered Si_3N_4 (with 2 w/o Al_2O_3 and 6 w/o Y_2O_3) was studied by acoustic emission method during indentation testing and during heating to and cooling from 920°C in air. The acoustic emission activity was discontinous and its magnitude and duration increased with load applied during indentation testing. Intense acoustic emission signals were detected on cooling Si_3N_4 ceramics to temperatures below 590°C, particularly below 450°C. X-ray diffraction studies showed the presence of SiO_2 (Quartz) after heating the sample to 920°C during acoustic emission study, compared to single phase β Si_3N_4 before the heat treatment.

INTRODUCTION

Silicon nitride (Si_3N_4) has been of great interest for high temperature engine components, war parts etc. due to its outstanding strength, oxidation resistance and stability.[1,2] A number of additives are used as sintering aids, such as Al_2O_3, Y_2O_3, MgO etc. Si_3N_4 contains varying amounts of oxygen as an impurity, in the form of amorphous silica at the grain boundaries and at triple points,[3,4] and sometimes as other compounds such as Si_2N_2O [5] or yttrium silicates. The amorphous phases are reported to crystallize during heat treatment or in service.

In the present study, acoustic emission techniques are employed to study microdeformations during indentation tests, which are often used for hardness and fracture toughness studies, and those arising from high temperature oxidation of Si_3N_4 ceramics. These studies are supported by x-ray diffraction examination.

EXPERIMENTAL

Specimens

The silicon nitride specimens, supplied by GTE, have the following characteristics:

Composition	:	2 w/o Al_2O_3, 6 w/o Y_2O_3
Archimedes Density	:	3.260 g cm^{-3} (99.98%)
Geometric Density	:	3.244 g cm^{-3} (99.53%)
Moduli: Young's	:	305 GPa
Shear	:	120 GPa
Bulk	:	225 GPa
Hardness: 1 Kg Knoop	:	13.41
1 Kg Vickers	:	12.92
Mean Strength (4 PT. Bend)	:	783 MPa
Phases (XRD) (Fig. 1)	:	βSi_3N_4

Heating

The Si_3N_4 specimen was heated in a resistance furnace to 920°C in air at the rate of 10°C/min, held at the peak temperature for five minutes and cooled to 225°C at the rate of 5°C/min, while acoustic emission data are collected on system 1.

---

* Permanent Address: Tata Research Development and Design Centre, 1 Mangaldas Road, Pune 411001, INDIA.

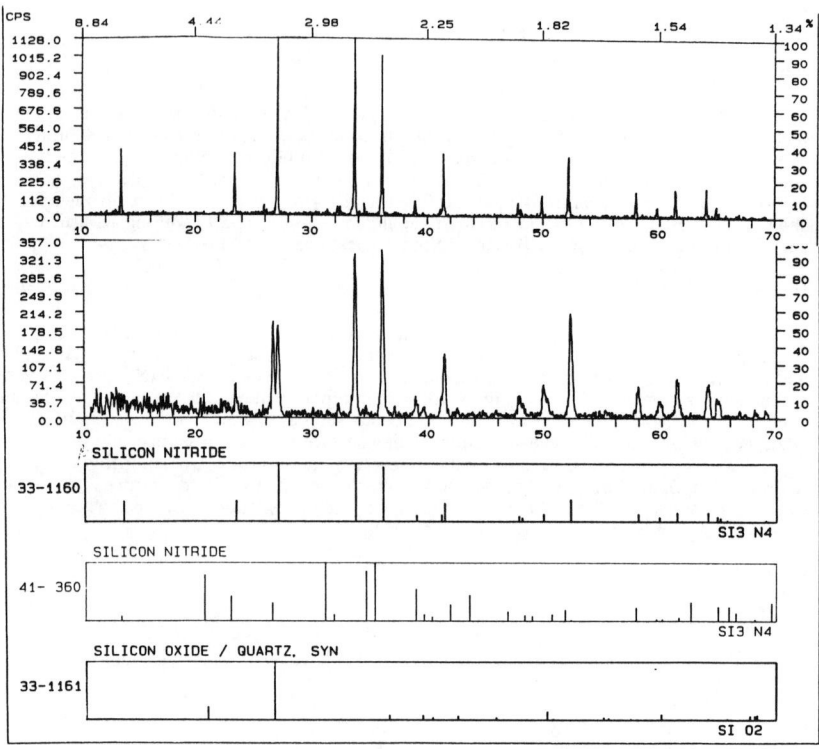

Figure 1 X-ray diffraction patterns of Si_3N_4 as received (top) and after heat treatment to 920°C in air for AE Study (bottom).

Indentation

Indentations with loads upto 500 g (5N) were made with a Leitz Hardness Tester. The transducer of the acoustic emission system 2, described below, was located close to the indenter and held in place by a clamp. Improved contact between the transducer and the specimen was achieved by the use of an ultrasonic couplant grease. Acoustic emission signals were monitored as a function of time after the application of load until no further signals were detected.

Acoustic Emission

Acoustic emissions are transient elastic waves arising from rapid release of energy within a material due to any microdeformation process such as microcracking, crystallographic phase transition etc. These can be detected by piezoelectric transducers mounted on the sample while it is subjected to mechanical stress, as in indentation testing, or while it is heated and cooled through a certain temperature range. Two kinds of acoustic emission equipment were used in the present work.

System 1, used for the study of Si_3N_4 while it is heated and cooled, makes use of a transducer with a center frequency of 500 kHz (in a range of 300 - 700 kHz). An alumina rod serves as a wave guide and is attached to the sample with a high temperature cement at one end and to the transducer with an ultrasonic couplant at the other. The electrical output from the transducer is amplified, filtered, and processed through a train of instrumentation, consisting of an amplifier/discriminator, totalizer and rate meter modules, to obtain "total or cumulative AE counts" and "count rate (counts/5°C)" data. The discriminator triggers a pulse whenever the amplifier output exceeds a certain adjustable threshold. The details of the system are described earlier.[6-8]

System 2, used for an indentation study of Si_3N_4, consists of a transducer which is placed on the sample near the indenter and held in place by a clamp. An ultrasonic couplant ensures good contact between the sample surface and the transducer. When the indentator is applied, simultaneously the electronic circuitry is turned on, which collects, filters and analyzes the acoustic emission signals and displays cumulative counts or count rate as a function of time. The equipment is Locan 320 System from Physical Acoustics Corp. and has been employed by the present authors to study the domain processes and microcracking during poling of piezoelectric lead zirconate titanate (PZT) ceramics.[9]

RESULTS AND DISCUSSION

Indentation Studies

No acoustic emission signals were detected upto loads of 200 g (2N). But at 300 g (3N) and 500 g (5N), indentations were formed and AE signals were recorded as a function of time (figure 2). Cumulative counts and count rates were obtained. The deformation process was discontinuous and discrete with the major acoustic activity occurring sometime (tens of seconds) after application of load. The total cumulative counts were more at a load of 5N compared to those at 3N, as may be expected. Under a load of 3N, the acoustic emission ceased after about 35 sec, whereas they continue upto 72 sec after the application of a load of 5N. Thus, the cumulative AE counts and the duration of AE activity are larger at a load of 5N than at 3N.

High Temperature Studies

During heating of Si_3N_4 a few AE counts (less than 600) were detected upto 550°C and none beyond (figure 3). On the other hand, during cooling of the same sample from 920°C, no AE signals were detected down to 550°C, but substantial AE activity (total cumulative counts of over 20,000) was evident below this temperature, particularly below 450°C. In order to detect any changes in the sample as a result of this heat treatment, x-ray diffraction pattern was obtained of the sample after the heating and cooling cycle. A significant amount of silica (α Quartz) was evident (figure 1). Since the sample was heated in air upto 920°C, oxidation appears to have taken place. Further, amorphus silica present in the grain boundaries and three grain junctions may also crystallize as a result of the heat treatment employed. The stresses developed during cooling due to the differential thermal expansion coefficients of quartz and βSi_3N_4 result in microdeformations, which appear as acoustic emission signals.

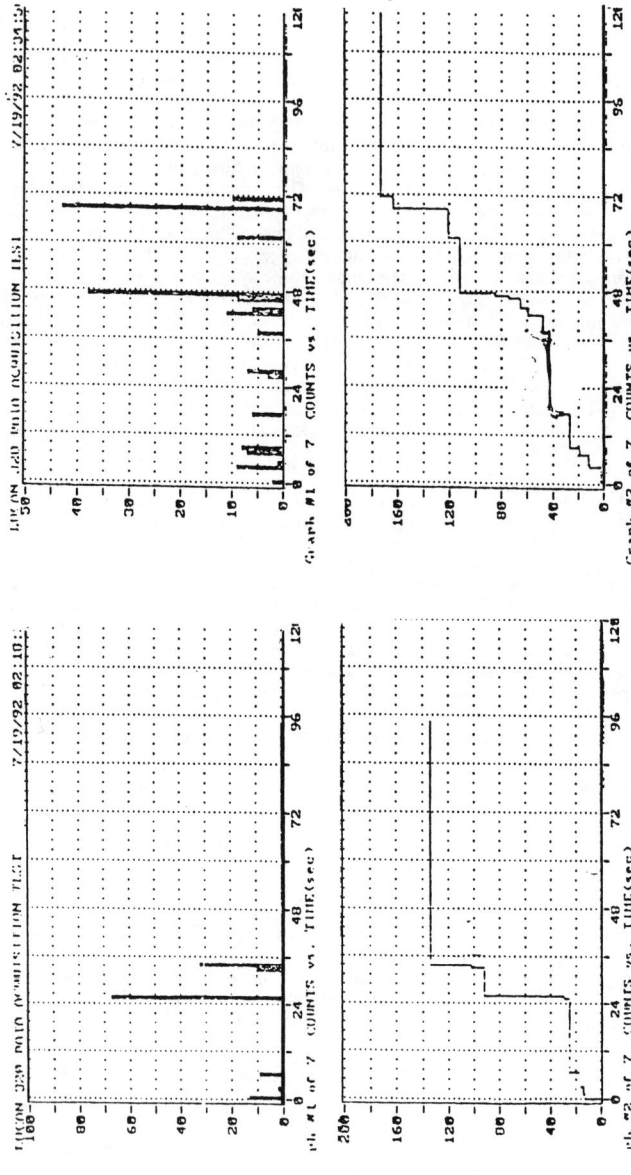

Figure 2 Acoustic emission data consisting of count rate (top) and cumulative counts (bottom) as a function of time after application of indentation load of 3N (left) and 5N (right) on Si_3N_4 ceramics.

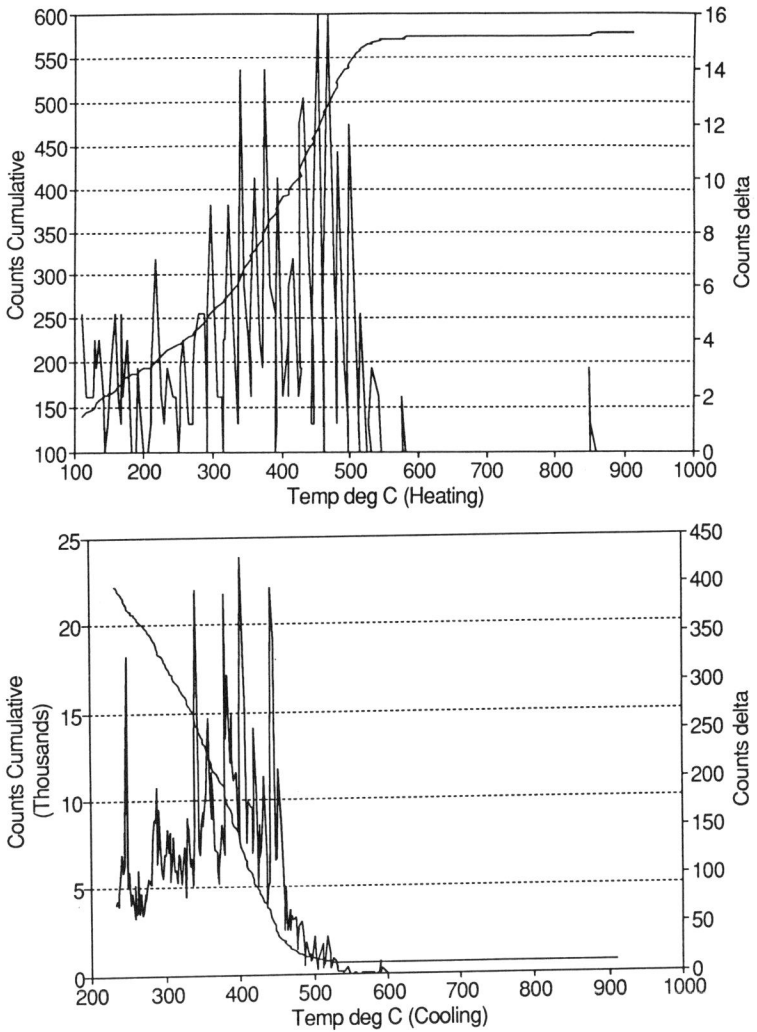

Figure 3   Acoustic emission data consisting of count rate and cumulative counts on heating (top)a sample of Si_3N_4 ceramics to 920°C and cooling (bottom) to 225°C.

CONCLUSIONS

Acoustics emission signals were detected during hardness testing. Differential contraction during cooling of multiphase Si_3N_4 ceramics leads to microcracking which again was detected AE techniques.

REFERENCES

1. G. Ziegler, J. Heinrich and G. Wotting, J. Mater. Sci. **22**, 3041 (1987).
2. J. Weiss, Ann. Rev. Mater. Sci. **11**, 381 (1981).
3. H. Du, R. E. Tressler, K. E. Spear and C. G. Pantano, J. Electrochem. Soc. **136**, 1527 (1989).
4. D. A. Bonnell, T. Y. Tien and M. Ruhle, J. Am. Ceram. Soc. **70**, 40 (1987).
5. F. F. Lange, S. C. Singhal, and R. C. Kuznicki, J. Am. Ceram. Soc. **60**, 249 (1977).
6. V. Srikanth and E. C. Subbarao, Ceram. Intl. **18**, 251 (1992).
7. V. Srikanth and E. C. Subbarao, Acta. Metall. **40**, 1091 (1992).
8. E. C. Subbarao and V. Srikanth, Physica **C171**, 449 (1990).
9. E. C. Subbarao, V. Srikanth, W. Cao and L. E. Cross, Submitted for Publication.

PART II D

Mechanical Performance

DEFORMATION AND TOUGHNESS OF α-SILICON NITRIDE SINGLE CRYSTALS

H. Suematsu, J. J. Petrovic* and T. E. Mitchell,
Center for Materials Science, Los Alamos National Laboratory, Los Alamos, NM 87545.
(*Materials Science and Technology Division)

ABSTRACT

Vickers and Knoop indentation methods were used to determine the deformation behavior and fracture toughness of single crystal α-Si_3N_4 at room temperature. The tests were performed on (0001), ($1\bar{1}00$) and ($11\bar{2}0$) faces; both hardness and toughness values were found to be independent of orientation within statistical variations. Thus, the mechanical properties of silicon nitride are essentially isotropic at room temperature.

Single crystals of α-Si_3N_4 were also compressed at 1760 and 1820°C at a strain rate of The sample tested at 1760°C showed yielding at a stress of 200MPa but it fractured before indicating a significant amount of plastic deformation. On the other hand, the sample compressed at 1820°C deformed to a strain of 2.7% at a stress of 40MPa before fracturing. A significant density of dislocations was observed by transmission electron microscopy. From conventional **g.b** analysis, the Burgers vector of the dislocations was determined to be 1/3<$11\bar{2}0$>. From the line direction of the dislocations, the primary slip system of α-Si_3N_4 is determined to be {$1\bar{1}01$}<$11\bar{2}0$>.

INTRODUCTION

Si_3N_4 is well known as a high-toughness and high-strength ceramic. The mechanical properties of Si_3N_4 depend on the sintering process, although the strength generally exceeds 700MPa and the toughness often reaches 7MPa.m$^{1/2}$ or more. The exhaust turbine of the turbo charger for an automobile engine is now often made from Si_3N_4 because of its high strength and light weight. It is also a candidate material for future structural applications such as in gas turbine generators operating above 1400°C.

There are a number of reasons to explain the high toughness of sintered Si_3N_4. A most probable explanation is that the β-Si_3N_4 grains tend to grow in an elongated morphology during the sintering process and then act as fibers to reinforce the sintered body. When a crack is propagating, the surface energy to by-pass the elongated grain and the friction to pull-out the grain requires the applied stress to spend more energy to fracture the material. Thus, the reason for the high toughness might be the shape of the grain rather than the toughness of the Si_3N_4 itself. However, this is not clear because the fracture toughness of Si_3N_4 itself has never been measured.

The strength of Si_3N_4 is almost constant up to 1200°C. At higher temperatures, the strength begins to drop because of grain boundary sliding. Additives such as Y_2O_3 are required to sinter the Si_3N_4 and they form a glassy second phase on the grain boundary which has a relatively low melting point. This grain boundary phase degrades the high temperature properties. Much work has been published on methods to remove or crystallize the harmful glassy phase. If such methods were successful, the maximum use temperature for Si_3N_4 might be increased and the deformation of Si_3N_4 would be governed by other factors, for example Coble creep or dislocation movement. Thus, the deformation behavior of Si_3N_4 itself is important to predict the high temperature mechanical properties of sintered Si_3N_4. However, knowledge of plastic deformation in Si_3N_4 by dislocation motion is limited[1-6].

In order to determine the mechanical properties of Si_3N_4 itself, single crystals are necessary. In the present paper, we describe the preparation of such crystals and their subjection to hardness tests and high temperature compression tests.

EXPERIMENTAL PROCEDURE

Single crystals of Si_3N_4 were made by a chemical vapor deposition process at Union Carbide Coating Service. $HSiCl_3$, NH_3 and H_2 gases were used as reactants and were introduced in a chamber at a pressure of 0.5 torr. A graphite bowl was placed in the chamber and heated to a temperature of 1300°C in order to act as a substrate to grow the crystals on.

Suitable single crystals were examined with a Laue back-reflection x-ray camera in order to determine their phase structure and orientation. All crystals were found to have the α structure (trigonal, space group P31c, a = 0.7758nm, c = 0.5623nm). Samples for hardness tests and compression tests(Fig.1) were cut from the crystal to the desired orientation using a diamond saw. The surfaces of the samples were ground and polished with diamond paste.

In order to determine fracture toughness and hardness, a micro hardness tester was used. Both Vickers and Knoop diamond indenters were attached to the hardness tester. During the hardness tests, the indenters were loaded to 300g. High temperature compression tests were performed on a mechanical testing machine with a vacuum furnace (Shimadzu, IS-10T). Samples were tested at temperatures of 1760 and 1820°C in a vacuum of 3×10^{-3}Pa. In order to inhibit the decomposition of Si_3N_4 at such high temperatures, the surfaces of the samples were covered with BN powder. The samples were compressed at a constant strain rate of 2.2×10^{-5}/s.

After the compression tests, specimens for transmission electron microscopy (TEM) were cut out of the deformed sample. They were ground with diamond paste to a thickness of 30mm and finally thinned in a Gatan ion milling machine with 5kV Ar ions to obtain foils for TEM observation. TEM observations were performed with a Philips CM30 operated at an accelerating voltage of 300kV.

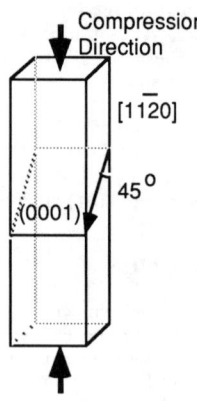

Fig.1 Sample orientation.

RESULTS AND DISCUSSION

Fracture toughness at room temperature

Figure 2 shows a typical Vickers indent on the (0001) plane of a Si_3N_4 single crystal. One of the diagonals of the square indent is parallel to $[1\bar{1}00]$ and the other is parallel to $[11\bar{2}0]$. Cracks initiated from the four corners of the indentation extend almost parallel to the diagonals and the lengths of the four cracks are much the same. This means that the fracture energy on the $(1\bar{1}00)$ and $(11\bar{2}0)$ planes is essentially the same. Table 1 shows the average and standard deviation of the toughness values on the $(10\bar{1}0)$, $(11\bar{2}0)$ and (0001) faces. The toughness value on $(11\bar{2}0)$ appears to be higher than that on the other planes, but this may be due to a difference in hardness. Comparing the length of the cracks on a given plane, it is seen that there is almost no difference. This fact also indicates that the toughness of single crystal α-Si_3N_4 is isotropic.

The fracture toughness values of single crystal Si_3N_4 are relatively low, 1.8-2.9MPa.m$^{1/2}$, compared with typical sintered Si_3N_4. As described above, sintered Si_3N_4 contains a whisker-like β phase and a secondary grain boundary phase. This means that pull-out and crack-bridging are the primary mechanisms for toughening in sintered Si_3N_4. HIPed Si_3N_4 without any additives also has a relatively low toughness[7]. Thus, it is concluded that α-Si_3N_4 itself has a low toughness but elongated β grains and secondary grain boundary phases make the Si_3N_4 tough.

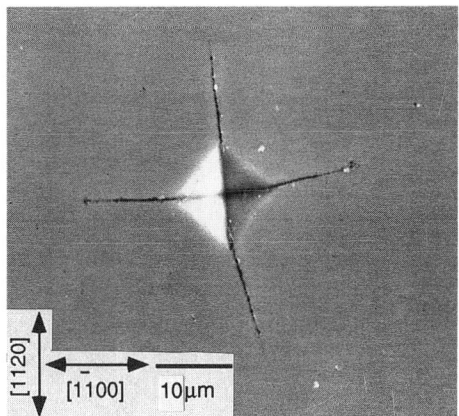

Fig.2 Optical micrograph of Vickers indentation on the (0001) plane of single crystal Si_3N_4.

Table 1 Vickers fracture toughness of Si_3N_4 single crystal.

Surface		($10\bar{1}0$)	($11\bar{2}0$)	(0001)
Vickers hardness (kg/mm^2)		2830±270	3050±130	2890±130
Crack length (μm)	// [$10\bar{1}0$]	–	29.6±3.0	38.0±3.7
"	// [$\bar{1}210$]	39.6±7.6	–	37.6±9.3
"	// [0001]	37.9±5.0	26.5±2.0	–
Average K_{IC} (MPa•m$^{1/2}$)		1.9±0.3	2.8±0.2	1.9±0.3

Knoop hardness anisotropy

Figure 3 shows the Knoop hardness of Si_3N_4 and indicates that the hardness has almost the same value on the three surfaces. Nevertheless, the variation of hardness with orientation on each face is limited to within 10%. There is a small regular variation with orientation for the ($11\bar{2}0$) face but it is not more than a standard deviation of the hardness value. These results are qualitatively consistent with those of Niihara and Hirai[8]. Daniels and Dunn[9] and Brooks et al.[10] proposed an equation to estimate hardness anisotropy from the geometry of the indenter and the primary slip system. The equations fit very well to most materials. If hardness anisotropy is apparent on the surface of a single crystal, the primary slip system can be identified from the equation. In most ceramics, it is quite rare that obvious plastic deformation occurs at low temperatures and so the Knoop hardness anisotropy method is the primary way to find the slip system particularly at room temperature. Since Si_3N_4 shows little anisotropy in Knoop hardness on any of the three surfaces tested, it is not possible to determine the primary slip system of Si_3N_4 at room temperature by this technique.

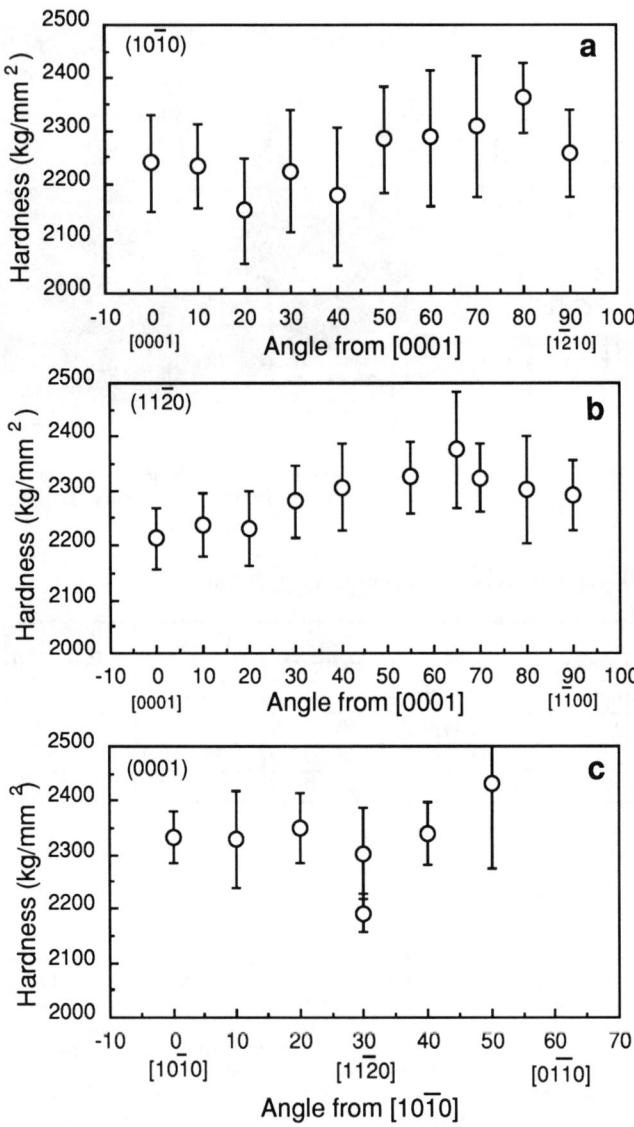

Fig.3 Knoop hardness anisotropy of single crystal Si3N4. (a) on (10$\bar{1}$0), (b) on (11$\bar{2}$0), (c) on (0001)

High temperature deformation

Stress-strain curves for crystals deformed in uniaxial compression at 1760°C and 1820°C are shown in Fig.4. For the test at 1760°C, the curve deviates from an elastic straight line at a stress of around 200MPa and shows evidence of plastic deformation. After yielding, the curve becomes a zig-zag line and fracture occurs at 270MPa. On the other hand, for the crystal compressed at 1820°C, yielding occurs at a relatively low stress(40MPa) and plastic deformation continued at about the same stress until the test was stopped at 2.7% strain.

A TEM sample was cut out from the crystal deformed at 1820°C. A typical deformation substructure is shown in Figs.5. Some short dislocation segments are seen and they are aligned almost parallel to each other, implying that they lie on the same slip plane. A **g·b** analysis was performed and the Burgers vector was found to be $1/3<11\bar{2}0>$ for all of these dislocations. A line direction analysis was carried out by taking TEM images of the dislocations viewed along several different zone axes and the dislocation line direction was found to be near $<1\bar{1}03>$. The closest low indices plane that is perpendicular to both the Burgers vector and the line direction is $\{1\bar{1}01\}$, so that the primary slip system at 1820°C is $\{1\bar{1}01\}<11\bar{2}0>$.

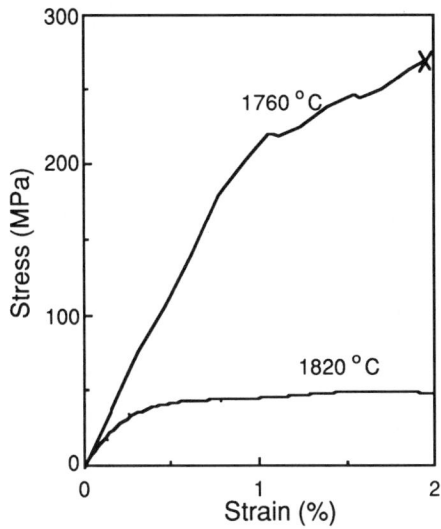

Fig.4 Stress strain curves of single crystal Si_3N_4

Some work has been published on dislocations in β-Si_3N_4. A Burgers vector of $<0001>$ was found for screw dislocations in as-sintered material[1-4]. However, the dislocations observed in the present work are formed during plastic deformation and are edge in character. α-Si_3N_4 has a structure formed by stscking two β-Si_3N_4 unit cells along the c-axis. Thus, the $<0001>$ Burgers vector is shorter in β-Si_3N_4(0.28nm) than in α(0.56nm) and has much smaller strain energy to form. Nevertheless, the $<0001>$ Burgers vector of α-Si_3N_4 is still shorter than $1/3<11\bar{2}0>$ and might be expected to be prefered for the compression axis chosen (Fig.1). Apparently, the observed $\{1\bar{1}01\}$ slip plane is relatively easy but this is not obvious from the crystal structure.

CONCLUSIONS

1) Hardness and fracture toughness were measured by the Vickers indentation method on the (0001) face of α-Si_3N_4 single crystals. The intrinsic fracture toughness of Si_3N_4 is found to be ~2.8MPa m$^{1/2}$, lower than in typical sintered Si_3N_4 containing elongated β grains and a secondary grain boundary phase.

2) Knoop hardness tests were conducted on (0001), (10$\bar{1}$0) and (11$\bar{2}$0) faces. The average hardness values were 2380kg/mm^2 and the variation was within 5%. No significant anisotropy was detected.

3) Si_3N_4 single crystals were deformed in compression at 1760 and 1820°C with a strain rate of 2.2x10^{-5}/s. Fracture occurred soon after yielding at 1760°C while plastic deformation occurred at relatively low stresses at 1820°C. TEM observations of dislocations showed that the primary slip system is $<11\bar{2}0>\{1\bar{1}01\}$.

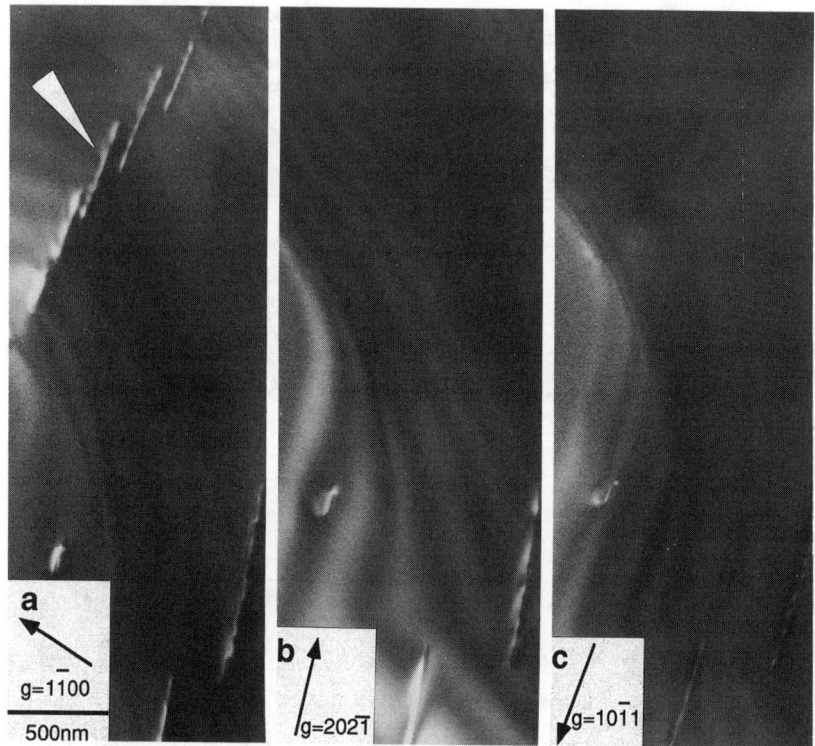

Fig.5 Dark field TEM image of dislocations in Si_3N_4 deformed at 1820°C to 2.7%. (a); $g=1\bar{1}00$, (b); $g=20\bar{2}1$, (c); $g=10\bar{1}1$. Dislocations shown by the arrow in (a) are out of contrast in both (b) and (c).

ACKNOWLEDGMENT

The high temperature compression tests were performed in the Research Laboratory for Nuclear Reactors, Tokyo Institute of Technology. The authors acknowledge Prof. T. Yano and Mr. M. Imai for help with the experiments. This work was supported by DOE-OBES.

REFERENCES

1. E. Butler, Phil. Mag., **21**, 829 (1971).
2. A. G. Evans and J. V. Sharp, J. Mater. Sci., **6**, 1291 (1971).
3. R. Kossowsky, J. Mater. Sci., 8 1603 (1973).
4. R. Kossowsky, D. G. Miller, and E. S. Diaz, J. Mater. Sci., **10**, 983 (1975).
5. R. A. Youngman and T. E. Mitchell, Rad. Effects, **74**, 267 (1983).
6. H. Suematsu, J. J. Petrovic and T. E. Mitchell, Proceedings of 50th Annual Meeting of the Electron Microscopy Society of America, pp.26-27 (1992).
7. H. Suematsu, Unpublished Data.
8. K. Niihara and T. Hirai, J. Mater Sci. Letters, **13**, 2276 (1978)
9. F. W. Daniels and C. G. Dunn, Trans. Am. Soc. Metals, **41**, 129 (1949).
10. C. A. Brooks, J. B. O'Neill and B. A. Redfern, Proc. Roy. Soc. London A., **322**, 73 (1971).

EVALUATION OF TENSILE STATIC, DYNAMIC, AND CYCLIC FATIGUE BEHAVIOR FOR A HIPed SILICON NITRIDE AT ELEVATED TEMPERATURES

CHIH-KUANG JACK LIN, MICHAEL G. JENKINS, and MATTISON K. FERBER
Metals and Ceramics Division, Oak Ridge National Laboratory, Oak Ridge, TN 37831-6064

ABSTRACT

Tensile fatigue behavior of a hot-isostatically-pressed (HIPed) silicon nitride was investigated over ranges of constant stresses, constant stress rates, and cyclic loading at 1150-1370°C. At 1150°C, static and dynamic fatigue failures were governed by a slow crack growth mechanism. Creep rupture was the dominant failure mechanism in static fatigue at 1260 and 1370°C. A transition of failure mechanism from slow crack growth to creep rupture appeared at stress rates $\leq 10^{-2}$ MPa/s for dynamic fatigue at 1260 and 1370°C. At 1150-1370°C, cyclic loading appeared to be less damaging than static loading as cyclic fatigue specimens displayed greater failure times than static fatigue specimens under the same maximum stresses.

INTRODUCTION

Ceramic materials offer significant potential for use as structural components in advanced heat engines. Designs with successful long-term performance require comprehensive information of the mechanical and thermal properties in these materials under various modes of loading at elevated temperatures where fatigue and creep resistances are major concerns. Creep rupture and slow crack growth (SCG) are typical failure mechanisms in ceramics under long-term loading at elevated temperatures [1]. For creep rupture, failure is usually associated with bulk deformation from extensive nucleation, growth, and coalescence of cavities and microcracks as a result of creep processes such as elemental diffusion or viscous flow. Alternatively, an SCG mechanism generally occurs by the extension of pre-existing flaws to final, critical sizes. However, localized creep damage in the vicinity of the crack tip may occur during SCG under certain conditions. The present study empirically characterizes the static, dynamic, and cyclic fatigue behavior of a HIPed silicon nitride in the elevated-temperature range of advanced heat engine designs. Direct comparisons of static, dynamic, and cyclic fatigue results are made to provide insight into the failure mechanisms in this material at elevated temperatures.

MATERIAL AND EXPERIMENTAL PROCEDURE

Test specimens were fabricated from a commercial HIPed silicon nitride designated PY6 (GTE Laboratories, Inc., Waltham, Mass., 1989-1990 vintage) which contained ~ 6 wt% yttrium oxide as the sintering aid. Uniaxial tension was employed for all the fatigue tests carried out at 1150, 1260, and 1370°C in ambient air. The geometry of the typical tensile specimen had an uniform gage section of 35 mm in length and 6.35 mm in diameter. In static fatigue tests, specimens were tested to failure under constant tensile stresses. In dynamic fatigue tests, specimens were failed at various stress rates: 3.7×10^1, 10^0, 10^{-1}, 10^{-2}, 10^{-3}, and 10^{-4} MPa/s. The stress-strain curve was also recorded for each dynamic fatigue test using a direct-contact extensometer which employed a remote capacitance sensor. Cyclic fatigue tests were conducted at a stress ratio R = 0.1 (R = minimum stress / maximum stress) with a symmetric trapezoidal wave form of 0.1 Hz (0.5 s ramp and 4.5 s dwell).

RESULTS AND DISCUSSION

Slow Crack Growth Model

In conventional ceramic fatigue analysis, an SCG model (with a power law relationship between crack growth rate and stress intensity factor) is often used to relate the static, dynamic, and cyclic fatigue behavior. With the assumption that all fatigue failures occur from the growth of initial flaws to final, critical sizes, the following relations can be obtained for uniform stress tests:

$$t_{sf} = \beta\sigma_s^{-N} \tag{1}$$

$$S_f^{N+1} = \beta(N+1)\dot{\sigma} \tag{2}$$

$$t_{df} = \beta(N+1)S_f^{-N} \tag{3}$$

$$t_{cf} = \beta\sigma_m^{-N}\tau\left\{\int_0^\tau [f(t)]^N dt\right\}^{-1} \tag{4}$$

In Eq. (1), t_{sf} is the time to failure for static fatigue under a constant stress, σ_s, β is a constant, and N is the crack growth exponent. Eq. (2) provides an indication of the stress rate ($\dot{\sigma}$) dependence of failure strength (S_f) in dynamic fatigue. Eq. (3) expresses the failure time for dynamic fatigue in terms of failure strength under a constant stress rate, $\dot{\sigma}$. Failure time in cyclic fatigue, t_{cf}, is given in Eq. (4) where σ_m is the maximum stress, τ is the cycle period, and f(t) is a function of time associated with the cyclic wave form such that the applied stress in a cycle can be described as $\sigma(t) = \sigma_m f(t)$. Details of the derivation of Eqs. (1)-(4) are given in Refs. 2-4.

The 'effective' times to failure for dynamic and cyclic fatigue can be defined as follows:

$$t_{df,eff} = t_{df}/(N+1) \tag{5}$$

$$t_{cf,eff} = t_{cf}\tau^{-1}\left\{\int_0^\tau [f(t)]^N dt\right\} \tag{6}$$

For equivalent values of S_f, σ_m, and σ_s, the 'effective' times to failure in dynamic and cyclic fatigue (Eqs. (5) and (6), respectively) are equal to the failure time in static fatigue (Eq. (1)) if slow crack growth is the primary mechanism for all the failures. Therefore, making comparison among the 'effective' times to failure in static, dynamic, and cyclic fatigue under the same maximum stresses will aid in assessing whether all fatigue failures are dominated by SCG.

Static Fatigue Behavior

Results of static fatigue tests are shown in a plot of log σ_s vs. log t_{sf} (Fig. 1). Linear regression analyses of the data (excluding the runout point) provided estimates of N as 22 at 1150°C, and 6.4 at 1260 and 1370°C. The static fatigue susceptibility of this HIPed silicon nitride increased with increasing temperature from 1150 to 1260°C as the value of N decreased from 22 to 6.4. However, static fatigue susceptibility did not change at temperatures greater than 1260°C because the values of N at 1260 and 1370°C were equal. The difference in N values implies a change of static fatigue failure mechanism between the temperatures of 1150 and 1260°C. Consistent with the high N value, SCG has been reported [5] as the dominant failure mechanism at 1150°C, while creep rupture was responsible for the low N values at 1260 and 1370°C. Furthermore, because the subcritical crack extension was presumably controlled by stress intensity factor, the failures at 1150°C were stress-controlled. On the other hand, at 1260 and 1370°C the creep-induced relaxation of crack-tip stresses precluded failure by a stress-controlled mechanism. In this case, failure resulted when the strain associated with creep damage reached a critical value in accordance with a Monkman-Grant relationship (i.e., strain-controlled failure). Detailed discussions concerning the stress and temperature sensitivities of the secondary (or minimum) creep rate and other creep behavior have been addressed previously [5].

Dynamic Fatigue Behavior

Fig. 2 illustrates the dynamic fatigue results in a log-log plot of S_f vs. $\dot{\sigma}$. These results show a continuous relationship of decreasing failure strength with decreasing stress rate at 1150°C, giving an N value of 98. At 1260 and 1370°C, this continuous relationship is interrupted at a stress rate between 10^{-2} and 10^{-1} MPa/s. The N values at stress rates >10^{-2} MPa/s were estimated as 24 at 1260°C and 7.1 at 1370°C. The N values at stress rates ≤10^{-2} MPa/s were estimated as 18 and 3.9 at 1260 and 1370°C, respectively. This shift of N values

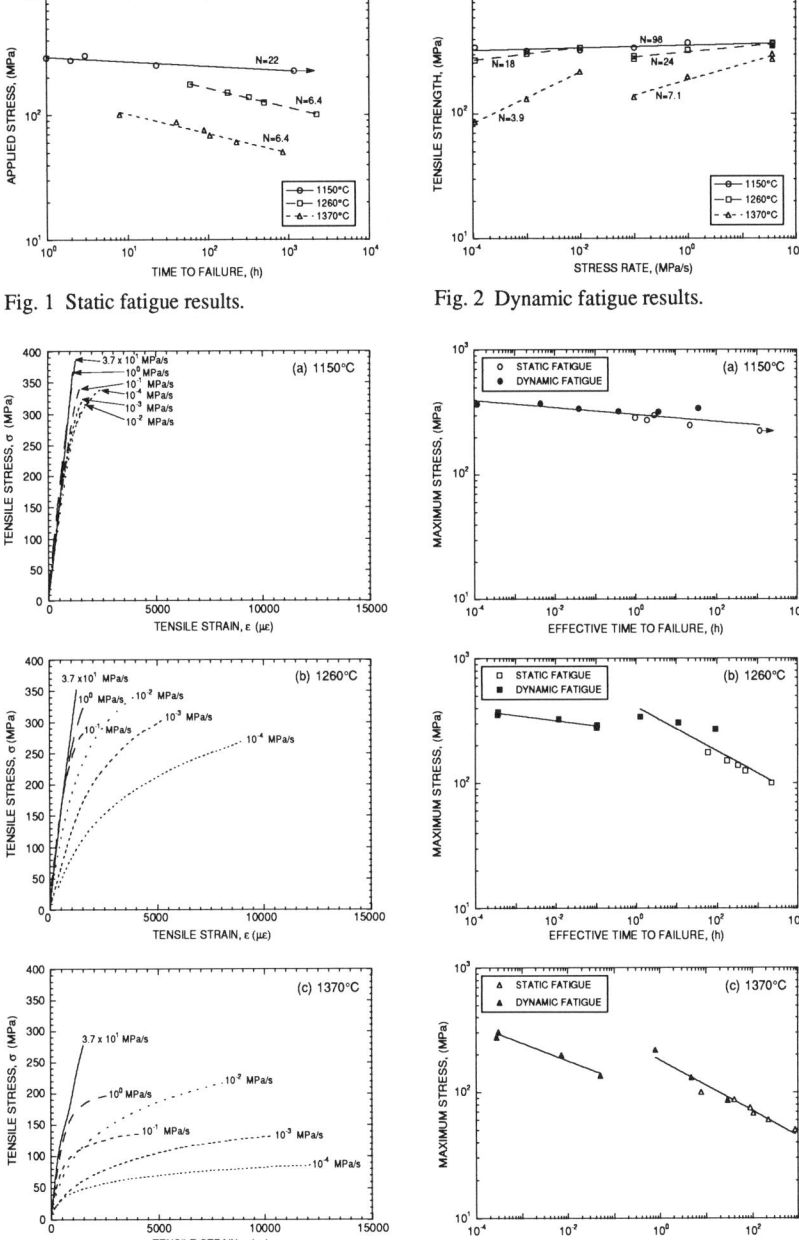

Fig. 1 Static fatigue results.

Fig. 2 Dynamic fatigue results.

Fig. 3 Stress-strain curves in dynamic fatigue tests.

Fig. 4 Comparison of static and dynamic fatigue results.

implies a change of dominant failure mechanism from SCG to creep rupture as stress rate decreased. This change in failure behavior is also evident in the stress-strain curves shown in Fig. 3. At 1150°C, all the stress-strain curves exhibit extensive linear behavior (with nearly identical slopes) becoming nonlinear just prior to failure reflecting compliance changes due to SCG. Similar stress-strain behavior was also present at 1260 and 1370°C for stress rates $>10^{-2}$ MPa/s. However, creep deformation (accumulation of creep cavities and microcracks) was apparent at 1260 and 1370°C for stress rates $\leq 10^{-2}$ MPa/s, as denoted by pronounced nonlinearity in the stress-strain curves.

Comparison of dynamic and static fatigue results is shown in log-log plots of maximum stress vs. 'effective' time to failure (Fig. 4). Good agreement between sets of static and dynamic fatigue data at 1150°C is illustrated in Fig. 4(a) by a single straight line. Thus, similar SCG failure mechanisms [5] are implied for dynamic and static fatigue at 1150°C. This is consistent with the assumption of the 'effective' time approach. At 1260 and 1370°C, all static and dynamic fatigue data could not be represented by a single straight line as shown in Figs. 4(b) and (c). Two straight lines were needed to reasonably fit all the data in two regions. The dynamic fatigue 'effective' failure times with stress rates $\leq 10^{-2}$ MPa/s were comparable with the static fatigue failure times (right sides of Figs. 4(b) and (c)) indicating similar failure mechanisms. Since failures of static fatigue specimens tested at 1260 and 1370°C were attributed to creep rupture [5], the dominant failure mechanism in dynamic fatigue with stress rates $\leq 10^{-2}$ MPa/s at 1260 and 1370°C is also creep rupture. These agreements imply that the growth of creep damage zone for this silicon nitride at 1260 and 1370°C might be of a power law relationship equivalent to that in the SCG model. Therefore, the above 'effective' time approach is applicable to provide first approximation for comparison of dynamic and static fatigue results. However, microstructural analyses (to be described) are necessary to verify these agreements in 'effective' time approach. The controlling failure mechanism in dynamic fatigue with stress rates $>10^{-2}$ MPa/s at 1260 and 1370°C was SCG as indicated by the greater N values for short failure times compared to those for long failure times. Although, unlike the results at 1370°C, all dynamic fatigue data at 1260°C could be fitted with a single straight line, neither the previously described macroscopic stress-strain relations nor the microstructural analyses results (to be described) could support this type of fit. Note that the stress-strain curves at 1260°C exhibited considerable nonlinearity for stress rates $\leq 10^{-2}$ MPa/s and remained nearly linear for stress rates $>10^{-2}$ MPa/s. Fractography also indicated extensive creep cavities in specimens tested at stress rates $\leq 10^{-2}$ MPa/s but not in those tested at stress rates $>10^{-2}$ MPa/s at both 1260°C and 1370°C (Figs. 5(a) and (b)). In general, 1260°C may be the transition temperature for the change of dynamic fatigue failure mechanism from a single mechanism (SCG) to dual mechanisms (SCG or creep rupture, depending on stress rate) for this HIPed silicon nitride.

Cyclic Fatigue Behavior

Results of cyclic fatigue testing are illustrated in a log-log plot of maximum stress vs. number of cycles to failure (Fig. 6). Tensile strength (obtained at a stress rate of 37 MPa/s) was plotted as a single cycle to failure. As seen in Fig. 6, number of cycles to failure increased with decreasing stress level and cyclic fatigue resistance decreased with increasing temperature. Cyclic and static fatigue results are compared in log-log plots of maximum stress vs. 'effective' time to failure (Fig. 7). Note that the levels of maximum applied stress required to generate failures within 1 to 1000 hours were greater in cyclic fatigue than in static fatigue. In other words, the 'effective' failure times under cyclic loading were longer than the static fatigue lifetimes with the same maximum applied stresses. This difference in lifetimes becomes greater when the comparison is made using the actual cyclic fatigue failure times. Figs. 7(a)-(c) show the difference between static and cyclic fatigue lifetimes decreasing with increasing stress level. Although a direct comparison could not be made at high applied stresses (failure time less than 1 hour) due to the absence of static fatigue data, both static and cyclic fatigue failures, at stress levels near the fast fracture strength, would result from SCG of a dominant inherent flaw. Therefore, the cyclic 'effective' lifetimes would be closer to the static fatigue lifetimes at higher stresses. Such a transition of dominant failure mechanism from creep rupture to SCG in static fatigue at elevated temperatures has been shown in some ceramics (e.g., Refs. 6-7). However, Figs. 7(a)-(c) indicate that cyclic and static fatigue mechanisms were different and that the SCG model could not describe the cyclic fatigue results based on the assumption that cyclic fatigue was the cyclic manifestation of static fatigue. In addition, indications are that cyclic fatigue mechanisms were not as damaging as static fatigue mechanisms for the given stress-temperature

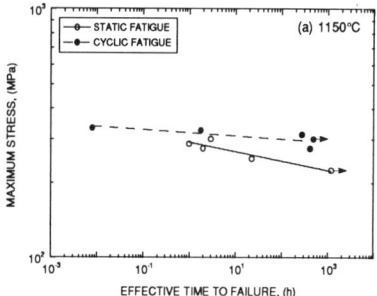
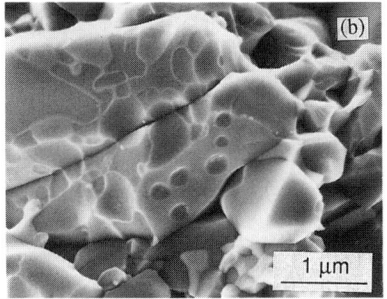

Fig. 5 SEM micrographs of dynamic fatigue specimens at 1370°C: (a) 10^{-1} MPa/s, (b) 10^{-3} MPa/s.

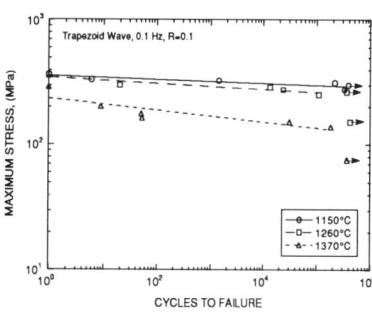

Fig. 6 Cyclic fatigue results.

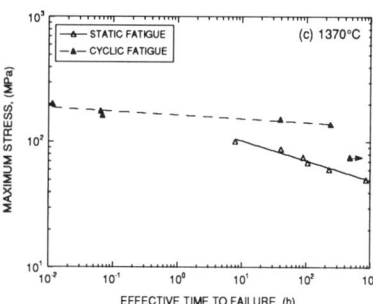

Fig. 7 Comparison of static and cyclic fatigue results.

conditions in this study. Such behavior at elevated temperatures is opposite of the room-temperature fatigue behavior of some ceramics where cyclic loading degrades strength more than static loading (e.g. Refs. 8-9).

Fractography of cyclic fatigue specimens did not display such extensive creep cavities as observed in the static fatigue specimens tested at 1260 and 1370°C. Failures of cyclic fatigue at 1150°C were related to pre-existing defects. Even though both static and cyclic fatigue failures at 1150°C were due to subcritical crack extension of pre-existing flaws, the stress-life plot implies that crack propagation rates were less in cyclic loading than in static loading. The lesser damage caused by cyclic fatigue relative to static fatigue in ceramics at elevated temperatures might be related to the viscous grain-boundary phases [10-13]. For example, the effects of bridging the crack surfaces by the viscous grain-boundary phases could be more pronounced in cyclic loading than in static loading under certain conditions [13]. Although the exact cyclic fatigue mechanisms at 1260 and 1370°C are not clear at present, unloading in each cycle may play an important role in either suppressing development of creep damage or retarding accumulation of creep deformation. Possible mechanisms include relaxation of local stress concentration at potential sites for creep damage development and recovery of creep deformation during unloading. More research, including measurement of strain-time relation, high-frequency testing, and microstructural analyses, is needed to identify the mechanisms and variables responsible for the cyclic fatigue behavior. The influences of other factors, such as the devitrification of secondary phases and the environmental effects (e.g., oxidation), on the fatigue behavior at different modes of loading also need to be examined.

CONCLUDING REMARKS

The tensile static, dynamic, and cyclic fatigue behavior of PY6 silicon nitride could not be well related by an SCG model, particularly at temperatures ≥ 1260°C. Caution is advised in applying the SCG model to predict stress-life relation for dynamic or cyclic fatigue based on static fatigue data. This is especially true at elevated temperatures where failure mechanisms other than SCG, such as creep processes and visco-elastic effects, may exist. Cyclic loading was apparently less damaging than static loading at 1150-1370°C, suggesting that static fatigue may be the major consideration in determining the long-term mechanical reliability of ceramics at elevated temperatures. However, more research is necessary to develop effective fatigue design methodologies for use in high performance ceramic structural components at elevated temperatures.

ACKNOWLEDGMENTS

Research sponsored by the Ceramic Technology Project, DOE Office of Transportation Technologies, under contract DE-AC05-84OR21400 with Martin Marietta Energy Systems, Inc. C.-K. J. Lin was supported in part by an appointment to the Oak Ridge National Laboratory Postdoctoral Research Program administered by the Oak Ridge Institute for Science and Education. The authors thank Drs. B. Gieseke and K. Liu for reviewing the manuscript.

REFERENCES

1. S.M. Wiederhorn and E.R. Fuller, Jr., Mater. Sci. Eng. **71**, 169 (1985).
2. S.M. Wiederhorn, in Fracture Mechanics of Ceramics, Vol. 2, edited by R.C. Bradt et al. (Plenum Press, New York, 1974), p. 623.
3. J.E. Ritter, in Fracture Mechanics of Ceramics, Vol. 4, edited by R.C. Bradt et al. (Plenum Press, New York, 1978), p. 667.
4. A.G. Evans and E.R. Fuller, Met. Trans. **5**, 27 (1974).
5. M.K. Ferber and M.G. Jenkins, J. Am. Ceram. Soc. **75**, 2453 (1992).
6. P.F. Becher, P. Angelini, W.H. Warwick, and T.N. Tiegs, J. Am. Ceram. Soc. **73**, 91 (1990).
7. G.D. Quinn, J. Mater. Sci. **25**, 4361 (1990).
8. S. Horibe, J. Eur. Ceram. Soc. **6**, 89 (1990).
9. R.H. Dauskardt, D.B. Marshall, and R.O. Ritchie, J. Am. Ceram. Soc. **73**, 893 (1990).
10. T. Fett, G. Himsolt, and D. Munz, Adv. Ceram. Mater. **1**, 179 (1986).
11. L.X. Han and S. Suresh, J. Am. Ceram. Soc. **72**, 1233 (1989).
12. C.-K.J. Lin and D.F. Socie, J. Am. Ceram. Soc. **74**, 1511 (1991).
13. C.-K.J. Lin, D.F. Socie, Y. Xu, and A. Zangvil, J. Am. Ceram. Soc. **75**, 637 (1992).

HIGH TEMPERATURE FATIGUE PROPERTIES OF SILICON NITRIDE IN NITROGEN ATMOSPHERE

YASUHIRO SHIGEGAKI, TAKASHI INAMURA, AKIHIKO SUZUKI AND TADASHI SASA
Ishikawajima-Harima Heavy Industries Co., Ltd., Research Institute, 3-1-15 Toyosu, Koto-ku, Tokyo 135 Japan

ABSTRACT

Cyclic and static fatigue properties of pressure-less sintered silicon nitride were evaluated at 1000°C in air and in nitrogen using four-point bending mode. The data of cyclic fatigue tests or static fatigue tests and the morphology of the fractured surfaces in nitrogen were compared with those in air. The cyclic fatigue behavior was remarkably influenced by the atmosphere, while the static fatigue was less influenced. Crack healing effect due to the oxidation around the crack are thought to be the most probable mechanism to affect the cyclic fatigue rate in air.

INTRODUCTION

Silicon nitride materials have excellent mechanical properties, such as, high strength, high fracture toughness and high thermal shock resistance. Active research and development have been conducted to apply these materials to high temperature machineries, such as gas turbines. To evaluate the reliability of the structural components, it is necessary to clarify the fatigue properties of the material. High temperature fatigue properties are especially important in silicon nitrides.

The high temperature fatigue properties of silicon nitrides are thought to be mainly controlled by slow crack growth or creep phenomena and the dominant degradative mechanism may change according to the applied stress and the temperature[1].

Slow crack growth phenomena usually starts from an inclusion, a pore or a surface flow which originates from the fabrication process, and is known to be influenced by the atmosphere, such as humidity or oxygen. In several ceramics, stress corrosion cracking phenomena around the crack tip due to water or oxygen has been reported to affect the crack propagation rate under the static or cyclic loading. Such phenomena can also take place in silicon nitrides. It has been suggested that oxidation of the grain or the grain boundary phase of silicon nitride at high temperature can generate both beneficial and degradative effects on the fatigue properties, depending on the degree of oxidation[2-5]. However, the evaluation of the effect of the oxidizing atmosphere has not yet been carried out satisfactorily, especially for both the static and the cyclic fatigue behavior in silicon nitrides.

The purpose of the present study is to evaluate the static and cyclic fatigue properties of a pressure-less sintered silicon nitride in air and in nitrogen at 1000°C along with the analysis of the effect of gas atmosphere on the slow crack growth fatigue mechanism.

EXPERIMENTAL PROCEDURE

Specimen

The typical mechanical properties of the pressure-less sintered silicon nitride are summarized in Table I. This material maintains its fast fracture strength up to 1200°C.

All fatigue tests were carried out by four-point bending mode. The self-aligned silicon carbide fixture was used with the inner span and the outer span of 30 and 50 mm, respectively. The dimensions of the specimen are shown in Figure 1. The tensile surface of the specimen has curvature to ensure the fracture at the central portion of the specimen. The stress concentration factor due to this curvature is 1.017, which can be ignored when the applied stress is calculated.

Table I Typical mechanical properties of the material

4-Point Bending Strength	RT	970 MPa
	1000°C	970 MPa
	1200°C	900 MPa
Fracture Toughness		6 MPam$^{1/2}$
Young's Modulus		300 GPa
Hardness		15 GPa

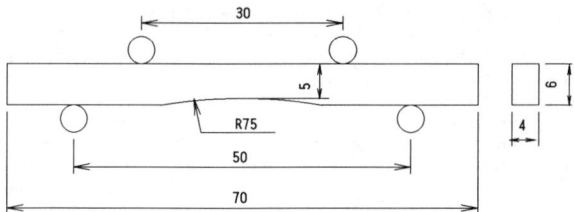

Figure 1 The dimensions of the specimen
(dimension in mm)

Testing Condition

The facilities for testing in nitrogen were equipped with a furnace system of controlled atmosphere, which consists of W-mesh type heater and vacuum or inert gas supplier system. The load was applied by a dead weight in the case of static fatigue tests and by a hydraulic actuator in the case of cyclic fatigue tests, respectively, after the temperature reached the desired constant value. The cyclic loading was applied with a sine wave of 40 Hz and the ratio of the minimum and maximum stress (R) was 0.1. Tests in air were also carried out under the same condition using similar facilities with a different type of heating device.

All tests were carried out at 1000°C because only the slow crack growth mechanism is dominant at this temperature for the material used in the present study (as shown in the fast fracture data (Table I)).

RESULTS AND DISCUSSION

Static Fatigue

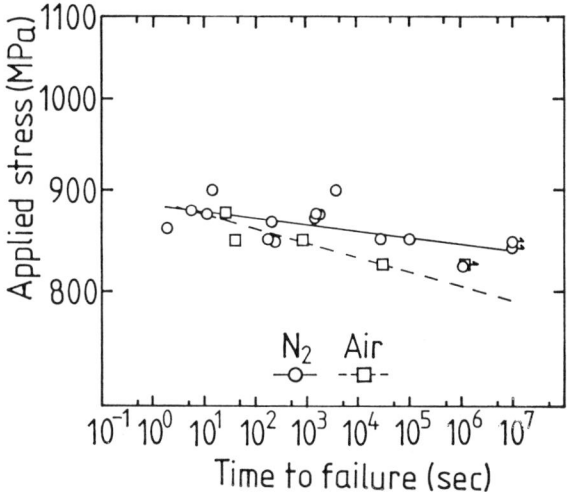

Figure 2 Static fatigue properties in air and in nitrogen at 1000°C

Figure 2 shows the plot of the time to failure with the applied stress in air and nitrogen at 1000°C, in which the arrows indicate the tests were terminated without failure. The lines represent the power-low relationship determined with the least square method.

The data in air and nitrogen are considered to be close through about 10^{-3} sec, but after prolonged time, the time to failure in nitrogen are longer than those in air at the same stress level. These facts are similar to the reported static fatigue results[3] of silicon nitride at 1400°C in the slow crack growth region as well as the stress corrosion effects, as seen in K_I-V diagrams for glass or alumina at room temperature.

Cyclic Fatigue

The relationship between the number of cycles to failure and the maximum stress in air and nitrogen at 1000°C is shown in Figure 3. The symbols are similar to those in Figure 2. In contrast to the static fatigue behavior, the cyclic fatigue specimens tested in air displayed longer failure times than those tested in nitrogen. These results indicate that the oxidizing environment produced a beneficial effect on the cyclic fatigue behavior.

Fracture Surface Morphology

SEM micrographs of the typical fracture surfaces of the static and cyclic fatigue specimen tested in nitrogen are shown in Figures 4 and 5, respectively. The micrographs of the specimens tested in air are not shown, but the morphology was similar to that in nitrogen. Almost all fracture origins were pores near (about 50 um below) the tensile surface. The flat

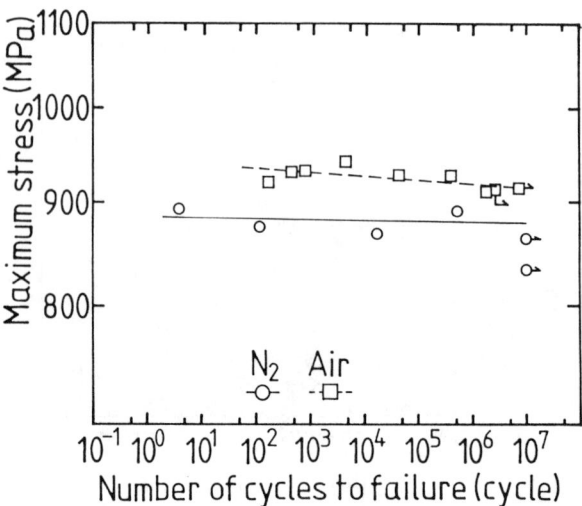

Figure 3 Cyclic fatigue properties in air and in nitrogen at 1000°C (40Hz, R=0.1)

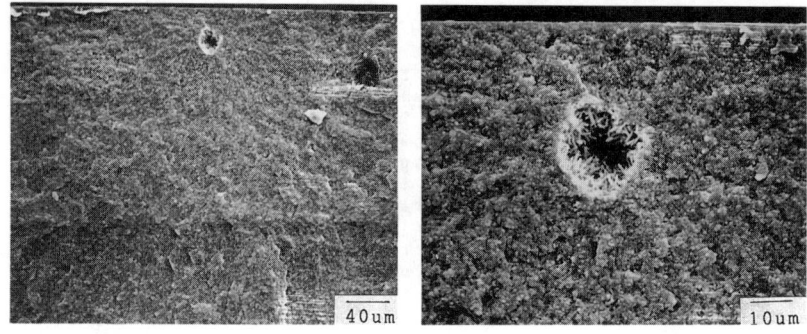

Figure 4 SEM photograph of the fracture surface of the static fatigue specimen in nitrogen (900MPa, 3780sec)

regions around the pores, which can be easily observed in both micrographs, were extended to the tensile surface. The high magnification micrograph indicated that cracks propagated intergranualy in both of the flat and the rough regions. Because of the detection of these flat area for crack propagation and no observation of creep damage in the survived specimens, this material demonstrated slow crack growth as the dominant failure mechanism at 1000°C in these stress regions, regardless of the loading manner and the atmosphere.

The Effect of Atmosphere

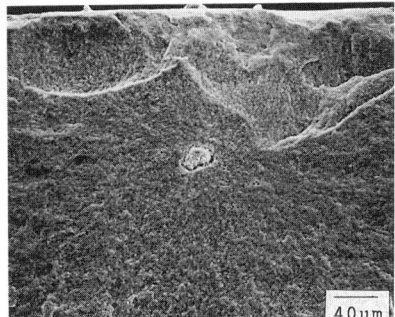 

Figure 5 SEM photograph of the fracture surface of the cyclic fatigue specimen in nitrogen (868MPa, 16759cycles)

To clarify the effect of the atmosphere on the cyclic fatigue behavior, the relations between the maximum stress and the cycle number in both air and nitrogen were calculated from the corresponding static fatigue data using Equation (1).

$$\sigma_f = B\, t_{eff}^{1/n} \qquad (1)$$

- $t_{eff} = t_f$ \hspace{2em} (static)

- $t_{eff} = \dfrac{t_f}{2\pi} \displaystyle\int_0^{2\pi} \left\{ \dfrac{1}{2}(1+R) + \dfrac{1}{2}(1-R)\sin\theta \right\}^n d\theta$ \hspace{1em} (cyclic)

The calculated results are shown in Figure 6 with the experimental data. The calculated results in nitrogen are considered to be close to the experimental data, while in air the estimated and experimental results are different at low stresses.

The oxidation rate of this material at 1000°C is quite low (typical weight gain is less than 0.1 mg/cm² even after 1000 hours). The crack propagates, however, along the intergranualar secondary phase, which can be more easily oxidized than the matrix. In such a case, oxidation is considered to affect the high temperature slow crack growth behavior with two kinetically competitive mechanisms. One is the stress corrosion effect[2], which promotes the propagation of the cracks, and the other is the crack healing effect due to the oxidation[5].

Generally the crack healing can take place only at lower stress intensity at the crack tip, especially below a certain threshold value, while at higher stress intensity only the crack growth take places. Therefore, the crack healing is only dominant in the case of the cyclic fatigue, because the cyclic loading provides the intervals of the low stress intensity which enables the crack healing. On the contrary, in the case of the static loading, the stress intensity is never decreased below the threshold. It is therefore considered that the cyclic fatigue rate in air was suppressed by the crack healing due to the oxidation on the subcritical crack surface.

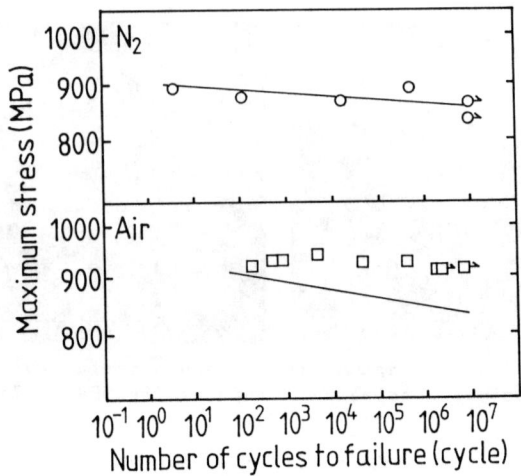

Figure 6 The comparison of the calculated cyclic fatigue curves with the experimental data

CONCLUSION

The static fatigue data in nitrogen were fairly consistent with those in air, but the cyclic fatigue data in nitrogen were positioned at lower stress side. The cyclic fatigue data in nitrogen agreed with the slow crack model prediction, but the data in air did not. The deviation in air is considered to be caused by the healing of the subcritical cracks.

ACKNOWLEDGEMENT

This work was performed under the management of the Engineering Research Association for High Performance Ceramics as a part of the R&D Project of Basic Technology for Future Industries supported by NEDO (New Energy and Industrial Technology Development Organization).

REFERENCE

1. T.Makino,Y.Nakasuji,M.Masuda and M.Matsui,in Proceedings of the 1st International Symposium on the Science of Engineering Ceramics, edited by S.Kimura and K.Niihara (Jap. Ceram.Soc.,Koda,Japan,1991),pp.189-194.
2. R.E.Tressler,N.Y.Jia,Z.Zheng and H.Takahashi,ibid,pp.167-176.
3. O.Van der Biest and C.Weber,in Ceramic Transactions, Vol.10,Corrosion and Corrosive Degradation of Ceramics, edited by R.E.Tressler and M.McNallan (Am.Ceram.Soc., Westerville,OH,1990),pp.129-139.
4. T.E.Easler,R.C.Bradt,R.E.Tressler,J.Am.Ceram.Soc. 64,12, 731(1981)
5. S.R.Choi,V.Tikare and R.Pawlik,Ceram.Eng.Sci.Proc. 12,9-10, 2190(1991)

CAVITY EVOLUTION DURING TENSILE CREEP OF Si_3N_4

WILLIAM LUECKE , S. M. WIEDERHORN , B. J. HOCKEY and G. G. LONG*
*National Institute of Standards and Technology, Gaithersburg, MD 20899

ABSTRACT

We have characterized the evolution of cavities during tensile creep of a Y_2O_3-hot isostatically pressed Si_3N_4, using precision density measurements, small-angle x-ray scattering (SAXS) and transmission electron microscopy (TEM). The cavities are bimodally distributed in size. Lenticular, 200 nm-size cavities are common, and lie primarily on two-grain boundaries. Irregularly shaped 500-1000 nm-size cavities are rare and lie at multi-grain junctions, but comprise approximately half of the total volume fraction of cavities. Although the material shows a continuous decrease in strain rate with strain, the cavity volume fraction evolves linearly with strain. Cavities account for approximately 85% of the total strain at any point during creep.

INTRODUCTION

Under low stress, high temperature deformation silicon nitride ceramics often fail by creep rupture [1, 2, 3]. The term creep rupture encompasses the generation and accumulation of creep damage, in the form of cavities. As the number and size of the cavities increase with strain, there comes a time when they coalesce to form a crack that rapidly (on the time-scale of the failure time) propagates, causing the specimen to fail [4, 5]. To design creep-rupture resistant ceramics, it is necessary to understand the microstructural origins of the creep damage that limit the lifetime under load. To this end, we have brought a number of analytical techniques to bear on a single silicon nitride, deformed in tension. We have characterized the creep response and lifetime of the material as a function of stress and temperature. Using a precision density measurement technique, we have measured the total cavity volume fraction in failed specimens, as well as that from specimens interrupted before failure. From small-angle x-ray scattering measurements (SAXS), we have obtained the size distribution of cavities in several representative specimens. Finally, we have done transmission electron microscopy to ascertain the location of the life-limiting cavities, as well as confirm the SAXS results. This note summarizes our observations on the creep rupture process of this silicon nitride.

EXPERIMENTAL DETAILS

This study tested a single, commercially-available silicon nitride[1] It is a hot-isostatically-pressed material, using 4 w/o yttria as a densifying aid. The microstructure is a bimodal size distribution of $\beta - Si_3N_4$ grains, with many equiaxed submicrometer-sized grains surrounding micrometer and larger acicular grains. Many of the acicular grains have aspect ratios greater than five. The densification aid is nearly fully crystallized in the as-received material; pockets of $Y_2Si_2O_7$ and $Y_5(SiO_4)N$ lie at multi-grain junctions. In contrast, however, a thin (0.5-1.5 nm) layer of amorphous material separates the two-grain boundaries.

Reference [6] describes the creep-testing apparatus and method in detail. A pneumatic bellows assembly applied the load to the specimen through SiC pull-rods. The creep specimens are dog-bone shaped, with a 2 x 2.5 mm cross-section and approximately 14 mm gauge length. They are attached to the pull-rods with SiC pins. A laser extensometry system[2] measures the

---

[1] Norton/TRW NT154. NT154 is a commercial designation for a material made by Norton/TRW Co. The use of commercial designations or company names does not indicate endorsement by the National Institute of Standards and Technology

[2] Zygo Corporation, Middlefield, CT 06455

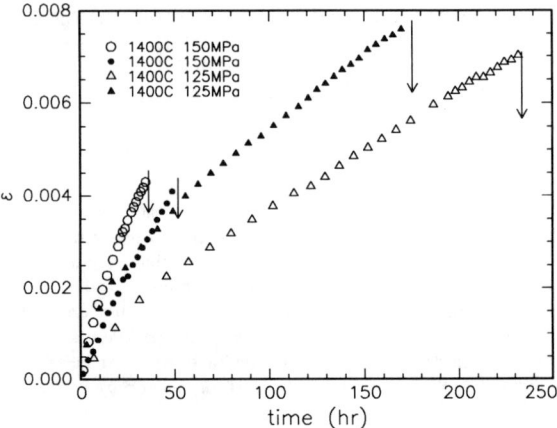

Figure 1: Creep curves for silicon nitride deformed at 1400°C.

displacement of two silicon carbide flags hung from the gauge length of the specimen. Although the extensometer system records the distance between the two flags very precisely, the technique introduces an unavoidable uncertainty of about 10% in the true gauge length of the specimen.

The stresses in this study lie in the range $100\text{MPa} < \sigma < 175\text{MPa}$ and the temperatures in the range $1370°C < T < 1430°C$. These conditions produce strain rates on the order of $10^{-8}s^{-1}$ and lifetimes between 1 and 1500 hours. Each experiment used only one deformation condition, and, unless noted, all specimens crept until failure.

During creep cavitation, the density of the gauge section decreases. Accordingly, we used a sink-float technique to measure the density change of the gauge length after creep, and indirectly, the volume fraction of cavities. From each creep specimen we cut a size-matched pair of samples: one from the gauge length and another from the grip end. Grinding 50 μm of material from each face, except the fracture surface, removed the oxide layer that would have interfered with accurate determination of the density. The density measurement apparatus is a modification of the one described by ASTM standard C729[7]. A constant temperature bath circulates water around a jacketed cylinder containing a mixture of methylene iodide and tetrabromoethane, whose density is approximately equal to that of the silicon nitride in question (approximately 3.2355g/cm^3). By increasing the bath temperature until the specimens sink, and then decreasing it until they float again, it is possible to bracket the neutral buoyancy temperature within 0.1°C. During each run we calibrated the coefficient of expansion of the fluid with a pair of standards of known density.[3] Over the limited range of temperatures the density of the solution changes linearly with temperature, so the density of the Si_3N_4 specimens can be calculated by interpolation. The technique renders densities accurate to ± 0.0005 g/cm^3. Under the assumption that any chemically-induced changes in density occur in both the grip and gauge sections equally, the volume fraction of cavities in the gauge section, f_v, is given by

$$f_v = \frac{\rho_{\text{grip}} - \rho_{\text{gauge}}}{\rho_{\text{grip}}} \quad (1)$$

RESULTS AND DISCUSSION

Figure 1 shows several creep curves for the material and illustrates a number of features generic to all the creep curves. The material shows a distinct, continuous decrease in strain rate

---

[3]Cargille Laboratories, Inc. Cedar Grove, NJ 07009

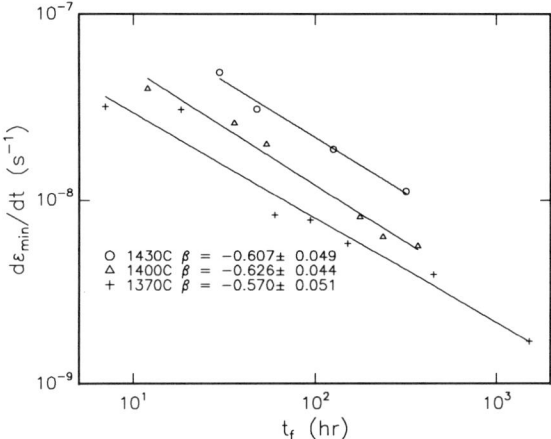

Figure 2: Monkman-Grant plot for silicon nitride

with time. Although one of the curves shows what might be a "secondary" or "steady-state" creep rate, this phenomenon is by no means universal for the material. The ratio of initial to final creep rates was generally around 10, but showed no systematic variation with either stress or temperature. Typical failure strains were 0.1% at the highest stresses and 1% at the lowest stresses. Plotting strain rate at failure as a function of stress revealed stress exponents of about 7. The temperature dependence of the creep rate, calculated for 125MPa and strain rate at failure, was 1230 ± 110 kJ/mol.

Failure occurs suddenly, without any evidence for tertiary creep. In all cases, the fracture surface indicated a large (5-25% of the total cross-sectional area) region of slow crack growth, the area decreasing with increasing stress. This region always extended in from the surface, usually from a corner. Most of the specimens failed at or near the end of the gauge length, probably due to the stress concentration caused by the small radius of curvature from the gauge length to the grip-end flange.

Figure 2 summarizes the creep data in the form of a modified Monkman-Grant [8] plot: minimum creep rate as a function of lifetime. Unlike many materials, which show a master curve independent of temperature on such a plot, the data for this silicon nitride fall into three distinct temperature strata. The equation,

$$\frac{d\epsilon_{\min}}{dt} = C_{\mathrm{MG}} t_f^\beta, \qquad (2)$$

fits the data well. The Monkman-Grant exponent, β, is independent of temperature. This result is in direct contradiction with results obtained on an earlier vintage NT154 [9], where no such temperature stratification occurred.

Figure 3 summarizes the results of the investigation into the level of creep damage at failure by plotting the volume fraction of cavities as function of failure strain. In the figure, the vertical error bars arise primarily from uncertainties in the densities of the calibration standards. The horizontal error bars estimate the uncertainty in the failure strain. The temperature stratification of the Monkman-Grant plot, Figure 2 disappears in this curve: the volume fraction of cavities at failure is a constant fraction of the failure strain regardless of temperature, stress or failure strain. From the volume fraction of cavities it is possible to calculate the strain due to the cavities. Assuming that all the material displaced from the cavities contributes to extension of the gauge length, the cavitation strain, ϵ_{cav}, is equivalent, for small strains, to the volume fraction of cavities, f_v. Given this relation, Figure 3 shows that about 85% of the strain to failure originates from the cavities. The dashed line in Figure 3 represents the case where all the strain

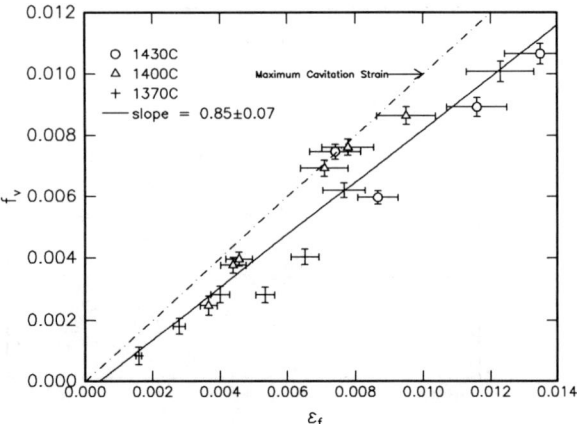

Figure 3: Volume fraction of cavities at failure as a function of failure strain

to failure originates from the cavitation.

To determine the nature of the evolution of creep damage with strain, we tested a suite of specimens at constant temperature and stress to different strains, interrupting the tests before failure. Figure 4 shows that the volume fraction of cavities (and hence the cavitation strain, as defined in the previous paragraph) evolves linearly with total strain. The solid data points in Figure 4 are the cavitation strains for specimens interrupted before failure. They lie on top of the best-fit line for the cavitation strain–failure strain plot, Figure 3. The open circles are the volume fractions of cavities for three specimens tested to failure under the same conditions. For clarity, the error bars are omitted The large primary creep rate in this material is, therefore, controlled by the cavitation process.

We determined the sized distribution of cavities in several specimens, previously tested to failure, by small-angle x-ray scattering (SAXS) at beamline X23A3 of the National Synchrotron Light Source. To calculate the volume fraction distribution of cavities in the gauge length, we first calculated the volume fraction distribution of all scatterers using a maximum entropy technique [10, 11][4] for each grip/gauge matched pair. The calculation assumes spherical scatterers, which is not an unreasonable approximation, as TEM results will show. The volume fraction distribution of cavities in the gauge section is the difference between the distributions of all scatterers in the grip and gauge sections. Figure 5 shows that there are two sizes of cavities in the material at failure. Numerically more prevalent are the 200 nm size cavities. The 600 nm size cavities occur only 1/30th as frequently, but comprise nearly half the total volume fraction.

Transmission electron microscopy confirms the results of the SAXS analysis. Figure 6 typifies the microstructure of specimens deformed at both 1370°C and 1430°C. The larger cavities indicated from the SAXS analysis occur at multi-grain junctions, and are irregularly shaped. The smaller cavities occur on two-grain boundaries, and are lenticularly shaped. A number of the small cavities cluster on any cavitated boundary, but many boundaries remain uncavitated.

SUMMARY

- The volume fraction of cavities evolves linearly with strain, even though the creep rate decreases continuously with strain.

---

[4]Using the computer program MAXE of UKAEA Harwell

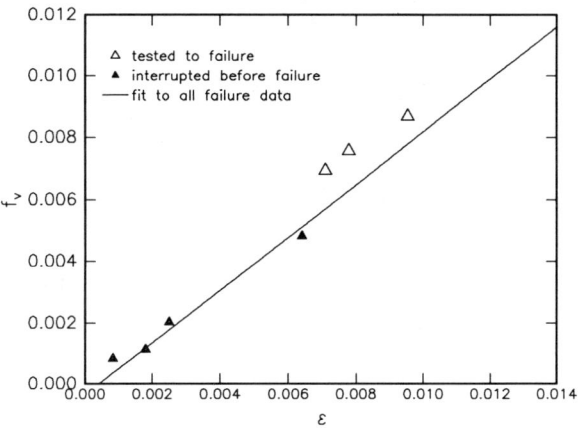

Figure 4: Cavity volume fraction evolution with strain for Si_3N_4 deformed at 1400° and 125MPa.

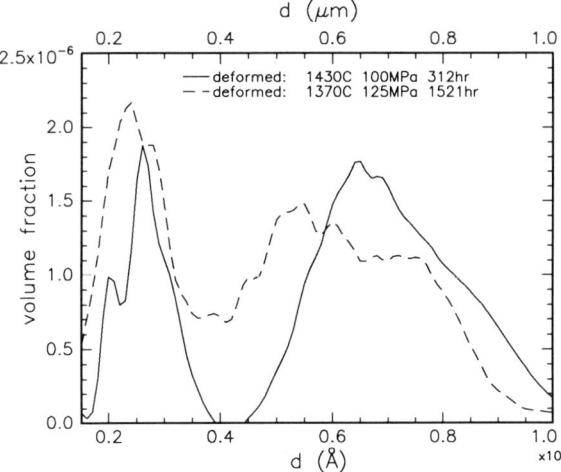

Figure 5: Distribution of cavities in two specimens tested to failure

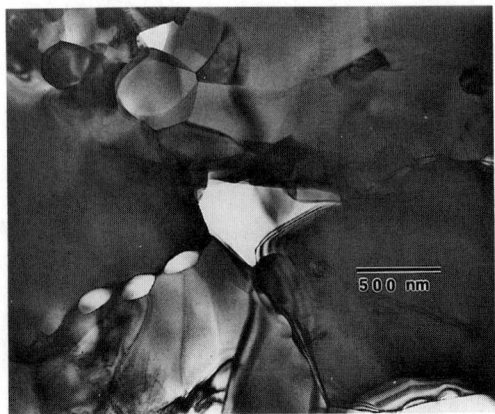

Figure 6: Transmission electron micrograph from the gauge length of the specimen that failed after 1521 hr under 125 MPa at 1370°C. The tensile axis is vertical.

- Cavitation accounts for approximately 85% of the strain at any point during creep.
- For this silicon nitride there are two types of cavities:
    1. Lenticular, 100-200 nm cavities located on two-grain boundaries
    2. Irregular, 500-1000 nm cavities located at multi-grain junctions

REFERENCES

1. M. K. Ferber and M. J. Jenkins, J. Am. Ceram. Soc. **75**, 2453 (1992).

2. S. M. Wiederhorn, B. J. Hockey, D. C. Cranmer, and R. Yeckley, submitted to J. Mater. Sci. (1991).

3. R. Kossowsky, D. G. Miller, and E. S. Diaz, J. Mater. Sci. **10**, 983 (1975).

4. A. G. Evans and A. Rana, Acta metall. **28**, 129 (1980).

5. S. M. Wiederhorn, B. J. Hockey, and T.-J. Chuang, Creep and Creep Rupture of Structural Ceramics, in *Toughening Mechanisms in Quasi-Brittle Materials*, edited by S. P. Shah, pages 555-576, Kluwer Academic Publishers, 1991.

6. Daniel F. Carroll and Sheldon M. Wiederhorn and D. E. Roberts, J. Am. Ceram. Soc. **72**, 1610 (1989).

7. "Standard Test Method for Density of Glass by the Sink-Float Comparator", Technical Report c729-75, American Society for Testing and Materials, 1990.

8. F. C. Monkman and N. J. Grant, Proc. ASTM **56**, 593 (1956).

9. D. C. Cranmer, B. J. Hockey, S. M. Wiederhorn, and R. Yeckley, to be published in Ceramic Engineering and Science Preceedings (1991).

10. G. G. Long et al., J. Appl. Cryst. **23**, 535 (1990).

11. J. A. Potton, G. J. Daniell, and D. Melville, J. Appl. Cryst. **21**, 663 (1988).

DAMAGE RESISTANCE OF *IN SITU* REINFORCED SILICON NITRIDE

Chien-Wei Li, Charles J. Gasdaska, Jeffrey Goldacker, and Siu-Ching Lui
Allied-Signal Inc., Research and Technology, Morristown, NJ 07962, USA

ABSTRACT

The room temperature fracture behavior for *in situ* reinforced (ISR) silicon nitride is correlated to its microstructure and *R*-curve behavior. The relation of strength to fracture origin suggests that stable growth of the intrinsic flaw precedes catastrophic fracture. Grain-bridging that generates a rising bridging stress behind the crack-tip has been proposed as the cause for stable crack growth, which in turn reduces the strength dependency on initial flaw size. As a result of strong bridging by the acicular β-Si_3N_4 grains, ISR Si_3N_4 is characterized for high Weibull modulus. At elevated temperatures, the material's tensile creep rupture behavior follows the Monkman-Grant type plot. A tensile creep rate of ~$10^{-9}s^{-1}$ at 1260°C/250 MPa, 1300°C/180 MPa, and 1350°C/90 MPa has been recorded. This relatively strong creep resistance is related to the sliding-resistance of the acicular grains and the properties of the amorphous film between the grains in ISR Si_3N_4.

I. Introduction

AS-800 is an *in situ* reinforced (ISR) silicon nitride containing rare-earth sintering aids produced by a gas pressure sintering process. This type of material consists of large populations of *in situ* formed acicular β-Si_3N_4 for toughening reported over a decade ago[1]. A summary of some of the key mechanical properties for a typical batch of AS-800 is presented in Table I.

Due to a grain-bridging mechanism ISR Si_3N_4 has a pronounced *R*-curve behavior[2,3], i.e., crack-resistance (toughness) increases with crack extension. How *R*-curve property relates to the microstructure, affects performance, and changes the design methodology are subjects that await further studies. This paper is a contribution to some of these issues, and deals with the correlation between the ISR's room temperature strength property and *R*-curve, and the high temperature tensile creep deformation behavior as it relates to microstructure[4].

II. Experiments

AS-800 samples labeled SN-1 and SN-2 of same composition were used for the RT property characterizations. Sample SN-3 of similar composition is used for the tensile creep measurement. Their main difference is in microstructure, with SN-1 having the finest and SN-2 and SN-3 having similar microstructures. The average grain widths are 0.6 and 0.7 µm for SN-1 and SN-2, respectively. The average apparent aspect ratio for both is ~2.1.

Fast-fracture strength was measured using bars with a 3 mm by 4 mm cross-section in 4-

Table I. Mechanical Properties for AS-800

4-Pt Flex. Strength*, MPa				4-Pt Flex Life >100 H, MPa		Fracture Toughness$ MPa.m$^{0.5}$
RT	1200°C	1315°C	1375°C	1000°C	1200°C	
840±46	648±35	610±18	559±30	480 (n#>100)	550 (n#>100)	8.5

*60 RT specimens, 10 specimen at each HT # Slow crack growth exponent $3-pt chevron-notch point bending with 20 mm inner and 40 mm outer spans. Fractography was conducted using an optical microscope under 500 or 1000x. The toughness property was evaluated using the indentation-strength in bending (ISB) technique[5]. Indentation loads ranging from 2 to 500 N were used to probe the R-curve property.

The tensile creep behavior was characterized using 5 cm-long flat (2.54 x 2.54 mm cross section) dogbone shaped specimens loaded with a pin and clevis fixture under constant load conditions. A scanning laser was used to measure the elongation between two silicon nitride flags mounted on the gauge section (2 cm long) of each specimen. Testing was performed for up to 408 hours. Density measurements using a sink/float technique in organic liquids were done to evaluate the contribution of cavitation to creep strain.

III. Results and Discussion

(1) Room Temperature Fracture

Indentation-strength data for SN-1 and SN-2 are plotted in Fig. 1(A). For comparison, published data[3] represented by the dotted line for another ISR Si_3N_4 having significantly coarser microstructure (average grain width=1.4μm) are included. The plot shows the typical features for materials with R-curve behavior[3,6], i.e., slope greater than -1/3; near-plateau strength at low indentation loads; and cross-over of the indentation-strength for materials of different scales of microstructure. These phenomena can be explained from bridging models[3,6]. Specifically, the model predicts a coarser-grained material to have lower strength but larger near-plateau zone at low indentation loads (small crack), but higher indentation-strength at high indentation loads (large crack). Figure 1(B) shows the log-normal plot of toughness calculated using the indentation-strength data[5], and the estimated critical crack sizes corresponding to different indentation loads. The critical crack sizes were estimated by multiplying the as-indented crack size by 2.5. Although this presumably underestimates the true critical crack size of an R-curve material, the error is not significant[3].

The trade-off between strength and large-crack toughness is apparent in Fig. 1. It may be questionable whether the toughness gain at a large crack size (~1 mm) for the coarsest-grained Si_3N_4, which has a chevron-notch toughness of 10 MPa.m$^{0.5}$, is significant. For instance, the difference in residual strength between SN-1 and the coarsest-grain Si_3N_4 for a 500-N indentation is about 10% (270 MPa vs. 300 MPa). Whether this noticeably improves

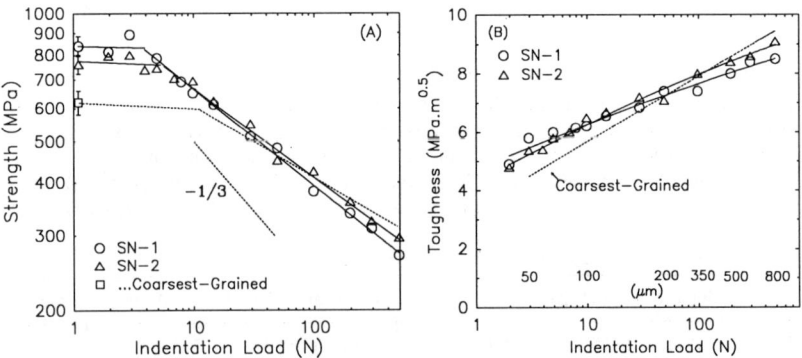

Fig. 1. (A) Effects of microstructure on indentation-strength (intrinsic strength arbitrarily located at 1.1 N); **(B)** corresponding estimated R-curve

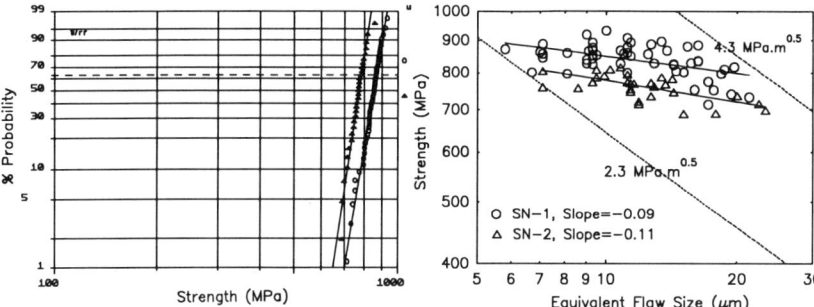

Fig. 2. Weibull Plots for SN-1(o) and SN-2(△); Weibull modulus=21.7 for SN-1 and 24.1 for SN-2

Fig. 3. Strength of ISR Si_3N_4 not sensitive to flaw size; dotted lines from LEFM calculation.

the lifetime of the material remains to be seen. In the mean time, SN-1's ~30% higher intrinsic strength, higher toughness for crack sizes up to ~300 μm, and respectable large-crack toughness would likely make it a better choice for applications.

The Weibull strength distributions for SN-1 and SN-2 are shown in Fig. 2. The data are censored and correspond to one type of fracture origin, i.e., large grain. The Weibull moduli are 21.7 for SN-1 and 24.1 for SN-2. Machining flaws are responsible for the failure of a few other bars (7 out of 60 for SN-1 and 5 out of 35 for SN-2). Inclusion of those data would have minor effects on the Weibull moduli.

The equivalent size of the fracture origin is plotted against strength in Fig. 3. This equivalent size is obtained by first converting the fracture origin into an ellipse, from which the maximum applied stress intensity K_a^m is calculated[7]. The equivalent flaw size, which is the radius of an embedded circular flaw experiencing the same applied stress and K_a^m as the ellipse, is then obtained. Notice that the two least-square-fitted lines in Fig. 3 have slopes close to -0.1. This is significantly different from -0.5, the value predicted by linear elastic fracture mechanics (LEFM). The two dotted lines in Fig. 3, representing the lower and upper bounds of the data points and having a slope of -0.5, correspond to applied stress intensity factors of 2.3 and 4.3 $MPa.m^{0.5}$, respectively. Behavior similar to that depicted by the data in Fig. 3 has been observed in other ceramics as well[8-13]. However, the lines in Fig. 3 appear to be much flatter than those reported for Y-TZP[8], SiC[9-11], Al_2O_3[11], and other Si_3N_4[11-13]. Indeed, the Weibull modulus for the ISR Si_3N_4 used in this study is much higher than the moduli reported for some of those materials. Work is in progress to examine the behavior of other types of intrinsic defects such as porosity in ISR Si_3N_4.

The data in Fig. 3 also show that for a given fracture origin size SN-1 has a higher average strength than SN-2. This is a clear indication that the local microstructure surrounding the fracture origin has played a significant role in the fracture process. There has been considerable discussion on the role of flaw size, grain size, residual stress, and R-curve on the strength of ceramics[6,14-16]. Singh et al.[14] first pointed out the possibility of stable crack growth in ceramics as the crack-tip meets the grain boundaries and transition from single crystal toughness to polycrystalline toughness takes place. This can explain the correlation shown in Fig. 3 since the initial flaw size is not the critical flaw size due to stable crack growth. This requires a steep R-curve rising faster than the applied stress intensity and until recently this issue has not been specifically addressed[3,17]. Modeling shows that stable crack growth in ceramics is possible if a residual stress acts on the intrinsic flaw as it does in the indentation crack[6]. Modeling also shows that, while residual stress exists, it is not a necessary factor for stable-crack-growth if there is a rising crack-closure (bridging) stress behind the crack-tip[3]. This is very likely to be the case for Si_3N_4

due to its characteristic grain-bridging mechanism.

Assuming a linear increase of bridging stress with crack-opening displacement and a crack profile not affected by the traction of the bridging element, it can be shown that the resultant R-curve is[3]

$$K_R = K_o \left(\frac{\lambda}{\lambda - \Delta c}\right)^{1/2} \quad (1)$$

where K_o is the intrinsic toughness, Δc is the crack extension, and λ is a characteristic length determined by the concentration, strength and the size of the bridging elements. The R-curve according to Eq. (1) has a concave shape and approaches K_o asymptotically, a behavior sought by Virkar et al[18]. This shape results in a crack that starts out with unstable growth and then changes to stable growth[18]. It is assumed that as the crack extension exceeds the size of the elastic-bridging zone D_b, the crack becomes unstable again because it extends into the region where grain pull-out controls and the bridging stress is no longer rising[3]. The applied stress intensity at that critical point should be considered equal to the "fracture toughness" K_c of the material.

Figure 3 indicates that $K_o <\sim 2.3$ and $K_c > \sim 4.3$ MPa.m$^{0.5}$. If $K_o = 2.3$ MPa.m$^{0.5}$, then for a flaw of 20 µm in radius, an applied stress as low as ~450 MPa would start unstable growth of the flaw. After this initiation, the quickly developing R-curve would stop the crack growth. With further increasing applied stress, the crack continues to grow until $\Delta c = D_b$ and critical fracture occurs. The critical stress, according to Fig. 3, is ~800 MPa for SN-1 and ~700 MPa for SN-2. The possibility that intrinsic flaws may grow unstably at a stress intensity as low as ~2 MPa.m$^{0.5}$ could explain part of the fatigue behavior of intrinsic flaws in Si_3N_4, which can grow at a stress intensity much lower than the fatigue threshold determined using large-cracks. Studies are being conducted to clarify these issues.

Since the critical crack size should be $\sim(c_o+D_b)$, then in order to calculate the critical stress σ_c one can replace Δc in Eq. (1) with D_b and equate Eq. (1) with the applied stress intensity. This results in the expression:

$$\log \sigma_c = \log \frac{1}{Y}[K_o(\lambda/(\lambda-D_b))]^{1/2} - \frac{1}{2}\log(c_o+D_b) \quad (2)$$

where Y is the crack geometry factor. Figure 4 re-plots the data shown in Fig. 3 with curves calculated according to Eq. (2). Good agreement is achieved by proper selection of D_b and K_c; their values determine λ[3]. Note that K_c and D_b for the coarser-grained SN-2 are larger than those for the SN-1, which is consistent with the microstructure. These selected parameters are also consistent with the values reported in the previous study[3].

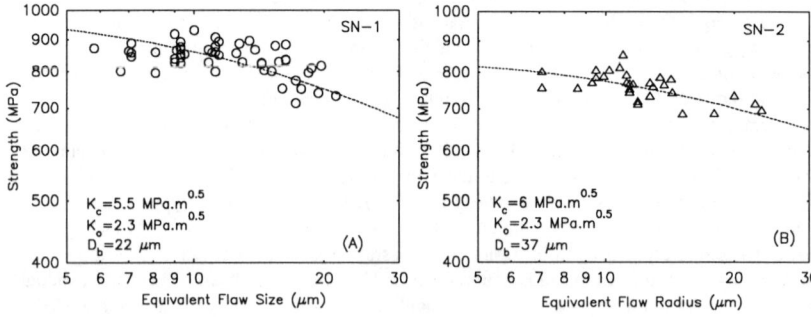

Fig. 4. Predicted fracture from Eq. (1) (dotted lines) for: **(A)** SN-1; **(B)** SN-2

(2) Tensile Creep Behavior

The tensile creep behavior for AS-800 SN-3 is summarized in Fig. 5. Creep curves typically exhibit a short (<24 hrs.) region of primary creep followed by a secondary or minimum creep rate. At higher temperatures and faster creep rates a region of tertiary creep may be observed. In Fig. 6 the minimum creep rate vs. stress has been plotted on a log-log plot. Notice that for the range of stresses and temperatures shown in Fig. 5 the stress exponents, n, are relatively large. In addition, apparent activation energies between 760 and >2000 KJ/mole are observed. As discussed below, we believe that the large stress exponents and activation energies rule out diffusional creep processes as the controlling mechanism, but they instead are related to the properties of the grain boundary "film" present between Si_3N_4 grains. It should be pointed out that the creep data in Fig. 5 represent a dramatic improvement in properties for Si_3N_4 compared to material available a few years ago. For example, at 1260°C AS-800 is capable of supporting stresses 4-5 times greater at creep rates identical to previous generations of silicon nitride.

The results for thirteen samples, eleven of which were tested to failure, are plotted in Fig. 6 as time to failure vs. minimum creep rate (Monkman-Grant plot[19]). The data in Fig. 6 encompass the entire range of stress and temperature which appear in Fig. 5. The results are well described by this treatment, as has also been demonstrated by Wiederhorn[20] for other silicon nitrides. Based on this failure prediction criterion, stresses for a 1000 hour lifetime are plotted in Fig. 7. Since the Monkman-Grant plot predicts a 1000 hour lifetime at a creep rate of $\sim 8 \times 10^{-10}$/sec these are the stresses which produce this creep rate. Since the longest creep run was 408 hours the 1000 hour prediction is the result of a factor of 2.5 extrapolation in time. Preliminary examination of fracture surfaces indicate that for T≤1300°C and stresses ≥200 MPa the failure mode is slow crack growth.

Stress exponents >4 in materials such as Si_3N_4 which do not deform by dislocation mechanisms are frequently attributed to cavitation. To determine the contribution of cavitation to the creep strain, precision density measurements were performed before and after creep testing. The results appear in Fig. 8 where the percent strain due to cavitation is plotted against the creep rate. The amount of strain due to cavitation was calculated as $3\varepsilon_{cav}=\Delta\rho/\rho$, where ρ is the initial density. At the highest strain rates where most strain accumulates cavitation contributes only around 20 percent of the total strain. This is consistent with estimates of cavitation based on TEM specimens. The cavities are present only in the grain boundary triple point interfaces and are not present at two-grain boundaries.

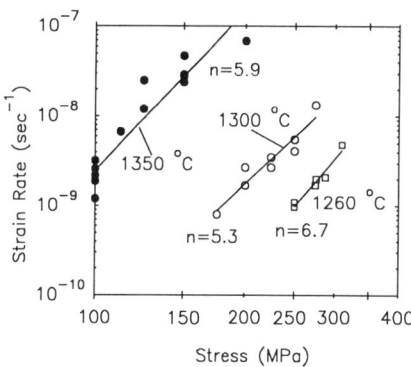

Fig. 5. Strain rate of ISR Si_3N_4 sensitive to stress and temperature changes

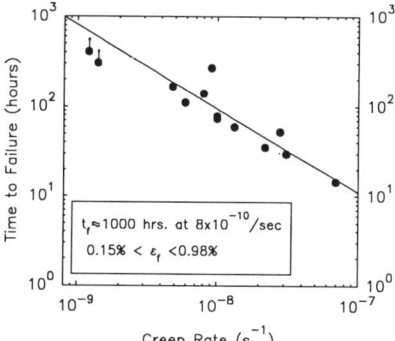

Fig. 6. Monkman-Grant plot predicts 1000-hour lifetime at a creep rate of 8×10^{-10} s^{-1}.

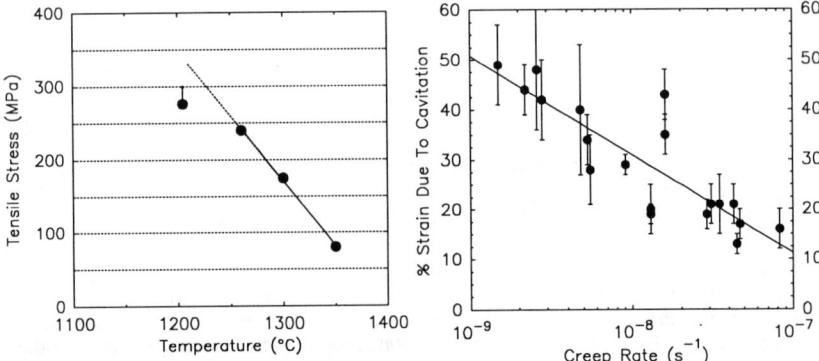

Fig. 7. Predicted maximum stresses for 1000-hour lifetime at indicated temperatures

Fig. 8. Cavitation's contribution to total strain decreases with increasing creep rate

The calculation of cavitational strain assumes that the volumetric strain due to cavitation is distributed evenly in all three coordinate directions[21]. However, even if all the volumetric strain is plated on boundaries perpendicular to the stress axis, cavitation would contribute only 60% of the total strain at the faster creep rates. In addition, this would not change the stress exponent since the general conclusion that cavitation contributes less strain as creep rates increase would still be valid. Since local tensile stresses exist at the majority of boundaries under creep conditions where grain boundary sliding and relaxation effects occur[22], it is unlikely that all the volume change contributes to the measured elongation. Indeed, no preferred orientation for cavitation is observed in TEM specimens so some material plates out on boundaries which are not perpendicular to the stress axis. Finally, from observation of modulus changes during stress change tests it does not appear that elastic compliance changes due to microcracking contribute significant strain.

While the triple point pockets are crystalline, an amorphous film 2-5 nm thick is present at all two-grain boundaries. The large stress exponents may result from the non-newtonian behavior of this film. It is known from experiments and theoretical modelling that glasses exhibit non-linear stress-strain behavior above a threshold stress[23,24]. For some glasses this threshold shear stress is ~200 MPa. The apparent stress exponents then can be very large. If it is assumed that the grain boundary film in Si_3N_4 exhibits similar behavior, then the large stress exponents could be explained. This assumes that the creep mechanism is one controlled by grain boundary sliding with accommodation by diffusion and cavitation. Activation energies of 760 KJ/mole are consistent with the temperature dependence of glass viscosities. The larger apparent activation energies (2000 KJ/mol) probably are a result of using a simple Arrhenius type of behavior to describe the complex properties of the grain boundary film whose properties are very sensitive to stress and temperature. Finally, the unusual microstructure of ISR Si_3N_4 undoubtedly changes the way sliding occurs and the development of the local boundary stresses compared to geometrically simple models of hexagonal tiles which are usually modelled. Further understanding of the relationship between sliding and ISR microstructure requires additional work.

IV. Conclusions

In summary, we have characterized the room temperature fast-fracture and high temperature (>1200°C) delayed-fracture behavior of the *in situ* reinforced (ISR) Si_3N_4. It is confirmed that a coarser-grained ISR, compared to a finer-grained, has a lower toughness in the small-crack region but higher toughness at large-cracks. The strength-initial flaw size relationship for the ISR Si_3N_4 does not follow the prediction of linear elastic fracture mechanics. Stable growth of original flaw due to a steep R-curve from grain-bridging can explain this discrepancy, and, as a result, reduces the strength dependency on initial flaw size. The strong bridging effects in ISR leads to characteristically high Weibull modulus for the material. The bridging model also explains that the lower strength for the coarser-grained ISR, when compared to the finer-grained at a given initial flaw size, is due to its larger bridging zone and less-steep R-curve, albeit its higher fracture toughness. High temperature tensile creep rupture behavior of the ISR follows the Monkman-Grant plot. The creep deformation has high apparent activation energies (≥760 KJ/mole). The relative contribution of creep-induced cavitation to the total strain drops as the creep rate increases. This leads us to believe that the shear-resistance of the amorphous film between Si_3N_4 grains has an important effect on the creep, and under high stresses the non-newtonian flow of the amorphous film causes the observed accelerating creep and large stress exponents.

Acknowledgements

The authors want to thank the support of managers John Yamanis and Cliff Ballard of the Ceramic Program, and the review of the manuscript by Dr. Yamanis. TEM characterization by Dr. Jordi Marti is appreciated. Part of the work is supported by the Department of Energy under the Advanced Heat Engine Project, Contract No. 85X-SH596C.

References

1. F. F. Lange, J. Am. Ceram. Soc. **62** (9-10), 428 (1979).
2. C.-W. Li and J. Yamanis, Ceram. Eng. Sci. Proc. **10** (7-8), 632 (1989).
3. C.-W. Li, D.-J. Lee, and S.-C. Lui, J. Am. Ceram. Soc. **75** (7), 1777 (1992).
4. P. J. Whalen, C. J. Gasdaska, and R. D. Silvers, Ceram. Eng. Sci. Proc. **11** (7-8), 633 (1990).
5. P. Chantikul, G. R. Anstis, B. R. Lawn, and D. B. Marshall, J. Am. Ceram. Soc. **64** (9), 539 (1981).
6. P. Chantikul, S. J. Bennison, and B. R. Lawn, J. Am. Ceram. Soc. **73** (8), 2419 (1990).
7. J. J. Petrovic and M. G. Mendiratta, in Fracture Mechanics Applied to Brittle Materials, ASTM STP 678, edited by S. W. Freiman (American Society for Testing and Materials, (1979), p. 83
8. J. Sung and P. S. Nicholson, J. Am. Ceram. Soc. **71** (9), 788 (1988).
9. S. G. Seshadri and M. Srinivasan, J. Am. Ceram. Soc. **64** (4), C-68 (1981).
10. K. Ikeda and H. Igaki, J. Am. Ceram. Soc. **70** (2) C-29 (1987).
11. H. P. Kirchner, R. M. Gruver, W. A. Sotter, Mater. Sci. Eng. **22** (2), 147 (1976).
12. S. Usami, H. Kimoto, I. Takahashi, and S. Shida, Eng. Fract. Mech. **26** (4), 745 (1986).
13. T. Hoshide, H. Furuya, Y. Nagase, and T. Yamada, Int. J. Fract. **26** (4), 229 (1984).
14. J. P. Singh, A. V. Virkar, D. K. Shetty, and R. S. Gordon, J. Am. Ceram. Soc. **62** (3-4), 179 (1979).
15. R. W. Rice, S. W. Freiman, and J. J. Mecholsky, J. Am. Ceram. Soc. **63** (3-4), 129 (1980).
16. S. J. Bennison and B. R. Lawn, J. Mater. Sci. **24** (9), 3169 (1989).
17. D. K. Shatty and J. S. Wang, J. Am. Ceram. Soc. **72** (7), 1158 (1989).
18. A. V. Virkar, D. K. Shetty, and A. G. Evans, J Am. Ceram. Soc. **64** (3-4), C-56 (1981).

19. F.C. Monkman and N.J. Grant, in <u>Deformation and Fracture at Elevated Temperatures</u>, eds., N.J. Grant and A.W. Mullendore, The M.I.T. Press Cambridge, MA (1965).
20. S.M. Wiederhorn, B.J. Hockey, D.C. Cranmer, D.E. Roberts and R. Krause, in <u>Preprints of the Ann. Auto. Technology Dev. Contractors' Coordination Meeting</u>, U.S. Dept. of Energy, 1991.
21. R. Raj, J. Am. Ceram. Soc. **65** (3), C-46 (1982).
22. W. Beeré, Phil. Trans. R. Soc. Lond. A. **288**, 177 (1978).
23. J.H. Li and D.R. Uhlmann, J. Non Crstall. Solids, **3**, 127 (1970).
24. J.H. Simmons, R.K. Mohr and C.J. Montrose, J. Appl. Phys. **53** (6), 4075 (1982).

CRACK GROWTH RESISTANCE OF CERAMIC COMPOSITE

SEIJIRO HAYASHI, H. BABA AND A. SUZUKI
Structure & Strength Department, Research Institute
Ishikawajima-Harima Heavy Industries Co., Ltd.
1-15 3-chome Toyosu, Koto-ku, Tokyo 135-91, JAPAN.

ABSTRACT

Fracture process zone in SiC_w/Si_3N_4 ceramic composite was studied by a hybrid experimental-numerical analysis employing moire interferometry and finite element analysis. A chevron-notched, wedge-loaded double cantilever beam specimen was used to obtain a stable crack growth. The relation between crack closure stress and crack opening displacement which govern fracture process zone was obtained.

INTRODUCTION

The fracture process zone characterizes the fracture of brittle materials. Although toughening models [1,2,3] have been proposed, these results have not led to development of tougher brittle materials or improved design methods for ceramic components to date. This is partly attributed to the lack of the information on the fracture process zone.

The objective of this paper is to estimate the fracture process zone in a SiC_w/Si_3N_4 ceramic composite. The major difficulty in studying fracture process zone is to determine constitutive relation (crack closure stress versus crack opening displacement relation). To overcome this difficulty, crack opening displacements along crack plane were obtained by moire interferometry and used as a boundary condition in a hybrid experimental-numerical analysis.

The ultimate purpose of this project is to develop a design code, in which the R-curve behavior is taken into account, for ceramic components. This study was conducted as the first step towards the goal.

EXPERIMENT

Specimen

The material used was a hot-pressed Si_3N_4 containing 20 wt% SiC whiskers, and 5 wt% MgO as a sintering aid. Whiskers were preferentially oriented perpendicular to the pressing axis. Figure 1 shows the distribution of whisker length and Fig. 2 shows that of whisker diameter. Elastic modulus and density of the material were 310 GPa and 3.19, respectively.

A chevron-notched, wedge-loaded double cantilever beam (CN-WL-DCB) specimen was machined from a 40-mm-diameter billet of which thickness is 6 mm. Figure 3 shows the specimen dimensions.

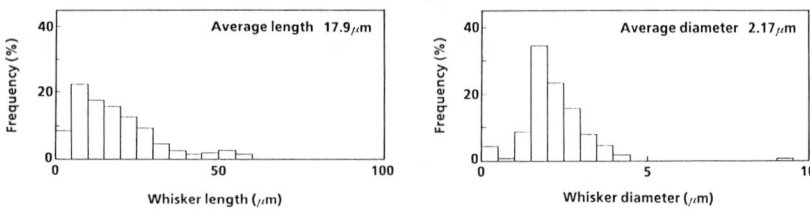

Fig. 1 Distribution of whisker length   Fig. 2 Distribution of whisker diameter

Loading procedure

The specimen was loaded in a rigid, displacement controlled testing machine whose capacity was 10 kN. Load was applied incrementally through a SiC loading pin. A crack stabilizer which avoids unstable fracture was attached to the specimen as indicated in Fig. 3. A compressive load of 300N was applied through a crack stabilizer.

Moire interferometry

Moire interferometry [4] was used to measure crack opening displacement associated with a stably growing crack. Optical set-up for the moire interferometry is shown schematically in Fig. 4. An argon laser with a coherent light of 514.5 nm wavelength was used as a light source.

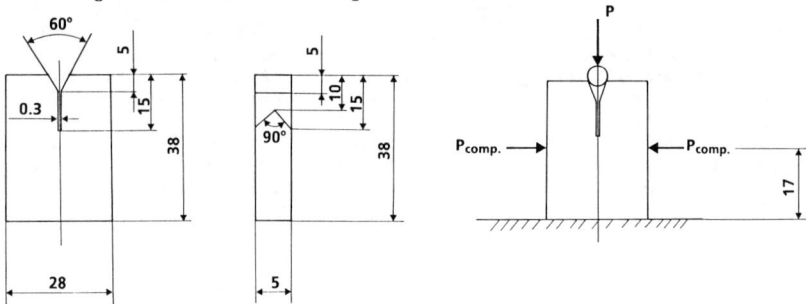

Fig. 3 CN-WL DCB Specimen and Loading condition (unit: mm)

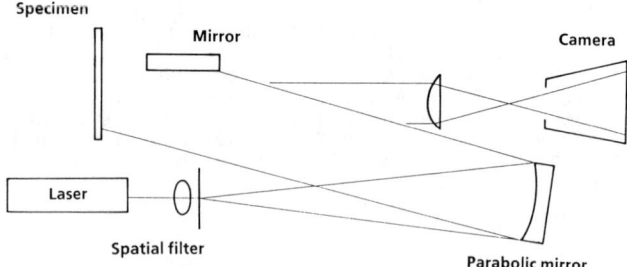

Fig. 4 Optical set-up for moire interferometry

A unidirectional diffraction grating of 1200 lines/mm was transferred to the surface of the specimen. By interfering ± first order diffractions from this specimen grating with a virtual grating of 2400 lines/mm, a fringe multiplication of 2 is obtained to produce moire fringes corresponding to the crack opening displacement (COD) on the surface. The COD is given by

$$COD = N/f \qquad (1)$$

where N is fringe order and f is the frequency of virtual reference grating. As can be seen from Eq. (1), the sensitivity of displacement is 1/2400 mm (=0.416 μm) per fringe order.

NUMERICAL ANALYSIS

A two dimensional finite element analysis was performed. Displacement at loading point determined from moire interferometry and half of the measured load obtained by load cell were prescribed at the loading point. The crack tip location was determined from moire fringe pattern and scanning electron microscopy observations (detail is described later). A trial and error procedure was employed to obtain a crack closure stress (CCS) versus crack opening displacement (COD) relation which provided numerical CODs that matched the measured CODs obtained from moire interferometry.

SCANNING ELECTRON MICROSCOPY OBSERVATION

Crack length measurement

Crack length is an important factor for the analysis of CCS versus COD relation. Some studies [5,6] determined crack tip from moire pattern. In this study, crack length of one identical specimen was measured by three methods. After moire pattern was recorded, the specimen was unloaded. The first method employed a photograph of moire pattern to measure the crack length. The second method used a microscope with measuring device. After measuring the crack length by the first and second methods, diffraction grating on the surface of the specimen was removed. The crack length was then measured by scanning electron microscopy (SEM) as the third method. Several colleagues measured the crack length by the three methods described above.

RESULTS AND DISCUSSION

Determination of crack length

The crack length measured by the first method is the same as that measured by the second method. However, SEM observation around crack tip (the third method) revealed that the crack tip located 0.3 mm ahead of the crack tip specified by moire pattern. The crack tip designated by moire pattern was labeled M and the crack tip which located 0.3 mm ahead of the moire-specified crack tip was labeled T in Fig. 5. No residual COD was observed around crack tip labeled T in the figure.

The preceeding observations indicate that moire detects the crack tip of a continuous crack opening within the sensitivity of moire. In other word, CODs less than the sensitivity of moire (0.416 μm) are not detected even though crack opening is continuous. Further studies concerning this point are being conducted.

In this study, we define the crack tip specified by SEM as the true crack tip.

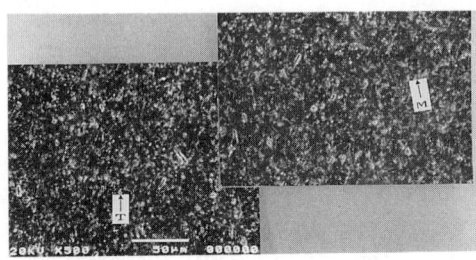

Fig. 5 Crack tip determined by moire pattern (labeled M) and SEM (labeled T)

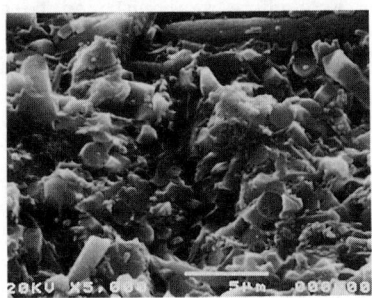

Fig. 6 SEM photograph of fracture surface

Fractography

SEM photograph of fracture surface is shown in Fig. 6. This reveals that inclined whiskers fracture on a plane normal to the whisker axis. Pullouts were rarely observed. These findings are consistent with the findings of Campbell et al [7]. The polished surface of the specimen used for crack length determination was also examined by SEM and showed that whiskers fractured on a plane normal to the whisker axis. This indicate that while being pulled out, whiskers were subjected to bending and failed. Therefore toghening was not provided by pullouts. However, a combination of debonding and bridging is a possible toughening mechanism.

Fracture process zone

Figure 7 shows a typical moire pattern. An asmmetry existed on the crack face because of slight crack deflection from the center line of the specimen.

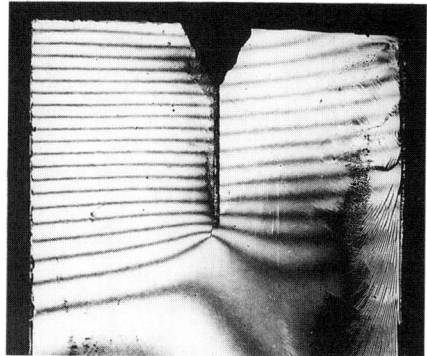

Fig. 7 Moire fringe pattern

The computed COD without the fracture process zone (FPZ) is compared with the measured COD in Fig. 8. Although difference between the computed COD without FPZ, dotted line in Fig. 8, and the measured COD is small, the former exceeds the latter. This implies that the FPZ exists behind the crack tip.
The relation between CCS and COD in Fig. 9 was used to match the computed COD with the measured COD. The computed CODs with FPZ is shown by a full line in Fig. 8.
Figure 10 shows the R-curve which was computed by the CCS acting on the COD in the FPZ of the specimen.

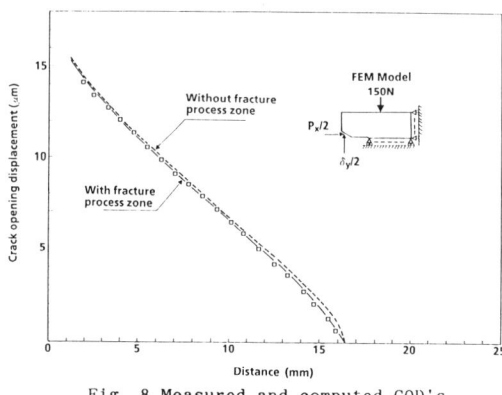

Fig. 8 Measured and computed COD's

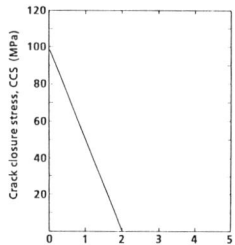

Fig. 9 Relationship between crack closure stress and crack opening displacement

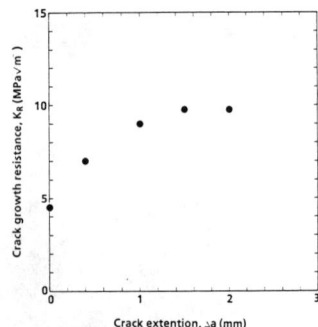

Fig. 10 Crack growth resistance curve

CONCLUSIONS

1. The crack tip obtained from the moire pattern located 0.3 mm behind the true crack tip. A possible explanation for the result is that the crack opening displacement ahead of the moire-specified-crack tip is less than the sensitivity (0.416μm) of moire interferometry.

2. Toughening of SiC_w/Si_3N_4 ceramic composite is possibly provided by a combination of debonding and bridging.

3. A quantitative relation between crack closure stress and crack opening displacement in the fracture process zone was obtained.

ACKNOWLEDGEMENTS

This work was performed under the management of Engineering Research Association for High Performance Ceramics as a part of The Fine Ceramics R & D Project of Basic Technology for Future Industries supported by the New Energy and Industrial Technology Development Organization. The authors are thankful to Mr. Hiroki Okuse and Mr. Hideyuki Kobayashi for their excellent technical contributions.

REFERENCES

1. D. R. Clarke, J. Am. Ceram. Soc. 63, 104 (1980).
2. J. W. Hutchinson, Acta Metall. 35, 1605 (1987).
3. B. Budiansky, J. W. Hutchinson and A. G. Evans, J. Mech. Phys. Solids 34, 167 (1986).
4. D. Post, in Handbook on Experimental Mechanics, edited by A. S. Kobayashi (Prentice Hall, Englewood Cliffs, NJ, 1987)p.314.
5. C. T. Yu and A. S. Kobayashi, in Experiments in Micromechanics of Failure Resistant Materials, edited by K. S. Kim (AMD- Vol. 130, ASME, 1991)pp. 79-83.
6. J. S. Epstein, G. B. May, K. E. Perry, J. E. Shull, H. Okada and S. N. Atluri, ibid (1991)pp.57-71.
7. G. H. Campbell, M. Rule, B. J. Dalgleish and A. G. Evans, J. Am. Ceram. Soc. 73, 521 (1990).

MECHANICAL PROPERTIES OF SiC WHISKER REINFORCED β-SiAlON COMPOSITES

C. YAMAGISHI, J. HAKOSHIMA, S. NAKAJOH, N. MIYATA, and K. TSUKAMOTO
Central Research Laboratory of Nihon Cement Co., Ltd, 1-2-23 Kiyosumi Koto-ku, Tokyo, Japan

ABSTRACT

The mechanical properties of SiC whisker, SiCw, reinforced β-Sialon composite were investigated. β-Sialon with 20 vol% SiCw showed excellent high temperature strength, 610 MPa at 1350°C, and fracture toughness, 5.3 MPa$\sqrt{m}$ (Chevron Notched Beam, CNB, method). This fracture toughness value was an improvement of 75% over that of β-Sialon (monolithic material). This composite also showed excellent oxidation-resistance; the weight gain due to oxidation after 200 hr at 1350°C in dry air was 0.3 mg/cm^2, and after oxidation the high temperature strength, 1350°C, was 683 MPa, an improvement over the pre-oxidation value. Also, in high temperature creep testing (200 hr at 1350°C and 300 MPa) deformation of the SiCw reinforced β-Sialon composite, SiCw/β-Sialon was 66% smaller than that of β-Sialon.

INTRODUCTION

Silicon nitride and Sialon[1-3] have excellent high temperature properties, so their application as high temperature structural materials for components of heat engines and Ceramic Gas Turbines, CGT, is of great interest. Utilization of the materials has begun to be examined. For example, the application of CGT for coal gas, operating at 1250°C, is under development.[4] Recently, due to the increasing scarcity of petroleum and the need for adaptable utilization of remaining resources, it has been proposed to develop a compact and efficient CGT for automobile engine applications. It has been postulated that such a CGT would operate at 1350°C, so materials capable of withstanding this temperature are required.

In this study, the high temperature (1350°C) mechanical properties of Sialon ceramics, prepared from high purity β-Sialon powder by carbo-thermal reduction, were investigated in order to apply β-Sialon to CGT components. SiC whisker reinforced β-Sialon composites, expected to enhance the mechanical properties of Sialon, were investigated as well.

EXPERIMENTAL PROCEDURES

Commercial SiCw (Tateho Chemical Industries, #1-S), β-Sialon powder (Nihon Cement Co. Ltd., $Si_{6-z}Al_zO_zN_{8-z}$, z = 0.5, average particle size: 0.4 μm) and Yb_2O_3 (Santoku Metal Industries) were used as sintering aids. The amount of SiCw additives was 20 vol%. The surface of the β-Sialon particles was covered with SiO_2. It is known from subsequent chemical analysis for oxygen that 2 mols SiO_2 per 13 mols β-Sialon were present in the β-Sialon powder. Four wt% AlN was added to the β-Sialon powder to allow the bulk composition to fall into the β-Sialon range (Eq.(1)).

$$13\ Si_{5.5}Al_{0.5}O_{0.5}N_{7.5} + 2\ SiO_2 + 4\ AlN \rightarrow 14\ Si_{5.25}Al_{0.75}O_{0.75}N_{7.25} \quad (1)$$
$$(z = 0.5) \hspace{4cm} (z = 0.75)$$

β-Sialon powder, Yb_2O_3 and AlN were mixed and then, with a sufficient amount of methanol, ball-milled for 24 hr. To produce SiCw/β-Sialon, 20 vol% SiCw was added to the slurry and dispersed using an ultrasonic homogenizer for 30 sec and ball-milled for 16 hr. Next, the slurry was dried in a rotary evaporator. The mixed powder was then hot-pressed for 1 hr at 1850°C, 26 MPa in a 0.1 MPa nitrogen atmosphere. The resulting hot-pressed discs (50 mm φ x 5 mm thick) were machined into specimens (3 mm x 4 mm x 40 mm) for mechanical testing.

Relative Density

Bulk density was measured by the Archimedes method and theoretical density was calculated from the true density of β-Sialon, SiCw and Yb_2O_3. Relative density was then calculated from the bulk density and the theoretical density.

Flexural Strength

Strength was tested using the three-point-bending method in which the span is 30 mm and the crosshead speed is 0.5 mm/sec, as per JIS R 1601. The test was carried out at 20°C and at 1350°C.

Fracture Toughness

Fracture toughness was tested by the CNB method. Details of the notch are shown in Fig. 1. After cutting the notch, the three-point-bending test was carried out. Again the span was 30 mm and the crosshead speed was 0.5 mm/sec. The fracture toughness is calculated using Eq. (2).[5]

$$K_{IC} = \frac{P_{max}}{BW^{1/2}} \times (3.08 + 5.00\, \alpha_0 + 8.33\, \alpha_0^2) \times \left(\frac{30}{W}\right) \times \left(\frac{\alpha_1 - \alpha_0}{1 - \alpha_0}\right) \times \left(\frac{1}{3.22}\right) \quad (2)$$

where P_{max} is maximum load, W is width, B is thickness, $\alpha_0 = a_0/W$ and $\alpha_1 = a_1/W$ (Fig. 1).

Oxidation Resistance

After an oxidizing environment exposure in flowing dry air (1 ℓ/min) at 1350°C for 200 hr, the specimen's weight gain and three-point-bending strength at 1350°C were measured.

High Temperature Creep Test

Deformations were measured after 200 hr of specimens subjected to 1350°C in air with a 300 MPa load applied by the three-point-bending method. (Fig. 2)

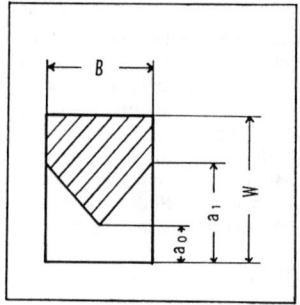

Fig. 1 Configuration of Chevron notched of specimen.

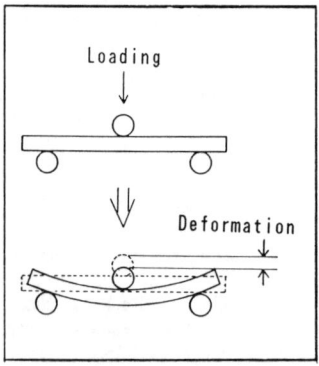

Fig. 2 Schematic illustration of creep test.

RESULTS AND DISCUSSION

Relative Density

The relative density of Sialon was 100% while that of SiCw/β-Sialon was 95% (Table 1). Since SiCw did not sinter at 1850°C, the porosity around the SiCw probably remained. Therefore, the density of the composite material was lower than that of the β-Sialon.

Flexural Strength

Table 1 shows the results of the three-point-bending test for Sialon and SiCw/β-Sialon at room temperature and at 1350°C. At room temperature, the flexural strength of β-Sialon was 668 MPa, and that of SiCw/β-Sialon was 705 MPa. The strength at room temperature increased with the addition of the SiCw; likely, the whiskers play a bridging role for the matrix. At 1350°C, the flexural strength of β-Sialon was 650 MPa, while that of the SiCw/β-Sialon was 617 MPa.

Up to 1350°C both materials maintained high levels of strength: greater than 600 MPa; therefore, it may be possible to utilize both materials in CGT components.

Table I: Mechanical Properties of β-Sialon and SiCw/β-Sialon.

Properties Materials	Flexural Strength (MPa)		Fracture Toughness ($MPa\sqrt{m}$)	Relative Density (%)
	at 20°C	at 1350°C		
β-Sialon	668	650	3.0	100
SiCw/β-Sialon	705	617	5.3	95

Fracture Toughness

Table 1 shows the fracture toughness results of β-Sialon and SiCw/β-Sialon. The fracture toughness of SiCw/β-Sialon was 5.3 $MPa\sqrt{m}$; a 75% improvement over Sialons 3.0 $MPa\sqrt{m}$. Fracture toughness of ceramics has been reported to improve upon addition of ceramic particles[6-7] or whiskers.[8-13] Since the major problem for Sialon is its low fracture toughness, the improvement in this respect is especially important. It seems plausible that the fracture toughness is improved by crack deflection and debonding around SiCw or pull-out of SiCw. Figure 3 is a SEM image of a fracture surface of SiCw/β-Sialon. The average grain size of β-Sialon is 1 μm and the shape is not elongated. The trace of a crack that caused debonding around a whisker is indicated by the arrow in Fig. 3.

3.0 μm

Fig. 3 SEM image of fracture surface of Si/β-Sialon.

Oxidation Test

Results of weight gain and three-point-bending test at 1350°C, after oxidation, are shown in Table 2. The weight gains for β-Sialon and SiCw/β-Sialon were 0.2 and 0.3 mg/cm^2, respectively. The notably low weight gain of Sialon was barely affected by the inclusion of SiCw.

After the oxidation exposure, the flexural strength at 1350°C for β-Sialon and the SiCw/β-Sialon were 785 and 683, respectively. At 1350°C, after oxidation, both materials had enough strength (greater than 600 MPa) to allow their application in CGT components.

Table II. Oxidation Resistance of β-Sialon and SiCw/β-Sialon.

Resistances Materials	Weight Gain (mg/cm^2)	Flexural Strength at 1350°C (MPa)	
		Post-oxidation	Pre-oxidation
β-Sialon	0.2	785	650
SiCw/β-Sialon	0.3	683	617

XRD

Figure 4 shows the XRD patterns for β-Sialon and SiCw/β-Sialon. In β-Sialon, a minute amount of garnet (Yb$_3$Al$_5$O$_{12}$) was detected but none in the SiCw/β-Sialon. Figure 5 shows the XRD patterns for the two materials after oxidation. In the β-Sialon, considerably more garnet was found than had been seen in the pre-oxidized material (Figure 5); and in the SiCw/β-Sialon some garnet was now present. During the oxidizing process, grain boundary glass phase was crystallized to garnet (which has excellent high temperature properties) so there was less glass to soften at high temperature. Therefore, the post-oxidation strengths at 1350°C were higher than before the oxidizing process.

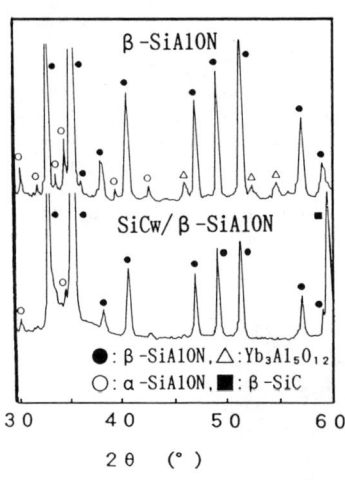

Fig. 4 XRD patterns before oxidation

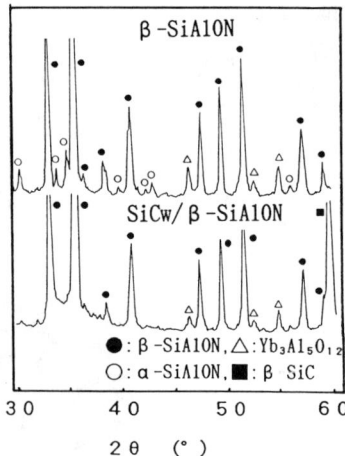

Fig. 5 XRD patterns after oxidation.

High Temperature

Figure 6 shows the deformation of the two materials after sustained loading at 1350°C. After 200 hr the β-Sialon and SiCw/β-Sialon had deformations of 637 and 421 µm, respectively. The SiCw/β-Sialon showed substantially less creep; only 66% of that for the non-reinforced Sialon.

Matsui et al.[11] reported that SiCw-reinforced silicon nitrite had better high temperature creep resistance than monolithic silicon nitrite, and suggested that the marked improvement resulted from suppressing grain boundary sliding due to the pinning effect of the SiCw.

Figure 7 shows a TEM image of SiC/β-Sialon. A grain boundary phase between β-Sialon and SiCw was not observed. With the grain boundary phase being glass that would soften at high temperature, it is predicted that the amount of sliding at the β-Sialon and SiCw boundary would be small if the glass phase was not present. For the SiCw/β-Sialon composite, it is suggested that with the pinning effect, where the SiCw is in contact with many β-Sialon grains and the glass boundary phase is absent, suppressed sliding contributes to the reduction of high temperature creep.

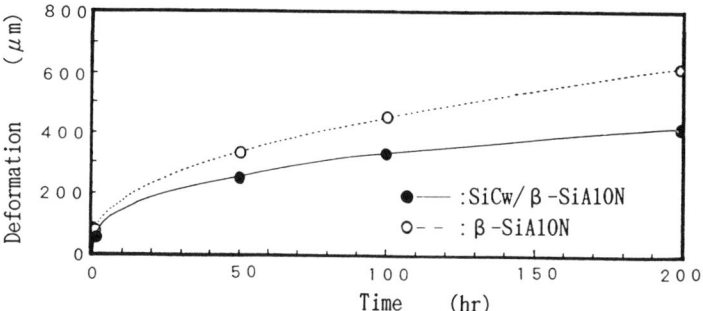

Fig. 6 Deformations of β-Sialon and SiCw/β-Sialon.

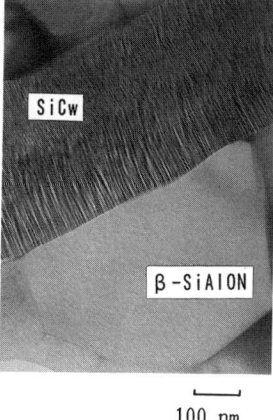

Fig. 7 Deformations of β-Sialon and SiCw/β-Sialon.

CONCLUSION

Sialon ceramics prepared from high purity β-Sialon powder by carbo-thermal reduction and SiCw/β-Sialon were produced by hot-pressing. The following results were obtained:

(1) High flexural strengths, greater than 600 MPa, of β-Sialon and SiCw/β-Sialon were maintained up to 1350°C.
(2) The fracture toughness of the SiCw-reinforced β-Sialon composite was 5.3 MPa$\sqrt{m}$; a 75% improvement over β-Sialon's value of 3.0 MPa$\sqrt{m}$. The improvement may be due in large part to debonding at the SiCw.
(3) SiCw/β-Sialon has inherited the excellent oxidation-resistance of β-Sialon. Weight gains for β-Sialon and SiCw/β-Sialon composite were 0.2 and 0.3 mg/cm^2, respectively. The flexural strengths of both materials at 1350°C, after oxidation, showed improvement over pre-oxidation values; 785 MPa for β-Sialon and 683 MPa for SiCw/β-Sialon. The improvements result from crystallization of the grain boundary glass phase.
(4) The creep deformations after 200 hr at 1350°C for β-Sialon and SiCw/β-Sialon composites were 637 and 421 μm, respectively. The SiCw/β-Sialon showed excellent creep resistance; the creep being only 66% of that for β-Sialon.

Sialon ceramics having high strength at 1350°C and excellent oxidation resistance were obtained. The use of SiCw in Sialon ceramics improves fracture toughness and high temperature creep performance resulting in a material that should have applications in CGT components.

ACKNOWLEDGEMENT

This work is a part of the Automotive Ceramics Gas Turbine Development Program by the Petroleum Energy Center.

REFERENCES

1. K.H. Jack, J. Mater. Sci., **11**, 1135 (1976).
2. Y. Oyama, Jpn. J. Appl. Phys., **11**, 760 (1972).
3. Y. Oyama and S. Kamigaito, Yokyo-Kyoukaisi, **80**, 327 (1972).
4. H. Tabata, (Proc. 9th Basic Technologies for Future Industries—Fine Ceramics, Tokyo, Japan, 1991) pp. 5.
5. D. Muntz, G. Himslt and J. Eschweiler, J. Am. Ceram. Soc., **63**, 341 (1980).
6. C. Yamagishi, J. Hakoshima, K. Tsukamoto and Y. Akiyama, J. Japan Soc. Powder and Powder Metall, **37**, 1056 (1990).
7. C. Yamagishi, K. Tsukamoto, J. Hakoshima, H. Shimojima and Y. Akiyama, J. Mat. Sci., **27**, 1908 (1992).
8. P.F. Becher, E.R. Fuller, Jr. and P. Angelinil, J. Am. Ceram. Soc., **74**, 2131 (1991).
9. G.H. Compbeel, M. Ruhle, B.J. Dalgleish and A.G. Evans, ibid., **73**, 521 (1990).
10. S.T. Burjan, J.G. Baldoni and M.L. Huckabee, J. Am. Ceram. Soc., **66**, 347 (1983).
11. T. Matsui, O. Komura and M. Miyake, J. Ceram. Soc. Jpn., **99**, 1103 (1991).
12. P. Sajgalik and J. Dusza, J. Mater. Sci., **10**, 776 (1991).
13. K. Ueno, S. Sodeoka and J. Hirooka, J. Ceram. Soc. Jpn., **100**, 525 (1992).

MECHANICAL PROPERTIES OF SILICON CARBIDE WHISKER-REINFORCED SILICON NITRIDE MATRIX COMPOSITES

YONG HUANG, HUIRONG LE, JIANBAO LI, LONGLIE ZHENG, JIANGUANG WU
Tsinghua University, Department of Materials Science and Engineering, Beijing 100084, P. R. China

ABSTRACT

In the present paper, the effect of La_2O_3-Y_2O_3-Al_2O_3 (LYA) and La_2O_3-Y_2O_3-MgO (LYM) additives on the mechanical properties of hot pressed SiC whisker-reinforced Si_3N_4 matrix composites and the relationship between whisker content and additive content were investigated in detail. The results show obvious strengthening in the case of LYA additives containing ≤ 2.4 wt% Al_2O_3. When the Al_2O_3 content of the additive was > 1.2 wt%, the fracture toughness decreased accompanied by decreased interfacial debonding and whisker pull-out. With LYM additives containing 2.4 wt% MgO, both the room temperature strength and toughness increased. A combination of Y_2O_3 + MgO (YM) with LYA additives produced substantial improvements in strength and toughness, up to 1050 MPa and ~11 MPa$\sqrt{m}$, respectively. The Weibull modulus of a SiC whisker-reinforced composite was found to be greater than that of either the unreinforced Si_3N_4 or Si_3N_4 containing SiC particles. The experiments also demonstrated that in order to achieve full density the additive content had to be increased with increased whisker content. Preliminary but encouraging results on strength retention at elevated temperatures were also reported.

INTRODUCTION

Si_3N_4 ceramics have excellent potential for a wide variety of engineering applications. However, low fracture toughness has limited applications involving high tensile stresses. Recently, SiC whisker-reinforced Si_3N_4 composites with improved fracture toughness have been described.[1,2] The ability of the SiC whiskers to toughen Si_3N_4 depends upon the degree of densification, interface debonding, and whisker orientation.[3] Si_3N_4 ceramics typically fracture along grain boundaries, presumably since the bonding is weak as compared to the strongly covalent bonding of the grains. In SiC whisker-reinforced composites, toughening contributions from interface debonding and whisker pull-out require control of interface bonding by selection of appropriate additives.

The effects of different additives and additive contents on the mechanical properties of SiC whisker-reinforced Si_3N_4 composites were investigated. At the same time, the relationship between whisker content and additive content was examined. This provided insight into how to optimize the toughness and strength by tailoring the compositions. Companion studies indicate that the improved toughness-strength properties were reflected in improved reliability with composites exhibiting increased Weibull moduli.

EXPERIMENTAL PROCEDURES

The raw materials consisted of Si_3N_4 powders (94% α-Si_3N_4; 1% free silicon; 0.7 μm average diameter) produced in this laboratory, commercial, chemically pure grades of La_2O_3, Y_2O_3, Al_2O_3, and MgO powders, and Tokai Carbon Ltd. SiC whiskers (0.6 μm average diameter; 10 to 80 μm long). The SiC whiskers were dispersed in ethanol with a surfactant. The dispersed whiskers were then added to an ethanol-based slurry of Si_3N_4 powder containing the appropriate additives and ball milled for 24 hours. The dried mixtures were hot pressed at temperatures of 1780 to 1830°C for 1 to 1.5 hours at 20 MPa applied pressure in a nitrogen atmosphere. Dense composites, as determined by the Archimedes method, were cut into test samples consisting of 4 x 3 x 30 mm flexure bars (with subsequently polished surfaces) and 6 x 3 x 30 mm bars for fracture toughness determinations. The strengths were determined in three point bending (span = 22 mm) and fracture toughness by the single edge notched beam method (notch radius = 0.1 mm) at room temperature. Microstructure was characterized by scanning electron microscopy SEM observations of fracture surfaces.

RESULTS AND DISCUSSION

Effect of Additive Composition

La_2O_3-Y_2O_3-Al_2O_3 (LYA) and La_2O_3-Y_2O_3-MgO (LYM) additives were used as densification aids with the amount of Al_2O_3 being varied in the LYA additive case. The influence of variation in additive type and composition on the mechanical properties of Si_3N_4 containing 10 wt% SiC whiskers and 20 wt% of the additives is shown in Fig. 1. When LYA additives are used, the strength of the composites increased from 880 to 1040 MPa as the alumina content in the LYA increased from 0 to 2.4 wt%. However, the fracture toughness remained unchanged with LYA containing up to 1.2 wt% alumina and decreased at higher alumina contents. In the LYM additive case, only LYM containing 2.4 wt% MgO was examined, but here the toughness, as well as strength, was raised substantially, 9.5 MPa√m and 1070 MPa, respectively. On the other hand, the use of 10 wt% MgO in the absence of La_2O_3 and Y_2O_3, resulted in composites (10 wt% SiC whiskers) with only modest improvements in strength and toughness, 850 MPa and 8.2 MPa√m, respectively.

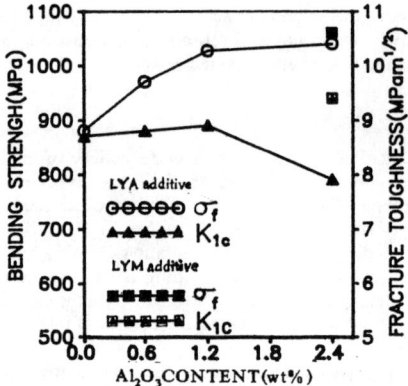

Fig. 1. Effect of Al_2O_3 and MgO content of additive on mechanical properties.

SEM observations of the fracture surfaces of the above composites revealed the following. In composites containing the LYA additives, the exposed whiskers exhibited very roughened surfaces, Figs. 2(a) and (b). However, the whiskers exposed on the fracture surfaces of the composite using the LYM additive were quite smooth, Fig. 2(c). These observations suggest chemical reaction between the SiC and the additives during densification when alumina is used as opposed to MgO. This may lead to stronger interfacial bonding in the LYA versus LYM additive case to limit debonding and pull-out. The rougher surfaces of the whiskers in the LYA additive composites would also tend to inhibit pull-out. These factors would contribute to the reduced toughening effects observed in the composites containing the LYA additives.

The influence of the amount of LYA (La_2O_3:Y_2O_3:Al_2O_3 = 6:6:1) added to the composite on strength and toughness is shown in Figure 3. While the toughness improvements appear to saturate at LYA contents of ≥ 15 wt%, a maximum in strength is observed for LYA contents of 25 wt%.

Densification additives consisting of a mixture of yttria plus magnesia YM and the LYA additive produced the greatest combined toughening-strengthening effect to date, Table I. As seen, equal amounts of the YM and LYA additives to composites containing 20 wt% SiC whiskers resulted in a toughness of 11 MPa√m and strength approaching 1100 MPa.

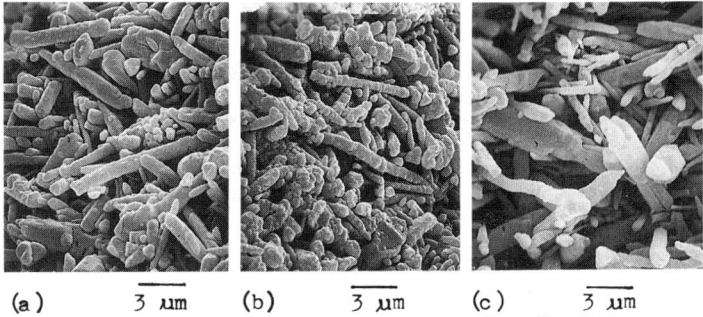

Fig. 2. Fracture surfaces of SiC whisker-reinforced Si_3N_4 composites containing LYA and LYM additives. LYA additive containing (a) 1 wt% alumina and (b) 2.4 wt% alumina and (c) LYM additive containing 2.4 wt% magnesia.

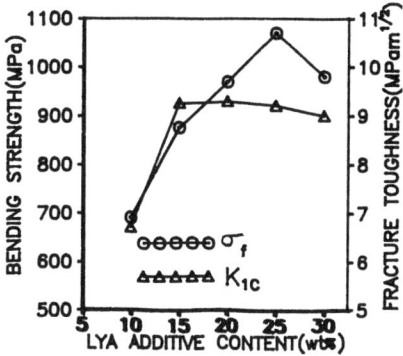

Fig. 3. Influence of densification additive additions on the strength and toughness. (10 wt% whisker)

Table I. Effect of a Combination of YM and LYA Additives on Mechanical Properties.

No.	Compositions(wt%)					σ_f (MPa)	K_{IC} (MPa$\sqrt{m}$)
	SiC(w)	Si_3N_4	$La_2O_3+Y_2O_3$	Al_2O_3	MgO		
LYA-0	10	90	20	0	0	880	8.7
LYA-4	10	90	17.6	2.4	0	1040	7.9
LYM	10	90	17.6	0	2.4	1070	9.5
M-1	10	90	0	0	10	850	8.2

Effect of SiC Whisker Content

Figures 4a and 4b reveal that the mechanical properties of composites are a strong function of the whisker content and the optimum whisker content can vary with densification (LYA) additive content. For example, when the (LYA) content is 20 wt%, the highest toughness (10 MPa√m) and strengths (1020 MPa) are achieved with whisker contents of 10 and 15 wt% respectively. Increasing the additive content to 25 wt % causes the maximum in toughness (10.5 MPa√m) and strength (1050 MPa) to shift to a whisker content of 20 wt %.

In addition, the SiC whisker-reinforced silicon nitride composite containing (10 wt% whiskers) exhibited a narrower distribution of strengths as compared to silicon nitride containing the same level of SiC particles approximately 0.1 μm in diameter. The Weibull modulus of the whisker reinforced composite was 23 versus 13 for the Si_3N_4-SiC particle composite.

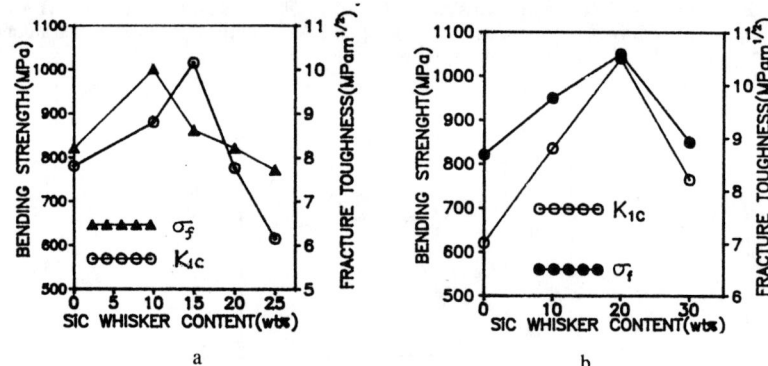

Fig. 4. Effect of whisker content in composites with different densification additive contents; a) 20 wt % additives and b) 25 wt % additives.

Strength Retention at Above 1300°C

To extend the present study to elevated temperatures, we have tested some (LY) composites with La_2O_3-Y_2O_3 additive composition (12.5 wt%) and 10 wt% SiC at 1300°C and 1370°C. In addition, heat treatment at 1425°C and 1500°C was conducted in an attempt to crystallize the grain boundary phase to improve high temperature properties. The results of this premliminary study were rather encouraging. As shown in Table II, these composites are able to retain strengths at elevated temperatures. Preliminary phase analysis by x-ray diffraction indicated that H-phase ($10R_2O_3$-$9SiO_2$-Si_3N_4) and J-phase ($4R_2O_3$-SiO_2-Si_3O_4) are crystallized after the 1450°C treatment, and similar solid solutions plus $LaYO_3$ formed after 1500°C heat treatment. Further progress of this will be reported elsewhere.

Table II. Effects of Heat Treatment on Properties of Composites.

heat-treatment \ properties	$\sigma_{R.T}$(MPa)	$\sigma_{1300°C}$(MPa)	$\sigma_{1370°C}$(MPa)
as hot-pressed	884 ± 64	734 ± 46	705 ± 36
1425°C / 2hr	839 ± 54	800 ± 88	780 ± 60
1500°C / 2hr	856 ± 21	827 ± 79	838 ± 11

CONCLUSIONS

(1) Silicon nitride matrix composites reinforced with 10 wt% SiC whiskers exhibit obvious strengthening when LYA densification aids containing up to 2.4 wt% alumina are employed. However, the accompanying toughness decreases when the LYA additive contains >1.2 wt% alumina.
(2) Substitution of the LYM additive for LYA in these same composites produced improvements in both the toughness and strength. This appears to be associated with changes in the whisker-matrix interfaces.
(3) The whisker content producing the optimum toughening and strengthening effects was a function of the additive content. This reflects, in part, the need for greater additive content with increase in whisker content.
(4) Combining LYA and YM additives produced the greatest combined improvements in toughness and strength, 11 MPa $\sqrt{m}$ and 1050 MPa, respectively.

ACKNOWLEDGMENTS

The authors wish to thank Professors Z. Z. Jiang and Z. D. Guan for their helpful comments and discussions.

REFERENCES

1. P.D. Shalek, J.J. Petrovic, G.F. Hurley, and F.D. Gac, Am. Ceram. Soc. Bull., **65**, (2) 352 (1986).

2. P.F. Becher, C.H. Hsueh, P. Angelini, and T.N. Tiegs, J. Am. Ceram. Soc., **71**, (12) 1050 (1988).

3. Y. Huang, J.B. Li et al., High Technology Letters, **1**, (7) 18 (1991).

THE MECHANICAL PROPERTIES OF A NOVEL Si_3N_4-AMORPHOUS Si_3N_4 COMPOSITE

IVAR E. REIMANIS*, J. J. PETROVIC*, H. SUEMATSU*, T. E. MITCHELL* AND O. S. LEUNG**
*Los Alamos National Laboratory, Los Alamos, NM 87545
**Materials Science and Engineering Department, University of California, Berkeley, CA 94720

ABSTRACT

The hardness and fracture toughness of a model two-phase composite consisting of crystalline Si_3N_4 particles in a matrix of amorphous Si_3N_4 are examined. The composite is created by heat treating high purity, partially amorphous CVD Si_3N_4 in N_2 for various times and temperatures in order to induce crystallization of the α phase. Microindentation tests at temperatures up to 1200 °C are conducted to evaluate the high temperature hardness and fracture toughness. The role of the microstructure is examined using optical and transmission electron microscopy. Finally, the relationship between the microstructure and the mechanical properties is discussed.

INTRODUCTION

Monolithic Si_3N_4 and Si_3N_4-based composites are well-known materials for high temperature structural components. Interfaces in these materials, including grain boundaries and matrix/particulate interfaces, control the high temperature mechanical properties, through mechanisms such as grain boundary sliding, crack deflection, microcrack toughening and fiber pullout [1, 2]. In this study, we examine the properties of a novel CVD-formed composite which consists of a matrix of amorphous Si_3N_4 containing particles of crystallized α-Si_3N_4. Provided that the CVD processing conditions are well-controlled, the interfaces in this composite contain no secondary chemical phases, thereby resulting in a unique model system to study the effect of internal interfaces and residual stress on stress-induced microcracking and crack deflection mechanisms. The aim of this work is to evaluate the role of the crystalline phase on the hardness and indentation fracture toughness of the composite.

EXPERIMENTAL

Partially amorphous Si_3N_4 was deposited on a cylindrical, high purity graphite mandrel using a $HSiCl_3$-NH_3-H_2 mixture at 1200 °C. Pressures ranging from 0.5- 2.0 Torr were used, producing coating thicknesses of 3 - 4 mm. After the CVD process, specimens were cut from the mandrel for anneal treatments at temperatures ranging from 1200-1600 °C for 1 h in a flowing N_2 environment. Results from the as-deposited specimens and the specimens annealed at 1400 °C are reported here.

X-ray diffraction analysis was used to identify the presence of the crystalline phase. The microstructures were revealed by immersing the specimens in concentrated HF acid, which preferentially etches amorphous Si_3N_4 over crystalline Si_3N_4. Crystalline regions could also be identified using dark field optical microscopy. Specimens for transmission electron microscopy (TEM) were prepared by thinning sections mechanically and milling with Ar ions. TEM in conjunction with electron energy loss spectroscopy (EELS) was conducted on the specimens to determine the nature of the amorphous matrix and whether or not it contained impurities.

Room temperature and high temperature (up to 1200 °C) hardness tests were performed using a Nikon model QM-2 hot hardness tester and loads of 1kg. The fracture toughness of the material was estimated using an indentation technique in which median type cracking was assumed [3]. A Young's modulus of 300 GPa was used for the Si_3N_4 [3].

RESULTS

Microstructure

The as-deposited Si_3N_4 contains approximately 20 vol % crystalline phase as determined by lineal analysis in the optical microscope. The x-ray diffraction results indicate that the crystalline regions are entirely α-Si_3N_4; no diffraction peaks corresponding to the β phase are observed, even in the annealed specimens. As seen in figure 1, the crystalline particles have a cylindrical morphology ranging from 20-150 μm in diameter, with the long axis parallel to the CVD direction. The textured nature of this composite is apparent upon comparing figure 1c) (viewing parallel to the CVD direction) with figure 1b) (viewing cross sectionally to the CVD direction). It was observed that particles beyond a critical size (20 - 45 μm in diameter with an aspect ratio of about 5) contain internal microcracks (figure 4); however, no decohesion of the particles from the matrix was observed. The TEM results showed that the regions between the crystalline particles are entirely amorphous. Furthermore, EELS analysis detected no elements other than Si and N.

A considerable microstructural change occurred as a result of the 1400 °C anneal. Firstly, the x-ray diffraction results indicate a much higher degree of crystallinity: the peak intensities increased by several orders of magnitude. Secondly, under the dark field aperture in the optical microscope, the entire specimen acquires a dark contrast, indicating that it is largely crystalline. While the large crystalline particles described above (figure 1) are no longer distinctly visible in the optical microscope, the HF etch reveals their grain boundaries. A greater number of microcracks were also observed. However, the main difference between the as-deposited and 1400 °C annealed specimens was observed in the matrix. TEM results showed that the matrix regions are mainly composed of small (~100 nm diameter) crystalline particles, determined to be α-Si_3N_4. Such a region may be seen in figure 2 where an adjacent amorphous region is also shown. Using EELS, elemental carbon was detected in the amorphous regions, but not in the crystalline regions. The source of the carbon was most likely the graphite mandrel used in the CVD process. It is probable that carbon diffuses through the Si_3N_4 during the anneal treatment. The observation that carbon was not detected in the crystalline regions corroborates evidence in the literature [e.g., 4] that carbon suppresses the crystallization of Si_3N_4.

Hardness and Fracture Toughness

The high temperature hardness of the as-deposited Si_3N_4 as well Si_3N_4 annealed at 1400°C decreased as a function of increasing temperature (figure 3). The hardness amd fracture toughness measurements were performed on sections with the orientations shown in figure 1c (viewing parallel to the CVD direction). The fracture toughness was measured to be 1.9 ± 0.4 MPa m$^{1/2}$ at room temperature, and 2.3 ± 0.3 MPa m$^{1/2}$ at 1200 °C. A Vickers indentation is shown in figure 5. Indentations placed in a region with a high volume fraction of particles (i.e., > 20 vol %) exhibited a greater degree of microcracking, as evidenced by the number of branches from the main crack. In general, a matrix crack which intercepts a particle penetrates the particle without any apparent crack deflection. However, in some instances, the crack deflects and propagates along the matrix/crystalline interface (figure 6) for a short distance before penetrating the particle. Finally, whenever the crack intercepts a particle at a glancing angle, the preferred crack path is along the matrix/crystalline interface.

The measured fracture toughness of the specimens which were annealed at 1400 °C is much higher: 4.4 ± 1.0 MPa m$^{1/2}$ at room temperature and 3.1 ± 0.5 MPa m$^{1/2}$ at 1200 °C. Furthermore, extensive microcracking and more tortuous crack paths were observed in the specimens annealed at 1400 °C.

DISCUSSION

From the results presented above it appears that stress-induced microcracking is a toughening mechanism in this composite. In stress-induced microcrack toughening, the reduced elastic modulus in the microcracked region is the primary contribution to the decreased local stress intensity at the crack tip. The modulus is a function of the microcrack density and the length of the microcracks [1], two parameters which usually require extensive TEM analysis for quantification. While we do not presently have enough information to estimate the toughening contribution, we can assess the microstructure with respect to the optimum toughness. To achieve an optimum toughness increase through microcracking, the particle size must be below the critical size for spontaneous microcracking, yet large enough for stress-induced microcracking to occur. The observations of spontaneous microcracking within large particles (figure 4) indicates that they experience residual tensile strain due to transformation stresses during crystallization. For spherical inclusions in residual tension surrounded by a matrix, the condition for internal spontaneous microcracking is that the sphere radius, R, must be greater than a critical size [5],

$$R_c = 2[K_c/E\varepsilon]^2$$

where K_c is the fracture toughness of the particle, E is its elastic modulus, and ε is the residual strain in the particle due to the transformation. To an order of magnitude, we can calculate the theoretical critical particle size. We will use $K_c = 3$ MPa m$^{1/2}$ for the particle [6] and E = 300 GPa [3]. ε is related to the volume change associated with crystallization. An estimation of ε requires specifying the density of the amorphous matrix. Assuming a value of $\varepsilon = 0.04$ based on the literature for CVD Si_3N_4 processed in a similar manner [7], we find that $R_c = 0.1$ μm. Comparing with our measured value (10 μm < R_c < 25 μm), this calculation suggests that spontaneous microcracking in the as-deposited material is much more extensive than is measurable in the optical microscope. It also indicates that a much finer microstructure is required for optimum toughness through microcracking.

CONCLUSIONS

We have reported the hardness and indentation fracture toughness, at room temperature and temperatures up to 1200 °C, for a novel Si_3N_4 composite. Annealing this composite at 1400 °C induces crystallization and increases the measured fracture toughness. The internal interfaces in these composites are quite strong compared with the particles, as evidenced by the lack of particle decohesion in the presence of tensile residual strains. However, even in the presence of strong interfaces, crack deflection does occur (figure 6). In contrast to toughness increases, the overall strength of the composite is expected to be low due to spontaneous microcracking within the particles. Suppressing spontaneous microcracking requires control over the particle size; to a first approximation, it appears that particle diameters less than about 200 nm are necessary to increase the strength and maximize toughening from stress-induced microcracking. The much finer matrix microstructure (particle sizes ~ 100 nm) produced by annealing the composite at 1400 °C appears to be a step in the right direction for increasing the strength and fracture toughness.

ACKNOWLEDGEMENTS

The authors acknowledge the Department of Energy's Office of Basic Energy Sciences, Division of Materials Science, for supporting this work. We thank Praxair Surface Technologies Inc., formerly known as Union Carbide Coatings Service Corporation, in Cleveland, Ohio, for supplying the CVD Si_3N_4.

REFERENCES

[1] M. Ruhle and A. G. Evans Progress in Materials Science Vol. 33, pp. 85 - 167 (1989).

[2] P. F. Becher J. Amer. Ceram. Soc. Vol. 74, pp255 - 269 (1991).
[3] G. R. Anstis, P. Chantikul, B. R. Lawn and D. B. Marshall J. Amer. Ceram. Soc. Vol. 64, No. 9, pp 533 - 538 (1981).
[4] M. Peuckert, T. Vaahs and M. Bruck Adv. Mater. 2, No. 9, pp. 398 - 404 (1990).
[5] D. J. Green in Fracture Mechanics of Ceramics, vol. 5, eds. R. C. Bradt, A. G. Evans, D. P. Hasselman and F. F. Lange, pp. 457-478, Plenum Press, NY (1983).
[6] H. Suematsu, T. E. Mitchell and J. J. Petrovic, in this symposium.
[7] K. Niihara and T. Hirai J. Mater. Sci. 11, pp. 593 - 603, (1976).

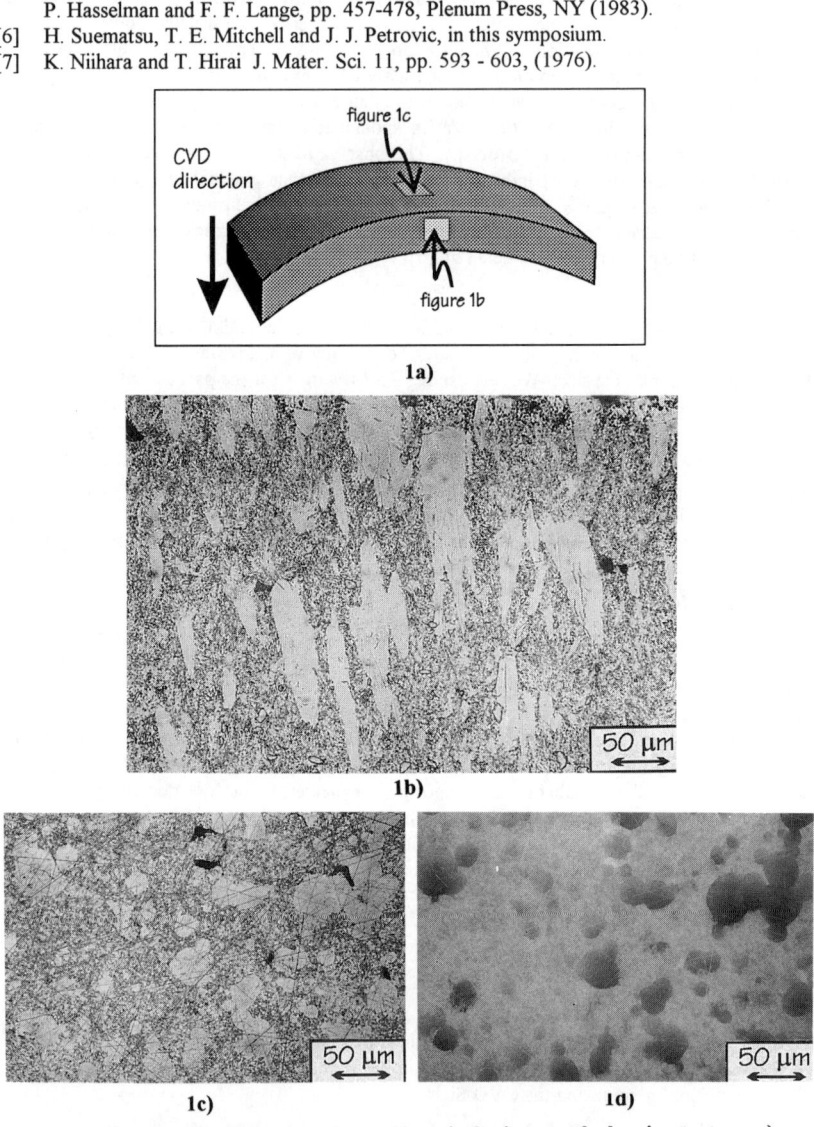

Figure 1 a) Schematic view of specimen; b) optical micrograph showing texture; c) optical micrograph viewing parallel to CVD direction; d) same region as in c), but under dark field conditions.

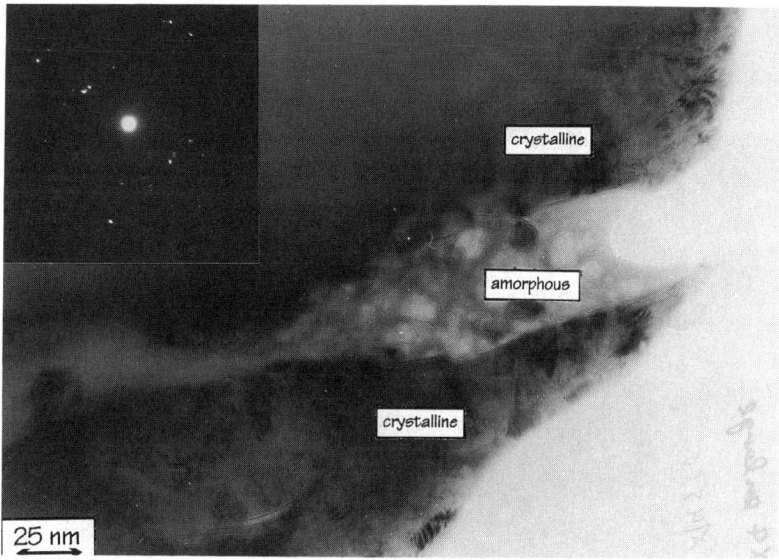

Figure 2 Transmission electron micrograph of specimen annealed at 1400°C for 1 h. Using EELS, elemental carbon was detected in the crystalline regions, but not in the amorphous matrix.

Hardness versus Temperature

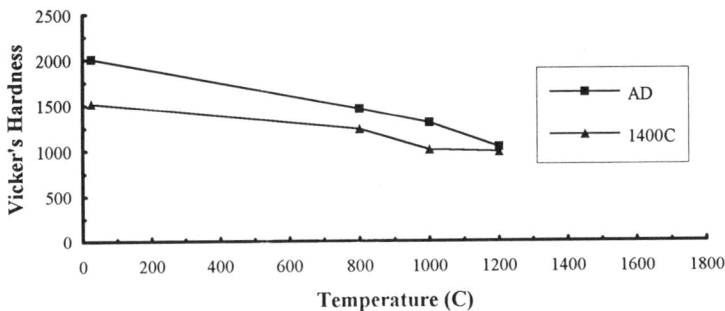

Figure 3 Hardness plotted as a function of temperature for the as deposited specimen (AD) and the specimen annealed at 1400°C for 1 h.

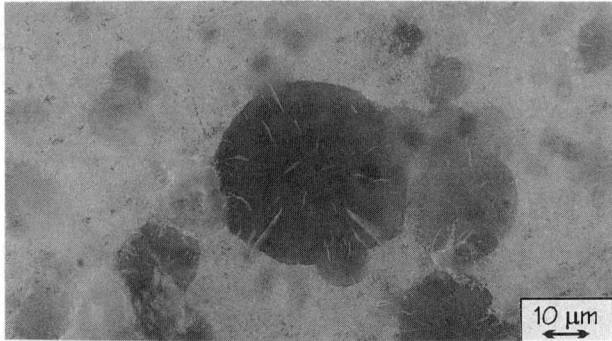

Figure 4 Spontaneous microcracking within α-Si_3N_4 particle.

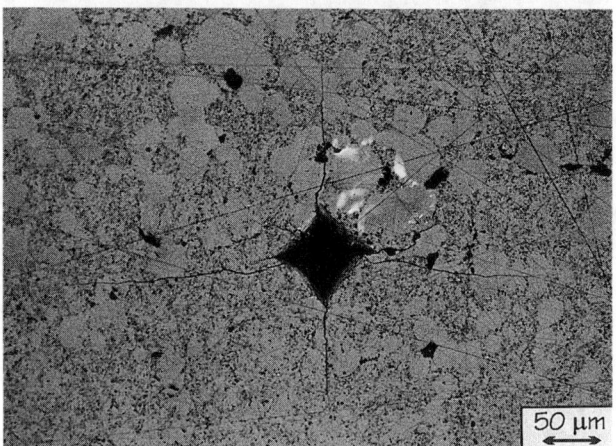

Figure 5 Vicker's indentation in as-deposited material.

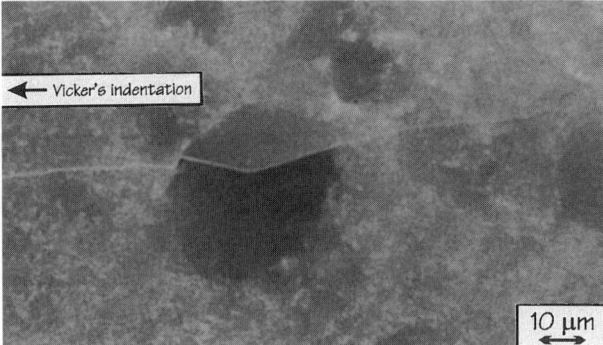

Figure 6 Crack/particle interaction.

PART II E

Application, Oxidation, and Corrosion

STUDIES ON Si₃N₄ CERAMIC CUTTING TOOL MATERIALS AND THEIR APPLICATIONS

HE-ZHUO MIAO, LONG-HAO QI, DE-JIN MA AND ZUO-ZHAO JIANG
Department of Materials Science & Engineering, Tsinghua University, Beijing 100084, China

ABSTRACT

Si_3N_4 ceramics are excellent metal cutting materials. The cutting behavior can be improved by hard particle dispersion phase hardening and heat treatment so as to reduce glass phase in the grain boundaries. Compared with cemented carbide tools, the cutting life of the composite Si_3N_4 cutting tool is about 10-100 times longer, and the optimum cutting speed is about 3-10 times faster. It performs well in cutting hardened tools, nickel based alloys, and other hard materials and can sustain shock loads in operations such as milling, planing and other types of interrupted cutting. Applications in various fields show that machining efficiency can be increased by 3-10 times resulting in savings of time, electricity, and machining of 30-70% or even more.

INTRODUCTION

Machine work is one of the most basic, extensive and important process in industrialized production. It directly affects the efficiency, cost and energy consumption of industrial production. Along with the development of modern science and technology, various high strength and high hardness engineering materials have been developed and employed. In most cases, these advanced materials are difficult to machine.

In recent years, ceramics have been increasingly utilized as cutting tools. As the first generation of ceramic cutting tools, the Al_2O_3 based ceramic cutting tool has high hardness, but its application is limited by its low toughness and low thermal shock resistance. Its properties can be improved by adding second phase particles or whiskers to the Al_2O_3 matrix and such composites have seen increasing use in industrial machining. The other family of ceramic cutting tools is based on Si_3N_4. The first development of Si_3N_4 cutting tools dates back to the middle of the 70's, by the group at Tsinghua University in China. This endeavor was successful because Si_3N_4 ceramic material has high hardness, high toughness, and a low thermal expansion coefficient. More recently, we have developed hard particle dispersion-reinforced Si_3N_4 ceramic cutting tools which prove to have improved resistance to wear, heat, thermal shock and mechanical shock. These tools have been put into practice in many industrial fields with encouraging results that show improved cutting efficiency and savings of 30-70% in manpower, electricity consumption, and machinery. This development will be reviewed below.

EXPERIMENTAL PROCEDURE AND RESULTS

It is very important to the cutting behavior of the ceramic cutting tool that the ceramic material have good heat resistance, fracture toughness, and low thermal expansion coefficient besides hardness. This was demonstrated during machining when the temperature rose quickly and the tip of the cutting tool wore due to fatigue microcracking caused by the effects of both mechanical and thermal stresses. Therefore, the cutting behavior of the ceramic cutting tool cannot be expressed simply by several indexes of material mechanical properties; it is the result of many factors. Generally, the cutting life (distance) serves as the main comparison standard according to the stipulation of ISO3685-1977 (E), i.e., at the same condition and same flank wear V_b, the cutting behavior of various tools are compared by the cutting distance. From our experiments, the behavior of ceramic cutting tools is relative to the factors described below.

Material Hardness

Hardness is the primary factor considered when choosing a cutting tool. The microhardness of Si_3N_4 ceramics at room temperature is about 1800 kg/mm²; not very high. If some dispersion particles with very high hardness, such as TiC and ZrC, are introduced into the Si_3N_4 matrix, the cutting behavior of the composite ceramic is improved.

Material Strength and Toughness

Since wear is caused by fatigue microcracks during machining, toughness K_{IC} is very important to the cutting behavior of the material, especially during high load machining. To prevent catastrophic failure of the insert, materials with high strength and high fracture toughness are needed.

Thermal Expansion Coefficient

Thermal stress during machining, especially in high speed cutting, may lead to the destruction of the insert. Therefore, a low thermal expansion coefficient of the Si_3N_4 matrix is helpful.

Heat Resistance

At the cutting temperature, the less the drop in strength and hardness of the ceramic insert, the better the result obtained. The types of additive (and the boundary phases formed by them) are also very important. The crystallizing treatment of the boundary glass phase will increase the high temperature hardness and strength of ceramic materials as well as improve the cutting behavior of the ceramic inserts.

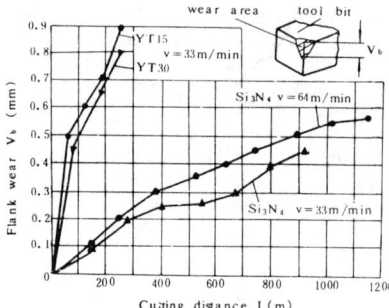

Fig. 1. Comparison between the tool life of Si_3N_4 ceramic and carbide tool in cutting hardened steel T10A (HRC60-62) Cutting condition: f = 0.1mm/rev; a = 03.mm.

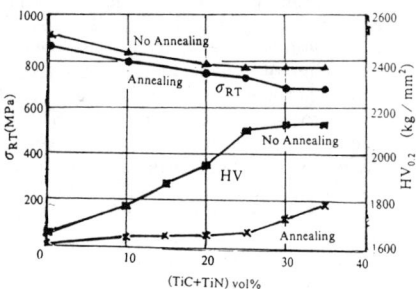

Fig. 2. The flexural strength and microhardness of composite ceramic Si_3N_4-TiC-TiN vary with TiC + TiN vol%.

Because the strength and fracture toughness of Si_3N_4 ceramics are high, the thermal expansion coefficient is low, and in the structural unit of Si-N_4 tetrahedra, the covalent bond between Si and N is very strong. These ensure that a Si_3N_4 ceramic cutting tool has good wear resistance.

When machining the hardened steel T10A (HRC60-62), if flank wear V_B = 0.4 mm, the cutting distance of a cemented carbide cutting tool (such as YT15, YT30, etc.) is only 50-70 meters (see Fig. 1). On the other hand, the cutting distance of Si_3N_4 cutting tool is 640 meters (cutting speed V = 64m/min) and 800 meters (cutting speed V = 33m/min), respectively. Thus, under this condition, the cutting life is about 8-10 times longer than that of a cemented carbide cutting tool. It also shows that the optimum cutting speed of a Si_3N_4 ceramic cutting tool is 4-10 times faster than that of a cemented carbide.

In order to improve the wear resistance of the cutting tool, the hard dispersion particles TiC were added to the Si_3N_4 matrix as follows: Si_3N_4, TiC and TiN powder were mixed with sintering aids such as Y_2O_3, MgO, Al_2O_3, AlN, etc. This was baked dry, hot pressed, annealed, cut, and machined.

Figure 2 shows the effects of TiC + TiN vol% dispersion phase on the flexural strength and microhardness of composite Si_3N_4 ceramic materials. As the content of TiC + TiN vol% increases, the R.T. flexural strength of the composite Si_3N_4 ceramic material decreases whereas the hardness increases. When TiC + TiN vol% is more than 15 vol%, there is a decrease in strength. The R.T. flexural strength and hardness of annealed specimens are slightly less than that of unannealed specimens. The material has better cutting behavior due to the increase in hardness by introducing hard dispersion particles and the decrease in the glass phase of the grain boundary by annealing treatment.

Figures 3 and 4 show experimental results of the cutting behavior of various composite Si_3N_4 ceramic cutting tool materials containing different volume percents of TiC + TiN where the cutting lives are different. Figure 3 shows the relationship between V_B and TiC + TiN vol% when the cutting distance is 730 meters. Figure 4 shows the relationship between V_B and TiC + TiN vol% when L = 1417 meters. The results show that the more TiC + TiN vol% content, the smaller flank wear V_B of composite Si_3N_4 cutting tool. Thus, when TiC + TiN vol% is less than 30%, the wear resistance of the cutting tool increases along with an increase in TiC + TiN vol%. Of course, too much TiC + TiN will make the material brittle.

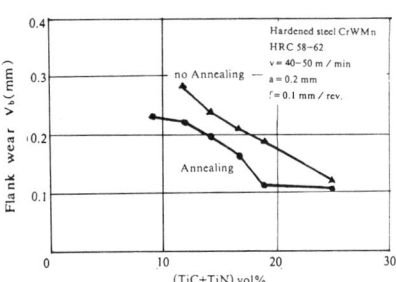
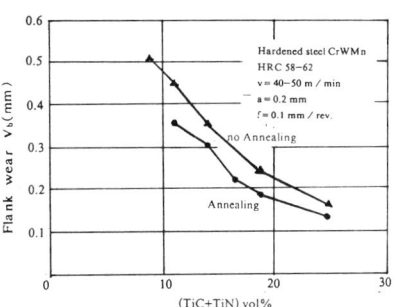

Fig. 3. When L = 730 m, the flank wear V_B vary with TiC + TiN vol%.

Fig. 4. When L = 1417 m, the flank wear V_B vary with TiC + TiN vol%.

As indicated above, increasing the heat resistance of the grain boundary phase will enhance behavior of the cutting tool (see Figs. 3 and 4).

When the specimens were treated by annealing for grain boundary glass phase crystallization, the V_B decreased, and thus, the wear resistance of the cutting tool increased. Although the room temperature flexural strength and hardness decreased little, the high temperature strength and high temperature hardness increased.

Figure 5 shows the flank wear curves of the Si_3N_4 ceramic cutting tool and the composite Si_3N_4 ceramic cutting tool when turning hardened tool steel CrWMn (HRC 59-62), and it indicates that the cutting life of the composite Si_3N_4 cutting tool is several times longer than that of the Si_3N_4 cutting tool.

Composite Si_3N_4 cutting tools possess good behavior in cutting hardened steel and better results can be obtained when turning chilled cast iron. Figure 6 and Table 1 show the results of turning chilled cast iron (HS 71-73). As shown in Table 1, the optimum cutting speed of the composite Si_3N_4 cutting tool was higher than the Si_3N_4 cutting tool and about 6 times higher than the cemented carbide tool.

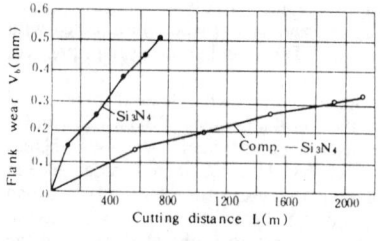
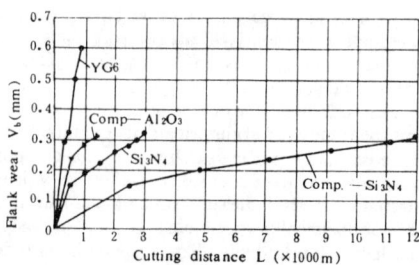

Fig. 5. Wear curves of Si$_3$N$_4$ cutting tool and composite Si$_3$N$_4$ cutting tool when turning hardened steel CrWMn.

Fig. 6. Wear curves of some cutting tools when turning chilled cast iron (HS 71-73).

Table I: Cutting Life Comparison of Some Cutting Tools When Cutting Chilled Cast Iron (HS 71-73)

Cutting Tool	Cutting Behavior V_0 (m/min)	Optimum Cutting Speed V (m/min)	Cutting Speed Used, $V_B = 0.3$ mm	Total Cutting Length L when Si$_3$N$_4$	Times of life of composite
Composite Si$_3$N$_4$	35	33.8	10,410	1	
Si$_3$N$_4$	29	31.4	2,690	3.87	
Composite Al$_2$O$_3$	29.6	30.5	2,508	4.15	
Cemented Carbide (YG6)	6	6	277	37.6	

When flank wear $V_B = 0.3$ mm, its cutting life (cutting distance) is 10,410 meters, or about 3.8 times the Si$_3$N$_4$ cutting tool, about 4 times that of the composite Al$_2$O$_3$ cutting tool made in China, and about 37 times that of cemented carbide tool YG6 (WC + Co).

The composite Si$_3$N$_4$ cutting tool can cut not only various hardened tool steels and chilled cast iron, but also nickel-based high temperature alloys, hard nickel-sprayed layers and their alloys, powder metallurgical sintered materials, pyrolitic graphite, glass fiber reinforced plastics and new engineering plastics. It not only does rough machining, semi-finish machining and finish machining, but also machining with great shock loads such as milling, planing and interrupted cutting.

APPLICATION IN PRODUCTION

This new type ceramic cutting tool has solved a series of production problems which were previously very difficult to solve. It has now been used in about 20 industrial fields such as machine making, metallurgy, automobile, vehicle, mine, energy, precise instrument, pump, and so on. Its usage has enabled substantial savings in working hours and electrical consumption to be realized. The following examples illustrate some of these practical applications.

Rough Machining of Superhard Cast Iron Shroud of Slurry Pump

In the past, annealing treatment was needed for machining because the workpiece was too hard to be cut by the traditional cutting tools and electrical usage was high. More recently, the new type of ceramic cutting tool has been used and the hard workpieces were cut directly without annealing which reduced the cutting time from 50 hours to 8 hours. Work hours and electrical consumption were also reduced significantly.

The cutting of Cr27 superhard cast iron shroud and Cr27 superhard cast iron side shroud (HRC 60) was also solved by using this cutting tool. Its cutting speed was 2-3 times faster than that of the Al_2O_3 cutting tool. It is especially useful for the cutting of high content Cr cast iron side shroud, which cannot be cut by other tools because of their high hardness, four pork ribs of the workpiece and the interrupted cutting with big shock. So far, only the composite Si_3N_4 cutting tool is qualified,. (Cutting speed used was 35 m/min; cutting depth: 5 mm; workpiece size: $\Phi 1020$ mm; workpiece material: Cr15Mo3; workpiece hardness: HRC 63-67.)

Interrupted Cutting of Hard Nickel Cast Iron Ring ($\Phi 3300$ mm) for Grinding Coal

The hardness of the workpiece was HRC 58-60, and there were six risers on the workpiece which formed in casting. Interrupted cutting and continuous cutting were tried but the cutting condition was so difficult that even the cemented carbide cutting tool could not be used. The composite Si_3N_4 tool, on the other hand, can resist the shock load, and its cutting life is 3-4 times longer than that of the composite Al_2O_3 tool, which is easily broken. The cutting parameters used were V = 39-48 m/min, a = 5-8 mm, f = 0.48 mm/rev.

Rough and Semi-Finish Turning of Chilled Cast Iron Roller

Very good results were obtained by using the composite Si_3N_4 cutting tool in the cutting of various chilled cast iron roller. The cutting efficiency increased 3-10 times, and 50-80% of the working hours and electrical consumption were saved.

Rough and Finish Machining of Superhard Roller Sleeve Spray Welded by Ni Alloy

This workpiece is used in a 2050 hot steel rolling mill. The spray welding material is 3309 Ni based alloy powder, and the hardness of the workpiece is HRCX 67. In the past, cemented carbide cutting tool was employed with a cutting speed of only 5.4 m/min (a = 0.5 mm, f = 0.1 mm/rev) and the tool wore easily. Fourteen tools were needed for a streth. The composite Si_3N_4 cutting tool however, is now used. With an increased cutting speed of 50 m/min, (a = 0.5 mm, f = 0.1 mm/rev), only one tool is needed for a streth because of the good wear resistance.

Semi-Finish Turning of 20 Cr Carbonized Hardened Steel (HRC 68)

The cemented carbide cutting tool cannot cut this hardened steel but the composite Si_3N_4 cutting tool can (V = 38 m/min, a = 0./25 mm, f = 0.15 mm/rev). The cutting life of the new tool is 66 times longer than that of the cemented carbide cutting tool YT 30.

Rough Machining of Chilled Cast Iron Tappet (HRC 54) Used in 190 Diesel Engine

One tool of cemented carbide could cut only 14 workpieces, and the wear width of the tool edge was up to 0.6 mm. Using the composite Si_3N_4 cutting tool, one tool can cut 122 workpieces, and the wear width of the tool edge is only 0.2 mm. The tool life increased 8 times, and the productive efficiency increased 8-12 times.

Interrupted Planing of High Carbon and High Chrome Wear-Resistant Cast Iron Jaw Plate in Stone-Breaking Machine

The hardness of the workpiece (600 x 350 x 200 mm) is HRC 62-64, and the shock force was very large during interrupted planing. When cemented carbide and an Al_2O_3 ceramic cutting tool were used, the edge of the inserts broke easily; this was not so for the composite Si_3N_4 cutting tool, which performed successfully. The planing speed was about 8 m/min, the cutting depth 3.5 mm, and the cutting efficiency increased by more than 5 times over that of other tools.

Machining of Nickel-based High Temperature Alloys (As Shown in Table 2)

Table II: Machining of Nickel-Base Alloys

Workpiece	Cutting Tool		Main Valuation
	Cemented Carbide	Composite Si_3N_4	
Front machine box; 169 (4880)	V = 31.8 m/min a = 1 mm f = 0.15 mm/rev coarseness 3.2μ	V = 193.5 m/min a = 2 mm f = 0.25 mm/rev coarseness 0.8μ	Cutting efficiency is GH increased by more than 10 times; surface smoothness is higher
Low pressure turbine axle; Incoloy 901 (F420)	V = 21 m/min a = 1 mm f = 0.1 mm/rev coarseness 3.2μ	V = 47.5 m/min a = 1.5-2 mm f = 0.2 mm/rev coarseness 1.6μ	Cutting efficiency is increased by 4-5 times

Rough and Finish Machining of Ti Alloy TCII Separate Ring of Gas Pressure

In the past, the cemented carbide cutting tool was used, and the cutting conditions were V = 47.5 m/min, a = 0.1-2 mm and f = 0.12 mm/rev. There were ripple and gnawed marks existing on the surface; coarseness was 6.3. After using the composite Si_3N_4 cutting tool, V = 105.6 m/min, a = 2 mm, and f = 0.1 mm/rev; the coarseness of the surface was lower than 1.6. The efficiency thus increased by more than 10 times.

It was proven by a series of applications in many factories that various benefits could be obtained by using the composite Si_3N_4 cutting tool. This new tool could cut superhard materials which could not be cut or were hard to cut with conventional cutting tools and the "turning, milling instead of grinding" process could be realized in the cutting of superhard materials. High speed cutting could be achieved on both hard-to-cut materials and ordinary metal materials. The machining process was simplified, cutting increased, and therefore manpower, electrical consumption, and the number of machines could be reduced by 30-70% or even more. Its popularization and application will bring great innovation to the field of machining and will bring great benefit to the users.

References

1. J-B. Zhou, H-Z. Miao et al., Science Report of Tsinghua University, TH78002, 1978.

2. J.K. Wu, H-Z. Miao, Z.Z. Jiang et al., (Proc. of Intl. Symp. on Factors in Densification and Sintering of Oxide and Non-Oxide Ceramics, Tokyo, Japan, 1978) pp. 443-457.

3. H-Z. Miao et al., Ceramurgia International, **6** (1) 36-39 (1980).

4. J-B. Zhou, H-Z. Miao et al., (Proc. of Intl. Conf. on Manufacturing Engineering, Melbourne, Australia, 1980).

5. Z-Z. Jiang et al., (Proc. of the First China-U.S. Bilateral Seminar on Inorganic Materials Research, Shanghai, May, 1983), pp. 184-191.

6. R.N. Katz, (Materials Sci. Monographs, Proc. of 6th CIMTEC, **38A**, 1987) pp. 145-161.

7. H-Z. Miao, Z.B. Luo and Z.Z. Jiang, (Proc. of the Second Intl. Metal Cutting Conference, Wuhan, May, 1985) pp. 583-595.

DEVELOPMENT OF ADVANCED SILICON NITRIDE VALVES
FOR COMBUSTION ENGINES AND SOME PRACTICAL EXPERIENCE ON THE ROAD

RAINER HAMMINGER* and JUERGEN HEINRICH**
* Hoechst AG, C 584, P.O. Box 80 03 20, 6230 Frankfurt 80, Germany
**Hoechst CeramTec AG, P.O. Box 13 60, 8672 Selb, Germany

ABSTRACT

One major goal of automotive engineers actually is to reduce fuel consumption and emissions of present internal combustion engines. First promissing results have been achieved by the substitution of conventional by silicon nitride valves. The weight reduction of about 60 % per valve is corresponding with lower inertia and spring forces, which lead to improved emissions, reduced fuel consumptions as well as high performances of the engines. Applying statistical methods, two advanced silicon nitride materials have been developed, concerning their application as "cool" inlet valves and "hot" exhaust valves, respectively. Using high quality powders and combined optimum processing parameters, 4-point-bending strength of more than 1200 MPa (room temperature) and 870 MPa (1000 °C), resp., have been realized. Weibull moduli of more than 30 have been reproduced. Grinding of valves produced from these materials has economically been demonstrated by drastically reducing the machining times. Some results of road tests as well as of simulated engine tests from cooperations with different automotive companies are presented in detail.

INTRODUCTION

Steadily growing requirements of environmetal protection are more and more decisive for the development trends in the automotive industry. Reduction of emissions and noise and low fuel consumption are on top of the engineers' specifications. An important approach towards the anti-pollution combustion engine has been made by Hoechst CeramTec AG (HCT) in close cooperation with the Hoechst research center (HOE) in developing and testing ceramic valves for reciprocating piston engines.

The valves control, on the one hand the inlet of the fuel air mix into the combustion chamber, and on the other the outlet of the emissions produced into the exhaust manifold by releasing (cam shaft controlled) the inlet and outlet channels. The valves have, thus, to be extremely wear and heat resistant, the more as they are (at a motor torque of 7000 revolutions per minut) abruptly opened and closed 3500 times every minute. While the valve stems remain rather cool the valve disks, as part of the combustion chambers, have to resist temperatures up to 900 °C. Low specific weight especially in comparison to the presently used metallic materials reduce the weight of the ceramic valves by more than two thirds against conventional valves. Consequently the size of the valve spring and the total length of the valve may be reduced. Further agreeable advantages of this new development are reduced overall heights and widths as well as a several kilograms lower cylinder head weight. This is achieved by using silicon nitride (Si_3N_4) valves. However, serial application of ceramic engine valves have to be discussed with respect to their mechanical performance and long term properties, their reliability and economy.

The present work deals with the development of materials and procedures to obtain suitable qualities of inlet and outlet valves, taking into consideration the cost-benefit factor (especially with regard to the final processing of the valves) and showing first practical experience in engine test rigs as well as in road trials.

SELECTION OF MATERIALS AND PROCEDURES

Characteristic requirements for ceramic valves

With regard to the characteristics required for ceramic valves both with the Otto and the Diesel engine the inlet and outlet valves have to be considered in different ways. The less strained inlet valve requires a medium room temperature strength (4-point bending, 3 x 4 x 45 mm) of > 800 MPa according to the latest state of knowledge while the medium service temperature of ceramic inlet valves is about 300 °C and the total temperature difference (from cold start to alpine performance) is supposed to be about 350 °C.

Much more complicated are the conditions with ceramic outlet valves. Here, a medium component temperature of approx. 850 °C is achieved while the maximum temperature difference is supposed to be about 950 °C. That is why the strength characteristics required are significantly higher (strength at 1000 °C > 900 MPa). The respective Weibull factor must be around m > 25. Moreover, a behaviour to subcritical cracking is required and may be characterized by a cracking parameter n > 50. These requirements for ceramic inlet and outlet valves are summarized in fig. 1.

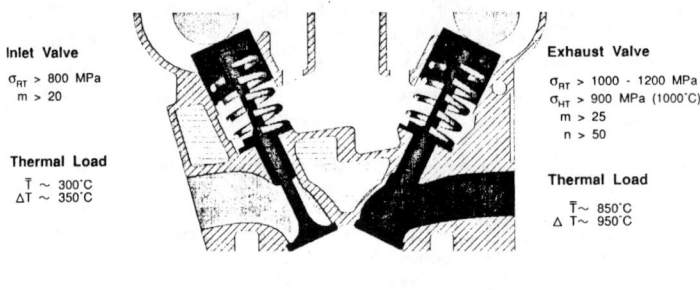

Fig. 1: Performance profile of ceramic engine valves

Materials development

To meet the requirements applicable for ceramic inlet and outlet valves HCT/HOE have developed two different material variations and have optimized them up to the prototype application in the engine. In order to achieve the specified requirements extensive tests have been run with regard to the influence of the starting powders and the sintering additives on the development of the microstructures and the resulting mechanical material characteristics. By the methods of statistical test planning the process parameters like processing, drying and shaping have been optimized to reproduce the component quality. Moreover, very extensive work has been dedicated to the optimization of the sintering technique. The connections are to be seen in detail from the literature |1-3|. The material HCT 90 meets the requirements applicable for the ceramic inlet valve (table I) by its medium bending strength (4-point bending, 3 x 4 x 45 mm) of > 920 MPa, a Weibull factor m > 19 and a K_{IC} factor of 7 MPa m$^{1/2}$. Meanwhile, also the requirements for ceramic outlet valves are met by the material HOE 120. The medium room temperature strength (4-point bending, 3 x 4 x 45 mm) is here exceeding 1200 MPa, whereas a 1000 °C high temperature strength is achieved by 870 MPa. The respective Weibull factor is around m = 23, the stress intensity factor K_{IC} has been increased to values between 9 and 11 MPa m$^{1/2}$ (table I).

Table I: Mechanical properties of Si_3N_4-components (samples out of valves)

	Materials		Preliminary Valve Specification
	HCT 90	HOE 120	
σ_{RT} (4-pt.) [MPa]	922	1202	> 900
$\sigma_{1000°C}$ (4-pt.) [MPa]	700	870	≈ 900
Weibull modulus	19	23	> 20
Density [g/cm³]	3,27	3,26	-
K_{IC} [MPa m$^{1/2}$]	7	9-11	-
Young's modulus [GPa]	320	310	-
Hardness (HV 10)	15	15	-

Selection of the procedure

For the production of ceramic inlet or outlet valves both Si_3N_4 and silicon powder may be used as starting materials |4|. Si_3N_4 powders are sintered at different gas pressures after being homogenized with sintering additives and after being shaped in nitrogen atmosphere. The gas pressure depends on the starting powder used, the concentration and the system of the sintering additives. Details on the production of material HOE 120 are to be seen from |5|. When using silicon as starting powder the components are converted into silicon nitride after their pre-homogenization with sintering additives and their shaping in nitrogen atmosphere. The subsequent densification is done in the same way as the sintering of Si_3N_4 pulverized bodies. Material HCT 90 is produced this way; the process is described in detail in |6|.

Shaping techniques for valves are depending on their geometry and the respective component size either by slip casting or by injection moulding. The valve geometries described have been injection moulded. In this process the starting powders are homogenized with the sintering and organic additives like low-molecular polyethylen resins and high-molecular polyethylen thermoplastics. These organic additives are burnt out before the sintering treatment in oxydizing atmosphere. A detailed description of this shaping process is to be seen from |7|.

VALVE SPRING RETAINER DESIGN

During the development work at HCT/HOE it turned out that the connection of the spring disk near the groove at the shaft end of the valves has repeatedly caused fractures. Due to the reduced shaft diameter stress peaks may occur near the groove that are specified by FE analysis as of the order of approx. 300 MPa. In case of irregular operation conditions, however, overturning moments may arise that result in significantly higher loads. Moreover, the grooving of the finished part causes additional costs. An alternative developed by HCT/HOE is to use a grooveless valve that is only friction-connected to the spring disk. Thus, the tensions arising in the connection area of the spring retainer may be reduced and tension peaks may be completely avoided |8|. The FE stress analysis of the conventional spring disk and of the grooveless valve spring retainer connection which has been successfully tested meanwhile by HCT/HOE is to be seen from Fig. 2. The grooveless version allows for maximum tensions of less than 100 MPa. Cycle load testing at a frequency of f = 100/s and a maximum force of F = 3000 N have been successfully passed. The new developed grooveless spring retainer connection is intended to be tested in the engine test rig shortly.

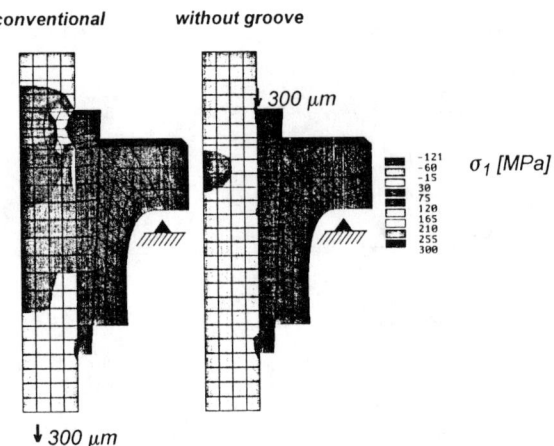

Fig. 2: Finite-element-analysis of valve spring retainer joint |8|

TEST BENCH AND ROAD TRIALS

After the development of the suitable materials and the selection of the right production method Hoechst started with the first vehicle equiped with ceramic inlet valves in October 1991. The outlet valves of the test model Daimler Benz 300 E/24 V remained conventional for this first trial on public roads. The results obtained after a running distance of approx. 30.000 street kilometers were so excellent that meanwhile a second vehicle test has been started with the same model, also equipped with ceramic inlet valves. Medium-term planning provides that up to 15 automobiles will be equiped with ceramic valves under Hoechst control to pass the road test.

In practical use on the road it is most obvious that the running behaviour of the modified engines is clearly smoother. As the inner friction moment is according to the speed lower by 5-30 % than that of engines with conventional valves. The noise level is reduced by a noticeable 18 decibels at a medium speed of 3000 rpm (fig. 3 and 4).

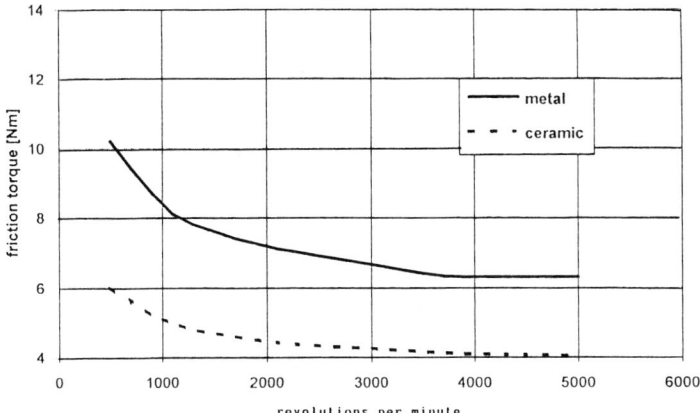

Fig. 3: Friction torque in the cylinder head
(Daimler Benz measurements |9|)

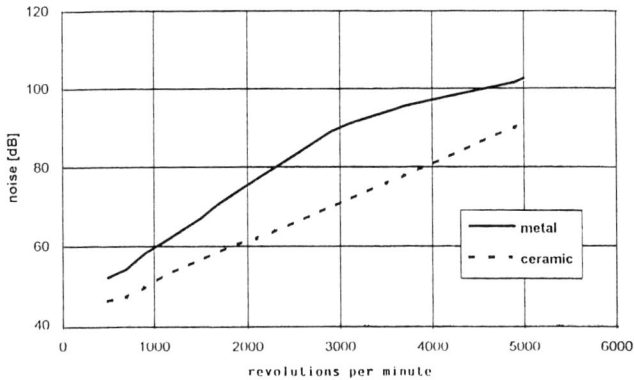

Fig. 4: Noise emission of the cylinder head
(Daimler Benz measurements |9|)

Also of great importance are the reductions in emission measured in the motor bench tests as per DIN cycles and the reductions in fuel consumption. While the emission of hydrocarbon (HC) and carbon monoxide (CO) was reduced by about 30 % and 20 % respectively, a dramatic reduction by approx. 80 % has been measured for the NO_x emission. Just as impressive is the reduction in fuel consumption by about 3-4 % |10|.

On the whole all results achieved so far are very promising due to the clearly reduced oscillating masses. Depending on the respective design even shorter opening and closing times of the valves as well as higher speed ratings may additionally be realized. For example, upon the development of a racing engine speed tests up to 16000 rpm have been run. Due to problems arising in the groove area of the racing valves different modifications have been made (for instance the groove geometry has been modified or copper sleeves have been used) and finally a long-time test of 14000 rpm has been successfully passed (fig. 5).

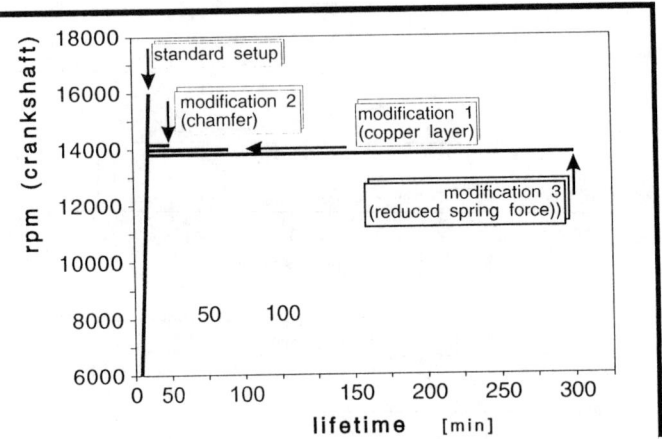

Fig. 5: Revolutions test of ceramic engine valves

COST-EFFECTIVENESS RATIO

From the actual point of view ceramic valves are first of all in price competition to Na-cooled metal outlet valves. Using Na-cooled valves in upper-class cars improves on the one hand the knock rating of the engines while it reduces on the other the fuel consumption under full load. Using ceramic outlet valves would in addition avoid the use of Na metal which is problematic with regard to environmental protection respective to recycling.

The costs involved for the production of unprocessed ceramic valve blanks may be roughly classified into the three categories of powder, shaping and sintering. With regard to a future serial production HCT/HOE is working intensively on an optimization of the individual processing steps in order to reduce the costs as far as possible. Fig. 6 shows how the total costs may be changed by the influence of the individual cost blocks required for the manufacture of the valve blanks. This graph shows that especially the powder prices are influencing the total valve costs to a great extent. Based on a given powder price of DM 50/kg a reduction of the powder costs by 20 % would reduce the total valve price by about 10 %, for example. In view of the large quantity production and the competitiveness with metal valves a target price of < DM 30/kg is aimed at for the powder. Compared to the powder costs the influence of the shaping and sintering cost blocks is relatively small. Nevertheless, experiments are being run to develop a continuous sintering process that works at atmospheric pressure in order to reduce the share of energy costs.

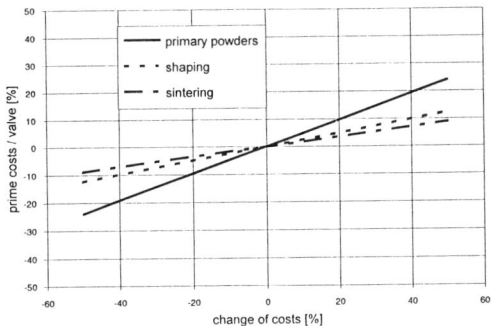

Fig. 6: Prime costs of ceramic engine valves

FINISHING

The principally required finishing of ceramic valves has always been an essential drag with regard to price competition. But meanwhile our development work has shown that finishing of ceramic valve blanks of an oversize shaft diameter of 1,5 mm can be done in 2-3 minutes. Inspite of the diamond tools required the processing costs are in the first approximation in the same range as the finishing costs of metal valves. Thus it seems near at hand that ceramic inlet and outlet valves - finishing included - may be produced at a competitive price approximately on the level of Na-cooled metal valves. At present HCT/HOE are trying to demonstrate within the scope of a large scale production trial on the finishing of ceramic valves that an economic realization is possible.

PROSPECTS

In the past two years ceramic valve prototypes made in pilot scale have proved on the one hand that it is possible to make them and on the other that they are able to function in the engine on the road. There is no doubt about it that till series production still many further problems will have to be solved. For example, a precondition for serial production is that a rather simple and reliable test method is on hand to detect micro-failures in the ceramic structure before the valves are finished, thus avoiding preliminary engine testing. Moreover, the reproducibility of the characteristics of ceramic components has to be proved in practical use for larger serial quantities. This proof, however, can be given only if ceramics and automotive manufacturers are cooperating very closely. The valves produced in large quantities have to be routine mounted and tested after having passed the quality assurance department. Fundamentally, this can be proved only by fleet trials of several hundreds of vehicles on the road. Within the field of electronic applications the ceramics manufacturers have already proved over the past years that they are in a position to manufacture components with failure rates in the ppm range. This knowledge and experience must be transmitted to components out of structural ceramics like the valve for example. To make use of all advantages offered by the ceramic valve it is in addition necessary to modify the cylinder head design of the engines at least partially. Progressive control periods would allow for a considerable reduction of the shaft length of the valve so that also the overall height of the cylinder head may be reduced. Simultaneously the engine output will be increased effectively by reducing the cylinder head height. One day, lower engine heights may even be advantageous for the automotive design and result in a further reduction of the drag coefficient c_w. Moreover, the results achieved so far point at a complete omission of hydraulic tappets in future due to non-wear of the valve seat and even at essential extensions of the mechanic control intervals for the air gap compensation. In view of the actually preferred multi-valve techniques (engine constructions comprising more than four valves per cylinder are already under development) in particular for high-power engines of small volumes with lowest emission values weight savings of several kilograms would be possible.

LITERATURE

1. P. Dötsch, M. Steiner, J. Heinrich: High Strength HIPRBSN - an Optimization by Statistical Methods. Proc. Third Int. Conf. on Hot Isostatic Pressing, June 10-14, 1991, Osaka, Japan

2. C. Boberski, H. Bestgen, R. Hamminger: Microstructural Development during Liquid Phase Sintering of Si_3N_4 Ceramics. J. Europ. Ceram. Soc. 9 (1992) 95-99

3. C. Boberski, R. Hamminger, M. Peuckert, F. Aldinger, R. Dillinger, J. Heinrich, J. Huber: High Performance Silicon Nitride Materials. Angew. Chem. Adv. Mater. 101 (1989) 1592-1601

4. J. Heinrich, E. Backer, M. Böhmer: Hot Isostatic Pressing of Si_3N_4 Powder Compacts and Reaction Bonded Si_3N_4. J. Am. Ceram. Soc. 71 (1988) C28-C31

5. C. Boberski, R. Hamminger: Mechanical Properties of Gas Pressure Sintered Silicon Nitride. Proc. 4. Int. Symp. Ceram. Mater. and Components for Engines (Göteborg), Elsevier, Amsterdam 1991

6. H. Gasthuber, J. Heinrich, J. Huber, M. Steiner, W. Bunk: Hot Isostatically Pressed Reaction Bonded Silicon Nitride Prechambers for the Diesel Engine. Am. Ceram. Soc. Bull. 68 (1989) 2104-2108

7. J. Heinrich: The Influence of Processing Conditions on Microstructure and Mechanical Properties of Reaction Bonded Silicon Nitride. DFVLR-FB 79-32, 1979

8. V. Olt, T. Thiemeier, O. Rosenfelder: Zuverlässigkeitsanalyse für keramische Bauteile: Numerische Spannungsbewertung oder Bauteilprüfung? Symp. Mechanical Properties of Ceramic Structural Materials. Sept. 1992, Karlsruhe, Germany, to be published

9. H. Gasthuber, R. Krebser: Strukturkeramik zur Reduzierung der dynamischen Massen im Ventiltrieb des Kolbenmotors. Haus der Technik e.V. Essen, T-30-911-056-1, Nov. 1991

10. H. Bühl, Daimler-Benz AG, Private Communication

OXIDATION KINETICS OF Si_2N_2O CERAMICS

JEANETTE PERSSON, PER-OLOV KÄLL AND MATS NYGREN
Department of Inorganic Chemistry, Arrhenius Laboratory, Stockholm University
S-106 91 Stockholm, Sweden

1. INTRODUCTION

Ceramics based on silicon nitride have long been considered promising candidates for high temperature engineering applications due to their excellent refractory properties combined with high strength and good resistance to oxidation. It is well known, however, that the performance of ceramics varies considerably with chemical composition, for example with type and amount of intergranular phase. Especially the evaluation of the high temperature behaviour of the ceramics under varying conditions is therefore important in increasing the knowledge of these materials.

In previous articles we have interpreted the oxidation behaviour of Si_2N_2O ceramics with use of the parabolic rate law, and a new rate law reading $\Delta w/A_o = a \cdot \arctan\sqrt{(bt)} + c\sqrt{t}$, [1,2]. In this article we further develop the arctan function, and also show that by an alternative mathematical approach the following rate equation can be obtained $(\Delta w/A_o)^2 = a_1 \ln(b_1 t + 1) + c_1 t$.

In the following, we compare the oxidation behaviour of two series of Si_2N_2O ceramics within the framework of the two rate laws.

2. EXPERIMENTAL

The preparation and characterization of the fully dense Si_2N_2O ceramics used in this study are described in Refs. 1 and 2.

Two series of Si_2N_2O samples were investigated: one series of essentially pure Si_2N_2O samples (here dentoted P), and a second series which contained traces of Ba on the surfaces of the samples (denoted Ba). In both series the matrix material was pure Si_2N_2O (i.e. no sintering additives were used in the preparation), but because one set of samples was insufficiently cleaned before the oxidation experiment the samples in this series contained minor amounts of Ba originating from the wax paste used in attaching the samples to the sample holder during polishing. Studies by SEM and EDX revealed the presence of Ba in the oxide scales of these samples.

The oxidation experiments were performed in a TG unit (SETARAM TAG 24). The oxidation was carried out isothermally with ceramic pieces of the approximate size 15x15x1 mm^3 for 20 hours in flowing dry oxygen of atmospheric pressure.

3. RESULTS AND DISCUSSION

3.1 Mathematical approach

Almost forty years ago Evans et al. [3] suggested that the logarithmic rate law behaviour frequently observed in low-temperature oxidation of certain metals such as iron, could be explained if one assumed that a reduction of the available cross-section area for diffusion occurred during the oxidation.

In our oxidation studies of Si_3N_4-based ceramics we have quite often observed a logarithmic or near-logarithmic rate law behaviour, rather than a pure parabolic one. We have also adopted the idea of a decreasing cross-section area for diffusion as a possible explanation of this non-parabolic behaviour. Such a reduction of the cross section might occur as a result of crystallization processes and formation of cracks and nitrogen bubbles within the oxide scale and/or at the scale/matrix interface. All these phenomena are regularly observed in the oxidation of Si_3N_4 ceramics.

On the other hand, to the extent that the growth of the oxide layer is governed by diffusion the parabolic rate law can be expected to hold, i.e

$$dX/dt = (1/2)\alpha X^{-1} \tag{1}$$

where X is the thickness of the formed oxide scale, t is time and α is the apparent diffusivity of the rate-determining species. Since, however, we are measuring the weight gain of the sample rather than the thickness of the oxide scale, the weight increase ought to be related to the thickness according to

$$\Delta w/A_o = \rho_o X \tag{2}$$

In (2) Δw is the weight increase of the oxidizing sample, A_o is the surface area of the sample, and the factor ρ_o represents the "effective density" of the growing oxide, as described in Ref. 1. Eq. (1) now can be re-written with use of (2) as

$$(2/A_o)(d\Delta w/dt) = (1/\rho_o)K_p A(t)X^{-1} \tag{3}$$

where K_p is the parabolic rate constant, and $A(t)$ is a function such that if $A(t)$ is constant $=1$ for all t, then the parabolic rate law will be a solution to (3), i.e.

$$(\Delta w/A_o)^2 = K_p t + B \tag{4}$$

In (4) B is an additive constant (ideally $=0$) and $K_p = \rho_o^2 \alpha$. However, if $A(t)$ in eq. (3) is not constant but monotonically decreasing, the solution will be clearly non-parabolic. We have chosen the following expression to account for the decreasing cross section area (see Ref. 1 for further details)

$$A(t) = (1 + (f\beta - t_o^{-1})t)/(1 + (\beta - t_o^{-1})t) \tag{5}$$

In (5) β is the constant that expresses the rate of decrease. The function $A(t)$ has the properties that $A(t)=1$ for all t if $f=1$, wheras if $0<f<1$ $A(t)=1$ when $t=0$ and decreases to $A(t)=f$ at $t=t_o$ (where t_o is assumed to be >0). In other words, f is the constant that expresses to what extent the observed oxidation curve diverges from "genuine" parabolic kinetics, implying that if f is close to unity, the oxidation curve will be essentially parabolic, while an f close to zero indicates that considerable changes (e.g. crystallization processes) have occurred during the oxidation, leading to a much reduced cross-section area for diffusion.

If a parabolic behaviour is observed at a later stage of the oxidation experiment, after an initial time lapse t_o ($t_o>0$), then one will have

$$(\Delta w/A_o)^2 = K_p^o t + B_o \tag{6}$$

In (6) K_p^o is the apparent parabolic rate constant for $t \geq t_o$, and B_o is an additive constant >0.

Regarding eq. (3), two different solutions exist if $0<f<1$. We may either substitute X in (3) for $(1/\rho_o)(K_p t+B)^{1/2}$, according to eq. (4), or for $\Delta w/(\rho_o A_o)$, according to eq. (2). In the former case integration of eq. (3) yields

$$\Delta w/A_o = a \cdot \arctan(bt+b')^{1/2} + c(t+b'b^{-1})^{1/2} \tag{7}$$

where a, b, b' and c are constants. In Ref. 1 we have assumed that B in eq. (4) is $=0$, yielding $b'=0$. Thus, the term $b'b^{-1}$ in (7) represents the zero-point error of the arctan function. A detailed discussion of eq. (7) is given in Ref. 1.

However, if the latter substitution is performed, the solution of (3) will be

$$(\Delta w/A_o)^2 = a_1 \ln(b_1 t+1) + c_1 t \tag{8}$$

where a_1, b_1 and c_1 are constants. It can be shown that the parabolic rate constant K_p is related to the constants of eqs. (7) and (8) by, respectively

$$K_p = (c + a\sqrt{b}/(1+b'))^2 \tag{9a}$$

and

$$K_p = a_1 b_1 + c_1 \tag{9b}$$

Also, the fraction f is related to these constants by

$$f = (c + a\sqrt{b}/(bt_o+b'+1))/\sqrt{K_p} \tag{10a}$$

for eq. (7), and by

$$f_1 = (c_1 + a_1 b_1/(b_1 t_o+1))/K_p \tag{10b}$$

for eq. (8). The constants f and f_1 usually have different values (except when both are =1), but are interrelated through the expression

$$f = (f_1(t_o+B/K_p)/(t_o+B_o/K_p^o))^{1/2} \tag{11}$$

The apparent parabolic rate constant K_p^o is related to K_p by

$$f^2 K_p = K_p^o(t_o+B/K_p)/(t_o+B_o/K_p^o) \tag{12a}$$

for the arctan function (7), and by

$$f_1 K_p = K_p^o \tag{12b}$$

for the logarithmic function (8). The relationships (11), (12a) and (12b) can be derived from the assumption that at $t=t_o$ the non-parabolic and the parabolic branches of the oxidation curve ought to have equal derivatives.

3.2 Experimental results

All oxidation curves of the Si_2N_2O samples except one exhibited a non-parabolic behaviour, either partly during the initial branch of the curve or extended throughout the curve. The duration of the non-parabolic part decreased with increasing temperature, however, and at higher temperatures a t_o value always could be determined (see Table I and II). After that point, i.e. for $t>t_o$, eq. (6) seemed to be valid. The non-parabolic parts of the curves could be fitted very well by both the arctan function (7) and the logarithmic function (8). The only sample which showed a parabolic behaviour during the whole oxidation was the one in the P series oxidized at 1300°C. The squared weight-gain curve for the P sample oxidized at 1300 °C together with the fitted function (4) is shown in Fig. 1a. The squared weight-gain curves for the P and Ba samples oxidized at 1350 and 1400 °C, respectively, together with the fitted arctan (7) and/or parabolic functions (6) are shown in Fig. 1b. The three curves represent three different types, namely "pure" parabolic (Fig. 1a), and "pure" arctan and mixed arctan-parabolic (Fig. 1b).

The calculated constants of the arctan function, including the calculated values of f, t_o and

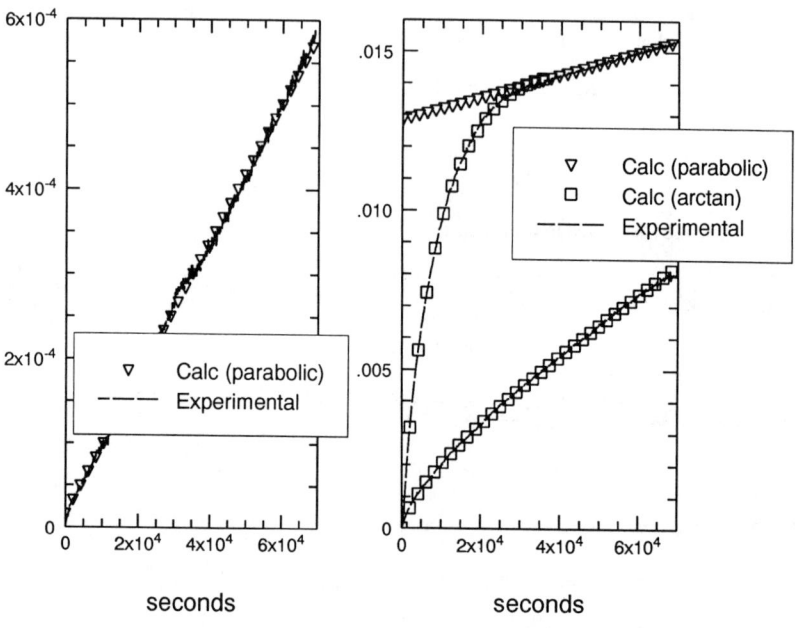

Fig. 1 a *Fig.1 b*

The squared weight gain ($mg^2 \cdot cm^{-4}$) plotted vs. time (s) for the oxidation of Si_2N_2O ceramics. The unbroken curves are the observed weight increases, whereas the squares represent the fitted arctan function (7) and the triangles the fitted parabolic functions (4) and (6). (a) pure Si_2N_2O at 1300 °C; (b) pure Si_2N_2O at 1350 (lower curve) and Ba-containing Si_2N_2O at 1400°C (upper curve).

K_p, for both series of oxynitrides are compiled in Table I, and the corresponding constants for the logarithmic function in Table II.

It is clear from Tables I and II that the K_p values are lower for the P series than for the Ba series, as might be expected. The f (and f_1) values are larger for the P series than for the Ba series, indicating that the decrease of the cross-section area for diffusion is somewhat less pronounced for the former samples than for the latter. In rough outline, the f and f_1 values decrease with increasing temperature for both series of samples. This observation appears to be consistent with the assumption that the reduction of the cross-section area is due to crystallization processes in the scale, as more extensive crystallization is expected to occur at higher temperatures. The oxygen diffusivity is reported to be much lower through crystalline than through amorphous SiO_2 [4]. In our previous articles [1,2] we have reported that extensive crystallization was actually observed in all samples, except for the one oxidized at 1300 °C in the P series. Accordingly, this sample exhibited pure parabolic oxidation behaviour.

A comparison between the arctan (Table I) and the logarithmic (Table II) functions shows that the t_o values in general are very similar for the two functions for both series of samples, and that the K_p values are always of the same magnitude. The differences between the K_p's

Table I

Temp [°C]	Sample Series	a [mg·cm⁻²] /10⁻²	b [s⁻¹] /10⁻⁴	b´ /10⁻³	c [mg·cm⁻²·s⁻¹/²] /10⁻⁵	t_o [s] /10³	f	K_p [mg²·cm⁻⁴·s⁻¹] /10⁻⁶
1300	P					0	1.00	0.00768
	Ba	9.99	0.528	-11.9	-0.804	57.7	0.24	0.528
1350	P	1.69	5.30	3.40	25.4	$>t_{max}$*	0.41	0.412
	Ba	6.61	1.83	-33.3	17.4	50.5	0.24	1.21
1400	P	6.29	1.55	-7.72	3.77	57.5	0.14	0.685
	Ba	15.5	1.16	-14.0	-28.4	33.6	0.04	1.98
1450	P	4.06	4.18	-4.45	8.02	34.9	0.15	0.837
	Ba	19.8	1.38	-11.7	-54.6	20.9	0.03	3.27
1500	P	6.15	7.10	-101.	5.64	14.4	0.11	3.53
	Ba	13.2	3.71	-38.5	-43.8	11.5	0.02	4.92
1550	P	4.28	13.3	-90.1	22.1	16.2	0.15	3.77
	Ba	8.21	8.76	-49.7	-9.13	10.3	0.06	6.08
1600	P	23.2	1.39	-18.6	-49.6	14.1	0.19	5.27
	Ba	22.2	4.04	-20.3	-123.	5.83	0.03	11.0

*) $t_{max} \geq 7 \cdot 10^4$ s

Table II

Temp [°C]	Sample Series	a_1 [mg²·cm⁻⁴] /10⁻³	b_1 [s⁻¹] /10⁻⁴	c_1 [mg²·cm⁻⁴·s⁻¹] /10⁻⁸	t_o [s] /10³	f_1	K_p [mg²·cm⁻⁴·s⁻¹] /10⁻⁶
1300	P				0	1.00	0.00768
	Ba	7.42	0.738	-2.96	58.2	0.14	0.518
1350	P	0.959	2.70	7.70	$>t_{max}$	0.25	0.416
	Ba	4.84	2.23	5.33	50.0	0.13	1.13
1400	P	3.11	2.26	-0.923	55.4	0.06	0.693
	Ba	11.6	1.99	-27.5	32.1	0.02	2.03
1450	P	1.41	6.10	0.840	33.3	0.06	0.868
	Ba	13.2	2.93	-46.9	22.3	0.01	3.41
1500	P	2.34	13.4	1.10	16.9	0.05	3.14
	Ba	6.00	9.10	-36.9	13.3	0.01	5.09
1550	P	1.72	18.2	9.85	16.7	0.06	3.22
	Ba	3.98	16.0	-26.9	9.15	0.02	6.10
1600	P	26.0	2.25	-78.2	14.6	0.12	5.06
	Ba	4.58	51.7	-8.09	13.8	0.01	23.6

obtained from the arctan and the logarithmic functions, do not exhibit any clear tendency but appear rather random. Also, the values of f and f_1 are consistent with eq. (11).

The zero-point corrections introduced for the arctan function ($b´b^{-1}$ in eq. (7)) in general were small, usually < 200 s. The introduction of such corrections, however, affected the K_p values significantly. For both series the K_p values increased, usually by 10-20%, and in one case by as much as 55% (the 1500 °C sample in the P series). So far we have not carried out the corresponding corrections for the logarithmic function (8). The Arrhenius plots for both series of samples are given in Fig. 2. The calculated E_a values for the P series are 279±40 kJ·mol⁻¹ for the arctan function, and 267±34 kJ·mol⁻¹ for the logarithmic. For the Ba series the values are 233±12 and 275±26 kJ·mol⁻¹ for the arctan and the logarithmic functions, respectively. The zero-point corrections discussed above did not significantly change the the E_a values obtained for the arctan function.

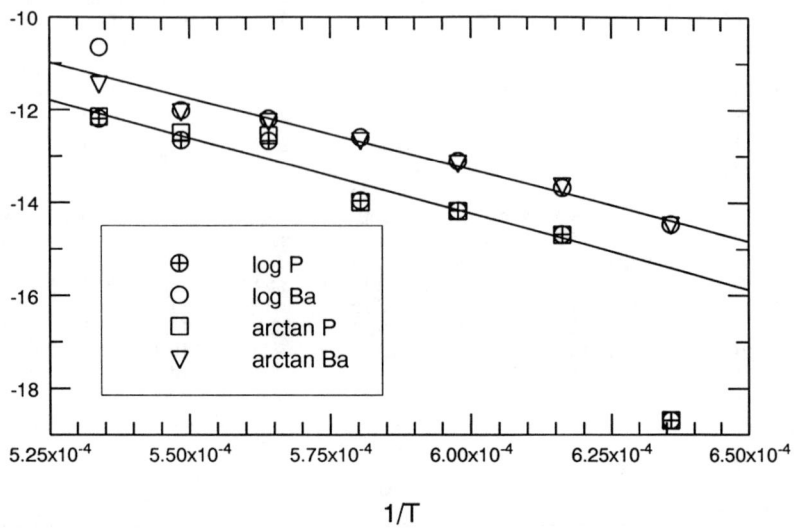

Fig. 2
Arrhenius plots of the parabolic rate constants K_p for the oxidation of pure (lower points) and Ba-containing (upper points) Si_2N_2O ceramics in the temperature range 1300-1600 °C. The squares and triangles are calculated from the arctan funtion (7); the circles from the logarithmic function (8). The slopes of the straight lines represent the calculated average activation energies for the two functions for the two series.

4. CONCLUSIONS

1. The arctan function (7) and the logarithmic function (8) yield simlar parabolic rate constants and activation energies for the oxidation of pure as well as of Ba-containing Si_2N_2O ceramics in the temperature range 1300-1600 °C.
2. Both functions seem to fit equally well to the non-parabolic branches of the observed oxidation curves.
3. The introduction of a zero-point correction for the arctan function in general increased the K_p values obtained for that function. The increases were in most cases of the order of 10-20%. However, the calculated activation energies remained essentially the same as if no zero-point corrections were undertaken.

REFERENCES

1. J. Persson, P.O. Käll and M. Nygren, *J. Am. Ceram. Soc.*, in print
2. J. Persson, P.O. Käll and M. Nygren, in *Proc. Irish Materials Forum Conference: Materials for Advanced Technology Applications*, edited by M. Buggy and S. Hampshire (Trans Tech Publications, Aedermannsdorf, 1992), p. 49
3. D.E. Davies, U.R. Evans and J.N. Agar, Proc. Roy. Soc., A225, 1954 (London), p. 29
4. G. H. Frischat, *Ionic Diffusion in Oxide Glasses*, (Trans Tech Publications, Aedermannsdorf, 1975), p. 129

OXIDATION BEHAVIOUR OF ZIRCONIA-SIALON COMPOSITES

YIBING CHENG* AND DEREK P.THOMPSON**
*Department of Materials Engineering, Monash University, Melbourne, Victoria 3168, Australia
**Department of Mechanical, Materials & Manufacturing Engineering, University of Newcastle upon Tyne, NE1 7RU, United Kingdom

ABSTRACT

In yttria densified ZrO_2-sialon composites, an increase in β':O'-sialon ratio results in a structural change in the zirconia phase because an increasing amount of nitrogen dissolves in the zirconia structure as the overall composition becomes increasingly nitrogen rich. Both ZrN and nitrogen-containing zirconia (N-ZrO_2) are observed in the β'-ZrO_2 composite and these phases show a poor oxidation performance above 600°C. However, the introduction of small amounts of O' phase into the β'-ZrO_2 composite has provided an effective oxygen-rich barrier to reduce the diffusion rate of nitrogen in zirconia and therefore O'-β'-ZrO_2 composites showed superior oxidation resistance compared with β'-ZrO_2 materials at temperatures between 600 and 1200°C.

INTRODUCTION

As possible candidates for engineering ceramic applications, zirconia-containing sialon composites have attracted wide attention. Zirconia is a readily available material with a melting point of 2680°C, and more importantly the t→m martensitic transformation provides a unique opportunity for toughening of high performance ceramics. Research activities concerned with zirconia-containing silicon nitride materials have covered the incorporation of zirconia for toughening[1-3], production of stabilized zirconia-sialon composites for refractory applications[4] and the use of zirconia as a sintering additive[5]. Unlike oxide ceramics, the structure of zirconia in nitrogen ceramics depends not only on the nature and amount of solute cations but also on the surrounding nitrogen environment because nitrogen can substitute for oxygen in the zirconia lattice and change its structure.

The reaction between nitrogen and zirconia has been widely observed in ZrO_2-containing sialon composites, producing zirconium oxynitrides, nitrogen-containing c-ZrO_2 or ZrN phase, depending on sintering conditions and composition[2,3,5,6]. Zr oxynitrides and ZrN oxidize readily at about 600°C[2,7,8], accompanied by a large volume expansion that often leads to a detrimental fragmentation of materials. Therefore different techniques, such as pre-reacting ZrO_2 with Y_2O_3, were used to eliminate the formation of the Zr oxynitride and ZrN phases and improve the oxidation properties of the materials[2,7,9,10]. However, another problem arises when the zirconia is doped with yttria, which is that nitrogen, acting as an anion stabilizer, can form a continuous solid solution with the yttria-containing zirconia in nitrogen ceramics[10,11]. The formation of nitrogen-containing zirconia (N-ZrO_2) in a nitrogen ceramic matrix is often unavoidable since nitrogen can be readily provided by both the matrix and the sintering atmosphere and thus the impact of any N-ZrO_2 produced on the oxidation properties of the material should also be carefully assessed. The present paper reports an oxidation study on some N-ZrO_2-containing zirconia-sialon composites.

EXPERIMENTAL

Four compositions were selected, in which the ZrO_2:Y_2O_3 ratio was always kept constant at 25:4 (in weight units) but the O':β' ratio was varied from 44:31 to 0:75 (by weight percetage) (Table 1). The designed Al_2O_3 contents of the O' and β' phases were x=0.15 and z=0.8 respectively and these compared to the upper limit of β'-sialon coexisting with O'-sialon[12]. The starting powders were Si_3N_4 (Starck, LC10), Al_2O_3 (Alcoa A17), Y_2O_3 (Rare Earth products, Ltd.) and m-ZrO_2 (Cookson). SiO_2 required to form O' was always introduced in the form of $ZrSiO_4$ (OPAZIR "S"). Sintering was carried out in a carbon resistance furnace in N_2. Bulk densities of the fired pellets were measured using the Archimedes method. Both Hägg-Guinier camera and diffractometer techniques were used to identify crystalline phases; phase identification was carried out mainly from the Hägg-Guinier photographs and the proportions of different phases were determined by measuring diffractometer profiles[13]. SEM microstructures were observed by scanning electron microscopy using a Camscan S4-80DV microscope.

Table I. Starting compositions (wt%)

Sample	O'	β'	ZrO_2	Y_2O_3
BOZ8	44	31	25	4
BOZ9	30	45	25	4
BOZ10	10	65	25	4
BOZ11	0	75	25	4

Note: O'—$Si_{1.85}Al_{0.15}O_{1.15}N_{1.85}$
β'—$Si_{5.2}Al_{0.8}O_{0.8}N_{7.2}$

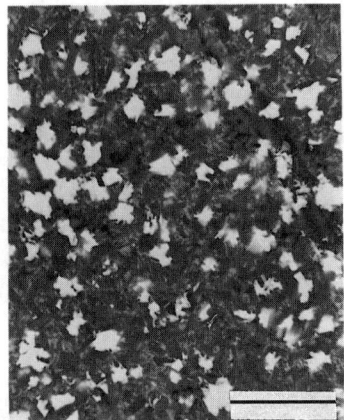

Fig.1 SEM back scattered image of Sample BOZ8 in which ZrO_2 are the white grains; O' the plate-like grains; β' the more equiaxial grains. marker bar = 10 μm.

RESULTS AND DISCUSSION

Formation of O'-β'-ZrO_2 composites

Samples were sintered at 1500°C for 0.5 hour followed by 1700°C for one hour. Under these conditions, the reaction is complete (Table 2) and the microstructure consists of O', β' and ZrO_2 phases (Fig.1). The increase in density with increasing β' sialon content is very obvious and is due to the β' phase possessing a higher density (3.2 gcm^{-3}) than O' (2.9 gcm^{-3}). Tetragonal zirconia in the material was quite stable and only a trace amount of m-ZrO_2 was observed in samples containing O'-sialon. It was noticed that the nitridation of ZrO_2 became thermodynamically favorable as the composition shifted from the O' to the β' region and eventually ZrN occurred in equilibrium with a β'-sialon matrix. A possible mechanism for this nitridation is

$$Si_3N_4(s) + 3ZrO_2(s) \rightarrow 3ZrN(s) + 3SiO_2(s) + \frac{1}{2}N_2(g) \quad (1)$$

The introduction of O' into the composites provides an oxygen rich environment which can affect the above reaction. As a result, even a small amount of O' in the sample effectively eliminated ZrN. It was found that the amount of O' in the final products was always noticeably lower than its designed figure. This may have been due to some of the SiO_2 not having reacted with Si_3N_4 and instead forming a glassy phase in the grain boundaries between ZrO_2 and the sialon grains. This oxygen-rich grain boundary then inhibits ZrN formation.

Table II. Density and phase identification of bulk samples after sintering (wt%)

Sample	D (gcm^{-3})	O'	β'	t-ZrO_2	m ZrO_2	ZrN
BOZ8	3.402	36.8	38.2	23.8	1.2	-
BOZ9	3.483	19.5	55.5	23.3	1.7	-
BOZ10	3.537	-	75	25.0	-	-
BOZ11*	3.556	-	75	>15.0	-	<10

* Amounts of ZrO_2 and ZrN in Sample BOZ11 were estimated.

There was no Zr oxynitride in these samples due to the presence of Y_2O_3 [2,10]. However, the reaction between nitrogen and the yttria-containing zirconia in the β'-rich matrix not only produced a ZrN phase but stabilized the remaining zirconia in a non-transformable tetragonal state as well. From the cell dimensions of the tetragonal zirconia in these samples after sintering (Table 3), it is seen that the c_t/a_t ratio of the phase gradually reduces as more β' is introduced although the $Y_2O_3:ZrO_2$ ratio in all the samples is constant. This structural change indicates that an increasing amount of nitrogen dissolves in the N-ZrO_2 solid solution as the overall composition becomes more nitrogen rich[11]. The tetragonal zirconia in the O'-rich matrix must have associated with much less nitrogen than its counterpart in the β'-rich material and consequently appears relatively more "transformable". It is suggested that the oxygen-rich environment in the O'-ZrO_2 microstructure has retarded the diffusion of nitrogen into the zirconia[14]. Thermodynamically Zr-O -containing compounds are much more stable than Zr-N ones in air and the replacement of nitrogen by oxygen in phases containing Zr-N bonds will spontaneously occur at an appropriate temperature (typically >600°C). Therefore a knowledge of the oxidation performance of both ZrN and N-ZrO_2 phases is imperative when considering applications for zirconia-sialon composites.

Table III. X-ray results for tetragonal zirconia in zirconia-sialon composites after sintering and oxidation respectively.

Sample	After Sintering*			After Oxidation**		
	a_t (Å)	c_t (Å)	c_t/a_t	a_t (Å)	c_t (Å)	c_t/a_t
BOZ8	5.108	5.189	1.0159	5.107	5.184	1.0151
BOZ9	5.111	5.184	1.0143	5.108	5.184	1.0149
BOZ10	5.116	5.189	1.0143	5.110	5.187	1.0151
BOZ11	5.121	5.172	1.0100	5.110	5.178	1.0133

* Samples were sintered at 1500°C/0.5 hour and then 1700°C/1 hour in N_2.
** Samples were heat-treated at 800°C for 2 hours in air.

Oxidation behaviour of nitrogen-containing tetragonal zirconia

ZrO_2 is a very stable oxide and shows excellent oxidation resistance at elevated temperature. However, it can be attacked by oxygen at high temperatures if nitrogen is dissolved in it. The sintered O'-β'-ZrO_2 samples were heat-treated at 800°C for two hours in air. The selection of such a low temperature for oxidation is to eliminate the possibility of β' or O'-sialon phase being oxidized and to only allow changes to occur in the N-ZrO_2 phase. After heat-treatment, it was found that the c_t/a_t ratio of the tetragonal zirconia significantly increased, particularly for Samples BOZ10 and BOZ11 in which β' sialon is the dominant phase (Table 3). The increase in tetragonality during this experiment can only be attributed to the substitution of oxygen for nitrogen in the tetragonal zirconia structure. This proves that the additional stabilization over oxide tetragonal zirconia with increasing β'-sialon content is largely due to an increase in nitrogen content of the zirconia although the yttria content in the zirconia may also vary slightly as a result of the different intake of yttria into grain boundary glass of different compositions.

The oxidation behaviour of the monolithic N-ZrO_2 phase was further investigated by heating of nitrogen-containing TZ-3Y ZrO_2 material in air at temperatures between 600 and 1300°C for 24 hours. Prior to oxidation, the TZ-3Y ZrO_2 (Toyo Soda, Japan) pellets were sintered at 1650°C for 2 hours in nitrogen (see ref. 11 for details of the experimental conditions). After sintering, the material contained about 0.6 wt% of nitrogen and the original t phase became a t' structure with a significant reduction in c_t/a_t ratio from 1.0149 to 1.0084, clearly suggesting nitrogen stabilization[11].

The oxidation results in Table 4 indicate a modest change in cell dimensions of the t' structure at 600°C. The substitution of oxygen for nitrogen becomes more apparent above 700°C and the tetragonality of the zirconia clearly increases with increasing oxidation temperature. After oxidation at 1300°C for 24 hours in air, the original cell dimensions of the 3Y-ZrO_2 were almost completely restored. It is clear that interchange of oxygen and nitrogen atoms in the zirconia lattice is reversible. Moreover, it was noticed that the colour of TZ-3Y ZrO_2 powder changed from white to grey after sintering in N_2 and became white again after oxidation above 800°C. Some volume expansion from this oxidation process is expected because the average size of the non-metal atoms is increasing. However, it was surprising to find that the magnitude of this expansion is very significant considering that only 0.6 wt% nitrogen has been introduced into the zirconia and as a result, severe spalling is observed after oxidation at 800°C (Fig. 2). Cracks were generally intergranular and may be initiated from a cleavage type of fracture occurring on a preferential crystallographic orientation of large crystals (Fig.2(b)). After oxidation at 1300°C in air, complete disintegration occurred and the densified N-ZrO_2 pellet turned to powder. From these results, it is clear that N-ZrO_2 is very detrimental to oxidation behaviour, particular if the amount of this phase in a composite is large; consequently N-ZrO_2, as well as Zr oxynitride and ZrN phases, should all be eliminated from the ziconia-containing nitrogen ceramics.

Table IV. X-ray results for TZ-3Y ZrO_2 samples sintered at 1650°C for 2 hours in N_2 and then oxidized at various temperatures for 24 hours

°C/hours	ΔW %**	a_t (Å)	c_t (Å)	a_t/c_t
TZ-3Y*	-	5.102	5.178	1.0149
1650/2, N_2	-1.73	5.110	5.153	1.0084
600/24, air	-0.6	5.110	5.160	1.0098
700/24, air	+1.2	5.105	5.173	1.0133
800/24, air	+1.0	5.103	5.173	1.0137
1300/24, air	+1.1	5.105	5.179	1.0145

* TZ-3Y ZrO_2 powder prior to sintering
** For weight changes, "-" means a weight loss; "+" a weight gain.

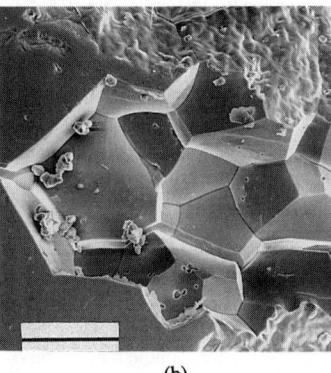

(a) (b)

Fig.2 SEM photographs of the TZ-3Y ZrO_2 (1650°C/2 hours, in N_2) after oxidation at 800°C for 24 hours in air. (a) marker bar = 300 μm; (b) marker bar = 30 μm;

Oxidation behaviour of ZrO_2-containing sialon composites

To examine the oxidation properties of O'-β'-ZrO_2 sialon composites, two large pieces of composition BOZ8 and BOZ11 were sintered at 1700°C for 2 hours. A number of test bars of size 10.7 x 4.5 x 2.2 mm were cut from the sintered samples for oxidation tests. Oxidation resistance was determined at temperatures ranging from 600°C to 1400°C in air for 48 hours respectively and the weight change was continuously monitored using a microbalance.

It can be seen that the ZrN-containing BOZ11 sample started to show a significant weight gain at 600°C (Fig.3). X-ray results in Table 5 indicate that no phase change takes place for both β' and O'-sialon phases but ZrN has disappeared after heat-treatment at 600°C, confirming that the weight gain at this temperature is mainly due to oxidation of ZrN in the material, corresponding to a reaction

$$ZrN(s) + O_2(g) \rightarrow ZrO_2(s) + \frac{1}{2}N_2(g) \qquad (2)$$

At 800°C, a violent increase in oxidation rate was recorded for Sample BOZ11 within the first five hours and the sample showed on the whole a significant weight gain, resulting in severe spalling (Fig.4(a)). Considering no obvious cracks were exhibited by the sample at 600°C, it is thought that a large volume expansion occurs during the transformation from ZrN to m-ZrO_2 at 800°C and in addition, the transition from N-ZrO_2 to ZrO_2, which takes place much more vigorously above 700°C, should also contribute to the overall volume expansion. Above 800°C, a glassy protective layer was formed on the surface, corresponding also to the appearance of cristobalite on the surface after cooling (Table 5), and a reduced rate of oxidation. Microcracks were discovered in large zirconium-containing grains at 1000°C, but the sample survived fragmentation (Fig.4(b)). Replacing 30 wt% of β' by O' in Composition BOZ8 successfully prevented the formation of ZrN and reduced the nitrogen content in the N-ZrO_2 and hence significantly improved the oxidation resistance, particularly at temperatures below 1200°C. It is thought that the oxygen rich silicon oxynitride phase plus grain boundary glass provide a shielding effect preventing ZrO_2 from nitriding to ZrN and retarding the diffusion of nitrogen into ZrO_2 even though the process may not have been completely inhibited. Moreover, at 1400°C, the oxidation of β' and O'-sialon phases became so severe that both materials exhibited a compareble poor oxidation resistance.

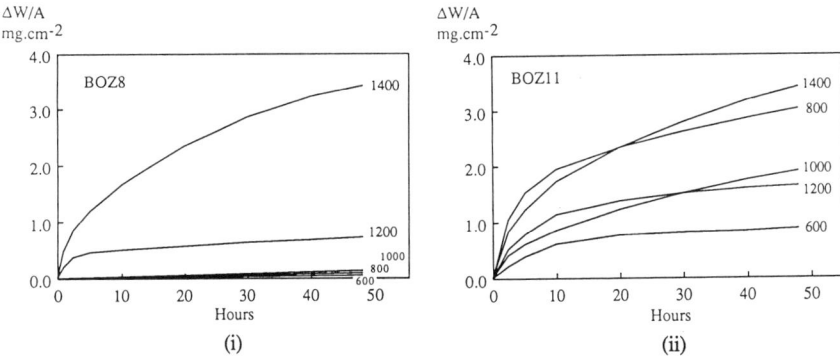

Fig. 3 Oxidation behaviour of O'-β'-ZrO_2 (i) and β'-ZrO_2 (ii) composites at different temperatures (°C) for 48 hours respectively.

Table V. Crystalline phases in the samples oxidized for 2 days at different temperatures

Sample	T°C	β'	O'	t	m	ZrN	SiO_2	mullite	$ZrSiO_4$	y-$Y_2Si_2O_7$
BOZ8	RT	ms	m	s	m	-	-	-	-	-
	600	ms	m	s	m	-	-	-	-	-
	800	ms	m	s	m	-	-	-	-	-
	1000	m	m	s	m	-	m	-	-	w
	1200	m	m	s	m	-	ms	-	m	m
	1400	-	mw	mw	-	-	w	-	vs	-
BOZ11	RT	vs	-	vs	vw	mw	-	-	-	-
	600	vs	-	vs	w	-	-	-	-	-
	800	vs	-	vs	w	-	-	-	-	-
	1000	ms	w	vs	mw	-	mw	-	-	w
	1200	ms	-	s	mw	-	ms	w	m	m
	1400	vw	-	w	vw	-	w	s	vs	-

CONCLUSIONS

The rate of nitrogen diffusion into zirconia is accelerated as the nitrogen content of the sialon matrix increases, resulting in the formation of N-ZrO_2 and eventually ZrN in β'-ZrO_2 sialon materials. Both ZrN and N-ZrO_2 oxidize above 600°C, and at 800°C the material experiences its worst oxidation performance, showing catastrophic spalling of β'-ZrO_2 composites. The formation of ZrN is depressed by the introduction of small amounts of O'-sialon phase into the β'-ZrO_2 composite because the oxygen rich silicon oxynitride phase and grain boundary glass provide a barrier retarding the diffusion of nitrogen into the zirconia and also the nitrogen content of the N-ZrO_2 phase is reduced by the same principle. Therefore O'-β'-ZrO_2 composites demonstrate superior oxidation resistance at temperatures between 600 and 1200°C.

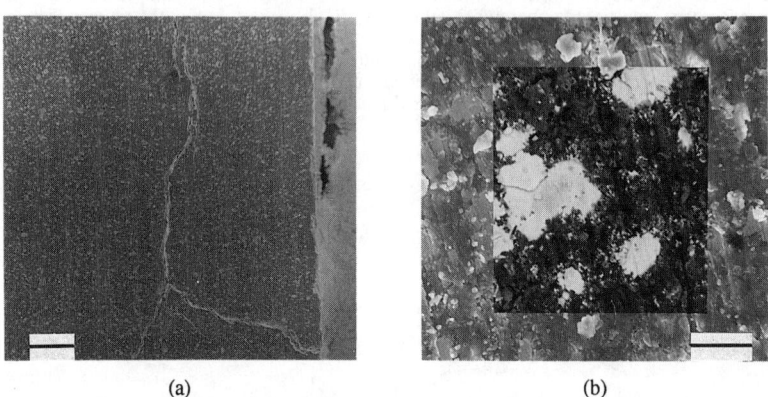

Fig. 4 SEM photographs of oxidized Sample BOZ11. (a) Catastrophic cracking after oxidation at 800°C for 2 days; marker bar = 300 µm. (b) a combination of SEM secondary and back scattered images (the dark square region) for Sample BOZ11 oxidized at 1000°C for 2 days. Notice cracks in the Zr-containing white grains; marker bar = 10 µm

ACKNOWLEDGEMENT

We sincerely thank the Cookson Group plc., U.K. for financial support during the period when the work was carried out.

REFERENCES

1. N.Claussen and J.Jahn, J.Am.Ceram.Soc. **61**(1-2), 94-5 (1978).
2. F.F.Lange, L.K.L.Falk and B.I.Davis, J.Mater.Res. 2(1), 66-76 (1987).
3. R.Pompe, in Structural Ceramics-Processing, Microstructure and Properties, ed. J.J.Bentzen et al., (Riso National Laboratory, Denmark, 1990), p.97.
4. D.B.Hoggard, H.K.Park, R.Morrison and S.Slasor, Am. Ceram. Bull. 69(7), 1163-6 (1990)
5. W.A.Sanders and D.M.Mieskowski, Adv. Ceram. Mater. 1(2), 166-73 (1986).
6. J.Weiss, L.J.Gauckler, H.L.Lukas, G.Petzow and T.Y.Tien, J.Mater.Sci. 16, 2997-3005 (1981).
7. G.N.Babini, A.Bellosi, P.Vincenzini, D.Dalle Fabbriche and R.Visani, in Science of Ceramics 12, ed by P.Vincenzini (Ceramurgica S.R.I., Faenza, 1984), p.471.
8. P.Vincenzini, A.Bellosi and G.N.Babini, Ceram. Inter. 12, 133-45 (1986).
9. A.K.Tjernlund, R.Pompe, M.Holmstrom and R.Carlsson, in Special Ceramics 8, ed. S.P.Howlett and D.Taylor (The Institute of Ceramics, Stoke-on-Trent, U.K., 1986), p.29.
10. Y.Cheng and D.P.Thompson, , J.Am.Ceram.Soc., in press.
11. Y.Cheng and D.P.Thompson, J.Am.Ceram.Soc., 74(5), 1135-8 (1991).
12. M.Y.H.L.Chan, Ph.D Thesis, University of Newcastle upon Tyne, (1987).
13. K.Liddell, M.Sc Thesis, University of Newcastle upon Tyne, (1980).
14. Y.Cheng and D.P.Thompson, Mater. Forum, in press

CORROSION OF SILICON NITRIDE CERAMICS BY NITRIC ACID

KUNIHIKO KANBARA,* N. UCHIDA,* K. UEMATSU,*T. KURITA,**
K. YOSHIMOTO** AND Y. SUZUKI
*Nagaoka University of Technology, Nagaoka, Niigata, Japan
**Power Reactor and Nuclear Fuel Development Co., Tokai-mura, Ibaraki, Japan

ABSTRACT

Corrosion of silicon nitride was studied in boiling nitric acid to examine its feasibility as a drying pan material in the reprocessing of nuclear fuel. Unlike stainless steel (a conventional drying pan material), the weight loss and strength degradation were negligible in the concentration of nitric acid. The corrosion increased with decreasing concentration of nitric acid. At the concentration of 1-6N, the maximum losses of weight and strength were 0.8% and over 40%, respectively, in 200 h. Ionic species dissolved in nitric acid were determined by ICP analysis and were found to be accurately correlated to the weight loss and thickness of the corrosion layer determined by micrography. In the corrosion layer, grain boundary glassy phase was selectively dissolved. Strength loss was correlated to the weight loss and was ascribed to the reduced load bearing area due to corrosion.

INTRODUCTION

In the reprocessing of nuclear fuel, the fuel dissolved in nitric acid is placed in a drying pan and heated by microwave to form an intermediate solid product. Non-uniform heating was a serious problem in the conventional drying pan made of stainless steel. It has a blocking feature against transmittance of microwaves which causes non-uniform heating. Silicon nitride ceramics have high transmittance of microwaves and excellent mechanical properties. These ceramics have a high potential for the present objectives. Recalling the extreme hazard associated with the fuel, however, it is crucial to fully understand the corrosion behavior of these ceramics in the acid environment. Corrosion behavior of silicon nitride ceramics has been intensively studied for the system sintered with yttria and alumina. Significant corrosion was reported in fluoric acid,[1] hydrochloric acid[2,3] and sulfuric acid,[4-6] etc. Corrosion has been the subject of much less study for the system sintered with magnesia and alumina. At present, no information is available on its corrosion characteristics in nitric acid.

This paper presents the corrosion characteristics of silicon nitride ceramics sintered with magnesia and alumina in nitric acid. This system contains no atomic species detrimental to fission and is better suited for the present objective than the system sintered with yttria and alumina. The specimen examined in this study is of commercial grade and was obtained from the same supplier as the drying pan. Following the conventional specifications, subjects were examined for weight and strength losses, ionic species dissolved in the nitric acid, and the thickness of the corrosion layer formed at the surface of the specimen. These results were found to be correlated to each other.

EXPERIMENT

The specimens were commercial products and were supplied in the form of test bars (3 mm x 4 mm x 40 mm). According to the supplier, they contain magnesia (0.17-0.83 wt%) and alumina (0.94-2.8 wt%) as sintering additives and iron oxide (0.26-0.9 wt%) as a major impurity. Before the corrosion experiments, they were cleaned in an ultrasonic bath (5 min) with acetone, ethanol and distilled water. Typically 5 specimens were used for each run. They were immersed in nitric acid (300 ml) and kept for various periods at the boiling temperature. An all glass apparatus consisting of flask and cooler was used. An electric heater was used for heating the flask and its output was adjusted to maintain a gentle boiling condition. With this apparatus, no measurable change occurred in the concentration of acid even after the prolonged corrosion run. After cooling and determining weight loss, specimens were subjected to strength

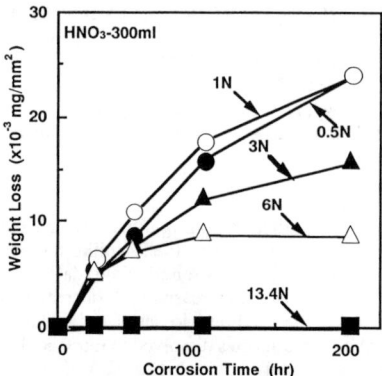

Fig. 1. Weight loss of specimen with corrosion time in nitric acid of various concentrations.

measurement. The 3-point flexural strength was measured with an Instron-type tester with a span length of 30 mm and a crosshead speed of 0.5 mm/min. The fracture surface was examined with SEM and optical microscope. The concentrations of ionic species dissolved in the nitric acid were determined by ICP analysis.

RESULTS AND DISCUSSION

Figure 1 shows the weight loss with corrosion time in nitric acid of various concentrations. Progressively increased weight loss was found as the nitric acid was diluted. For corrosion time under 25 h, the weight loss was approximately the same for all concentrations under 6N. Thereafter, the weight loss tended to be saturated with time. The time for saturation was shortened with increasing concentration of the acid. At the concentration of 6N, the weight change ceased after 100 h. The rate of weight loss for the period over 25 h decreased with increasing concentration of nitric acid. In concentrated nitric acid (13.4N), no measurable weight loss was found at any period.

Figure 2 shows the flexural strength of specimens after corrosion. At concentration 3N, the strength clearly decreased with corrosion time. The rate of strength loss decreased with increasing time. After 100 h, the strength loss tended to level off. At corrosion time of 200 h, the strength was reduced to 60-70% of the initial value. Similar strength was found for other specimens corroded in dilute nitric acid of concentration under 6N. In concentrated nitric acid, however, no significant change was found in strength after corrosion; the strength after corrosion was approximately the same as the original value. The similar dependence of strength loss on acid concentration was reported for corrosion in hydrochloric[2,3] and sulfuric acids.[4-6]

Figure 3 shows the optical micrograph of the fracture surface of a corroded specimen. A light color layer is clearly noted at the surface region. Separate SEM examination of this layer showed that individual particles of silicon nitride were clearly visible with void spaces between them. For the inner region of the specimen, particles of silicon nitride were less clearly observed; the space between particles was filled with continuous phase, and the microstructure was the same as that of the virgin specimen. The layer with light color clearly corresponds to the region from which the grain boundary phases were selectively dissolved.

Figure 4 shows the change of concentration with corrosion time for various ionic species dissolved in nitric acid. Ions of silicon, aluminum and magnesium were found and their concentrations increased monotonically with time. The concentrations of aluminum and magnesium were approximately equal for all corrosion periods, whereas the concentration of silicon was approximately three times larger. Iron, which was present in the ceramic was not

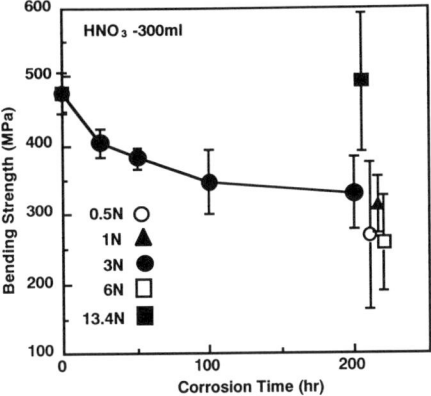

Fig. 2. Change of flexural strength with corrosion time in nitric acid of various concentrations.

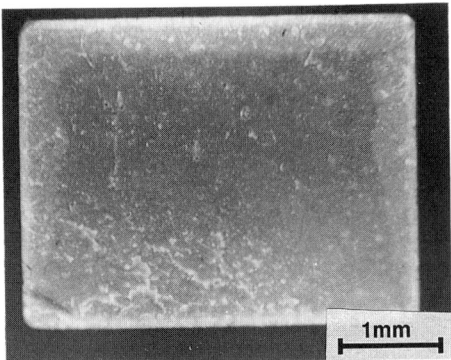

Fig. 3. Microstructure of fracture surface.

found. The sources of ions dissolved in nitric acid must be the corresponding oxides in the ceramics; very likely the grain boundary glassy phase. The approximate equal concentration for magnesium and aluminum in the acid suggests that only a fraction of alumina added was dissolved in the acid. Alumina is known to dissolve in silicon nitride in the Sialon phase. In the glassy phase, there seems to be only a fraction of total aluminum.

Figure 5 shows the relation between the weight loss and the thickness of the corroded layer. These linear relations again show that the weight loss occurs by the dissolutions of glassy phase which contains oxides of silicon, aluminum and magnesium. Comparison of weight loss and thickness shows that 5.3 wt% of the total mass was dissolved in the corroded layer. This value is understandable if homogeneous dissolution of grain boundary glassy phase only occurred in corrosion. With the chemical composition of the specimen, the total content of alumina and magnesia is 1.1-3.6 wt%. The ICP analysis of this study shows that the ratio of soluble silicon to aluminum or magnesium was approximately three. This value corresponds to the silica content of about 0.7-10 wt%. With these values, the total amount of glassy phase is estimated to be 1.8-14 wt% of the total mass in the present specimen. This value is close to the glassy phase soluble in the acid, 5.3 wt%.

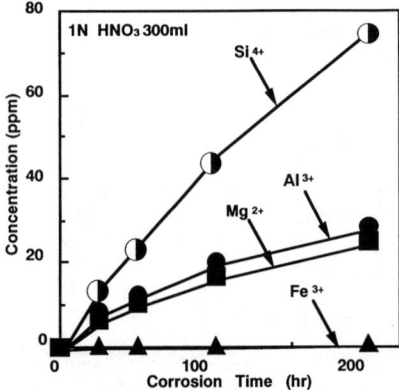

Fig. 4. Change of concentration with corrosion time for various ionic species dissolved in nitric acid.

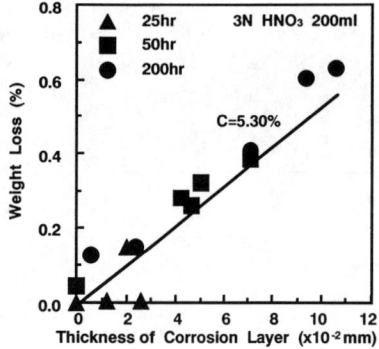

Fig. 5. Relation between thickness of corrosion layer and weight loss.

Figure 6 shows the relation between the thickness of corrosion layer and the flexural strength. The strength decreased with increasing thickness of corrosion layer. The solid lines in the figure show the strength calculated with the assumption that the strength reduction was due to the reduction of the load bearing area caused by corrosion. The numbers on lines show the assumed virgin strengths of this material. In the region of the rather thick corrosion layer, good agreement was found between the measured and calculated strengths. This result suggests that the reduction of load bearing area associated with corrosion is responsible for the decreased flexural strength. The presence of a surface compressive layer may explain the strength being lower than the calculated value for the region of thin corrosion layer. Presence of surface compressive layer is the subject of a further study.

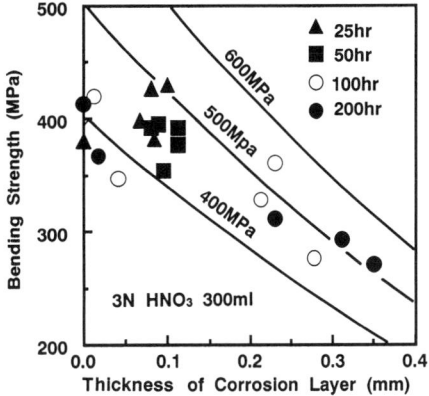

Fig. 6. Relation between thickness of corrosion layer and bending strength.

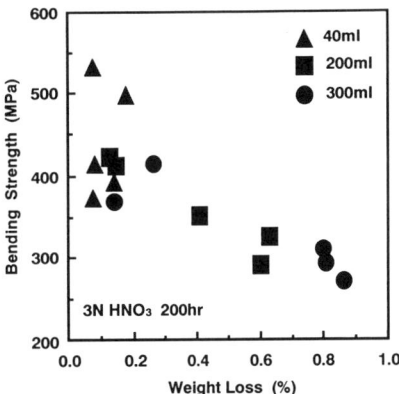

Fig. 7. Relation between weight loss and bending strength.

Figure 7 shows the relation between weight loss and flexural strength at various amounts of nitric acid. Although the strength varied significantly even with the same condition of corrosion, there is a clear correlation between strength and weight loss. The strength decreases with increasing weight loss. This result shows that the parameter which affects the strength more significantly is the weight loss rather than the amount of nitric acid used in the corrosion experiment. This result is readily understandable if one recalls the correlation between weight loss and thickness of corrosion layer (Fig. 5); weight loss is directly related to the thickness of corrosion layer, which is related to strength loss (Fig. 6). More quantitatively, take the data for weight loss of 0.8% as an example. This value corresponds to the dissolution of 15% of total glassy phase or the 15% loss of load bearing area in the ceramic. Considering the cross-sectional size of a specimen 3 mm x 4 mm, the load bearing area is reduced to 2.73 mm x 3.73 mm through corrosion. The corresponding reduction in strength is 23%, provided the material characteristics are the same for specimens before and after corrosion except the load bearing area. If the virgin strength of the ceramic was 500 MPa, as determined above, the strength expected for the weight loss 0.8% would be 370 MPa, which is in good agreement with the measured value 310 MPa.

CONCLUSION

Corrosion behavior of silicon nitride sintered with magnesia and alumina in various concentrations of nitric acid was studied. Conclusions reached are:
(1) Corrosion was significant in dilute nitric acid under 6N, but was negligible in concentrated nitric acid.
(2) In corrosion, the grain boundary glassy phase was dissolved selectively and uniformly and a corrosion layer was formed on the surface of the specimen.
(3) Good correlation exists between weight loss, ions dissolved in nitric acid, thickness of corrosion layer and strength loss.
(4) The reduction of strength with corrosion can be ascribed to the loss of a load bearing area with corrosion.

REFERENCES

1. T. Yoshio and K. Oda, Yogyo-Kyokai-Shi, **94** (2) 116 (1986).

2. T. Sato, Y. Tokunaga, T. Endo, M. Shimada, K. Komeya, K. Nishida, M. Komatsu and T. Kameda, J. Mater. Sci., **23**, 3440 (1988).

3. T. Sato, Y. Tokunaga, T. Endo, M. Shimada, K. Komeya, K. Nishida, M. Komatsu and T. Kameda, J. Am. Ceram. Soc., **71**, 1074 (1988).

4. A. Okada, S. Iio, T. Asano, M. Yoshimura, J. Japanese Ceram. Soc., **99**, 1260 (1991).

5. A. Okada, S. Iio, T. Asano, M. Yoshimura, J. Japanese Ceram. Soc., **100**, 80 (1992).

6. A. Okada, S. Iio, T. Asano, M. Yoshimura, J. Japanese Ceram. Soc., **100**, 965 (1992).

STRUCTURE AND CORROSION PROPERTIES OF PVD COATINGS IN THE SYSTEM SI-AL-O-N

OTTO KNOTEK*, FRANK LÖFFLER* AND WOLFRAM BEELE*
*Institut fuer Werkstoffkunde B, Aachen University of Technology, Templergraben 55, W-5100 Aachen, Germany

ABSTRACT

Si-Al-O-N ceramics, especially ß'-Sialon, are well known as a material offering a very good resistance against aggressive media combined with a good thermal shock behaviour. This combination recommends the Si-Al-O-N ceramics also as coating for products used in applications with high thermal gradients plus corrosion attack.

Therefore this paper presents new developments in the deposition of Si-Al-ON, especially in combination with the aircraft alloy TiAl6V4 as substrate. Investigations of the coating structure in dependence on the deposition parameters were done and the resistance against oxidation and attack of H_2SO_4 containing media were examined.

INTRODUCTION

Si-Al-O-N ceramics have been in development since the early 70's[1,2]. This system has the possibility of developing complex ceramics with advanced thermal properties in comparison to conventional binary ceramics. The chemical production of Sialon (Sialon is the trade name for ß'-Si-Al-O-N, Pat. by LUCAS) can take place in a number of ways. In most cases, known hard materials such as Al_2O_3 and Si_3N_4 are used as the primary products. However, sialon materials also include other components. Apart from Al_2O_3 and Si_3N_4, the materials formed are essentially the binary phases AlN and SiO_2 and the ternary phases Al_3O_3N and Si_2N_2O in the Si-Al-O-N system[3] (see Fig. 1).

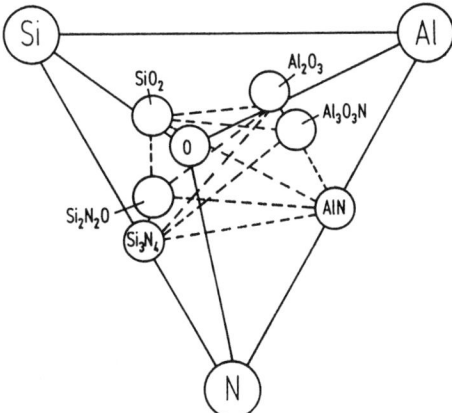

Fig. 1: Tetraeder showing the binary and ternary substructures in the system Si-Al-O-N[1]

These are supplemented by a number of special sialon phases. These compounds have not yet been fully researched, however. The most important are shown in Figure 2. The ß'-Sialon phase attains particularly good properties. Good values have been recorded[4], particularly regarding thermal stability to liquid steel and thermal shock resistance in this application. The ß'-Sialon phase has the composition $Si_{6-z}Al_zO_zN_{8-z}$ with z lower than 4.2[2]. Since silicon nitrides possess the atomic order of silicate structures, they belong to the silicates, in which nitrogen is partially replaced by oxygen. Silicate structures are characterized by almost infinitely extended combinations, in which for example an Si atom is also linked with four O atoms by atomic bonding. The system includes both amorphous and crystalline structures, and the number of degrees of freedom appears to permit a hitherto unsuspected range of possible technical applications. Of interest is the atomic bonding in this system, allowing many new types of crystal structure[4]. Sialon is therefore of particular interest for ceramic coatings.

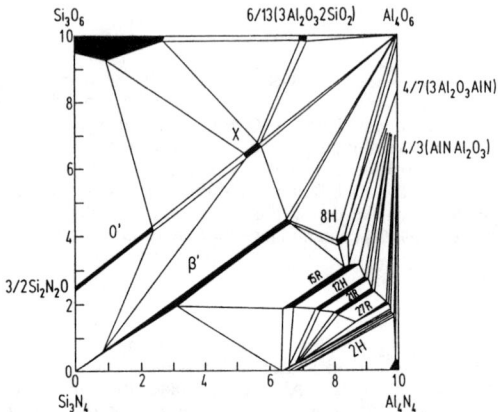

β'-Sialon: $Si_{6-Z}Al_ZO_ZN_{8-Z}$; Z <= 4,2

Fig. 2: Phase diagramm of the system Si-Al-O-N[1]

Early papers reported the deposition of amorphous Si-Al-O-N coatings using an reactive R.F. magnetron sputter process with aluminium-silicon targets and nitrogen plus oxygen as reactive gases[5,6,7]. With the knowledge of the bulk materials it is expected that crystalline Si-Al-O-N coatings should have better thermal properties combined with good wear properties.

In the work reported here, a β'-Si-Al-O-N target was used to achieve the deposition of crystalline coatings. An R.F. magnetron sputter process with 13.56 MHz R.F. frequency and 1 Pa argon pressure was used. The nitrogen gas flow was varied between 0% and 16% p_{N2}/p_{Ar}, the oxygen gas flow was varied between 0% and 4% p_{O2}/p_{Ar}.

The aircraft alloy TiAl6V4 was chosen as substrate material because of the need to develop high temperature corrosion protective thin layers with good mechanical properties for new applications. The Si-Al-O-N coating was deposited in a first step without a bondlayer on the alloy, in the second step a Ti-interlayer with a thickness below 1μm was sputtered first.

COATING STRUCTURE

Figure 3 shows the findings from the X-ray diffractometry. The deposition on the Ti-alloy gives, depending on the reactive gas flow, in some cases amorphous structures whereas all coatings deposited on the Ti-interlayer are crystalline. β'-Sialon was not detected. The coatings deposited with 4% to 6 % p_{N2}/p_{Ar} and without added oxygen have Si-Al-O-N phases (four element complex structures) together with some substructures, like alumina phases or Si_2ON_2. In comparison to the earlier publication mentioned above, one can see that crystalline Si-Al-O-N coatings are possible.

Figure 4 shows two exemplary SEM photos of the Si-Al-O-N rupture structure. No grain structure is detectable, but the surface is rougher than that examined on amorphous coatings. The Si-Al-O-N coatings have a dense morphology so that good mechanical and corrosion properties were expected.

Although a β'-Sialon target was used, no β'-phase was detected. This indicates a decomposition during the deposition. The small fields of stable phases shown in Fig. 2 are reason for the small gas range in which complex Si-Al-O-N phases were formed. The high cooling gra dient of the PVD process is another reason why the typical bulk material structures were not formed.

Phase	pN2/pAr=0% pO2/pAr=0%	pN2/pAr=4% pO2/pAr=0%	pN2/pAr=6% pO2/pAr=0%	pN2/pAr=10% pO2/pAr=0%	pN2/pAr=0% pO2/pAr=2%	pN2/pAr=0% pO2/pAr=4%	pN2/pAr=4% pO2/pAr=2%
χ-Sialon	amorphous	x				amorphous	amorphous
X-Sialon			x				
Si2ON2			x	x			
Al2O3 (sev.)			x				
Si3N4 (sev.)							
SiO2					x		

Phases formed on the Ti alloy

Phase	pN2/pAr=0% pO2/pAr=0%	pN2/pAr=4% pO2/pAr=0%	pN2/pAr=6% pO2/pAr=0%	pN2/pAr=10% pO2/pAr=0%	pN2/pAr=0% pO2/pAr=2%	pN2/pAr=0% pO2/pAr=4%	pN2/pAr=4% pO2/pAr=2%
χ-Sialon		x	x				
X-Sialon							
Si2ON2	x				x		x
Al2O3 (sev.)	x	x	x	x	x	x	
Si3N4 (sev.)				x			
SiO2	x		x	x	x	x	x

Phases formed on the Ti-interlayer

Fig. 3: X-ray examination of the coating structure dependent on the reactive gas flow

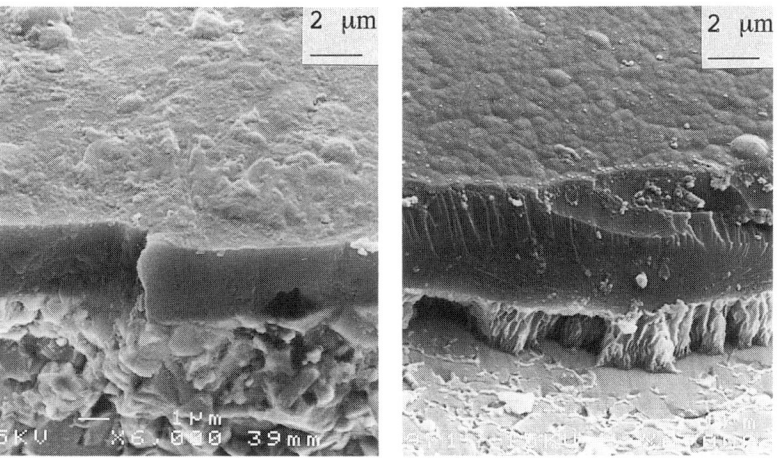

Fig. 4: SEM photos of crystalline Si-Al-O-N on TiAl6V4 with (right) and without (left) a Ti-interlayer

MECHANICAL PROPERTIES

The evaluation of the mechanical properties was done by the measurement of the microhardness and the adhesive strength. Both criteria together give an idea about the possibility to use these coatings in an tribological application.

The microhardness

The highest microhardness was achieved by the coatings deposited without added oxygen. Figure 5 gives the Vickers-microhardness values for the coatings deposited without a Ti-interlayer, but there was no major difference measured on the coatings with Ti-interlayer. The coatings deposited with a nitrogen gas content between 0% and 6% p_{N2}/p_{Ar} have a microhardness of 1500 HV0.05 and higher. Higher nitrogen contents gave a hardness reduction down to 800 HV0.05. The comparison with a cemented carbide substrate shows in general the same behaviour. High values were examined for 2% to 6% p_{N2}/p_{Ar}, whereas microhardnesses in the range of the cemented carbide itself were measured for higher nitrogen contents. The reason for this behaviour is the reduction of the deposition rate. The coatings deposited with high nitrogen gas contents were thinner than 2 µm so that the measurement was influenced by the substrate hardness. The coatings deposited with a nitrogen gas content below 6% p_{N2}/p_{Ar} had a thickness between 3 and 3.5 µm so that the vickers indentation was set in an area where no major substrate influence takes place.

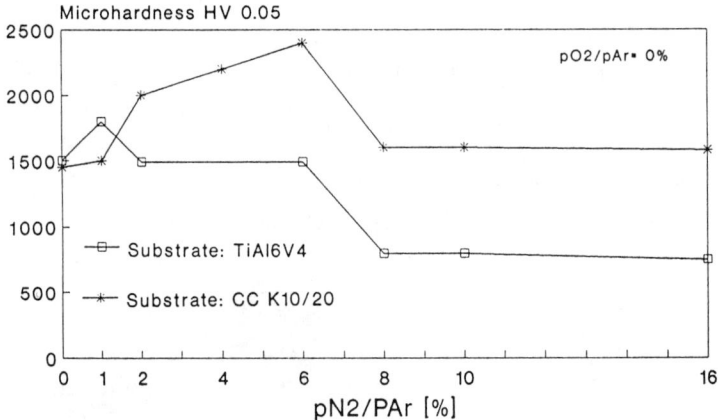

Fig. 5: Microhardness of Si-Al-O-N coatings deposited with various nitrogen gas contents

If oxygen is added as reactive gas, the deposition rate is extremely reduced. Therefore it was impossible to measure the microhardness. A more sensible unit, e.g. a nano indenter, would help to measure the correct hardness values on these coatings. However, the very low deposition rate for these deposition parameters seems to be too far away from an economical application.

The adhesion

The adhesive strength was measured with a scratch test unit. The most important result of this investigation is the effect of the Ti-interlayer on the adhesion. Whereas the critical load of Si-Al-O-N was always below 8 N on the Ti-alloy, considerable higher loads were possible with the Ti-interlayer. Figure 6 shows that the best results are achieved again by the coatings deposited with a nitrogen gas content between 2% and 6% p_{N2}/p_{Ar} and without any added oxygen gas.

These oxygen added coatings performed comparable to the coatings with higher nitrogen gas content so that no improvement was achieved.

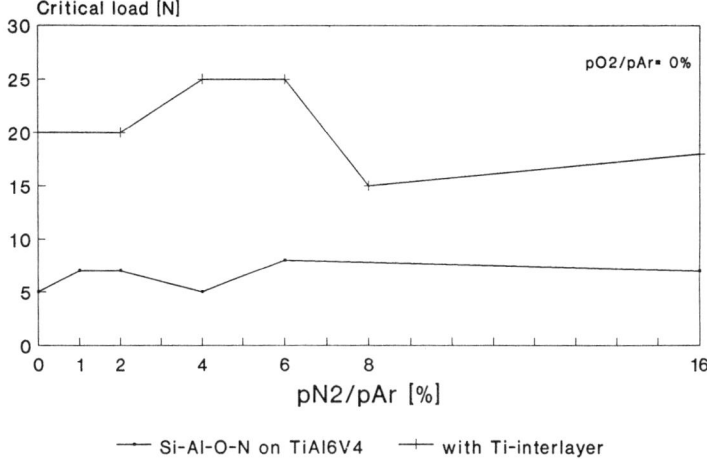

Fig. 6: Critical load of Si-Al-O-N coatings (with/ without Ti-interlayer)

PROTECTIVE EFFECT OF SI-AL-O-N ON TIAL6V4 AT 800°C IN AIR

The Si-Al-O-N coated TiAl6V4 specimens were annealed in an atmosphere furnace. The temperature was held at 800°C for 100h. Uncoated TiAl6V4 shows at this temperature a rapid oxidation. The driving forces for the oxidation are so high that the alloy is not able to passivate the surface. The oxidation was characterized by cross-sections prepared by the colour etching after WECK[8]. It is therefore impossible to reproduce the photos in black and white. No oxidized zone was detected on specimens coated with Si-Al-O-N coatings thicker than 2 μm. In comparison, the cross sections of uncoated specimens had a 200 to 300 μm thick oxide layer. Below this brittle oxide layer, the grain structure of the substrate changed, too; the α-Ti disappears and oxides are destroying the original grain structure. This active zone is more than 300 μm thick so that one can say that, in minimum, a 500 μm thick surface zone is destroyed after 100h in this test. The difference in the oxidation behaviour with and without Si-Al-O-N coating gives absolutely convincing results for this coating as a protective layer against high temperature oxidation.

The thermal shock resistance was also proven. The specimens were heated up to 450°C and in cycles shocked down in 1 minute to 50°C by pressurized air. 50 cycles were performed. This rough test had no influence on all specimens with Ti-interlayer. Only some specimens with Si-Al-O-N coated on the Ti-alloy showed small local defects.

RESISTANCE AGAINST SULPHIDATION

The specimens with Si-Al-O-N coated together with a Ti-interlayer were tested in a very aggressive fluid. The specimen surface (10 mm²) was placed 24 h in contact with an H_2SO_4 containing media. The corrosion attack was measured by the weight decrease. All specimens had no weight change after 24 h tests in 1%, 2% and 4% H_2SO_4 containing media. Finally the specimens were tested in **80% H_2SO_4**. An uncoated TiAl6V4 reference had 60 mg weight decrease. All coated specimens performed much better than the reference (see Figure 7), especially the specimen deposited without additional reactive gas. Again, the one deposited with 6% p_{N2}/p_{Ar} performed excellent; the weight decrease was lower than 1 mg which is, in comparison to the reference, 60 times better.

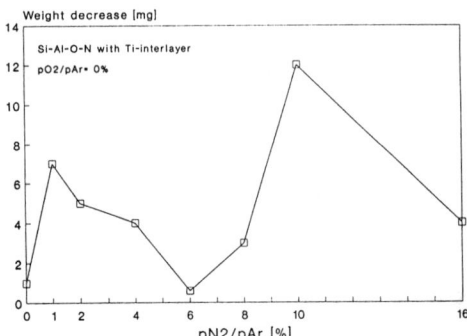

Fig. 7: Weight decrease of Si-Al-O-N coated TiAl6V4 (with Ti-interlayer) after 24 h in a 80% H_2SO_4 containing medium. Test conditions: 23°C temperature, 10 mm² contact surface, specimen weight around 2.5 g.

CONCLUSIONS

Using a ß'-Sialon target, depositing crystalline Si-Al-O-N coatings with the R.F. magnetron sputter process was achieved. Complex phases, κ-Sialon, X-Sialon and Si_2ON_2 were formed. The Ti-interlayer supports the formation of crystalline structures.

The evaluation of the mechanical properties showed under consideration of the deposition rate that the best coatings were deposited on a Ti-interlayer with a nitrogen gas content between 2% and 6% p_{N2}/p_{Ar}, without additional oxygen gas. These are also the parameters where the complex phase structures were formed. Interpreting these results, one can say that the best mechanical properties are connected with the formation of complex Si-Al-O-N phases, although the ß'-phase was not detected.

The oxidation and the sulphidation tests showed clearly that Si-Al-O-N coatings are advisable on TiAl6V4 as an oxidation barrier for high temperature applications and as corrosion resistant layer in an absolute aggressive media. These are clear advantages in comparison to "conventional" protective oxide coatings, e.g. Al_2O_3, ZrO_3 or SiO_2. PVD-Al_2O_3 modifies its crystalline structure at 800°C which causes cracks, ZrO_3 shows no protection against oxygen diffusion and SiO_2 has no functional mechanical properties.

Together with the expected advantages in the erosion behaviour, Si-Al-O-N coatings will be able to improve the properties of TiAl6V4 parts in many of its applications.

ACKNOWLEDGEMENT

This work was only possible with the support of Dr. M. Böhmer and his crew at SHM in Aachen who produced the target material. The high purity and porousfree HIP-material was an excellent base for the investigation of the sputter parameters reported.

REFERENCES

1. L.J. Gauckler, H.L. Lucas and G. Petzow, Presentation at the 2nd Powder Met. Sem., Max Planck Inst., Stuttgart 1974, J. Amer. Ceram. Soc. 58, 346 ff., 1975.
2. K.H. Jack, Sialons and related nitrogen ceramics, J. of Mat. Sci. 11, 1135 ff., 1976.
3. K.H. Jack, Phase Diagramms, Refractory Mat., Vol. 5, 241 ff., edited by Allen a. Alper, Academic Press (1978).

4. O. Knotek, A. Barimani and F. Löffler, On the formation of amorphous and metastable PVD-coatings, Mat. Res. Soc. Symp. Proc. Vol. 187 (1990).
5. O. Knotek, M. Atzor and F. Löffler, On magnetron sputtered multilayer coatings, Proc. 7th Int. Conf. on "Ion & Plasma Assisted Techniques", Geneve 1989.
6. O. Knotek and F. Löffler, On reactively sputtered Si-Al-O-N coatings, Surface Mod. Tech. III, edited by T.S. Sudarshan and D.G. Bhat, TMS 1990.
7. O. Knotek, F. Löffler, W. Beele and L. Wolkers, Properties and stability of Al- and Si-based ceramic PVD thin films, Proc. 8th SIMCER, Rimini 1992, in the framework of CERMAT'92, Nov. 1992.
8. E. Weck and E. Leistner, Metallographic instructions for colour etching by immersion, Vol. 77/III, Deutscher Verlag f. Schweißtech. (1986).

Author Index

Adlerborn, J.E., 423
Akimune, Yoshio, 405
Ampuero, Silvia, 227
Avella, F., 295

Baba, H., 481
Babonneau, F., 239, 245
Becher, P.F., 147
Beele, Wolfram, 539
Bell, Nelson, 359
Bickmore, Clint R., 251
Björklund, H., 423
Blanchard, Cheryl R., 303
Boskovic, S., 373
Bowen, Paul, 227
Braue, W., 329, 347
Bruley, J., 65

Cannon, R.M., 65
Carpenter, R.W., 329
Carroll, Daniel F., 411
Chen, I-Wei, 147, 209
Chen, L., 399
Cheng, Yibing, 527
Cinibulk, M.K., 65
Clarke, D.R., 65

Danforth, S.C., 341
Das Chowdhury, K., 329
Davidson, David L., 303

Ekström, Thommy, 121
Engström, E.U., 323
Estry, Hal W., 251

Falk, L.K.L., 323, 423
Ferber, Mattison K., 455
Feuer, H., 133
Fujita, Teruaki, 335

Gasdaska, Charles J., 473
Gazza, George, 257
Ge, C., 399
Gerardin, Corine M., 233
Gilde, Gary, 257
Goldacker, Jeffrey, 473
Graef, Renee C., 303
Grathwohl, Georg, 417
Gugel E., 133

Hakoshima, J., 487
Hamminger, Rainer, 513
Hampshire, Stuart, 93, 393
Hayashi, Seijiro, 481
Heinrich, Juergen, 513
Hiraga, Kenji, 335
Hirosaki, Naoto, 405
Hockey, B.J., 467
Hoffmann, M.J., 3, 65, 147

Hofmann, Heinrich, 105
Huang, Yong, 493
Hwang, C. James, 411
Hwang, Shyh-Lung, 147, 209

Inamura, Takashi, 461
Ishikawa, K., 435
Iwamoto, Y., 435

Jack, Kenneth H., 15
Jenkins, Michael G., 455
Jennings, Hamlin M., 277
Jiang, Zuo-Zhao, 507
Johnson, D. Lynn, 277

Käll, Per-Olov, 521
Kamiya, H., 435
Kanbara, Kunihiko, 533
Katz, R. Nathan, 197
Kerber, A., 105
Kiggans, Jr., James O., 283
Kim, Young-Wook, 265
Kleebe, H.-J., 65
Knotek, Otto, 539
Komeya, K., 29
Kühne, Arnd, 417
Kurita, T., 533

Laine, Richard M., 251
Larker, H.T., 423
Lauten, Frederick S., 315
Le, Huirong, 493
Lee, June-Gunn, 265
Lee, K.J., 373
Leigh, M., 393
Leung, O.S., 499
Lewis, M.H., 159
Li, Chien-Wei, 473
Li, Jianbao, 493
Lin, Chih-Kuang Jack, 455
Lin, H.T., 147
Liu, J., 329
Livage, J., 233
Löffler, Frank, 539
Long, G.G., 467
Lu, Ping, 341
Luecke, William, 467
Lui, Siu-Ching, 473

Ma, De-Jin, 507
Matsui, Minoru, 173
Menchoffer, Paul A., 353
Messier, Donald R., 365
Miao, He-Zhuo, 507
Mitchell, T.E., 449, 499
Mitomo, Mamoru, 405
Miyata, N., 487
Mueller, Brian L., 251

Nakajoh, S., 487
Natansohn, S., 295
Niihara, Koichi, 335
Nomura, H., 435
Nunn, Stephen D., 353, 359
Nygren, Mats, 521

Oberacker, Rainer, 417
Okumiya, M., 435
O'Reilly, K.P.J., 393

Pan, Yi-Ming, 303
Pasto, Arvid E., 295
Patel, Parimal J., 365
Persson, Jeanette, 521
Petrovic, J.J., 449, 499
Petzow, G., 3
Ploetz, Kristin L., 283, 353, 359
Plucknett, Kevin P., 289, 381
Pyzik, Aleksander J., 411

Qi, Long-Hao, 507
Quinn, G.D., 347

Raman, R.V., 309
Rankin, Janet, 315
Redington, M., 393
Reimanis, Ivar E., 499
Rele, S.V., 309
Ring, Terry A., 227
Rourke, W.J., 295
Rühle, M., 65
Rundgren, K., 323

Sakurai, C., 429
Sasa, Tadashi, 461
Sasaki, Gen, 335
Sato, Y., 429
Schwab, Stuart T., 303
Sglavo, V.M., 245
Sheldon, Brian W., 315
Shigegaki, Yasuhiro, 461
Shinohara, N., 435
Soraru, G.D., 239, 245
Srikanth, V., 441
Subbarao, E.C., 441
Suematsu, H., 449, 499

Suganuma, Katsuaki, 335
Sugita, K., 429
Sugiura, I., 435
Sun, W.Y., 39, 387
Suzuki, Akihiko, 461, 481
Suzuki, Y., 533
Symons, W.T., 341

Tajima, Yo, 189
Takahashi, H., 435
Tanaka, I., 65
Tarry, C.A., 441
Taulelle, F., 233
Thomas, Jeffrey J., 277
Thompson, Derek P., 79, 527
Tiegs, Terry N., 283, 353, 359
Tien, Tseng-Ying, 51, 373
Toro, Patricio, 271
Tsubaki, J., 435
Tsukamoto, K., 487

Uchida, N., 533
Ueki, M., 429
Uematsu, K., 435, 533

van Dijen, F. 105
Vogt, U., 105
Vulcan, F., 245

Walck, J.C., 441
Waldner, Kurt F., 251
Walls, Claudia A., 353, 359
Wang, P.L., 387
Wasén, J., 423
Wiederhorn, S.M., 467
Wilkinson, David S., 289, 381
Wittke, Oscar, 271
Woetting, G., 133
Wu, Jianguang, 493

Xia, Y., 399

Yamada, Y., 435
Yamagishi, C., 487
Yen, T.S., 39, 387
Yoshimoto, K., 533

Zheng, Longlie, 493

Subject Index

acoustic emission, 441
alloy design, 39
applications, 15, 29, 173, 189, 197

bearing (balls), 133, 197

carbothermal reaction, 105, 271
cavitation, 467, 473
characterization, 65, 323, 329, 335, 423, 429, 435
composites, 303, 309, 329, 335, 423, 481, 487, 493, 499, 507
corrosion, 533, 539
cost, 29, 105, 189, 513
creep, 3, 51, 159, 173, 209, 257, 417, 455, 467, 473, 487
crystallization, 3, 51, 93, 159, 323, 341, 347, 493, 499
cutting tools, 189, 197, 399, 507
CVD, 189, 315

deformation, 209, 449
densification, 283, 387
design, 173
dielectric properties, 197
dislocation, 449
dispersant, 257, 265, 271

engine (components), 15, 29, 133, 189, 197, 513
electronic (components), 197

FTIR, 227, 271, 315
fabrication, 245
fatigue
  cyclic, 147, 173, 455, 461
  dynamic, 173, 455
  rolling contact, 197
  static, 147, 173, 347, 455, 461
fibers, 197, 303, 365, 423
fracture map, 173

gas
  pressure sintering, 29, 189, 257, 353, 359
  turbine, 15, 173
glass ceramic, 93
grain
  aspect ratio, 51, 209, 411, 417
  boundary
    composition, 323, 329
    glass, 65, 121, 209, 323, 341, 365, 507
    phases, 39, 51, 79, 159, 347, 353, 359, 399, 493
  size, 51, 147, 209, 289, 309, 411, 417
green body, 257, 265, 277

hardness, 39, 121, 159, 373, 411, 441, 449, 499, 539

heat treatment, 79, 93, 493
HIP, 159, 189, 289, 309, 411, 417, 423, 455, 513

infiltration, 303
in-situ composites, 29, 405, 411, 473
interface, 329, 335, 411, 473

joining, 15

liquid phase sintering, 399, 435

M-Si-Al-O-N, 79, 121, 159, 393
market, 197
metallic defects, 365
microcracking, 159, 335, 449
microstructure
  design, 3, 39, 159, 189, 209
  development, 3, 29, 51, 133, 147, 209, 289, 359, 411, 417, 423, 435, 473, 499, 539
microwave, 277, 283, 289
moire interferometry, 481
multiaxial stresses, 173

NMR, 159, 233, 239
nucleation, 93, 209, 315

oxidation, 29, 121, 173, 295, 323, 359, 381, 441, 461, 487, 521, 527, 539
oxygen content, 105, 295
oxynitride glass, 79, 93, 197, 365

phase
  relationship, 3, 39, 51
  transformation, 209, 323, 393, 429, 441
polycarbosilane, 233, 239, 245, 303
polymer precursors, 233, 245, 251, 303
polysilazane, 233, 303
porosity, 105, 417, 435
powder, 29, 105, 133, 227, 257, 295, 405, 429
PVD film (coatings), 539
pyrolysis, 233, 303

R curve (behavior), 3, 147, 473, 481
RBSN, 277, 283
reliability, 3, 29, 133, 147, 173, 189, 257, 405, 473
road test, 513

S_2N_2O, 381, 521
SAXS, 467
Si-Al-O-N, 539
SiH_4, 315
$Si(NH)_2$, 227

Sialon
  composites, 15, 39, 79, 93, 209, 373
  α′, 51, 121, 209, 373, 387, 393
  β′, 121, 151, 209, 245, 487
  o′, 121
SiC, 309, 329, 335, 423, 481, 487, 493
silicon carbonitride, 233
single crystal, 449
sintering, 39, 105, 121, 133, 189, 265, 277, 381, 387, 393, 417, 435
  additives
    Y_2O_3, 121, 159, 189, 257, 265, 283, 323, 347, 359, 429, 493
    Al_2O_3, 189, 265, 283, 323, 347, 493
    SiO_2, 189, 257, 341
    ZrO_2, 323
    MgO, 429, 493
    Cr_2O_3, 189
    V_2O_5, 189
    Sc_2O_3, 29
    HfO_2, 29
    metal nitrides, 399
    rare earth oxides, 121, 159, 283, 359, 387, 393, 429, 493
slip casting, 257, 271
slow crack growth, 147, 173, 455
spin test, 173, 189

strength
  high temperature, 3, 29, 39, 51, 147, 159, 173, 295, 309, 353, 359, 493
  room temperature, 29, 39, 51, 133, 147, 159, 173, 189, 257, 295, 309, 353, 359, 365, 411, 417, 423, 429, 473, 493, 533

Taguchi method, 295
tape casting, 265
TEM, 65, 289, 329, 335, 341, 347, 467, 499
thermal
  decomposition, 227
  shock, 3
TiAl6V4, 539
TiN, 159, 507
TiB_2, 159
TiC, 507
toughness, 29, 39, 51, 121, 147, 159, 173, 189, 309, 335, 353, 373, 405, 411, 417, 429, 449, 481, 487, 493, 499
transient phases, 209, 381

wear, 29, 197
whisker, 271, 423

x-ray absorption spectroscopy, 239

Y-Si-Al-O-N, 15, 39, 51, 121, 209, 347, 373

ZrO_2, 323, 527